SOUTHA

(Unl

PHOTOINDUCED ELECTRON TRANSFER

PART D. Photoinduced Electron Transfer Reactions: Inorganic Substrates and Applications

PHOTOINDUCED ELECTRON TRANSFER

PART D. Photoinduced Electron Transfer Reactions: Inorganic Substrates and Applications

edited by

MARYE ANNE FOX

Department of Chemistry, The University of Texas at Austin, Austin, TX 78712-1167, U.S.A.

and

MICHEL CHANON

Laboratoire de Chimie Inorganique Moléculaire, Université d'Aix-Marseille, rue Henri Poincaré, 13397 Marseille Cédex 13, France

ELSEVIER
Amsterdam — Oxford — New York — Tokyo 1988

ELSEVIER SCIENCE PUBLISHERS B.V.
Sara Burgerhartstraat 25
P.O. Box 211, 1000 AE Amsterdam, The Netherlands

Distributors for the United States and Canada:

ELSEVIER SCIENCE PUBLISHING COMPANY INC.
655, Avenue of the Americas
New York, NY 10010, U.S.A.

ISBN 0-444-87125-X (Part D)
ISBN 0-444-87121-7 (Set)

Printed in The Netherlands

PREFACE

Electron transfer reactions are of great importance to nearly every subdiscipline of chemistry. The simple transfer of a single electron has been shown repeatedly to be a common activating mode for organic, inorganic, and biological molecules, and the very ubiquity of such reactions has guaranteed that their investigation would involve the most fundamental questions of modern chemistry. The fact that photoexcitation induces enhanced redox reactivity via electron transfer also provides a convenient method for experimentally testing theoretical predictions regarding structural and energetic effects.

The very bulk of this work will quickly convince even the most casual reader that a great deal is known about photoinduced electron transfer reactions. In assembling the manuscripts contained herein, we have tried to capture in print some of the diversity and excitement inherent to this broad field. One will find in these chapters contributions from theorists and experimentalists, from organic and inorganic chemists, from the perspective of the synthetic and mechanistic viewpoint. Some contributions are fundamental basic research; others clearly show practical applications of these principles. These volumes are intended, then, to serve a joint purpose: as a reference resource and an introductory overview to the diverse research accomplished via photoexcitation of electron donor - acceptor systems.

We have organized this work into four parts. The first deals with the theoretical and conceptual factors which influence electron transfer. The second covers experimental methodology and medium effects. The third and fourth deal with reactivity, with most organic transformations being addressed in Part C and most inorganic reactions covered in Part D. Neither part should be thought of as providing exhaustive coverage, but each should provide an overview of typical reactions observed for these classes of compounds. Part D also provides examples of photoinduced electron transfer in current use in important applications.

Obviously, there will be significant interdependence among the parts. Subject, chemical, and author citation indices appear at the end of the individual Parts A–C, and comprehensive cumulative indices are included in Part D. In using these indices to locate a particular citation, the reader is advised to read a few pages before and after the listed citation as the topic of interest is often discussed from several viewpoints, and for brevity only a single citation is usually given for a specific section. In locating a particular compound, one might also wish to search for structurally analogous materials and to scan a few pages before and after the citation since a family of compounds may have been indexed with a single entry. Furthermore, compounds with very complex names, especially those referred to in the text by structure numbers, are not included in the chemical indices. Often a reactant will be included in the chemical index without identifying all the products formed in a given chemical transformation.

Despite the obvious magnitude of what is known about photoinduced electron transfer, many unanswered questions persist. This general field is an extremely active area of current chemical research, and we presume that an enhanced scientific sophistication will continue to develop even as these chapters are in press. It is our hope that some of the fascination of such studies is conveyed in these chapters and that the individual manuscripts will be an effective means for stimulating both provocative discussions in the scientific community and for suggesting possible venues for further investigation.

In our own cases, the work described herein has been made possible by the generous support of several scientific funding agencies: to MC: the CNRS and the Scientific and Technic UER of Marseille III University; and to MAF: the U.S. National Science Foundation, the Office of Basic Energy Research of the U.S. Department of Energy, the U.S. Army Research Office, the Robert A. Welch Foundation, and the Texas Advanced Technology Research Program. We also gratefully acknowledge a travel grant sponsored jointly by NSF (U.S.) and by CNRS (France) which made possible the joint chapter written by one of us (MAF) with Professor Pierre Pichat in Lyon.

Only one who has worked as editor of such a large opus can appreciate the magnitude of the work involved in preparing the indices found at the end of each volume. Our colleague in Austin, Professor James Whitesell, gave us indispensable aid in managing the truly unwieldy data base compiled herein. Without his personal help and encouragement, not to mention the innumerable hours of programming and patient instruction on the vagaries of file manipulation, this work would have been impossible. Dr. J.L. Larice wrote the program allowing the direct printing of Authors indices in camera-ready form. We are also grateful to Robert Fox who invested half of a summer in bringing this project to fruition. Cheryl Melrose and Kathy Blum also helped with data entry. The correspondence relevant to the book was handled entirely by Susie Pruett, without whom, as she has proved for over a decade, progress in Austin would grind to a halt. Mrs. V. Nicolas and V. Milliat took care of entering the names of authors of the 8000 references covered by this work.

Much of the collaborative work on this project was done during the summer of 1987 while one of us (M.C.) visited Austin via a French-American cultural agreement between the University of Law, Economics, and Science of Aix-Marseille and the University of Texas. This was a fruitful time, both personally and scientifically.

We also gratefully acknowledge the cheerful assistance of the editorial staff of Elsevier Science Publishers in Amsterdam who assembled the chapters and worried about some significant technical details.

Our best wishes for an amusing trip through the diverse area of chemistry which can be addressed by photoinduced electron transfer!

Marye Anne Fox, Austin
Michel Chanon, Marseille

CONTENTS

Volume D. Photoinduced electron transfer reactions: inorganic substrates and applications

CONTENTS OF OTHER PARTS

Part A. Conceptual basis

Part B. Experimental techniques and medium effects

Part C. Photoinduced electron transfer reactions: organic substrates

LIST OF CONTRIBUTORS (Parts A–D)

Angelo Albini, Dipartimento di Chimica Organica, Universitá di Pavia, 27100 Pavia, Italy

J.V. Alfimov, Institute of Chemical Physics, USSR Academy of Sciences, 142432, Chernogolovka, Moscow Region, USSR

Oded Anner, Department of Physical Chemistry, The Hebrew University of Jerusalem, Jerusalem 91904, Israel

J.P. Astruc, Laboratoire de Physique des Lasers, Université Paris-Nord, 93430 Villetaneuse Cédex, France

James E. Baggott, Department of Chemistry, The University, Whitenights, P.O. Box 224, Reading RG6 2AD, England

Vincenzo Balzani, Dipartimento di Chimica "G. Ciamician", Universitá di Bologna, and Istituto FRAE-CNR, Bologna, Italy

S. Baral, Department of Chemistry, Syracuse University, Syracuse, NY 13244-1200, USA

R. Barbé, Laboratoire de Physique des Lasers, Université Paris-Nord, 93430 Villetaneuse Cédex, France

B. Blaive, Ecole Supérieur d'Ingenierie de Pétroléochemie et de Synthese Organique Industrielle, Faculté des Sciences et Techniques de Saint-Jérôme, Marseille 13397, Cédex 13, France

James R. Bolton, Photochemistry Unit, Department of Chemistry, The University of Western Ontario, London, Ontario N6A 5B67, Canada

W. Russel Bowman, Department of Chemistry, Loughborough University of Technology, Loughborough, Leicestershire LE11 3TU, England

Steven G. Boxer, Department of Chemistry, Stanford University, Stanford, CA 94305, USA

Louis Brus, AT&T Bell Laboratories, Murray Hill, NJ 07974, USA

Daniel F. Calef, Lawrence Livermore National Laboratory, University of California, P.O. Box 808, Livermore, CA 94550, USA

Michel Chanon, Laboratoire de Chimie Inorganique Moléculaire, Université de Droit, d'Économie et des Sciences d'Aix-Marseille, Faculté des Sciences et Techniques de Saint-Jérôme 13397, Cédex 13, France

Xiaohong Ci, Department of Chemistry, University of Rochester, Rochester, NY 14627, USA

Maurice Comtat, Laboratoire de Génie Chimique, Centre National Recherche Scientifique UA 192, Université Paul Sabatier, 118, Route de Narbonne, 31062 Toulouse Cédex, France

John S. Connolly, Photoconversion Research Branch, Solar Energy Institute, 1617 Cole Blvd., Golden, CO 80401, USA

J.A. Delaire, Physico-Chimie des Rayonnements, Université de Paris-Sud, Bâtiment 350 - Centre d'Orsay, 91405 Orsay, France

C. Desfrançois, Laboratoire de Physique des Lasers, Université Paris-Nord, 93430 Villetaneuse Cédex, France

Lennart Eberson, Division of Organic Chemistry I, Chemical Center, University of Lund, P.O. Box 740, S-220 07, Lund, Sweden

Jens Eriksen, Department of Chemistry, Sultan Qaboos University, P.O. Box 32486, Al Khod, Sultanate of Oman

J. Faure, Physico-Chimie des Rayonnements, Université de Paris-Sud, Bâtiment 350 - Centre d'Orsay, 91405 Orsay, France

Janos H. Fendler, Department of Chemistry, Syracuse University, Syracuse, NY 13244-1200, USA

Marye Anne Fox, Department of Chemistry, University of Texas, Austin, TX 78712, USA

Stefan Franzen, Department of Chemistry, Stanford University, Stanford, CA 94305, USA

Shunichi Fukuzumi, Department of Applied Chemistry, Faculty of Engineering, Osaka University, Suita, Osaka 565, Japan

S. Gaspard, Institut de Chimie des Substances Naturelles, Centre National de la Recherche Scientifique, 91190 Gif-sur-Yvette, France

Paul G. Gassman, Department of Chemistry, University of Minnesota, Minneapolis, MN 55455, USA

Charles Giannotti, Institut de Chimie des Substances Naturelles, Centre National de la Recherche Scientifique, 91190 Gif-sur-Yvette, France

L. Giral, Department of Chemistry, Faculté des Sciences de Montpellier, Place E. Bataillon, Montpellier, France

Richard A. Goldstein, Department of Chemistry, Stanford University, Stanford, CA 94305, USA

Michael Grätzel, Departement de Chimie, Ecole Polytechnique Federale de Lausanne, EPFL-Ecublens, CH-1015 Lausanne, Switzerland

Yehuda Haas, Department of Physical Chemistry, The Hebrew University of Jerusalem, Jerusalem 91904, Israel

Anthony Harriman, Davy-Faraday Research Laboratory, The Royal Institution, 21 Albemarle St., London W1X 4BS, England; current address: Center for Fast Kinetics Research, University of Texas, Austin, TX 78712, USA

M. Dale Hawley, Department of Chemistry, Willard Hall, Kansas State University, Manhattan, KS 66506, USA

Mikio Hoshino, The Institute of Physical and Chemical Research, Wako, Saitama 351-01, Japan

Guilford Jones, II, Department of Chemistry, Boston University, Boston, MA 02215, USA

Michel Julliard, Laboratoire de Chimie Inorganique Moléculaire, Université de Droit, d'Économie et des Sciences d'Aix-Marseille, Faculté des Sciences et Techniques de Saint-Jérôme, Marseille 13397, Cédex 13, France

Tomoji Kawai, The Institute of Scientific and Industrial Research, Osaka University, 8-1 Mihogaoka, Ibaraki, Osaka 567, Japan

Larry Kevan, Department of Chemistry, University of Houston, Houston, TX 77004, USA

A.D. Kirk, Department of Chemistry, University of Victoria, Victoria, British Columbia V8W 2Y2, Canada

P. Krausz, Institut de Chimie des Substances Naturelles, Centre National de la Recherche Scientifique, 91190 Gif-sur-Yvette, France

Alain Lablache-Combier, Laboratoire de Chimie Organique Physique, Université des Sciences et Techniques de Lille, 59655 Villeneuve d'Ascq Cédex, France

Cooper H. Langford, Department of Chemistry, Concordia University, 1455 de Maisonneuve Blvd. West, Montreal, Quebec H3G 1M8, Canada

Frederick D. Lewis, Department of Chemistry, Northwestern University, Evanston, IL 60201, USA

H.R. Mäcke, Abteilung Nuklearmedizin, Kantonsspital Basel, CH-4031 Basel, Switzerland

Patrick S. Mariano, Department of Chemistry, University of Maryland, College Park, MD 20742, USA

David C. Mauzerall, The Rockefeller University, New York, NY 10021, USA

C. Montginoul, Department of Chemistry, Faculté des Sciences de Montpellier, Place E. Bataillon, Montpellier, France

Carol Moralejo, Department of Chemistry, Concordia University, 1455 de Maisonneuve Blvd. West, Montreal, Quebec H3G 1M8, Canada

Seiichiro Nakabayashi, The Institute of Physical and Chemical Research, Wako, Saitama 351-01, Japan

P. Neta, Chemical Kinetics Division, Center for Chemical Physics, National Bureau of Standards, Gaithersburg, MD 20899, USA

Nicole Paillous, Laboratoire des Interactions Moléculaires et Réactivité Chimique et Photochimique, Centre National de la Recherche Scientifique UA 470, Université Paul Sabatier, 118, Route de Narbonne, 31062 Toulouse Cédex, France

Pierre Pichat, Equipe Photocatalyse du Centre National de la Recherche Scientifique, Ecole Centrale de Lyon, B.P. 163, 69131 Ecully Cédex, France

Norbert J. Pienta, Department of Chemistry and Biochemistry, University of Arkansas, Fayetteville, AR 72701, USA

Keith F. Purcell, Department of Chemistry, Willard Hall, Kansas State University, Manhattan, KS 66506, USA

Joseph Rabani, Department of Physical Chemistry and Energy Research Center, The Hebrew University of Jerusalem, Jerusalem 91904, Israel

J. Santamaria, Laboratoire de Recherches Organiques de l'E.S.P.C.I., Université Pierre et Marie Curie, 10 Rue Vauquelin, 75231 Paris Cédex, France

V.A. Sazhnikov, Institute of Chemical Physics, USSR Academy of Sciences, 142431, Chernogolovka, Moscow Region, USSR

Franco Scandola, Dipartimento di Chimica, Universitá di Ferrara, and Centro di Fotochimica del CNR, Ferrara, Italy

J.P. Schermann, Laboratoire de Physique des Lasers, Université Paris-Nord, 93430 Villetaneuse Cédex, France

F. Schue, Department of Chemistry, Faculté des Sciences de Montpellier, Place E. Bataillon, Montpellier, France

Nick Serpone, Department of Chemistry, Concordia University, 1455 de Maisonneuve Blvd. West, Montreal, Quebec H3G 1M8, Canada

B. Serre, Department of Chemistry, Faculté des Sciences de Montpellier, Place E. Bataillon, Montpellier, France

Yasuhiko Shirota, Department of Applied Chemistry, Faculty of Engineering, Osaka University, Yamadaoka, Suita, Osaka 565, Japan

Haruo Shizuka, Department of Chemistry, Gunma University, Kiryu, Gunma 376, Japan

Ada Sulpizio, Dipartimento di Chimica Organica, Université di Pavia, 27100 Pavia, Italy

Toshio Tanaka, Department of Applied Chemistry, Faculty of Engineering, Osaka University, Suita, Osaka 565, Japan

Arnd Vogler, Institut für Anorganische Chemie, Universität Regensburg, Universitätsstraße, 8400 Regensburg, West Germany

William L. Waltz, Department of Chemistry and the Saskatchewan Accelerator Laboratory, Saskatoon, Saskatchewan S7N 0W0, Canada

Michael R. Wasielewski, Chemistry Division, Argonne National Laboratory, Argonne, IL 60439, USA

David G. Whitten, Department of Chemistry, University of Rochester, Rochester, NY 14627, USA

Frank Wilkinson, Department of Chemistry, Loughborough University of Technology, Loughborough, Leicestershire LE11 3TU, England

Alan F. Williams, Departement de Chimie Minérale, Analytique, et Appliquée, Université de Genève, CH-1211 Genève 4, Switzerland

5. PHOTOINDUCED ELECTRON TRANSFER REACTIONS: INORGANIC SUBSTRATES

Chapter 5.1

Activation of Oxygen by Photoinduced Electron Transfer

M.A. Fox

1. INTRODUCTION

1.1. Possible Oxygenation Pathways

Seemingly paradoxically, oxygen is both extremely stable and highly reactive. It is essential, of course, for metabolism and hence for life. Furthermore, its incorporation into organic molecules makes possible many synthetically important chemical transformations, in part because of the ease with which the oxidation level of oxygen-containing compounds can be altered.

Oxygen is unusual in that it exists as a ground state triplet so that its reactivity with most molecules, which are ground state singlets, is inhibited by spin restriction. It can be easily converted, however, by interaction with electrons, triplet sensitizers, or reactive intermediates to any of several highly reactive species. For example, as shown in Scheme 1, addition of an electron generates a doublet O_2^-, superoxide. Superoxide is both paramagnetic

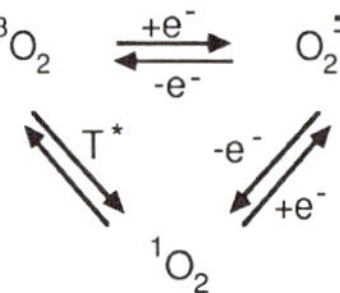

Scheme 1. Interconversion of Ground State (Triplet) Oxygen with Superoxide and Singlet Oxygen. (T is a triplet sensitizer.)

and negatively charged (by virtue of the extra electron) and is highly reactive. In addition to its chemical reactivity, superoxide has been implicated as a reactive intermediate in many biological systems (ref. 1) involving oxidoreductases (refs. 2-4), cytochromes (refs. 6-9), flavins (ref. 10), or proteins (ref. 11).

Superoxide, as a doublet, does not face the spin restriction of ground state oxygen, and, upon re-oxidation, can be converted either to an activated singlet or back to the ground state triplet. Thus, sequential forward and reverse electron transfer to ground state oxygen can provide an alternate route to singlet oxygen besides the conventional triplet sensitization pathway.

Singlet oxygen is highly reactive, and its extensive involvement in cycloadditions and ene reactions of unsaturated molecules is well-established (ref. 12). The lowest lying singlet state, $^1\delta g$, is approximately 23 kcal/mol more energetic than its triplet precursor and it is only because of the spin forbiddenness of the intersystem crossing required for its relaxation that $^1\delta gO_2$ lives long enough for efficient chemical reactivity. Although solvent dependent, singlet oxygen lifetimes of microseconds have been observed in common organic solvents at room temperature.

Both ground state oxygen and superoxide are attacked by reactive intermediates, Scheme 2.

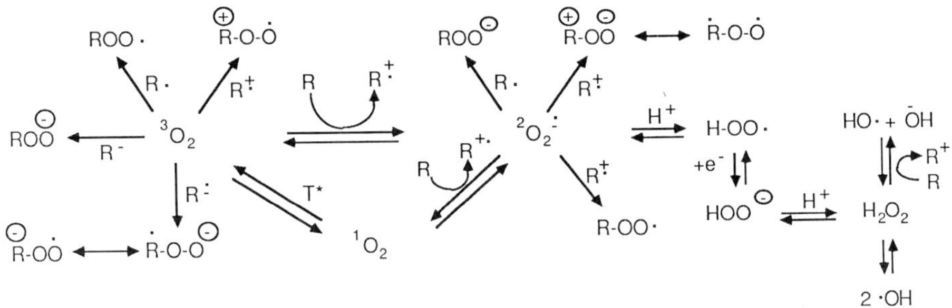

Scheme 2. Reactive Intermediates Derived from Superoxide and Triplet Oxygen

If a reactant R acts as an electron donor toward oxygen (right horizontal path), the formation of superoxide will be accompanied by formation of a donor cation radical. Triplet oxygen can be reduced more easily (by about 0.34 V) than singlet oxygen, but the latter species can indeed also be converted to O_2^{-}. In acidic media, superoxide will be protonated, forming the hydroperoxy radical $HO_2^{\bullet}$. Further reduction would then convert the radical to the anion, protonation of which gives hydrogen peroxide. With an appropriate catalyst, hydrogen peroxide can be converted to hydroxy radicals, either by direct decomposition or through a reductive pathway which also produces hydroxide.

Both $^3O_2$ and $^2O_2^-$ are rapidly attacked by radicals and radical cations (upward vertical paths, Scheme 2). Respectively, these routes produce alkylhydroperoxy radicals, radical cations, or anions, or biradical intermediates. In addition, triplet oxygen can trap anionic intermediates, producing alkylhydroperoxy anions (left horizontal) or alkylhydroperoxy radical anions (downward vertical). Charge repulsion makes the analogous routes unfavorable with superoxide.

Given the complexity of the intermediates which can be rationally envisioned to derive from oxygen, it is not surprising that much mechanistic speculation accompanies studies of oxygenation. This dynamic interplay will be clearly reflected in the mechanistic discussions of the chapter.

1.2. Photoactivation of Oxygen

Lacking an effective chromophore, oxygen is itself transparent throughout the visible and much of the ultraviolet spectrum. Its photoactivation must therefore be sensitized. Triplet sensitization by an absorptive sensitizer produces singlet oxygen via energy transfer from suitably energetic excited triplets, eqn. 1.

$$S \xrightarrow{h\nu} {}^1S^* \longrightarrow {}^3T^* \xrightarrow{O_2} {}^1O_2 \qquad (+T_0) \tag{1}$$

Singlet oxygen chemistry is extensive and is beyond the scope of this treatment (refs. 12-14).

More germane to the topic of the chapter is its sensitized photoactivation via electron transfer. Photoexcitation renders a molecule both more easily oxidized (by population of an energetically higher lying orbital) and more easily reduced (by creation of a vacancy in what was a filled orbital in the ground state). Since oxygen can act as a reasonable electron acceptor, many photoexcited states can transfer a single electron to produce a radical ion pair, eqn. 2.

$$S \xrightarrow{h\nu} S^* + O_2 \longrightarrow S^{\dot{+}} + O_2^{\dot{-}} \tag{2}$$

The positively and negatively charged radicals can remain associated or participate in ion pairing equilibria, eqn. 3, which are analogous,

$$S^{\dot{+}}O_2^{\dot{-}} \rightleftharpoons S^{\dot{+}}//O_2^{\dot{-}} \rightleftharpoons S^{\dot{+}} + O_2^{\dot{-}} \quad (3)$$

respectively, to the contact ion pairs, solvent separated ion pairs and free ions whose existence in solvolysis chemistry has been well established. The reactivity of each member of the ion pair will vary according to its degree of association. Clearly the photoinduced electron transfer of eqn. 2 opens many of the possible reaction channels of Scheme 2. The efficiency of the photoreaction will depend, however, on the relative rates of reaction of the ion radicals compared with that of thermal reverse electron transfer, eqn. 4. (Note, however, that the reaction of two doublets can produce either triplet

$$S^{\dot{+}} + O_2^{\dot{-}} \longrightarrow S + O_2 \quad (4)$$

or singlet oxygen, provided that the energetics for electron transfer are favorable.)

1.3. Chapter Organization

This general background having been established, we will look in more detail at the redox chemistry of oxygen. We will then consider pathways converting superoxide to singlet oxygen and other oxidizing radicals and will survey recent developments regarding the mechanism of photoinduced oxygenations. Since relatively recent reviews are available in both homogeneous solution (refs. 15-18) and in heterogeneous systems (refs. 19-22), this section will constitute a mechanistic overview rather than a bibliographic compilation.

2. REDOX CHEMISTRY OF OXYGEN

2.1. Redox Properties of Oxygen

The oxidation-reduction chemistry of dioxygen in aqueous solution can be summarized as in Scheme 3, for the standard state of O_2 at unit activity (ref. 23).

$$O_2 \xrightarrow{-0.16\ V} HO_2^{\bullet}/O_2^{\bullet-} \xrightarrow{+0.89\ V} H_2O_2 \xrightarrow{+0.51\ V} HO\cdot \xrightarrow{+218\ V} H_2O$$

+.36 V (O_2 → H_2O_2) +1.35 V (H_2O_2 → H_2O)

+0.86 V (O_2 → H_2O)

Scheme 3. Redox Chemistry of Oxygen

Hydrogen peroxide, which can be generated from O_2 reduction or water oxidation, is formed by proton-induced disproportionation of superoxide.

The redox properties of O_2 will be highly dependent on solvent, however (refs. 24-26). The direct oxidation is only quasi-reversible and indirect methods have been adopted for its characterization. The oxidation of H_2O_2 under anhydrous conditions, for example, occurs by single electron transfer to $HO_2^{\bullet}$ which either disproportionates to H_2O_2 and O_2 (at a rate of $10^7\ M^{-1}S^{-1}$ in acetonitrile) (ref. 23) or is further oxidized to O_2 by a second electron transfer. The redox potentials of superoxide have also been determined via pulse radiolysis (ref. 27) and kinetic absorption spectrophotometry (refs. 27-30). Quantitative agreement between laboratories has proven to be difficult, Table 1.

Table 1. Redox Potentials (E_o) Estimated for Dioxygen and Superoxide

	-0.7	70
	+0.57V	29
$O_2/O_2^{\bullet-}$	-0.33V	27,30
	+0.07V	28
$O_2^{\bullet-}/H_2O_2$	+1.3	70
	+0.36V	28
$^{\bullet}HO_2$	>1.0V	29

Superoxide oxidation was reversible, although its reduction was not (ref. 28).

Studies of the kinetics of electron transfer between aqueous superoxide and Co(III) amine complexes establish that the initial rate limiting step involves outer sphere electron transfer. The widely variant values observed upon altering the redox potential of the complex did not correlate well with Marcus predictions, an observation attributed to very different solvation requirements by dioxygen and superoxide (ref. 31). The rate data for electron transfer to organic substrates can be correlated well with Marcus theory if a reasonable reorganization parameter (18 kcal/mole) is used (ref. 32). That solvation is important in controlling these redox potentials is evident from the laser photodetachment threshold of O_2^-, which shifts from the infrared to the visible region in non-polar liquids (ref. 33).

Special mention should be made of single electron reductions of oxygen involving hydroquinones and semiquinones, since natural quinones are effective traps for $O_2^{\dot{-}}$ in living cells (ref. 34) and since quinones have often been involved in non-oxygenated photoinduced electron transfer studies. Oxygen is found to complex with quinones (ref. 17), establishing the equilibrium shown in eqn. 5 (ref. 35).

$$O_2^{\dot{-}} + Q \rightleftharpoons O_2 + Q^{\dot{-}} \qquad (5)$$

In oxygenated solution, the semiquinone half-life is thus shortened and the mean diffusion distance of the semiquinone is less than 0.6 A (refs. 36,37). This observation then implies that oxygen can trap a photogenerated electron formed in a primary donor(D)-to-quinone(Q) electron transfer, eqn. 6, forming

$$D + Q \xrightarrow{h\nu} D^{\dot{+}}Q^{\dot{-}} \xrightarrow{O_2} D^{\dot{+}} + O_2^{\dot{-}} \qquad (6)$$

a new radical pair, at least in competition with back electron transfer. Superoxide can then act both as an electron carrier and as a powerful nucleophile (ref. 38). Dicyanoanthracene (DCA) can function in parallel fashion, oxidation products having been formed via superoxide and singlet oxygen generated by photoexcitation of a donor-DCA charge transfer complex (ref. 39).

Charge transfer complexation involving oxygen and organic compounds is probably best indicated by the shift of the long wavelength edge of the absorption bonds of many organic compounds in the presence of oxygen (ref. 40). At least on solid supports, oxygen-aromatic charge transfer complexes are thought to form stable complexes with fixed geometries (ref. 41), with oxygen acting as the acceptor and the organic molecule as the donor.

The partial electron transfer characteristic of charge transfer bands can be driven to completion upon photoexcitation. For example, flash photolysis of aryl amines in oxygenated solution forms a transient radical cation which can be spectroscopically observed, eqn. 7 (refs. 42-44). The biphotonic

$$PhN(CH_3)_2 \;\; O_2 \xrightarrow{h\nu} PhN^{+\cdot}(CH_3)_2 + O_2^{\cdot -} \qquad (7)$$

ionization observed in degassed solution becomes monophotonic in air. The ion radicals formed probably exist as contact ion pairs in non-polar solvents and as free ions in polar solvents (ref. 44). Such reactions are also possible routes for dye sensitization (ref. 45).

Similar electron transfer consequences are also observed with inorganic complexes (refs. 46-51). Both superoxide and singlet oxygen are sometimes formed via excited state quenching (refs. 51,52).

2.2. Singlet Oxygen

Analogous electron transfer pathways equilibrating superoxide and singlet oxygen have also been suggested, eqn. 8. The electrophilicity of

$$^1O_2 + D \rightleftharpoons D^{+\cdot} + O_2^{\cdot -} \qquad (8)$$

singlet oxygen is well-documented (refs. 13,14). The relative reaction rates of quenching of singlet oxygen by organic donor arenes correlate well with their half wave oxidation potentials (ref. 14). Aryl amines (refs. 53-55), phenols (ref. 56), hydroxytryptophan (ref. 57), and sulfides (refs. 58-60) have been shown to form charge transfer complexes with singlet oxygen, producing superoxide. The yield of observable radical ions depends in all cases on the ease of ionization of the substrate and on the solvation of the resulting ions. Excited triplets (ref. 61) or singlets have been used to generate superoxide. Spin traps used in parallel with laser flash photolysis have established that singlet oxygen quenching by 1,3-diphenylisobenzofuran or by rubrene is substantially a reversible electron transfer, eqn. 9, which

$$D + {}^1O_2 \longrightarrow D^{\dot{+}} \cdots O_2^{\dot{-}} \longrightarrow D + {}^3O_2 \qquad (9)$$

occurs about 80 times faster than direct chemical reaction with these purported singlet oxygen traps (ref. 62).

The reverse process, formation of singlet oxygen from superoxide, also has chemical precedent. Superoxide and various cation radicals annihilate, for example, to produce variable yields of singlet oxygen (refs. 63,64). The efficiency of the disproportionation of superoxide itself to generate singlet oxygen, eqn. 10, was found to depend on solvation,

$$2O_2^{\dot{-}} + 2H^+ \rightleftharpoons H_2O_2 + {}^1O_2 \qquad (10)$$

especially when water was present (refs. 65-67). Under other conditions, no evidence for formation of singlet oxygen from superoxide could be obtained (ref. 68).

Because of the apparent ease of interconversion between singlet oxygen and superoxide, a clear experimental test to differentiate these primary photogenerated oxidants would be useful. One such test involves microemulsion-dispersed furfuryl alcohol, which forms an endo peroxide with singlet oxygen but is inert toward superoxide, eqn. 11 (ref. 69). This control

$$\text{2-hydroxyfuran} \xrightarrow[\text{microemulsion}]{^1O_2} \text{endoperoxide (–O–O–) bearing OH}; \qquad \text{2-hydroxyfuran} \xrightarrow[\text{microemulsion}]{O_2^{\cdot-}} \text{NR} \qquad (11)$$

is achieved by the selective penetration of singlet oxygen through the surfactant aggregate interface where it is detected in the hydrophobic core. (Superoxide is excluded by the anionic surface charge.) Care must be taken, however, since hydrated electrons may penetrate the interface depending on the initial pulse dose.

Two other tricks that have proved useful for this goal involve the different regiochemistries and decay rates observed in reactions involving these reagents. For example, Mattes and Farid have described an electron-transfer sensitized oxygenation of 1,1-dimethylindene in which opposite regiochemistry is observed in a cation radical - superoxide pathway from that seen with singlet oxygen (ref. 70). That the lifetime of singlet oxygen is ten times longer in CD_3CN than in CH_3CN (ref. 71) has also been used experimentally to distinguish these two reactive species.

2.3. Radicals as Sources for Superoxide

Superoxide has been detected in the oxidative quenching of radicals. The interaction of hydroxy radical with trialkylamines, for example, has been shown in kinetic studies to form superoxide, eqn. 12 (ref. 72). Adducts

$$\cdot OH + NMe_3 \longrightarrow H_2O + Me_2N\text{-}\dot{C}H_2 \xrightarrow{O_2} O_2^{\cdot-} + CH_2{=}\overset{\oplus}{N}Me_2 \qquad (12)$$

formed by the attack of hydroxy radical on substituted uracils are converted, in the presence of oxygen, to peroxy radicals which oxidize tetramethyl-p-phenylenediamine with a rate constant of $10^8\ M^{-1}s^{-1}$ (ref. 73). Again pH effects are important in governing the positions of the electron transfer equilibria.

3. MECHANISTIC PATHWAYS FOR PHOTOOXYGENATION OF CATION RADICALS

3.1. Possible Routes

From the reaction diversity discussed above, it is probably not surprising that many pathways have been proposed for oxygen incorporation into organic substrates. Although electron transfer catalysis has been suggested in the formation of peroxidic carbon-oxygen bonds at carbanionic centers, eqn. 13 (refs. 74,75), this type of oxygenation usually does not require light activation. In this section we will focus our attention on photoinduced electron transfer oxidations, most of which proceed via cation radicals. We

$$R^- + O_2 \longrightarrow R\cdot + O_2^{\cdot-}; \quad R\cdot \xrightarrow{O_2} RO_2\cdot; \quad RO_2\cdot \xrightarrow{R^-} RO_2^- \ (+R\cdot); \quad RO_2^- \xrightarrow{H_2O} ROOH \qquad (13)$$

recognize that such intermediates $D^{\cdot+}$ are usually generated in situ by electron transfer from a donor D to an appropriate acceptor, either by photochemical excitation of a donor-acceptor system, eqns. 14 and 15, or of a charge transfer complex, eqn. 16, or by electrochemical, eqn. 17, or photoelectrochemical, eqn. 18, methods.

$$D^* + A \longrightarrow D^{\cdot+} + A^{\cdot-} \qquad (14)$$

$$D + A^* \longrightarrow D^{\cdot+} + A^{\cdot-} \qquad (15)$$

$$[D\cdots A]^* \xrightarrow{h\nu} D^{\cdot+} + A^- \qquad (16)$$

$$D \xrightarrow{-e} D^{\cdot+} \qquad (17)$$

$$D \underset{h+}{\xrightarrow{h\nu}} D^{\cdot+} \qquad (18)$$

The resulting cation radical can then react with either superoxide, eqn. 19, or triplet oxygen, eqn. 20, or by back transfer to generate singlet oxygen which then attacks, e.g., by an ene insertion, eqn 21.

$$D^{+\cdot} + {}^3O_2^{\cdot\cdot} \longrightarrow \overset{\oplus}{D}\text{-O-O}^{\ominus} \longleftrightarrow \dot{D}\text{-O-O}\cdot \qquad (19)$$

$$D^{+\cdot} + {}^3O_2 \longrightarrow \overset{+}{D}\text{-O-O}^{\cdot} \qquad (20)$$

$$H\text{-}D^{+\cdot} + {}^3O_2^{\cdot\cdot} \longrightarrow H\text{-}D + {}^1O_2 \longrightarrow D\text{-O-O-H} \qquad (21)$$

Examples will be presented which address each of these routes. Methods which have proved to be successful to divert the competition between these routes[13,14] will also be briefly discussed. Since these reactions have been reviewed elsewhere (refs. 17,18), we provide only representative examples.

3.2. Olefins

3.2.1. Oxidative Cleavage via Dioxetanes

By far, the most extensively studied oxygenation routes have involved photosensitized oxidation of alkenes. A characterizable 1,2-dioxetane product can be isolated upon dicyanoanthracene sensitization of an electron rich alkene in the presence of oxygen, eqn. 22 (ref. 76).

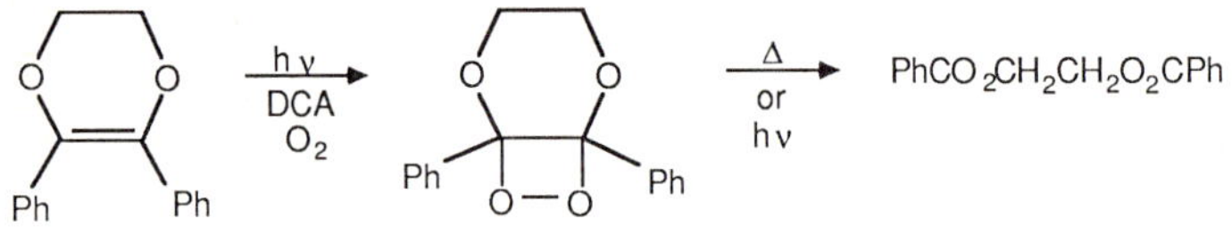

$$\xrightarrow[\text{DCA, } O_2]{h\nu} \qquad \xrightarrow[\text{or } h\nu]{\Delta} PhCO_2CH_2CH_2O_2CPh \qquad (22)$$

Ph Ph O O ... Ph O—O Ph

Ultimately the diester shown is obtained in high yield. Trapping of a cation radical by superoxide, either directly or through two single bond-forming steps, eqn. 23, is a reasonable route to this precursor, although the

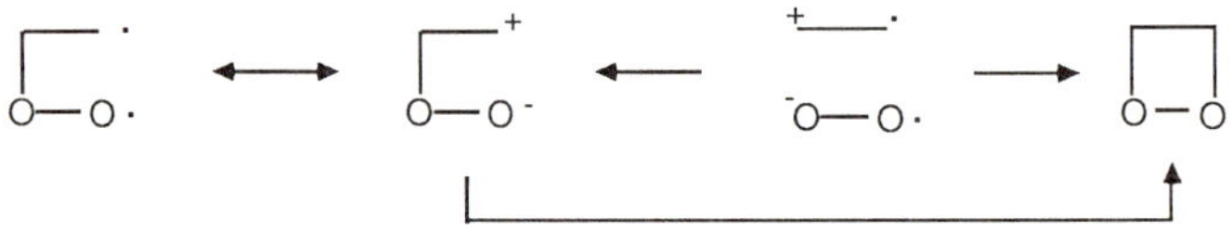

(23)

same product is formed by allowing the starting material to interact with singlet oxygen. When generated in the dark, the cation radical was shown to react faster with superoxide than with singlet oxygen, so a kinetic differentiation of the applicable routes could be made (ref. 77).

Similar intermediates have been suggested in the photoinduced oxygenation of other olefins, although they have not proved to be isolable (refs. 78-86).

With trans-stilbene, for example, oxidative cleavage to benzaldehyde, eqn. 24 (ref. 78),

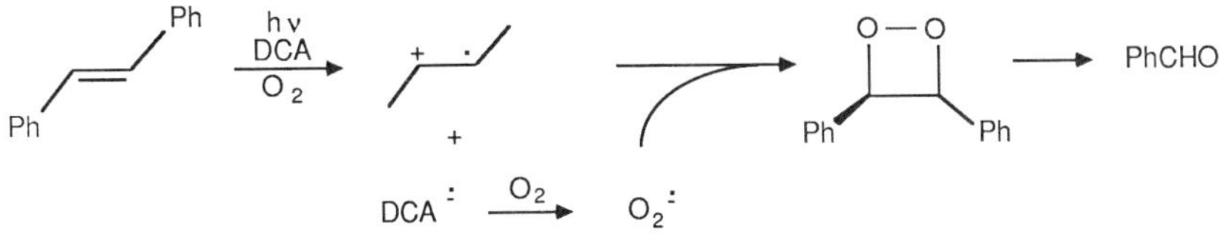

Ph–CH=CH–Ph $\xrightarrow[O_2]{h\nu,\ DCA}$ [Ph–CH(+)–CH(·)–Ph] + $DCA^{\bar{\cdot}}$ $\xrightarrow{O_2}$ $O_2^{\bar{\cdot}}$ ⟶ dioxetane (Ph, Ph) ⟶ PhCHO (24)

is thought to proceed through a dioxetane formed by collapse of the cation radical-superoxide pair, since the ion radicals can be detected spectroscopically (refs. 76,86-88). Furthermore, 1,1-diphenylethylene can be converted to benzophenone and epoxide by heating it as a ground state complex with tetrafluoro-tetracyanoquinodimethane in the presence of oxygen (ref. 89). The preparation of dioxetanes[90] and oxidative cleavage products (refs. 91,92) via singlet oxygen, however, is also precedented.

Photoexcitation of donor-acceptor pairs or a charge transfer complexes provides a route for formation of the olefin cation radical in the presence of superoxide. A similar result can also be achieved by heterogeneous sensitization by photoexcited semiconductor particles suspended in an oxygenated solution of the olefin (refs. 19-22). With a wide band gap semiconductor, where appreciable energy is stored in the photogenerated electron-hole pair, simultaneous formation of an olefin cation radical and superoxide is possible (refs. 84,85). As shown in Scheme 4, irradiation of

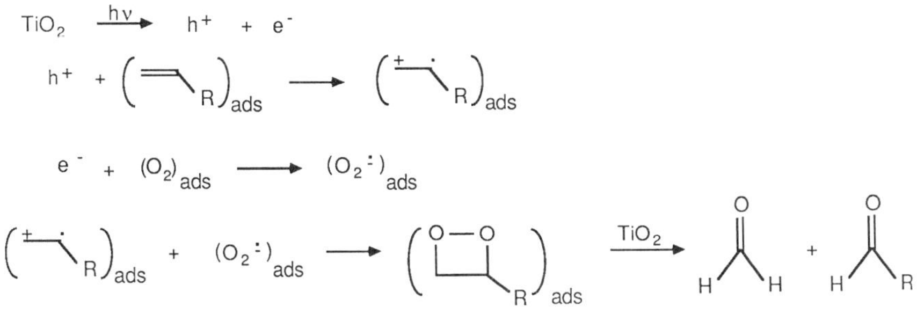

$TiO_2 \xrightarrow{h\nu} h^+ + e^-$

$h^+ + (\text{olefin-R})_{ads} \longrightarrow (\text{olefin}^{+\cdot}\text{-R})_{ads}$

$e^- + (O_2)_{ads} \longrightarrow (O_2^{\bar{\cdot}})_{ads}$

$(\text{olefin}^{+\cdot}\text{-R})_{ads} + (O_2^{\bar{\cdot}})_{ads} \longrightarrow (\text{dioxetane-R})_{ads} \xrightarrow{TiO_2}$ HCHO + RCHO

Scheme 4. Photelectrochemical Formation of a Surface-Adsorbed Radical Ion Pair

TiO_2 generates an electron-hole pair poised at the band edges of the semiconductor. These potentials are sufficiently large to form olefin cation radicals and superoxide on a common surface. Diffusion together along this metal oxide surface then allows for chemistry parallel to that observed in reactions occurring via cation radical-superoxide collapse in homogeneous solution. Since this topic is covered elsewhere (refs. 19-22,93), no further consideration is given here.

Besides cation radical trapping with superoxide, we must also consider whether the photogenerated cation radical might also react with ground state (triplet) oxygen. The relative importance of oxygenation deriving from superoxide and from triplet oxygen is difficult to sort in the photochemical and photoelectrochemical routes in which both reagents are present. Conventional electrochemical routes for olefin oxidation do not conveniently allow for simultaneous superoxide formation, since the poised electrode surface is highly oxidizing and, hence, would not allow for the presence of appreciable quantities of superoxide. Oxygen can coexist with the electrogenerated cation radical, however, and mechanistic testing for the occurrence of oxygen trapping of photogenerated cation radicals may perhaps be best accompished in the dark by electrochemical generation of the intermediate.

Precedent for dioxetane formation via triplet oxygen trapping of an olefin cation radical is available for highly substituted and/or strained olefins, e.g., adamantylideneadamantane (refs. 94-97) and other tetraalkyl olefins (refs. 98-102). The sequence shown in Scheme 5 has been suggested as the probable route for such conversions. Evidence for the open intermediate formed by single bond (non-concerted) trapping of the cation radical derives both from theoretical calculations (ref. 103) and from the double-bond rotational isomerization of a bicyclo<3.2.1> octylidene (ref. 98). It is interesting to note that this route initiates chain oxygenation, with chain lengths increasing appreciably upon cooling (ref. 101). With

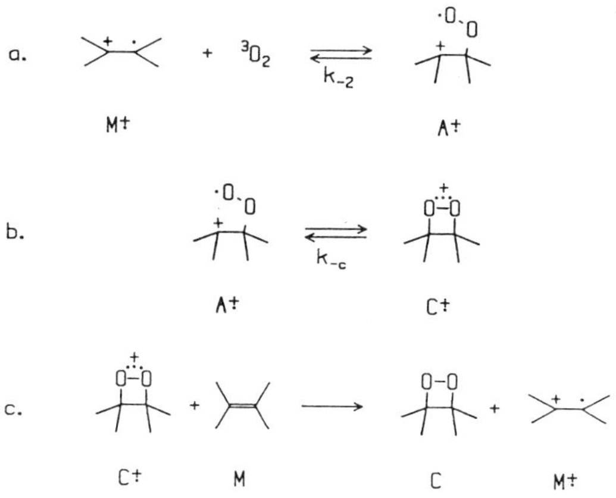

Scheme 5. Oxygen Trapping of Radical Cations

biadamantylidene, for example, the chain length of about 10 observed in methylene chloride at room temperature increases to over 800 at $-78^{\circ}C$. The operation of such a cation radical chain would amplify greatly the charge separation caused by the initial photoinduced electron transfer if this route were operative as well in the photochemical routes for cation radical generation.

3.2.2. Oxetane Formation

Net oxidative cleavage is also often accompanied by oxetane formation (refs. 78,80,84,85,104-106), e.g., eqn. 25.

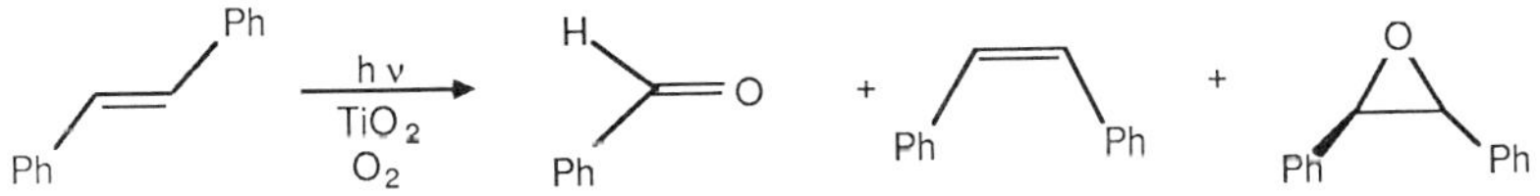

(25)

The mechanism for epoxide formation has not been unambiguously established, but autooxidation routes have been suggested, occurring either through a peroxy intermediate, Scheme 6 (ref. 107), or through a tetroxy intermediate, Scheme 7.

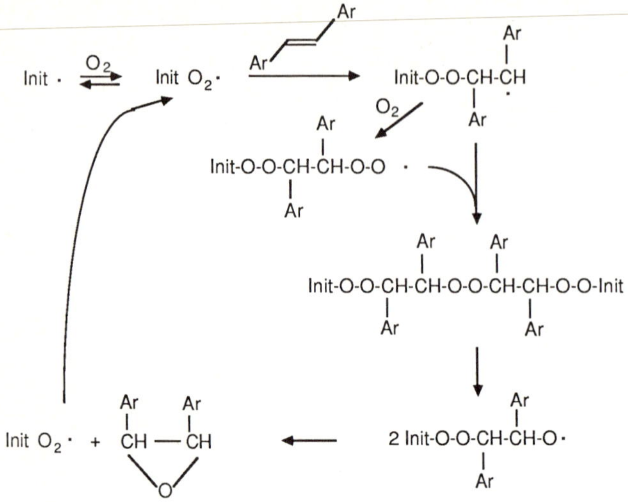

Scheme 6. Proposed Route for Oxetane Formation via Peroxy Intermediates

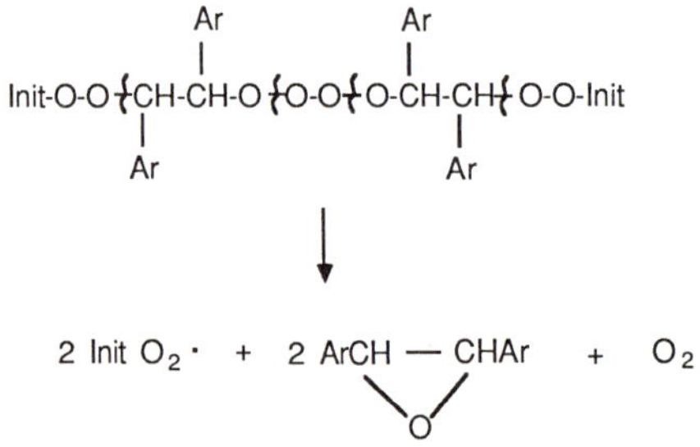

Scheme 7. Proposed Route for Oxetane Formation via Tetroxy Intermediates

The pathways are analogous to those suggested by Bartlett and coworkers in the photochemical formation of epoxides from aryl olefins upon direct irradiation (ref. 108) or sensitization by ketones or diketones (ref. 109). In these reactions, neither singlet oxygen nor simple radical autooxidation could explain the observed product distributions.

In metal oxide semiconductor-induced reactions, direct oxygen transfer from the surface also remains a mechanistic possibility (ref. 104).

3.2.3. Peroxidation

The epoxides formed in this way are themselves subject to further oxygenation. For example, in the diphenylethylene-sensitized photooxygenation of 1,2-diphenylcyclobutene, ozonides are produced, eqn 26 (refs. 90,105).

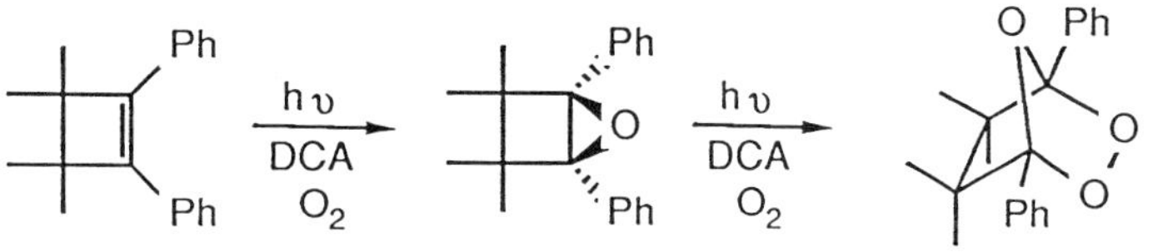

(26)

With the more highly stained cyclopropenes, cleavage, rather than than oxygenated rings, are formed, eqn. 27 (ref. 110). Quantum yields substantially

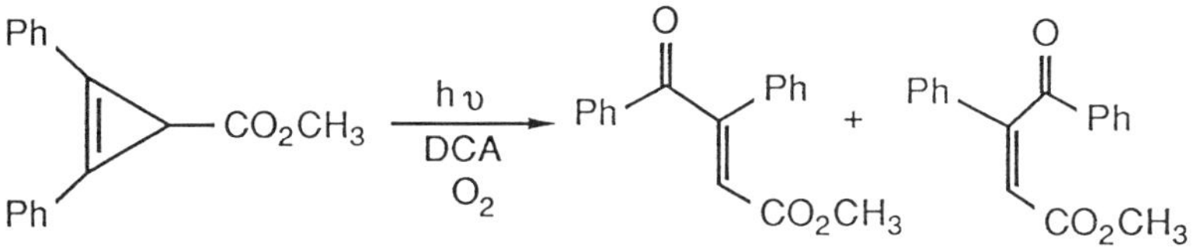

(27)

over unity are observed. These same products are also formed upon reaction with singlet oxygen (sensitization with rose bengal) and by chemical peroxidation, so the mechanism is ambiguous.

Ozonides are also produced directly from epoxides when biphenyl is employed as a co-sensitizer, eqn 28 (refs. 90,111-113). Photooxygenation

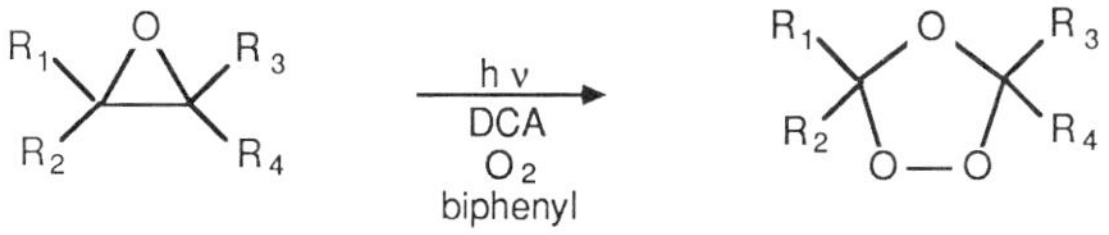

(28)

is often more efficient in the presence of a co-sensitizer, i.e., another oxidizable olefin whose function is to more effectively trap the singlet excited state of the sensitizer or to separate into free radical ions the radical ion pair formed initially in the electron transfer event. Although the reaction can proceed with reasonable efficiency without biphenyl if the oxirane is electron-rich, its presence is required for most of the epoxides, because direct electron transfer from most oxiranes to excited dicyanoanthracene is inefficient. The initial stereochemical relationship of 2- and 3-substituents in the epoxide is retained in the ozonide, suggesting that the ring-opened radical cation is trapped more rapidly than it can equilibrate geometrically.

Analogous conversions in aziridines have also been reported, eqn. 29 (ref. 114).

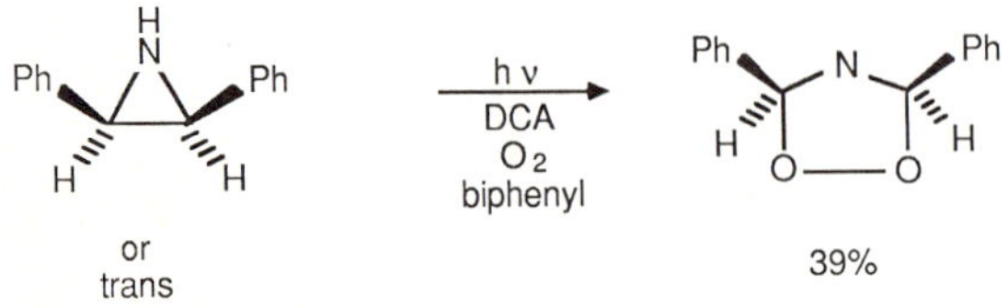

(29)

Here, co-sensitization is not compulsory, because of the higher ring electron density, but the reaction is no longer stereospecific. The authors could exclude the direct involvement of singlet oxygen.

The photooxygenation of methylenecyclopropanes follows a similar route, eqn. 30 (refs. 115,116).

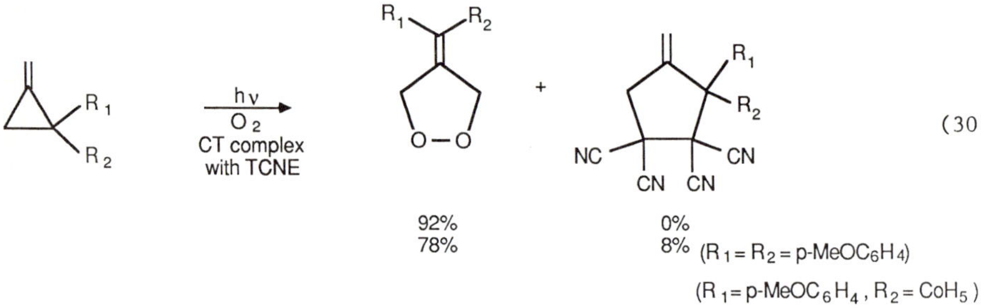

(30)

Delocalization must occur in the ring-opened intermediate, for rearranged products are also isolated with asymmetrically substituted methylenecyclopropanes (ref. 116). The stability of the ring-opened cation radical is critical in controlling the course of the reaction, for with less electron-rich rings trapping by the TCNE radical anion within the contact ion pair or the excited charge transfer complex begins to compete. The formation of these dioxolanes is accompanied by the suppression of the ring-opening isomerization observed in degassed solution (refs. 115,116).

Co-sensitized photooxygenation of vinylcyclopropanes also occurs, eqn. 31 (ref. 117), in competition with oxidative cleavage, although the operative

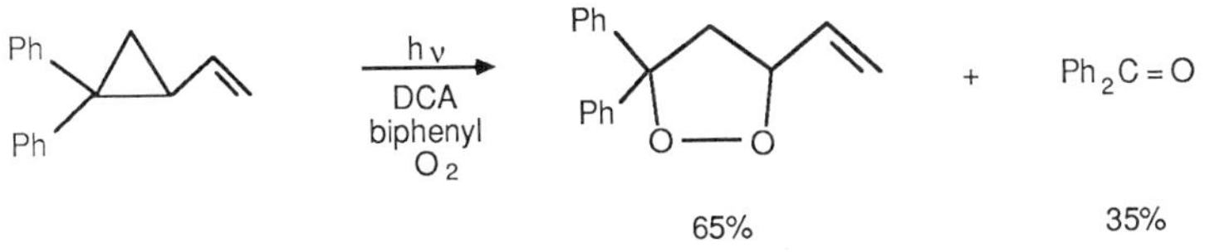

Ph₂C=O (31)

65% 35%

sequence of ring-opening and trapping reactions remains unclear.

3.2.4. Oxygenation Consequent to Dimerization

Since the rate of trapping of an olefin cation radical by its neutral precursor is diffusion controlled (refs. 118-121), oxygenation routes involving intermediate formation of the dimer cation radical and its derived peroxy cation radical must also be considered, Scheme 8.

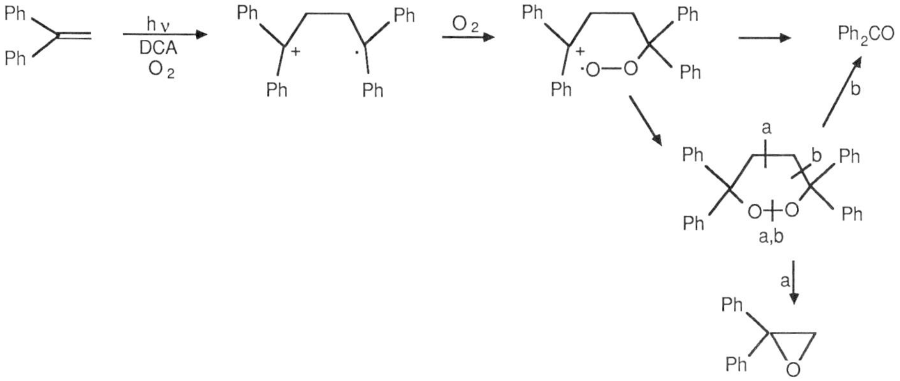

Scheme 8. Oxygenation via a Dimeric Cation Radical

Precedent for dimer cation radical trapping by triplet oxygen is available from studies of the electrochemical oxidation of olefins discussed earlier. Dioxanes have been isolated when diarylethylenes are irradiated in the presence of oxygen and catalytic amounts of antimony(V) chloride in dichloromethane (ref. 122) or by sensitization with tetracyanoanthracene (ref. 18). Cleavage of the dioxane can be envisioned as a possible route for either epoxidation, path a, Scheme 8, or oxidative clevage, path b, Scheme 8.

3.2.5. Nucleophilic Solvent Assisted Oxygenation

Trapping of a photogenerated olefin cation radical by nucleophilic solvent is also extremely rapid. When diphenylethylene, for example, is irradiated with tetracyanoethylene in the presence of methanol, the methoxy hydroperoxide is a major product, eqn 32 (ref. 119).

$$Ph_2C{=}CH_2 \xrightarrow[\text{TCA},\ O_2,\ CH_3CN/MeOH]{h\nu} Ph_2C(OOH)CH_2OMe \qquad (32)$$

Presumably, trapping of the cation radical by methanol forms a carbon radical which is subsequently trapped in a slower pathway by oxygen. 1,1-Dimethylindene photooxygenation is similarly assisted by solvent (ref. 70).

3.3. Dienes

3.3.1. Oxygenation

Since dienes usually have lower oxidation potentials than the structurally related olefins, formation of diene cation radicals by parallel photoinduced electron transfer pathways is quite likely. The chain oxygenation of diene cation radicals has been suggested by Bartlett (ref. 123), Barton (ref. 124), Haynes (ref. 125), and Tang (ref. 126). Subsequent electrochemical study has shown that when a diene cation radical is held in an s-cis conformation rapid oxygen addition occurs, forming endoperoxide in high chemical yield, eqn 33 (ref. 127).

$$\xrightarrow[O_2]{-e^-} \qquad (33)$$

O-O

The diene does not give significant peroxidation from singlet oxygen, and, as with olefin cation radicals, a mechanism is suggested in which sequential C-O bond formation occurs. With the cation radical of biadamantylideneethylene, an s-trans diene, oxygenation was not observed, possibly because of reversible oxygen trapping.

This route to cyclic peroxides can also be catalyzed by Lewis acids (refs. 122,124-126,128,129). Although transition metal Lewis acids can initiate this chemistry in the dark, trityl cation and non-transition metal acids require irradiation. With bicyclohexenyl, for example, only the cis-peroxide is formed, eqn 34 (ref. 124),

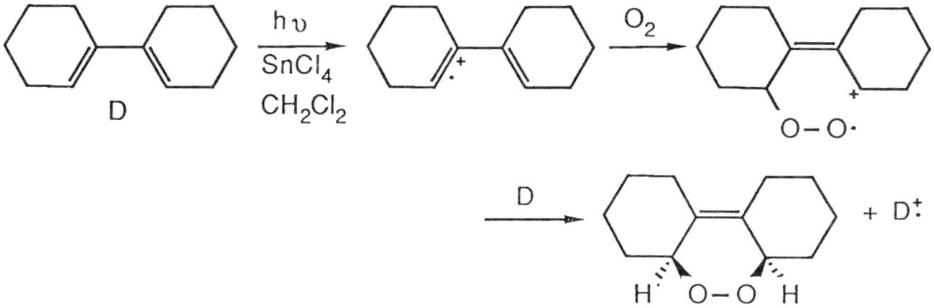

(34)

whereas a mixture of cis and trans-isomers is formed via free radical oxidation. Singlet oxygen is not involved (ref. 125), and the diene cation radical can also be formed by irradiation of a ground state complex (ref. 129).

3.3.2. Oxygenation Consequent to Dimerization

As with olefins, trapping of the diene cation radical by its neutral diene precursor can initiate endo-peroxide formation. Such an oxygenation product is formed, for example, when a mixture of t-butylcyclohexadiene and trityl cation are irradiated in oxygenated solution, eqn 35 (ref. 130).

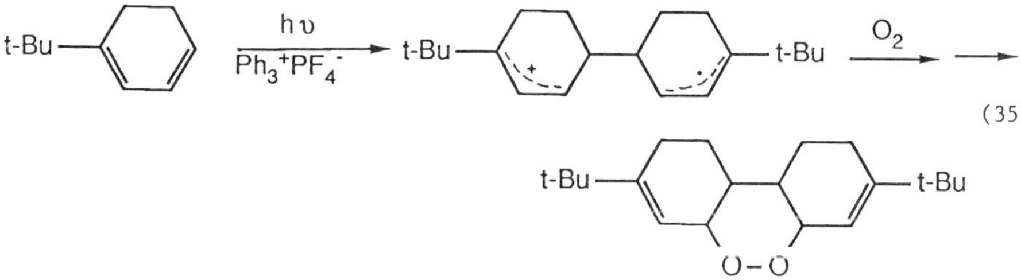

(35)

3.4. Alkynes

If a parallel pathway occurs in the photooxygenation of alkynes as is followed with alkenes, the formation of fragmentation products derived from a dioxetene might be expected, eqn 36.

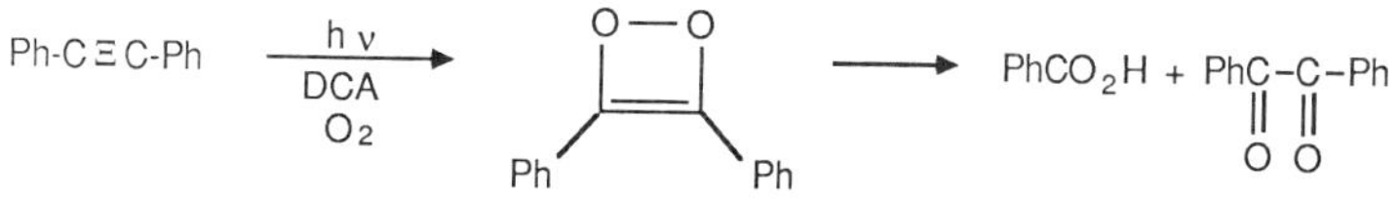

$$\text{Ph-C}\equiv\text{C-Ph} \xrightarrow[\text{DCA, } O_2]{h\nu} \text{dioxetene} \longrightarrow PhCO_2H + PhC(=O)\text{-}C(=O)\text{-}Ph \quad (36)$$

With diphenylacetylene or its ring methylated derivatives, benzils and benzoic acids consistent with this pathway are indeed isolated (refs. 119,131). Alkyl- and phenylacetylenes are not oxidized since the initial electron transfer from dicyanoanthracene would be endothermic.

3.5. Alkylarenes

Aromatic rings bearing benzyllic hydrogen are often oxidatively cleaved by photoinduced electron transfer. The yield of oxidative cleavage of bis-p-methoxybibenzyl, eqn 37 (ref. 132),

$$(\text{MeO-C}_6\text{H}_4\text{-CH}_2)_2 \xrightarrow[\text{DCA, } O_2]{h\nu} \text{MeO-C}_6\text{H}_4\text{-CHO} \quad (\sim 83\%) \qquad (37)$$

for example, drops to only traces when the electron-rich substituent is replaced by hydrogen. A time-resolved flash study of analogous reactions reveals transient formation of the arene cation radical (ref. 133). The possible involvement of cation radicals in the photo-Fenton reaction has also been proposed (refs. 134,135), Analogous products are obtained with more highly methylated benzenes (refs. 136,137), and even oxidative cleavage of biphenyl can be observed (ref. 138).

With methylnaphthalenes, divergent chemistry is observed upon electron transfer sensitization and oxygenation with singlet oxygen, eqn 38 (ref. 139).

$$\text{1-methylnaphthalene} \xrightarrow[O_2]{h\nu} \text{1-naphthaldehyde} + \text{1-methyl-1,4-epidioxy-1,4-dihydronaphthalene} \qquad (38)$$

	1-naphthaldehyde	endoperoxide
(DCA/CH_3CN)	75%	20%
($^1O_2$)		100%

Presumably, the photogenerated cation radical deprotonates, generating a benzyllic radical which is trapped by oxygen. Side chain deprotonation can be inhibited if the cation radical is generated on an irradiated semiconductor suspension, where nearly exclusive oxidative cleavage of the ring is observed (ref. 140). The formation of endoperoxide from 1,4-dimethylnaphthalene has been attributed to spin reversal assistance by the electron transfer sensitization (ref. 141).

3.6. Miscellaneous Reactions

Although mechanistic details have not been nearly as extensively examined as in the photooxygenation of olefins, photoinduced electron transfer routes also provide methods for oxidizing alcohols, phenols and sulfides. Both simple alkyl alcohols (ref. 142) and benzyl alcohols (ref. 143) can be so sensitized. Structural effects on the reaction efficiency suggest that a balance must be met between electron transfer and proton transfer rates in hydroxylated compounds.

Sensitized formation of semiquinones and hydroperoxysemiquinones occurs upon sensitization of ortho- and para-alkylated phenols, eqn 39 (ref. 78).

$$\text{2,4,6-}R_3C_6H_2OH \xrightarrow[\text{DCA}]{h\nu} \text{4-}R\text{-4-OOH-2,6-}R_2\text{-cyclohexa-2,5-dienone} + \text{4-}R\text{-4-OH-2,6-}R_2\text{-cyclohexa-2,5-dienone} \qquad (39)$$

In the electron transfer initiated oxidation of dialkyl or diarylsulfides, the corresponding sulfoxide is obtained (ref. 145). Diphenyl sulfide was more reactive than diethyl sulfide, in contrast to the relative reactivity observed with singlet oxygen. Analogous conversions can also be accomplished on irradiated TiO_2 suspensions (refs. 145,146).

4. CONCLUSIONS

Photoinduced electron transfer can be initiated by irradiating a donor-acceptor pair, a charge transfer complex, or a light-sensitive semiconductor surface. The radical ion pair formed in this primary photochemical event is highly reactive, and the radical cation formed by this method can be oxygenated by triplet oxygen, superoxide, or other oxy radicals formed in the reaction medium.

The radical anion can act as an electron source to reduce dissolved oxygen to superoxide. Superoxide can also be formed electrochemically at modest negative potentials. In addition to directly attacking the geminate cation radical (or electron poor neutral substrates), superoxide can participate in protonation equilibria or back electron transfer. The former route provides access to hydroperoxy radicals, while the latter opens a route, as an alternate to the conventional triplet sensitization, for the formation of singlet oxygen.

A rich variety of competing chemical reactivity is thus available from photoinitiated activation of molecular oxygen. A challenge in this area remains in predicting the relative efficiencies of the divergent oxygenation reactions available from the myriad of activated oxygen species discussed herein and in controlling the course of such oxygenation pathways.

5. ACKNOWLEDGEMENTS

Our research program on radical ion chemistry has been generously supported by the U.S. National Science Foundation and by the Robert A. Welch Foundation.

6. REFERENCES

1. A. Shafferman, G. Stein, Biochim. Biophys. Acta, 416 (1975) 287.
2. P. Wardman, E.D. Clarke, Biochem. Biophys. Res. Commun., 69 (1976) 942.
3. K. Inoue, T. Matsuura, I. Saito, J. Photochem., 25 (1984) 511.
4. E.J. Land, A.J. Swallow, Biochim. Biophys. Acta, 234 (1971) 34.
5. Y. Sawada, I. Yamazaki, Biochim. Biophys. Acta, 327 (1973) 257.
6. M.G. Simic, I. A. Taub, J. Tocci, P.A. Hurwitz, Biochem. Biophys. Res. Commun., 62 (1975) 161.
7. K. Tajima, M. Sakamoto, K. Okada, K. Mukai, K. Ishizu, H. Sakurai, H. Mori, Biochem. Biophys. Res. Commun., 115 (1983) 1002.
8. H. Seki, Y.A. Ilan, Y. Ilan, G. Stein, Biochim. Biophys. Acta, 440 (1976) 573.
9. L.R. deAlvare, K. Goda, T. Kimura, Biochem. Biophys. Res. Commun., 69 (1976) 687.
10. A. Yoshimura, S. Kato, Bull. Chem. Soc. Jpn., 53 (1980) 1877.
11. J.A. Fee, R.L. Ward, Biochem. Biophys. Res. Commun., 71 (1976) 427.
12. H.E. Wasserman, R.W. Murray, Singlet Oxygen, Academic Press, New York, 1979.

13. D.R. Kearns, Chem. Rev., 71 (1971) 395.
14. B. Stevens, Acc. Chem. Res., 6 (1973) 90.
15. C.S. Foote, Tetrahedron, 25 (1985) 2089.
16. K. Gollnick, A. Schnatterer, Photochem. Photobio. 43 (1986) 365.
17. F.D. Lewis, Chapter 4.1, this work.
18. S.L. Mattes, S. Farid, Org. Photochem., 6 (1983) 233.
19. M.A. Fox, Top. Curr. Chem., 142 (1987) 72.
20. M.A. Fox, Org. Electrochem., 1 (1985) 177.
21. M.A. Fox, Accts. Chem. Res., 16 (1983) 314.
22. M.A. Fox, New Trends and Applications of Photocatalysis and Photoelectrochemistry in Environmental Problems, M. Schiavello, (Ed.), NATO Adv.Sci. Ser., 1988, in press.
23. P. Cofre, D.T. Sawyer, Inorg. Chem., 25 (1986) 2089.
24. J. Wilshire, D.T. Sawyer, Accts. Chem. Res., 12 (1979) 205.
25. D.T. Sawyer, J.S. Valentine, Accts. Chem. Res., 14 (1981) 393.
26. D.H. Chin, G. Chiericato, E.J. Nanni, D.T. Sawyer, J. Am. Chem. Soc., 104 (1982) 1296.
27. Y.A. Ilan, G. Czapski, D. Meisel, Biochim. Biophys. Acta, 430 (1976) 209.
28. P.S. Rao, E. Hayon, J. Phys. Chem. 79 (1975) 397.
29. P.S. Rao, E. Hayon, Biochem. Biophys. Res. Commun., 51 (1973) 468.
30. Y.A. Ilan, D. Meisel, G. Czapski, Isr. J. Chem., 12 (1974) 891.
31. M.S. McDowell, J.H. Espenson, A. Bakac, Inorg. Chem., 23 (1984) 2232.
32. D. Meisel, Chem. Phys. Lett., 34 (1975) 263.
33. U. Sowada, R.A. Holroyd, J. Chem. Phys., 70 (1979) 3586.
34. I.B. Afanas'ev, N.I. Polozova, Zhur. Ubsch. Khim., 15 (1979) 1802.
35. R.L. Willson, Biochem. Soc. Trans., (1974) 1082.
36. B.A. Svingen, G. Powis, Arch. Biochem. Biophys., 209 (1982) 119.
37. M. Simic, E. Hayon, Biochem. Biophys. Res. Commun., 50 (1973) 364.
38. I.B. Afanas'ev, S.V. Prigoda, T.Ya. Mal'tseva, G.I. Samokhvalov, Int. J. Chem. Kinetics, 6 (1974) 643.
39. J. Santamaria, R. Ouchabane, Tetrahedron Lett., 42 (1986) 5559.
40. H. Tsubomura, R.S. Mulliken, J. Am. Chem. Soc., 82 (1960) 5966.
41. H. Ishida, H. Takahashi, H. Sato, H. Tsubomura, J. Am. Chem. Soc., 92 (1970) 275.
42. M. Hori, H. Itoi, H. Tsubomura, Bull. Chem. Soc. Jpn., 43 (1970) 3765.
43. H. Shizuka, Y. Sawaguri, T. Morita, Bull. Chem. Soc. Jpn., 45 (1972) 24.
44. Y. Taniguchi, A. Miyoshi, N. Mataga, H. Shizuka, T. Morita, Chem. Phys. Lett., 15 (1972) 223.
45. V. Kasche, L. Lindqvist, J. Phys. Chem., 68 (1964) 817.
46. D.G. Whitten, Accts. Chem. Res., 13 (1980) 83.
47. A. Tkac, Int. J. Radiat. Phys. Chem., 7 (1975) 457.
48. C.P. Anderson, D.J. Salmon, T.J. Meyer, R.C. Young, J. Am. Chem. Soc., 99 (1977) 1980.
49. A. Bakac, J.H. Espenson, I.I. Creaser, A.M. Sargeson, J. Am. Chem. Soc., 105 (1983) 7624.
50. J.R. Darwent, K. Kalyanasundaram, J. Chem. Soc., Faraday Trans. 2, 77 (1981) 373.
51. G.S. Cox, D.G. Whitten, C. Giannotti, Chem. Phys. Lett., 67 (1979) 511.
52. G.S. Cox, C. Bobillier, D.G. Whitten, Photochem. Photobiol., 36 (1982) 401.
53. I. Saito, T. Matsuura, K. Inoue, J. Am. Chem. Soc., 105 (1983) 3200.
54. L.E. Manring, C.S. Foote, J. Phys. Chem., 86 (1982) 1257.
55. I. Saito, T. Matsuura, K. Inoue, J. Am. Chem. Soc., 103 (1981) 188.
56. M.J. Thomas, C.S. Foote, Photochem. Photobio., 24 (1978) 683.
57. K. Inoue, T. Matsuura, I. Saito, J. Photochem., 25 (1984) 511.
58. K. Inoue, T. Matsuura, I. Saito, Tetrahedron, 41 (1985) 2177.
59. M.L. Kacher, C.S. Foote, Photochem. Photobio., 29 (1979) 765.
60. C.S. Foote, J.W. Peters, J. Am. Chem. Soc., 93 (1971) 3795.
61. S.K. Chattopadhyay, C.V. Kumar, P.K. Das, J. Photochem., 30 (1985) 81.
62. A.P. Darmayan, G. Moger, J. Photochem., 26 (1984) 269.

63. E.A. Mayeda, A.J. Bard, J. Am. Chem. Soc., 95 (1973) 6223.
64. N.D. Martin, C.W. Jefford, Tetrahedron Lett., (1982) 3949.
65. A.U. Khan, Photochem. Photobio. 28 (1978) 615.
66. A.U. Khan, J. Am. Chem. Soc., 103 (1981) 6516.
67. A.U. Khan, J. Am. Chem. Soc., 99 (1977) 370.
68. R. Poupko, I. Rosenthal, J. Phys. Chem., 77 (1973) 1722.
69. M.-T. Maurette, E. Oliveros, P.P. Infelta, K. Ramsteiner, A.M. Braun, Helv. Chim. Acta, 66 (1983) 722.
70. S.L. Mattes, S. Farid, J. Am. Chem. Soc., 104 (1982) 1454.
71. P. Ogilby, C.S. Foote, J. Am. Chem. Soc., 103 (1981) 1219.
72. S. Das, M.N. Schuckmann, H.P. Schuckmann, C. von Sonntag, Chem. Ber., 120 (1987) 319.
73. S. Fujita, S. Steenken, J. Am. Chem. Soc., 103 (1981) 2540.
74. G.A. Russell, J. Am. Chem. Soc., 76 (1954) 1595.
75. R.D. Guthrie, Carbanions in Organic Synthesis, J.C. Stowell, (Ed.), Wiley, New York, 1979.
76. A.P. Schaap, K.A. Zaklika, B. Kaskar, L.W.M. Fung, J. Am. Chem. Soc., 102 (1980) 389.
77. C.S. Foote, Tetrahedron, 41 (1985) 2221.
78. J. Eriksen, C.S. Foote, T.L. Parker, J. Am. Chem. Soc., 99 (1977) 6645.
79. M. Kajima, H. Sakuragi, K. Tokumaru, Tetrahedron Lett., (1981) 2889.
80. J. Eriksen, C.S. Foote, J. Am. Chem. Soc., 102 (1980) 6083.
81. L.E. Manring, M.K. Kramer, C.S. Foote, Tetrahedron Lett., 25 (1984) 2523.
82. L.E. Manring, J. Eriksen, C.S. Foote, J. Am. Chem. Soc., 102 (1980) 4275.
83. J. Koo, G.B. Schuster, J. Am. Chem. Soc., 99 (1977) 5403.
84. M.A. Fox, C.C. Chen, J. Am. Chem. Soc., 103 (1981) 6757.
85. T. Kanno, T. Oguchi, H. Sakuragi, K. Tokumaru, Tetrahedron Lett., 21 (1980) 467.
86. L.T. Spada, C.S. Foote, J. Am. Chem. Soc., 102 (1980) 391.
87. M.A. Fox, B. Lindig, C.C. Chen, J. Am. Chem. Soc., 104 (1982) 5828.
88. W. Hub, U. Kluter, S. Schneider, F. Dorr, J.D. Oxman, F.D. Lewis, J. Phys. Chem., 88 (1984) 2308.
89. N.J. Peacock, G.B. Schuster, J. Org. Chem., 49 (1984) 3356.
90. S. Futamura, S. Kusunose, H. Ohta, Y. Kamiya, J. Chem. Soc., Perkin I, (1984) 15.
91. D.S. Steichen, C.S. Foote, J. Am. Chem. Soc., 103 (1981) 1855.
92. A. Albini, G.F. Bettinetti, Chem. Ind. (Milan), 62 (1980) 615.
93. M.A. Fox, P. Pichat, Chapter 6.1, this work.
94. S.F. Nelsen, R. Akaba, J. Am. Chem. Soc., 103 (1981) 2096
95. E.L. Clennan, W. Simmons, C.W. Almgren, J. Am. Chem. Soc., 103 (1981) 2098.
96. R. Akaba, S. Aihara, H. Sakuragi, K. Tokumaru, J. Chem. Soc., Chem. Commun., (1987) 1262.
97. S.F. Nelsen, Accts. Chem. Res., 20 (1987) 269.
98. S.F. Nelsen, D.L. Kapp, J. Am. Chem. Soc., 108 (1986) 1265.
99. S.F. Nelsen, M.F. Teasley, J. Org. Chem. 51 (1986) 3474.
100. S.F. Nelsen, D.L. Kapp, R. Akaba, D.H Evans, J. Am. Chem. Soc., 108 (1986) 6853.
101. S.F. Nelsen, D.L. Kapp, M.F. Teasley, J. Org. Chem., 49 (1984) 579.
102. S.F. Nelsen, M.F. Teasley, J. Org. Chem., 51 (1986) 3221.
103. C.C. Chen, M.A. Fox, J. Comput. Chem., 4 (1983) 488.
104. D.A. Sackett, M.A. Fox, J. Phys. Org. Chem., (1988) in press.
105. G.W. Griffin, G.P. Kirschenheuter, C. Vaz, P.P. Umrigar, D.C. Lakin, S. Christensen, Tetrahedron Lett., 41 (1985) 2069.
106. G.P. Kirschenheuter, G.W. Griffin, J. Chem. Soc., Chem. Commun., (1983) 596.
107. T. Kanno, M. Hisaoko, H. Sakuragi, K. Tokumaru, Bull. Chem. Soc. Jpn., 54 (1981) 2330.
108. P.D. Bartlett, M.E. Landis, J. Am. Chem. Soc., 99 (1977) 3033.
109. N. Shimizu, P.D. Bartlett, J. Am. Chem. Soc., 98 (1976) 4193.

110. K.A. Brown-Wensley, S.L. Mattes, S. Farid, J. Am. Chem. Soc., 100 (1978) 4162.
111. A.P. Schaap, S. Siddiqui, P. Balakrishnan, L. Lopez, S.D. Gagnon, Isr. J. Chem., 23 (1983) 415.
112. S. Futamura, S. Kusunose, H. Ohta, Y. Kamiya, J. Chem. Soc., Chem. Commun., (1982) 1223.
113. A.P. Schaap, L. Lopez, S.D. Gagnon, J. Am. Chem. Soc., 105 (1983) 664.
114. A.P. Schaap, G. Prasad, S. Siddiqui, Tetrahedron Lett., 25 (1984) 3035.
115. T. Miyashi, M. Kamata, T. Mukai, J. Am. Chem. Soc., 108 (1986) 2755.
116. Y. Takahashi, T. Miyashi, T. Mukai, J. Am. Chem. Soc., 105 (1983) 6511.
117. S.C. Shim, J.S. Song, J. Org. Chem., 51 (1986) 2817.
118. S. Farid, S.E. Hartman, T.R. Evans, The Exciplex, M. Gordon, W.R. Ware((Ed.), Academic Press, New York, 1975, p.327.
119. S.L. Mattes, S. Farid, J. Am. Chem. Soc., 108 (1986) 7356.
120. S.L. Mattes, S. Farid, J. Chem. Soc., Chem. Commun., (1980) 126.
121. G.C. Calhoun, G.B. Schuster, J. Am. Chem. Soc., 106 (1984) 6870.
122. R.K. Haynes, M.K.S. Probert, I.D. Wilmot, Austr. J. Chem., 31 (1978) 1737.
123. P.D. Bartlett, Amer. Chem. Soc. Sympos. Ser., 69 (1978) 15.
124. D.H.R. Barton, G. LeClerc, P.D. Magnus, I.D. Menzies, J. Chem. Soc., Perkin I, (1975) 2055.
125. R.K. Haynes, Austr. J. Chem., 31 (1978) 121, 131.
126. T. Tang, H.J. Yue, J.F. Wolf, F. Mares, J. Am. Chem. Soc., 100 (1978) 5248.
127. S.F. Nelsen, M.F. Teasley, J. Am. Chem. Soc., 108 (1986) 5505.
128. D.H.R. Barton, G. LeClerc, P.D. Magnus, I.D. Menzies, J. Chem. Soc., Chem. Commun., (1972) 447; ibid., (1974) 511.
129. S.F. Nelsen, M.F. Teasley, D.C. Kapp, J. Am. Chem. Soc., 108 (1986) 5503.
130. M.F. Arain, R.K. Haynes, S.C. Vonwillwe, T.W. Hambley, J. Am. Chem. Soc., 107 (1985) 4582.
131. S.L. Mattes, S. Farid, J. Chem. Soc., Chem. Commun., (1980) 457.
132. A. Albini, S. Spreti, J. Chem. Soc., Perkin II, (1988) in press.
133. L.W. Reichel, G.W. Griffin, A.J. Muller, P.K. Das, S.N. Ege, Can. J. Chem., 62 (1984) 424.
134. M. Fujihara, Y. Satoh, T. Osa, Nature, 293 (1981) 206.
135. M. Fujihara, Y. Satoh, T. Osa, Bull. Chem. Soc. Jpn., 55 (1982) 666
136. A. Albini, S. Spreti, Z. Naturforsch., 41b (1986) 1286.
137. I. Saito, K. Tamoto, T. Matsuura, Tetrahedron Lett., (1979) 2889.
138. K. Mizuno, N. Ichinose, T. Tamai, Y. Otsuji, Tetrahedron Lett., 26 (1985) 5823.
139. J. Santamaria, P. Gabillet, L. Bokobza, Tetrahedron Lett., (1984) 2139.
140. M.A. Fox, J.N. Younathan, C.C. Chen, J. Org. Chem., 99 (1984) 1969.
141. J. Santamaria, Tetrahedron Lett., (1981) 4511.
142. K. Hatano, K. Usui, Y. Ishida, Bull. Chem. Soc. Jpn., 54 (1981) 413.
143. S. Fukuzumi, S. Kuroda, T. Tanaka, J. Chem. Soc., Chem. Commun. (1987) 120.
144. S. Futamura, K. Yamazaki, K. Ohta, Y. Kamiya, Bull. Chem. Soc. Jpn., 55 (1982) 3852.
145. R.S. Davidson, J.E. Pratt, Tetrahedron Lett., 24 (1983) 547.
146. M.A. Fox, A.A. Wahab, submitted for publication.

Chapter 5.2

Dioxygen-Transition Metal Complexes

H.R. Mäcke and A.F. Williams

1. INTRODUCTION

Virtually all the transition metals form complexes in which a dioxygen ligand is bound to one or two transition metal atoms, and these complexes have been intensively studied for their properties as possible catalysts, as oxygen carriers, and as models for biological systems (refs. 1-2). These complexes are frequently strongly colored, and this property has been used for colorimetric analysis; the coloration is generally due to low-lying charge-transfer transitions (refs. 3-5), and there is therefore a potential for photo-induced electron transfer chemistry. It is our intention in this chapter to review the spectroscopic and photochemical data concerning these complexes and to relate these results to the electronic structures, giving particular attention to processes involving electron transfer to or from dioxygen.

Dioxygen complexes are unusual in two respects: (i) the variety of coordination modes encountered (see Fig. 1) and (ii) the varying formal electronic charge associated with the dioxygen ligand. Thus the dioxygen ligand may be regarded as coordinated neutral O_2, superoxide (O_2^-) or peroxide (O_2^{2-}), and this has sometimes led to controversy over the assignment of the formal oxidation state of the ligand (ref. 6): it is for this reason that we prefer the general name 'dioxygen complex'.

Structure type	Designation	Example
M–O–O	η^1	$[Co(CN)_5O_2]^{3-}$
M<(O–O)	η^2	$[(Ph_3P)_2PtO_2]$
M–O–O–M	$\eta^1:\eta^1$	$[(H_3N)_5CoO_2Co(NH_3)_5]^{5+}$
M<(O–O)>M	$\eta^2:\eta^2$	$[(UO_2Cl_3)_2O_2]^{4-}$
M<(O–O)–M	$\eta^1:\eta^2$	$[(Ph_3P)_2ClRhO_2]_2$

Fig. 1. The classification of structural types of dioxygen complex.

We will discuss separately the results for the three structural types η^1, η^2 and η^1:η^1 which are most commonly encountered. The other structures are rare and no photochemical or spectral data are available. Detailed discussions of the bonding have been given in two recent reviews (refs. 1 & 7), and the electronic spectra have been reviewed (refs. 3-5).

Two general points remain to be made before discussing specific structural types. Firstly, the bonding of the dioxygen ligand involves essentially the π^* orbitals of the dioxygen molecule. It is useful to separate these orbitals into the π^*_h lying in the plane of the M-O-O unit, and the π^*_v which is perpendicular to it. As shown in Figure 2, the π^*_h orbital interacts strongly with the metal d-orbitals in both geometries, whereas the π^*_v orbital has a π overlap with the d-orbitals in η^1 geometry and a very weak δ overlap with the d-orbitals in η^2 geometry. For this reason the π^*_v orbital is generally more or less non-bonding, while the π^*_h orbital is strongly mixed with the d-orbitals, and gives a bonding and an antibonding orbital.

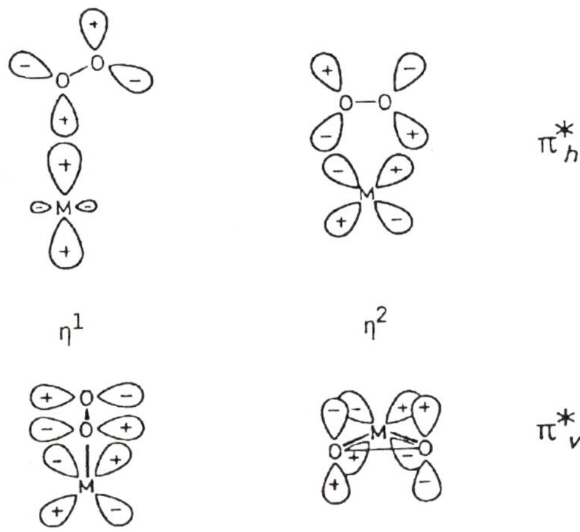

Fig. 2. Interactions between the π^* orbitals of dioxygen and metal d- orbitals in η^1 and η^2 geometries.

Although the general features of the electronic spectra of dioxygen complexes are well established, in some cases confirmation of the assignment of transitions is desirable. A more serious point is that the great majority of dioxygen complexes contain ligands other than dioxygen and these ligands may give rise to other electronic transitions; an obvious example is the field of metalloporphyrin chemistry. The strongest absorptions may have little direct relationship to the metal-dioxygen chromophore, and any photochemical activity is then difficult to relate to the metal-dioxygen bonding.

2. η^1 COMPLEXES

These complexes are generally pseudo-octahedral, with the composition $L_5M(O_2)$, where M is either iron or cobalt. Figure 3 shows a simple molecular orbital diagram for such a complex.In oxyhemoglobin and related compounds, the π^*_h orbital is doubly occupied and acts as a donor to the iron atom, and the π^*_v orbital is empty and can act as a π-acceptor from the d_π orbital (one of the t_{2g}

subset). This arrangement may be regarded formally as singlet dioxygen complexed to low spin Fe(II). The η^1 cobalt complexes, which have been widely studied as oxygen carriers (ref. 8), contain one more electron, and the π^*_v orbital is singly occupied and may be studied by E.P.R.; these complexes are regarded formally as complexes of superoxide with Co(III). The final well-studied η^1 complex is oxyhemerythrin (ref. 9), which is thought to contain an η^1 peroxide ligand bound to high spin Fe(III).

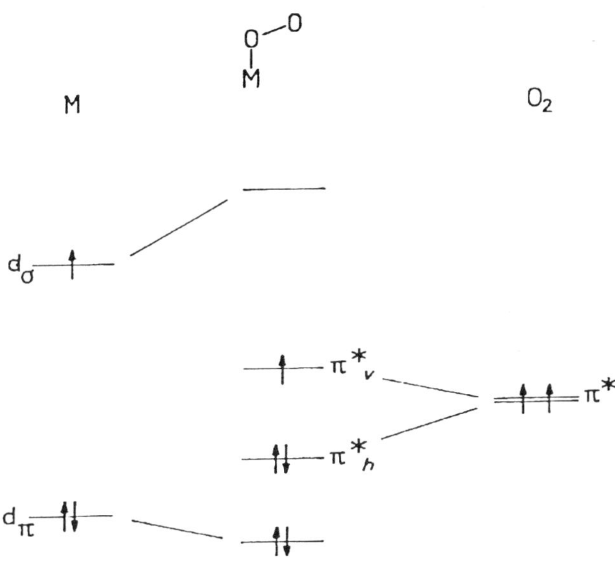

Fig. 3. Molecular orbital diagram for an η^1 cobalt dioxygen complex. In iron complexes the π^*_v orbital is unoccupied.

The electronic spectra of oxyhemoglobin and oxymyoglobin have been studied in detail (refs. 10 - 12), and show a multitude of transitions. Thus, for oxyhemoglobin, Eaton *et al.* (ref. 10) found seven transitions other than porphyrin π-π^* transitions. Using Extended Hückel calculations, they made the following assignments: Bands I to IV (7700, 8700, 10200, and 12700 cm^{-1}) involve transitions to the lowest unoccupied M.O. composed of iron d_{xz} and dioxygen π^*_v orbitals; Bands V and VI (18400 and 22000 cm^{-1}) are d-d transitions; Band VII (31000 cm^{-1}) is a strongly electric dipole allowed transition associated with a transition from a dioxygen π bonding orbital to the lowest unoccupied M.O. mentioned above. MS X-α and Pariser-Parr-Pople calculations (ref. 13) were in general agreement with these assignments.

Photochemical studies of oxyhemoglobin (HbO_2) and oxymyoglobin (MbO_2) have concentrated on the kinetics of recombination of photodetached dioxygen, and on the nature of the hemoglobin or myoglobin formed after photodissociation. Steady state irradiation of HbO_2 and MbO_2 leads to dissociation of molecular oxygen, and it has been known for some time that the quantum yields for this process are rather low (HbO_2 $\phi = 0.03$; MbO_2 $\phi = 0.05$) compared with the photodissociation of CO from carboxyhemoglobin ($\phi = 0.5$) and carboxymyoglobin ($\phi = 1$) (ref. 14). Although recent studies using fast, time-resolved spectroscopy (refs. 15-19) have used a variety of experimental conditions and time windows, there is a general agreement on the following points:

(i) The photodissociation of dioxygen or of carbon monoxide leads to intermediates with identical spectra which are, however, broader and red-shifted in comparison with ground state Mb and Hb.

(ii) The recombination of dioxygen may take place by two mechanisms: a rapid, geminate recombination involving dioxygen that has not diffused from the active site; and a slower recombination involving the diffusion of dioxygen to the binding site. The rapid recombination appears to be much faster for dioxygen than for CO, and half lives of the order of 100 picoseconds have been reported.

Photodissociation has been observed upon irradiation with 307 nm, 353 nm, and 530 nm light, but the results obtained do show a dependence on irradiation wavelength. In general, a d-d excited state is held to be responsible for the photoactivity; in HbO_2 and MbO_2 the lowest lying excited states are iron to dioxygen charge transfer states (*vide supra*) which lie below the d-d excited states, and might thereby offer a means of excited state relaxation which is absent in the carbon monoxide complexes. Recent work on hemoglobin, myoglobin, and synthetic heme complexes (ref. 19) suggests that geminate recombination may occur *via* two routes involving different intermediates, and has postulated the existence of a number of intermediates. In conclusion, we may say that the lower quantum yields for photodissociation of dioxygen than for carbon monoxide are at least partially due to the more rapid recombination of dioxygen; the lower quantum efficiency of the initial photodissociation step has yet to be definitely established.

There has been one report that irradiation of oxyhemoglobin with light of wavelength below 300 nm leads to elimination of superoxide anion with formation of methemoglobin (ref. 20), the expected result of metal to ligand charge transfer. A recent investigation of the action of white light on hematoporphyrin derivative (ref. 21) concluded that the cytotoxic species produced on irradiation is not superoxide anion, but may be singlet oxygen, $^1O_2$. Photolysis of oxygenated cytochrome o leads to dissociation of dioxygen (ref. 22). Gamma-irradiation of oxyhemoglobin or oxymyoglobin leads to electron capture and the formation of a $Fe(II)$-O_2^- complex (ref. 23). No photochemical data have been reported for hemerythrin, but resonance Raman spectra have shown the absorption band at 500 nm to be associated with peroxide to iron(III) charge transfer (ref. 9); a similar band is observed at 454 nm in the complex formed by hydrogen peroxide with $[Fe(III)(EDTA)]^-$ which is thought to be isostructural with hemerythrin (ref. 24).

The cobalt η^1 dioxygen complexes have been studied in some detail (ref. 8), and their spectra discussed (refs. 3-5). Lever (ref. 5) assigns a weak band ($\varepsilon = 100\ M^{-1}cm^{-1}$) near 500 nm to charge transfer from the metal t_{2g} orbitals to the singly occupied ligand π_v^* orbital , and a strong band ($\varepsilon = 5000\ M^{-1} cm^{-1}$) near 300 nm to transfer from the doubly occupied π_h^* orbital to metal d^*_σ orbitals. Resonance Raman spectra show enhancement of the O-O vibrations on irradiation in the 500 nm region (refs. 25,26), although Nakamoto (ref. 26) attributes this to ligand to metal charge transfer. This absorption appears to be photochemically active, since exchange of coordinated $^{18}O_2$ with free $^{16}O_2$ is observed during the irradiation, and irradiation of the complex $[Co(TPP)O_2]$ at 532 nm leads to dissociation of

O_2 (ref. 27).

3. η^2 COMPLEXES

η^2 complexes are found with virtually all transition metals, and the interaction of dioxygen with the metal atom is shown in Fig. 4. One metal d-orbital interacts strongly with the π^*_h orbital, but the π^*_v orbital interacts only very weakly with the metal d-orbitals and is essentially non-bonding. In all known compounds, both π^* orbitals are occupied, and the ligand is generally regarded as peroxide O_2^{2-}, in agreement with the observed bond lengths and O - O stretching frequencies, although in many cases the compound is prepared from molecular oxygen or superoxide rather than from H_2O_2. The possible electronic transitions are: (i) π^*_h to d^*, at high energy and rarely observed, and (ii) π^*_v to d^* which is frequently observed as a weak ($\varepsilon = 10^2$-10^3 M^{-1} cm^{-1}) absorption in the visible or U.V. region. Lever and Gray (refs. 3-5) argue that the low intensity arises from the symmetry of the π^*_v orbitals. Resonance Raman spectra of the complex $CrO(O_2)_2.H_2O$ (ref. 28) show enhancement of the O - O stretch on irradiation of the charge transfer band; for the complex formed by Ti(IV) in acid solution, which is also thought to contain η^2 dioxygen, the pre-resonance Raman spectrum shows enhancement of the Ti - O stretch but not of the O - O stretch (ref. 29).

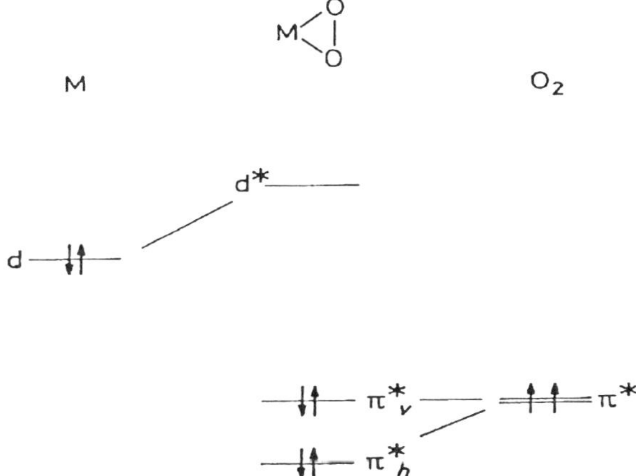

Fig. 4. Molecular orbital diagram for an η^2 dioxygen complex.

The photochemical activity of these complexes is quite well documented, and essentially involves the loss of molecular oxygen. The photolysis of complexes Ti(porph)(O_2) (porph = porphyrin) with a high pressure mercury lamp (ref.30) or at 404 nm (ref. 31) leads to formation of the complex Ti(porph)O. The proposed mechanism is

$$\text{Ti(porph)}(O_2) \xrightarrow{h\nu} \text{Ti(porph)} + O_2 \qquad (1)$$

$$\text{Ti(porph)} + \text{Ti(porph)}(O_2) \longrightarrow 2\ \text{Ti(porph)O} \qquad (2)$$

The photo-ejected dioxygen may be in the form of $^1O_2$ since it has been shown to oxidise cyclohexene, and produces a characteristic radical with the singlet oxygen quencher 2,2,6,6-tetramethylpiperidine. The Mo(IV) complex Mo(porph)(O_2) gives a similar reaction on photolysis, and this makes an interesting contrast with the bis-dioxygen complex Mo(porph)$(O_2)_2$which yields the *cis* dioxo complex Mo(porph)O_2 upon irradiation (ref. 32). The Ti(IV) and Mo(IV) complexes show ligand to metal charge transfer, the reduced metal species reacting in a second step with a second dioxygen complex, whereas the bis(dioxygen) complex does not give the stable complex Mo(porph)(O_2) expected from ligand-to-metal transfer, but shows ligand to ligand charge transfer, one peroxo group being oxidized, and the other reduced.

A series of η^2 complexes of iridium formed by oxidative addition of dioxygen to square planar iridium(I) complexes have been investigated (ref. 33), and found to give O_2 and the iridium(I) complex on irradiation at 366 nm, although a nitrogen purge is required for the complexes which bind dioxygen irreversibly. A test for the possible formation of singlet oxygen with tetramethylethylene proved negative. Geoffroy, Hammond, and Gray conclude that the transition responsible for the photoactivity is an iridium(III)-to-phosphine charge transfer, and the dioxygen is then lost from the iridium(IV) like excited state.

Somewhat similar behavior is observed for the complex $(Ph_3P)_2Pt(O_2)$ which is stable to irradiation at wavelengths longer than 300 nm in the presence of air, but which decomposes slowly under a nitrogen purge (ref. 34). The products of the reaction were identified as $(Ph_3P)_2Pt$ (which gives a characteristic emission spectrum at low temperatures, and may react further with the solvent) and singlet oxygen (shown by reaction with 1,3 diphenylisobenzylfuran and tetramethylpiperidine). A calculation has suggested that the electronic transition responsible for this reaction has π^*_v to metal charge transfer character (ref. 35). Photolysis of solutions of the square planar d^8 complex $PdL_2(O_2)$ (L = tertiary phosphine) in dichloromethane or dichloroethane using 366 or 405 nm light gives PdL_2Cl_2 provided that there is an excess of phosphine present; the phosphine is transformed to phosphine oxide. In this case, however, a test for singlet oxygen (with 1-methylcyclohexene) proved negative. The complex PdL_2Cl_2 is thought to be formed by oxidative addition of the solvent to a PdL_2 intermediate. The quantum yields for irradiation at 405 nm were quite high (ϕ = 0.5) but no reaction was observed for wavelengths greater than 500 nm (ref. 36).

There are one or two other references to reactions of η^2 complexes: complexes of the type $MoO(O_2)_2LL'$ have been reported to be photosensitive in methanol solution (ref. 37), and complexes of the type $Cr(O_2)_2$(diethylenetriamine) have been found to hydroxylate organic substrates more efficiently in the presence of light (ref. 38).

In summary, η^2 dioxygen complexes do exhibit photoinduced electron transfer reactions, although the transition responsible is not always the O_2-to-metal charge transfer. There is a fair amount of evidence to suggest that the molecular oxygen liberated may be in the singlet ($^1\Delta_g$) state, although the variety of different methods used to identify the singlet oxygen leads to some ambiguity. It is perhaps the d^0 systems, which react readily with hydrogen peroxide, which are the most promising for the generation of singlet oxygen or the photogeneration of molecular oxygen from hydrogen peroxide.

4. η^1:η^1 COMPLEXES

Although stable η^1:η^1 complexes are known for Pt(II) and Rh(III), and transient binuclear iron complexes have been reported, the vast majority of these complexes contain two cobalt atoms and have the composition $L_5Co(O_2)CoL_5$ or $L_4Co(O_2;X)CoL_4$ where X is a second bridging group, either OH^-

or NH_2^-. The formal oxidation state of the cobalt atoms in these complexes is always +III, but the dioxygen ligand may have one of two oxidation states, either peroxo- (O_2^{2-}) or superoxo- (O_2^-); E.P.R. spectra have shown clearly that in the superoxo-complexes the unpaired electron is localized on the dioxygen ligand. The chemistry of these complexes has been reviewed recently (ref. 39).

The cobalt atoms in these complexes have pseudo-octahedral coordination, and we may imagine the bonding as involving the interaction of the d_{z^2} orbitals of each CoL_5 fragment with the π^* orbitals of dioxygen: this is shown in Figure 5. In the planar case ($\theta = 0$ or 180°), one of the π^* orbitals (π^*_v) has π symmetry with respect to the plane and interacts weakly with the metal ions, whereas the other (π^*_h) lies in the plane and interacts strongly with the metals; in the non-planar case, this distinction no longer exists, and the energy difference between the two π^* orbitals is less important. The superoxo-complexes have structures which are planar or very nearly so, but a wide variety of dihedral angles is found for peroxo- complexes, suggesting that the rotational barrier about the O-O bond is not very high.

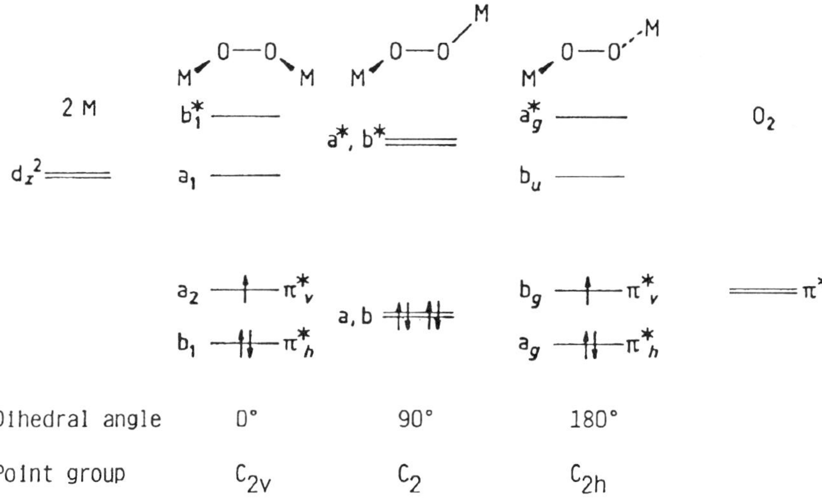

Fig. 5. The interaction of the metal d_{z^2} orbitals with the π^* orbitals in a η^1:η^1 complex as a function of the dihedral angle θ. The orbital occupations correspond to superoxo (θ = 0°, 180°) and peroxo (θ = 90°) complexes respectively.

4.1. Electronic spectra

The absorption spectra of these compounds have been studied in some detail and have been reviewed (refs. 3-5); although the general features are well established, further work is necessary, especially if the photochemical activity of these compounds is to be understood.

4.1.1. η^1:η^1 superoxo complexes

It is convenient to begin with a discussion of some simple μ-superoxo species for which a detailed study of the solution spectra has been made (ref. 40) and has recently been supported by a polarised single crystal investigation of $[(H_3N)_5Co(O_2)Co(NH_3)_5](NO_3)_2Cl_3.2H_2O$ which shows planar C_{2h} symmetry (Fig. 5; ref. 41). The solution spectrum of $[(H_3N)_5Co(O_2)Co(NH_3)_5]^{5+}$ is shown below, and the bands of interest indicated by Roman numerals. The discussion follows the assignments of

Miskowski *et al.* (ref. 40) unless otherwise stated.

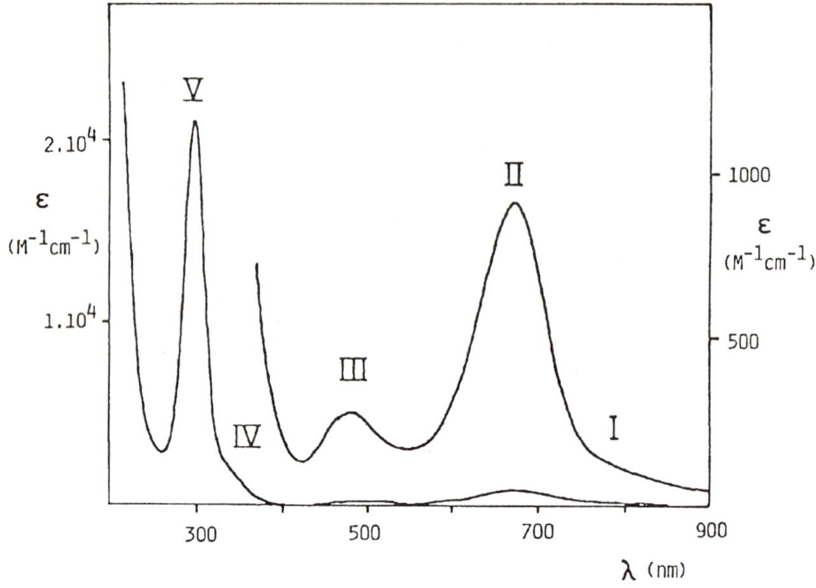

Fig. 6. The spectrum of $[(H_3N)_5Co(O_2)Co(NH_3)_5]^{5+}$ in 0.1M HCl.

Band I (800 nm, $\varepsilon = 110\ M^{-1}cm^{-1}$) is a weak π^*_h-to-π^*_v transition localized on the dioxygen ligand. Band II (671.5 nm, $\varepsilon = 927\ M^{-1}cm^{-1}$) is a charge transfer transition from the fully occupied cobalt t_{2g} orbitals to the singly occupied π^*_v orbital; this assignment has been confirmed by resonance Raman spectroscopy (ref. 42) and the polarization and vibrational structure of the single crystal spectrum. Band III (480 nm, $\varepsilon = 309\ M^{-1}cm^{-1}$) is a d-d transition assigned as the $^1A_{1g}$ - $^1T_{1g}$ transition of an octahedral complex. Band IV (345 nm, $= 3450\ M^{-1}cm^{-1}$) has been assigned to a second d-d transition ($^1A_{1g}$ - $^1T_{2g}$), but Lever (ref. 5) suggested an alternative assignment to a π^*_v-to-d_σ charge transfer. It is generally accepted that such a transition would be weak, and lie at lower energy than the π^*_h-to-d_σ charge transfer; it has further been suggested that it might give rise to an observable circular dichroism feature, and this has been observed in certain cases (refs. 3 - 5). Band V (302 nm, $\varepsilon = 20160\ M^{-1}cm^{-1}$) is assigned to the π^*_h-to-d_σ charge transfer. Finally a strong absorption at 225 nm ($\varepsilon = 21000\ M^{-1}cm^{-1}$) is associated with an interligand π - π^*_v transition; this transition has also been seen in a mononuclear η^1 cobalt dioxygen complex (ref. 43)

The spectrum of the decacyano complex $[(NC)_5Co(O_2)Co(CN)_5]^{5-}$ is in accord with these assignments: the MLCT transition (Band II) and the d-d bands (Band III) are blue shifted. The LMCT transition is at 310.5 nm ($\varepsilon = 17500\ M^{-1}cm^{-1}$), but a second transition is seen at 268 nm ($\varepsilon = 5230\ M^{-1}cm^{-1}$); Miskowski *et al* (ref. 40) suggest that this arises from the asymmetry of the ligand field in the $(NC)_5Co(O_2)$ chromophore, and the consequent large splitting of the $d_{x^2-y^2}$ and d_{z^2} orbitals: the 268 nm band is thus due to π^* - $d_{x^2-y^2}$ charge transfer. Resonance Raman spectra have been used to confirm the assignment of the MLCT band (refs. 42,44).

4.1.2. η^1:η^1 peroxo complexes

The spectra of the μ-peroxo complexes should in principle be simpler since bands I,II, and VI of the μ-superoxo spectra are absent as a result of the double occupation of the π^*_v orbital. In practice, the interpretation of the dioxygen-to-metal charge transfer spectra has proved to be difficult and no wholly satisfactory explanation has yet been given. In their review of η^1:η^1 complexes, Fallab and Mitchell (ref. 39) have classified the charge transfer spectra into three groups:

(i) Singly bridged complexes showing one absorption band near 300 nm (ε = 11000 $M^{-1}cm^{-1}$) sometimes with a much weaker band near 400 nm. X-ray structure determination shows that these complexes have *trans* planar geometry. An example is given in Figure 7.

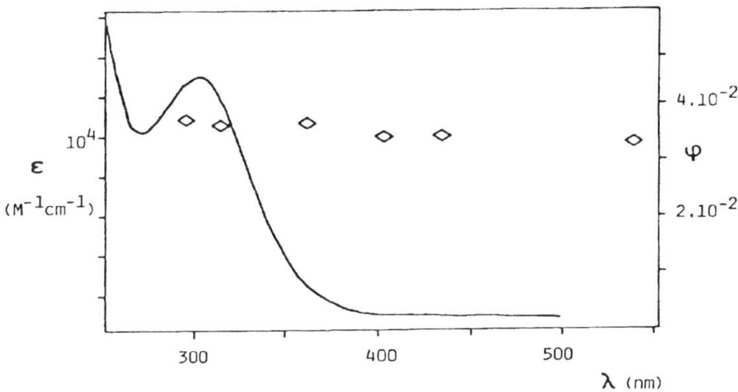

Fig. 7. The spectrum of $[(papd)CoO_2Co(papd)]^{4+}$. The lozenges indicate the quantum yield for loss of dioxygen as a function of excitation wavelength (see section 4.2.1).

(ii) Complexes with an OH or NH_2 group forming a second bridge between the cobalt atoms: these show two bands (ε = 6000 $M^{-1}cm^{-1}$) of which the second lies below 300 nm. X-ray structures show these complexes to be strongly distorted from planarity.

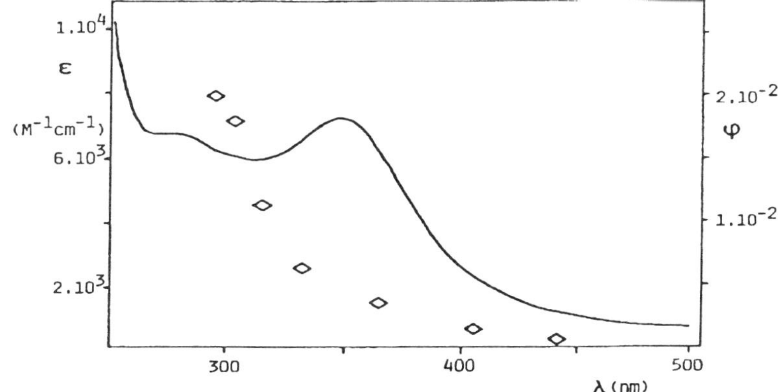

Fig. 8. The spectrum of $[(tren)Co(OH,O_2)Co(tren)]^{3+}$. The lozenges indicate the quantum yield for loss of dioxygen as a function of excitation wavelength (see section 4.2.1).

(iii) Complexes showing two bands of roughly equal intensity ($\varepsilon = 5000\ M^{-1}cm^{-1}$) between 300 and 400 nm. In these complexes the ligand field around the cobalt atom deviates strongly from octahedral symmetry, and splitting of the e^*_g levels is likely. Examples of this group include $[(tren)(CN)Co(O_2)Co(CN)(tren)]^{2+}$ whose spectrum is shown in Figure 9, and complexes with amino acids such as $[(his)_2Co(O_2)Co(his)_2]$.

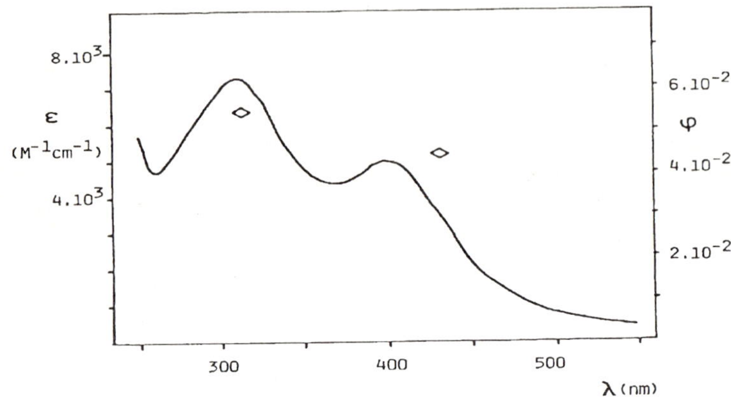

Fig. 9. The spectrum of $[(tren)(CN)Co(O_2)Co(CN)(tren)]^{2+}$. The lozenges indicate the quantum yield for loss of dioxygen as a function of excitation wavelength (see section 4.2.1).

Lever and Gray have discussed this behavior in terms of the dihedral angle θ of the Co-O-O-Co unit (Fig.5, refs 3-5). For planar systems ($\theta = 0°$ or $180°$) the π^*_h - d_σ transition is expected to be intense, and the π^*_v - d_σ transition to be weak; the first transition is therefore associated with the strong band at 300 nm, and the second with the weaker feature near 400 nm. In non-planar systems, both π^* orbitals of dioxygen may interact with d_σ and the 'in plane' or 'out of plane' distinction is lost; two π^* - d_σ transitions are therefore expected, corresponding to transfer from orbitals a and b in the $\theta = 90°$ case shown in Figure 5. Lever and Gray's model explains well the class (i) spectra , and the observation of two bands in the doubly-bridged systems which are known to be non-planar, but two problems remain. Firstly, the X-ray structure of the class (iii) complex $[(tren)(CN)Co(O_2)Co(CN)(tren)]^{2+}$ shows it to be very nearly planar (ref. 45), and the spectra in solution and in a KBr pellet show two charge transfer bands of approximately equal intensity (Fig.9); a possible explanation of this would be the splitting of the e_g orbitals in the asymmetric ligand field as proposed for $[(NC)_5Co(O_2)Co(CN)_5]^{5-}$ above. Secondly, according to Lever and Gray's model, the distortion from planarity results in the π^*_h orbital losing, and the π^*_v orbital gaining, bonding character. The π^*_h - d_σ, should therefore be red shifted; in fact the charge transfer band at higher energy in the doubly bridged complexes is blue shifted in comparison with the mononuclear complexes. The possibility that this band may in fact involve charge transfer from the second bridging group cannot be discounted.

Resonance Raman spectroscopy has been used to study the nature of the transitions in amine, amino-acid, and cyanide complexes (refs. 44,46,47). Enhancement of the O-O stretching mode is observed as the exciting wavelength is reduced and approaches the dioxygen-to-cobalt charge transitions in the region 320 - 360 nm. In complexes where the cobalt is complexed by Schiff bases or dimethylglyoxime, the binding of dioxygen is reversible, and the O-O stretching frequencies are much higher, implying much less transfer of charge from cobalt to dioxygen in the ground state. Resonance Raman spectra of these complexes (refs. 26, 48-51) show the dioxygen-to-metal charge transfer band to lie much lower, typically around 500 nm, and the measurement of the resonance Raman spectrum is frequently accompanied by the decomposition of the sample or exchange with free molecular O_2 (ref. 51). A similar relation between the dioxygen-to-cobalt charge transfer frequency and the stability of the μ-peroxo complex has been observed for polydentate amine complexes (ref. 52).

In closing this discussion of the spectra of μ-peroxo complexes, it is interesting to note the results of Solomon and his collaborators on the oxygen transporting protein hemocyanin (ref. 53). The structure of the oxygenated form of hemocyanin is thought to involve a peroxide group bridging two Cu(II) ions with a dihedral angle close to zero. Extensive spectroscopic, resonance Raman, and circular dichroism studies have identified three transitions associated with dioxygen-to-copper charge transfer: (i) a weak band at 580 nm ($\varepsilon = 1000\ M^{-1}cm^{-1}$) assigned to the transition $a_2 - b_1$ in Fig. 5 (i.e. $\pi^*_v - d_\sigma$); (ii) a strong CD feature at 480 nm assigned to the magnetic dipole allowed transition $a_2 - a_1$ (also $\pi^*_v - d_\sigma$); (iii) a strong absorption at 345 nm ($\varepsilon = 20000\ M^{-1}cm^{-1}$) assigned to a $b_1 - b_1^*$ ($\pi^*_h - d_\sigma$) transition. A fourth transition ($b_1 - a_1$) is thought to be masked by UV absorption by the protein. The arguments advanced by Solomon *et al.* are powerful and imply that up to four dioxygen-to-cobalt charge transfer transitions could be observed for binuclear cobalt peroxo complexes, even if the π^* to $d_{x^2-y^2}$ transitions are neglected. Further experimental work is clearly necessary in this subject.

4.2. Photochemistry

4.2.1. η^1:η^1 peroxo complexes

Very little work has been published on the photochemistry of these complexes, but unpublished work by one of the authors (H.R.M.) suggests the existence of a very varied photochemistry (ref. 54); two preliminary communications of this work have appeared (ref. 55). Steady state irradiation of thermodynamically stable μ-peroxo complexes produces little change: there is slow formation of mononuclear Co(III) complexes with a quantum yield in the range $10^{-4} - 10^{-5} ME^{-1}$. If, however, the experiment is repeated using a μ-peroxo complex labelled with ^{57}Co in the presence of unlabelled Co^{2+} an exchange of cobalt with a quantum yield of $5.10^{-2} ME^{-1}$ is observed:

$$[(papd)^{57}CoO_2{}^{57}Co(papd)]^{4+} + 2\ Co^{2+} \xrightarrow{h\nu} [(papd)CoO_2Co(papd)]^{4+} + 2\ ^{57}Co^{2+} \quad (3)$$

It will be recalled that photoinitiated exchange of O_2 was observed during the measurement of the resonance Raman spectrum of [(salen)CoO_2Co(salen)] (ref. 51). In the presence of EDTA as a scavenger for Co^{2+}, the quantum yield for photodecomposition rises very considerably, and release of dioxygen is observed. The results are consistent with the formation of a radical pair as the primary photoproduct, followed by recombination or dissociation with subsequent scavenging of Co^{2+}:

$$Co^{III}O_2^{2-}Co^{III} \xrightarrow{h\nu} Co^{III}O_2^{-}Co^{II} \quad (4)$$

$$Co^{III}O_2^{-}Co^{II} \longrightarrow Co^{III}O_2^{2-}Co^{III} \quad (5)$$

$$Co^{III}O_2^{-}Co^{II} \longrightarrow Co^{III}O_2^{-} + Co^{II} \quad (6)$$

$$Co^{III}O_2^{-} \longrightarrow Co^{II} + O_2 \quad (7)$$

Support for this mechanism comes from the observation that irradiation of the complex $[(papd)CoO_2Co(papd)]^{4+}$ in an oxygen saturated solution gives $[(papd)CoO_2]^{2+}$, identified by its UV and EPR spectra and its decomposition kinetics. This work has been extended to study the dependence of the photochemical behavior on the excitation wavelength and on the presence of a second bridging group. The results are summarised in Table 1:

TABLE 1.
Quantum yields ($10^{-3}ME^{-1}$) for loss of dioxygen from μ-peroxo compounds.

Compound	Excitation Wavelength (nm)							
	297	302	313	334	366	405	435	546
$[(papd)CoO_2Co(papd)]^{4+}$			38		37	33	34	32
$[(tren)(CN)CoO_2Co(CN)(tren)]^{2+}$			55				40	
$[(tren)Co(OH,O_2)Co(tren)]^{3+}$	20	18	11	6.2	3.3	1.5	0.6	
$[(en)_2Co(OH,O_2)Co(en)_2]^{3+}$	19		14	7.7	4.8			
"					2.2[a]			
$[(dmtrien)Co(OH,O_2)Co(dmtrien)]^{3+}$	35		25					

[a] value from ref. 56 at 5 °C; all other values from ref. 54 at 25 °C

The monobridged complex $[(papd)CoO_2Co(papd)]^{4+}$ shows a clean photochemical decomposition in presence of EDTA. The quantum yield is essentially independent of the exciting wavelength below 546 nm (see Fig. 7). This suggests that the excited state responsible for photochemical activity is not that associated with the absorption at 300 nm (π^*_h - d_σ). Possible alternatives are a ligand field excited state, or the π^*_v - d_σ excited state which Solomon's work on copper complexes suggests to lie well below the π^*_h - d_σ excited state (ref. 53), and which, on the grounds of orbital symmetry, would be more likely to lead to a stable $Co^{III}O_2^-$ species.

The other monobridged species studied is $[(tren)(CN)CoO_2Co(CN)(tren)]^{2+}$ which gives a type (iii) spectrum (Fig. 9). The quantum yields are comparable to $[(papd)CoO_2Co(papd)]^{4+}$ but the decomposition is not so clean, and Co(III) complexes are also formed.

The dibridged species show clean photochemically induced elimination of dioxygen, with a strong dependence of quantum yield on excitation wavelength. At low excitation energy the quantum yields are much smaller than for the monobridged species, but when the excitation reaches the higher energy charge transfer band, the quantum yields are comparable with the monobridged species. A full explanation of these results must await a satisfactory description of the spectra of these complexes, but it would seem that the second bridging group hinders the photodecomposition at low energies, presumably by hindering reaction (6). The increased quantum yield at low excitation wavelengths may arise from a weakening of the second bridge in the excited state.

A published study (ref. 56) of the photolysis at 366 nm of $[(en)_2Co(OH,O_2)Co(en)_2]^{3+}$ using EDTA to scavenge Co^{2+} or the proton to scavenge ethylenediamine gives quantum yields in reasonable agreement with the results presented here. The quantum yield on irradiation at 515 nm was too small to be measured. The authors also report an accelerated decomposition of $[(dien)(en)CoO_2Co(en)(dien)]^{4+}$ upon irradiation, but that $[(en)_2Co(NH_2,O_2)Co(en)_2]^{3+}$ shows much less photochemical sensitivity.

Finally we may note that the postulated primary photoproduct $Co^{III}O_2^-Co^{II}$ has superoxide character, and might therefore be expected to show some oxidizing character. Preliminary results obtained by one of us (H.R.M.) show that reducing agents such as sulphite, iodide and Fe^{2+}

do indeed act as efficient scavengers of the excited state.

4.2.2. η^1:η^1 superoxo-complexes

The photochemical activity of these complexes has long been recognized, and several papers have appeared. A summary of results is given in Table 2:

TABLE 2.
Quantum yields (ME^{-1}) for photolysis of μ-superoxo complexes.

Compound	Excitation Wavelength (nm)							pH	ref.
	254	313	320	350	365	405	480		
$[(H_3N)_5Co(O_2)Co(NH_3)_5]^{5+}$	0.3	0.85		0.65				0.7	57
"			0.30	0.24		0.1	<0.01	1	58
$[(H_3N)_4Co(NH_2,O_2)Co(NH_3)_4]^{4+}$			0.49	0.40		0.1		1	58
$[(en)_2Co(NH_2,O_2)Co(en)_2]^{4+}$			.065	.040		.005		1	59
$[(tetren)Co(O_2)Co(tetren)]^{5+}$			<.005					1	59
$[(NC)_5Co(O_2)Co(CN)_5]^{5-}$			0.2	0.2		0.05	0.01	1	58
"		.094		.06				7	40
"	0.2	0.07						>4	60
"	0.2	0.12						<2	60
$[(NC)_4Co(NH_2,O_2)Co(CN)_4]^{4-}$			<.005					1	58

The first paper to appear (ref. 57) studied the decammine system, and showed the stoichiometry of photolysis to be

$$[(H_3N)_5CoOOCo(NH_3)_5]^{5+} \xrightarrow{h\nu} [Co(NH_3)_5(OH_2)]^{3+} + Co^{2+} + O_2 + 5\ NH_3 \qquad (8)$$

The photochemical activity was associated with the 302 nm absorption assigned earlier to a π^*_h - d_σ charge transfer. These results were confirmed by the second paper to appear (ref. 58) which also studied the doubly bridged complex $[(H_3N)_4Co(NH_2,O_2)Co(NH_3)_4]^{4+}$. The products of photolysis of this compound were identical to those of reaction (8), showing the bridging amide ion to be held on the cobalt(III) center. In the presence of chloride, $[Co(NH_3)_5Cl]^{2+}$ is formed, and it was deduced from the relative amounts of chloro and aquocobaltpentammines formed that the species $[Co(NH_3)_5]^{3+}$ is a probable intermediate. A test for singlet oxygen with a 1,3 cyclohexadiene/benzene mixture proved negative.

These results are consistent with a mechanism of the type

$$Co^{III}O_2^-Co^{III} \xrightarrow{h\nu} Co^{III}O_2Co^{II} \qquad (9)$$

$$Co^{III}O_2Co^{II} \longrightarrow Co^{III} + O_2 + Co^{II} \qquad (10)$$

Although the photolysis of both peroxo and superoxo species is thought to involve charge transfer excited states, there are a number of significant differences. The quantum yields of superoxo complexes are often an order of magnitude greater than those found for peroxo complexes, but become vanishingly small at higher wavelengths where the μ-peroxo complex $[(papd)CoO_2Co(papd)]^{4+}$ is still active. For the superoxo complexes, the excited state will contain dioxygen in a $(\pi^*_h)^1(\pi^*_v)^1$ configuration favorable for dissociation to triplet molecular oxygen. For the peroxo complexes, the excited state resulting from π^*_h - d_σ charge transfer has the configuration $(\pi^*_h)^1(\pi^*_v)^2$ which is not that required for the formation of a stable $Co^{III}O_2^-$ complex postulated as an intermediate $((\pi^*_h)^2(\pi^*_v)^1)$. Lower quantum efficiency for the peroxo complexes may therefore be understood, and the lower energy of the π^*_v - d_σ charge transfer state would explain why photoactivity is still observed at longer wavelengths.

The effect of the other ligands present is also different in peroxo and superoxo photochemistry: the presence of a second bridge in the peroxo complexes drastically reduces the quantum yield at long wavelengths, but has virtually no effect on the quantum yield of the superoxo complexes. On the other hand, the quantum yield for the superoxo complexes drops very markedly if the ammines are substituted by ethylenediamine or by the pentadentante ligand tetren (ref. 59) whereas no such effect is seen with the peroxo complexes; this effect should however be verified by study of photolysis of superoxo complexes in the presence of a scavenger for cobalt(II) such as EDTA, since the only reasonable explanation for the effect is the greater thermodynamic and kinetic stability of the Co(II) complexes with the chelating ligands.

The decacyano superoxide complex $[(NC)_5Co(O_2)Co(CN)_5]^{5-}$ shows a different stoichiometry for photolysis:

$$2\,[(NC)_5CoOOCo(CN)_5]^{5-} + 2\,H^+ + 4\,H_2O \xrightarrow{h\nu} 4\,[Co(CN)_5(H_2O)]^{2-} + H_2O_2 + O_2 \qquad (11)$$

The first results (refs. 58,40) showed a dependence of quantum yield on pH, and a recent study reports the influence of excitation wavelength and pH (ref. 60). As is usual for superoxo complexes, no photochemical activity is observed for the MLCT transition at 480 nm. For the d-d transition at 373 nm, a quantum yield of 0.05 ME^{-1} is found. At 300 nm (assigned to a π^*_h - d_{z^2} transition) the quantum yield is 0.07 ME^{-1} above pH 4, but rises to 0.12 at pH < 2. The 268 nm band, which was assigned to a π^*_h - $d_{x^2-y^2}$ transition, shows a quantum yield of 0.2 ME^{-1}, independent of pH.

If $[(NC)_5Co(O_2)Co(CN)_5]^{5-}$ were to decompose in a similar way to the ammines, the expected reaction would be:

$$[(NC)_5Co(O_2)Co(CN)_5]^{5-} \xrightarrow{h\nu} [(NC)_5Co(OH_2)]^{2-} + O_2 + [Co(CN)_5]^{3-} \qquad (12)$$

In this case however the cobalt(II) complex produced is a low spin, highly reducing complex, and may not show the same lability as the Co(II) ammines; $[Co(CN)_5]^{3-}$ reacts rapidly with O_2 to give $[(NC)_5Co(O_2)]^{3-}$ which decomposes to $[(NC)_5Co(OH_2)]^{2-}$, O_2, and H_2O_2 in acid solution (ref. 61). The stoichiometry of reaction (11) may therefore arise from a secondary thermal reaction following the photochemical reaction (12). Hoshino et al. (ref. 60) observe the reduction of cyctochrome(III) c during the photolysis of $[(NC)_5Co(O_2)Co(CN)_5]^{5-}$, and attribute this to the formation of superoxide anion, but this could equally be due to the presence of $[Co(CN)_5]^{3-}$. Finally the lower lability of the cobalt(II) cyanide complex could equally explain the apparent lack of activity of $[(NC)_4Co(NH_2,O_2)Co(CN)_4]^{4-}$ (ref. 58), in which the presence of a second bridging group confers photochemical stability on the complex.

CONCLUSIONS

The most frequently observed photochemical reaction of dioxygen complexes is the loss of molecular oxygen. Since the dioxygen ligand in a dioxygen complex can almost invariably be regarded as carrying negative charge, this may be regarded as a photoinduced electron transfer, and is indeed frequently produced by excitation of a dioxygen-to-metal charge transfer band; however, there are exceptions, as seen for Ir(III), where the transition responsible is thought to be an iridium-to-phosphine charge transfer, and in oxyhemoglobin, where d-d excitation is apparently involved. The product of the reaction is a coordinatively unsaturated complex, which in many cases is highly reducing (for example, the Ti(II)(porph) complex), and may therefore be a useful reactive intermediate. Although the results available are still somewhat contradictory, there is a fair amount of evidence that the dioxygen can in certain cases be liberated in the form of excited $^1\Delta_g$ singlet dioxygen, although the results seem to be rather dependent on the method of detection used. A general feature of the photochemistry of these complexes is the rapidity of recombination of the liberated dioxygen: to observe photochemical activity, inert gas purges or scavengers for the initial products are frequently required. It seems probable that future research will develop in this direction, and that the photochemical decomposition of dioxygen complexes may develop as a useful source of reactive intermediates.

There is currently little evidence for the production of superoxide anion by photochemical means, and none at all for liberation of peroxide. The final possibility, the reduction of the dioxygen to oxide, has been reported only for the bis(dioxygen) complex [Mo(porph)$(O_2)_2$], where the electrons are transferred from one dioxygen to another. The rarity of this reaction is surprizing, given the oxidizing nature of hydrogen peroxide, and the facility with which dioxygen can be completely reduced by certain metal ions; there is, however, no sign in the electronic spectra of a transition corresponding to charge transfer from metal to dioxygen σ^* orbitals.

ABBREVIATIONS AND STRUCTURAL FORMULAE

dien	1,4,7 triazaheptane (diethylenetriamine), $H_2NCH_2CH_2NHCH_2CH_2NH_2$
dmtrien	4,7-dimethyl-1,4,7,10 tetraazadecane
EDTA	ethylenediamine tetraacetate, $[(O_2CCH_2)_2NCH_2CH_2N(CH_2CO_2)_2]^{4-}$
en	1,4 diazabutane (ethylenediamine), $H_2NCH_2CH_2NH_2$
HbO_2	oxyhemoglobin
his	histidinate
LMCT	ligand to metal charge transfer
MbO_2	oxymyoglobin
MLCT	metal to ligand charge transfer
papd	1,13-diamino-4,7,10-triazatridecane
porph	porphyrinate anion
tetren	tetraethylenepentamine
TPP	tetraphenylporphyrinate
tren	tris(2-aminoethyl)amine, $N(CH_2CH_2NH_2)_3$
ε	molar extinction coefficient
ϕ	quantum yield (mole^{-1} einstein^{-1})

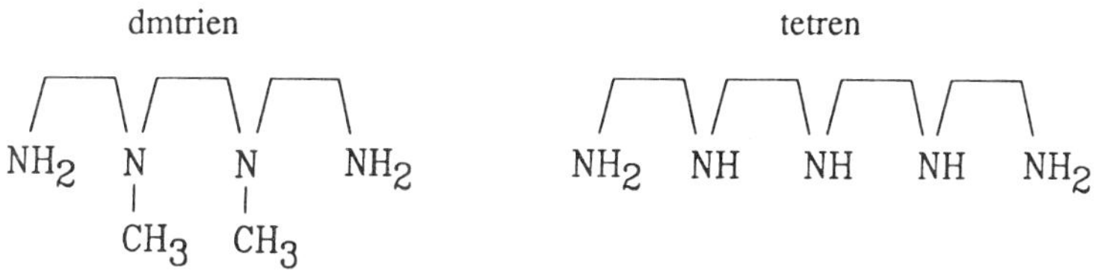

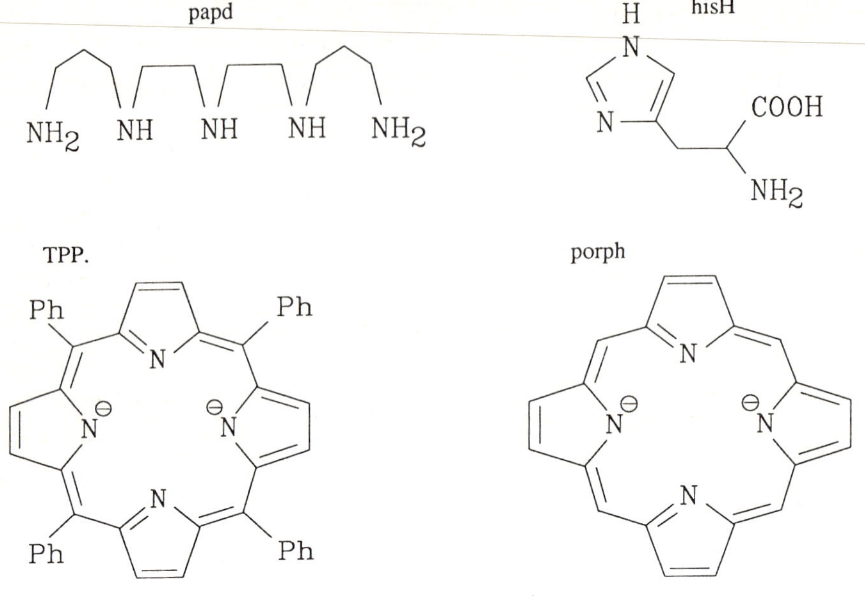

1,3 diphenylisobenzylfuran

Ph

O

Ph

The coordination shown below is found in the oxygenated forms of hemoglobin, myoglobin, cytochrome o, and hematoporphyrin, with iron in the +II oxidation state and coordinated by a porphyrin. In carboxyhemoglobin and carboxymyoglobin the dioxygen is replaced by carbon monoxide. In methemoglobin, the iron is in the +III oxidation state with an aquo ligand bound in place of dioxygen.

O—O

N—Fe—N

L

ACKNOWLEDGEMENTS

We gratefully acknowledge the help of Prof. J.M.J. Tronchet, Mme C. Gratiot, and M. G. Moret of the University of Geneva in the preparation of the camera-ready text.

REFERENCES

1 M.H. Gubelmann and A.F. Williams, Structure and Bonding, (Berlin), 55 (1983) 1.
2 L. Vaska, Acc. Chem. Res. 9 (1976) 175.
3 A.B.P. Lever and H.B. Gray, Acc. Chem. Res. 11 (1978) 348.
4 A.B.P. Lever, J. Mol. Structure 58 (1980) 123.
5 A.B.P. Lever, Inorganic Electronic Spectroscopy, 2nd edn., Elsevier, Amsterdam, 1984 pp. 285-296.
6 D.A. Summerville, R.D. Jones, B.M. Hoffmann, and F. Basolo, J. Chem. Ed. 56 (1979) 157; R.B. VanAtta, C.E. Strouse, L.K. Hanson, and J.S. Valentine, J. Amer. Chem. Soc. 109 (1987) 1425.
7 R. Boca, Coord. Chem. Rev. 50 (1983) 1.
8 R.D.Jones, D.A. Summerville and F. Basolo, Chem. Rev. 79 (1979) 139.
9 I.M. Klotz and D.M. Kurtz, Jr., Acc. Chem. Res. 17 (1984) 16.
10 W.A. Eaton, L.K. Hansen, P.J. Stephens, J.C. Sutherland and J.B.R. Dunn, J. Amer. Chem. Soc. 100 (1978) 4991.
11 A.K. Churg and M.W. Makinen, J. Chem. Phys. 68 (1978) 1913.
12 M. W. Makinen, A.K. Churg, and H.A. Glick, Proc. Natl. Acad. Sci. USA 75 (1978) 2291.
13 D.A. Case, B.H. Huynh, and M. Karplus, J. Amer. Chem. Soc. 101 (1979) 4433.
14 Q.H. Gibson and S. Ainsworth, Nature 180 (1957) 1416.
15 B.I. Greene, R.M. Hochstrasser, R.B. Weissman and W.A. Eaton, Proc. Natl. Acad. Sci. USA 75 (1978) 5255.
16 D.A. Chernoff, R.M. Hochstrasser, and A.W. Steele, Proc. Natl. Acad. Sci. USA 77 (1980) 5606.
17 J.L. Martin, A. Migus, C. Poyart, Y. Lecarpentier, A. Antonetti and A. Orszag, Biochem. Biophys. Res. Commun. 107 (1982) 803.
18 D.A. Duddel, R.J. Morris and J.T. Richards, Biochim. Biophys. Acta 621 (1980) 1.
19 J.A. Hutchinson and L.J. Noe, IEEE J. Quantum Electronics QE-20 (1984) 1353; K. Caldwell, L.J. Noe, J.D. Ciccone, and T.G. Traylor, J. Amer. Chem. Soc. 108 (1986) 6150.
20 L.S. Demma and J.M. Salhany, J. Biol. Chem. 252 (1977) 1226.
21 S.L. Gibson, H.J. Cohen and R. Hilf, Photochem. Photobiol. 40 (1984) 441.
22 Y. Orii and D.A. Webster, J. Biol. Chem. 261 (1986) 3544.
23 M.C.R. Symons and R.L. Petersen, Proc. Roy. Soc B201 (1978) 285.
24 R.E. Hester and E.M. Nour, J. Raman Spectrosc. 11 (1981) 35.
25 T. Szymanski, T.W. Cape, R.P. Van Duyne and F. Basolo, J. Chem. Soc. Chem. Comm. (1979) 5.
26 K. Nakamoto, Y. Nonaka, T. Ishiguro, M.W. Urban, M. Suzuki, M. Kozaka, Y. Nishida, and S. Kida, J. Amer. Chem. Soc. 104 (1982) 3386.
27 M. Hoshino, Chem. Phys. Lett. 120 (1985) 50.
28 R.E. Hester and E.M. Nour, J. Raman Spectrosc. 11 (1981) 39.
29 V. Peruzzo and R.E. Hester, J. Raman Spectrosc. 5 (1976) 115.
30 C.J. Boreham, J.M. Latour, J.C. Marchon, B. Boisselier-Cocolios and R. Guilard, Inorg. Chim. Acta 45 (1980) L69.
31 P. Bergamini, S. Sostero, O. Traverso, P. Deplano and L.J. Wilson, J. Chem. Soc., Dalton Trans. (1986) 2311.
32 H. Ledon, M. Bonnet and J.Y. Lallemand, J. Chem. Soc. Chem. Comm. (1980) 702.

33 G.L. Geoffroy, G.S. Hammond and H.B. Gray, J. Amer. Chem. Soc. 97 (1975) 3933.
34 A. Vogler and H. Kunkely, J. Amer. Chem. Soc. 103 (1981) 6222.
35 J.G. Norman, Inorg. Chem. 16 (1977) 1328.
36 D. Deal and J.I. Zink, Inorg. Chem. 20 (1981) 3995.
37 A.D. Westland, F. Haque and J.M. Bouchard, Inorg. Chem. 19 (1980) 2255.
38 C.K. Ranganathan, T. Ramasami, D. Ramaswamy and M. Santappa, Leather Sci. (Madras) 28 (1981) 351.
39 S. Fallab and P.R. Mitchell, Adv. Inorg. Bioinorganic Mechanisms 3 (1984) 311.
40 V.M. Miskowski, J.L. Robbins, I.M. Treitel, and H.B. Gray, Inorg. Chem. 14 (1975) 2318.
41 V. Miskowski, B.D. Santarsiero, W.P. Schaefer, G.E. Ansok and H.B. Gray, Inorg. Chem. 23 (1984) 172.
42 T.C. Strekas and T.G. Spiro, Inorg. Chem. 14 (1975) 1421.
43 A.B.P. Lever and S.R. Pickens, Inorg. Chem. Acta 45 (1980) L185.
44 R.E. Hester and E.M. Nour, J. Raman Spectrosc. 11 (1981) 43.
45 R. Heiniger, Ph.D. Thesis, University of Basel, 1984; M. Zehnder and S. Fallab, Helv. Chim. Acta 87 (1984) 392.
46 T.B. Freedman, C.M. Yoshida, and T.M. Loehr, J. Chem. Soc. Chem. Comm. (1974) 1017.
47 R.E. Hester and E.M. Nour, J. Raman Spectrosc. 11 (1981) 64.
48 R.E. Hester and E.M. Nour, J. Raman Spectrosc. 11 (1981) 59.
49 R.E. Hester and E.M. Nour, J. Raman Spectrosc. 11 (1981) 49.
50 K. Nakamoto, M. Suzuki, T. Ishiguro, M. Kozuka, Y. Nishida and S.Kida, Inorg. Chem. 19 (1980) 2822.
51 M.Suzuki, T. Ishiguro, M. Kozuka and K. Nakamoto, Inorg. Chem. 20 (1981) 1993.
52 S.R. Pickens and A.E.Martell, Inorg. Chem. 19 (1980) 15.
53 E.I. Solomon, Pure Appl. Chem. 55 (1983) 1069; E.I. Solomon, K.W. Penfield and D.E. Wilcox, Struct. Bonding (Berlin), 53 (1983) 1.
54 H.R. Mäcke, unpublished results.
55 H.R. Mäcke, Abstracts of the XXII ICCC, Budapest, Hungary, 1982; Abstracts of the XXIII ICCC, Boulder, Colorado, USA, 1984.
56 M. Kikkawa, Y. Sasaki, S. Kawata, Y.Hatakeyama, F.B. Ueno, and K. Saito, Inorg. Chem. 24 (1985) 4096.
57 J.E. Barnes, J. Barrett, R.W. Brett and J.J. Brown, J. Inorg. Nucl. Chem. 30 (1968) 2207.
58 J.S. Valentine and D.S. Valentine, J. Amer. Chem. Soc. 93 (1971) 1111.
59 J.S. Valentine and D.S. Valentine, Inorg. Chem., 10 (1971) 393.
60 M. Hoshino, M. Nakajima, M. Takakubo and M. Imamura, J. Phys. Chem. 86 (1982) 221.
61 M.H. Gubelmann, B. Bocquet, S. Rüttimann, and A.F. Williams, unpublished results.

Chapter 5.3

Photoinduced Electron Transfer in Hexacoordinate Inorganic Complexes

N. Serpone

1.0 INTRODUCTION

No other area of inorganic chemistry has experienced a more dramatic and dynamic growth than light-induced electron transfer from (or to) excited states of transition metal complexes to (or from) some suitable electron acceptor (or donor). The search for alternative new fuels to substitute the 'declining' world supply of fossil fuels has given the impetus to this field which has seen no parallels in the last decade (1975-1986). Sunlight was to provide the photons to assist in the generation of these new fuels (e.g., hydrogen from the photosplitting of water). While photoinduced electron transfer processes are not novel, the demonstration of electron transfer from an excited state of a transition metal complex, e.g.,$^*Ru(bpy)_3^{2+}$, to a ground state cobalt(III) centre as in $Co(NH_3)_5Br^{2+}$ was first reported by Gafney and Adamson in 1972 [1]. Light irradiated organic donors had been known [2,3] to senistize the decompositions of Co(III) ammines and the substitution reactions of Cr(III) ammines. The novel development at the time was to discover that a coordination compound such as $Ru(bpy)_3^{2+}$ (bpy = 2,2'-bipyridine) could act as a sensitizer for both types of reactions [1,4,5]. The sensitized redox decomposition of Co(III) ammines by excited $^*Ru(bpy)_3^{2+}$, acting as a one-electron reducing agent, was not without controversy; however, the proposal was later confirmed in other laboratories [6].

The excited state of a molecule may be considered an <u>electronic isomer</u> of the ground state molecule and has all the kinetic and thermodynamic properties available to the latter species. The main difference lies in the rather short lifetime (10^{-12} to 10^{-3}s) and the greater gree energy <u>vis-a-vis</u> the ground state. It is pertinent to consider this electronic isomer in more detail with emphasis placed on those originating from transition metal complexes, and in particular hexacoordinated complexes. Several excellent reviews and books have been published and these may be consulted for greater details [6-17]. We herein explore (albeit not exhaustively) and critically examine the photoinduced intermolecular electron transfer processes with specific reference to those between excited octahedral transition metal complexes and ground-state six-coordinate metal complexes and, where relevant, other metal ions and complexes. Complexes containing porphyrin and related ligands, and organometallic compounds are not covered. Intermolecular electron transfer processes involving organic electron acceptors and donors are covered in another section of this book [18]; intramolecular charge transfer photochemistry that leads to photoredox products in hexacoordinate complexes are also discussed elsewhere in connection with wavelength-dependent electron transfer phenomena [19]. Much attention will be devoted to coordination compounds containing 2,2'-bipyridine (I) and 1,10-phenanthroline (II) or their derivatives.

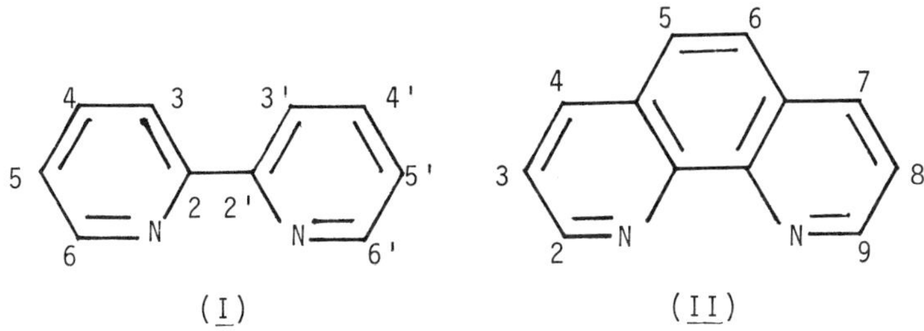

2.0 NATURE AND PROPERTIES OF EXCITED STATES OF TRANSITION METAL COMPLEXES

2.1 Configuration and Nomenclature of Excited States

The nature of the excited states of transition metal (TM) complexes is best understood from a consideration of the various molecular orbitals that originate from metal-ligand interactions. Figure 1 depicts a simplified energy level diagram and defines the relevant orbitals of the metal and of the ligands; where the ligand possesses π-type orbitals, antibonding ligand π* are also included. The ensuing molecular orbitals (M.O.'s) are classified as either σ or π and further denoted by a subscript (M or L) to indicate whether the M.O. is predominantly metal (σ_M, π_M) or ligand (σ_L, π_L) in character. This notation leads to such terms as "metal-localized, metal-centred, or ligand field" and "ligand-localized or ligand-centred"; it will be useful in recognizing the type of electronic

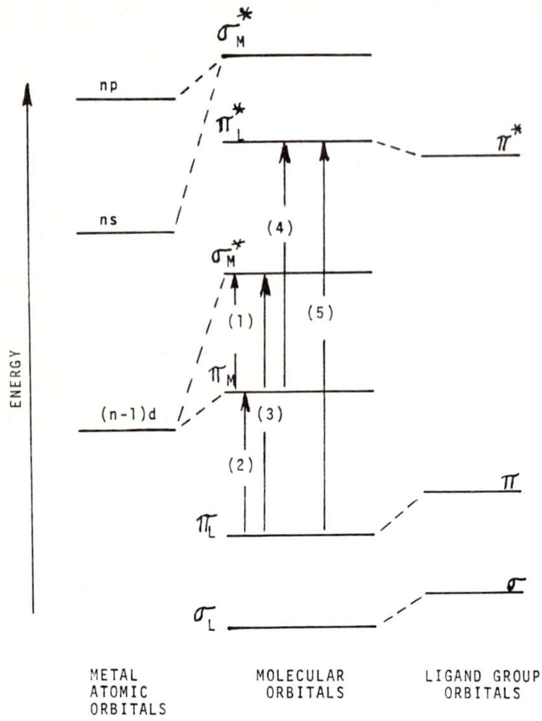

Figure 1.- Molecular orbital energy level diagram for a hexacoordinate transition metal complex of octahedral or near-octahedral symmetry.

transitions on absorption of light and the nature of the resulting excited state(s). Four basic transitions (and excited states) may be described:

(1) metal-centred or ligand field (MC or LF) transitions in which the electron is simply redistributed amongst the metal orbitals;

(2) and (3) ligand-to-metal charge transfer transitions (LMCT) in which the electron is formally transferred from the ligand to the metal (the metal is formally reduced and the ligand is formally oxidized);

(4) metal-to-ligand charge transfer (MLCT) transitions in which the electron is transferred from the metal to the ligand orbitals (the ligand is formally reduced and the metal is formally oxidized);

(5) ligand-centred transitions (LC) in which the electronic

transition occurs between predominantly ligand molecular orbitals.

In complexes of the first row transition metals, the ligand field spitting (10Dq) is such that MC transitions are relatively low energy transitions and the resulting states are low energy states. By contrast, the charge-transfer (LMCT or MLCT) transitions are not so influenced by the ligand field, but are dependent on the ionization energies and electron affinities of the metal and of the ligands. The energies (in fact the properties) of excited states can be tuned by an appropriate selection of ligands and variation of oxidation states of the metals. The nature of the lowest energy excited state will thus depend primarily on the nature of the metal the nature of the ligand, and the oxidation state of the metal. The above calls attention to the fact that energy ordering of excited states can be altered to meet some desired specific goal. Figure 2 illustrates examples [15,20] of fine-tuning the nature of the low

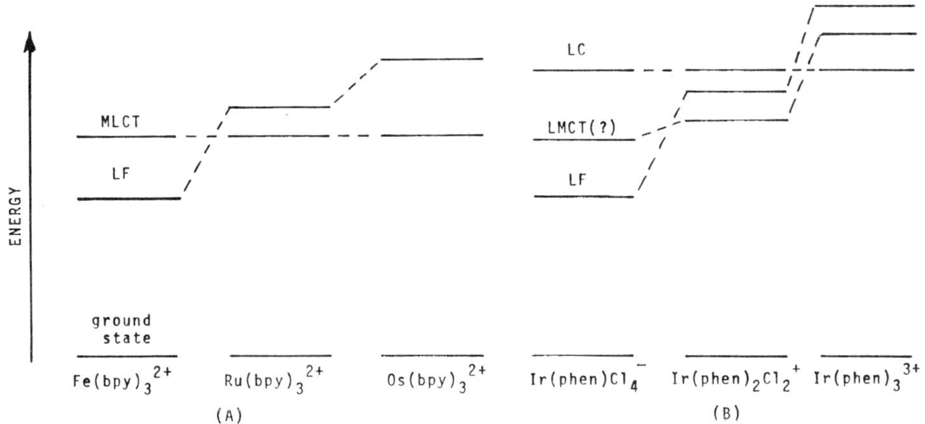

Figure 2.- Energy ordering and fine-tuning the nature of the lowest energy excited states as a function of the nature of the metal and ligands. Adapted from refs. 15 and 20.

lying excited state of a TM complex. In Figure 2A, changing the metal from Fe(II) to Ru(II) to Os(II) leads to a progressive increase in the energy of the LF state such that the $Ru(bpy)_3^{2+}$ and $Os(bpy)_3^{2+}$ have low-lying MLCT excited states. For a given metal ion, such as Ir(III), the progressive substitution of phenanthroline ligands (phen) for the Cl^- leads to a change of the low energy excited state from LF for $Ir(phen)Cl_4^-$, to LMCT for $Ir(phen)_2Cl_2^+$, to LC for $Ir(phen)_3^{3+}$. These variations lead to a rich and varied photochemistry of transition metal complexes, a subject beyond the scope of this work. It suffices to note that since electron transfer occurs between a donor and an acceptor in a bimolecular process the electron transfer events that implicate excited states will be dictated greatly by the lifetime of the excited state, and specifically by the other competing deactivation pathways.

2.2 Factors Affecting the Lifetime of Excited States

As an example we consider $Cr(bpy)_3^{3+}$ which has been investigated extensively in our laboratories [17]. The events occurring under light excitation and in the absence of quencher are summarized in Scheme I and portrayed in the Jablonski diagram of Figure 3.

Scheme I:

$$(^4A_2)Cr(bpy)_3^{3+} \xrightarrow{h\nu} (^4T_2, a^4T_1, b^4T_1)Cr(bpy)_3^{3+} \qquad (1)$$

$$(^4T_2, a^4T_1, b^4T_1)Cr(bpy)_3^{3+} \longrightarrow (^4T_2)Cr(bpy)_3^{3+} \qquad (2)$$

$$(^4T_2)Cr(bpy)_3^{3+} \xrightarrow{^4k_{nr}} (^4A_2)Cr(bpy)_3^{3+} \qquad (3)$$

$$\xrightarrow{^4k_{isc}} (^2E)Cr(bpy)_3^{3+} \qquad (4)$$

$$\xrightarrow{^4k_{rad}} (^4A_2)Cr(bpy)_3^{3+} + h\nu' \qquad (5)$$

$$\xrightarrow[H_2O]{^4k_{rx}} \text{products} \qquad (6)$$

$$(^2E)Cr(bpy)_3^{3+} \xrightarrow{^2k_{nr}} (^4A_2)Cr(bpy)_3^{3+} \qquad (7)$$

$$\xrightarrow{^2k_{rad}} (^4A_2)Cr(bpy)_3^{3+} + h\nu'' \qquad (8)$$

$$\xrightarrow[H_2O]{^2k_{rx}} \text{products} \qquad (9)$$

$$(^2E)Cr(bpy)_3^{3+} + (^4A_2)Cr(bpy)_3^{3+} \xrightarrow[Cl^-]{^2k_g} 2(^4A_2)Cr(bpy)_3^{3+} \qquad (10)$$

The low-energy excited state of a given spin multiplicity for $Cr(bpy)_3^{3+}$ is a MC (or LF) state. The intrinsic factors responsible for deactivation of a given excited state, in the absence of a chemical reaction are: (i) vibrational relaxation within an electronic manifold, (ii) non-radiative transitions to another excited

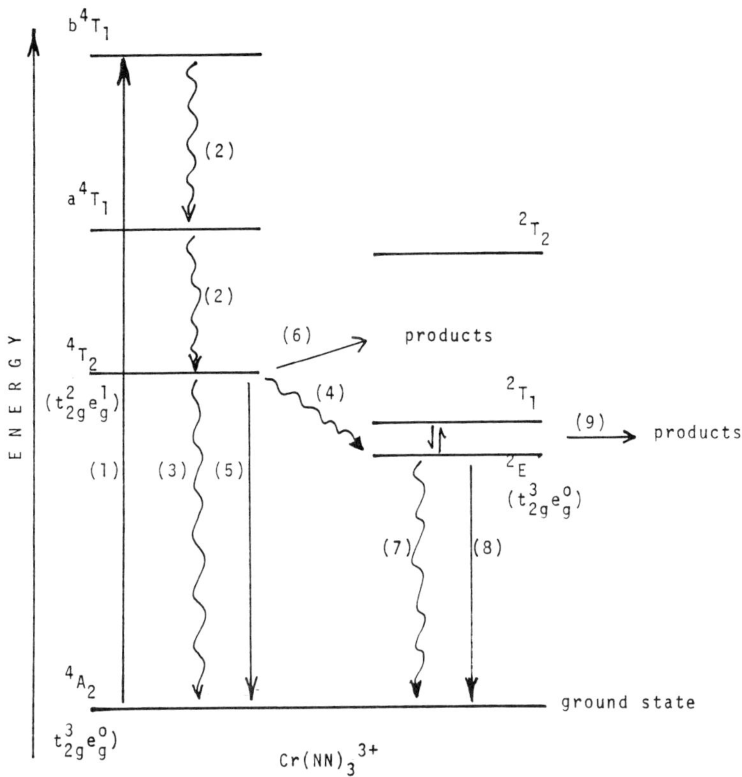

Figure 3.- Jablonski diagram for $Cr(bpy)_3{}^{3+}$. Adapted from ref.17.

state (intersystem crossing, isc) or to the ground state, and (iii) radiative transitions (emission: fluorescence and/or phosphorescence). The lifetime and fate of the excited state thus depends on the kinetic competitiveness of the various deactivation modes; $\tau^{-1}(^4T_2) = {}^4k_{nr} + {}^4k_{rad} + {}^4k_{isc}$ and $\tau^{-1}(^2E) = {}^2k_{nr} + {}^2k_{rad}$ in the absence of a reactive step. Bimolecular events may occur if the lifetime of the excited state is $> 10^{-9}$-10^{-10}s. Rates of excited state electron transfer reactions typically cover a range of 10^5-$10^{10}M^{-1}s^{-1}$, and if electron transfer is to compete with other deactivation modes, the rates must be comparable to, or faster than the rate of decay of the excited state by other modes.

2.3 Redox Properties of Excited States

Electronic excited states may be looked upon as new chemical species with different physical and chemical properties from the ground state. Redox properties are an example. Excited states are simultaneously more powerful oxidants and reductants than is the ground state. The origin of this enhanced redox behaviour is understood considering the one-electron orbital diagram of Figure 4. The redox ability of the excited state depends on the ionization potential and the electron affinity of this new species. Light

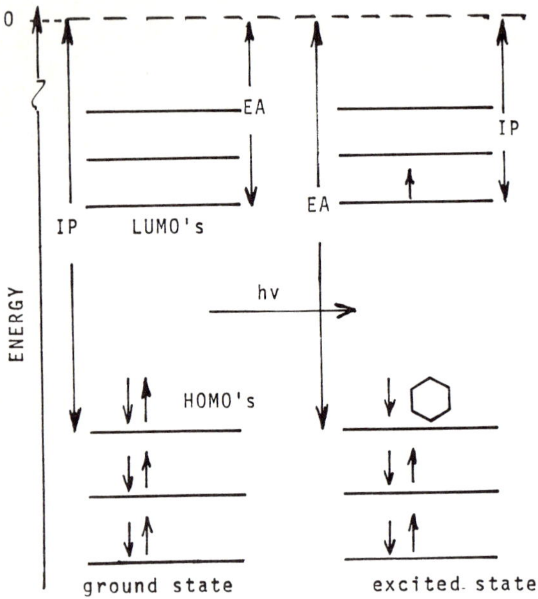

Figure 4.- One-electron orbital diagram.

excitation of an electron from a HOMO to a LUMO orbital decreases the ionization potential; the presence of a "hole" in the HOMO orbital of the excited state increases the electron affinity. An example of these changes is depicted in Figure 5 for $Cr(polypyridine)_3^{3+}$ [21].

Where the Stokes shift between ground-state absorption and excited-state emission is very small, and the changes in shape, size, and solvation between the two states are small, the entropy difference between the ground and excited state is negligibly small. The redox potentials of the excited state may be estimated from

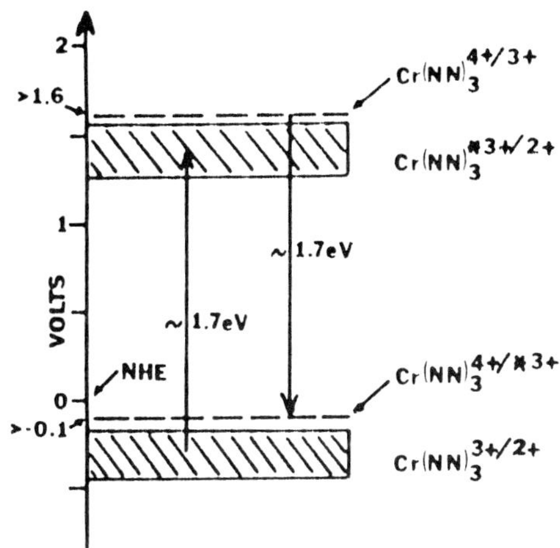

Figure 5.- Redox potential scheme for $Cr(polypyridine)_3^{3+}$ complexes. Reproduced with permission from ref. 21.

eqn's 11 and 12 for $Cr(NN)_3^{3+}$ species (see Table 1):

$$E^o(Cr^{4+}/Cr^{*3+}) = E^o(Cr^{4+}/Cr^{3+}) - E_{oo}(*Cr^{3+}/Cr^{3+}) \quad (11)$$

$$E^o(*Cr^{3+}/Cr^{2+}) = E^o(Cr^{3+}/Cr^{2+}) + E_{oo}(*Cr^{3+}/Cr^{3+}) \quad (12)$$

where $E_{oo}(*Cr^{3+}/Cr^{3+})$ is the one-electron potential corresponding to the 0-0 spectroscopic energy of the excited state (ca. 1.7 eV). The validity of this procedure has been verified for the MLCT state of $Ru(bpy)_3^{2+}$ [22,23]; the $E^o(Ru^{3+}/*Ru^{2+})$ is estimated as -0.84 V in accord with -0.81 V for the $Ru(bpy)_3^{3+}/*Ru(bpy)_3^{2+}$ couple in acetonitrile [24]. This was taken [22] as good evidence that the entropy difference between ground and excited state is negligible for $Ru(bpy)_3^{2+}$. For some $Cr(NN)_3^{3+}$ complexes, Balzani and coworkers [25] have experimentally estimated redox potentials using a series of quenchers of graded potentials. Some redox potentials of ground and excited states of $Cr(NN)_3^{3+}$ and $Ru(NN)_3^{2+}$ are summarized in Table 1; the E_{oo} value for the Ru(II) complexes is ca. 2.12 eV [17,26].

TABLE 1
Redox potentials (vs.NHE) of $Cr(NN)_3^{3+}$ and $Ru(NN)_3^{2+}$ species

NN	$E^o(Cr^{3+/2+})$[a] V	$E^o(Cr^{*3+/2+})$[a] V	$E^o(Ru^{3+/2+})$[b] V	$E^o(Ru^{3+/*2+})$[b] V	$E^o(Ru^{2+/+})$[b] V	$E^o(Ru^{*2+/+})$[b] V
bpy	-0.26	+1.44	+1.26	-0.86	-1.28	+0.84
4,4'-Me_2bpy	-0.45	+1.25	+1.10	-0.94	-1.44	+0.69
4,4'-Ph_2bpy	-0.28	+1.39	+1.17	-0.85	-	-
phen	-0.28	+1.42	+1.26	-0.87	-1.36	+0.82
5-Mephen	-0.30	+1.39	+1.23	-0.92	-1.31	+1.00
5-Clphen	-0.17	+1.53	+1.36	-0.77	-1.15	-
5-Brphen	-0.15	+1.55	+1.37	-	-0.76	-
4,7-Me_2phen	-0.45	+1.23	+1.09	-0.94	-1.47	+0.67
5,6-Me_2phen	-0.29	+1.40	+1.20	-0.93	-1.34	+0.80
3,4,7,8-Me_4-phen	-0.57	+1.11	+1.02	-	-1.11	-
5-Phphen	-0.21	+1.49	-	-	-	-
4,7-Ph_2phen	-0.26	+1.41	+1.22	-	-1.31	-

[a] reference 17; [b] reference 26.

3.0 BIMOLECULAR ELECTRON TRANSFER

Transition metal complexes may undergo redox reactions via outer-sphere and inner-sphere pathways. Inner-sphere redox reactions are not expected to occur within the lifetime of the excited state species since ligand substitution processes are generally slow, and thus cannot compete with excited-state decay. By contrast, an outer-sphere redox pathway forms an activated complex in which the number and type of ligands on the two reactants remain unchanged. This pathway may be very fast, and can compete with other deactivation modes of the excited-state species. The quenching rates are governed by both intrinsic (self-exchange rates of the reactants) and extrinsic (overall free energy change of the reaction) factors [6].

3.1 Quenching by Electron Transfer

Energy transfer and electron transfer are two of the major dynamic quenching pathways for deactivating excited states. The latter consists of a vectorial electron displacement from an occupied donor orbital to an unoccupied acceptor orbital. The process requires close approach (< 10 A) of the reaction pair for effective orbital overlap. Theoretical descriptions of electron transfer quenching of excited states find origin from the work of Marcus [27], Hush [28], and others [29]; they were recently extended by Sutin [30] and Balzani [9,31] to quenching excited states of transition metal complexes.

Where the excited state of a TM complex (sensitizer, S) is relatively long-lived, it may undergo a bimolecular energy transfer reaction (eqn 13) or electron transfer with a quencher molecule, Q. The sensitizer may be oxidatively quenched (S* acts as a reductant; eqn 14) or may be reductively quenched (S* acts as an oxidant; eqn 15). The quenching reactions 14 and 15 may subsequently be followed by back electron transfer to yield the ground-state reactants.

$$S^* + Q \xrightarrow{k_{en.tr.}} S + Q^* \quad \text{(energy transfer)} \quad (13)$$

$$S^* + Q \xrightarrow{k_{ox}} S^+ + Q^- \xrightarrow{k_{bet}} S + Q \quad \text{(oxidative } e^- \text{ transfer)} \quad (14)$$

$$S^* + Q \xrightarrow{k_{red}} S^- + Q^+ \xrightarrow{k_{bet}} S + Q \quad \text{(reductive } e^- \text{ transfer)} \quad (15)$$

If all three reactions contribute to the quenching of S*, then $k_q(obs) = k_{en.tr.} + k_{ox} + k_{red}$. The path for electron transfer quenching is summarized in Scheme II (adapted from ref. 16), where the first step in quenching is the formation of an en-

Scheme II:

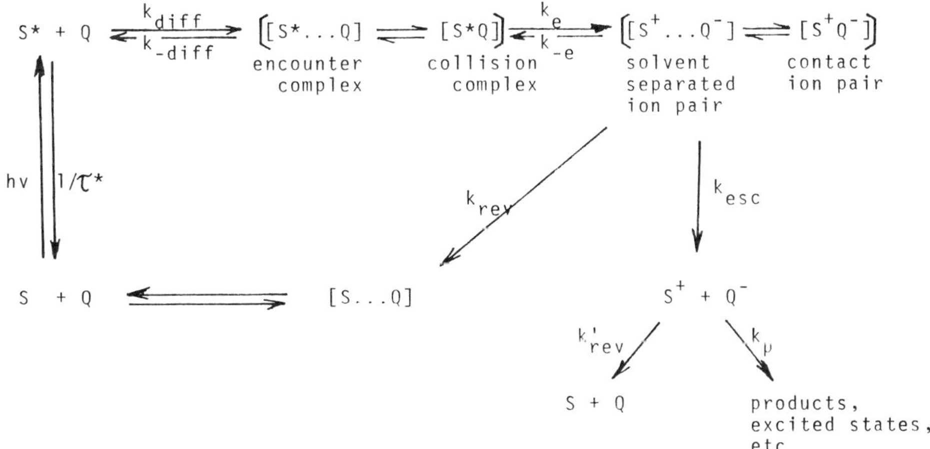

where k_{diff} and k_{-diff} are diffusion-controlled rate constants k_e and k_{-e} are the bimolecular rate constants for the activated electron transfer step, k_{rev} is the rate constant for the reversible electron transfer back to the ground state and k_{esc} is the rate constant for cage escape by S^+ and Q^- ions.

counter complex which consists [16,32] of an intermolecular ensemble of an excited-state and ground-state species typically ca. 5-10 A apart and solvated by several shells of solvent molecules with the innermost forming the solvent cage, defined by [].

The occurrence of electron transfer between S* and Q will depend on the overall free energy change, ΔG, which accompanies the reaction [29]. It will be thermodynamically favoured if $\Delta G < 0$; this is a critical requirement if electron transfer is to compete with other deactivation modes of S*. The rate constant for the quenching process, k_q, may be expressed in terms of the rate constants (eqn 16) in Scheme II:

$$k_q = \frac{k_{diff}}{1 + \dfrac{k_{-diff}}{k_e} + \dfrac{k_{-diff}k_{-e}}{[k_{rev} + (k'_{rev} + k_p)k_{esc}]k_e}} \tag{16}$$

If $k_e >> k_{-e}$, k_q is given by equation 17 which further reduces to

$$k_q = \frac{k_{diff}k_e}{k_e + k_{-diff}} \tag{17}$$

equation 18 for $k_{diff} = k_{-diff}$. For slow electron transfer react-

$$1/k_q = 1/k_e + 1/k_{diff} \tag{18}$$

ions ($k_e << k_{diff}$), the rate of quenching of S* is the activated rate of electron transfer; $k_q \sim k_e$. Under the assumption that $k_{esc} << k_{rev}$ and that

$$k_e = Zk_{et} \tag{19}$$

and

$$k_{et} = \kappa \exp(-\Delta G_e^{\#}/RT) \tag{20}$$

equation 16 becomes

$$k_q = \frac{k_{diff}}{1 + \dfrac{k_{-diff}}{Z}[\exp\{\Delta G_e^{\#}/RT\} + \exp\{\Delta G_e^{o}/RT\}]} \tag{21}$$

which may be used to estimate the dependence of k_q on the overall free energy change ΔG_e^o, if the free energy of activation for the electron transfer step, $\Delta G_e^{\#}$, can be related to ΔG_e^o. Marcus [27,33] suggested the classical relationship 22 to relate $\Delta G_e^{\#}$ to ΔG_e^o; Rehm and Weller [29,34] proposed the empirical relationship 23.

$$\Delta G_e^{\#} = (\lambda/4)[1 + \Delta G_e^o/\lambda]^2 \quad (22)$$

$$\Delta G_e^{\#} = [(\Delta G_e^o/2)^2 + (\lambda/4)^2]^{1/2} \quad (23)$$

where $\lambda = \lambda_i + \lambda_o$ and represent the intrinsic barriers to electron transfer corresponding to changes in bond lengths (λ_i) and in solvent reorganization (λ_o).

A detailed kinetic analysis and critical review of these two relationships have been presented by Balzani and coworkers [9]. It is instructive to consider briefly the terms in eqns 19 and 20: Z is the collision frequency at unit molar concentration of reactants within the encounter complex; κ is a constant reflecting the electronic barrier of the process, and $\Delta G_e^{\#}$ represents contributions from nuclear barriers [16]. Thus, if $\kappa << 1$ and $\Delta G_e^{\#} = 0$, the rate of electron transfer is primarily dependent on electronic barriers; where $\kappa \sim 1$ and $\Delta G_e^{\#} > 0$, the nuclear barriers become rate-determining. Moreover, where κ approaches unity the reaction is said to be adiabatic and where κ approaches 0, the reaction potential surfaces do not cross and the electron transfer reaction is then defined as nonadiabatic [16]. Hence, κ represents the probability that the electron jump from reactant surface to the product surface will occur. The question of adiabatic vs. nonadiabatic reactions has caused some controversies recently, and a good discussion may be found in recent accounts by Sutin [35], Balzani and Scandola [31], and by Karvanos and Turro [16].

Treatments of electron transfer pathways by Marcus [27,36] and Hush [37] predict that plots of log k_q vs. driving force, ΔG^o, and for constant reorganization energy should be parabolic (see Figure 6). Clearly, for positive and slightly negative values of ΔG_e^o, the dependence of log k_q on ΔG_e^o is virtually identical. However, in the region where ΔG_e^o becomes increasingly more negative, the Marcus theory predicts a dramatic decrease in k_q (the Marcus inverted region); by contrast, the Rehm-Weller empirical treatment predicts that k_q reaches a plateau value under these conditions. Several attempts have been made to locate cases where the inverted Marcus behaviour is observed; these attempts have not been without

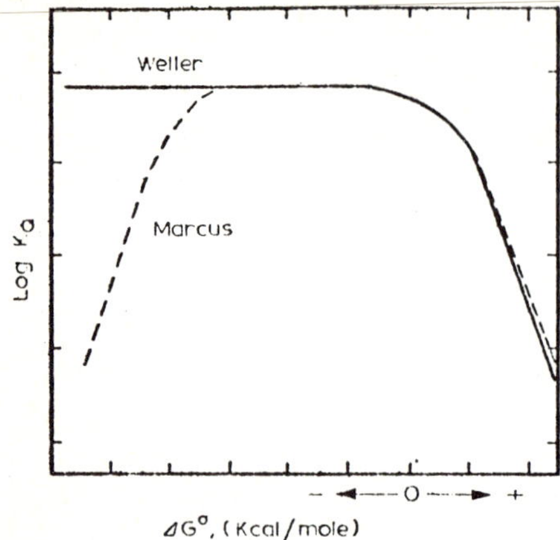

Figure 6.- Dependence of log k_q on ΔG_e^o calculated from eqn 21, where ΔG_e^o is given by the Marcus and Rehm-Weller relationships. Reproduced with permission from ref.17.

criticisms. Vestiges of the Marcus inverted region in electron transfer quenching of a transition metal complex excited state have been found in the quenching of $(^3MLCT)Ru(bpy)_3^{2+}$ by other polypyridyl metal complexes; there is a small decrease in k_q with increasing exergonic character of the quenching reaction (ΔG_e^o becomes increasingly more negative) [38].

The occurrence of an electron transfer quenching pathway can be confirmed if the Marcus [36] relationship 24 for electron transfer reactions is satisfied:

$$k_q = k_{12} = (k_{11}k_{22}K_{12}f_{12})^{1/2} \tag{24}$$

where $\log f_{12} = (\log K_{12})^2/4 \log(k_{11}k_{22}/Z^2)$, and k_{11} and k_{22} are the rate constants for the self-exchange reactions 25 and 26, respectively. K_{12} is the equilibrium constant for the cross-reaction 27, and Z is taken as $10^{11}M^{-1}s^{-1}$; eqn 27 holds when $\Delta G_{12} \sim 0$.

$$M + M^+ \underset{}{\overset{k_{11}}{\rightleftharpoons}} M^+ + M \tag{25}$$

$$*Cr(NN)_3^{3+} + Cr(NN)_3^{2+} \overset{k_{22}}{\rightleftharpoons} Cr(NN)_3^{2+} + *Cr(NN)_3^{3+} \tag{26}$$

$$*Cr(NN)_3^{3+} + M \overset{K_{12}}{\rightleftharpoons} Cr(NN)_3^{2+} + M^+ \tag{27}$$

Also,

$$\log k_{12} = 0.5 \log k_{11}k_{22} + 0.5(1 + \alpha)\log K_{12} \quad (28)$$

where $\alpha = (\log K_{12})/4\log(k_{11}k_{22}/Z^2)$. A linear relationship in a plot of $\log k_q$ (= $\log k_{12}$) vs. $(1 + \alpha) \log K_{12}$ with slope 0.5 and intercept 0.5 $\log k_{11}k_{22}$ would be indicative of electron transfer quenching. Moreover, if k_{11} is known, the value of k_{22} can be estimated by fitting the experimental k_{12} (= k_{et}) data assuming various values for k_{22}. Log K_{12} is a measure of the driving force for electron transfer and varies as ΔE^o, the difference in redox potentials of the reacting partners.

The effect of the driving force on electron transfer reactions has come under critical scrutiny [25,35]. Creutz and Sutin [38] have pointed out that electron transfer rates should begin to decrease at high enough driving forces; solid evidence where k_q's span several orders of magnitude is still lacking in transition metal complexes. However, intramolecular quenching by electron transfer of an organic donor excited state, chemically coupled to an acceptor molecule by spacer molecules, has recently been demonstrated to exhibit an inverted Marcus-type behaviour [39].

A last point worth emphasizing is that the lifetime of an excited state species imposes a restriction on nuclear reorganization and electronic coupling [16]. Changes in bond lengths, solvent reorganization, and dynamic molecular motion necessary for effective orbital overlap must take place during the lifetime of the excited state of the sensitizer S*. Electron transfer events, therefore, strongly depend on the photophysical properties of S*.

We now turn to actual investigations into excited-state electron transfer events. Polypyridyl transition metal complexes have been widely studied as they provide a wide range of excited state properties, which can be varied by altering the metal centre, altering the type of ligands, or simply by addition of substituents to the polypyridine framework. It is this "excited state tuning" that allows the transition metal complexes to serve as both reductants and oxidants, an attribute not possible for the metal ion or the free ligand molecule alone.

3.2 Discriminating Between Electron Transfer and Energy Transfer Quenching Pathways.

Having shown that quenching of an excited state occurs by a dynamic mechanism, it remains to determine whether it is by energy transfer or by electron transfer. This is not trivial, and establishing which of these is operative requires consideration of both kinetic and thermodynamic factors. Kinetically, the energy of activation, needed to reorganize inner and outer shells of the reactant pair prior to energy and/or electron transfer, will be important. The thermodynamic feasibility of an energy transfer pathway will be related to the 0-0 spectroscopic energy levels of the donor /acceptor pair; that for electron transfer will be a function of the redox potentials of the donor/acceptor pair.

The prediction of efficient quenching by energy transfer requires knowledge of the excited state energies of the donor and acceptor. In many cases, these are known for many transition metal complexes from emission studies, or from the onset of the lowest energy absorption band of the complex. Where the energy of the acceptor state is $\gtrsim$ 1000 cm^{-1} than that of the donor state, quenching by energy transfer will be inefficient [40]. Spin must be preserved in this pathway. For example, while energy transfer from a donor singlet to an acceptor triplet is spin forbidden, the presence of spin-orbit coupling in transition metal complexes may relax the spin selection rules and "singlet to triplet" energy transfer becomes possible when the donor singlet is at higher energy than the acceptor triplet. Naturally, the lifetimes of the two states must also be considered (see above). Experimentally, this quenching pathway may be confirmed from observations of luminescence from the acceptor state(s) (see reaction 13) or from the growth of a transient absorption (flash photolysis) corresponding to an acceptor excited state in a time scale commensurate with the decay of the donor state; electron transfer quenching is by no means precluded.

Quenching by electron transfer is indicated if a relationship between the quenching rate constants, k_q, and the driving force (ΔG), obtained from redox potentials can be demonstrated. Thus, a linear plot of log k_q vs. ΔG (slope: 1/2.303RT) for small values of ΔG and which subsequently reaches a plateau (the Rehm-Weller expression) at diffusion-controlled rates, or shows a decrease for very negative ΔG (the Marcus behaviour; see Figure 6) is strong evidence for an electron transfer pathway [41]. Proof that the emitting state is acting as an electron donor (*S; oxidative quenching) requires the use of a quencher substrate (Q) to which energy transfer cannot occur readily and which accepts electrons to yield a reduced product that is not rapidly oxidized by S^+; likewise, demonstration of reductive quenching requires the use of a substrate to which energy transfer is inefficient and which can donate electrons to give an oxidized product that is not rapidly reduced by S^- [21]. Direct evidence of the formation of the oxidized or reduced forms of the acceptor or donor (reactions 14 and 15) using a spectroscopic flash photolysis technique for reversible redox systems, or from a continuous irradiation study on irreversible redox systems may confirm electron transfer. However, the interpretation of such experiments is not free of ambiguities [21,42,43]. Less direct evidence for electron transfer quenching mechanisms may also be obtained from steady-state [43], relative rate [44], and from spectroscopic considerations [44,45]. Lin and Sutin [43] have exploited the "tuning" of excited-state redox potentials in several Ru(II) complexes to distinguish between energy transfer and electron transfer pathways. As well, Meyer and coworkers [46,47] cleverly exploited the differential flash photolysis technique as a relaxation method in this regard. Several additional methods are available to establish the quenching pathway and/or to confirm the nature of the intermediates following quenching of the donor excited states [16].

4.0 ELECTRON TRANSFER INVOLVING EXCITED STATES OF Ru(II) AND Os(II) OCTAHEDRAL TRANSITION METAL COMPLEXES

4.1 Quenching of Ru(II) and Os(II) Polypyridyl Complexes

4.1.1 By Metal Ions In Aqueous Media

Since the first reports [1,48-50] that $Ru(bpy)_3^{2+}$ (electronic configuration $t_{2g}^6e_g^0$) sensitizes the redox decomposition of such metal complexes as $Co(NH_3)_5Cl^{2+}$ and $PtCl_4^{2-}$, this complex has proven to be the most versatile photosensitizer amongst transition metal complexes [16]. The many studies done on $Ru(bpy)_3^{2+}$ show the lowest excited state to be a MLCT state, $^3(d,\pi^*)$, formed by promotion of a t_{2g} electron to a π^*-orbital of the polypyridyl ligand; the electron appears localized on only one of the bpy ligands [9, 15,26]. This strongly emitting excited state is produced with unitary efficiency, is relatively long-lived ($\tau \sim 0.6$ us), and is relatively undistorted with respect to the ground state (Stokes shift is small). $Ru(bpy)_3^{2+}$ appears inert to photosubstitution at 25°C in water; at higher temperatures, photolabilization occurs from a LF excited state rather than from $(^3MLCT)Ru(bpy)_3^{2+}$ [16,26]. In this respect, $*Ru(bpy)_3^{2+}$ behaves like $Ru(bpy)_3^{2+}$ and $Ru(bpy)_3^{3+}$, also inert to substitution. Hence the chemistry of $*Ru(bpy)_3^{2+}$ is expected to be restricted to outer-sphere electron- and energy-transfer processes.

The quenching of $*Ru(bpy)_3^{2+}$ by a variety of aquated metal ions {Fe^{3+}(aq), Cu^{2+}(aq), Eu^{3+}(aq), Tl^{3+}(aq), Ag^{+}(aq), and Hg^{2+}(aq)} has been investigated by several groups. Table 2 summarizes [26] the rate constants for quenching by electron transfer (reactions 14 or 15) and identifies the process as either oxidative or reductive quenching. The above inorganic ions quench oxidatively to give $Ru(bpy)_3^{3+}$. Thermodynamic and kinetic considerations as well as flash photolysis studies support electron transfer quenching of $*Ru(bpy)_3^{2+}$ by Eu^{3+}(aq) and Tl^{3+}(aq).

TABLE 2
Kinetics and redox quantum yields in the quenching of $^{*}Ru(bpy)_3^{2+}$ by inorganic ions and molecules

Quencher	Medium	$10^{-7}k_q(M^{-1}s^{-1})$	Mechanism	$\Phi(Ru^{n+})$	Ref.
Fe^{3+}	H_2SO_4(0.5M)	270	Oxidative	0.83	43,45,46,52-55
Fe^{2+}	$HClO_4$(0.5M)	1.6			45,52
Cu^{2+}	H_2SO_4(0.5M)	6.2	Oxidative	0.56	56-58
Eu^{2+}	H_2O(u = 0.5M)	2.8	Reductive	1.0	45
Eu^{3+}	H_2O(u = 0.5M)	$\leq$0.008	Oxidative	>0	59,60
Tl^{3+}	$HClO_4$(0.5M)	11	Oxidative	2.0	52,61
Tl^{3+}	HCl (1.0M)	330			61
Tl^{+}	$HClO_4$(0.5M)	<0.1			52
Ag^{+}	H_2O	0.35	Oxidative		62,63
Ag^{+}	CH_3CN	0.011	Oxidative		62
Hg^{2+}	$HClO_4$(0.1M)	15	Oxidative	>0.6	64
$HgCl_2$	$HClO_4$(0.1M)	38		<0.01	64

Several features of Table 2 are worth noting. The rate constant k_q for the $*Ru(bpy)_3^{2+}/Eu^{3+}$ couple, $\lesssim 0.8 \times 10^5 M^{-1}s^{-1}$, is very much lower than diffusion-controlled rates. This has been taken [51] as evidence for a nonadiabatic mechanism, controlled by electronic rather than nuclear factors. The outer 5s and 5d orbitals of the Eu^{3+} acceptor shield the deeply buried 4f orbital, thus preventing orbital overlap, a prerequisite for an adiabatic reaction [16]. Doubt has been expressed on the assignment of an electron transfer mechanism in the reductive quenching of $*Ru(bpy)_3^{2+}$ by Eu^{2+}(aq) inasmuch as the excited-state energies of $*Eu^{2+}$ are not known with certainty [45]. However, elaborate kinetic arguments would seem to indicate that quenching does take place by electron transfer [60]. This has recently been demonstrated by Connolly and coworkers (reaction 29) [65].

$$*Ru(bpy)_3^{2+} + Eu^{2+}(aq) \longrightarrow Ru(bpy)_3^{+} + Eu^{3+}(aq) \qquad (29)$$

Rate constants for electron transfer quenching of $*Ru(NN)_3^{2+}$ (NN = a polypyridine ligand) by Eu^{3+} and Eu^{2+} encapsulated in a cryptate molecule, $[Eu \subset 2.2.1]^{3+}$ and $[Eu \subset 2.2.1]^{2+}$, in aqueous media are summarized in Table 3 (reactions 30 and 31) [66]. Compari-

$$*Ru(NN)_3^{2+} + [Eu \subset 2.2.1]^{3+} \longrightarrow Ru(NN)_3^{3+} + [Eu \subset 2.2.1]^{2+} \qquad (30)$$

$$*Ru(NN)_3^{2+} + [Eu \subset 2.2.1]^{2+} \longrightarrow Ru(NN)_3^{+} + [Eu \subset 2.2.1]^{3+} \qquad (31)$$

TABLE 3
Rate constants for electron transfer quenching of $*Ru(NN)_3^{2+}$ by europium(II) and (III) cryptates in aqueous media (1 M KCl)[a]

Quencher	*Ru(II) complex	$10^{-7}k_q(M^{-1}s^{-1})$
$[Eu \subset 2.2.1]^{3+}$	$*Ru(bpy)_2(4,4'\text{-}Cl_2bpy)^{2+}$	0.56
	$*Ru(bpy)_2(4\text{-}NO_2bpy)^{2+}$	2.1
	$*Ru(bpy)_2(4,4'\text{-}Me_2bpy)^{2+}$	3.3
	$*Ru(bpy)_3^{2+}$	4.9
	$*Ru(bpy)(4,4'\text{-}Me_2bpy)_2^{2+}$	7.1
	$*Ru(4,4'\text{-}Me_2bpy)_3^{2+}$	5.6
	$*Ru(5,6\text{-}Me_2phen)_3^{2+}$	6.0
	$*Ru(3,4,7,8\text{-}Me_4phen)_3^{2+}$	3.4
$[Eu \subset 2.2.1]^{2+}$	$*Ru(4,4'\text{-}Me_2bpy)_3^{2+}$	130
	$*Ru(bpy)(isobiq)_2^{2+}$	130
	$*Ru(bpy)_2(DMCH)^{2+}$	100
	$*Ru(bpy)_2(isobiq)^{2+}$	120
	$*Ru(bpy)_3^{2+}$	130
	$*Ru(bpy)(biq)_2^{2+}$	170
	$*Ru(bpy)_2(4\text{-}NO_2phen)^{2+}$	230

[a] reference 66.

son of the k_q values in Table 3 with k_q values for electron transfer quenching by $Eu^{3+}(aq)$ and $Eu^{2+}(aq)$ shows that cryptation decreases the intrinsic barrier and/or increases the adiabatic coefficient of the electron transfer reaction. However, the rate constants dependence on the free energy changes (driving force) of the electron transfer process demonstrate that the data for oxidative quenching by $[Eu\subset 2.2.1]^{3+}$ do not correlate with the data for reductive quenching of $*Ru(NN)_3^{2+}$ by $[Eu\subset 2.2.1]^{2+}$. This asymmetric behaviour may result from several factors: (i) different shapes of the potential energy wells for $[Eu\subset 2.2.1]^{3+}$ and $[Eu\subset 2.2.1]^{2+}$; (ii) different work terms for the formation of the precursor complex; and (iii) different distances of closest approach of $[Eu\subset 2.2.1]^{3+}$ and $[Eu\subset 2.2.1]^{2+}$ with the hydrophobic $Ru(NN)_3^{2+}$ complexes [66].

Oxidative quenching by $Fe^{3+}(aq)$ has been investigated extensively as it is one of the rare cases where the back electron transfer is very inefficient (reaction 32) [43,45,46,52-55]. This

$$*Ru(bpy)_3^{2+} + Fe^{3+}(aq) \longrightarrow Ru(bpy)_3^{3+} + Fe^{2+}(aq) \quad (32)$$

$*Ru(bpy)_3^{2+}/Fe^{3+}$ couple affords an example where misinterpretation of data may occur when relying exclusively on flash photolysis techniques [23]. In 0.5M H_2SO_4, $k_q = 2.7 \times 10^9 M^{-1}s^{-1}$ and $k_{bet} = 5.2 \times 10^6 M^{-1}s^{-1}$. The $Ru(bpy)_3^{3+}$ product of reaction 32 is formed in high yields ($\gtrsim$ 80%) in aqueous media as determined by flash spectroscopic techniques [23,43,53]. The observation of the oxidized form $Ru(bpy)_3^{3+}$ would normally be taken as evidence for a primary electron transfer process in reaction 32. Unfortunately, the existence of a low-lying excited state of Fe^{3+} at 1.60 eV [Note that the $*Ru(bpy)_3^{2+}$ emitting state is at 2.12 eV] could also lead to collisional energy transfer; $Ru(bpy)_3^{3+}$ could, in principle, be formed via reactions 33 and 34. Excited $*Fe^{3+}$ may be quenched

$$*Ru(bpy)_3^{2+} + Fe^{3+} \longrightarrow Ru(bpy)_3^{2+} + *Fe^{3+} \quad (33)$$

$$Ru(bpy)_3^{2+} + *Fe^{3+} \longrightarrow Ru(bpy)_3^{3+} + Fe^{2+} \quad (34)$$

(eqn 34 is exothermic: $\Delta G = -0.8$ eV) by the Ru(II) complex to yield the oxidized Ru(III) substrate. Kinetic arguements seem to negate the importance of reactions 33 and 34, and a primary electron transfer process seems indicated for reaction 32. $Ru(terpy)_2^{3+}$ and $Fe^{2+}(aq)$ are produced by flash photolysis of solutions containing $Ru(terpy)_2^{2+}$ and $Fe^{3+}(aq)$; $k_q < 2.9 \times 10^9 M^{-1}s^{-1}$ and terpy is 2,2',2"-terpyridine [67]. The solution medium can also have considerable influence on the rates of electron transfer owing to the different quencher species that may prevail in various media. For example, in Cl^- and SO_4^{2-} solutions containing $Fe^{3+}(aq)$, such species as $FeCl_4^-$ and $Fe(SO_4)_2^-$ may be present. Ionic strength also influences rates of electron transfer: in 0.11 M $HClO_4$, k_q of reaction 32 is $1.5 \pm 0.1 \times 10^9 M^{-1}s^{-1}$ but in 0.050 M $HClO_4$, k_q is $2.3 \pm 0.1 \times 10^9 M^{-1}s^{-1}$ [21].

Quenching of $*Ru(bpy)_3^{2+}$ by $Fe^{2+}(aq)$ is less effective than by $Fe^{3+}(aq)$ or $Tl^{3+}(aq)$. An electron transfer quenching mechanism seems not possible in this case [52], but the k_q pattern for several quenchers parallels that for outer-sphere redox reactions of $Os(bpy)_3^{3+}$ with several reductants (including $Fe^{2+}(aq)$) and electron transfer is not precluded [45]. Quenching of $*Ru(bpy)_3^{2+}$ by Tl^+ (aq) is not efficient; $k_q < 10^6 M^{-1}s^{-1}$ [52].

Solution medium control of k_q is demonstrated for Tl^{3+} in HCl and H_2SO_4 solutions: $k_q(HCl) = 3.3 \times 10^9 M^{-1}s^{-1}$ compared to k_q $(H_2SO_4) = 1.1 \times 10^8 M^{-1}s^{-1}$. The anionic species $TlCl_4^-$ interact efficiently with the Ru(II) complex but the radical ion pairs do not escape (Scheme II) the primary solvent cage $[\Phi Ru^{3+}) < 0.01]$ [26]. Where the polypyridine ligands possess substituents with acid/base properties, the pH of the medium becomes an important parameter in the magnitude of the electron transfer quenching rate constant. For $*Ru(bpy)_2(dcbpy)^{2+}$, where dcbpy is 4,4'-dicarboxy-2,2'-bipyridine, quenching by $Cu^{2+}(aq)$ yields $k_q(\text{pH } 1.5) = 8.21 \times 10^7 M^{-1}s^{-1}$ and $k_q(\text{pH } 4.3) = 1.3 \times 10^9 M^{-1}s^{-1}$ [68]. The increase in k_q at the higher pH demonstrates the importance of electrostatic interactions (work terms in the Marcus theory) between the net charge on the excited state (pK_{a2}^* 4.1) and the cupric ion. Particularly significant, in terms of the nature of the solution medium, is the electron transfer quenching of some $*Ru(NN)_3^{2+}$ complexes by $Cu^{2+}(aq)$ and chlorocopper(II) complexes in chloride media [69]. The observed rate constants for decay of the charge-transfer excited state is given by equation 35:

$$k_{obs} = k_o + k_q[Cu^{2+}(aq)] + k_q'[CuCl^+] + k_q''[CuCl_2] \qquad (35)$$

where k_o is the rate constant for spontaneous decay. The quenching rate constants are collected in Table 4. The efficiency of quenching by the $CuCl_n^{(2-n)+}$ increases with n. The weaker than expected dependence of k_q's on ΔG^o may be due to possible restraints imposed by the stereochemistry of the encounter-pair under condition of

TABLE 4

Rate constants for electron transfer quenching of $*Ru(NN)_3^{2+}$ by $Cu^{2+}(aq)$ and chlorocopper(II) complexes.[a]

NN	$Cu^{2+}(aq)$ $10^{-7}k_q(M^{-1}s^{-1})$	$CuCl^+$ $10^{-8}k'_q(M^{-1}s^{-1})$	$CuCl_2$ $10^{-9}k''_q(M^{-1}s^{-1})$
bpy	5.67	1.05	2.04
phen	6.14	2.22	2.05
4,7-Me_2phen	7.81	3.16	2.55
3,4,7,8-Me_4phen	4.83	2.31	2.15

[a] reference 69.

of high driving force for the reaction. In particular, the substitution of methyl groups on the ligands could lead to an increased encounter distance and to a consequent decrease in the overlap of donor and acceptor orbitals [69].

Aqueous Cu^{2+} also quenches $*Ru(NN)_2(CN)_2$ (NN = bpy, phen) at least in part by oxidative quenching to yield Cu^+ as does $Cu(acac)_2$; k_q's are, respectively, $3.9 \times 10^8 M^{-1}s^{-1}$ and $4.8 \times 10^8 M^{-1}s^{-1}$ by Cu^{2+}, and $1.1 \times 10^9 M^{-1}s^{-1}$ and $1.5 \times 10^9 M^{-1}s^{-1}$ by $Cu(acac)_2$ [70].

$Ru(bpy)_3^{3+}$ is formed irreversibly from quenching of $*Ru(bpy)_3^{2+}$ by Tl^{3+} [reactions 36-38; $\Phi(Ru^{3+}) = 2.0$] [52,61,71,72]. The high yield of Ru^{3+} arises from the dark oxidation of the ground

$$*Ru(bpy)_3^{2+} + Tl(III) \longrightarrow Ru(bpy)_3^{3+} + Tl(II) \quad (36)$$

$$Ru(bpy)_3^{2+} + Tl(II) \longrightarrow Ru(bpy)_3^{3+} + Tl(I) \quad (37)$$

state species $Ru(bpy)_3^{2+}$ by the Tl(II) intermediate. The overall reaction is:

$$2Ru(bpy)_3^{2+} + Tl(III) \xrightarrow[\Phi\ =\ 2]{h\nu} 2Ru(bpy)_3^{3+} + Tl(I) \quad (38)$$

and the whole oxidative quenching process is illustrated in Scheme III [52]. Irradiation leads to the ^{1}MLCT state (step 1) which can decay to the ground state (step 2) or can intersystem cross to the triplet charge-transfer state (^{3}MLCT; step 3). This ^{3}CT can also decay to the ground state (step 4) or can emit phosphorescence (step 5) whose quenching is accompanied by conversion of $Ru(bpy)_3^{2+}$

Scheme III:

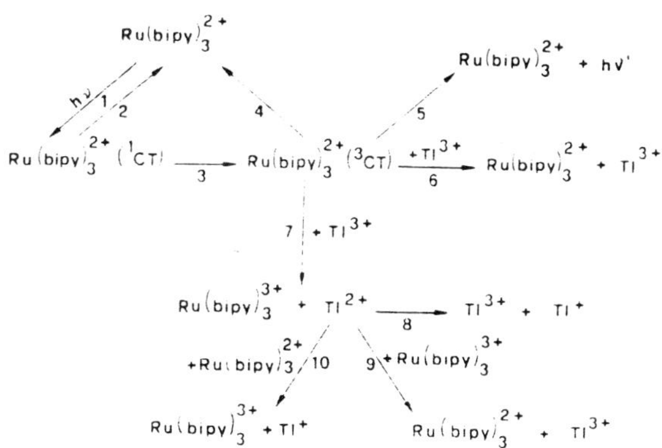

to $Ru(bpy)_3^{3+}$ (electron transfer step 7; reaction 36), and Tl(III) to Tl(II). This intermediate can disproportionate (step 8), reduce $Ru(bpy)_3^{3+}$ (step 9) or oxidize $Ru(bpy)_3^{2+}$ (step 10). The observation that 1 M methanol does not alter $\Phi(Ru^{3+})$ supports the mechanism of quenching by direct electron transfer from the ^3CT state.

The relative importance of electron-transfer and energy-transfer quenching mechanisms has been ascertained from an investigation on a series of luminescent $Ru(NN)_3^{2+}$ complexes [22]. The absorption and emission spectra of this series (Table 5) are nearly identical making it likely that the rates of energy transfer quenching with a given acceptor Q will be constant for this series. By contrast, redox potentials of $Ru(NN)_3^{2+}$ may be altered by changing the NN ligand substituents and electron transfer quenching rates with a given acceptor should reflect the differences in the driving force. The k_q values for quenching by Fe^{3+}(aq) are large and close to the diffusion-controlled value but have not yet reached the diffusion-controlled limit. For energy transfer quenching to lead to electron transfer products requires oxidation of $Ru(bpy)_3^{2+}$ by excited $*Fe^{3+}$(aq) to be more rapid ($k_4 > k_{-5}$ in Scheme IV) than dissociation of $[Ru(bpy)_3^{2+}|*Fe^{3+}]$ into $Ru(bpy)_3^{2+}$ and $*Fe^{3+}$ [22]. Further, for energy transfer to be important in Scheme IV necessitates that oxidation of $Ru(NN)_3^{2+}$ by $*Fe^{3+}$ be more rapid than oxidation of $*Ru(NN)_3^{2+}$ by Fe^{3+} ($k_4 > k_2$). Such a reactivity order is unlikely [22] because (i) oxidation of $*Ru(NN)_3^{2+}$ by Fe^{3+} has 0.5V larger driving force and (ii) the reorganization energy for this reaction involving Fe^{3+} is smaller since the added electron to Fe^{3+} enters a t_{2g} orbital (reaction 39). By contrast, when $*Fe^{3+}$ is the

TABLE 5

Rate constants for quenching $*Ru(NN)_3^{2+}$ by several metal ions in aqueous media [a]

NN	$E^o(Ru^{*2+/3+})$ V	$Fe^{3+}(aq)$ [b] $10^{-9}k_q(M^{-1}s^{-1})$	$Eu^{3+}(aq)$ [c] $10^{-5}k_q(M^{-1}s^{-1})$	$Cr^{3+}(aq)$ [d] $10^{-7}k_q(M^{-1}s^{-1})$	$Cu^{2+}(aq)$ [e] $10^{-7}k_q(M^{-1}s^{-1})$
4,4'-Me_2bpy	+0.94	2.9	~5	1.0	9.7
bpy	+0.84	2.7	≤0.8	1.2	6.2
3,4,7,8-Me_4phen	+1.11	3.4	-	-	10.0
3,5,6,8-Me_4phen	+1.04	2.5	-	-	-
4,7-Me_2phen	+1.01	3.0	21	1.3	8.6
5,6-Me_2phen	+0.93	2.6	7.5	1.4	7.6
5-Mephen	+0.90	2.6	4.2	1.3	6.0
5-Phphen	+0.87	2.7	-	-	5.7
phen	+0.87	2.8	~1	1.4	7.5
5-Brphen	+0.76	2.3	≤0.5	1.3	4.2
5-Clphen	+0.77	2.3	≤0.5	1.2	4.0

[a] reference 22. [b] 0.5M H_2SO_4. [c] 0.025M HCl; major species is $EuCl^{2+}$ and u = 2.7M. [d] u = 1M; 0.04M H_2SO_4. [e] in 0.5M H_2SO_4; reference 56.

Scheme IV:

$$*Ru(NN)_3^{2+} + Fe^{3+} \underset{k_{-1}}{\overset{k_1}{\rightleftharpoons}} [*Ru(NN)_3^{2+}|Fe^{3+}]$$

$$[*Ru(NN)_3^{2+}|Fe^{3+}] \xrightarrow{k_3} [Ru(NN)_3^{2+}|*Fe^{3+}]$$

$$Ru(NN)_3^{2+} + *Fe^{3+} \underset{k_{-5}}{\overset{k_5}{\rightleftharpoons}} [Ru(NN)_3^{2+}|*Fe^{3+}]$$

$$[*Ru(NN)_3^{2+}|Fe^{3+}] \xrightarrow{k_2} [Ru(NN)_3^{3+}|Fe^{2+}]$$

$$[Ru(NN)_3^{2+}|*Fe^{3+}] \xrightarrow{k_4} [Ru(NN)_3^{3+}|Fe^{2+}]$$

$$[Ru(NN)_3^{3+}|Fe^{2+}] \rightleftharpoons Ru(NN)_3^{3+} + Fe^{2+}$$

$$\underset{(t_{2g}^5e_g^0)(\pi^*)^1}{*Ru(NN)_3^{2+}} + \underset{(t_{2g}^3e_g^2)}{Fe^{3+}(aq)} \longrightarrow \underset{(t_{2g}^5e_g^0)}{Ru(NN)_3^{3+}} + \underset{(t_{2g}^4e_g^2)}{Fe^{2+}(aq)} \quad (39)$$

oxidant, electron transfer occurs between orbitals of mismatched symmetry. The larger distortions required about the iron for energy transfer and the high yield of electron transfer products are not consistent with energy transfer as a major quenching pathway [22]. $*Ru(phen)_3^{2+}$ and $*Ru(terpy)(bpy)(NH_3)^{2+}$ are quenched by Fe^{2+} (aq) via electron transfer in 1M $HClO_4$: k_q are 2.5 x $10^9M^{-1}s^{-1}$ and 3.1 x $10^9M^{-1}s^{-1}$, respectively [73].

The dependence of the quenching rate constants on the oxidation potentials (Table 5) of the emitting states is consistent with an electron transfer quenching mechanism. For $Eu^{3+}(aq)$ quenching of $*Ru(NN)_3^{2+}$, k_q's vary by more than two orders of magnitude. A plot of log k_q vs. log K_q (see eqn 28) is linear with slope 0.49 and intercept 1.55; this confirms electron transfer quenching [22]. For $Cr^{3+}(aq)$ quenching, k_q is insensitive to the nature of the ligand substituents in $Ru(NN)_3^{2+}$ and to the differences in redox potentials. An energy transfer pathway is thus indicated for $Cr^{3+}(aq)$ [22]. These conclusions are illustrated in Figure 7 which depicts plots of log k_q against $-E^0(Ru^{3+}/*Ru^{2+})$ [56,74]. The free energy

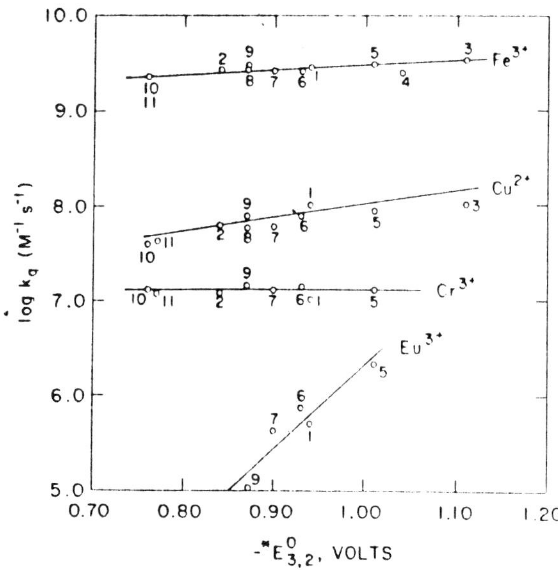

Figure 7.- Plot of the logarithm of the rate constant for quenching of $*Ru(NN)_3^{2+}$ by $Eu^{3+}(aq)$, $Cr^{3+}(aq)$, $Cu^{2+}(aq)$, and $Fe^{3+}(aq)$ vs. the $Ru(NN)_3^{3+}/*Ru(NN)_3^{2+}$ reduction potential of the $Ru(NN)_3^{2+}$ complexes: 1, NN = 4,4'-Me_2bpy; 2, bpy; 3, 3,4,7,8-Me_4phen; 4, 3,5,6,-8-Me_4phen; 5, 4,7-Me_2phen; 6, 5,6-Me_2phen; 7, 5-Mephen; 8, 5-Ph-phen; 9, phen; 10, 5-Brphen; 11, 5-Clphen. Reproduced with permission from ref.56.

dependence of the quenching reaction between $*Ru(NN)_3^{2+}$ and $Cu^{2+}(aq)$ is smaller than expected for an electron transfer mechanism [56]. A possible reason is that quenching proceeds by parallel energy-transfer and electron-transfer pathways. However, while the former is thermodynamically possible, kinetic arguments suggest quenching to be predominantly by electron transfer. Additional reasons for the small k_q dependence on ΔG^0 have invoked a special form of Cu^{2+}(aq) in the medium used, a Jahn-Teller distorted Cu^{2+} (aq), and the transfer of a π^* electron in the redox step. It may be that the equivalent of eqn 28 (eqn 40) overestimates the rates of electron transfer reactions with large driving forces. The origin of this effect lies in the detailed shapes of the potential

$$\log k_{12} = 0.55 \log k_{11} + 0.50 \log (k_{22}K_{12}f_{12}) \tag{40}$$

energy surfaces, in the departure from adiabaticity, in nuclear tunneling, and other quantum mechanical contributions [56]. It would seem that the simple Marcus model breaks-down at large driving forces.

The rate constants for quenching the charge-transfer states of $*Ru(bp\text{-}4,4'\text{-}ds)_3^{4-}$ and $*Ru(bpy)(bp\text{-}4,4'\text{-}ds)_2^{2-}$, where bp-4,4'-ds is III, by Cu^{2+}(aq) in neutral aqueous solution show unusual

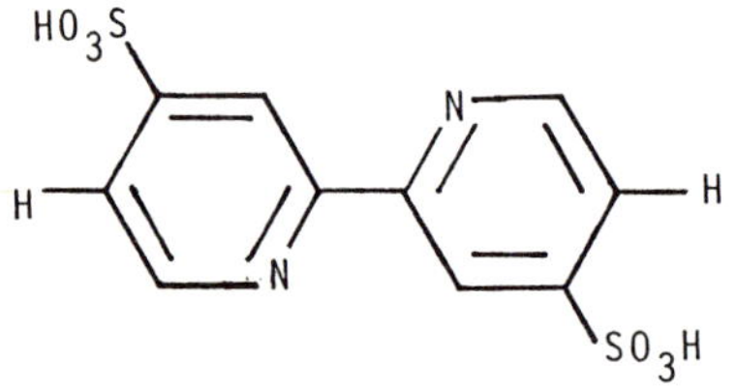

III

(non-Arrhenius) temperature dependences [75]. This anomalous behaviour is related to the presence of SO_3- substituents on the bpy ligand. Electron transfer quenching involves parallel inner-sphere and outer-sphere pathways (Scheme V); k_d and k_{-d} are the forward and reverse rate constants, respectively, for formation of the encounter complex, k_e is the rate constant for outer-sphere electron transfer to form products, k_i and k_{-i} are the forward and reverse rate constants for formation and dissociation of the inner-sphere complex, and k_s is the rate constant for formation of the electron transfer products from the inner-sphere complex. The kinetic scheme V depicts the interactions between SO_3- groups and aqueous Cu^{2+} ions only in the Ru(II) complex excited state; this represents a second dynamic quenching process operating in parallel with outer-sphere electron transfer. That such interactions occur only in the excited state is in keeping with the view that excitation increases the basicity of the complex. The activation parameters for the outer-sphere and inner-sphere pathways are collected in Table 6

Scheme V:

$[Ru(bp\text{-}4,4'\text{-}ds)_3]^{4-} \underset{k_0}{\overset{h\nu}{\rightleftharpoons}} \{[Ru(bp\text{-}4,4'\text{-}ds)_3]^{4-}\}^*$

$\{[Ru(bp\text{-}4,4'\text{-}ds)_3]^{4-}\}^* + Cu^{2+}_{aq} \underset{k_{-d}}{\overset{k_d}{\rightleftharpoons}} \{[Ru(bp\text{-}4,4'\text{-}ds)_3]^{4-}\cdot Cu^{2+}_{aq}\}^*$

$\{[Ru(bp\text{-}4,4'\text{-}ds)_3]^{4-}\cdot Cu^{2+}_{aq}\}^* \underset{k_{-i}}{\overset{k_i}{\rightleftharpoons}} [(bp\text{-}4,4'\text{-}ds)_2Ru(bp\text{-}SO_3,SO_3Cu_{aq})]^{2-}$

$\{[Ru(bp\text{-}4,4'\text{-}ds)_3]^{4-}\cdot Cu^{2+}_{aq}\}^* \xrightarrow{k_e} [Ru(bp\text{-}4,4'\text{-}ds)_3]^{3-} + Cu^{+}_{aq}$

$[(bp\text{-}4,4'\text{-}ds)_2Ru(bp\text{-}SO_3,SO_3Cu_{aq})]^{2-} \xrightarrow{k_s} [Ru(bp\text{-}4,4'\text{-}ds)_3]^{3-} + Cu^{+}_{aq}$

TABLE 6

Rate constants and activation parameters for the quenching of $^*Ru(bp\text{-}4,4'\text{-}ds)_3{}^{4-}$ emission by Cu^{2+} at 25°C.[a]

Parameters	Aqueous Media[b] pH 5-6	D_2O[b]	0.25M H_2SO_4[c]
$10^{-7}k_q(M^{-1}s^{-1})$	18.7	16.2	7.53
$10^{-8}A_o(M^{-1}s^{-1})$	4.09	2.94	8.37
$10^{-14}A_i(M^{-1}s^{-1})$	1.75	5.71	17.1
ΔE_o (kJ/mol)	27.4	19.8	6.4
ΔE_i (kJ/mol)	38.0	41.5	46.5
$\Delta G^{\#}_o$ (kJ/mol)	26.6	26.7	28.5
$\Delta G^{\#}_i$ (kJ/mol)	29.7	30.2	32.6
$\Delta H^{\#}_o$ (kJ/mol)	0.2	-0.5	3.9
$\Delta H^{\#}_i$ (kJ/mol)	35.5	38.9	44.0
$\Delta S^{\#}_o$ (J/K/mol)	-87	-91	-83
$\Delta S^{\#}_i$ (J/K/mol)	+20	+29	+38

[a] Reference 75; the subscripts 'o' and 'i' refer to outer and inner sphere electron transfer. [b] Ionic strength = 0.2 M with $NaClO_4$. [c] Ionic strength = 0.95 M with $CuSO_4$

[75]. It is intriguing to note that entropies of activation for the inner-sphere path are positive. This is reasonable if water molecules from the primary coordination sphere of Cu^{2+}(aq) were being substituted by charged groups. Also, both inner- and outer-sphere paths involve reaction between species of opposite charge and thus large pre-exponential factors are to be expected [75]. The small differences between the energetics of the quenching reaction in H_2O and in D_2O result from the lack of (or negligible) hydrogen-bonding interactions.

In photoinduced electron transfer involving $*Ru(bpy)_3^{2+}$ excited states and Ag^+ or Hg^{2+} ions as quenchers, the reduced form of the quencher species Ag^o or Hg^+ dimerize (or aggregate with Ag^+ or Hg^+, respectively) to give Ag_2^+ or Hg_2^{2+} [62-64,76]. The back electron transfer between the electron-transfer product $Ru(bpy)_3^{3+}$ and these Ag_2^+ or Hg_2^{2+} is slow. Under such conditions, $Ru(bpy)_3^{3+}$ lives long enough that it can be intercepted by such other reductants as Et_3N [76] or triethanolamine (TEOA). These systems show significant photopotentials and applications of these to other than energy storage, such as generation of useful and interesting species under mild conditions, have been explored [76]. For example, photopotentials in the range 200-300 mV have been produced from a photogalvanic cell based on $*Ru(bpy)_3^{2+}$ and Hg^{2+} [64]. The Hg_2^{2+} trap species is formed with high overall efficiency. $HgCl_2$ was also used as a quencher [77] to test the possible formation of Hg_2Cl_2 trap species and because quenching by the neutral $HgCl_2$ might not be influenced by ionic strength effects. The results are summarized in Table 7. The dependence of k_q on the redox potential (Weller plots) of the diimine donor $*Ru(NN)_3^{2+}$ leaves little doubt that all three mercury(II) species of Table 7 react by electron transfer. Contrary to the expectations from simple electrostatic considerations ($HgCl_2 < HgCl_3^- < HgCl_4^{2-}$), the experimental order of reactivity of the mercury(II) species is $HgCl_4^{2-} << HgCl_2 < HgCl_3^-$. The low reactivity of $HgCl_4^{2-}$ arises from a larger inherent barrier to electron transfer [$\Delta G^{\#}(0)$]; they are 6.6 ± 0.4 kcal/mol, 4.7 ± 0.5 kcal/mol, and 4.3 ± 0.8 kcal/mol for $HgCl_4^{2-}$, $HgCl_2$, and $HgCl_3^-$, respectively, [77]. A factor contributing to the larger value of $\Delta G^{\#}(0)$ for $HgCl_4^{2-}$ may be steric; the size of the anionic species prevents an encounter complex to form at an optimum distance for electron transfer. Orbital overlap would therefore be less effective. The greater reactivity of phenanthroline (phen) derivatives is caused by the presence of a cen-

TABLE 7
Electron transfer quenching constants of $*Ru(NN)_3^{2+}$ by mercury(II) species.[a]

NN	$E^o(Ru^{*2+/3+})$ V	$HgCl_2$ $10^{-9}k_q$[b] $(M^{-1}s^{-1})$	$HgCl_2$ $10^{-9}k_q$[c] $(M^{-1}s^{-1})$	$HgCl_3^-$ $10^{-9}k_q$[d] $(M^{-1}s^{-1})$	$HgCl_4^{2-}$ $10^{-9}k_q$[d] $(M^{-1}s^{-1})$
3,4,7,8-Me_4phen	+1.11	2.95	-	-	-
4,7-Me_2phen	+1.01	2.82	2.82	8.51	1.02
4,4'-Me_2bpy	+0.94	1.51	1.91	5.13	0.724
5,6-Me_2phen	+0.93	2.88	2.57	5.01	0.776
5-Mephen	+0.90	1.91	2.63	6.31	0.813
phen	+0.87	1.32	1.86	5.50	0.324
bpy	+0.84	0.170	0.229	1.66	0.0372
5-Clphen	+0.77	0.389	0.575	2.51	0.112
5-Brphen	+0.76	0.468	0.575	1.82	0.251

[a] Reference 77. [b] Ionic strength = 0.09 M $NaNO_3$. [c] Ionic strength = 0.99 M $NaNO_3$. [d] Ionic strength = 0.99 M ($NaNO_3$ + NaCl).

tral aromatic ring in the phen structure, absent in the bpy ligands. Mercury(II) substrates, such as $HgCl_2$, are known to form complexes with aromatic molecules; similar interactions in the substances of Table 7 lead to stabilization of the encounter complexes and enhancement of k_q's in the Ru(II)-phen derivatives. This interaction <u>does not occur</u> in the ground state $Ru(NN)_3^{2+}$ species [77].

The influence of electrolytes on the photoredox kinetics and cage escape yields of redox products in the quenching of $*Ru(bpy)_3^{2+}$ by Tl(III) and Hg(II) substrates has been investigated to define parameters for efficient photoelectrochemical cells [78]. One parameter, which characterizes the redox quenching process, is the quantum yield for the cage escape of the redox products [Φ(redox)] from the geminate pair (eqn 41). A high cage escape

$$*Ru(bpy)_3^{2+} + Q \longrightarrow [Ru(bpy)_3^{3+}\ldots Q^-] \xrightleftharpoons{\Phi(redox)} Ru(bpy)_3^{3+} + Q^- \quad (41)$$

yield is essential in any photoelectrochemical energy conversion device. The k_q and Φ(redox) data are collected in Table 8. Differ-

TABLE 8

Summary of k_q, back electron transfer rate constants k_{bet}, and cage escape quantum yields Φ_{redox} in the photoredox quenching of $*Ru(bpy)_3^{2+}$.[a]

Quencher	Medium	u,[b] M	$10^{-9}k_q(M^{-1}s^{-1})$	$k_{bet}(M^{-1}s^{-1})$	Φ_{redox}
Tl(III)	0.05M H_2SO_4	1.5	0.10	irreversible	2.00
	1.0M HCl	1.0	3.4	-	-
	3.0M HCl	3.0	4.3	2.9×10^{10}	0.14
Hg(II)	3.0M HCl	3.0	0.19	-	<0.01
	0.5M H_2SO_4	1.5	0.15	part.irrev. (25%)	0.32
	0.5M $HClO_4$	0.5	0.15	part.irrev. (25%)	-

[a] reference 78. [b] Ionic strength.

ences in the two acids for Tl(III) quenching arises from a change in the chemical nature of the quenching species (see above); this affects the driving force of the reaction. For Hg(II) quenching in H_2SO_4 or $HClO_4$, the geminate pair separates to yield moderate amounts of photoproducts [Φ(redox) $\sim$ 0.3] and the reaction is partially irreversible. In HCl, hardly any redox products are formed (< 0.01). The partial irreversibility of the photoredox process in H_2SO_4 arises from the formation of Hg_2^{2+} trap species (see above) which are slow to react with the Ru(III) product of reaction 41.

4.1.2 By Metal Ions in Micellar Solutions

It was noted earlier that $HgCl_2$ might be a superior quencher to test because the insolubility of the anticipated Hg_2Cl_2 trap species would render the energy-wasting back electron transfer step slower than found for Hg_2^{2+} species (< 10 $M^{-1}s^{-1}$) [77]. Unfortunately, no redox products of $HgCl_x^{2-x}$ quenchers have been observed in non-micellar media even under flash photolysis conditions. Evidently, the rate of cage escape in reactions 42-44 is slower than the energetically favourable back electron transfer reaction 43. To im-

$$*Ru(NN)_3^{2+} + HgCl_2 \rightarrow [*Ru(NN)_3^{2+}|HgCl_2] \rightarrow [Ru(NN)_3^{3+}|HgCl_2^{\cdot-}] \quad (42)$$

$$[Ru(NN)_3^{3+}|HgCl_2^{\cdot -}] \longrightarrow [Ru(NN)_3^{2+}|HgCl_2] \longrightarrow Ru(NN)_3^{2+} + HgCl_2 \quad (43)$$

$$[Ru(NN)_3^{3+}|HgCl_2^{\cdot -}] \longrightarrow Ru(NN)_3^{3+} + HgCl_2^{\cdot -} \quad (44)$$

prove the cage escape of the energetically more valuable products of reaction 44, excited-state interactions of $Ru(NN)_3^{2+}$ photosensitizers with $HgCl_x^{2-x}$ ($x = 2,3,4$) have been studied in sodium lauryl sulfate micellar solutions. The $Ru(NN)_3^{2+}$ species are tightly bound to the surface of the micelles (<1% may still be free in solution) and the quenchers appear to remain entirely in the aqueous phase (Figure 8) [79]. Quenching proceeds via oxidative electron

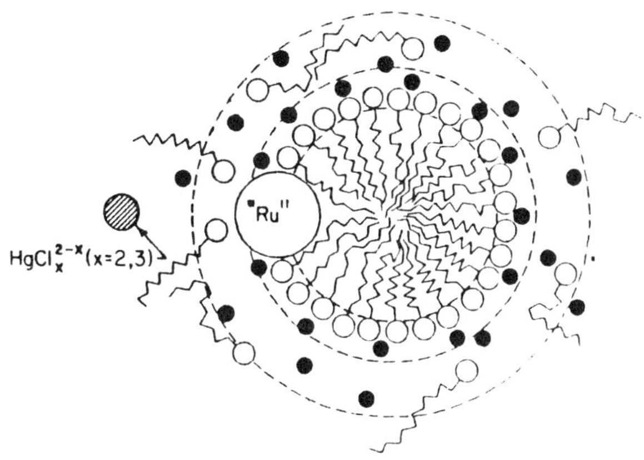

Figure 8.- Representation of $Ru(NN)_3^{2+}$ imbedded in a micelle. Dashed lines represent the boundaries of (from the interior out):the inner core, the Stern layer, and the Gouy-Chapman layer. The solid circles are Na^+ ions, and the open circles with tails are LS^-. Reproduced with permission from ref. 79.

transfer to yield $Ru(NN)_3^{3+}$ and $HgCl_2^{\cdot -}$ free radicals; the latter species was not detected owing to the inefficient ejection of the radical from the anionic micelle. This is a result of the very fast reaction 45. The quenching data taken under various conditions are shown in Table 9. According to the model used [79], the quen-

$$HgCl_2^{\cdot -} \longrightarrow HgCl\cdot + Cl^- \quad (45)$$

ching rate is determined by the rate of collision between solution quenchers and surface-bound photosensitizer(s). Only a small fraction of such collisions leads to quenching; the 2-3 orders of magnitude decrease in k_q (compare the data of Table 9 with those of

TABLE 9

Rate constants for electron transfer quenching of a series of luminescent $*Ru(NN)_3^{2+}$ complexes with Hg(II) species in N_2-purged Sodium Lauryl Sulfate (NaLS) solutions.[a]

NN	$E^o(Ru^{*2+/3+})$ V	$HgCl_2$[b] $10^{-6}k_q$ $(M^{-1}s^{-1})$	$HgCl_2$[d] $10^{-6}k_q$ $(M^{-1}s^{-1})$	$HgCl_3^-$[e] $10^{-6}k_q$ $(M^{-1}s^{-1})$
3,4,7,8-Me_4phen	+1.11	74.8	60.9	26.0
4,7-Me_2phen	+1.01	30.4	24.0	11.5
		22.9[c]	-	18.2[c]
4,4'-Me_2bpy	+0.94	18.8	12.0	5.78
5,6-Me_2phen	+0.93	16.7	12.5	3.79
5-Mephen	+0.90	13.2	10.5	3.36
4,7-Ph_2phen	+0.90	0.95	0.80	-
phen	+0.87	11.7	7.16	4.32
		7.14[c]	-	5.93[c]
bpy	+0.84	3.79	1.84	<0.06
5-Clphen	+0.77	2.45	1.44	<0.09
5-Brphen	+0.76	2.88	1.14	0.45

[a] reference 79. [b] u = 0.0M. [c] u = 0.40M with $NaNO_3$ + NaCl. [d] u = 0.10M with $NaNO_3$. [e] u = 0.10M with NaCl.

Table 7) is ascribed (in part) to a geometric shielding factor (the effectiveness of encounters between quencher and micellized photosensitizer is decreased by a factor of ∿ 30) and to the negative micellar charge which influences the $\Delta G^{\#}(0)$'s. These represent the energy necessary to reorganize the nuclear configurations of the precursor complex prior to the electron transfer event. For $HgCl_2$ and $HgCl_3^-$ quenchers in 0.1 M NO_3^- and Cl^- media, respectively, $\Delta G^{\#}(0)$ are 7.0 ± 0.5 kcal/mol and 7.6 ± 0.5 kcal/mol [79] (compare with the values in aqueous media). These larger $\Delta G^{\#}(0)$'s reflect the highly hydrophobic nature of the micellar environment, and thus the work necessary to reorganize the media around the photosensitizer, in order to accomodate an increase in sensitizer change, is expectedly greater than a similar rearrangement in non-micellar media. Moreover, the observed reactivity is $HgCl_2 > HgCl_3^- > HgCl_4^{2-}$ in accord with electrostatic expectations.

Contrary to the above, the enhanced luminescence quenching of $*Ru(bpy)_3^{2+}$ by Cu^{2+} in micellar sodium dodecyl sulfate media is quantitatively accountable in terms of a completely micellized photosensitizer and rapid equilibration of the distribution of the quencher between the aqueous and micellar phases: $k_q = 2.1 \times 10^5$ s^{-1}; below the c.m.c., $k_q = 9{,}5 \times 10^6 s^{-1}$ [80]. The quenching is slow relative to equilibration. For $*Ru(bpy)_2(CN)_2$, emission quenching by Cu^{2+} is faster than equilibration. Apparently, Cu^{2+} has a particular affinity for micelles containing this photosensitizer (quasi-static quenching by $\sim$ 70%).

4.1.3 By Cobalt(III) Complexes

Since the report [1] that $*Ru(bpy)_3^{2+}$ can act as a one-electron reducing agent, numerous quenching experiments have been carried out; the quencher molecules have included a wide variety of inorganic transition metal complexes: oxalates [5], bipyridyls [81], phenanthrolines [45,82], ammines [1,45], cyanides [83-85], acetylacetonates [58], ethylenediaminetetraacetate EDTA complexes [86,87], and dimeric u-oxo bridged cobalt complexes [88-90].

The photodecomposition of cobalt(III) complexes has been investigated in detail; these were the first inorganic complexes used to test whether quenching of the $*Ru(bpy)_3^{2+}$ emission occurs via energy transfer or electron transfer [1]. The details of the mechanisms of the decomposition reactions have been controversial. The pentaammine complexes $Co(NH_3)_5X^{2+}$ (X = F, Cl, Br) and Co(HEDTA) Cl^- decompose upon photoreduction yielding $Co^{2+}(aq)$ (reactions 46 and 47) [1]. The quantum yields of $Co^{2+}(aq)$ formation, $\Phi(Co^{2+})$,

$$*Ru(bpy)_3^{2+} + Co^{III}(NH_3)_5X^{2+} \longrightarrow Ru(bpy)_3^{3+} + Co^{II}(NH_3)_5X^+ \qquad (46)$$

$$Co^{II}(NH_3)_5X^+ \longrightarrow Co^{2+}(aq) + 5NH_3 + X^- \qquad (47)$$

are 0.0003 (F^-), 0.063 (Cl^-), 0.104 (Br^-), and $\sim$ 0.2 for the $HEDTA^{3-}$ complex. Gafney and Adamson [1] concluded that the mechanism of the reaction "appears to be one of electron transfer from $*Ru(bpy)_3^{2+}$ to $Co(NH_3)_5X^{2+}$ during an encounter"; they did not exclude the possibility that ordinary photosensitization occurred, followed by oxidation of $Ru(bpy)_3^{2+}$ in a secondary reaction involving radical species. In a later paper [5], the potential for

$*Ru(bpy)_3^{2+}$ to act as a one-electron photoredox sensitizer was tested with the oxalate $Co(C_2O_4)_3^{3-}$ complex (eqn 48); $k_q = 6.7 \times 10^9 M^{-1}s^{-1}$ and $\Phi(Co^{2+}) = 0.85$ in 0.1N H_2SO_4. By contrast, Natarajan and Endicott [42,86,87,91] argued, from photolysis and scavenging studies on cobalt-EDTA and -ammine complexes, that the mechanism involves triplet-triplet energy transfer, followed by an intramolecular oxidation of one of the ligands of the Co(III) substrate (eqns 49-52; for X = Br). The details of the mechanism were subse-

$$*Ru(bpy)_3^{2+} + Co(C_2O_4)_3^{3-} \longrightarrow Ru(bpy)_3^{3+} + Co^{2+}(aq) + 3C_2O_4^{2-} \quad (48)$$

quently re-examined by Navon and Sutin [44] who concluded that the sensitized decompositions of $Co(NH_3)_5X^{2+}$ ($X = Br^-, Cl^-, H_2O, NH_3$) occur via direct electron transfer as outlined in reactions 46 and 47; the quenching rates increase in the order $NH_3 < H_2O < Cl^- < Br^-$ with the latter being diffusion-controlled. To the extent that

$$*Ru(bpy)_3^{2+} + Co(NH_3)_5Br^{2+} \longrightarrow Ru(bpy)_3^{2+} + *Co(NH_3)_5Br^{2+} \quad (49)$$

$$*Co(NH_3)_5Br^{2+} \longrightarrow [Co^{II}(NH_3)_5;Br^\cdot]^{3+} \longrightarrow Co^{2+}(aq) + 5NH_3 + Br^\cdot \quad (50)$$

$$Br^\cdot + Br^- \longrightarrow Br_2^{\overline{\cdot}} \quad (51)$$

$$Br_2^{\overline{\cdot}} + Ru(bpy)_3^{2+} \longrightarrow Ru(bpy)_3^{3+} + 2Br^- \quad (52)$$

$Ru(bpy)_3^{3+}$ is produced in yields of ∿100% for all halides, the Endicott energy-transfer mechanism would suggest that the oxidizing radicals ($X^\cdot$ and $X_2^{\overline{\cdot}}$) are formed from excited states of the *Co (III) complexes with close to unitary efficiency, followed by a quantitative reaction with $Ru(bpy)_3^{2+}$. These two conditions are not satisfied by the cobalt complexes [44]. None of the Co(III) complexes investigated possess low-lying reactive excited states that could act as energy acceptors. The results on the variations in the yield of $Ru(bpy)_3^{3+}$ and $Co^{2+}(aq)$ in 50% 2-propanol can also be interpreted [44] without invoking competitive radical reactions.

Substantial differences exist between the excited states $*Ru(bpy)_3^{2+}$ and $*Os(bpy)_3^{2+}$. The lifetime of the latter (∿19 ns) is considerably less than that of $*Ru(bpy)_3^{2+}$ under similar conditions (∿600 ns); the reduction potentials of $Os(bpy)_3^{3+}$ (0.88 V in 1M acid) is also less than that of $Ru(bpy)_3^{3+}$ (1.24 V in 1M acid). Such differences have been exploited to obtain information on the quenching mechanism of $*Os(bpy)_3^{2+}$ by a number of acidopentaammine cobalt(III) complexes [92]. If an energy transfer path were operative, the lower energy and shorter lifetime of $*Os(bpy)_3^{2+}$ would predict this donor to be less efficient than the *Ru analogue. The consistency of the reaction stoichiometry [$Os(bpy)_3^{3+}$ and Co^{2+}(aq) are produced in equimolar amounts] along with the high limiting quantum yields of Co^{2+}(aq) [$\Phi \sim 1.09 \pm 0.18$ vs. 0.52 ± 0.21 for $*Ru(bpy)_3^{2+}$ as the donor under identical conditions] for this low energy *Os(II) donor indicates that reduction of $Co^{III}(NH_3)_5Cl^{2+}$ occurs via oxidative electron transfer. The efficiency of electron transfer per quenching encounter is essentially 100%, and the overall efficiency of the photoredox reaction is determined by the rate of the back reaction [Co(II) + Os(III) $\rightarrow$ Co(III) + Os(II)] relative to the rate of the reactions which convert the reduced cobalt complex to a form not readily oxidizable by the $M(bpy)_3^{3+}$ complex [92].

Table 10 summarizes the oxidative quenching rate constants for the EDTA, ammine, and oxalato cobalt(III) complexes noted above. Of the tris-oxalato complexes of Co(III), Cr(III), and Fe(III) investigated, only the Co complex quenches $*Ru(bpy)_3^{2+}$ by electron transfer efficiently. Oxidative quenching by $Co(HEDTA)X^-$ complexes

TABLE 10
Rate constants for oxidative electron transfer and redox quantum yields following quenching of $*Ru(bpy)_3^{2+}$ by some Co(III) complexes.

Quencher	Medium	$\Phi(M^{n+})$	$10^{-9}k_q(M^{-1}s^{-1})$	Ref.
$Co(NH_3)_6^{3+}$	0.5M H_2SO_4	small(Co^{2+})	0.01	1,44
$Co(NH_3)_5Cl^{2+}$	0.5M H_2SO_4	1.0(Ru^{3+})	0.93	1,44
	u = 0.02-0.22M acetate buffer	-	0.22-1.4	93
$Co(NH_3)_5H_2O^{3+}$	0.5M H_2SO_4	1.0(Ru^{3+})	0.15	44
$Co(NH_3)_5Br^{2+}$	0.5M H_2SO_4	1.0(Ru^{3+})	2.5	1,42,44,91
$Co(C_2O_4)_3^{3-}$	0.5M H_2SO_4	0.85(Co^{2+})	6.7	4,5
$Co(EDTA)NO_2^-$	H_2O, pH 3.0	0.71(Co^{2+})	1.3	86,87
$Co(HEDTA)Br^-$	H_2O, pH 3.0	0.80(Co^{2+})	3.0	86,87
$Co(HEDTA)Cl^-$	H_2O, pH 3.0	0.24(Co^{2+})	0.8	86,87
$Co(EDTA)H^{2-}$	H_2O, pH 3.0	0.10(Co^{2+})	0.3	86,87

produce Co^{2+}(aq) with relatively high quantum yields depending on the nature of X. Recently, Kimura and coworkers [93] used the redox chemistry inherent in the oxidative quenching of $*Ru(bpy)_3^{2+}$ by $Co(NH_3)_5Cl^{2+}$ to produce $Co(EDTA)^-$ very efficiently in acetate buffer. The ligand substitution reaction between the substitution-inert $Co(NH_3)_5Cl^{2+}$ complex and $EDTA^{4-}$ is mediated by small amounts of $Ru(bpy)_3^{2+}$ acting as a photocatalyst (Scheme VI).

Scheme VI:

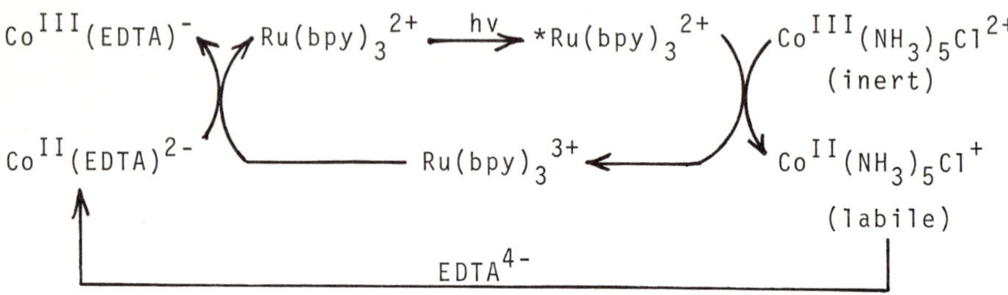

When L is pyridine (py), 1,2-bis(4-pyridyl)ethane, 4,4'-bipyridine, 1-methyl-4,4'-bipyridyl, and 1,2-bis(4-pyridyl)ethene, net photochemistry is observed on electron transfer quenching of $*Ru(bpy)_3^{2+}$ with $Co^{III}(NH_3)_5L^{n+}$ [94]. The k_q's and Φ(redox) data are given in Table 11 [94,95]. Flash photolysis experiments failed to reveal the existence of radical intermediates. Quantum yield measurements show that, in the $Ru(bpy)_3^{2+}$-$Co(NH_3)_5(NC_5\text{-}C_5N)^{3+}$ and -$Co(NH_3)_5(NC_5\text{-}C_5NCH_3)^{4+}$ systems, a Co^{2+}(aq) is formed for every quenching event; this rules out a cage recombination reaction between the primary one-electron reduction product of the Co(III) complex and the $Ru(bpy)_3^{3+}$. Scheme VII summarizes the quenching events: k_{sep} is the rate constant for cage escape ($10^{10} s^{-1}$), k_{et} and k'_{et} are the rate constants for intramolecular electron transfer from the ligand radical to the cobalt(III) centre in the solvent cage and in the cage-free ions, respectively; L· is the one-electron reduction product radical. For L = 4,4'-bipyridine, 1-methyl-4,4'-bipyridyl, and 1,2-bis(4-pyridyl)ethene, the details of electron transfer quenching proceeds via a chemical mechanism, whereby the electron is transferred to the pyridine ligand and, in a subsequent step, from the bound radical to the cobalt(III) centre. When electron transfer occurs within the solvent cage $k_{et} > 10^{11} s^{-1}$; if the caged ions escape $k'_{et} > 4 \times 10^6 s^{-1}$. Hence, a chem-

TABLE 11

Rate constants and quantum yields of Co^{2+} production in the oxidative electron transfer quenching of $*Ru(bpy)_3^{2+}$ by $Co(NH_3)_5L^{n+}$ complexes.[a]

L	$10^{-8}k_q$ ($M^{-1}s^{-1}$)	$\Phi(Co^{2+})$	$10^{-9}k_d$ ($M^{-1}s^{-1}$)	$10^{-9}k_{-d}$ (s^{-1})	$10^{-10}k''_{et}$ (s^{-1})
pyridine	1.72[b]	-	1.8	6.8	0.07
NC_5-C_5N	5.67[b]	0.17	-	-	-
	12.0[c]	0.15[d]	-	-	-
	7.25[d]	-	-	-	-
$NC_5-C_5NCH_3^+$	10.9[b]	0.14[d]	1.4	6.8	2.5
	7.23[d]	-	-	-	-
$NC_5-CH_2CH_2-C_5N$	2.02[b]	-	-	-	-
$NC_5-CH=CH-C_5N$	11.5[b]	-	-	-	-
$p-NO_2-C_6H_4-COO-$	24	0.011	3.0	4.2	1.7
$o-NO_2-C_6H_4-COO-$	13	0.051	2.9	4.4	0.4
C_6H_5-COO-	1.5	0.45	2.9	4.4	0.02
CH_3-COO-	2.1	-	2.8	5.3	0.06

[a] references 94,95. [b] pH 5.6, 0.01M phosphate buffer, u = 0.10M $LiClO_4$. [c] 0.0944M $HClO_4$, u = 0.10M. [d] 0.50M H_2SO_4.

Scheme VII:

$$*Ru(bpy)_3^{2+} + Co^{III}(NH_3)_5L^{n+} \xrightarrow{k_q} [Ru(bpy)_3^{3+};Co^{III}(NH_3)_5L^{n\dot{+}}]$$

$$[Ru(bpy)_3^{3+};Co^{III}(NH_3)_5L^{n\dot{+}}] \xrightarrow{k_{esc}} Ru(bpy)_3^{3+} + Co^{III}(NH_3)_5L^{n\dot{+}}$$

$$[Ru(bpy)_3^{3+};Co^{III}(NH_3)_5L^{n\dot{+}}] \xrightarrow{k_{et}} [Ru(bpy)_3^{3+};Co^{II}(NH_3)_5L^{n+}]$$

$$Ru(bpy)_3^{3+} + Co^{III}(NH_3)_5L^{n\dot{+}} \xrightarrow{k'_{et}} Ru(bpy)_3^{3+} + Co^{II}(NH_3)_5L^{n+}$$

ical mechanism obtains in this case when a ligand is implicated in the electron transfer [94]. Rate comparisons have also been used [96] in support of a chemical mechanism in the quenching of $*Ru(bpy)_3^{2+}$ by $Fe(CN)_5(pyrCH_3)^{2-}$ and $Co(CN)_5(pyrCOOH)^{2-}$ by an electron transfer pathway whereby the pyrazine (pyr) ligands are reduced to the corresponding radicals.

Evidence for cage recombination reactions has recently been demonstrated for $Co(NH_3)_5L^{n+}$ quenching of $(Ru(bpy)_3{}^{2+}$; L is p-nitro benzoate, o-nitrobenzoate, benzoate, and acetate [95]. Quenching of $*Ru(bpy)_3{}^{2+}$ by these cobalt complexes occurs by electron transfer to the unique ligand of the quencher (L = nitrobenzoates) or to the cobalt(III) centre of the quencher (L = benzoate and acetate) to produce ligand-centred or cobalt-centred successor complexes in the solvent cage (see Scheme VII). For the latter complexes, cage dissociation leading to the production of cobalt(II) and cage recombination resulting in the regeneration of the reactants are competitive; substantial yields of Co(II) are observed. For ligand-centred complexes, cage recombination predominates and the $\Phi(Co^{2+})$ are low (Table 11) [95]. It is instructive to consider the k_q step in Scheme VII in greater detail; this is shown in reactions 53 and 54. It is clear that for the (L =) p-nitrobenzoate, o-nitrobenzoate, and bipyridine complexes, the k_q values approach the calculated values of k_d and hence every encounter [or every other encounter

$$*Ru(bpy)_3{}^{2+} \quad + \quad Q \; \underset{k_{-d}}{\overset{k_d}{\rightleftharpoons}} \; [*Ru(bpy)_3{}^{2+}...Q] \tag{53}$$

$$[*Ru(bpy)_3{}^{2+}...Q] \xrightarrow{k''_{et}} [Ru(bpy)_3{}^{3+}...Q^-] \tag{54}$$

for $Co(NH_3)_5(o\text{-}NO_2C_6H_4COO)^{2+}$] results in electron transfer. In contrast, k_q values for the benzoate, acetate, and pyridine complexes are at least one order of magnitude smaller than diffusion-controlled limits; <10% of the encounters result in deactivation of $*Ru(bpy)_3{}^{2+}$ [95].

Flash photolysis of the systems $Ru(bpy)_3{}^{2+}$-$Co(phen)_3{}^{3+}$ or -$Co(bpy)_3{}^{3+}$ generates the corresponding Ru(III) and Co(II) complexes in relatively high yields (1.1 ± 0.3 and 0.52 ± 0.02, respectively); the respective k_q's are $2.18 \times 10^9 M^{-1}s^{-1}$ and $2.27 \times 10^9 M^{-1} s^{-1}$ [97]. The quenching reaction is relatively exothermic, $\Delta G \sim -1.2$ eV; however, no evidence of a Co(II) intermediate containing a coordinated radical anion species is observed. Electron transfer may in part involve nuclear tunneling. The back reactions (eqns 55 and 56) k_{bet} ($\sim 2 \times 10^8 M^{-1}s^{-1}$) are about an order of magnitude smaller than the diffusion-corrected rate constants obtained from the Marcus model (eqn 24), $k_{calc} = 2 \times 10^9 M^{-1}s^{-1}$. The difference

$$Ru(bpy)_3^{3+} + Co(phen)_3^{2+} \longrightarrow Ru(bpy)_3^{2+} + Co(phen)_3^{3+} \qquad (55)$$

$$Ru(bpy)_3^{3+} + Co(bpy)_3^{2+} \longrightarrow Ru(bpy)_3^{2+} + Co(bpy)_3^{3+} \qquad (56)$$

between k_{calc} and k_{bet} indicates that non-adiabaticities associated with the spin multiplicities, or a mixed d-π* pathway for electron transfer have only a slight retarding effect on the rates of these two reactions.

Rate constants for the quenching of polypyridine-ruthenium (II), $Ru(NN)_3^{2+}$ excited states by caged cobalt(III) amine complexes range from 2 x 10^8 to 1 x $10^9 M^{-1}s^{-1}$ at 25°C (Table 12) [98]. The quenching process involves parallel energy transfer ($k \sim 1 \times 10^8 M^{-1}s^{-1}$) and electron transfer ($k_e \sim 0.1$-$1 \times 10^9 M^{-1}s^{-1}$) from $*Ru(NN)_3^{2+}$ to $Co(cage)^{3+}$ substrates. The latter mechanism predominates at high driving force. The free energy dependence of the quenching and back reaction rate constants is accounted for in terms of a weak-interaction semi-classical model; the quenching reactions are adiabatic, $\kappa \sim 1$, while the back reactions are non-adiabatic, $\kappa \sim 10^{-3}$-10^{-4}. The relatively large yields of electron transfer products (Y_{el} in Table 12 range from 0.3 to 1.0) observed for these systems are a consequence of this non-adiabatic character of the back reaction and result from the poor coupling of the $Ru(NN)_3^{3+}$ and $Co(cage)^{2+}$orbitals [98].

The continuing studies of outer-sphere electron transfer reactions are aimed (i) at understanding and possibly delineating the underlying nuclear and electronic factors that influence the rate constants, and (ii) at correlating these factors to some fundamental molecular parameter (chemical composition, geometric structure, ionic charge,...). Cobalt(III) and cobalt(II) complexes play a major role in these efforts because rate constants are typically less than the diffusion-controlled limit [14,30,99-106] for various reasons: (1) electron transfer events implicate $\sigma^*(e_g)$ antibonding orbitals resulting in major changes in nuclear coordinates, (2) the electron transfer event is spin-forbidden and thus can only take place by mixing excited- and ground-state wave functions, and (3) poor orbital overlap between excited donor and ground state acceptor orbitals [107]. A detailed quenching study of the luminescence from $*Ru(NN)_3^{2+}$ and $*Os(phen)_3^{2+}$ by several cobalt(III) complexes has recently been reported by Sandrini and coworkers [107]. The k_q's correlate with the rate constants of electron transfer between the ground states of the same Co(III) subs-

TABLE 12
Rate constants for the quenching of $*Ru(NN)_3^{2+}$ emission by various caged cobalt(III) ammine complexes.[a]

Quencher	Excited Redox Partner (NN)	$E^o_{Co}-*E^o_{Ru}$ V	$10^{-8}k_q(M^{-1}s^{-1})$	$10^{-8}k_{bet}$[b] $(M^{-1}s^{-1})$	Y_{el}
$Co(sep)^{3+}$	bpy	0.52	2.8(2.9)	5.5(1.58)[c]	0.51
(u=0.50M H_2SO_4)	4,4'-Me_2bpy	0.62	4.5	-	-
(0.20M LiCl)	5-Brphen	0.44	2.2(2.8)	-	-
	5-Clphen	0.45	2.0(2.7)	8.2(1.68)	0.35
	phen	0.55	5.5(7.0)	6.4(1.58)	0.53
	5-Mephen	0.58	5.5(7.3)	-	-
	5,6-Me_2phen	0.61	5.7(7.2)	-	-
	4,7-Me_2phen	0.69	7.0(9.5)	5.4(1.41)	0.64
	3,5,6,8-Me_4phen	0.72	3.0	-	-
	3,4,7,8-Me_4phen	0.79	6.9	-	-
$Co(azamesar)^{3+}$	5-Brphen	0.42	2.0	-	-
(u = 0.2M LiCl)	phen	0.53	4.0	-	-
	4,7-Me_2phen	0.67	8.4	-	-
$Co(ammesarH)^{4+}$	5-Clphen	0.58	2.3	-	-
(0.10M HCl +	phen	0.68	5.2	-	-
0.10M LiCl)	4,7-Me_2phen	0.82	7.4	-	-
$Co(diamsar)^{3+}$	5-Brphen	0.41	1.5	-	-
(0.16M LiCl +	5-Clphen	0.42	1.4	1.7(1.71)	0.30
0.05M N-ethyl morpholine)	phen	0.52	3.4	0.96(1.61)	0.65
	4,7-Me_2phen	0.66	5.8	0.31(1.44)	0.90
$Co(diamsarH_2)^{5+}$	5-Brphen	0.75	5.9	-	-
(0.10M HCl +	phen	0.86	8.6	0.079(1.27)	1.0
0.10M LiCl)	4,7-Me_2phen	1.00	10.6	-	-

[a] reference 98. [b] u = 0.20 M with LiCl. [c] $E^o_{Ru}-E^o_{Co}$ in V for back reaction is equivalent to reactions 55 and 56: $Ru(NN)_3^{3+}$ + Co-(cage)$^{2+}$ $\longrightarrow$ $Ru(NN)_3^{2+}$ + $Co(cage)^{3+}$, where cage denotes cryptand type ligands such as sepulchrate (sep; see above).

trates and $Ru(NH_3)_6^{2+}$ or $Cr(bpy)_3^{2+}$ as reductants; no such correlation is observed [108] with excited $*Cr(bpy)_3^{3+}$, which is completely quenched via energy transfer. In the case of $*M(NN)_3^{2+}$ (M = Ru, or Os) the major quenching pathways (k_q's are > $10^7 M^{-1}s^{-1}$; Table 13) are the energy- (reaction 57) and electron-transfer processes which are spin allowed and thermodynamically allowed, respectively.

TABLE 13

Quenching rate constants for the oxidative redox reaction of $*M(NN)_3^{2+}$ with cobalt(III) complexes.[a]

No.	Quencher	Excited Redox Partner	$10^{-8}k_q(obs)$[b] $(M^{-1}s^{-1})$	ΔG[e] eV	$10^{-8}k_q(corr)$ $(M^{-1}s^{-1})$
1.	$Co(NH_3)_6^{3+}$	$*Ru(bpy)_3^{2+}$	0.40	-0.96	1.2
		$*Ru(phen)_3^{2+}$	0.12	-1.02	0.67
		$*Ru(DM\text{-}bpy)_3^{2+}$	0.087	-1.04	0.49
		$*Os(phen)_3^{2+}$	1.0	-1.06	5.1
		$*Ru(DM\text{-}phen)_3^{2+}$	0.23	-1.13	1.3
2.	$Co(en)_3^{3+}$	$*Ru(bpy)_3^{2+}$	0.20	-0.68	0.61
		$*Os(phen)_3^{2+}$	1.0	-0.78	2.9
3.	$Co(phen)_3^{3+}$	$*Ru(bpy)_3^{2+}$	19	-1.26	19
		$*Os(bpy)_3^{2+}$	57[c]	-1.42	45
4.	$Co(NH_3)_5H_2O^{3+}$	$*Ru(bpy)_3^{2+}$	0.75	-1.23	3.9
		$*Os(phen)_3^{2+}$	3.7	-1.33	15
		$*Os(bpy)_3^{2+}$	2.1[c]	-1.39	5.3
5.	$Co(NH_3)_5F^{2+}$	$*Ru(bpy)_3^{2+}$	0.80		
		$*Os(phen)_3^{2+}$	2.9		
6.	$Co(NH_3)_5Cl^{2+}$	$*Ru(bpy)_3^{2+}$	5.8		
		$*Os(phen)_3^{2+}$	9.3		
		$*Os(terpy)_2^{2+}$	8.7[f]		
7.	$Co(NH_3)_5NCS^{2+}$	$*Ru(bpy)_3^{2+}$	3.0		
		$*Os(phen)_3^{2+}$	6.8		
8.	$Co(NH_3)_5(OOCH)^{2+}$	$*Ru(bpy)_3^{2+}$	0.47		
		$*Os(phen)_3^{2+}$	4.1		
9.	$Co(NH_3)_5NO_2^{2+}$	$*Ru(bpy)_3^{2+}$	1.0		
10.	cis-$Co(en)_2(H_2O)Cl^{2+}$	$*Ru(bpy)_3^{2+}$	6.0		
		$*Os(phen)_3^{2+}$	9.6		
11.	cis-$Co(en)_2(NH_3)Cl^{2+}$	$*Ru(bpy)_3^{2+}$	4.2		
		$*Os(phen)_3^{2+}$	9.3		
12.	cis-$Co(en)_2Cl_2^{+}$	$*Ru(bpy)_3^{2+}$	14.3		
		$*Os(phen)_3^{2+}$	22.5		
13.	trans-$Co(en)_2Cl_2^{+}$	$*Ru(bpy)_3^{2+}$	20.0		
		$*Os(phen)_3^{2+}$	26.5		
14.	cis-$Co(en)_2(NCS)Cl^{+}$	$*Ru(bpy)_3^{2+}$	15.5		
		$*Os(phen)_3^{2+}$	21.2		
15.	trans-$Co(en)_2(NCS)Cl^{+}$	$*Ru(bpy)_3^{2+}$	10.7		
		$*Os(phen)_3^{2+}$	16.0		
16.	cis-$Co(en)_2(NCS)_2^{+}$	$*Ru(bpy)_3^{2+}$	7.6		
		$*Os(phen)_3^{2+}$	19.0		
17.	trans-$Co(en)_2(NCS)_2^{+}$	$*Ru(bpy)_3^{2+}$	4.8		
		$*Os(phen)_3^{2+}$	17.0		
18.	$Co(bpy)_3^{3+}$	$*Ru(bpy)_3^{2+}$	23[d]	-1.20	13

[a] reference 107. [b] in 0.1M H_2SO_4 (u = 0.12M), 22°C. [c] u = 0.5M H_2SO_4. [d] u = 1.0M H_2SO_4. [e] free energy change calculated from standard redox potentials, see ref. 107. [f] reference 109.

$$*Ru(NN)_3^{2+} + (^1A_{1g})Co^{III} \longrightarrow Ru(NN)_3^{2+} + (^3T_{1g}, ^5T_{2g})Co^{III} \quad (57)$$

The correlation of the quenching data of Table 13, illustrated in Figure 9, has been taken as evidence [107] that reaction 57 is of little or of no consequence and quenching occurs by electron transfer. An examination of the pertinent data by the approach [31,51, 110] based on log k_q^{corr} against ΔG plots suggests [107] the following conclusions. (i) The electron transfer reactions of cobalt com-

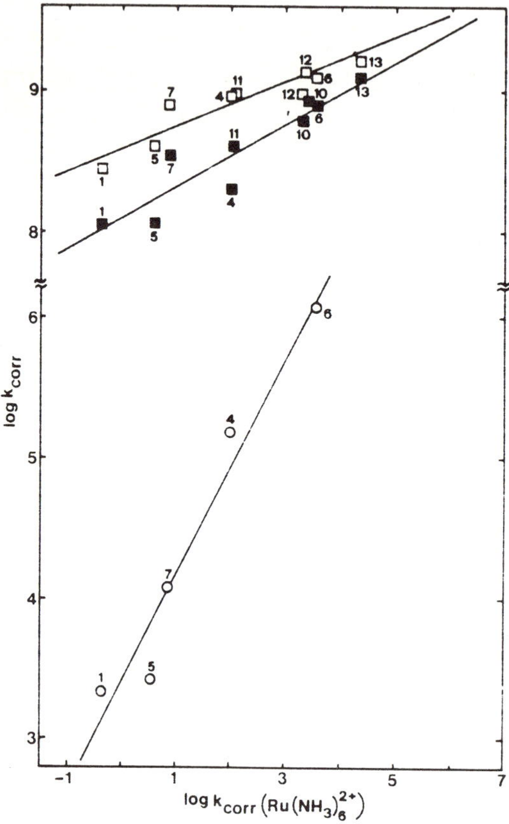

Figure 9.- Plot of the logarithm of the rate constant for the quenching of $*Ru(bpy)_3^{2+}$ (■) and $*Os(phen)_3^{2+}$ (□) and the oxidation of $Cr(bpy)_3^{2+}$ (o) by cobalt(III) complexes vs. the logarithm of the rate constants for the oxidation of $Ru(NH_3)_6^{2+}$ by the same Co(III) complexes. The cobalt complexes are numbered as in Table 13. Reproduced with permission from ref. 107.

plexes are intrinsically non-adiabatic; the electronic transmission coefficient κ and $\Delta G^{\#}$, the intrinsic barrier of the self-exchange reactions, are: 0.01-1 and 24 ± 2 kcal/mol for $Co(NH_3)_6^{3+}$ and $Co(NH_3)_5(H_2O)^{3+}$; 10^{-2}-10^{-4} and 17 ± 2 kcal/mol for $Co(en)_3^{3+}$; 10^{-4}-10^{-6} and 13 ± 2 kcal/mol for $Co(bpy)_3^{3+}$ and $Co(phen)_3^{3+}$ [107]. (ii) The decrease in κ from $Co(NH_3)_6^{3+/2+}$ to $Co(polypyridine)_3^{3+/2+}$ results from the involvement of a $\sigma^*(e_g)$ cobalt orbital in the electron transfer event and in a decrease in overlap between this $\sigma^*(e_g)$ orbital and the donor orbital as the size of the ligands increases.

Quantum yields of production of Co^{2+}(aq) from quenching of $*Ru(bpy)_3^{2+}$ with $Co(NH_3)_6^{3+}$, $Co(NH_3)_5Cl^{2+}$, and $Co(en)_3^{3+}$, and quenching rate constants have been determined recently; the results are summarized in Table 14 [111]. These data indicate that parallel

TABLE 14
Rate constants and quantum yields of Co(II) formation in the quenching of $*Ru(NN)_3^{2+}$ by cobalt(III) complexes.[a]

Quencher	Excited State (NN)	$10^{-7}k_q(M^{-1}s^{-1})$	$\Phi(Co^{2+})$	$\Phi_{Co^{2+}}(corr)$[b]
$Co(NH_3)_6^{3+}$	bpy	1.5	0.0045-0.0046	0.45
	5-Clphen	3.2		
	phen	4.6		
	5,6-Me_2phen	5.3		
	4,7-Me_2phen	7.7		
$Co(NH_3)_5Cl^{2+}$	bpy	38	0.18-0.19	1.05
$Co(en)_3^{3+}$	bpy	3.4	0.0021-0.0024	0.11
	5-Clphen	4.8		
	phen	6.5		
	5,6-Me_2phen	9.1		
	4,7-Me_2phen	11		

[a] reference 111. [b] quantum yields of formation of Co^{2+}(aq) corrected for unquenched $*Ru(bpy)_3^{2+}$.

energy-transfer and electron-transfer pathways are in fact operative for $Co(NH_3)_6^{3+}$ and $Co(en)_3^{3+}$. When corrected for the energy transfer contribution, the resulting electron transfer k_q's are 6.8 x $10^6 M^{-1}s^{-1}$ and 3.7 x $10^6 M^{-1}s^{-1}$, respectively. However, this finding of an energy transfer contribution does not imply that the other Co(III) complexes of Table 13 also react by energy transfer.

The corrected $\Phi(Co^{2+})$ show that for every quenching act, an electron transfer event occurs 11% and 45% of the time for $Co(en)_3^{3+}$ and $Co(NH_3)_6^{3+}$, respectively. As noted earlier, every quenching act implicates electron transfer for quenching of $*Ru(bpy)_3^{2+}$ by $Co(NH_3)_5Cl^{2+}$. Zahir and coworkers [111] have questioned the reliability of the derived values of the electronic transmission coefficient, κ, and the free energy of activation reported by Sandrini et al. [107] on the basis that the rate constants of the self-exchange reactions (e.g., reaction 58) estimated from eqn 24 are incompatible with the experimentally measured self-exchange rate constants;

$$Co(phen)_3^{3+} + Co'(phen)_3^{2+} \longrightarrow Co(phen)_3^{2+} + Co'(phen)_3^{3+} \qquad (58)$$

$k_{calc}(Co^{3+/2+}) = 0.038\ M^{-1}s^{-1}$ against a $k_{expt}(Co^{3+/2+})$ of $42\ M^{-1}s^{-1}$ [111]. Moreover, rate constants for quenching of $*Ru(bpy)_3^{2+}$ by $Co(NH_3)_6^{3+}$, $Co(en)_3^{3+}$, and $Co(phen)_3^{3+}$ include energy transfer contributions and unless these are subtracted, any treatment of k_q's as electron transfer rate constants may be erroneous [111]. This calls attention to the importance of measuring redox quantum yields to assess the relative contributions of energy transfer and electron transfer events in the deactivation of an excited state.

Additional criticisms by Furholz and Haim [112] for the above approach have appeared. These authors have presented an alternate treatment which makes use of the measured properties of the individual couples and of Marcus' cross relationship, modified [113] to take into account non-adiabaticity. It would seem that in the approach of Sandrini et al [107], (a) the fittings of the experimental data are insensitive to the value(s) of k_{11} of the self-exchange reaction of the quenchers, at least in the mildly non-adiabatic regime ($\kappa_{12}>10^{-3}$), and (b) the homogenized rate constants (k_q^{cor} in Table 13) can be fitted to various values of κ_{11} (from 1.4×10^{-4} to 1) and $\Delta G^{\#}_{11}$ (hence, no unique solution) [112]. The two treatments lead to different conclusions as to the outer-sphere electron reactions between $M(NN)_3^{3+/2+}$ and $Fe^{3+/2+}$ couples, for example. In the Sandrini et al. interpretation [107], the self-exchange of the iron(III)/iron(II) couple would be taken as non-adiabatic and the cross reactions also as non-adiabatic; in the Furholz and Haim's approach [112], the self-exchange of the $Fe^{3+/2+}$ would be adiabatic and the cross reactions non-adiabatic. Balzani and Scandola [114] have replied to these and additional critiques; they suggest in fact that the two treatments are not "alternate" of one another, but rather "complement" one another, and both may be used in different contexts. For instance, when kinetic differences between A/A^- couples in outer-sphere electron transfer reactions are sought, a comparison between parameters derived from their cross reactions, with the same series of homogeneous partners, may likely prove more instructive than one between parameters derived from their self-exchange reactions. It is clear that the adiabatic vs. non-adiabatic debate [115] continues unabated, and appears to depend mostly on the different approaches taken to obtain values of the electronic transmission coefficients κ.

Quenching of $*Ru(bpy)_2(CN)_2$ and $*Ru(phen)_2(CN)_2$ by $Co(NH_3)_6^{3+}$ ($1.39 \times 10^9 M^{-1}s^{-1}$ and $2.3 \times 10^9 M^{-1}s^{-1}$), $Co(NH_3)_5Cl^{2+}$ ($3.35 \times 10^9 M^{-1}s^{-1}$ and $3.9 \times 10^9 M^{-1}s^{-1}$), and $Co(C_2O_4)_3^{3-}$ ($2.98 \times 10^9 M^{-1}s^{-1}$ and $3.24 \times 10^9 M^{-1}s^{-1}$) occurs with much higher k_q's than most complexes which quench only by energy transfer ($\sim 10^7 M^{-1}s^{-1}$) [70]. In keeping with the quenching of $*Ru(bpy)_3^{2+}$, the major component is oxidative electron transfer to yield Co^{2+}(aq), but some quenching by energy transfer is not ruled out [70].

Both the dinuclear superoxo-bridged Co(III)-ammine complexes IV and V quench $*Ru(bpy)_3^{2+}$ predominantly by oxidative electron transfer to the superoxide group, in accord with thermal reduction of these cobalt(III) complexes by powerful reducing agents [89].

$$[(NH_3)_4Co\overset{O-O}{\underset{NH_2}{\diamond}}Co(NH_3)_4]^{4+} \qquad [(en)_2Co\overset{O-O}{\underset{NH_2}{\diamond}}Co(en)_2]^{4+}$$

IV V

More detailed studies suggest that quenching occurs by parallel energy-transfer (20%) and electron-transfer (80%) mechanisms (Scheme VIII) [88,90]; the quenching data are presented in Table 15. The back thermal reaction k_{bet} is smaller ($10^7 M^{-1}s^{-1}$) than k_q. The electron transfer pathway <u>does not lead</u> to formation of the cobaltous ions Co^{2+}(aq) nor to the observed Ru(III) substrate. Formation of these ions takes place via the energy-transfer pathway [90].

TABLE 15

Results from quenching $*Ru(bpy)_3^{2+}$ by superoxo-bridged dicobalt(III) complexes.[a]

Quencher, Q	Medium (u, M)	$10^{-9}k_q(M^{-1}s^{-1})$[b]	$\Phi(Co^{2+})$[c]	$\Phi(Ru^{3+})$[c]
$[(NH_3)_5Co\text{-}O\text{-}O\text{-}Co(NH_3)_5]^{5+}$	1.0M HCl (1.0)	1.91	0.021	0.022
	0.5M H_2SO_4 (1.5)	4.62		
	1.0M $HClO_4$(1.0)	4.30		
$[(NH_3)_4Co(\mu\text{-}O\text{-}O)(\mu\text{-}NH_2)Co(NH_3)_4]^{4+}$	1.0M HCl (1.0)	4.00	0.039	0.041
	1.0M $HClO_4$ (1.0)	8.84		
$[(en)_2Co(\mu\text{-}O\text{-}O)(\mu\text{-}NH_2)Co(en)_2]^{4+}$	1.0M HCl (1.0)	3.20	0.012	0.076
$[(bpy)_2Co(\mu\text{-}O\text{-}O)(\mu\text{-}NH_2)Co(bpy)_2]^{4+}$	0.81M HCl (0.81)	4.47	0.031	0.015
$[(phen)_2Co(\mu\text{-}O\text{-}O)(\mu\text{-}NH_2)Co(phen)_2]^{4+}$	1.0M HCl (1.0)	8.10	-	-

[a] reference 90. [b] k_q of scheme VIII and comprises energy- and electron-transfer contributions. [c] for energy transfer contribution (see scheme VIII).

Scheme VIII:

$$[Ru(bpy)_3^{2+};{*Q}] \rightarrow Ru(bpy)_3^{3+} + 2Co^{2+} + radical + \ldots$$

$$Ru(bpy)_3^{2+} + radical \rightarrow Ru(bpy)_3^{3+} + \ldots$$

20% (upper path)

$${*Ru(bpy)_3^{2+}} + Q$$

80% (lower path)

$$[Ru(bpy)_3^{3+};Q^-] \xrightarrow{k_{bet}} Ru(bpy)_3^{2+} + Q$$

(Q^- are the corresponding peroxo complexes)

4.1.4 By Osmium- and Ruthenium-Ammines

Quenching of $*Ru(NN)_3^{2+}$ by substituted bipyridinium cations rhodium(III) polypyridyls (discussed later), and osmium(III)-ammine complexes shows no evidence of an inverted Marcus behaviour [116]. After correction for an energy-transfer component ($k_q = k_{el} + k_{en}$), the rate constants for reduction of $Os^{III}(NH_3)_5X^{2+}$ (k_{el} = $2 \times 10^6 - 2 \times 10^8 M^{-1}s^{-1}$ for X = Cl, Br, I, H_2O, and NH_3) and for oxidation of $Os(NH_3)_5N_2^{2+}$ are in accord with expectations and consistent with $Os^{3+/2+}$ exchange rates (10^2-$10^5 M^{-1}s^{-1}$ range); the data are collected in Table 16 [116]. Variations in k_q's reflect changes in the driving force, a function of the redox potential of $*Ru(NN)_3^{2+}$ and the nature of NN as in outer-sphere electron transfer reactions of ground state complexes. Additionally, k_q's for redox quenching are affected by the redox potentials of the Q^+/Q or Q/Q^- couples and the intrinsic electron transfer barriers of the $Q/*Ru(NN)_3^{2+}$ couples [116]. The self-exchange reactions of the $*Ru(NN)_3^{2+}/Ru(NN)_3^{3+}$ (eqn 59) and $Ru(NN)_3^+/*Ru(NN)_3^{2+}$ (eqn 60) couples possess very small electron transfer barriers [15].

$$*\underline{Ru}(NN)_3^{2+} + Ru(NN)_3^{3+} \xrightarrow{k \gtrsim 10^8 M^{-1}s^{-1}} \underline{Ru}(NN)_3^{3+} + *Ru(NN)_3^{2+} \quad (59)$$

$$\underline{Ru}(NN)_3^{+} + *Ru(NN)_3^{2+} \xrightarrow{k \sim 10^9 M^{-1}s^{-1}} *\underline{Ru}(NN)_3^{2+} + Ru(NN)_3^{+} \quad (60)$$

The rate constants for quenching $*Ru(NN)_3^{2+}$ by Os^{III}-ammine complexes increase with the reducing power of $*Ru(NN)_3^{2+}$

TABLE 16

k_q's for quenching $*Ru(NN)_3^{2+}$ by osmium(II) and osmium(III) ammine complexes in 0.5M sulfuric acid.[a]

Quencher (E^o, V)	$10^{-8}k_q(M^{-1}s^{-1})$					
	5-Clphen	bpy	5,6-Me_2phen	4,7-Me_2phen	3,5,6,8-Me_4phen	3,4,7,8-Me_4phen
$Os^{II}(NH_3)_5N_2^{2+}$ (+0.56)	5.2	0.9	-	$\lesssim$0.03	-	-
$Os(NH_3)_5Cl^{2+}$ (-0.85)	$\lesssim$0.2	$\lesssim$0.4	-	3.0	-	4.2
$Os(NH_3)_6^{3+}$ (-0.78)	$\lesssim$0.09	$\lesssim$0.14	$\lesssim$0.3	0.50	-	0.79
$Os(NH_3)_5I^{2+}$ (-0.77)	3.2	3.2	6.6	-	6.6	8.5
$Os(NH_3)_5(H_2O)^{3+}$ (-0.73)	$\lesssim$0.05	$\lesssim$0.07	$\lesssim$0.2	0.87	0.49	1.4

[a] reference 116

[i.e., as $E^o(Ru(NN)_3^{3+/2+})$ becomes more negative]; this is expected for oxidative quenching of $*Ru(NN)_3^{2+}$ [116]. It is thermodynamically favoured (cf. E^o in Table 16) for all of the $Os(NH_3)_6^{3+}$ and $Os(NH_3)_5(H_2O)^{3+}$ systems and for two of the four $Os(NH_3)_5Cl^{2+}$ systems. The iodopentaammine-osmium(III) complex seems unique in this series of quenchers; quenching occurs predominantly by energy transfer inasmuch as as this complex manifests LMCT absorptions at 24,500 cm^{-1} and near-IR absorption at 4,000 cm^{-1}. No redox products were detected in the quenching by osmium(III) complexes: cage escape yields < 0.05 and k(cage escape) $\sim 10^9 M^{-1}s^{-1}$ [116]. By contrast, the k_q's for the $Os(NH_3)_5N_2^{2+}$ complex increase with the oxidizing power of $*Ru(NN)_3^{2+}$, indicative of reductive quenching which in light of the Os(III)/Os(II) redox potential (+0.56 V) is thermodynamically favoured for all the $Ru(NN)_3^{2+}$ systems examined. A parallel is found [117] between the relative quenching efficiencies and the susceptibilities of the complexes trans-$OsO_2(en)_2^{2+}$, cis-$Os(en)_2H_2^{2+}$, and cis-$Os(en)_2Cl_2^+$ to oxidation or reduction by $*Ru(bpy)_3^{2+}$; k_q's are respectively, $1.67 \times 10^9 M^{-1}s^{-1}$, $2.28 \times 10^8 M^{-1}s^{-1}$, and $5.21 \times 10^7 M^{-1}s^{-1}$ in 0.5 M (HCl + NaCl) solution. The low value of k_q for the dihydride complex reflects the mechanistic tendency of this complex to act as a two-electron reducing agent. While the $Os(en)_2Cl_2^+$ species is subject to oxidation and possibly to reduction by $*Ru(bpy)_3^{2+}$ but with a substantial lower driving force than the dioxo and dihydrido species, energy transfer may also be implicated [117].

Quenching of $*Ru(bpy)_3^{2+}$ emission by $Ru(NH_3)_5X^{3+}$ ($X = Cl^-$, NH_3) and $Ru(NH_3)_6^{2+}$ has been studied [22,44,45,53]; k_q's are 2.7 x $10^9 M^{-1}s^{-1}$ ($X = Cl^-$), 1.2-2.7 x $10^9 M^{-1}s^{-1}$ ($X = NH_3$), and 2.5 x $10^9 M^{-1}s^{-1}$ for $Ru(NH_3)_6^{2+}$. The oxidative quenching rate constant for the $*Os(bpy)_3^{2+}/Ru(NH_3)_6^{3+}$ system is 4.8 x $10^9 M^{-1}s^{-1}$ [118]. An oxidative electron transfer mechanism has also been implicated in the quenching of $*Ru(bpy)_3^{2+}$ by $Ru(NH_3)_5X^{3+}$ complexes (eqn 61), followed by rapid back electron transfer (eqn 62) to reform the reactants in

$$*Ru(bpy)_3^{2+} + Ru(NH_3)_6^{3+} \longrightarrow Ru(bpy)_3^{3+} + Ru(NH_3)_6^{2+} \quad (61)$$

$$Ru(bpy)_3^{3+} + Ru(NH_3)_6^{2+} \longrightarrow Ru(bpy)_3^{2+} + Ru(NH_3)_6^{3+} \quad (62)$$

their electronic ground states; no net chemical change occurs [44]. Moreover, the $Ru(NH_3)_5X^{3+}$ substrates quench more efficiently than the corresponding $Co(NH_3)_5X^{3+}$ complexes (cf. Table 14). Reactions with $Co(NH_3)_6^{3+}$ are slow because of the height of the intrinsic barrier; the self-exchange of $Co(NH_3)_6^{2+}$-$Co(NH_3)_6^{3+}$ is $10^{-9}M^{-1}s^{-1}$ compared to 8.2 x $10^2 M^{-1}s^{-1}$ for the $Ru(NH_3)_6^{2+}$-$Ru(NH_3)_6^{3+}$ electron exchange. The slow rate for quenching by $Co(NH_3)_6^{3+}$ cannot be due to an unfavourable driving force, since the redox potentials of $Co(NH_3)_6^{3+}$ and $Ru(NH_3)_6^{3+}$ are similar (+0.1 V) [44]. Spectroscopic considerations do not support energy transfer for $Ru(NH_3)_6^{3+}$ quenching of $*Ru(bpy)_3^{2+}$.

4.1.5 By Transition Metal Cyanide Complexes

Quenching of electronically excited states of $M(NN)_3^{2+}$ (M = Ru or Os) by transition metal cyanide complexes has been investigated extensively [22,45,58,70,84,85,96] to delineate between the various quenching pathways: diffusional vs. static and energy vs. electron transfer. Table 17 summarizes the data in which the predominant mechanism is electron transfer. The cyanide complexes used (Table 17) span a wide range of excited state energies and redox potentials, thus making a delineation of quenching mechanisms possible. Thus far in this work, the predominant quenching mechanism has been oxidative electron transfer from $*Ru(NN)_3^{2+}$ to the quencher. By contrast, reductive quenching by inorganic complexes appears scarce; it is more common with organic quenchers [16,26]. Both $Mo(CN)_8^{4-}$ and $Os(CN)_6^{4-}$ display reductive quenching; a small fraction of the oxidized $Mo(CN)_8^{3-}$ and $Os(CN)_6^{3-}$ escapes the back

TABLE 17

Quenching of $*Ru(NN)_3^{2+}$ (NN = bpy, phen) and other S* by various transition metal cyanide complexes via electron transfer.

Quencher	Ionic Strength	$10^{-8}k_q(M^{-1}s^{-1})$ bpy(path)	$10^{-8}k_q(M^{-1}s^{-1})$ phen(path)	Ref.
$Mo(CN)_8^{4-}$	0.50M NaCl	3.4(reductive)		84
$Os(CN)_6^{4-}$	0.50M NaCl	12 (reductive)		84
$Ru(CN)_6^{4-}$	0.50M NaCl	~0.1(reductive)		85
$Fe(CN)_6^{4-}$	0.50M NaCl	33 (reductive)		84
	-	730 (energy)	730 (energy)	58
	0.02M acetate buffer	35 (reductive)		45
$Co(CN)_6^{3-}$	0.50M NaCl	< 0.01(oxidative)		85
$Fe(CN)_6^{3-}$	0.50M NaCl	65 (oxidative)		22,84
	-	510 (energy)	440 (energy)	58
	0.50M H_2SO_4	≤73		21
$Ni(CN)_4^{2-}$	0.50M NaCl	56 (red./energy)		84
	-	28 (energy)		58
$Fe(CN)_5CO^{3-}$	0.50M NaCl	0.06(reductive)		96
$Fe(CN)_5(Me_2SO)^{3-}$	0.50M NaCl	0.56(reductive)		96
$Fe(CN)_5(pzCO_2)^{4-}$	0.50M NaCl	19 (reductive)		96
$Fe(CN)_5(imid)^{3-}$	0.50M NaCl	38 (reductive)		96
$Fe(CN)_5py^{3-}$	0.50M NaCl	42 (reductive)		96
$Fe(CN)_5(NMPz)^{2-}$	0.50M NaCl	53 (see text)		96
$Fe(CN)_5NO^{2-}$	0.50M NaCl	60		96

S*	Q	$10^{-9}k_q(M^{-1}s^{-1})$	Ref.
$*Os(bpy)_3^{2+}$	$Fe(CN)_6^{3-}$	13.0	22
		9.4(oxidative)	85
	$Mo(CN)_8^{4-}$	0.28(reductive)	85
	$Fe(CN)_6^{4-}$	0.38(reductive)	85
	$Ru(CN)_6^{4-}$	0.13(reductive)	85
	$Co(CN)_6^{3-}$	0.13(oxidative)	85
	$Ni(CN)_4^{2-}$	0.09(oxidative)	85
$*Ru(bpy)_2(CN)_2$	$Fe(CN)_6^{3-}$	6.9(ox./energy)	70
	$Fe(CN)_6^{4-}$	0.61(red./energy)	70
$*Ru(phen)_2(CN)_2$	$Fe(CN)_6^{3-}$	6.54(ox./energy)	70
	$Fe(CN)_6^{4-}$	0.51(red./energy)	70

reduction by $Ru(bpy)_3^+$ [84]. Energy transfer and oxidative electron transfer are not thermodynamically plausible for $Fe(CN)_6^{4-}$ and reductive quenching is indicated; with $Fe(CN)_6^{3-}$, oxidative electron transfer is the major quenching path for $*Ru(bpy)_3^{2+}$ and $*Os(bpy)_3^{2+}$. Support for reductive quenching by $Fe(CN)_6^{4-}$ comes from the observation that $Ru(bpy)_3^+$ is formed in the $*Ru(bpy)_3^{2+}/Fe(CN)_6^{4-}$ system [59]. The suggestions of energy transfer quenching by Demas and Addington [58] for this system finds no experimental support. These authors also suggested energy transfer quenching in $*Ru(bpy)_3^{2+}/Fe(CN)_6^{3-}$ on the basis of an absorption band observed by Kiss et al.[119] at 18,200 cm^{-1} in the spectrum of $Fe(CN)_6^{3-}$. Such a band, however, has not been seen by others [120, 121] and is absent [85] in freshly prepared solutions of pure $Fe(CN)_6^{3-}$, although an absorption feature at 19,000 cm^{-1} does show up on aging under the spectrophotometer light. Energy transfer, therefore, is not likely unless it involves "non-spectroscopic" excited states. Quenching of the two dicyano complexes by $Fe(CN)_6^{3-}$ and $Fe(CN)_6^{4-}$ (Table 17) appears to occur via parallel energy transfer and electron transfer [70].

For $Co(CN)_6^{3-}$, $Pd(CN)_4^{2-}$, and $Pt(CN)_4^{2-}$ complexes, $k_q < 10^6 M^{-1}s^{-1}$, an energy transfer pathway is safely ruled out for the Pd(II) and Pt(II) species [84], but may occur for $Co(CN)_6^{3-}$ [58]. The latter does quench $*Os(bpy)_3^{2+}$ rather efficiently ($k_q = 1.3 \times 10^8 M^{-1}s^{-1}$) by oxidative electron transfer [85].

The validity of the above inferences is tested in Figure 10 which shows a plot of log k_q against the free energy change for reductive and oxidative quenching [85]. The data points corresponding to reductive or oxidative quenching lie on a Marcus-type band. The scatter from a common curve in the plot arises from differences in the inherent barriers to electron transfer by the excited states and more so by the quenchers. Where energy transfer is the more plausible pathway, the point corresponding to reductive and oxidative electron transfer lie far from the Marcus-type curve [85].

The quenching rate constants of $*Ru(bpy)_3^{2+}$ by the first five pentacyanoferrates in Table 17 parallel the increasing reducing power of these complexes; k_q's approach the diffusion-controlled limit when E^0 becomes < 0.5 V [96]. This rate dependence indicates a dominant reductive quenching mechanism (eqn 63). For the other

$$*Ru(bpy)_3^{2+} + Fe(CN)_5L^{x-} \longrightarrow Ru(bpy)_3^{+} + Fe(CN)_5L^{(x-1)-} \qquad (63)$$

$Fe(CN)_5L^{x-}$ complexes, quenching occurs via a different mechanism. In the case of the N-methylpyrazinium complex, $Fe(CN)_5(NMPz)^{2-}$, a

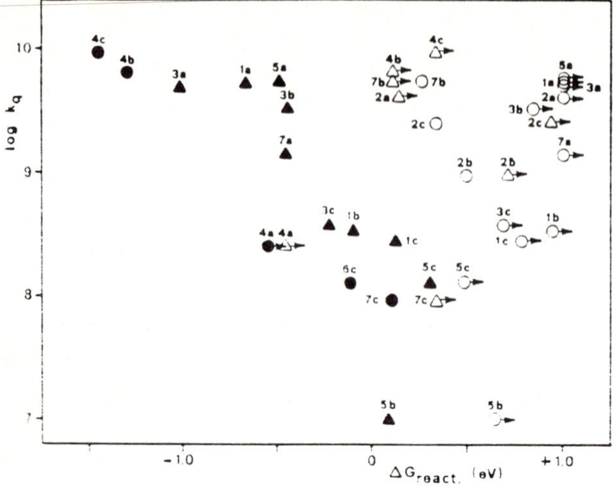

Figure 10.- Relation between k_q and the free energy change for reductive (Δ) and oxidative (o) quenching. Full points correspond to the quenching mechanism that is thought to be more probable. For each point, the $M(bpy)_3{}^{n+}$-quencher system is identified by a number indicating the quencher (see below) and a letter indicating the excited complex [a, b, and c are for $*Cr(bpy)_3{}^{3+}$, $*Ru(bpy)_3{}^{2+}$, and $*Os(bpy)_3{}^{2+}$, respectively]. The number indicates the following quencher: (1) $Mo(CN)_8{}^{4-}$, (2) $Cr(CN)_6{}^{3-}$, (3) $Fe(CN)_6{}^{4-}$, (4) $Fe(CN)_6{}^{3-}$, (5) $Ru(CN)_6{}^{4-}$, (6) $Co(CN)_6{}^{3-}$, (7) $Ni(CN)_4{}^{2-}$. Reproduced with permission from ref. 85.

quenching path featuring reduction of the $*Ru(bpy)_3{}^{2+}$ excited state by iron(II) is unlikely since k_q is an order of magnitude larger than observed for similar complexes of comparable E^o's [96]. It would seem that the bound NMPz ligand in $Fe(CN)_5(NMPz)^{2-}$ quenches the excited state oxidatively to produce $Ru(bpy)_3{}^{3+}$ and the neutral N-methylpyrazinium radical bound to iron(II) (eqn 64). Quenching by

$$*Ru(bpy)_3{}^{2+} + Fe^{II}(CN)_5(NMPz^+)^{2-} \xrightarrow[(ox.)]{k_q} Ru(bpy)_3{}^{3+} + Fe^{II}(CN)_5(NMPz\cdot)^{3-} \qquad (64)$$

the iron(II) complex via parallel pathways with more than one process contributing to the larger value of k_q cannot be ruled out [96].

4.1.6 By Transition Metal Polypyridyl Complexes: Vestiges of the Marcus Inverted Region

The quenching of the ^{3}MLCT state of $*Ru(bpy)_3^{2+}$ by $Cr(bpy)_3^{3+}$ occurs via oxidative electron transfer (k_q = 3.3 x $10^9 M^{-1} s^{-1}$) [122]. An interesting aspect of this $Ru(bpy)_3^{2+}/Cr(bpy)_3^{3+}$ system is that the metal-centred doublet state of $*Cr(bpy)_3^{3+}$ can, in turn, be quenched by $Ru(bpy)_3^{2+}$ by reductive electron transfer, k_q = 4.0 x $10^8 M^{-1} s^{-1}$; this was the first example of a bimolecular redox process to and involving metal-centred excited states. The redox process was later confirmed by a very elegant flash photolysis experiment [83]. Both $Ru(bpy)_3^{2+}$ and $Cr(bpy)_3^{3+}$ absorb a significant fraction of the exciting light, leading to formation of $Cr(bpy)_3^{2+}$ via (i) reaction 65 followed by reaction 66, (ii) reaction 67 followed by reaction 68, or (iii) reaction 69 followed by 66. Addition of I^-, which quenches $*Cr(bpy)_3^{3+}$, to the system in the

$$Cr(bpy)_3^{3+} \xrightarrow{h\nu} \longrightarrow *Cr(bpy)_3^{3+} \quad (65)$$

$$*Cr(bpy)_3^{3+} + Ru(bpy)_3^{2+} \xrightarrow{red.} Cr(bpy)_3^{2+} + Ru(bpy)_3^{3+} \quad (66)$$

$$Ru(bpy)_3^{2+} \xrightarrow{h\nu} \longrightarrow *Ru(bpy)_3^{2+} \quad (67)$$

$$*Ru(bpy)_3^{2+} + Cr(bpy)_3^{3+} \xrightarrow{ox.} Ru(bpy)_3^{3+} + Cr(bpy)_3^{2+} \quad (68)$$

$$*Ru(bpy)_3^{2+} + Cr(bpy)_3^{3+} \xrightarrow{en.tr.} Ru(bpy)_3^{2+} + *Cr(bpy)_3^{3+} \quad (69)$$

flash experiment shows that complete scavenging of $*Cr(bpy)_3^{3+}$ by I^- still gives $Cr(bpy)_3^{2+}$ on flash excitation (but less, ~50%); this suggests that both electron transfer pathways (eqns 65-66 and 67-68) take place in the $Ru(bpy)_3^{2+}/Cr(bpy)_3^{3+}$ system. $Cr(bpy)_3^{2+}$ decays via oxidation by $Ru(bpy)_3^{3+}$, k = 2.6 x $10^9 M^{-1} s^{-1}$ [83]. Such a system is remarkable in that the same pair of high-energy products is obtained irrespective of the absorbing species; the energetics are summarized in Scheme IX [118] which illustrates the conversion of light to chemical energy (1.51 eV).

A remarkable feature of the Marcus theory is its prediction (Section 3.1) of the 'inverted' region when $-\Delta G^o_{12}$ for the cross reaction is > $2(\Delta G^{\#}_{11} + \Delta G^{\#}_{22})$; crossover into the inverted region (see Figure 6) occurs when log K_{12} = 2 log $(Z^2/k_{11}k_{22})$ and is favoured when both K_{12} and $k_{11}k_{22}$ are large [38]. Rehm and Weller have sought evidence for this inverted region in the electron transfer quenching of hydrocarbon fluorescence; unfortunately, k_q's lie in the diffusion-controlled limit in the range of ΔG^o_{12} from -10 to

Scheme IX:

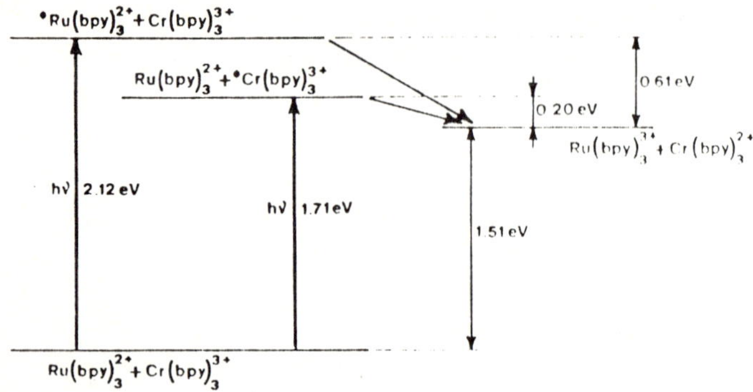

-62 kcal/mol with the onset of crossover expected at ∿ -15 kcal/mol [29]. Creutz and Sutin [38] have also sought evidence for the inverted region in the quenching of $*Ru(bpy)_3^{2+}$ and $*Ru(4,4'\text{-}Me_2bpy)_3^{2+}$ by $Ru(bpy)_3^{3+}$, $Ru(4,4'\text{-}Me_2bpy)_3^{3+}$, $Os(bpy)_3^{3+}$, and $Cr(bpy)_3^{3+}$, systems with small reorganization energies. The driving force ranges from ΔG^o_{12} = -10 to -46 kcal/mol with inversion expected at - G^o_{12} 16 kcal/mol. For the seven systems investigated, k_q's (0.5M H_2SO_4, 25°C) are 3-4 x $10^9 M^{-1}s^{-1}$. The data are illustrated in Figure 11 which shows vestiges of the inverted Marcus behaviour; the conclusion that inversion is observed is inescapable [38]. Effects by excited state self-exchange reactions (eqn 70) on the quenching reaction that might lead to lower k_q's were ruled out. As well, the possible discontinuity of the data for Os(III) and Ru(III) quenchers in Figure 11 as might occur if quenching also took place by other mechanisms was not considered significant; indeed the conclusions would be strengthened because if k_q (= k_{en} + k_{ox}) were a composite, then the k_{ox} component would be even smaller. Additional evidence for the inverted region in such systems arises from 71;

$$*Ru(bpy)_3^{2+} + Ru(NN)_3^{3+} \rightleftharpoons Ru(bpy)_3^{3+} + *Ru(NN)_3^{2+} \qquad (70)$$

$$2*Ru(bpy)_3^{2+} \longrightarrow Ru(bpy)_3^{+} + Ru(bpy)_3^{3+} \qquad (71)$$

an upper limit of $10^7 M^{-1}s^{-1}$ has been estimated [38]. Table 18 summarizes the data for electron transfer quenching of $*M(NN)_3^{2+}$ complexes by several transition metal polypyridyl complexes [82,109, 116,122]. Rate constants in the quenching of $*Ru(NN)_3^{2+}$ by

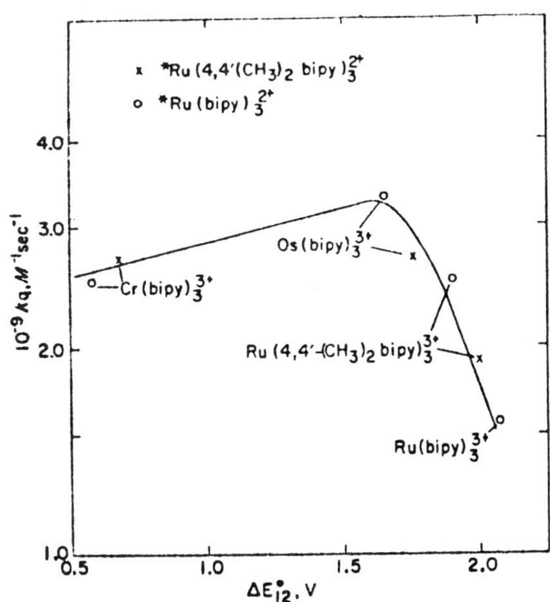

Figure 11.- Plot of the logarithm of k_q, the quenching rate constant ($0.5M\ H_2SO_4$, 25°C) vs. E^0_{12}, the driving force for electron transfer quenching. Reproduced with permission from ref. 38.

$Rh(4,4'-Me_2bpy)_3^{3+}$ are $0.001-1.0 \times 10^9 M^{-1}s^{-1}$, and are highly sensitive to the reducing power of $*Ru(NN)_3^{2+}$ [116]. Results of continuous and flash photolysis experiments have afforded a detailed mechanistic scheme (Scheme X) to be deduced [81] in the quenching of $*Ru(bpy)_3^{2+}$ by $Rh(bpy)_3^{3+}$. Light absorption yields $*Ru(bpy)_3^{2+}$

Scheme X:

$$*Ru(bpy)_3^{2+} + Rh(bpy)_3^{3+} \rightleftharpoons [*Ru(bpy)_3^{2+} | Rh(bpy)_3^{3+}]$$

$$\downarrow\uparrow$$

$$Ru(bpy)_3^{3+} + Rh(bpy)_3^{2+} \underset{}{\overset{k_{esc}}{\leftrightharpoons}} [Ru(bpy)_3^{3+} | Rh(bpy)_3^{2+}]$$

$$\downarrow k_{-e}$$

$$Ru(bpy)_3^{2+} + Rh(bpy)_3^{3+} \leftrightharpoons [Ru(bpy)_3^{2+} | Rh(bpy)_3^{3+}]$$

which is oxidized by $Rh(bpy)_3^{3+}$ ($k = 3.9 \times 10^8 M^{-1}s^{-1}$) to give $Ru(bpy)_3^{3+}$ and $Rh(bpy)_3^{2+}$; cage escape yield is 0.15. The back reaction of $Ru(bpy)_3^{3+}$ with $Rh(bpy)_3^{2+}$ ($k = 3 \times 10^9 M^{-1}s^{-1}$) can be pre-

TABLE 18
Rate constants for electron transfer quenching of excited states of polypyridyl complexes in aqueous solutions.

Quencher	Ionic Strength (Medium)	$10^{-9}k_q(M^{-1}s^{-1})$	Ref.
Donor = $*Ru(bpy)_3^{2+}$			
$Cr(bpy)_3^{3+}$	0.2M (NaCl)	3.3 (oxidative)	122
$Os(bpy)_3^{2+}$	0.1M (NaCl)	1.5 (red./energy)	82
$Ru(5\text{-}NO_2phen)_3^{2+}$	0.1M (NaCl)	2.0 (ox./energy)	82
$Ru(TPTZ)_2^{2+}$	0.5M (NaCl)	1.2 (ox./energy)	82
$Rh(5\text{-}Brphen)_3^{3+}$	0.5M (Na_2SO_4)	1.5	116
$Rh(5\text{-}Clphen)_3^{3+}$	0.5M (Na_2SO_4)	1.5	116
$Rh(5\text{-}Phphen)_3^{3+}$	0.5M (Na_2SO_4)	1.6	116
$Rh(phen)_3^{3+}$	0.5M (Na_2SO_4)	0.4	116
	0.5M (H_2SO_4)	0.68(oxidative)	81
$Rh(bpy)_3^{3+}$	0.5M (Na_2SO_4)	0.6 (oxidative)	116
	0.5M (NaOH)	0.37(oxidative)	81
$Rh(4,7\text{-}Me_2phen)_3^{3+}$	0.5M (Na_2SO_4)	1.73	116
$Rh(5,6\text{-}Me_2phen)_3^{3+}$	0.5M (Na_2SO_4)	0.3	116
$Rh(5\text{-}Mephen)_3^{3+}$	0.5M (Na_2SO_4)	0.4	116
$Rh(4,4'\text{-}Me_2bpy)_3^{3+}$	0.03M(Na_2SO_4)	0.03	116
$Co(bpy)_3^{+}$	0.10M(Et_4NClO_4)	6.6	109
$Co(4,4'\text{-}Me_2bpy)_3^{+}$	0.10M(Et_4NClO_4)	2.7	109
$Co(terpy)_2^{+}$	0.10M(Et_4NClO_4)	2.0	109
$Co(bpy)_3^{2+}$	0.10M(Et_4NClO_4)	0.11	109
$Co(4,4'\text{-}Me_2bpy)_3^{2+}$	0.10M(Et_4NClO_4)	0.16	109
$Co(terpy)_2^{2+}$	0.10M(Et_4NClO_4)	3.3	109
$Ru(bpy)_3^{3+}$	0.10M(Et_4NClO_4)	3.2	109
$Rh(bpy)_2(H_2O)_2^{3+}$	0.50M(H_2SO_4)	0.44	81
$Rh(bpy)_2(OH)_2^{+}$	0.50M(NaOH)	0.012(energy?)	81
$Rh(bpy)_2(H_2O)Cl^{2+}$	0.50M(H_2SO_4)	1.0	81
$Rh(bpy)_2(OH)Cl^{+}$	0.50M(NaOH)	0.4	81
Donor = $*Ru(phen)_3^{2+}$			
$Rh(bpy)_3^{3+}$	0.5M (Na_2SO_4)	0.6	116
	0.50M(H_2SO_4)	0.99	81
	0.50M(NaOH)	0.71	81

TABLE 18 (cont'd)

Quencher	Ionic Strength (Medium)	$10^{-9}k_q(M^{-1}s^{-1})$	Ref.
$Rh(5,6\text{-}Me_2phen)_3^{3+}$	0.5M (Na_2SO_4)	0.95	116
$Rh(5\text{-}Mephen)_3^{3+}$	0.5M (Na_2SO_4)	0.98	116
$Rh(phen)_3^{3+}$	0.50M(H_2SO_4)	1.16	81
$Rh(4,4'\text{-}Me_2bpy)_3^{3+}$	0.5M (Na_2SO_4)	0.2	116
$Rh(bpy)_2(H_2O)_2^{3+}$	0.50M(H_2SO_4)	0.65	81
$Rh(bpy)_2(OH)_2^{+}$	0.50M(NaOH)	0.012	81
Donor = $*Ru(5\text{-}Clphen)_3^{2+}$			
$Os(bpy)_3^{2+}$	1.0M (Na_2SO_4)	2.5(red./energy)	82
$Rh(bpy)_3^{3+}$	0.5M (Na_2SO_4)	0.21	116
	0.50M(NaOH)	0.28	81
	0.50M(H_2SO_4)	0.41	81
$Rh(5\text{-}Mephen)_3^{3+}$	0.5M (Na_2SO_4)	0.16	116
$Rh(4,4'\text{-}Me_2bpy)_3^{3+}$	0.5M (Na_2SO_4)	0.001	116
$Rh(bpy)_2(H_2O)_2^{3+}$	0.50M(H_2SO_4)	0.29	81
$Rh(bpy)_2(OH)_2^{+}$	0.50M (NaOH)	0.011	81
Donor = $*Ru(5\text{-}Mephen)_3^{2+}$			
$Rh(bpy)_3^{3+}$	0.5M (Na_2SO_4)	0.7	116
$Rh(5,6\text{-}Me_2phen)_3^{3+}$	0.5M (Na_2SO_4)	1.3	116
$Rh(5\text{-}Mephen)_3^{3+}$	0.5M (Na_2SO_4)	1.0	116
$Rh(4,4'Me_2bpy)_3^{3+}$	0.5M (Na_2SO_4)	0.49	116
Donor = $*Ru(4,7\text{-}Me_2phen)_3^{2+}$			
$Rh(bpy)_3^{3+}$	0.5M (Na_2SO_4)	0.9	116
	0.50M(H_2SO_4)	1.25	81
	0.50M(NaOH)	0.99	81
$Rh(5,6\text{-}Me_2phen)_3^{3+}$	0.5M (Na_2SO_4)	1.6	116
$Rh(5\text{-}Mephen)_3^{3+}$	0.5M (Na_2SO_4)	1.3	116
$Rh(4,4'\text{-}Me_2bpy)_3^{3+}$	0.5M (Na_2SO_4)	1.2	116
$Co(bpy)_3^{+}$	0.10M(Et_4NClO_4)	3.8	109
$Co(bpy)_3^{2+}$	0.10M(Et_4NClO_4)	0.19	109

TABLE 18 (cont'd)

Quencher	Ionic Strength (Medium)	$10^{-9}k_q(M^{-1}s^{-1})$	Ref.
$Ru(4,7\text{-}Me_2phen)_3^{3+}$	0.10M(Et_4NClO_4)	2.5	109
$Rh(bpy)_2(H_2O)_2^{3+}$	0.50M(H_2SO_4)	1.39	81
$Rh(bpy)_2(OH)_2^{+}$	0.50M (NaOH)	0.025	81
Donor = $*Ru(4,4'\text{-}Me_2bpy)_3^{2+}$			
$Rh(bpy)_3^{3+}$	0.50M (H_2SO_4)	1.14	81
	0.50M (NaOH)	0.77	81
$Rh(phen)_3^{3+}$	0.5M (H_2SO_4)	1.37	81
$Rh(bpy)_2(H_2O)_2^{3+}$	0.5M (H_2SO_4)	0.78	81
$Rh(bpy)_2(OH)_2^{+}$	0.5M (NaOH)	0.018	81
Donor = $*Os(5\text{-}Clphen)_3^{2+}$			
$Fe(phen)_3^{2+}$	0.50M (NaCl)	1.4 (ox./energy)	82
$Ru(TPTZ)_2^{2+}$	0.50M (NaCl)	2.6 (oxidative)	82

vented by scavenging the Ru(III) with triethanolamine (k = 0.2 x $10^8M^{-1}s^{-1}$) [81]. Quenching of $*Ru(bpy)_3^{2+}$ by $Rh(bpy)_3^{3+}$ via energy transfer and/or reductive electron transfer is an uphill process; oxidative quenching is thermodynamically favourable.

Contrary to these earlier conclusions [81,123] that oxidative quenching of $*Ru(bpy)_3^{2+}$ by $Rh(bpy)_3^{3+}$ produces $Rh^{II}(bpy)_3^{2+}$ rather than $Rh^{III}(bpy)_2(bpy^-)^{2+}$, the more recent quenching data (Table 18) observed for $Rh(NN)_3^{3+}$ complexes do not appear consistent with the rate-determining formation of $Rh^{II}(NN)_3^{2+}$. Rather, the large $k_{11}k_{22}$ required to fit the data [116] to a Marcus treatment implicates $Rh^{III}(NN)_2(NN^-)^{2+}$ as the immediate quenching products in these systems.

4.1.7 By Transition Metal 1,2-Dithiolene Complexes

Transition metal 1,2-dithiolene complexes (VI) can undergo several sequential one-electron redox processes. For the simple dithiolenes of the type $MS_4C_4R_4^{n-}$ (M = Ni, Pt), two one-electron redox steps (n = 0, 1, and n = 1, 2) are typically observed [124]. Quenching of $*Ru(bpy)_3^{2+}$ by these complexes (for R = CN, n = 2; for

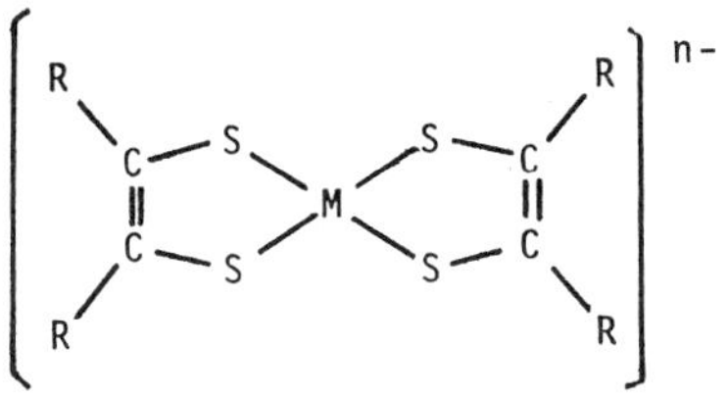

VI

R = Ph, n = 0) occurs predominantly by electron transfer [125-127]. The nature of the quenching process in non-micellar solutions has been determined by laser flash photolysis and radical scavenging experiments [127]. Quenching by $MS_4C_4(CN)_4^{2-}$ quenchers, the pathway may be energy transfer and/or reductive electron transfer, followed by very efficient cage recombination, particularly in solvents of low dielectric constant (k_q for M = Ni is 12.0 x $10^{10} M^{-1} s^{-1}$ in CH_3CN and 4.1 x $10^{10} M^{-1} s^{-1}$ in DMF; for M = Pt, k_q = 2.0 x 10^{10} $M^{-1} s^{-1}$ in DMF) [127]. The unimolecular rate constants for the quenching process within the encounter complex are strikingly different for $MS_4C_4(C_6H_5)_4$ (k > 2.2 x $10^{10} s^{-1}$) and $MS_4C_4(CN)_4^{2-}$ (k = 8 x 10^8 s^{-1}). The difference in these two rates result from differences in the electronic transmission coefficients. With structurally similar quenchers, oxidative electron transfer of $*Ru(bpy)_3^{2+}$ may be more adiabatic than reductive quenching; this could be related to the orbitals involved in the two types of electron transfer [127]. In oxidative quenching, an external ligand-localized $\pi*$ orbital donates the electron, while in reductive quenching the electron enters an internal, metal-centred t_{2g} orbital of the *Ru(II) complex.

4.1.8 <u>By Biologically Important Molecules</u>

Reduced blue copper proteins quench $*Ru(bpy)_3^{2+}$ by reductive electron transfer; k_q's (aqueous solution, u = 0.1M phosphate, pH 7.0) are 4.2 x $10^8 M^{-1} s^{-1}$ (stellacyanin), 1.6 x $10^9 M^{-1} s^{-1}$ (plastocyanin), and 6.9 x $10^8 M^{-1} s^{-1}$ (azurin) [128]. The mechanism has been confirmed from observation of transient absorption attributed to the oxidized copper protein PCu(II) and to $Ru(bpy)_3^+$ in the $PCu(I)/*Ru(bpy)_3^{2+}$ system. By contrast, cytochrome-c oxidatively quenches $*M(bpy)_3^{2+}$ and $*M(phen)_3^{2+}$ (M = Ru, Os); the lower limit of the bimolecular electron transfer rate between *M(II) and cyt-c^{3+} are 2 x $10^8 M^{-1} s^{-1}$ (Ru/bpy), 8 x $10^7 M^{-1} s^{-1}$ (Ru/phen), 5 x $10^9 M^{-1}$ s^{-1} (Os/bpy), and 7 x $10^8 M^{-1} s^{-1}$ (Os/phen) [129].

4.2 Pressure Effects on Photoinduced Electron Transfer Reactions: $*Ru(bpy)_3^{2+}$ and Various Metal Complexes

Electron transfer between photoexcited $Ru(bpy)_3^{2+}$ and Mo-$(CN)_8^{4-}$ or $Eu^{2+}(aq)$ yields large positive ($\Delta V^{\#}$ = +24.7 cm^3/mol) and negative (-11 cm^3/mol) volumes of activation (Table 19), respectively in experiments to to 300 MPa. Pressure dependence studies of this kind are scarce. It must be stressed that k_q's must be obtained under conditions identical to those of $\Delta V^{\#}$; this is not the case for the data of Table 19 [130], and no conclusions are possible in comparing $\Delta V^{\#}$'s with k_q's.

4.3 Effect of Molecular Structure on Quenching of $*Ru(NN)_3^{2+}$

In the energy transfer quenching of $*Ru(bpy)_3^{2+}$ by several cis- and trans-Co(III) and -Cr(III) β-diketonate complexes [131], little or no discrimination occurs during the quenching encounter for the Co(III) complexes (k_q's are near the diffusion limit) and for $Cr(bzac)_3$ (bzac = benzoylacetonate); with the chromiumtrifluoroacetylacetonates, the cis isomer is ∿ 40% more efficient in quenching than trans-$Cr(tfac)_3$ [131]. No extensive systematic studies have been carried out where the major quenching pathways is oxidative or reductive electron transfer. In the five pairs of cis- and trans-$Co(en)_2XY$ complexes (cf. Table 13), which quench $*Ru(bpy)_3^{2+}$ by oxidative electron transfer, it would generally appear that the cis isomers are slightly more efficient (k_q's cis $\gtrsim$ k_q's trans); but the differences are too small.

Photoinduced oxidative electron transfer between optical isomers has recently been studied by Kobayasi and coworkers [132] in the luminescence quenching of $(-)_D$-$*Ru(bpy)_3^{2+}$ by $(+)_D$-, rac-, and $(-)_D$-$Co(EDTA)^-$ in aqueous and methanol/water solutions. The rate constant for quenching of $(-)_D$-$*Ru(bpy)_3^{2+}$ (Table 20) in mixtures of methanol/water by $(-)_D$-$Co(EDTA)^-$ [$k_q(\Phi)$ = 1.38 ± 0.03 x $10^{10}M^{-1}s^{-1}$] is greater than observed with $(+)_D$-$Co(EDTA)^-$ (1.22 ± 0.01 x $10^{10}M^{-1}s^{-1}$); the mean k_q value 1.29 ± 0.03 x $10^{10}M^{-1}s^{-1}$ is obtained with rac-$Co(EDTA)^-$. In water (natural ionic strength), only very small differences are evident between k_q's of $(-)_D$-*Ru-$(bpy)_3^{2+}$ by $(+)_D$-$Co(EDTA)^-$, $(-)_D$-$Co(EDTA)^-$, and rac-$Co(EDTA)^-$; in methanol/water solutions, the $(-)_D$-isomer is more efficient than the $(+)_D$-enantiomer. Also, k_q's for rac-$Co(EDTA)^-$ vary with the concentration of MeOH in methanolic aqueous media and correlate with the medium's viscosity. Electron transfer essential to the

TABLE 19

Volumes of activation and quenching rate constants for $*Ru(bpy)_3^{2+}$ with various quenchers.[a]

Quencher	Electron Transfer Mechanism	$\Delta V^{\#}$ (cm^3/mol)	$10^{-9}k_q$ ($M^{-1}s^{-1}$)	Ionic Strength (Medium)
$Fe(CN)_6^{3-}$	oxidative	0.0 ± 0.1	5.6	0.25M(NaCl+HCl)
$[(en)_2Co(u_2\text{-}NH_2,O_2)Co(en)_2]^{4+}$	oxidative	0 to +1.0	1.9	0.25M(NaCl+H_2SO_4)
$[(NH_3)_5Co(u\text{-}O_2)Co(NH_3)_5]^{5+}$	oxidative	$+1 \pm 2$	1.9	0.25M(NaCl+H_2SO_4)
$[(CN)_5Co(u\text{-}O_2)Co(CN)_5]^{5-}$	oxidative	0 to +1.2	4.8	0.25M(NaCl+HCl)
Tl^{3+}	oxidative	$+0.2 \pm 0.1$	1.02	4.75M ($HClO_4$)
$Fe(CN)_6^{4-}$	reductive	0.0 ± 0.5	3.0	0.5M (NaCl)
$Mo(CN)_8^{4-}$	reductive	$+24.7 \pm 0.6$	0.63	0.25M(NaCl+HCl)
$Os(CN)_6^{4-}$	reductive	$+6.8 \pm 2.0$	0.84	0.25M(NaCl+HCl)
$IrCl_6^{3-}$	reductive	$+1.1 \pm 0.5$	0.83	0.25M(NaCl+HCl)
Eu^{2+}(aq)	reductive	-11.0 ± 1.0	0.0082	0.5M ($HClO_4$)
Fe^{2+}(aq)	reductive	-0.6 ± 0.6	0.01	0.25M(NaCl+H_2SO_4)

[a] reference 130.

TABLE 20
Rate constants in the quenching of $(-)_D$-$*Ru(bpy)_3^{2+}$ by optical isomers of Co(EDTA) determined from lifetime measurements.[a]

Quencher	Medium	$10^{-9}k_q$, $M^{-1}s^{-1}$
rac-$Co(EDTA)^-$	H_2O	6.65
	50% MeOH	3.89
	81% MeOH	8.48
	90% MeOH	12.7
$(+)_D$-$Co(EDTA)^-$	50% MeOH	3.50
$(-)_D$-$Co(EDTA)^-$	50% MeOH	4.18
$(+)_D$-$Co(EDTA)^-$	81% MeOH	8.20
$(-)_D$-$Co(EDTA)^-$	81% MeOH	9.14
$(+)_D$-$Co(EDTA)^-$	90% MeOH	11.8
$(-)_D$-$Co(EDTA)^-$	90% MeOH	14.1

[a] reference 132.

quenching process arises only in a contact ion-pair; any difference in the quenching cross-section between the $(+)_D$- and $(-)_D$-$Co(EDTA)^-$ enantiomers occurs from a difference in the rate of electron transfer within this ion-pair. The Δ-$(-)_D$-$*Ru(bpy)_3^{2+}$ leads to preference for Λ-$(-)_D$-$Co(EDTA)^-$ in the oxidative deactivation. In the back electron transfer reaction, a rather slow oxidation of rac-$Co(EDTA)^{2-}$ by Λ-$(+)_D$-$Ru(bpy)_3^{3+}$ yields a slight excess of Δ-$(+)_D$-$Co(EDTA)^-$ in aqueous solution [133]. The rate constants for quenching of $(-)_D$-$*Ru(bpy)_3^{2+}$ in 90% MeOH by $(+)_D$-$Co(EDTA)^-$ and $(-)_D$-Co-$(EDTA)^-$ are temperature dependent (15-45°C); the activation parameters are, respectively, $\Delta G^{\#}$ (kcal/mol) = 3.70 $\pm$ 0.02 and 3.59 $\pm$ 0.05, $\Delta H^{\#}$ (kcal/mol) = 2.7 and 3.4, and $T\Delta S^{\#}$ (kcal/mol) = -1.0 and -0.2 [132]. The electron transfer rate increases whenever reorientation of the counter ions in the precursor ion-pair yields closer contact of the reactant pair or provides a more effective path for electron transfer; the $(-)_D$-isomer makes closer contact with $(-)_D$-$*Ru(bpy)_3^{2+}$ than does $(+)_D$-$Co(EDTA)^-$. The preference for $(-)_D$-Co-$(EDTA)^-$ is achieved for a higher activation enthalpic barrier compensated in turn by the higher activation entropy, a result of a greater solvent reorientation accompanying the effective electron transfer [132].

4.4 Outer-Sphere and Inner-Sphere Intramolecular Electron Transfer in Excited Binuclear Complexes and Ion-Pairs

Highly charged transition metal complex cations form ion pairs with most anions, and where these act as reducing agent, an outer-sphere ligand-to-metal charge-transfer (OSLMCT) absorption band(s) may appear in the visible or near-UV spectra region. This area has recently been reviewed by Balzani et al. [134].

When solutions containing $Ru(bpy)_3^{2+}$ and SO_3^{2-} ions are flash photolyzed at moderate to high laser intensities, the reduced $Ru(bpy)_3^+$ is seen to form during (fast) and after (slow) the 20-ns laser pulse [135]. The slow contribution to reductive quenching of $*Ru(bpy)_3^{2+}$ by sulfite is due to a dynamic process (k_q = 3 x $10^5 M^{-1} s^{-1}$); the yields of $Ru(bpy)_3^+$ in the fast component (reactions 72, 73) is a sensitive function of laser intensity. The biphotonic pro-

$$*Ru(bpy)_3^{2+} + SO_3^{2-} \underset{}{\overset{*K}{\rightleftharpoons}} *Ru(bpy)_3^{2+}/SO_3^{2-} \qquad (72)$$

$$*Ru(bpy)_3^{2+}/SO_3^{2-} \xrightarrow{*k_2} Ru(bpy)_3^{+} + SO_3^{-} \qquad (73)$$

duction of $Ru(bpy)_3^+$ may have two origins: (i) the ion-paired sulfite acts as an efficient scavenger for the shorter-lived second excited state, $**Ru(bpy)_3^{2+}$ or (ii) alternatively, association of $*Ru(bpy)_3^{2+}$ with SO_3^{2-} gives rise to an outer-sphere complex displaying an OSLMCT band estimated at $\lambda \sim$ 550 nm; the latter interpretation appears more likely [135].

The $Ru(NH_3)_5py^{3+}$ complex in solutions of various MX salts forms ion pairs, $Ru(NH_3)_5py^{3+}/X^-$, characterized by OSLMCT absorptions for X = Cl^-, Br^-, I^-, and $C_2O_4^{2-}$ [136]. Flash photolysis excitation into the OSLMCT band (X = I^-; Figure 12) yields the electron transfer metastable products $Ru(NH_3)_5py^{2+}$ and I_3^- which relax to the starting materials via a slow second order reaction. When X is an irreversible reductant like $C_2O_4^{2-}$, continuous photolysis leads to substantial net photoreduction of the metal centre. Analogous results are obtained with $Os(NH_3)_5Cl^{2+}$ and $M(CN)_6^{4-}$ (M = Fe, Ru, Os) [137]. Cationic and anionic complexes of the same metal also form ion pairs that exhibit outer-sphere (also known as intervalence) CT bands [138]; for example, $Ru^{III}(NH_3)_5Cl^{2+}/Ru^{II}(CN)_6^{4-}$

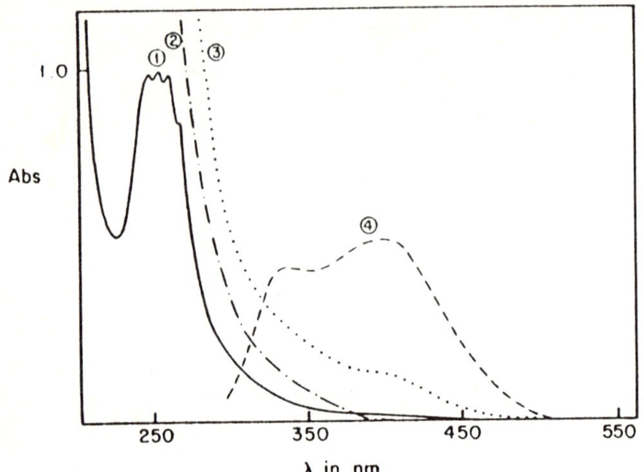

Figure 12.- Spectra of aqueous $Ru(NH_3)_5py^{3+}$ in the presence and absence of I^-: (1) $[Ru(NH_3)_5py]_2(S_2O_6)_3$ (2.2×10^{-4}M) aqueous solution, pH 3; (2) 1.0M NaI in pH 3 aqueous solution; (3) $[Ru(NH_3)_5py]_2(S_2O_6)_3$ (2.2×10^{-4}M) in 1.0M NaI solution; (4) difference of spectrum 3 minus spectrum 2 plus spectrum 1. Reproduced with permission from ref. 136.

has a band at 510 nm (Figure 13). Photolysis in this band yields the binuclear complex $[(NH_3)_5Ru^{III}NCRu^{II}(CN)_5]^-$ (Φ = 0.002) via ra-

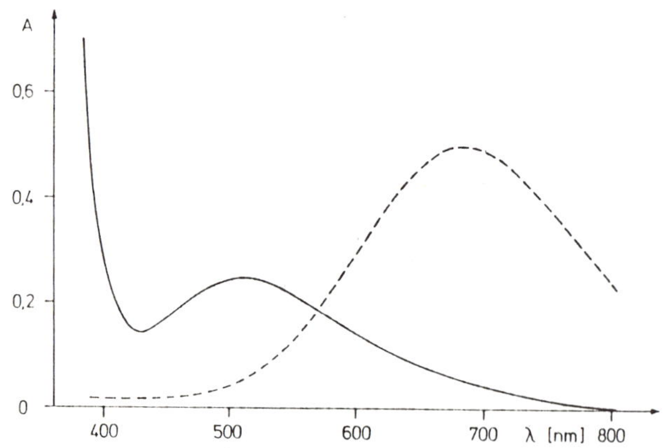

Figure 13.- Electronic absorption spectra of aqueous solutions of equimolar $[Ru(NH_3)_5Cl]Cl_2$ and $K_4[Ru(CN)_6]$ (solid line) and $Na[(NH_3)_5Ru(NC)Ru(CN)_5]\cdot 3H_2O$ (dashed line) at 298°K. Reproduced with permission from ref. 138.

pid aquation in competition with back electron transfer (Scheme XI). In the complex $[(NH_3)_5Co^{III}(NC)Ru^{II}(CN)_5]^-$, light excitation

Scheme XI:

$$Ru^{III}(NH_3)_5Cl^{2+}/Ru^{II}(CN)_6^{4-} \xrightarrow{h\nu} Ru^{II}(NH_3)_5Cl^{+}/Ru^{III}(CN)_6^{3-}$$

$$\longrightarrow Ru^{II}(NH_3)_5Cl^{+} + Ru^{III}(CN)_6^{3-}$$

$$Ru^{II}(NH_3)_5Cl^{+} \xrightarrow{-Cl^-,\ +H_2O} Ru^{II}(NH_3)_5H_2O^{2+}$$

$$Ru^{II}(NH_3)_5H_2O^{2+} \xrightarrow[+Ru^{II}(CN)_6^{4-}]{} [(NH_3)_5Ru^{II}NCRu^{II}(CN)_5]^{2-}$$

$$[(NH_3)_5Ru^{II}NCRu^{II}(CN)_5]^{2-} + Ru^{III}(CN)_6^{3-} \longrightarrow Ru^{II}(CN)_6^{4-} + [(NH_3)_5Ru^{III}NCRu^{II}(CN)_5]^{-}$$

into the intervalence charge transfer band at 375 nm [Ru^{II} to Co^{III}] leads to an inner-sphere photoredox reaction to give Ru-$(CN)_6^{3-}$ and Co^{2+})aq); quantum yield Φ= 0.46 [139]. Similarly, excitation into the outer-sphere intervalence CT band (λ = 360 nm) of the $Co(NH_3)_6^{3+}/Ru(CN)_6^{4-}$ ion pair also yields photoredox products but the quantum yield is substantially smaller (Φ = 0.034) [140].

Luminescence quenching of $*Ru(bpy)_3^{2+}$ by $S_2O_8^{2-}$ in aqueous and mixed CH_3CN/water solutions occur by both unimolecular and bimolecular electron transfer; the rate-determining step in both processes is solvent reorganization of one or both reactants [141]. The non-diffusional pathway results from the ground state association to form the $Ru(bpy)_3^{2+}/S_2O_8^{2-}$ ion pair. The diffusional (dynamic) electron transfer quenching mechanism in this system has been discussed thoroughly by Bolletta et al.[71].

Advantage can be taken of the substantial association between ions of high and opposite charge in outer-sphere systems to determine rate constants for the thermally activated "intramolecular" electron transfer process, and to assess whether dynamic quenching process is diffusion-controlled or activation-controlled [142]. Stern-Volmer plots of the emission quenching in $*Os(5\text{-}Cl\text{-}phen)_3^{2+}$-$Fe(CN)_6^{4-}$ and $*Ru(bpy)_3^{2+}$-$Fe(CN)_6^{4-}$ systems exhibit upward curvature indicative of a static process. The initial slopes yield the second-order quenching rate constants (25°C) at two ionic strengths u = 0.001M and 0.10M, respectively: $k_q = 1.7 \times 10^{10} M^{-1}s^{-1}$ and $3.2 \times 10^9 M^{-1}s^{-1}$ (Os^{II} system), and $5.1 \times 10^{10} M^{-1}s^{-1}$ and $7.2 \times 10^9 M^{-1}s^{-1}$ (Ru^{II} system). In the latter, quenching is diffusion-controlled and is activation-controlled for the Os(II) system (Scheme XII) [142]. The unimolecular electron transfer rate const-

<u>Scheme XII:</u>

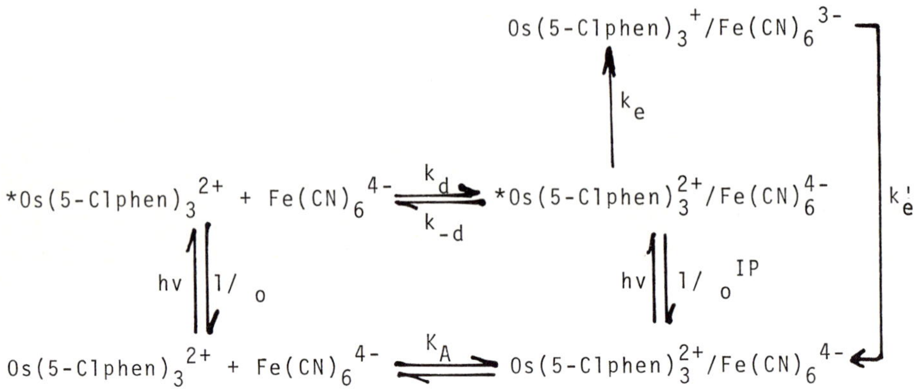

$Os(5\text{-}Clphen)_3^{+}/Fe(CN)_6^{3-}$

$\uparrow k_e$; k_e' (from $Os(5\text{-}Clphen)_3^{+}/Fe(CN)_6^{3-}$ to $Os(5\text{-}Clphen)_3^{2+}/Fe(CN)_6^{4-}$)

$*Os(5\text{-}Clphen)_3^{2+} + Fe(CN)_6^{4-} \underset{k_{-d}}{\overset{k_d}{\rightleftharpoons}} *Os(5\text{-}Clphen)_3^{2+}/Fe(CN)_6^{4-}$

hv $\uparrow\downarrow$ 1/ o ; hv $\uparrow\downarrow$ 1/ $_o^{IP}$

$Os(5\text{-}Clphen)_3^{2+} + Fe(CN)_6^{4-} \overset{K_A}{\rightleftharpoons} Os(5\text{-}Clphen)_3^{2+}/Fe(CN)_6^{4-}$

ant k_e for $*Ru(bpy)_3^{2+}$-$Fe(CN)_6^{4-}$ is larger than k_e for $*Os(5\text{-}Cl\text{-}phen)_3^{2+}$-$Fe(CN)_6^{4-}$ ($>1.0 \times 10^8 s^{-1}$) in accord with the larger reduction potential of $*Ru(bpy)_3^{2+}$ (+0.84 V) against +0.72 V for $*Os\text{-}(5\text{-}Clphen)_3^{2+}$. Also, k_e is rate-limiting in Scheme XII; k_e' is >> k_e as expected on the basis of the considerably higher exoergonicity of the Os(I)-Fe(III) reactions (ΔE^o = 1.5 V) versus E^o of 0.3 V in the *Os(II)-Fe(II) reaction. For the $*Os(5\text{-}Clphen)_3^{2+}$-$Fe(CN)_6^{4-}$ system, $k_e = 8.7 \times 10^7 s^{-1}$ (u = 0.001M) and $1.6 \times 10^8 s^{-1}$ (u = 0.10 M) [142].

An exciting photophysical study of the unsymmetrical ligand-bridged Os(II,II) species, $[(bpy)_2(CO)Os^{II}(L)Os^{II}(phen)(dppe)\text{-}Cl]^{3+}$ and the mixed valence Os(II,III) dimer $[(bpy)_2(CO)Os^{II}(L)\text{-}Os^{III}(phen)(dppe)Cl]^{4+}$ (where dppe = 1,2-bis(diphenylphosphino)-cis-ethene) has recently been reported [143]. The II,II dimers are characterized by two independent MLCT chromophores: $d\pi(Os) \rightarrow$ phen

L:

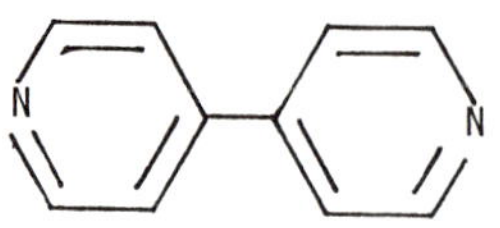

4,4'-bpy

<u>VII</u>

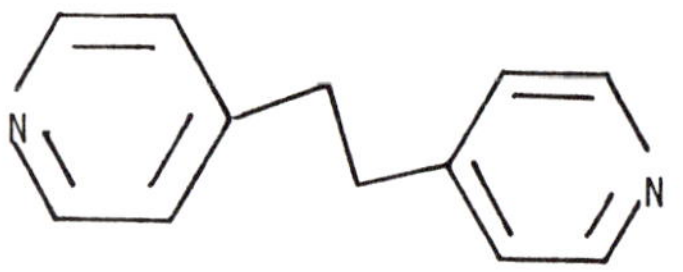

bpa

<u>VIII</u>

and $d\pi(Os) \rightarrow$ bpy. Quenching of the higher, Os(bpy)-localized MLCT state does not occur via intramolecular energy transfer to the Os-(phen) chromophore though such a process is energetically favourable; rather it takes place via reductive electron transfer from the adjacent Os^{II}(phen) site (Scheme XIII). Interestingly, the k_1/k_2 step represents electron transfer within a photogenerated mixed-valence dimer. In addition, the intramolecular reductive quenching

Scheme XIII:

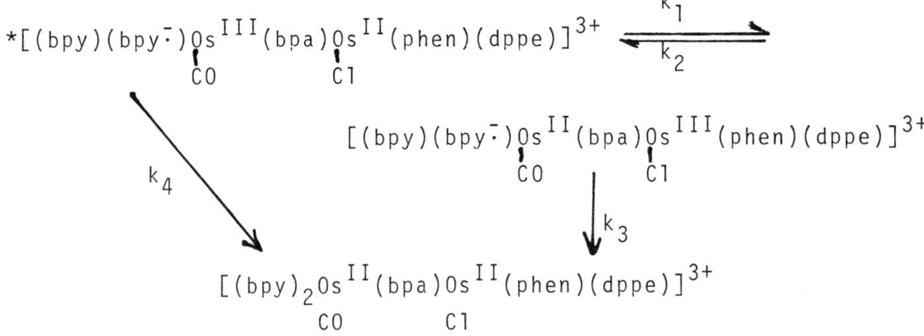

step produces a charge-separated, remote MLCT excited state [(bpy)-$(bpy^-)(CO)Os^{II}(bpa)Os^{III}(phen)(dppe)Cl]^{3+}$, in which the 'electron/hole' pair is separated by the intervening Os(II) centre and the bpa-bridging ligand. Photoinduced charge separation of this type raises the possibility of providing the basis for relatively long lived intramolecular charge storage given the relative isolation of the oxidized (hole) and reduced (electron) sites. Disappointingly, this remote MLCT state is unexpectedly short-lived ($\sim$ 100 ns) [143], probably a result of the rotational flexibility of the bpa bridging ligand which may lead to "through-space contact" of the two redox sites; decay may therefore occur via outer-sphere electron transfer. By contrast, intramolecular electron transfer quenching seems unimportant for quenching the Os(bpy)-localized MLCT state in the mixed-valence dimers $[(bpy)(bpy^{\cdot -})(CO)Os^{III}(L)$-$Os^{III}(phen)(dppe)Cl]^{4+}$. Several factors may be responsible for the apparent lack of quenching by the nearby Os(III) centre [143]; (i) the potential oxidative electron transfer step is highly exergonic ($\sim$ -1.4 eV), falls into the inverted Marcus region, and is subject to the vibrational overlap problems associated with the energy gap law; (ii) k_q may be strongly affected by the relatively large distance between the bpy-localized electron and the Os(III) acceptor site. Likewise, no significant intramolecular electron transfer quenching occurs during the lifetime of the Ru(bpy) MLCT state (eqn 74) following light excitation into the $d\pi(Ru) \rightarrow$ bpy chromophore of the Ru(II,III) mixed-valence dimer [144,145]. The rapid decay of

$$*[(NH_3)_5Ru^{III}(4,4'\text{-bpy})\underset{\displaystyle Cl}{Ru}^{III}(bpy)(bpy^{\cdot -})]^{4+} \longrightarrow$$

$$[(NH_3)_5Ru^{II}(4,4'\text{-bpy})\underset{\displaystyle Cl}{Ru}^{III}(bpy)_2]^{4+} \quad (74)$$

$(bpy^{\cdot -})Os^{II}(bpa)Os^{III}(phen)$ to $(bpy)Os^{II}(bpa)Os^{II}(phen)$ is enigmatic inasmuch as the process involves a rather exothermic charge recombination which should relatively slow owing to unfavourable Franck-Condon terms, and moreover, involves electron transfer over a considerable distance. Picosecond laser, time-resolved spectroscopic studies of the Os(II,II) dimer might reveal additional information regarding the kinetics of the ultra-fast events; the parallel exists between these charge-separated species and charge separation in semiconductors [146].

Theoretical considerations into electron transfer processes in binuclear $(L_1)_nM_1\text{-B-}M_2(L_2)_m$ complexes with emphasis on stabilizing the charge displacement over a long period against a potential gradient have been presented in some details [147]. This may have significance in photosynthetic oxygen evolution.

4.5 Solar Energy Conversion Involving Photoinduced Electron Transfer Reactions of Transition Metal Complexes

The extensive photoredox chemistry demonstrated by $*Ru(bpy)_3^{2+}$ and related complexes (Section 4.1) has led to numerous investigations into the potential utility of these complexes as photosensitizers towards the photochemical conversion and storage of solar energy of which the photodissociation of H_2O is a pragmatic example; in retrospect, this has proven a formidable task [26]. Formation of a molecule of dihydrogen and dioxygen implicates multi-electron transfers (reactions 75,76); E^0 0.0 V (vs. NHE) for reaction 75 and 1.23 V for reaction 76, respectively, at pH 0.

$$2\ e^- + 2\ H_2O \longrightarrow 2\ OH^- + H_2 \quad (75)$$

$$2\ H_2O \longrightarrow 4\ H^+ + O_2 + 4\ e^- \quad (76)$$

Though $*Ru(bpy)_3^{2+}$ has the appropriate thermodynamic driving force to induce reaction 75, the kinetic barriers have precluded direct photoreduction of water. Several redox processes have been studied [26,148,149] involving suitable acceptor relay species (reaction 77) formed in an oxidative quenching step; the generation of oxygen (reaction 78) may implicate the oxidized sensitizer S^+, or an oxidized electron donor (D^+) formed in a reductive quenching step.

$$2\ H_2O + 2\ A^- \xrightarrow{cat.} 2\ A + 2\ OH^- + H_2 \qquad (77)$$

$$4\ D^+ + 2\ H_2O \xrightarrow{cat.} 4\ D + 4\ H^+ + O_2 \qquad (78)$$

Several reports have appeared on the efficient photogeneration of powerful reductants and oxidants as well as coupling of these to some cyclic process of water photocleavage. Such multi-electron transfer events as depicted by reactions 75 and 76 are rather inefficient, and redox catalysts are imperative in driving the reactions to a successful outcome. Balzani and coworkers [150] have recently enunciated some requirements needed for ideal electron relays and photosensitizer species, with particular reference to the needs for a high turnover number for a relay and for a high turnover number and a long lifetime for the photosensitizer. Unfortunately, processes in pure homogeneous media have required "sacrificial" donors and/or acceptors to intercept the separated photogenerated redox products that have escaped the solvent cage before back reaction occurs. We now consider some systems which have utilized transition metal complexes as photosensitizers and as electron relays. Important results are also available on photoelectrochemical cells employing transition metal complexes [26,151]; these are beyond the scope of this work.

Visible light photolysis of a system consisting of $Ru(bpy)_3^{2+}$ as the photosensitizer, $Rh(bpy)_3^{3+}$ as the electron relay, and triethanolamine (TEOA) as the sacrificial electron donor generates H_2 efficiently [152-154] in aqueous media in the presence of $PtCl_6^{2-}$ which turns into an active colloidal Pt^0 (thus heterogeneous) catalyst during irradiation. The outline of the reaction pathways leading to formation of H_2 and other Rh species is illustrated in Scheme XIV [81]. At about the same time, Sutin and coworkers [81,123] demonstrated that H_2 evolves mainly from Pt^0-catalyzed reactions of $Rh(bpy)_3^{2+}$, contrary to the suggestion by Lehn et al. [152,153] that the $Rh(bpy)_2^+$ species is involved. Following oxida-

Scheme XIV:

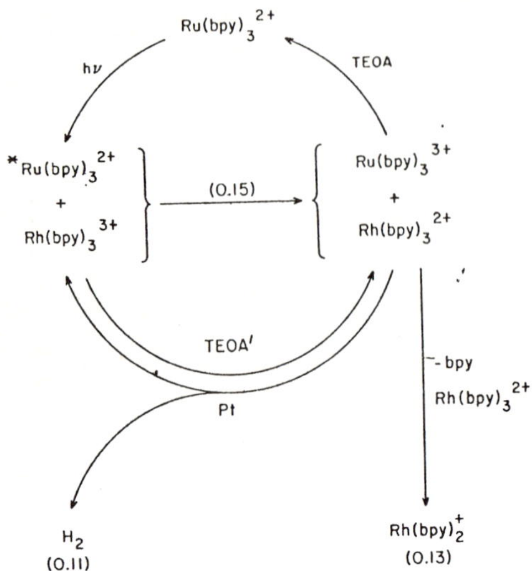

tive electron transfer from $^{*}Ru(bpy)_3^{2+}$, the oxidized $Ru(bpy)_3^{3+}$ is reduced by the electron donor TEOA ($k \sim 0.3 \times 10^7 M^{-1}s^{-1}$) back to the Ru(II) ground state; the oxidized deprotonated $TEOA^+$ radical cation reduces an additional molecule of $Rh(bpy)_3^{3+}$ to give a net yield of Rh(II) of 0.3 ± 0.1. In the absence of Pt^0, the Rh(II) species undergoes ligand loss and subsequent disproportionation to give $Rh(bpy)_2^+$ which can exist in various forms but none of which yield H_2 in the dark. The cage escape yield for the Rh(II) substrate (Scheme XIV) is 0.15 and the quantum yield of H_2 formation is 0.11, a value close to $\Phi(Rh^I)$ in the absence of Pt catalysis [81].

Reductive quenching of $^{*}Ru(bpy)_3^{2+}$ yields the strongly reducing, long-lived $Ru(bpy)_3^+$ which can give H_2 directly with a Pt catalyst or can reduce a metal complex to yield unstable metal (eg. Co) hydrides [26]. Systems employing Co(II) either as a macrocycle [155] or as a labile bpy complex, formed in situ, and $Ru(bpy)_3^+$ (formed by reductive quenching with Eu^{2+} ions) reduce Co(II) to Co(I). This latter species oxidatively adds H_3O^+ to form an unstable d^6 hydride which rapidly undergoes decomposition to yield the parent complex and H_2 (Φ up to $\sim$ 13%). Other systems that have been employed and the relevant results are depicted in Table 21 for the photoreduction and photooxidation of water.

Photochemically produced $Ru(bpy)_3^{3+}$ has received extensive

TABLE 21

Photoredox systems in the photoproduction of H_2 and O_2 from water using transition metal complexes.[a]

Photosensitizer	Electron Relay	Sacrificial Donor	Catalyst	Φ_{cage}	Φ_{H_2}	Ref.
photoreduction of water						
$*Ru(bpy)_3^{2+}$	$Rh(bpy)_3^{3+}$	TEOA	Pt^o (in situ)	0.15	0.11	81,156
		EDTA	Pt^o (in situ)	0.15	0.04	81
	$Eu^{2+}(aq)$	-	$Co^{II}L$	1.00	0.05	155
	$[Fe_4S_4(SBzl)_4]^{2-}$		-	-	?	157
photooxidation of water						
$*Ru(bpy)_3^{2+}$	Q = $Co(NH_3)_5Cl^{2+}$	k_q ($M^{-1}s^{-1}$) =	$9.3x10^8$	cat.:	RuO_2 powder	158-160
					RuO_2 zeolite	161
					RuO_2 colloid	158,159,162
					RuO_2 electr.	72,163
					IrO_2 powder	160,161
	$Co(C_2O_4)_3^{3-}$		$6.7x10^9$		RuO_2 colloid	158
					RuO_2 powder	158
	$Tl^{3+}(aq)$		$4.9x10^8$ (1N H_2SO_4)		RuO_2 colloid	158
					RuO_2 electr.	72,163
	$Ag^+(aq)$		$3.5x10^6$		RuO_2 zeolite	63

[a] reference 26.

study towards its possible role as an oxidant for reaction 78. Creutz and Sutin [164] demonstrated that this Ru(III) complex can oxidize water to molecular oxygen and itself is partially decomposed through a very complicated and obscure pathway; the process is catalyzed by RuO_2 [160,165]. Irradiation of $Ru(bpy)_3^{2+}$ in the presence of a sacrificial electron acceptor ($S_2O_8^{2-}$ or $Co(NH_3)_5Cl^{2+}$) and RuO_2 generates O_2 continuously (eqns 79,80). Lehn and coworkers [166] were the first to utilize cobalt(III) complexes as A; where

$$*Ru(bpy)_3^{2+} + A \longrightarrow Ru(bpy)_3^{3+} + A^- \quad (79)$$

$$4\ Ru(bpy)_3^{3+} + 2\ H_2O \longrightarrow 4\ Ru(bpy)_3^{2+} + 4\ H^+ + O_2 \quad (80)$$

A is $Co(NH_3)_5Cl^{2+}$, k_q = 9.3 x $10^8 M^{-1}s^{-1}$ and Φ of formation of the oxidized Ru(III) complex is 0.31. The back reaction between the Co^{2+}(aq) cation and $Ru(bpy)_3^{3+}$ is prohibited on thermodynamic grounds (ΔG = +0.54 eV). Shafirovich et al.[167] showed that simple Co(II) complexes can effectively catalyze the reduction of one-electron oxidants such as $Ru(bpy)_3^{3+}$, $Fe(bpy)_3^{3+}$, and $IrCl_6^{2-}$ with these reduction reactions leading to oxidation of water to O_2. They claim [167] that irradiation of $Ru(bpy)_3^{2+}$ in aqueous media (pH 7) in the presence of $Co(NH_3)_5Cl^{2+}$ yields O_2 with Φ = 0.025. Studies by Harriman et al.[168] confirm that both colloidal RuO_2 and simple cobalt(II) ions are effective catalysts for O_2 production; $\Phi(O_2) \sim$ 12% with $Co(NH_3)_5Cl^{2+}$ as the sacrificial electron acceptor. However, when A is Fe^{3+} (which is a reversible redox couple Fe^{3+}/Fe^{2+}) no O_2 is observed on prolonged irradiation; k_q for the oxidative quenching of $*Ru(bpy)_3^{2+}$ is 5.5 x $10^8 M^{-1}s^{-1}$ at pH 5 and the back reaction has k $\sim$ 3.4 x $10^8 M^{-1}s^{-1}$ so that the reverse electron transfer step ensures the lifetime of $Ru(bpy)_3^{3+}$ to be very low under these conditions of maximum $\Phi(O_2)$. The importance of a catalyst (RuO_2 or $CoSO_4$) in producing O_2 is demonstrated in Figure 14 which clearly shows that in the absence of the catalyst, no O_2 is formed; $\Phi(O_2)$ = 0.030 (colloidal RuO_2), 0.003 (RuO_2 powder), and 0.020 ($CoSO_4$) at pH 5. The significance of $CoSO_4$ is interesting; both the rate and yield of O_2 formation are dependent on [$CoSO_4$] and on pH. The probable mechanism resulting in O_2 is summarized in the series of reactions of Scheme XV [168].

Other acceptors which upon reduction undergo irreversible change and hence do not back react with $Ru(bpy)_3^{3+}$ are summarized in Table 21.

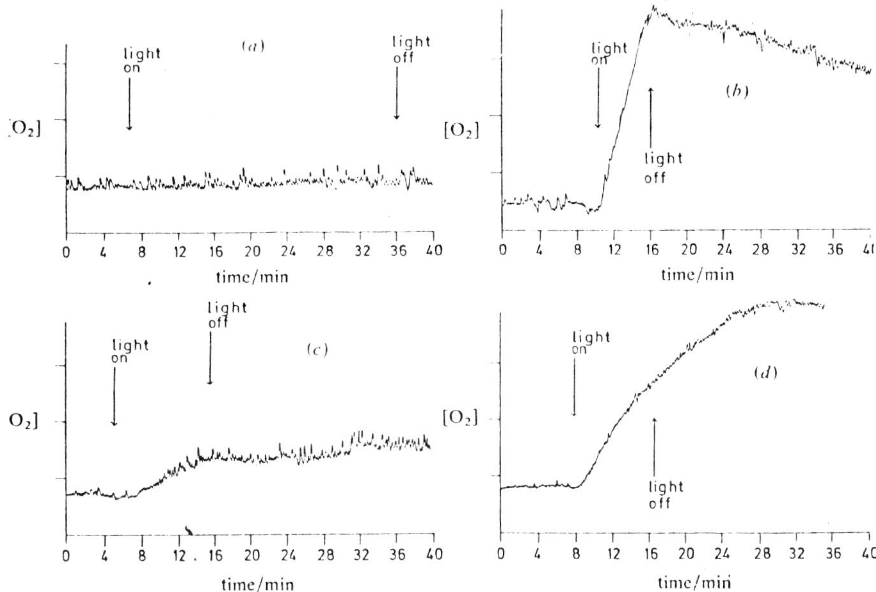

Figure 14.- Typical traces for formation of O_2 from the irradiation of $Ru(bpy)_3^{2+}$ in aqueous solution at pH 5 containing $Co(NH_3)_5Cl^{2+}$; (a) no added catalyst; (b) colloidal RuO_2; (c) RuO_2 powder; (d) $CoSO_4$. Reproduced with permission from ref. 168.

Scheme XV:

$$Ru(bpy)_3{}^{2+} + Co(NH_3)_5Cl^{2+} \xrightarrow{h\nu} Ru(bpy)_3{}^{3+} + Co^{2+} + \ldots$$

$$Co^{2+}(cat.) + Ru(bpy)_3{}^{3+} \longrightarrow Co^{3+} + Ru(bpy)_3{}^{2+}$$

$$Co^{3+} + Ru(bpy)_3{}^{3+} \longrightarrow Co^{4+} + Ru(bpy)_3{}^{2+}$$

$$Co^{4+} + 2\ H_2O \longrightarrow Co^{2+} + H_2O_2 + 2\ H^+$$

$$H_2O_2 + 2\ Ru(bpy)_3{}^{3+} \longrightarrow O_2 + 2\ H_2 + 2\ Ru(bpy)_3{}^{2+}$$

$$Co^{3+} \longrightarrow \text{dimers (?)}$$

5.0 ELECTRON TRANSFER REACTIONS WITH EXCITED STATES OF Rh(III) COMPLEXES

Except for their use as electron relays in solar energy conversion schemes (Section 4.5), Rh(III)-polypyridyl complexes have received little attention until recently [169]. The scarcity of studies originates from an almost non-emitting excited *Rh(III) species at ambient temperatures in fluid media [170]. Excited-state absorption (ESA) from flash photolysis of $Rh(phen)_3^{3+}$ provides a precious handle in monitoring the otherwise elusive $*Rh(phen)_3^{3+}$ excited state(s) of this powerful oxidant. At room temperature, the ligand-centred π-π* triplet is in fast equilibrium with a metal-centred d-d triplet state. Very weak ($\Phi < 10^{-5}$) dual emission from $*Rh(phen)_3^{3+}$ originates from the $^3(\pi\pi^*)$ state (450 nm) and from the ^{3}(dd) state (580 nm); the dd emission is broad and structureless (τ of both states $\sim$ 250 $\pm$ 20 ns) [169,170]. When flash photolyzed, $*Rh(phen)_3^{3+}$ is reductively quenched by Fe^{2+}(aq) (reaction 81) in the presence of methylviologen which quantitatively scavenges the Rh(II) product; the extent of quenching is followed by observing the formation of MV^+ radical cations (absorption band at 603 nm). The formation of MV^+ in a 1:1 ratio with respect to the number of absorbed photons by $Rh(phen)_3^{3+}$ demonstrates that the cage escape yield is unity and that the intersystem crossing from the singlet to the triplet manifold is unity at room temperature [169].

$$*Rh(phen)_3^{3+} + Fe^{2+}(aq) \xrightarrow[u\,=\,1\,M]{4.0 \times 10^9 M^{-1}s^{-1}} Rh(phen)_3^{2+} + Fe^{3+}(aq)$$

$$Rh(phen)_3^{2+} \xrightarrow{MV^{2+}} Rh(phen)_3^{3+} + MV^{\dot{+}} \qquad (81)$$

6.0 ELECTRON TRANSFER REACTIONS WITH EXCITED STATES OF Ir(III) COMPLEXES

The popularity of ruthenium(II)-polypyridyl complexes as photosensitizers has led others to seek new transition metal complexes with more favourable redox properties and longer lifetimes for their excited states, in order to explore and possibly exploit photoinduced energy and/or electron transfer processes. In this regard, Bergeron and Watts [171] have studied quenching processes of excited $Ir(bpy)_2(H_2O)(bpy)^{3+}$ and $Ir(bpy)_2(OH)(bpy)^{2+}$ by a variety of metal complexes. The nature of these two Ir(III) complexes has since been determined by x-ray crystallography by Serpone et al. [172] to be the ortho-metallated species IX, $(bpy)_2Ir^{III}(C^3,N'$-$bpyH)^{3+}$, and X, $(bpy)_2Ir^{III}(C^3,N'$-$bpy)^{2+}$; it has recently been confirmed by others [173] by a variety of techniques. The reduction

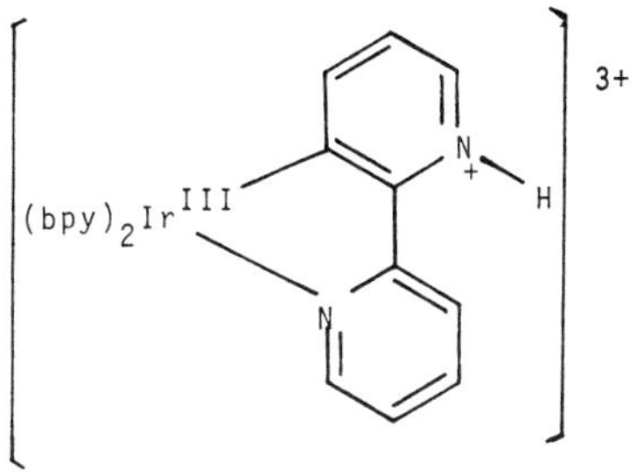

$[(bpy)_2Ir^{III}(C^3,N'\text{-bpyH})]^{3+}$ IX $[(bpy)_2Ir^{III}(C^3,N'\text{-bpy})]^{2+}$ X

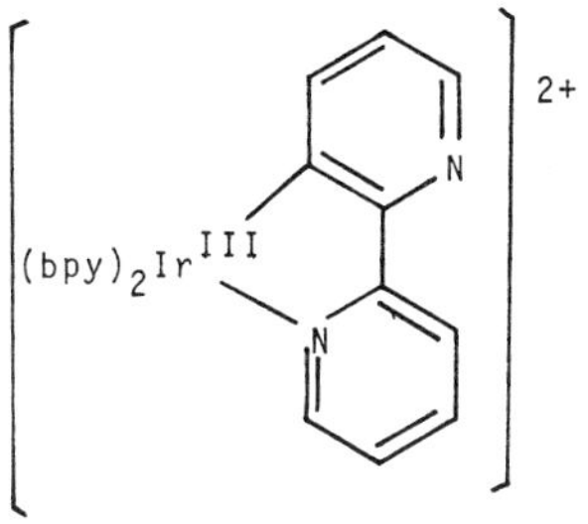

potential of the deprotonated form *X is +1.84 V; for the $Ir(bpy)_3^{3+}$ complex, $^*E^0$(redox) = +2.26 V [171]. The excited iridium complexes are weaker reducing agents than the related $^*Ru(bpy)_3^{2+}$. However, the excited state donor energies, 259 kJ/mol (X) and 246 kJ/mol (IX), considerably exceed the donor energy of $^*Ru(bpy)_3^{2+}$ (205 kJ/mol) and are slightly below that of $^*Ir(bpy)_3^{3+}$ (271 kJ/mol). Hence energy transfer and reductive electron transfer pathways are expected to be major quenching processes for *IX and *X, except for the quenchers Fe^{3+}, Tl^{3+}, $Tl(OH)^{2+}$, and Eu^{3+} (Table 22) where oxidative electron transfer is possibly the only quenching pathway to contribute to the deactivation of the excited state(s). The Eu^{3+} ion does not quench *IX and *X but Tl^{3+} quenches somewhat in accord with expectations from the corresponding redox potentials (-0.43 V and -0.37 V, respectively). By contrast, oxidative quenching of $^*Ru(bpy)_3^{2+}$ by Tl^{3+} is efficient owing to the more favourable oxidation potential of $^*Ru(bpy)_3^{2+}$ (10^8 - $10^9 M^{-1} s^{-1}$; Table 8) compared to $^*(bpy)_2Ir(C^3,N'$-$bpy)^{2+}$. Reductive quenching of this species by Eu^{2+} is more efficient than for the $^*Ru(bpy)_3^{2+}$ complex. No net photochemical changes occur in the $^*(bpy)_2Ir(C^3,N'$-$bpy)^{2+}$-Eu^{2+} system. Delineation of the contributions from energy transfer from electron transfer in the quenching events has not been addressed [171]. Bromide ion quenches *X by formation of exciplexes [174].

TABLE 22
Luminescence quenching results of *IX and *X by various metal ions.[a]

Quencher	$10^{-7}k_q$, $M^{-1}s^{-1}$ [b] $(bpy)_2Ir^{III}(C^3,N^1\text{-}bpyH)^{3+}$	Quencher	$10^{-7}k_q$, $M^{-1}s^{-1}$ [d] $(bpy)_2Ir^{III}(C^3,N^1\text{-}bpy)^{2+}$
Fe^{3+}	3.7	$Fe(OH)^{2+}$	31
Tl^{3+}	0.006	$Tl(OH)^{2+}$	1.6
Co^{2+}	0.61	Co^{2+}	1.5
Ni^{2+}	0.28	Ni^{2+}	0.42
Eu^{3+}	0	Eu^{3+}	0
Eu^{2+}	20[c]	-	-

[a] reference 171. [b] in 0.5M $HClO_4$. [c] in 0.5M HNO_3. [d] in acetate buffer pH 4.66, 0.1M acetic acid + 0.08M sodium acetate.

7.0 ELECTRON TRANSFER REACTIONS WITH EXCITED STATES OF EUROPIUM (III) ENCAPSULATED IN A CRYPTATE

The $[Eu\subset 2.2.1]^{3+}$ encapsulated complex exhibits a bright luminescence in aqueous solutions at ambient temperature; its lowest excited state ($\tau \sim 215$ ns) is a $^5D_0$ metal-centred state which is quenched by $M(CN)_6^{4-}$ (M = Fe, Ru, Os) complexes via a (dynamic) reductive electron transfer process (reaction 82) [134,175]. The

$$*[Eu\subset 2.2.1]^{3+} + M(CN)_6^{4-} \longrightarrow [Eu\subset 2.2.1]^{2+} + M(CN)_6^{3-} \qquad (82)$$

bimolecular quenching constants (aqueous media, 1 M KCl, room temperature) are 7.5×10^8, 2.2×10^8, and $7.8 \times 10^8 M^{-1}s^{-1}$ for the Fe, Ru, and Os cyanides, respectively. In more concentrated $M(CN)_6^{4-}$ solutions, $[Eu\subset 2.2.1]^{3+}$ is strongly associated to $M(CN)_6^{4-}$. Irradiation into the outer-sphere, intervalence charge-transfer (OSCT) band leads to an ion-paired species containing europium(II) and $M(CN)_6^{3-}$ (eqn 83). Reductive quenching (eqn 82) leads to a success-

$$[Eu^{III}\subset 2.2.1]^{3+}/M^{II}(CN)_6^{4-} \xrightarrow[(OSCT)]{h\nu} [Eu^{II}\subset 2.2.1]^{2+}/M^{III}(CN)_6^{3-} \qquad (83)$$

or complex which is identical to that obtained upon OSCT excitation of the ground-state ion-pair $[Eu\subset 2.2.1]^{3+}/M(CN)_6^{4-}$. The processes are illustrated in Scheme XVI [175]. The thermal and the two light-

Scheme XVI:

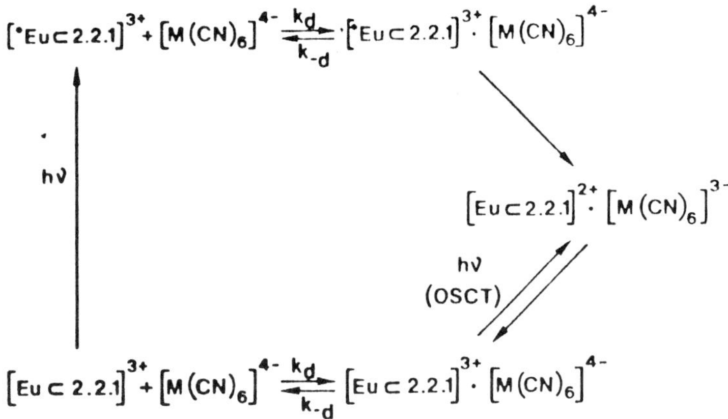

induced electron transfer steps exhibit the same intrinsic barrier and orbital overlap since the $^5D_0$ excited state possesses the same equilibrium geometry and orbital configuration as the ground state $^7F_0$ [175]. Reductive quenching, however, is spin forbidden, but the influence of the spin factor on the quenching event(s) is probably small, owing to the presence of heavy atoms.

8.0 ELECTRON TRANSFER REACTIONS WITH EXCITED STATES OF VANADIUM COMPLEXES

$V(bpy)_3^{2+}$ and $V(phen)_3^{2+}$ have lowest energy excited states which are probably either $^2E/^2T_1$ or quartet ^{4}MLCT (d^3 system) states. The excited state lifetimes are ∿0.5 ns and ∿1.8 ns, respectively [176]. Irradiation of $V(phen)_3^{2+}$ in its lowest energy absorption band (∿ 640 nm), in the presence of such electron acceptors as MV^{2+}, Eu^{3+}, or $VO(phen)_2^{2+}$, leads to photooxidation of the $*V(phen)_3^{2+}$ excited state and ligand loss. Where the electron acceptor is MV^{2+} (k_q ∿ $1.2 \times 10^9 M^{-1}s^{-1}$) or Eu^{3+}(aq), the photooxidation follows reactions 84-86 [176]. With $VO(phen)_2^{2+}$ as the oxidant, the sequence is different (eqns 87,88) and yields the same V(III,III)

$$V(phen)_3^{2+}(^4A_2) \xrightarrow[< 35\ ps]{h\nu} *V(phen)_3^{2+}(^2E \text{ or } ^4MLCT) \qquad (84)$$

$$*V(phen)_3^{2+} + A \longrightarrow V(phen)_3^{3+} + A^- \qquad (85)$$

$$2\ V(phen)_3{}^{3+} + 2\ H_2O \longrightarrow [(phen)_2V^{III}(u\text{-}OH)_2V^{III}(phen)_2]^{4+} \quad (86)$$

$$*V(phen)_3{}^{2+} + VO(phen)_2{}^{2+} \longrightarrow V(phen)_3{}^{3+} + VO(phen)_2{}^{+} \quad (87)$$

$$V(phen)_3{}^{3+} + VO(phen)_2{}^{+} + H_2O \longrightarrow phen + [(phen)_2V^{III}(u\text{-}OH)_2V^{III}(phen)_2]^{4+} \quad (88)$$

dimer as in reaction 86. Analogous events occur with $V(bpy)_3{}^{2+}$;however, the rates of the reaction are only ∿ 1/4 of those with V-$(phen)_3{}^{2+}$ under comparable conditions. The short lifetime of the lowest energy excited states makes photoinduced oxidative electron transfer relatively inefficient, but the photoredox reactions are irreversible owing to the hydrolytic dimerization noted in the above photochemistry.

9.0 ELECTRON TRANSFER REACTIONS WITH EXCITED STATES OF IRON COMPLEXES

2-Pyrazinecarboxylatobis(ethylenediamine)cobalt(III) reacts rapidly in aqueous media with $(CN)_5Fe^{II}(OH_2)^{3-}$(aq) to form XI; with the tetraammine analogue, XII forms [177]. These are of the

$[(CN)_5Fe^{II}\text{-}N$(pyrazine-carboxylate)$N\text{-}Co^{III}(en)_2]^-$

XI

$[(CN)_5Fe^{II}\text{-}N$(pyrazine-carboxylate)$N\text{-}Co^{III}(NH_3)_4]^-$

XII

type of 'precursor' complexes that occur in bimolecular events; this precursor contains an electron source and an electron sink, joined by a common bond system. Differences in the photoreactivities of XI and XII are understood if the mechanism involves, in each case, an excited state in which iron(III) is linked to cobalt (III) via a short-lived, bridging pyrazine radical. In the MLCT excitation (Fe^{II}(d) → pyrazine(π*)) of XII, the tetraamminecobalt(III) moiety is sufficiently reactive to oxidize the bridging radical after virtually every excitation event; for the photoinduced electron transfer to the $(NH_3)_4Co(III)$ moiety, k is > $10^9 s^{-1}$. The $(en)_2$-Co(III) moiety in XI is considerably less oxidizing in competing for the electron in the MLCT excited state.

$*Fe(bpy)_3{}^{2+}$ is oxidatively quenched by Fe^{3+}(aq); k_q ∿ 10^9 $M^{-1}s^{-1}$ [178].

10.0 BIMOLECULAR ELECTRON TRANSFER REACTIONS INVOLVING Cr(III) POLYPYRIDYL COMPLEXES

A comparison of the redox potentials in Table 1 between the ground-state and excited-state species shows the latter to be better oxidants. The reduction potentials of the Cr(III) complexes parallel those of the corresponding ruthenium(III) complexes [23], with the former potentials ∿1.5 V more negative. The potentials for reduction of the ^2E excited states are also ∿ 1.7 V more positive than the corresponding ground-state reduction potentials. This uniform shift of the excited state redox potentials is due to the similarity in the wavelengths for maximum emission in the Cr-$(NN)_3^{3+}$ complexes.

$Cr(bpy)_3^{3+}$ has proven a good candidate in studies of bimolecular excited-state redox reactions, inasmuch as its lowest MC excited state(s), $^2E/^2T_1$, is relatively long-lived in aqueous media ($^2\tau$ ∿ 0.063 ms) [179], is photostable in acidic media [180], and exhibits an efficient emission [181] which can easily be monitored. $Ru(bpy)_3^{2+}$ quenches this emission via reductive electron transfer [83,122]. Flash photolysis experiments carried out in the presence of $Ru(bpy)_3^{2+}$ and iodide ion, which selectively quenches the ^2E state of $Cr(bpy)_3^{3+}$, have unequivocally demonstrated the nature of the quenching mechanism (reactions 65 and 66) [83]; k_q = $4.0 \times 10^8 M^{-1}s^{-1}$ [122]. The products of reaction 66 are not observed because of a fast thermal back electron transfer reaction to give the original ground-state reactants.

Transition metal cyanide complexes quench the emission from $(^2E)Cr(bpy)_3^{3+}$ [85]; Table 23 summarizes the electron transfer quenching constants. Quenching of $*Cr(bpy)_3^{3+}$ occurs via reductive electron transfer with $Mo(CN)_8^{4-}$, $Fe(CN)_6^{4-}$, $Ru(CN)_6^{4-}$, and Ni-$(CN)_4^{2-}$, and via oxidative electron transfer with $Fe(CN)_6^{3-}$ and Co-$(CN)_6^{3-}$. The relationship between k_q's and the redox potentials of the excited state $*Cr(bpy)_3^{3+}$ and quenchers provides some insights into the thermodynamic favourability of either a reductive or oxidative electron transfer pathway. The quenching rate constant k_q of the cyanide complexes show no Marcus inverted region (see Figure 10) [85].

$Cr(bpy)_3^{2+}$ and $Fe^{3+}(aq)$ result from reductive quenching of $*Cr(bpy)_3^{3+}$ by $Fe^{2+}(aq)$ at pH 3 [83,182,183]. These electron transfer products could also be produced if an energy transfer step occurred prior to electron transfer (eqns 89,90) [182]. For such a

TABLE 23
Reductive electron transfer rate constants (k_q) for quenching $*Cr(NN)_3^{3+}$ complexes.

NN	Quencher	$10^{-8}k_q(M^{-1}s^{-1})$	Conditions	Ref.
bpy	Fe^{2+}	0.16	a	182
		0.37	b	21
4,4'-Me_2bpy	Fe^{2+}	0.0082	a	182
		0.022	b	21
4,4'-Ph_2bpy	Fe^{2+}	0.20	b	21
5-Clphen	Fe^{2+}	0.48	a	182
		0.12	b	21
5-Brphen	Fe^{2+}	0.84	b	183
phen	Fe^{2+}	0.15	a	182
		0.32	b	21
5-Mephen	Fe^{2+}	0.10	a	182
		0.34	b	183
5,6-Me_2phen	Fe^{2+}	0.26	b	183
4,7-Me_2phen	Fe^{2+}	0.0089	a	182
		0.060	a	21
3,4,7,8-Me_4phen	Fe^{2+}	0.0092	b	21
5-Phphen	Fe^{2+}	0.45	b	183
4,7-Ph_2phen	Fe^{2+}	0.25	b	21
bpy	$Mo(CN)_8^{4-}$	52	c	85
	$Fe(CN)_6^{4-}$	49	c	85
	$Ru(CN)_6^{4-}$	55	c	85
	$Ni(CN)_4^{2-}$	14	c	85
	$Ru(bpy)_3^{2+}$	6.1	a	182
		4.0	d	122
4,4'-Me_2bpy	$Ru(bpy)_3^{2+}$	2.0	a	182
phen	$Ru(bpy)_3^{2+}$	8.3	a	182
bpy	$Ru(4,4'\text{-}Me_2bpy)_3^{2+}$	11	a	182
	$Ru(3,4,7,8\text{-}Me_4phen)_3^{2+}$	13	a	182
	$Ru(4,7\text{-}Me_2phen)_3^{2+}$	11	a	182
phen	$Ru(4,7\text{-}Me_2phen)_3^{2+}$	14	a	182
bpy	$Ru(5\text{-}Mephen)_3^{2+}$	9.9	a	182

TABLE 23 (cont'd)

NN	Quencher	$10^{-8}k_q(M^{-1}s^{-1})$	Conditions	Ref.
4,7-Me_2phen	$Ru(5\text{-}Mephen)_3^{2+}$	11	a	182
bpy	$Ru(phen)_3^{2+}$	9.1	a	182
	$Ru(5\text{-}Clphen)_3^{2+}$	6.1	a	182
phen	$Ru(5\text{-}Clphen)_3^{2+}$	8.1	a	182
5-Mephen	$Ru(5\text{-}Clphen)_3^{2+}$	8.4	a	182
bpy	$Ru(5\text{-}NO_2phen)_3^{2+}$	1.8	a	182
phen	$Ru(5\text{-}NO_2phen)_3^{2+}$	2.6	a	182
bpy	$Os(bpy)_3^{2+}$	15	a	182
phen	$Os(bpy)_3^{2+}$	15	a	182
bpy	$Ru(bpy)_3^{3+}$	0.06	a,e	182
	$Os(bpy)_3^{3+}$	<0.01	a,e	182
	$Fe(CN)_6^{3-}$	2.6	c,e	85
	$Co(CN)_6^{3-}$	<0.002	c,e	85

[a] 1M H_2SO_4 at 25°C. [b] 1M HCl at 25°C. [c] water, u = 0.5M NaCl at ca. 20°C. [d] water, u = 0.2M at 25°C. [e] oxidative electron transfer (?).

pathway to be feasible, however, requires reaction 90 to be faster than electron transfer involving the encounter complex (reaction 89); that is, faster than reaction 91. The involvement of energy

$$*Cr(NN)_3^{3+} + Fe^{2+}(aq) \longrightarrow [*Fe^{2+} | Cr(NN)_3^{3+}] \quad (89)$$

$$[*Fe^{2+} | Cr(NN)_3^{3+}] \longrightarrow [Fe^{3+} | Cr(NN)_3^{2+}] \quad (90)$$

$$[Fe^{2+} | *Cr(NN)_3^{3+}] \longrightarrow [Fe^{3+} | Cr(NN)_3^{2+}] \quad (91)$$

transfer is unlikely inasmuch as reaction 90 requires more reorganizational energy than reaction 91; also, the orbital symmetry for reaction 90 is less favourable than for reaction 91. Other studies [83,183] are in agreement with a reductive electron transfer mechanism for quenching $*Cr(NN)_3^{3+}$ by $Fe^{2+}(aq)$. To the extent that Fe^{2+} has available low-lying excited states at 10,400 cm^{-1} and 14,400 cm^{-1} above the ground state [184], energy transfer is energetically feasible. The determination of redox quantum yields and thus the extent of electron transfer has not been addressed [182,183] (but see below). An oxidative path is unlikely since E^0 > 1.6 V (NHE) for the $Cr(NN)_3^{4+}/Cr(NN)_3^{3+}$ couple and $*E^o > -0.1$ V for the $Cr(NN)_3^{4+}/*Cr(NN)_3^{3+}$ couple (see Figure 5) [21].

Quenching of various $*Cr(NN)_3^{3+}$ by $Ru(NN)_3^{2+}$ complexes takes place via reductive electron transfer; variations in the quenching rates parallel changes in the driving force for electron transfer, ΔE^o. The $(^3CT)Ru(NN)_3^{2+}$ state is higher in energy than $(^2E)Cr(NN)_3^{3+}$ (2.1 eV vs. 1.7 eV) [182], ruling out an energy transfer quenching process; the rate constants for reaction 92 are collected in Table 23. The values of k_q increase with increasing

$$*Cr(NN)_3^{3+} + Ru(NN)_3^{2+} \xrightarrow{k_q} Cr(NN)_3^{2+} + Ru(NN)_3^{3+} \qquad (92)$$

driving force of the reaction, ΔE^o, but much less dramatically than predicted by the Marcus model [36,38].

Quenching of $*Cr(NN)_3^{3+}$ excited states by quencher species capable of oxidizing chromium(III) to chromium(IV) is inconclusive as regards the operating quenching mechanism(s) [182]. Of the two quenchers examined, $Ru(bpy)_3^{3+}$ and $Os(bpy)_3^{3+}$, the former is expected to quench more efficiently than the Os(III) complex if oxidative electron transfer occurred, since the driving force ΔE^o for the Ru(III) reaction is more favourable (1.25 V vs. 0.82 V) [182]. Indeed, $Os(bpy)_3^{3+}$ is an inefficient quencher compared to $Ru(bpy)_3^{3+}$; $k_q \lesssim 1 \times 10^6 M^{-1}s^{-1}$. The corresponding k_q for the $Ru(bpy)_3^{3+}$ reaction is $6 \times 10^6 M^{-1}s^{-1}$ (1 M H_2SO_4, 25°C). An energy transfer pathway cannot be discounted for these two M(III)-polypyridyl quenchers; the Ru(III) complex has an absorption band at 674 nm [24] which is in close proximity to the emission band at 695 nm [185] of $*Cr(bpy)_3^{3+}$. Both the Ru(III) and Os(III) complexes also have spin-forbidden bands at longer wavelengths, which may be close to, or overlap with the 728 nm emission of $*Cr(bpy)_3^{3+}$ [21]. Inconclusive results also obtain with $Fe(CN)_6^{3-}$ and $Co(CN)_6^{3-}$ as oxidative quenchers of $*Cr(bpy)_3^{3+}$ [85].

The Marcus model for outer-sphere electron transfer reactions, in the form of equations 24 and 28, affords an estimation of the self-exchange rate constants k_{22} for the couples involved in reaction 26. Determination of these self-exchange rate constants permits a comparison of the excited state reaction 26 with the corresponding ground state reaction 93, inasmuch as the two

$$\underline{Cr}(bpy)_3^{3+} + Cr(bpy)_3^{2+} \rightleftharpoons \underline{Cr}(bpy)_3^{2+} + Cr(bpy)_3^{3+} \qquad (93)$$

states may differ in shape, size and/or dipole moment. Any difference should be reflected in the reorganizational parameter λ (see eqns 22 and 23) for the ground- and excited-state reactions. The Stokes shift between ground-state absorption and excited-state emission is related to the nuclear coordinates of these states, and, expectedly also to the differences in λ of the ground and excited states. Thus, an observed Stokes shift of 0 cm^{-1} implies similar reorganizational parameters for the two states.

Several groups [182,183,186] have investigated the $*Cr(NN)_3^{3+}/Cr(NN)_3^{2+}$ self-exchange reactions in reductive electron transfer quenching experiments. Both Brunschwig and Sutin [182] and Serpone et al. [183] have determined the self-exchange rate using $Fe^{2+}(aq)$ as the quencher (reaction 94). The self-exchange reactions corresponding to this cross-reaction 94 are noted as re-

$$*Cr(NN)_3^{3+} + Fe^{2+}(aq) \xrightarrow{k_q} Cr(NN)_3^{2+} + Fe^{3+}(aq) \qquad (94)$$

actions 26 and 95, with k_{22} and k_{11} as the self-exchange rate contants, respectively. For the Fe^{2+}/Fe^{3+} couple, k_{11} = 4.0 $M^{-1}s^{-1}$ [187]. Serpone et al.[183] report an apparent k_{22} of $10^5 M^{-1}s^{-1}$

$$Fe^{2+}(aq) + Fe^{3+}(aq) \rightleftharpoons Fe^{3+}(aq) + Fe^{2+}(aq) \qquad (95)$$

from the intercept of a log k_{12} vs. log K_{12} plot (Figure 15). Endicott and Ferraudi [188] estimate k_{22} for the $*Cr(bpy)_3^{3+}/Cr(bpy)_3^{2+}$ couple as 3 x $10^5 M^{-1}s^{-1}$, based on data for the $Ru(bpy)_3^{2+}/*Cr(bpy)_3^{3+}$ cross-reaction and employing a self-exchange rate constant of $\sim$ 4 x $10^9 M^{-1}s^{-1}$ [23,38] for the $Ru(bpy)_3^{3+}/Ru(bpy)_3^{2+}$ couple. Brunschwig and Sutin [182] estimate $k_{22} \sim 1 \times 10^8 M^{-1}s^{-1}$ from a series of $Ru(NN)_3^{2+}/*Cr(NN)_3^{3+}$ reactions (eqn 26) in 1 M H_2SO_4 at 25^oC. However, an experimental value of k_{ex} = 600 $M^{-1}s^{-1}$ was obtained based on $Fe^{2+}/*Cr(NN)_3^{3+}$ reactions. The discrepancy between the experimental and theoretical estimations of k_{22} may be due to the non-adiabaticity of the cross-reaction 94; the self-exchange reactions 26 and 95 are probably adiabatic. If the Marcus relationship 24 is modified to include a term for the probability (p) of electron transfer in the respective activated complex, eqn 26 becomes

$$k_{12} = p_{12}(k_{11}k_{22}K_{12}f_{12})/p_{11}p_{22})^{1/2} \qquad (96)$$

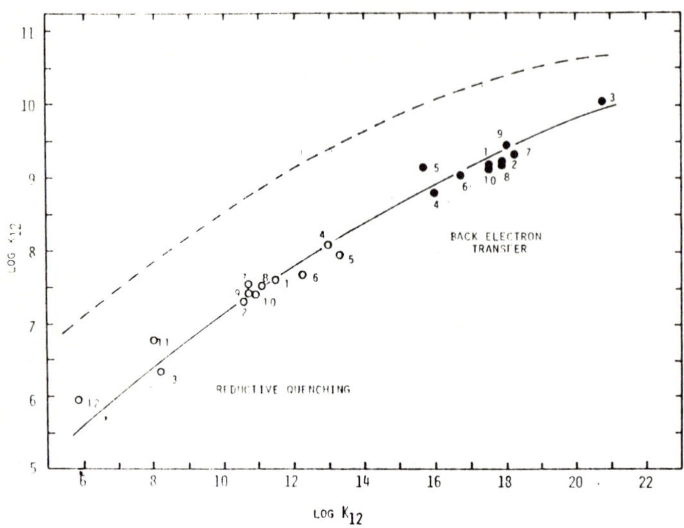

Figure 15.- Log k_q vs. log K_{12} plot (●) for the quenching of *Cr-$(NN)_3^{3+}$ excited states by Fe^{2+}(aq) in air-equilibrated, 1 M HCl aqueous solutions; (O) log k_{12} vs. log K_{12} plot for reduction of Fe^{3+}(aq) by $Cr(NN)_3^{2+}$ transients. The dashed line is the calculated curve for $k_{22} = 10^8 M^{-1}s^{-1}$ and the solid line, calculated curve for $k_{22} = 10^5 M^{-1}s^{-1}$. The numbers indicate the following NN: (1) bpy, (2) 4,4'-Ph_2bpy, (3) 4,4'-Me_2bpy, (4) 5-Clphen, (5) 5-Brphen, (6) 5-Phphen, (7) 5-Mephen, (8) phen, (9) 5,6-Me_2phen, (10) 4,7-Ph_2phen, (11) 4,7-Me_2phen. Reproduced with permission from ref. 183.

In the adiabatic limit, $p_{12} = p_{11} = p_{22} = 1$, and eqn 96 reduces to eqn 26. However, taking the cross-reaction 94 as adiabatic, $p_{11} = p_{22} = 1$ and $p_{12} << 1$; the intercept in a log k_{12} vs. $(1 + \alpha)$ log K_{12} plot will yield $k_{22}p_{12}$ and not simply k_{22}. Since $p_{12} << 1$, k_{22} for reaction 26 will be $> 10^5 M^{-1}s^{-1}$, the apparent value obtained by Serpone et al. [183], and is probably ca. 10^8-$10^9 M^{-1}s^{-1}$ [182]. The rate constant for the ground-state self-exchange (reaction 93) is $\sim 10^8$-$10^9 M^{-1}s^{-1}$ [183]. Ferraudi and Endicott [186] report $k_{ex} \sim 4 \times 10^9 M^{-1}s^{-1}$ for reaction 93 from flash photolysis studies of $Cr(bpy)_3^{3+}$ in the presence of $Cr(Me_2phen)_3^{3+}$ in methanolic solution (25^oC, u = 0.1 M).

These results demonstrate the similarities in the excited state and ground state self-exchange reactions. This is not unexpected; the reactants in both reactions have the same charges, and the $^4A_2$ and ^2E states are virtually undistorted with respect to each other. These similarities suggest that the ground- and excited-state reactions only differ with respect to their inner-sphere barriers, $*\Delta G^{\#}_{in}$ [182], if both reactions are adiabatic. For metal-centred states, the differences in these barriers for the ground-state, $\Delta G^{\#}_{in}$, and excited-state, $*\Delta G^{\#}_{in}$, exchange reactions are approximated [182] by eqn 97; $\underline{f}$ is a breathing force constant, $\underline{a}$ is

$$\Delta(\Delta G^{\#}_{in}) = *\Delta G^{\#}_{in} - \Delta G^{\#}_{in}$$

$$= \frac{3f_2}{f_2 + f_3}\left[(f_3^* - f_3)a_2^2 + \frac{E_s}{6} \cdot \frac{(a_3^* + a_3 + 2a_2)}{(a_3^* + a_3)}\right] \qquad (97)$$

the Cr-N bond distance and subscripts 2 and 3 denote the charge on the complex. The $3f_2/(f_2 + f_3)$ term is [189] small since the $^4A_2$ and ^2E states have the same electron configuration. The second term in eqn 97 has five possible fates [182,183]. To the extent that the t_{2g} orbitals in both the $^4A_2$ and ^2E states are equally occupied, $a_3^* = a_3$, and $E_s = 0$; thus $\Delta(\Delta G^{\#}_{in})$ would necessarily be 0, and the inner-sphere barriers for the ground- and excited-state reactions would be identical. This reasoning suggests that the exchange rate for the two reactions should be nearly identical as observed.

The lifetime of the $^4T_2$ state of $Cr(bpy)_3^{3+}$ is $\sim$ 50 ps [190]. This state cannot be involved in reaction 26 since $a_3^* > a_3$ and because its lifetime is too short for bimolecular encounters. The other possibility is the $^2T_1$ state, in thermal equilibrium with the ^2E state. These latter two states differ only in the t_{2g} orbital population. Unfortunately, the available data and the uncertainties inherent in the Marcus relationships have precluded a delineation as to the reactive excited state in the self-exchange reactions involving the $^2E/^2T_1$ pair.

The ^2E state does not appear to be implicated in the reductive quenching of $*Cr(bpy)_3^{3+}$ in alcoholic media [186]. This

$$*Cr(bpy)_3^{3+} + ROH \longrightarrow Cr(bpy)_3^{2+} + ROH^+ \qquad (98)$$

suggestion is based on observations that both the $(^2E)Cr(bpy)_3^{3+}$ transient and $Cr(bpy)_3^{2+}$ are detected in aqueous alkaline and alcoholic media, and that the $(^2E)Cr(bpy)_3^{3+}$ lifetime is relatively independent of solution media while the $(^{2}E)Cr(bpy)_3^{3+}$ yield ($^{4}\eta_{isc}$) is medium-dependent. The suggestion is [186] that the solvent dependent process leading to the chromium(II) species is a prompt, upper-state process that occurs in competition with excited state relaxation to form the ^{2}E transient. The investigation of Ferraudi and Endicott [186] implies the involvement of either the $^{4}T_2$ or $^{2}T_2$, or some higher excited state. Reductive quenching by solvent (DMF) has also been implicated in the photosolvation of $Cr(bpy)_3^{3+}$ in argon-purged solutions (Scheme XVII) [191]; however, in aqueous solutions, the intervention of Cr(II) species is not evident. Cage escape is rapid, occurring by self exchange of the

Scheme XVII:

$$Cr(bpy)_3^{3+} \xrightarrow{h\nu} {}^{*}Cr(bpy)_3^{3+}$$

$$\downarrow DMF$$

$$Cr(bpy)_3^{3+} + DMF \nleftarrow [Cr(bpy)_3^{2+}...DMF^{+}]$$

$$\downarrow \text{cage escape}$$

$$Cr(bpy)_2(DMF)_2^{2+} + bpy \xleftarrow{DMF} Cr(bpy)_3^{2+} + DMF^{+}$$

$$Cr(bpy)_2(DMF)_2^{2+} + Cr(bpy)_3^{3+} \longrightarrow Cr(bpy)_3^{2+} + Cr(bpy)_2(DMF)_2^{3+}$$

DMF^{+} radical cation with molecules in the outer solvation sphere and eventually with bulk solvent.

$(^{2}E)Cr(bpy)_3^{3+}$ is also reductively quenched by ferrocene and ferrocene carboxylic acid in 70% CH_3CN/H_2O [192]. Fe^{2+} in 1 M $HClO_4$ and $Fe^{II}Cl_x^{(2-x)+}$ in 1 M HCl also quench reductively, but quenching is more efficient in 1 M HCl than in 1 M $HClO_4$. The fraction F_1 of Cr(II) redox products, k_q's, rate constants for the electron transfer quenching component, and the reverse electron transfer reactions are presented in Table 24. The efficiencies of formation of electron transfer products is < 20%. Two other mechanisms contribute to the deactivation of $(^{2}E)Cr(bpy)_3^{3+}$: (i) energy transfer, and (ii) paramagnetic ion-induced intersystem crossing from ^{2}E to the ground state $^{4}A_2$ [192]. The smaller F_1 value for $Fe^{II}Cl_x^{(2-x)+}$ is due to a faster reverse electron transfer reaction in 1 M HCl (probably via an inner-sphere pathway), while the lower value for ferrocene-COOH acid is consistent with the greater

TABLE 24
Reductive electron transfer kinetics in the quenching of $(^2E)Cr(bpy)_3^{3+}$.[a]

Quencher	$k_q(M^{-1}s^{-1})$	$k_e(M^{-1}s^{-1})$	$k_{-e}(M^{-1}s^{-1})$	Fraction of Cr(II) Products (F_1)
Fe^{2+} [b]	7×10^6	1.2×10^6	1.0×10^9	0.17
$Fe^{II}Cl_x^{(2-x)+}$ [c]	4×10^7	3.2×10^6	1.4×10^9	0.081
Ferrocene	6.2×10^9	1.1×10^9	-	0.18
Ferrocene-COOH	7.3×10^9	0.5×10^9	-	0.069

[a] reference 192. [b] in 1M $HClO_4$. [c] in 1M HCl.

driving force for the reverse electron transfer in the geminate pair (in the solvent cage), ΔG = -1.26 eV compared to -0.81 eV for the same reaction with oxidized ferrocene.

11.0 ELECTRON TRANSFER REACTIONS IN EXCITED COBALT(III) COMPLEXES

The intramolecular charge-transfer photochemistry of cobalt(III) complexes has been thoroughly studied; it has been reviewed extensively by Balzani and Carassiti [7], Endicott [2], and by Ford and coworkers [193].The picture emerging from these studies is the strong tendency of cobalt(III) complexes to undergo intramolecular photoredox decomposition when irradiated in their LMCT absorption bands.

Efforts have also been devoted to investigating outer-sphere or ion-pair charge-transfer (IPCT) processes [194-199]. Irradiation of $Co^{III}(phen)_2(C_2O_4)^+/RCOO^-$ and $Co(NH_3)_5X^{n+}/B(C_6H_5)_4^-$ in their IPCT absorption bands leads to formation of Co(II) complexes (reaction 99) [196] and $Co^{2+}(aq)$ (reaction 100) [197,199].

$$Co^{III}(phen)_2(C_2O_4)^+/RCOO^- \xrightarrow[CTTM]{h\nu} Co^{II}(phen)_2(C_2O_4) + RCOO^{\cdot} \longrightarrow R^{\cdot} + CO_2 \qquad (99)$$

R = H, CH_3, C_2H_5, $C(CH_3)_3$, CCl_3, CF_3, C_6H_5

$$Co(NH_3)_5X^{n+}/B(C_6H_5)_4^- \xrightarrow[\lambda > 350\ nm]{h\nu} Co^{2+}(aq) + 5NH_3 + X^- + {}^{\cdot}B(C_6H_5)_4 \qquad (100)$$

$X = NH_3, F^-, Cl^-, Br^-, NO_3^-, NO_2^-, N_3^-, CH_3COO^-$

The quantum yield of Co^{2+}(aq) formation is dependent on the excitation wavelength, on the nature of the ligand X, and on the nature of the solution medium; $\Phi(Co^{II})$ decreases with increase in the concentration of methanol in the solvent mixture $MeOH/CH_2Cl_2$ [199]. While $Co(NH_3)_5(H_2O)^{3+}$, $Co(NH_3)_4(H_2O)_2^{3+}$, and $Co(en)_2(H_2O)Cl^{2+}$ show no or negligible intramolecular photoredox reaction on irradiation at $\lambda > 350$ nm, the addition of chromate(VI) to the aqueous/alcoholic media leads to an efficient formation of Co(II) together with reduction of chromium(VI) to chromium(III). Chromium(V) is an important intermediate in the primary step of the photoinduced event [198]. The photochemical formation of $[Cl(en)_2Co\text{-}NC\text{-}Fe(CN)_5]^{2-}$ from irradiating a mixture of $Co(en)_3^{3+}$ and $Fe(CN)_6^{4-}$ proceeds via electron transfer [200].

In section 4.5 we noted that some electron relay substances (eg., $Co(NH_3)_5Cl^{2+}$ and $Rh(bpy)_3^{3+}$) undergo irreversible decomposition with consequent decreased usefulness of the relay in a light energy conversion cycle. Cobalt(III)-sepulchrate (XIII) seems an ideal candidate for utilization as an electron relay species. This complex exhibits some remarkable properties [201]:

$Co(sep)^{3+}$:

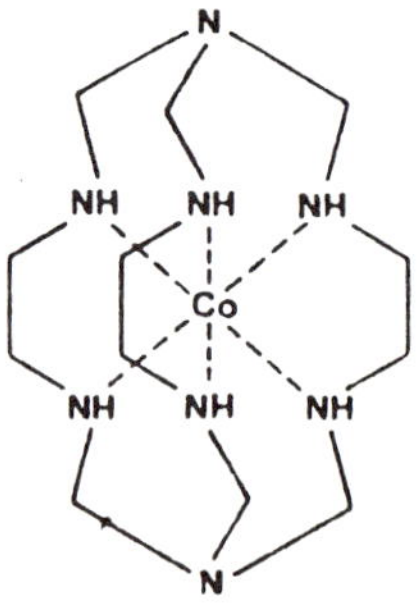

(XIII)

(i) it undergoes a reversible reduction at -0.54 V (vs. SCE), (ii) the reduced cobalt(II) complex is appreciably inert to ligand substitution and to metal exchange in view of the cage nature of the sepulchrate ligand, and (iii) the self-exchange electron transfer rate is relatively high (5.1 $M^{-1}s^{-1}$; u = 0.2M). Other relay species have been considered: for example, Eu^{3+}(aq), $V(H_2O)_6^{3+}$, $Co(en)_3^{3+}$. However, their usefulness is limited owing to (i) low self-exchange rates arising from high intrinsic barriers and/or non-adiabaticity, and (ii) to the irreversible redox behaviour exhibited by some of these metal complexes.

In aqueous solutions, $Co(sep)^{3+}$ quenches $*Ru(bpy)_3^{2+}$ by oxidative electron transfer: $k_q = 3.6 \times 10^8 M^{-1}s^{-1}$ in 1 M H_2SO_4 [201] ($2.6 \times 10^8 M^{-1}s^{-1}$; u = 0.1 M NaCl [202]) and is larger than the k_q ($6.2 \times 10^7 M^{-1}s^{-1}$) for $Co(en)_3^{3+}$ under the same conditions, even though the reduction potential for $Co(sep)^{3+}$ is more negative [-0.54 V vs. -0.45 V (SCE)]. The net amount of electron transfer is ∿ 90% [202]. Houlding and coworkers [202] also investigated the cobaltocene-like complex $Co(C_5H_4COOH)_2^+$ as a potential relay species; k_q for quenching $*Ru(bpy)_3^{2+}$ is $1.5 \times 10^9 M^{-1}s^{-1}$ over the pH range 1-10. The quenching pathway is either energy transfer and/or oxidative electron transfer followed by rapid back electron transfer within the cage. The photochemistry of $Co(sep)^{3+}/X$ ions pairs ($X = I^-$ and $C_2O_4^{2-}$) has been investigated recently by Pina and coworkers [203-205].

Light irradiation of $Co(sep)^{3+}/X^{n-}$ ion pairs in their IPCT bands (near-UV spectra region) leads to an electron displacement from the anion to the cation yielding the reduced $Co^{II}(sep)^{2+}$ species via Scheme XVIII for $X^{n-} = I^-$ [204]. In neutral aqueous

<u>Scheme XVIII</u>:

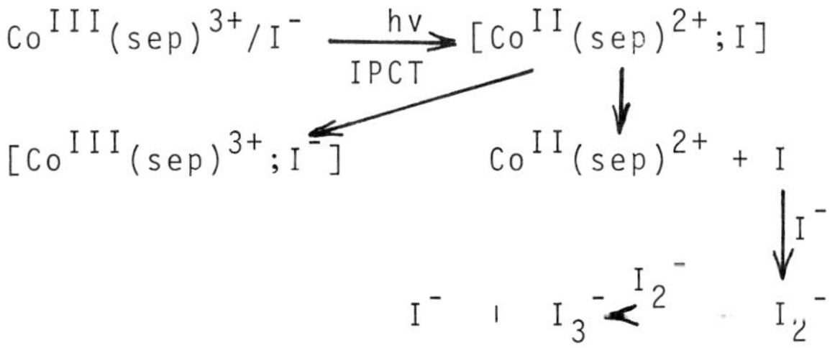

$$Co^{III}(sep)^{3+}/I^- \xrightarrow[IPCT]{h\nu} [Co^{II}(sep)^{2+};I]$$

$[Co^{II}(sep)^{2+};I] \longrightarrow [Co^{III}(sep)^{3+};I^-]$

$[Co^{II}(sep)^{2+};I] \longrightarrow Co^{II}(sep)^{2+} + I$

$I + I^- \longrightarrow I_2^-$

$I_2^- + I_2^- \longrightarrow I^- + I_3^-$

$$Co^{II}(sep)^{2+} + I_2^- \longrightarrow Co^{III}(sep)^{3+} + 2I^- \qquad (101)$$

media, I_2^- (and/or I^- and/or I_3^-) oxidizes $Co(sep)^{2+}$; no net photochemistry is observed. However, in acidic media, decomposition of this cobalt(II) species (reaction 102; k ∿ 0.012 $M^{-1}s^{-1}$) competes with the re-oxidation reaction 101. When X^{n-} is $C_2O_4^{2-}$ and

$$Co^{II}(sep)^{2+} \xrightarrow{H^+} Co^{2+}(aq) + \text{other products} \qquad (102)$$

the pH is 3.0, light excitation into the IPCT band leads to electron transfer from $HC_2O_4^-$ to $Co(sep)^{3+}$ to give the reduced Co(II) substrate and CO_2 according to Scheme XIX:

Scheme XIX:

$$Co^{III}(sep)^{3+}/HC_2O_4^- \xrightarrow{h\nu\ (IPCT)} [Co^{II}(sep)^{2+};HC_2O_4]$$

$$[Co^{II}(sep)^{2+};HC_2O_4] \longrightarrow [Co^{III}(sep)^{3+};HC_2O_4^-]$$

$$[Co^{II}(sep)^{2+};HC_2O_4] \longrightarrow Co^{II}(sep)^{2+} + CO_2 + CO_2^- + H^+$$

$$CO_2^- \xrightarrow{Co^{III}(sep)^{3+}} Co^{II}(sep)^{2+} + CO_2$$

The potential utility of these ion pairs to the design of photosensitizers for cyclic redox processes has been discussed by Pina et al. [202-205]. Reactions 103 and 104 are typical examples where $Co(sep)^{3+}$ plays the role of a photosensitizer (Scheme

$$4\ I^- \ +\ O_2\ \ 4\ H^+ \xrightarrow[Co(sep)^{3+}]{h\nu} 2\ I_2\ +\ 2\ H_2O \qquad (103)$$

$$HC_2O_4^- \ +\ H^+ \xrightarrow[Co(sep)^{3+}]{h\nu} 2\ CO_2\ +\ H_2 \qquad (104)$$

XX) [205]; in this scheme, the oxalate ion acts as the sacrificial

Scheme XX:

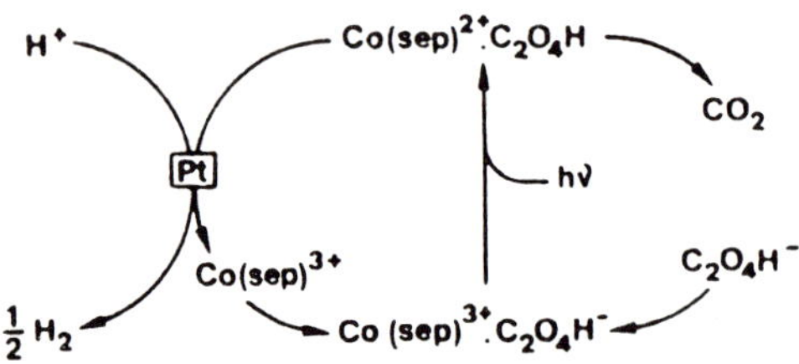

electron donor. The parallel with cyclic schemes (cf. Scheme XIV) involving $*Ru(bpy)_3^{2+}$ for the generation of H_2 is evident. Unfortunately, the lack of strong absorption of light energy in the visible spectral region, coupled with the very short lifetime of

$Co(sep)^{3+}$ [203-205] precludes the utilization of this species in a light energy conversion cycle. Its role in other electron transfer processes needs to be explored.

12.0 CONCLUDING REMARKS

Clearly the last decade has witnessed a dramatic evolution in studies aimed at electron transfer processes, particularly from excited state species. The advent of excited state electron transfer processes is now well established. These studies, however, have not been without controversy when it comes to distinguishing between energy transfer quenching and electron transfer quenching. Often, the quenching rate constants have been taken to reflect completely the kinetics of electron transfer events. It appears, more careful examinations are required inasmuch as quenching often implicates both energy and electron transfer. Here the redox quantum yields need be examined.

Demonstration of the so-called 'inverted' Marcus region by excited state complexes in their electron transfer reactions, which are more exergonic than those of the related ground state species, has been somewhat disappointing. Nevertheless, the existence of the 'inverted' region has recently been demonstrated in donor/acceptor molecules joined by spacer molecules. The various factors (Franck Condon and electronic) affecting rates of electron transfer and models have been examined actively by several groups. Electron transfer from excited states in various organized media has received much attention. At present, much of the focus of photochemists is directed at electron transfer events in heme-proteins using transition metal complexes as probes.

Were it not for the oil crisis of 1973 and the subsequent euphoria in solar energy conversion and storage investigations with water splitting as a principal theme, much of the work examined in this article would not have received the attention it has. Though the 'crisis' is over, the efforts continue unabated to probe electron transfer events with novel and different molecules.

13.0 ACKNOWLEDGEMENTS

We are grateful to Dr. Mary Jamieson for her assistance in the literature search and for useful discussions. We are also grateful to the American Chemical Society, the Royal Society of Chemistry of the UK for permission to reproduce some work. Our research is supported by NSERC (Canada) and by NATO (Grant No. 843/84).

14.0 Appendix

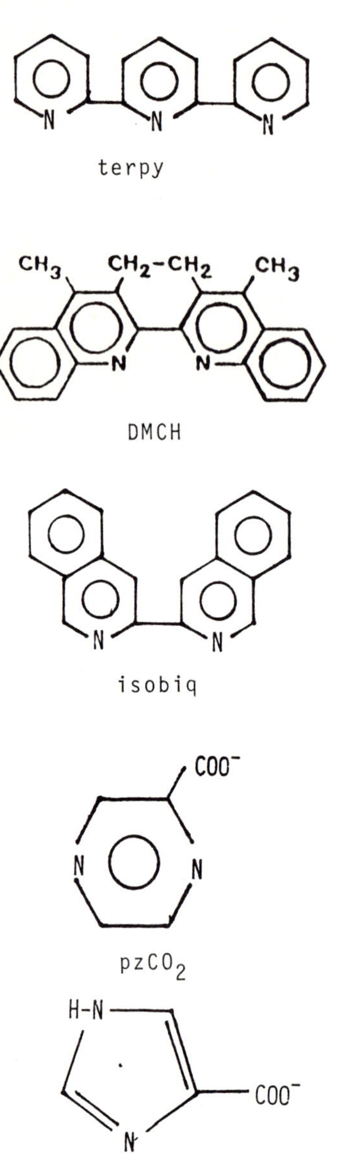

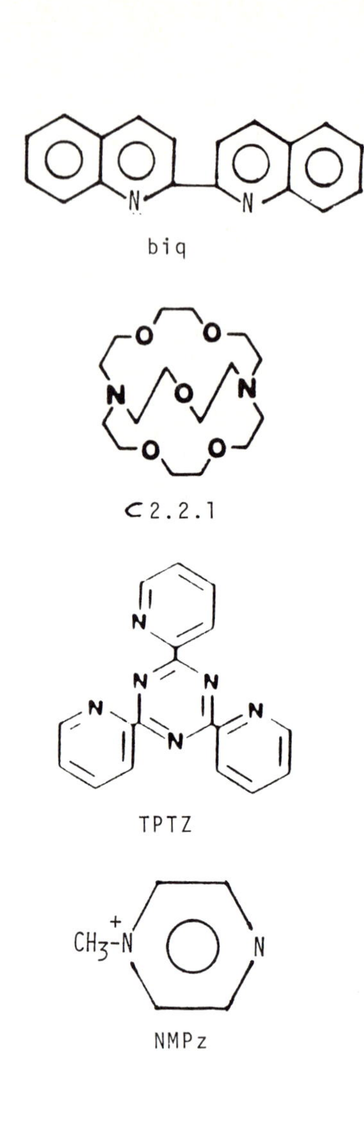

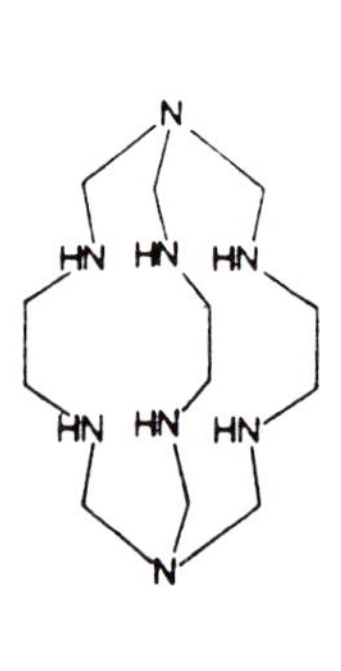

sep

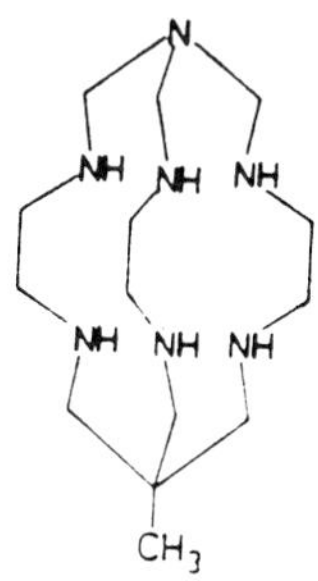

azamesar

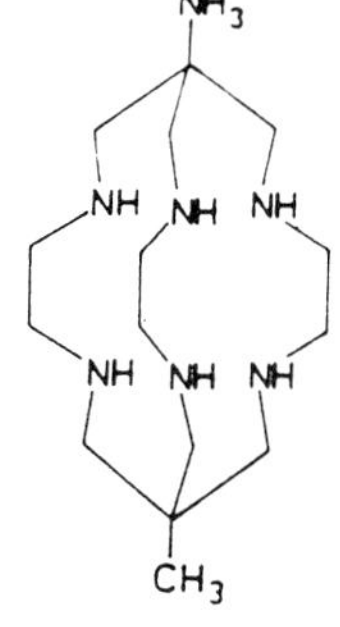

ammesar H

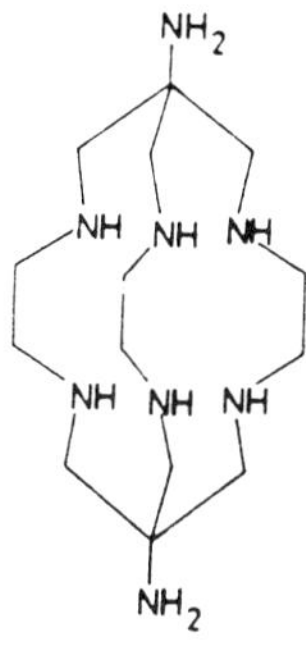

diamsar

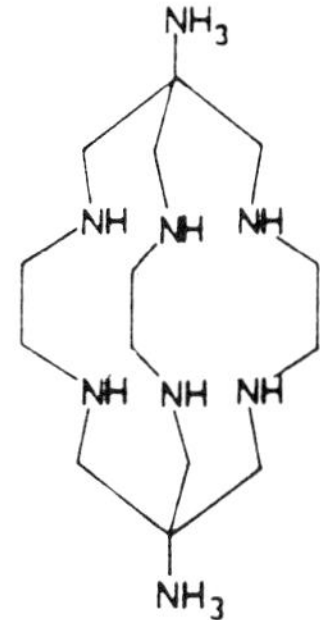

diamsar H_2

15.0 REFERENCES

1 H.D. Gafney and A.W. Adamson, J.Am.Chem.Soc., 94 (1972) 8238.

2 J.F. Endicott, in "Concepts of Inorganic Photochemistry", A.W. Adamson and P.D. Fleischauer, Eds., Wiley-Interscience, New York, 1975, chapter 3, p. 81; and references therein.

3 E. Zinato, in "Concepts of Inorganic Photochemistry", A.W. Adamson and P.D. Fleischauer, Eds., Wiley-Interscience, New York, 1975, chapter 4, p. 143; and references therein.

4 J.N. Demas and A.W. Adamson, J.Am.Chem.Soc., 93 (1971) 1800.

5 J.N. Demas and A.W. Adamson, J.Am.Chem.Soc., 95 (1973) 5159.

6 See e.g., V. Balzani, L. Moggi, M.F. Manfrin, and F. Bolletta, Coord.Chem.Rev., 15 (1975) 321; and references therein.

7 V. Balzani and V. Carassiti, "Photochemistry of Coordination Compounds", Academic Press, New York, 1970.

8 A.W. Adamson and P.D. Fleischauer, Eds., "Concepts of Inorganic Photochemistry", Wiley-Interscience, New York, 1975.

9 V. Balzani, F. Bolletta, M.T. Gandolfi, and M. Maestri, Topics Curr.Chem., 75 (1978) 1.

10 A.W. Adamson, Pure Appl.Chem., 51 (1979) 313.

11 M.S. Wrighton, Ed., "Inorganic and Organometallic Photochemistry", Adv.Chem.Ser., vol. 168, American Chemical Society, Washington, D.C., 1978.

12 M.Gratzel, Ed., "Energy Resources through Photochemistry and Catalysis", Academic Press, New York, 1983.

13 F. Scandola, in "Rearrangements in Ground and Excited States", Academic Press, New York, vol. 3, 1980, p. 59.

14 N. Sutin, Prog.Inorg.Chem., 30 (1983) 441.

15 N. Sutin and C. Creutz, Pure Appl.Chem., 52 (1980) 2717.

16 G.J. Karvanos and N.J. Turro, Chem.Rev., 86 (1986) 401.

17 M.A. Jamieson, N. Serpone, and M.Z. Hoffman, Coord.Chem.Rev., 39 (1981) 121.

18 V. Balzani and F. Scandola, in "Photoinduced Electron Transfer" M.A. Fox and M. Chanon, Eds., Elsevier, Amsterdam, The Netherlands, vol. D, chapter 5.4, 1988.

19 C.H. Langford, C. Moralejo, and M. Hogben, in "Photoinduced Electron Transfer", M.A. Fox and M. Chanon, Eds., Elsevier, Amsterdam, The Netherlands, vol. B, chapter 2.9, 1988.

20 L. Moggi, in "Photoelectrochemistry, Photocatalysis, and Photoreactors", M. Schiavello, Ed., NATO-ASI Series, vol. 146, D. Reidel Publ.Co., Dordrecht, The Netherlands, 1985, p. 197.

21 N. Serpone, M.A. Jamieson, M.S. Henry, M.Z. Hoffman, F. Bolletta, and M. Maestri, J.Am.Chem.Soc., 101 (1979) 2907.

22 C.-T. Lin and N. Sutin, J.Phys.Chem., 80 (1976) 97.

23 C.-T. Lin, W. Bottcher, M. Chou, C. Creutz, and N. Sutin, J. Am.Chem.Soc., 98 (1976) 6536.

24 C.R. Bock, T.J. Meyer, and D.G. Whitten, J.Am.Chem.Soc., 97 (1975) 2909.

25 R. Ballardini, G. Varani, M.T. Indelli, F. Scandola, and V. Balzani, J.Am.Chem.Soc., 100 (1978) 7219.

26 K. Kalyanasundaram, Coord.Chem.Rev., 46 (1982) 159; and references therein.

27 R.A. Marcus, Annu.Rev.Phys.Chem., 15 (1964) 155.

28 N.S. Hush, in "Mechanistic Aspects of Inorganic Reactions", ACS Symposium Series 198, D.B. Rorabacker and J.F. Endicott, Eds., American Chemical Society, Washington, D.C., 1982, chapter 13.

29 D. Rehm and A. Weller, Israel J.Chem., 8 (1970) 259.

30 N. Sutin, in "Bioinorganic Chemistry", G.L. Eichorn, Ed., vol. 2, Elsevier, New York, 1973, chapter 19, p. 611.

31 V. Balzani and F. Scandola, in "Energy Resources through Photochemistry and Catalysis", M. Gratzel, Ed., Academic Press, New York, 1983, chapter 1, p. 1.

32 N.J. Turro, "Modern Molecular Photochemistry", Benjamin/Cummings, Menlo Park, CA., 1978, chapter 19.

33 R.A. Marcus, J.Chem.Phys., 24 (1956) 966.

34 R. Rehm and A. Weller, Ber.Bunsenges.Phys.Chem., 73 (1969) 834.

35 N. Sutin, in "Tunneling in Biological Systems", B. Chance, D. C. DeVault, H. Frauenfelder, R.A. Marcus, J.R. Schrieffer, and N. Sutin, Eds., Academic Press, New York, 1979, p. 201.

36 R.A. Marcus, J.Chem.Phys., 43 (1965) 2654; J.Chem.Phys., 43 (1965) 679.

37 N.S. Hush, Prog.Inorg.Chem., 8 (1967) 391; Electrochim.Acta, 13 (1968) 1005.

38 C. Creutz and N. Sutin, J.Am.Chem.Soc., 99 (1977) 241.

39 J.R. Miller, L.T. Calcaterra, and G.L. Closs, J.Am.Chem.Soc., 106 (1984) 3047; see also M.R. Wasielewski, M.P. Niemczyk, W. A. Svec, and E.B. Pewitt, J.Am.Chem.Soc., 107 (1985) 1080.

40 N.J. Turro, Pure Appl.Chem., 49 (1977) 405.

41 V. Balzani and F. Scandola, in "Photochemical Conversion and Storage of Solar Energy", J.S. Connolly, Ed., Academic Press, New York, 1981, chapter 4.

42 P. Natarajan and J.F. Endicott, J.Phys.Chem., 77 (1973) 971.

43 C.-T. Lin and N. Sutin, J.Am.Chem.Soc., 97 (1975) 3543.

44 G. Navon and N. Sutin, Inorg.Chem., 13 (1974) 2159.

45 C. Creutz and N. Sutin, Inorg.Chem., 15 (1976) 496.

46 R.C. Young, F.R. Keene, and T.J. Meyer, J.Am.Chem.Soc., 99 (1977) 2468.

47 C.P. Anserson, D.J. Salmon, R.C. Young, and T.J. Meyer, J.Am. Chem.Soc., 99 (1977) 1980.

48 I. Fujita and H. Kobayashi, J.Chem.Phys., 52 (1970) 4904.

49 I. Fujita and H. Kobayashi, J.Chem.Phys., 59 (1972) 2902.

50 I. Fujita and H. Kobayashi, Ber.Bunsenges.Phys.Chem., 76 (1972) 115.

51 V. Balzani, F. Scandola, G. Orlandi, N. Sabbatini, and M.T. Indelli, J.Am.Chem.Soc., 103 (1981) 3370.

52 G.S. Laurence and V. Balzani, Inorg.Chem., 13 (1974) 2976.

53 C.R. Bock, T.J. Meyer, and D.G. Whitten, J.Am.Chem.Soc., 96 (1974) 4710.

54 M.I.C. Ferreira and A. Harriman, J.Chem.Soc.Faraday Trans.2, 75 (1979) 874.

55 D.G. Taylor and J.N. Demas, J.Chem.Phys., 71 (1979) 1032.

56 M.A. Hoselton, C.-T. Lin, H.A. Schwarz, and N. Sutin, J.Am. Chem.Soc., 100 (1978) 2383.

57 J.E. Baggott and M.J. Pilling, J.Phys.Chem., 84 (1980) 3012.

58 J.N. Demas and J.W. Addington, J.Am.Chem.Soc., 98 (1976) 5800.

59 C. Creutz and N. Sutin, J.Am.Chem.Soc., 98 (1976) 6384.

60 C. Creutz, Inorg.Chem., 17 (1979) 1046.

61 E. Sutcliffe, M. Neumann-Spallart, and K. Kalyanasundaram, (1980) unpublished results. Quoted in ref. 26.

62 T.K. Foreman, C. Giannotti, and D.G. Whitten, J.Am.Chem.Soc., 102 (1980) 1170.

63 K. Chandrasekaran, T.K. Foreman, and D.G. Whitten, Nouv.J.Chim. 5 (1981) 275.

64 B.A. DeGraff and J.N. Demas, J.Am.Chem.Soc., 102 (1980) 6169.

65 P. Connolly, J.H. Espenson, and A. Bakac, Inorg.Chem., 25 (1986) 2169.

66 N. Sabbatini, S. Dellonte, A. Bonazzi, M. Ciano, and V. Balzani Inorg.Chem., 25 (1986) 1738.

67 R.C. Young, J.K. Nagle, T.J. Meyer, and D.G. Whitten, J.Am. Chem.Soc., 100 (1978) 4773.

68 T. Shimidzu, T. Iyoda, and K. Izaki, J.Phys.Chem., 89 (1985) 642.

69 J.E. Baggott and M.J. Pilling, J.Chem.Soc.Fraraday Trans.1, 77 (1981) 261.

70 J.N. Demas, J.W. Addington, S.H. Peterson, and E.W. Harris, J. Phys.Chem., 81 (1977) 1039.

71 F. Bolletta, A. Juris, M. Maestri, and D. Sandrini, Inorg.Chim. Acta, 44 (1980) L175.

72 M. Neumann-Spallart, K. Kalyanasundaram, C. Gratzel, and M. Gratzel, Helv.Chim.Acta, 62 (1980) 1111.

73 R.C. Young, T.J. Meyer, and D.G. Whitten, J.Am.Chem.Soc., 98 (1976) 286.

74 N. Sutin and C. Creutz, Adv.Chem.Ser., 168 (1978) 1.

75 S. Anderson, E.C. Constable, K.R. Seddon, J.E. Turp, J.E. Baggott, and M.J. Pilling, J.Chem.Soc.Datlton Trans., (1985) 2247.

76 D.G. Whitten, Acc.Chem.Res., 13 (1980) 83; and references therein.

77 B.L. Hauenstein, Jr., K. Mandal, J.N. Demas, and B.A. DeGraff, Inorg.Chem., 23 (1984) 1101.

78 K. Kalyanasundaram and M. Neumann-Spallart, Chem.Phys.Lett., 88 (1982) 7.

79 W.J. Dressick, B.L. Hauenstein, Jr., J.N. Demas, and B.A. DeGraff, Inorg.Chem., 23 (1984) 1107.

80 S.J. Atherton, J.H. Baxendale, and B.M. Hoey, J.Chem.Soc.Faraday Trans.1, 78 (1982) 2167.

81 S.-F. Chan, M. Chou, C. Creutz, M. Matsubara, and N. Sutin, J. Am.Chem.Soc., 103 (1981) 369.

82 C. Creutz, M. Chou, T.L. Netzel, M. Okumura, and N. Sutin, J. Am.Chem.Soc., 102 (1980) 1309.

83 R. Ballardini, G. Varani, M.A. Scandola, and V. Balzani, J.Am. Chem.Soc., 98 (1976) 7432.

84 A. Juris, M.T. Gandolfi, M.F. Manfrin, and V. Balzani, J.Am. Chem.Soc., 98 (1976) 1047.

85 A. Juris, M.F. Manfrin, M. Maestri, and N. Serpone, Inorg.Chem. 17 (1978) 2258.

86 P. Natarajan and J.F. Endicott, J.Am.Chem.Soc., 94 (1972) 3635.

87 P. Natarajan and J.F. Endicott, J.Am.Chem.Soc., 95 (1973) 2470.

88 K. Chandrasekaran and P. Natarajan, J.Chem.Soc.Dalton Trans., (1981) 774.

89 K. Chandrasekaran and P. Natarajan, J.Chem.Soc.Chem.Commun., (1977) 774.

90 K. Chandrasekaran and P. Natarajan, Inorg.Chem., 19 (1980) 1714.

91 P. Natarajan and J.F. Endicott, J.Phys.Chem., 77 (1973) 1823.

92 E. Finkenberg, P. Fisher, S.-M.Y. Huang, and H.D. Gafney, J. Phys.Chem., 82 (1978) 526.

93 M. Kimura, M. Yamashita, and S. Nishida, Inorg.Chem., 24 (1985) 1527.

94 K.R. Leopold and A. Haim, Inorg.Chem., 17 (1978) 1753.

95 W. Bottcher and A. Haim, J.Am.Chem.Soc., 102 (1980) 1564.

96 H.E. Toma and C. Creutz, Inorg.Chem., 16 (1977) 545.

97 R. Berkoff, K. Krist, and H.D. Gafney, Inorg.Chem., 19 (1980) 1.

98 C.-Y. Mok, A.W. Zanella, C. Creutz, and N. Sutin, Inorg.Chem., 23 (1984) 2891.

99 R.D. Cannon, "Electron Transfer Reactions", Butterworths, London, 1980.

100 N. Sutin, Acc.Chem.Res., 15 (1982) 275.

101 D.B. Rorabacker and J.F. Endicott, Eds., "Mechanistic Aspects of Inorganic Reactions", ACS Symposium Series, vol. 198, American Chemical Society, Washington, D.C., 1982.

102 J.F. Endicott, K. Kumar, T. Ramasani, and F.P. Rotzinger, Prog. Inorg.Chem., 30 (1983) 141.

103 M. Chou, C. Creutz, and N. Sutin, J.Am.Chem.Soc., 99 (1977) 5615.

104 E. Buhks, M. Bixon, J. Jortner, and G. Navon, Inorg.Chem., 18 (1979) 2014.

105 J.F. Endicott, B. Durham, D.M. Glick, T.J. Anderson, J.M. Kuszai, W.G. Schmonsees, and K.P. Balakrishnan, J.Am.Chem.Soc., 103 (1981) 1431.

106 J.F. Endicott, B. Durham, and K. Kumar, Inorg.Chem., 21 (1982) 2437.

107 D. Sandrini, M.T. Gandolfi, M. Maestri, F. Bolletta, and V. Balzani, Inorg.Chem., 23 (1984) 3017.

108 M.T. Gandolfi, M. Maestri, D. Sandrini, and V. Balzani, Inorg. Chem., 22 (1983) 3435.

109 D.K. Liu, B.S. Brunschwig, C. Creutz, and N. Sutin, J.Am.Chem. Soc., 108 (1986) 1749.

110 V. Balzani, F. Bolletta, and F. Scandola, J.Am.Chem.Soc., 102 (1980) 2152.

111 K. Zahir, W. Bottcher, and A. Haim, Inorg.Chem., 24 (1985) 1968.

112 U. Furholz and A. Haim, Inorg.Chem., 24 (1985) 3091.

113 N. Sutin, in "Inorganic Reactions and Methods", J.J. Zuckerman, Ed., VCH Verlagsgesellschaft, Weinheim, West Germany, vol. 15, 1986.

114 V. Balzani and F. Scandola, Inorg.Chem., 25 (1986) 4457.

115 Open Discussions (N. Sutin and W. Siebrand) at the 4th. Microsymposium on the Photochemistry and Photophysics of Coordination Compounds, Mont Gabriel, Quebec, Canada, July 1980.

116 C. Creutz, A.D. Keller, N. Sutin, and A.P. Zipp, J.Am.Chem.Soc. 104 (1982) 3618.

117 J.M. Malin, Inorg.Chim.Acta, 45 (1980) L87.

118 V. Balzani, F. Bolletta, M. Ciano, and M. Maestri, J.Chem.Educ. 60 (1983) 447.

119 A.V. Kiss, J. Abraham, and I. Hegedus, Z.Anorg.Allg.Chem., 244 (1940) 98.

120 J. Brigando, Bull.Soc.Chim.Fr., (1957) 503.

121 J.J. Alexander and H.B. Gray, J.Am.Chem.Soc., 90 (1968) 4260.

122 F. Bolletta, M. Maestri, L. Moggi, and V. Balzani, J.Chem.Soc. Chem.Commun., (1975) 901.

123 G.M. Brown, S.-F. Chan, C. Creutz, H.A. Schwarz, and N. Sutin, J.Am.Chem.Soc., 101 (1979) 7638.

124 R.P. Burns and C.A. McAuliffe, Adv.Inorg.Radiochem., 22 (1979) 303.

125 R. Frank and H. Rau, Z.Naturforsch., 37a (1982) 1253.

126 R. Frank and H. Rau, J.Phys.Chem., 87 (1983) 5181.

127 C. Chiorboli, F. Scandola, and H. Kisch, J.Phys.Chem., 90 (1986) 2211.

128 A. English, V.R. Lum, P.J. DeLaive, and H.B. Gray, J.Am.Chem. Soc., 104 (1982) 870.

129 K.C. Cho, C.M. Che, F.C. Cheng, and C.L. Choy, J.Am.Chem.Soc., 106 (1984) 6843.

130 F.B. Ueno, Y. Sasaki, T. Ito, and K. Saito, J.Chem.Soc.Chem. Commun., (1982) 328.

131 S.-M.Y. Huang and H.D. Gafney, J.Phys.Chem., 81 (1977) 2682.

132 Y. Kaizu, T. Mori, and H. Kobayashi, J.Phys.Chem., 89 (1985) 332.

133 D.A. Geselowitz and H. Taube, J.Am.Chem.Soc., 102 (1980) 4525.

134 V. Balzani, N. Sabbatini, and F. Scandola, Chem.Rev., 86 (1986) 319.

135 C. Creutz, N. Sutin, and B.S. Brunschwig, J.Am.Chem.Soc., 101 (1979) 1297.

136 D.A. Sexton, J.C. Curtis, H. Cohen, and P.C. Ford, Inorg.Chem., 23 (1984) 49.

137 A. Vogler, A.H. Osman, and H. Kunkely, Coord.Chem.Rev., in press.

138 A. Vogler and J. Kisslinger, J.Am.Chem.Soc., 104 (1982) 2311.

139 A. Vogler and H. Kunkely, Ber.Bunsenges.Phys.Chem., (Weinheim) 79 (1975) 83.

140 A. Vogler and J. Kisslinger, Angew.Chem.Int.Ed.Eng., 21 (1982) 77.

141 H.S. White, W.G. Becker, and A.J. Bard, J.Phys.Chem., 88 (1984) 1840.

142 W. Rybak, A. Haim, T.L. Netzel, and N. Sutin, J.Phys.Chem., 85 (1981) 2856.

143 K.S. Schanze, G.A. Neyhart, and T.J. Meyer, J.Phys.Chem., 90 (1986) 2182; Inorg.Chem., 24 (1985) 2121.

144 J.C. Curtis, J.S. Bernstein, R.H. Schmehl, and T.J. Meyer, Chem.Phys.Lett., 81 (1981) 48.

145 J.C. Curtis, J.S. Bernstein, and T.J. Meyer, Inorg.Chem., 24 (1985) 385.

146 G. Rothenberger, J. Moser, M. Gratzel, N. Serpone, and D.K. Sharma, J.Am.Chem.Soc., 107 (1985) 8054.

147 G. Calzaferri, in "Photosynthetic Oxygen Evolution", H. Metzler Ed., Academic Press, New York, 1978, p. 31.

148 J. Rabani, in "Energy Storage", J. Silverman, Ed., Pergamon Press, Oxford, 1979, p. 46.

149 O.I. Micic and M.T. Nenadovic, in "Energy Storage", J. Silverman, Ed., Pergamon Press, Oxford, 1979, p. 445.

150 V. Balzani, A. Juris, and F. Scandola, in "Homogeneous and Heterogeneous Photocatalysis", E. Pelizzetti and N. Serpone, Eds. NATO-ASI Series, vol. C174, D.Reidel Publ.Co., Dordrecht, The Netherlands, 1986, p. 1.

151 W.J. Dressick, T.J. Meyer, B. Durham, and D.P. Rillema, Inorg. Chem., 21 (1982) 3451; W.J. Dressick and T.J. Meyer, Israel J. Chem., 22 (1982) 153.

152 M. Kirsch, J.-M. Lehn, and J.-P. Sauvage, Helv.Chim.Acta, 62 (1979) 1345.

153 J.-M. Lehn, J.-P. Sauvage, and R. Ziessel, Nouv.J.Chim., 5 (1981) 291; J.-M. Lehn and J.-P. Sauvage, Nouv.J.Chim., 1 (1977) 449.

154 R. Ballardini, G. Varani, and V. Balzani, J.Am.Chem.Soc., 102 (1980) 1719.

155 G.M. Brown, B.S. Brunschwig, C. Creutz, J.F. Endicott, and N. Sutin, J.Am.Chem.Soc., 101 (1979) 1298.

156 G.M. Brown, S.-F. Chan, C. Creutz, H.A. Schwarz, and N. Sutin, J.Am.Chem.Soc., 101 (1979) 7638.

157 Y. Okuno and O. Yonemitsu, Chem.Lett., (1980) 959.

158 K. Kalyanasundaram and M. Gratzel, Angew.Chem.Int.Ed.Eng., 18 (1979) 701.

159 K. Kalyanasundaram, O.I. Micic, E. Pramauro, and M. Gratzel, Helv.Chim.Acta, 62 (1979) 2532.

160 J. Kiwi and M. Gratzel, Chimia, 33 (1979) 289.

161 J.-M. Lehn, J.-P. Sauvage, and R. Ziessel, Nouv.J.Chim., 4 (1980) 355.

162 A.L. Elizasova, L.A. Matvienko, V.N. Parmon, and K.I. Zamaraev, Dokl.Akad.Nauk.SSSR., 249 (1979) 863.

163 M. Neumann-Spallart and K. Kalyanasundaram, Ber.Bunsenges.Phys. Chem., 85 (1981) 704.

164 C. Creutz and N. Sutin, Proc.Natl.Acad.Sci.USA., 72 (1975) 2858.

165 J. Kiwi and M. Gratzel, Angew.Chem.Int.Ed.Eng., 17 (1978) 860.

166 J.-M. Lehn, J.-P. Sauvage, and R. Ziessel, Nouv.J.Chim., 3 (1979) 423.

167 V.Ya. Shafirovich, N.K. Khanmov, and V.V. Strelets, Nouv.J. Chim., 4 (1980) 81.

168 A. Harriman, G. Porter, and P. Walters, J.Chem.Soc.Faraday Trans.2, 77 (1981) 2373.

169 M.T. Indelli, A. Carioli, and F. Scandola, J.Phys.Chem., 88 (1984) 2685.

170 F. Bolletta, A. Rossi, F. Barigelletti, S. Dellonte, and V. Balzani, Gazz.Chim.Ital., 111 (1981) 155.

171 S.F. Bergeron and R.J. Watts, J.Am.Chem.Soc., 101 (1979) 3151.

172 W.A. Wickramasinghe, P.H. Bird, and N. Serpone, J.Chem.Soc.Chem. Commun., (1981) 1284.

173 N. Serpone, G. Ponterini, M.A. Jamieson, F. Bolletta, and M. Maestri, Coord.Chem.Rev., 50 (1983) 209; R.J. Watts, personal communication to N. Serpone, March 1982; K.R. Seddon, personal communication to N. Serpone, March 1982; N. Serpone, W.A. Wickramasinghe, and P.H. Bird, unpublished results; G. Nord, A.C. Hazell, R.G. Hazell, and O. Farver, Inorg.Chem., 22 (1983) 3429; R.J. Spellane, R.J. Watts, and J.C. Curtis, Inorg.Chem., 22 (1983) 4060.

174 A. Shalma-Schvok, S. Gershumi, J. Rabani, H. Cohen, and D. Meyerstein, J.Phys.Chem., 89 (1985) 2460.

175 N. Sabbatini, A. Bonazzi, M. Ciano, and V. Balzani, J.Am.Chem. Soc., 106 (1984) 4055.

176 S. Sadiq. Shah and A.W. Maverick, Inorg.Chem., 25 (1986) 1867.

177 D.A. Piering and J.M. Malin, J.Am.Chem.Soc., 98 (1976) 6045.

178 J. Phillips, J.A. Koningstein, C.H. Langford, and R. Sasseville J.Phys.Chem., 82 (1978) 622.

179 R. Sriram, M.Z. Hoffman, M.A. Jamieson, and N. Serpone, J.Am. Chem.Soc., 102 (1980) 1754.

180 M.S. Henry and M.Z. Hoffman, Adv.Chem.Ser., 168 (1978) 91.

181 N.A.P. Kane-Maquire, J. Conway, and C.H. Langford, J.Chem.Soc. Chem.Commun., (1974) 801.

182 B.S. Brunschwig and N. Sutin, J.Am.Chem.Soc., 100 (1978) 7568.

183 N. Serpone, M.A. Jamieson, S. Emmi, P.G. Fuochi, Q.G. Mulazzani and M.Z. Hoffman, J.Am.Chem.Soc., 103 (1981) 1091.

184 C.K. Jorgensen, Adv.Chem.Phys., 5 (1963) 33.

185 N.A.P. Kane-Maguire and C.H. Langford, J.Chem.Soc.Chem.Commun., (1971) 875.

186 G.J. Ferraudi and J.F. Endicott, Inorg.Chim.Acta, 37 (1979) 219.

187 J. Silverman and R.W. Dodson, J.Phys.Chem., 56 (1952) 846.

188 J.F. Endicott and G.J. Ferraudi, J.Am.Chem.Soc., 99 (1977) 5812.

189 Y. Saito, J. Takemoto, B. Hutchinson, and K. Nakamoto, Inorg. Chem., 11 (1972) 2003.

190 N. Serpone and M.Z. Hoffman, J.Phys.Chem., 91 (1987) in press.

191 G.B. Porter and J. VanHouten, Inorg.Chem., 19 (1980) 2903; 18 (1979) 2053.

192 T. Ohno and S. Kato, Bull.Chem.Soc.Jpn., 57 (1984) 1528.

193 P.C. Ford, D. Wink, and J. DiBenedetto, Prog.Inorg.Chem., 30 (1983) 213.

194 H. Hennig, K. Jurdeczka, and P. Thomas, Z.Chem., 16 (1976) 161.

195 H. Hennig, P. Scheibler, R. Wagener, and D. Rekorek, J.Signal AM., 8 (1980) 383.

196 D. Rehorek and H. Hennig, Z.Chem., 20 (1980) 109.

197 D. Rehorek, D. Schmidt, and H. Hennig, Z.Chem., 20 (1980) 223.

198 H. Hennig, P. Scheibler, R. Wagener, P. Thomas, and D. Rehorek, J.Prakt.Chem., 324 (1982) 279.

199 H. Heenig, D. Walther, and P. Thomas, Z.Chem., 23 (1983) 446.

200 C.H. Langford and R. Sasseville, Can.J.Chem., 59 (1981) 647.

201 M.A. Rampi-Scandola, F. Scandola, A. Indelli, and V. Balzani, Inorg.Chim.Acta, 76 (1983) L67.

202 V. Houlding, T. Geiger, E. Kolle, and M. Gratzel, J.Chem.Soc. Chem.Commun., (1982) 681.

203 F. Pina, M. Ciano, Q.G. Mulazzani, M. Venturi, V. Balzani, and L. Moggi, Sci.Papers I.P.C.R., 78 (1984) 166.

204 F. Pina, M. Ciano, L. Moggi, and V. Balzani, Inorg.Chem., 24 (1985) 844.

205 F. Pina, Q.G. Mulazzani, M. Venturi, M. Ciano, and V. Balzani, Inorg.Chem., 24 (1985) 848.

Chapter 5.4

Photoinduced Electron Transfer in Coordination Compounds: Ion-Pairs and Supramolecular Systems

Vincenzo Balzani and Franco Scandola

1 INTRODUCTION

1.1 Scope and limitations

Photo-induced electron transfer processes involving transition metal complexes have been extensively studied in recent years for fundamental reasons (refs. 1-2) as well as in view of practical applications (ref. 3). In most cases these studies have been concerned with bimolecular processes in fluid solution. The kinetic treatment of such processes is based cn the separation between diffusive and reactive steps (Fig. 1). The latter ones are considered as unimolecular reactions taking place in the so-called precursor or successor complexes. In most cases such complexes are simple encounter complexes whose

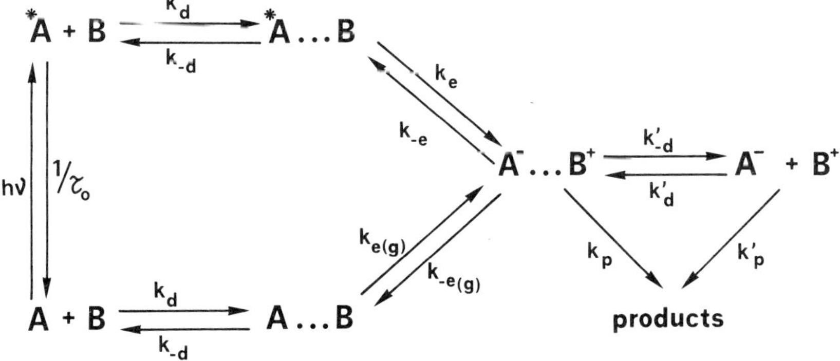

Fig. 1. Kinetic scheme for intermolecular photoinduced electron transfer processes in fluid solution.

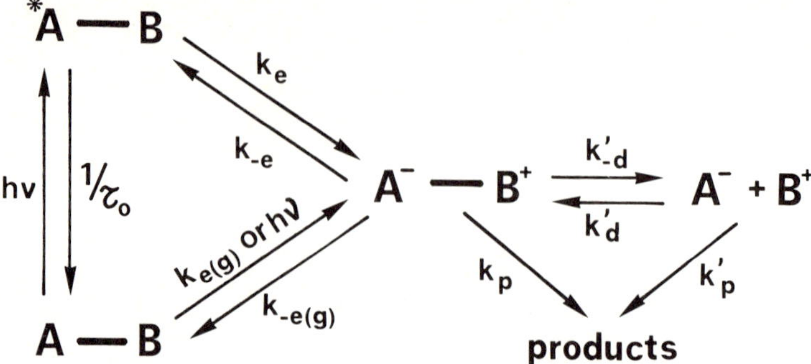

Fig. 2. Kinetic scheme for intramolecular photoinduced electron transfer.

concentrations are too small to be directly monitored. Direct excitation of the ground state precursor complex A...B is also negligible in these cases. Under such conditions, the rate of the unimolecular reactive step is either completely hidden by diffusion or, at best, it can be indirectly inferred from the overall second order rate constant.

It is possible to design systems in which, by virtue of some kind of specific interaction, the stationary concentrations of the precursor and/or successor complexes are sufficiently high to allow direct observation of these species. Fig. 2 shows the limiting case of a system where the A and B species are associated but maintain their chemical identities. This goal can be achieved via either electrostatic interactions or suitable covalent linking between the two reaction partners. In the first case one is dealing with what is usually called an (outer-sphere) ion-pair, while in the latter case the system can be considered as a true supermolecule. In these systems, light excitation can lead to substantially localized excited levels of A (or B), or to "charge-tranfer" levels (Fig. 2). From the view point of Fig. 1, localized and charge transfer excitations would correspond to formation of the precursor and successor complexes, respectively, of the photoinduced electron transfer process. In principle, systems where the two partners are associated offer a unique opportunity to monitor unimolecular electron transfer steps directly. In this chapter we will review and discuss recent studies dealing with the photochemistry and photophysics of ion-pairs and supermolecules made up by coordination compounds. Some fundamental principles of non-radiative and radiative electron transfer processes will first be recalled.

1.2 Electron transfer kinetics

The theoretical treatment of the kinetics of electron transfer reactions (both as bimolecular and unimolecular processes) has been one of the major

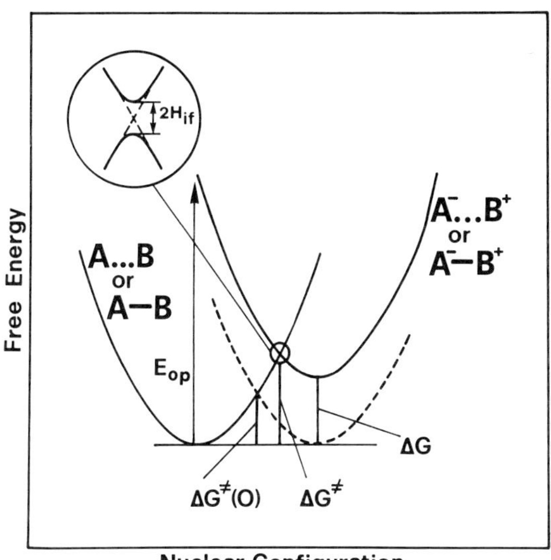

Fig. 3. Schematic representation of thermal and photoinduced electron transfer processes. For details, see text.

topics in physical inorganic chemistry for the past twenty years and is reviewed in a number of excellent reviews (refs. 1-2,4-7). In this section we will simply recall, within the framework of a classical nonadiabatic approach (refs. 2,7), the various factors which are expected to affect the rate of unimolecular electron transfer processes.

The unimolecular rate constant of an electron transfer reaction (e.g. eq. (1)),

$$A...B \xrightarrow{k_e} A^{-}...B^{+} \qquad (1)$$

is given by

$$k_e = \nu_n \kappa \exp(-\Delta G^{\neq}/RT) \qquad (2)$$

The quantities involved in eq. (2) can be discussed referring to Fig. 3, which depicts the electron transfer process in terms of free energy surfaces for reactants and products.

As far as the exponential term of eq. (2) is concerned, the free energy of activation $\Delta G^{\neq}$ is related to the energy required to reach the distorted nuclear configuration (crossing point) at which the Franck-Condon principle is satisfied for the electron transfer process. This energy is classically related by means of a free energy relationship (eq. 3) to: (i) an "intrinsic barrier" parameter, $\Delta G^{\neq}(0)$,

$$\Delta G^{\neq} = \Delta G^{\neq}(0)\ [1 + \Delta G/4\Delta G^{\neq}(0)]^2 \tag{3}$$

which represents the degree of distortion (horizontal displacement in Fig. 3) between the reactant and product curves; (ii) the actual free energy change (vertical displacement in Fig. 3) of the electron transfer process. The intrinsic barrier $\Delta G^{\neq}(0)$ can be viewed as the sum of distinct contributions from solvent reorganization, $\Delta G^{\neq}(0)_{out}$, and reorganization in the inner coordination sphere, $\Delta G^{\neq}(0)_{in}$:

$$\Delta G^{\neq}(0) = \Delta G^{\neq}(0)_{out} + \Delta G^{\neq}(0)_{in} \tag{4}$$

Appropriate equations are available to evaluate the outer and inner contributions in terms of molecular parameters and solvent dielectric behavior (ref. 2). For a given degree of distortion, the classical model predicts that $\Delta G^{\neq}$ initially decreases with increasing exergonicity, goes to zero for $\Delta G = -4\Delta G^{\neq}(0)$, and then increases again for larger exergonicities. This last region is the so called "Marcus inverted region" of electron transfer reactions (ref. 4). According to more rigorous, but less practical, quantum mechanical models (refs. 8-10), an inverted behaviour (rate decreasing with increasing driving force) is still predicted to occur, although with some quantitative differences with respect to the classical model (linear instead of quadratic dependence, usually referred to as "energy gap law") (ref. 11). For reasons which are not yet completely clear (refs. 12,13), common bimolecular electron transfer reactions in fluid solution do not behave experimentally as expected in this regard, since no conclusive example of inverted behavior has been reported to date for these systems. On the other hand, electron transfer processes in rigid matrices (ref. 14) or within covalently bound supramolecular systems (refs. 15,16), offer clear examples of the inverted behavior. The energy gap law is also obeyed for a number of electronically excited states whose decay can be viewed as an intramolecular electron transfer (refs. 17,18).

The ν_n term in the pre-exponential part of the rate constant (eq. 2) is an effective nuclear frequency for motion on the parabolic energy surfaces of Fig. 3 and can be expressed as a function of both the outer solvent reorganizational frequency, ν_{out}, and the inner vibrational frequencies of the reactant and product species, ν_{in} (ref. 12):

$$\nu_n^2 = \frac{\nu^2_{out}\,\Delta G^{\neq}(0)_{out} + \nu^2_{in}\,\Delta G^{\neq}(0)_{in}}{\Delta G^{\neq}(0)_{out} + \Delta G^{\neq}(0)_{in}} \tag{5}$$

When $\Delta G^{\neq}(0)_{out} >> \Delta G^{\neq}(0)_{in}$, ν_n is dominated by the solvent reorganizational frequency. Examples of electron transfer reactions which appear to be controlled by the dielectric relaxation time of the solvent have been reported (for a recent example, see ref. 19).

The κ term in the pre-exponential part of the rate constant is the transmission coefficient, related to the probability of conversion from reactant to product in the avoided crossing region, which can be expressed as a function of the electronic coupling matrix element H_{if} between the initial and final states of the system (eqs. (6) and (7)),

$$\kappa = \frac{2\,[1-\exp(-\nu_{el}/2\nu_n)]}{2 - \exp(-\nu_{el}/2\nu_n)} \tag{6}$$

$$\nu_{el} = 2H_{if}^2/h\,[\pi^3/4\Delta G^{\neq}(0)RT]^{1/2} \tag{7}$$

The electronic interaction H_{if} depends essentially on the overlap between the donor and acceptor orbitals. In the usual bimolecular electron transfer processes, the geometric configuration of the encounter is not fixed, a number of different configurations being explored within the lifetime of the encounter complex. This may lead to some kind of optimization of the transmission coefficient, which is actually considered to be unitary (adiabatic behavior) for many bimolecular electron transfer reactions (refs. 1,4,6). In intramolecular processes, such as those involving ion-pairs or supermolecules, the geometry is fixed or at least restricted and the actual magnitude of H_{if} could play a more important role. It should be noted that according to eqs. 6 and 7, $\kappa \approx \nu_{el}/\nu_n$ for small H_{if} and $\kappa \approx 1$ when H_{if} is large. Thus, the preexponential factor is proportional to $(H_{if})^2$ for small H_{if} (non-adiabatic regime) and saturates to ν_n for large H_{if} (adiabatic regime).

1.3 Optical electron transfer

In a chemical system such as an ion-pair or supermolecule containing several potential redox sites, an electron can be transferred from one site to another not only thermally, but also optically (ref. 20). These optical transitions are usually called charge transfer (CT) or intervalence transfer (IT) transitions. The latter terminology is particularly used for binuclear metal complexes containing the same metal ion in two different valence states. The theory of IT spectra, which has been originally worked out by Hush and developed more recently by other authors, is discussed in a number of comprehensive reviews (refs. 21-23). In Hush's classical model (ref. 20), the properties of an optical IT transition (energy, intensity, band shape) are related in a simple way to the kinetic and thermodynamic factors for the corresponding thermal electron transfer process and can be conveniently discussed in terms of the same free energy diagram (Fig. 3) used for thermal electron transfer processes (for a detailed discussion of the relationships between the energy and free energy scales used to describe these phenomena, see ref. 24). According to the Hush model, the optical IT energy ("vertical" electron transfer energy or $\bar{\nu}_{max}$ for the IT band) E_{op}, is related by eq. (8) to both the intrinsic barrier parameter $\Delta G^{\neq}(0)$ and the thermodynamic driving force of the thermal reaction, ΔG, and, by virtue of eq. 3

$$E_{op} = 4\Delta G^{\neq}(0) + \Delta G \qquad (8)$$

also to the actual activation free energy, $\Delta G^{\neq}$, of the corresponding thermal process. On the other hand, the halfwidth of the IT band, $\Delta\bar{\nu}_{1/2}$, is related to the optical energy and to the thermodynamic driving force of the thermal process by the following equation:

$$\Delta\bar{\nu}_{1/2}(cm^{-1}) = 48.06\ (E_{op} - \Delta G)^{1/2} \qquad (9)$$

Finally the "integrated" IT intensity, $\varepsilon_{max}\Delta\bar{\nu}_{1/2}$, is related to the optical energy, to the distance, r, between the two redox centers involved, and to the electronic matrix element, H_{if}, between the initial and final states of the transition (or of the thermal electron transfer process):

$$\varepsilon_{max}\Delta\bar{\nu}_{1/2} = (r(\text{Å})^2\ H_{if}^{\,2})/(4.20\times10^{-4}\ E_{op}) \qquad (10)$$

The above expressions clearly show that, at least in principle, the relevant kinetic parameters of a unimolecular thermal electron transfer process (i.e., the degree of adiabaticity, κ , and the activation free energy, $\Delta G^{\neq}$) should be calculable from the spectroscopic parameters of the corresponding IT transition. More practically, a very quick estimate of the relative "degree of adiabaticity" of various unimolecular electron transfer paths in a complex molecule can be made by looking at the relative intensities of the corresponding IT spectral transitions.

Although the Hush model was originally conceived for metal-to-metal charge transfer transitions in mixed-valence bi- or polynuclear complexes, in principle it can be extended to any kind of transition between localized redox

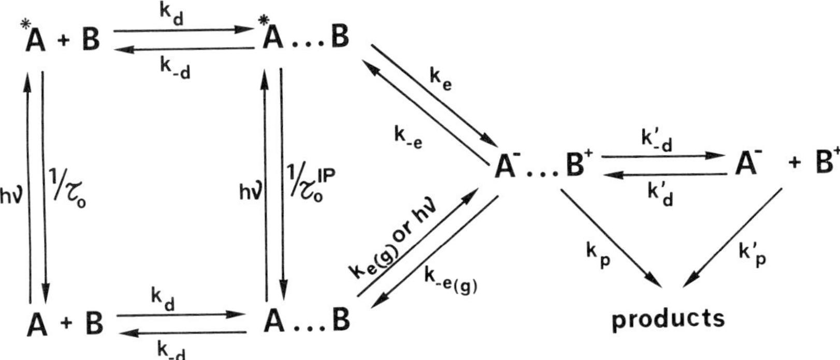

Fig. 4. Kinetic scheme for photoinduced electron transfer processes between species that give rise to outer-sphere ion-pairs.

centers. As a matter of fact, it is entirely possible to use this model to discuss metal-to-ligand charge transfer (MLCT) (ref. 25), ligand-to-ligand charge transfer (LLCT) (refs. 26-28), and ion-pair charge transfer (IPCT) (ref. 29) bands.

2. ION-PAIR SYSTEMS

Coordination compounds are usually charged species that can give rise to ion-pairs. If the complexes involved in an ion pair maintain their chemical identities (i.e., when their coordination spheres remain intact), the system is called an outer-sphere ion-pair and the interaction is generally weak. The photoinduced phenomena taking place in an outer-sphere ion-pair system can be dealt with neither with the kinetic scheme of Fig. 1 nor with that of Fig. 2. The association constants, in fact, are generally in the range 10^1-10^3 M^{-1}, which means that under the usual experimental conditions substantial amounts of both free and paired species may be present. In these cases, a more complete kinetic scheme (Fig. 4) must be used (vide infra).

2.1 Electronic levels

In an outer-sphere ion-pair system, besides electronic transitions within each parent ion, charge transfer (CT) transitions (sometimes called Outer Sphere Charge Transfer (OSCT), Ion Pair Charge Transfer (IPCT), Intervalence Transfer (IT), or Second Sphere Charge Transfer (SSCT)) can also occur. In most cases these charge transfer transitions promote an electron from a molecular orbital predominantly localized on a metal to a molecular orbital predominantly localized on another metal (metal-to-metal charge transfer, MMCT) of the ion pair. Recently, Vogler and Kunkely (refs. 30, 31) have reported examples of outer-sphere charge transfer transitions from a metal of a partner to the ligands of the other partner (MLCT) and also from the ligands of one partner to the ligands of the other partner (LLCT) of ion pairs.

Depending on the specific nature of the parent ions, the CT levels of the ion pair may lie at high or low energy. As discussed elsewhere (ref. 32), the three limiting cases of interest are as follows.

(i) The CT levels lie at very high energy. In such a case, the absorption spectrum of the ion pair is essentially the summation of the spectra of the two isolated ions and no CT photochemical reaction can take place. However, the weak interaction may induce radiationless transitions between levels centered on the different partners. This may cause a cascade nonradiative deactivation to the lowest excited state of the system, with quenching of the excited state manifestations (photochemical reactions and/or luminescence) of one of the two partners and sensitization of those of the other one.

(ii) The CT levels lie in the energy range of the spectroscopic levels of the two isolated partners. In such a case the ion pair shows, in some spectral region, a somewhat stronger absorption than that corresponding to the summation of the absorption spectra of the two isolated ions. Excitation of the ion pair in the CT band or in higher energy bands may cause either a photoreaction of CT origin or radiationless deactivation to the lowest excited state of the system. Excitation below the CT levels may be followed by radiationless deactivation to the lowest energy excited state, which is essentially localized on one of the two ions. From this level luminescence and/or reaction can take place as in case (i).

(iii) A CT level is the lowest excited state of the system. In such a case, a new, broad absorption band may appear at low energy. Radiationless deactivation to the lowest excited state can take place, with complete quenching of the photochemical and luminescence properties of the two partners. New photochemical properties resulting from the lowest CT excited state may or may not appear depending on the stability of the primary electron transfer products (vide infra).

In an energy diagram, a CT transition can be represented as E_{op} in Figs. 3 and 5. Light excitation in the CT band leads to $A^{-}...B^{+}$, an electronic isomer of the starting species. Excitation is generally followed by rapid back electron transfer to the A...B species, as shown in Fig. 5a (in photophysical terms, the back electron transfer reaction is described as a radiationless deactivation). Dissociation and reassociation of the two partners of the $A^{-}...B^{+}$ species can also occur, as shown in Fig. 4. Irreversible formation of stable photoproducts can only take place when A^{-} and/or B^{+} undergo a fast

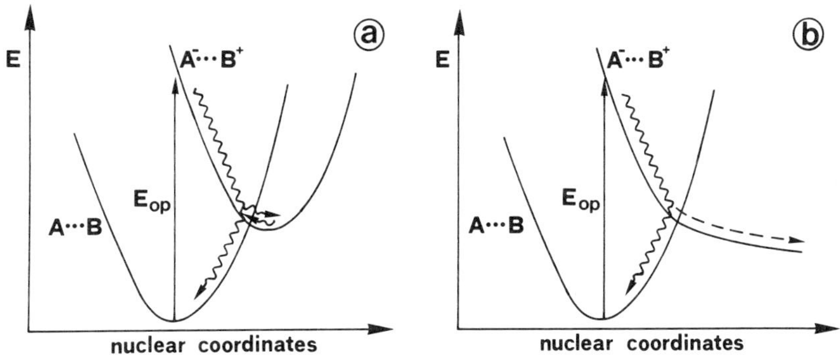

Fig. 5. Schematic representation of photoinduced reversible (a) and irreversible (b) electron transfer processes in ion-pair systems.

irreversible reaction such as ligand dissociation or reaction with a scavenger (Fig. 5b).

2.2 Monitoring the system

In photoinduced electron transfer experiments involving ion pairs, the behavior of the system can be studied by monitoring the luminescence of the excited states and/or the chemical (transient or net) changes produced by light absorption. Continuous or pulsed light excitation can be used.

2.2.1 Dynamic and static quenching. Electron transfer quenching of excited states in a system involving ion-pair formation can be discussed on the basis of the scheme shown in Fig. 4 (refs. 33-35) which differs from the simpler scheme proposed by Rehm and Weller (ref. 36) for dynamic quenching (Fig. 1) because it also takes into account the light absorption and emission by the ion pairs. In the scheme of Fig. 4, k_d and k_{-d} are the diffusion controlled rate constants for the formation and dissociation of the precursor complex, k_{-e}, $k_{e(g)}$ and $k_{-e(g)}$ are the rate constants for electron transfer in the precursor and successor complex, and τ_o and τ_o^{IP} are the lifetimes due to radiative and nonradiative (excepting electron transfer) deactivation of the "free" and "paired" excited state. In the case of weak interaction, τ_o^{IP} can be taken to be equal to τ_o and the extinction coefficients of the "free" and "paired" species can also be taken to be the same. It should be noted that an implicit assumption of the scheme of Fig. 4 is that the precursor complex of the dynamic quenching coincides with the excited preformed ion pair. It has been pointed out by several authors (refs. 37-40) that this last assumption may be not correct, especially for non-spherical systems and for species that can undergo copenetration. In the case of weak interaction, the free and paired excited species, $^*A$ and $^*A...B$, emit at the same wavelength and the luminescence quantities that can be monitored are the total luminescence intensity and the decays of the $^*A$ and $^*A...B$ species. For the simple case of electron transfer quenching without ion-pair formation, the well known relationship (ref. 41) between luminescence intensity and lifetime in the absence (I_o and τ_o) and in the presence (I and τ) of the quencher are obeyed

$$I_o/I = 1 + k_q\tau_o[B] \qquad (11)$$

$$\tau_o/\tau = 1 + k_q\tau_o[B] \qquad (12)$$

and thus the intensity and lifetime Stern-Volmer plots coincide.

When ion pairs are formed, the scheme of Fig. 4 yields the following complicated relationship for I_o/I vs [B] (refs. 33,34)

$$I_o/I = \frac{(1+k_q\tau_o[B])(1+K[B])}{1+\alpha[B]} \qquad (13)$$

where

$$\alpha = \frac{k_{-d}K+(k_{em}^{IP}/k_{em})\{(1/\tau_o+k_d[B])K+k_d\}}{1/\tau_o^{IP}+k_e+k_{-d}}$$

In eq. 13, K is the ion-pair formation constant, taken as k_d/k_{-d}, k_{em}^{IP} and k_{em} are the rate constants for emission by the ion pair and by the free $^*A$ species, respectively, and k_q is given by

$$k_q = \frac{k_dk_e}{k_e+k_{-d}} \cdot \frac{1+1/(\tau_o^{IP}k_e)}{1+1/\{\tau_o^{IP}(k_e+k_{-d})\}} \qquad (14)$$

Eq. 13 obviously reduces to the very simple eq. 11 when ion-pairs are not present, but it is apparent that the kinetic behavior of a photoinduced electron transfer process between species giving rise to ion-pairs may be quite complicated (refs. 33-35,40). Some interesting limiting cases are as follows.

(a) When $k_e \ll 1/\tau_o$ the trivial case of no intensity and lifetime quenching obtains.

(b) When $k_e \ll k_{-d}$, the presence of ion pairs is not relevant for the quenching process because the precursor complexes $^*A...B$ obtained by direct excitation of the A...B ion pairs undergo dissociation before being able to react. Under such conditions, the quenching process obeys the laws of dynamic quenching (eqs. 11 and 12) and it may result efficient or inefficient depending on whether k_e is larger or smaller than $1/\tau_o$.

(c) When k_e is much larger than $1/\tau_o^{IP} + k_{-d}$, the A molecules that are excited while they are paired with B undergo a fast static quenching with rate constant k_e, whereas the excited free A molecules can be quenched at a diffusion controlled rate via dynamic formation of encounters with B. The observable behavior of the system will depend on the relative values of k_e, $1/\tau_o + k_d[B]$, and $1/\tau_{res}$ (where τ_{res} is the time resolution of the equipment used to monitor the luminescence lifetime). Two interesting sub-cases are as follows:

(c-1) When $1/k_e$ is shorter than τ_{res}, the decay of the excited ion pairs cannot be followed and the luminescence emission decreases with time according to a single exponential decay which is related to the dynamic quenching ($1/\tau = 1/\tau_o + k_d[B]$). On the other hand, under stationary excitation conditions (i.e., for measurements of emission intensities with a spectrofluorimeter) both the dynamic and static quenching processes are effective. It follows that the I_o/I and τ_o/τ versus [B] Stern-Volmer plots are no longer coincident, with the intensity plot showing an upward curvature (at constant ionic strength) caused by the increasing importance of the static component as [B] increases. Under such conditions, $\alpha[B]$ in eq. 13 may be much smaller than unity even for sufficiently high B concentrations, so that eq. 13 reduces to the following quadratic equation:

$$I_o/I = (1+k_d\tau_o[B])(1+K[B]) = 1 + (k_d\tau_o+K)[B] + k_d\tau_o K[B]^2 \qquad (15)$$

This case has often been observed (refs. 40-43) and the luminescence data have been used to evaluate the ion pair association constant, K.

(c-2). When $1/k_e$ is longer than τ_{res} but still much shorter than $\tau_o+k_d[B]$, the intensity of the luminescence emission decreases with time according to two distinct exponentials related to the decays of $^*A$ and $^*A...B$. Such conditions, however, are very difficult to obtain because (i) the time window between τ_o and τ_{res} may be narrow, (ii) when k_e becomes relatively small it will be hardly larger than $1/\tau_o^{IP}+k_{-d}$, and (iii) sufficiently high concentrations of both free $^*A$ and paired $^*A...B$ need to be present to obtain observable signals. In particular, for small K values a large amount of B has to be used to have enough ion pairs, but this increases the rate of the dynamic quenching (whose rate constant is $\tau_o+k_d[B]$) which can thus approach the rate of the static quenching, k_e. When the two rates are close enough, the luminescence decay may appear to be exponential within the experimental uncertainty and its value may decrease with increasing [B] up to a saturation value given by the rate of the intramolecular electron transfer step, k_e. Clear examples of double exponential decay and of saturation of the luminescence decay with increasing [B] have recently been reported (ref. 35).

(d) When $k_e > 1/\tau_o + k_{-d}$ (so as to ensure static quenching) and $k_d[B]\tau_o \ll 1$ (so as to make negligible the dynamic quenching), eqs. 15 reduces to

$$I_o/I = (1 + K[B]) \quad (16)$$

and the lifetime of the free $^*A$ species is not quenched. Whether or not the decay of the excited ion pairs can be seen will again depend on the relative values of $1/k_e$ and τ_{res}. The above conditions, however, are difficult to obtain because $k_e > 1/\tau_o + k_{-d}$ means that τ_o cannot be too small and the dynamic quenching is diffusion controlled. Thus, $k_d[B]\tau_o$ will be much smaller than unity only when the concentration of free B is small, but at the same time there should be a non negligible amount of ion pairs, which implies a large K. Quenching through this sort of purely static mechanism has been observed for flavin compounds in aqueous solution (ref. 44).

Before closing this section we would like to recall that the ionic strength of the solution affects, of course, both the ion-pair association constants and the diffusion and dissociation rate constants, with profound effects on the I_o/I and τ_o/τ vs [B] plots (refs. 40-43).

2.2.2 Photoinduced chemical reactions. Light absorption by an ion-pair system can lead to excited levels of the isolated partners or to CT levels. The subsequent behavior of the system will depend on the relative position of the various levels, as discussed in section 2.1. The most interesting and quite common case is that corresponding to Fig. 4, where the CT level $A^-...B^+$ is lower in energy than the localized excited states and $^*A$ may or may not be luminescent. Light excitation of these systems leads, directly or indirectly, to the formation of species (that can still be ion pairs) containing reduced A and oxidized B, $A^-...B^+$. Unless the energy of $A^-...B^+$ is lower than that of A...B (and, in such a case, the reaction would also have occurred thermally), the $A^-...B^+$ species will appear as a more or less short lived transient since it must either go back to A...B (in the solvent cage or via dissociation and reassociation) or undergo some competitive reaction (usually, ligand dissociation or reaction with scavenger) leading to permanent photoproducts. Continuous photolysis experiments can only reveal, of course, permanent products while flash experiments with appropriate time resolution may allow us to observe the formation and decay of the $A^-...B^+$ transients.

2.3 Selected examples

Table 1 shows a selected collections of data concerning photoinduced electron transfer processes in ion pairs. As one can see, CT bands have been observed in a large number of cases. Even when a CT band has not been observed, the occurrence of $^*A$ quenching (by an electron transfer mechanism) assures that a low lying CT level must be present although its spectroscopic observation is prevented by the high intensity tails of the bands corresponding to the localized transitions. The observed photoreaction refers in most cases to the formation of permanent products deriving from the ligand dissociation of the reduced partner.

We will now discuss in more detail some cases of particular interest.

2.3.1. Correlations between optical and thermal quantities.

As we have seen in section 1.2, the rate constant for a thermal electron transfer reaction is given by:

$$k_e = \nu_n \kappa \exp(-\Delta G^{\neq}/RT) \qquad (2)$$

where $\Delta G^{\neq}$ can be classically expressed by the Marcus quadratic equation (ref. 4)

$$\Delta G^{\neq} = \Delta G^{\neq}(0)\ [(1+\Delta G)/4\Delta G^{\neq}(0)]^2 \qquad (3)$$

as a function of the free energy change ΔG and of the intrinsic barrier $\Delta G^{\neq}(0)$ (Fig. 3). We have also seen that in the classical Hush model (ref. 5) the energy of the optical electron transfer transition is related to ΔG and $\Delta G^{\neq}(0)$ by the equation

$$E_{op} = 4\Delta G^{\neq}(0) + \Delta G \qquad (8)$$

that taken together with eq. 3 yields the following equation:

$$\Delta G = E_{op}{}^2/4(E_{op} - \Delta G) \qquad (17)$$

Furthermore, the integrated intensity of the optical transition is related to the interaction energy H_{if} (eq. 10) whose value governs, through eqs. 6 and 7, the preexponential factor of the expression of the rate constant for the thermal process (eq. 2). The above relationships between thermal and optical electron transfer parameters have so far been applied and discussed in a few experimental cases (ref. 38).

TABLE 1.

Data on some selected outer-sphere ion-pairs between coordination compounds.[a]

cation	anion	CT band λ_{max}, nm(ε_{max})	Luminescence quenching	Photoreaction	Ref.
$Co(NH_3)_6^{3+}$ [b]	$Fe(CN)_6^{4-}$	440(300)	-	-	38
$Co(en)_3^{3+}$	$Fe(CN)_6^{4-}$	430(105)	-	yes	38,45
$Co(NH_3)_6^{3+}$ [c,d]	$Ru(CN)_6^{4-}$	342(243)	-	yes	46,47
$Ni(tim)^{2+}$	$Ni(mnt)^{2-}$ [e]	840[f]($\sim 10^4$)	-	-	31
$Rh(bpy)_3^{3+}$	$Fe(CN)_6^{4-}$ [g]	480(61)	-	yes	30
$Ru(NH_3)_6^{3+}$	$Fe(CN)_6^{4-}$ [h]	744[f]	-	-	48
$Ru(NH_3)_6^{3+}$	$Fe(CN)_5py^{3-}$ [i]	725[f]	-	-	49
$Ru(NH_3)_5py^{3+}$ [j]	$Fe(CN)_6^{4-}$ [k]	910(33)	-	yes	29,50
$Ru(NH_3)_5Cl^{2+}$	$Ru(CN)_6^{4-}$	510(20)	-	yes	51
$Ru(bpy)_3^{2+}$ [l,m]	$Fe(CN)_6^{4-}$	-	yes	-	33,43
$Ru(bpy)_3^{2+}$	$NiS_4C_4(CN)_4^{2-}$	-	yes	-	34
$Ru(bpy)_3^{2+}$ [n,o]	$Mn(OH)PW_{11}O_{39}^{6-}$	-	yes	-	40
$Ru(bpy)_3^{2+}$ [n]	$Co(H_2O)SiW_{11}O_{39}^{6-}$	-	yes[p]	-	35
$Ru(bpy)_2(4,4'\text{-}Cl_2bpy)^{3+}$ [n]	$Co(H_2O)SiW_{11}O_{39}^{6-}$	-	yes[q]	-	40
$Os(NH_3)_5Cl^{2+}$	$Fe(CN)_6^{4-}$ [r]	438	-	yes	47
$Os(5\text{-}Clphen)_3^{2+}$	$Fe(CN)_6^{4-}$	-	yes[s]	-	33
$[Eu \subset 2.2.1]^{3+}$	$Fe(CN)_6^{4-}$ [t]	530(110)	yes	-	52,53

a) in aqueous solution, unless otherwise noted. For all systems, except otherwise noted, the cation is the oxidant species.

b) for the ion-pair involving $Co(sep)^{3+}$, see the same reference.

c) in DMSO, λ_{max}=360 nm (ref. 46).

d) for the ion-pairs involving $Co(en)_3^{3+}$, $Co(pn)_3^{3+}$, and $Co(sep)^{3+}$, see the same reference.

e) for the ion-pairs involving $Pd(MNT)^{2-}$ and $Pt(MNT)^{2-}$, see the same reference.

f) solid state.

g) for the ion-pairs involving $Ru(CN)_6^{4-}$ and $Os(CN)_6^{4-}$, see the same reference.

h) for the ion-pair involving $Ru(CN)_6^{4-}$, see the same reference.

i) for ion pairs involving similar complexes containing pyridine substituted ligands, see the same reference.

j) for ion-pairs involving similar complexes containing pyridine substituted ligands, see ref. 29.

k) for ion-pairs involving $Ru(CN)_6^{4-}$ and $Os(CN)_6^{4-}$, see ref. 29.

l) for the ion-pairs involving $Ru(phen)_3^{2+}$, see ref. 43.

m) for the ion-pair involving $Mo(CN)_8^{4-}$, see ref. 42; for the ion-pair involving $Co(C_2O_4)_3^{3-}$, see ref. 43.

n) the oxidant species is the anion.

o) for the ion pair involving $Ru(bpy)_2(4,4'-Cl_2bpy)^{2+}$, see the same reference.

p) two exponential decays corresponding to the dynamic and static components.

q) rate saturation corresponding to the emission of the ion pair.

r) for the ion-pairs involving the $Ru(CN)_6^{4-}$ and $Os(CN)_6^{4-}$ complexes, see the same reference.

s) emission from both free and ion paired $Os(5\text{-}Cl\ phen)_3^{2+}$ species.

t) for the ion pairs with $Ru(CN)_6^{4-}$ and $Os(CN)_6^{4-}$, see the same reference.

The most extensive collection of homogeneous experimental data is that reported by Curtis and Meyer (ref. 29) who have studied the ion-pairs formed by $M(CN)_6^{4-}$ anions (M = Fe, Ru, Os) with $Ru(NH_3)_5L^{3+}$ cations (L = pyridine or substituted pyridine). For these systems, the intrinsic barrier $\Delta G^{\neq}(0)$ should be constant and the optical energy is thus expected to increase linearly with increasing ΔG (eq. 8). This expectation is fully confirmed (note, however, that in the original paper (ref. 29) E_{op} was plotted against ΔH) and the intercept of the linear plot is in fair agreement with the value of $4\Delta G^{\neq}(0)$ obtained from the intrinsic barriers of the correspondent self-exchange electron transfer reactions (ref. 38). Linear plots of E_{op} vs ΔG have also been obtained for the $[Eu\subset 2.2.1]^{3+}$-$M(CN)_6^{4-}$ (M = Fe, Ru, Os) ion pairs in aqueous solution (refs. 52,53) and for the solid salts $[Ru(NH_3)_6][Fe(CN)_5L]$ (L is CO, $(CH_3)_2SO$, or a nitrogen heterocycle) (ref. 49).

Eqs. 2 and 17 allow one to calculate the rate constant for the thermal electron transfer reaction from the free energy change, the optical transition energy, and the assumption of an adiabatic behavior (κ = 1). When such an exercise is applied (ref. 38) to the ion pairs between $M(CN)_6^{4-}$ (M = Fe, Ru) and $Ru(NH_3)_5Py^{3+}$, $Co(NH_3)_6^{3+}$, and $[Eu\subset 2.2.1]^{3+}$, the ratios of the rate constants for the reactions involving $Fe(CN)_6^{4-}$ or $Ru(CN)_6^{4-}$ are 1.4×10^6, 2.0×10^6, 6.7×10^6, i.e. in reasonable agreement with the constancy predicted by the Marcus relationship for cross-reactions. However, a ratio of 2×10^6 for the reactivities of $Fe(CN)_6^{4-}$ and $Ru(CN)_6^{4-}$ would imply that the rate constant for the $Fe(CN)_6^{4-/3-}$ self-exchange is about 10^4 times higher than that of the $Ru(CN)_6^{4-/3-}$ self-exchange (ref. 38), while other results as well as theoretical expectation indicate that the rate constants of the two self-exchange reactions must be equal (or nearly so).

As pointed out by Haim (ref. 38), there are also problems with the absolute values of the rate constants. In the single case where a measured value is available, i.e. for the ion pair $Ru(NH_3)_5Py^{3+}$-$Fe(CN)_6^{4-}$, the calculated rate constant (1×10^6 s^{-1}) is about 10^3 times higher than the measured rate constant (1.7×10^3 s^{-1}). As a possible reason for this disagreement, it has been suggested that the ion-pairs for optical and thermal electron transfer may have different intimate structures (ref. 38). A similar problem is also emerging when the results obtained in static and dynamic quenching processes are compared (ref. 54).

It should be noted that for both the $M(CN)_6^{4-}$-$Ru(NH_3)_5L^{3+}$ (ref. 29) and $M(CN)_6^{4-}$-$[Eu\subset 2.2.1]^{3+}$ (ref 53) systems the value of H_{if} is calculated (eq. 10) to be larger than 100 cm^{-1}, indicating that the electron transfer reactions (in the optical ion-pair configurations) are in the adiabatic regime.

Table 2.

Some selected data concerning the dynamic and static electron transfer quenching of $^{*}RuL_3^{2+}$ complexes by $Mn(OH)PW_{11}O_{39}^{6-}$ (MnW_{11}^{6-}) and $Co(H_2O)SiW_{11}O_{39}^{6-}$ (CoW_{11}^{6-}).[a]

$^{*}RuL_3^{2+}$-MW_{11}^{6-}	$\Delta G(eV)^{b}$	$k_q(M^{-1}s^{-1})^{c}$	$k_d(M^{-1}s^{-1})^{d}$	$k_e(s^{-1})^{e}$	Remarks
$^{*}Ru(bpy)_3^{2+}$-MnW_{11}^{6-}	-0.34	$2.1x10^{10}$	$2.2x10^{10}$	($\geqslant 5x10^{8}$)	Static and dynamic quenching. One exponential decay.
$^{*}Ru(bpy)_3^{2+}$-CoW_{11}^{6-}	-0.20	$2.3x10^{10}$	$2.2x10^{10}$	$8.5x10^{7}$	Static and dynamic quenching. Two exponential decays.
$^{*}Ru(bpy)_2(4,4'Cl_2bpy)^{2+}$-$CoW_{11}^{6-}$	-0.09	f	-	$4x10^{6}$	τ_o/τ reaches a plateau as $[CoW_{11}^{6-}]$ increases.
$^{*}Ru(bpy)_2(biq)^{2+}$-CoW_{11}^{6-}	+0.35	$<3x10^{8}$	$1.1x10^{10}$	($<5x10^{-5}$)	No observable quenching.

a) From refs. 35,40. Aqueous solution, room temperature. For the ionic strengths, see the original papers.

b) Free energy change for the photoinduced electron transfer process comprehensive of the work term. RuL_3^{2+} is oxidized.

c) Experimental quenching rate constant obtained from the τ_o/τ Stern Volmer plots. In the case of two exponential decays, the slower one was used.

d) Calculated diffusion rate constant.

e) Rate constant for intramolecular electron transfer.

f) Non linear τ_o/τ Stern-Volmer plots.

2.3.2. Direct measurements of intramolecular electron transfer rates. As mentioned in section 1.1, one of the goals in studying photoinduced electron transfer reactions in ion pairs is the possibility to measure the rate constant of the intrinsic (intramolecular) act of electron transfer. This goal, however, is not at all easy to reach, and there are only a few cases described in the literature (refs. 33,35) where intramolecular rate constants of electron transfer reactions involving excited states have been directly measured or evaluated.

The best example is probably that involving the RuL_3^{2+}-MW_{11}^{6-} ion pairs in aqueous solutions, where RuL_3^{2+} is $Ru(bpy)_3^{2+}$, $Ru(bpy)_2(4,4'\text{-}Cl_2bpy)^{2+}$, or $Ru(bpy)_2(biq)^{2+}$ (bpy = 2,2'-bipyridine, biq = 2,2'-biquinoline) and MW_{11}^{6-} is $Mn(OH)PW_{11}O_{39}^{6-}$ or $Co(H_2O)SiW_{11}O_{39}^{6-}$ (ref. 35,40). The RuL_3^{2+} complexes have excited state oxidation potentials in the range -0.87/-0.37 V, while the MW_{11}^{6-} species are reduced at -0.88 and -1.02 V. Thus, by an appropriate choice of the excited RuL_3^{2+} complex and of the MW_{11}^{6-} quencher it has been possible to tune the rate of the electron transfer quenching process. Other favorable conditions were the relatively long lifetime of the excited states (125-400 ns), the relatively short time resolution of the available equipment for lifetime measurements (2 ns), and the high charge product (-12) which implies high association constants (of the order of $10^4\ M^{-1}$) and small k_{-d} values (of the order of $10^6\ s^{-1}$). The most interesting results have been obtained for the ion pairs shown in Table 2 and can be described on the basis of the scheme of Fig. 4. The $Ru(bpy)_3^{2+}$-MnW_{11}^{6-} ion pair corresponds to the case of the most exergonic electron transfer reaction. The luminescence decay was strictly exponential (within the time resolution of the equipment) and the τ_o/τ versus $[MW_{11}^{6-}]$ plot was linear, yielding a rate constant for the dynamic quenching practically equal to the calculated diffusion rate constant. The I_o/I versus $[MW_{11}^{6-}]$ plot exhibited an upward curvature and was fitted by eq. 11. For the less exergonic $Ru(bpy)_3^{2+}$-CoW_{11}^{6-} system, a double exponential decay was observed. Taking the longer decay, the τ_o/τ versus $[CoW_{11}^{6-}]$ plot was linear yielding again a dynamic quenching constant practically equal to the diffusion rate constant. The I_o/I versus $[CoW_{11}^{6-}]$ plot showed an upward curvature and was fitted by eq. 13. The double exponential decay yielded experimental decay constants $-m_1 = 8.9\text{x}10^7\ s^{-1}$ and $-m_2 = 9.6\text{x}10^6\ s^{-1}$, which correspond to the static and dynamic quenching paths. For the $Ru(bpy)_2(4,4'\text{-}Cl_2bpy)^{2+}$-$CoW_{11}^{6-}$ system, the rate of the electron transfer step had to be smaller because of the smaller exergonicity. The luminescence emission was found to decay according to a single exponential, but the τ_o/τ vs $[CoW_{11}^{6-}]$ plot exhibited a downward curvature reaching a plateau for $\tau_o/\tau = 1.7$. This shows that the quenching limiting factor at high CoW_{11}^{6-} concentrations becomes the rate of the intramolecular electron transfer step, k_e, that from the lifetime of the plateau value was estimated to be $\sim 4\times10^{6}$ s^{-1}. Finally, for the $Ru(bpy)_2(biq)^{2+}$-$CoW_{11}{}^{6-}$ system the electron transfer step is noticeably endergonic and using the Marcus relationship its rate constant is estimated to be much smaller than $\tau_o{}^{IP}$. This explains why there is no quenching effect.

2.3.3. Photoinduced reactions. As shown in Table 1, photoinduced reactions in ion-pair systems are rather common. In most cases, only the formation of permanent products deriving from secondary reactions has been observed. The importance of secondary reactions to obtain a net photochemical effect (Fig. 5) is well illustrated by the behavior of the ion pairs of $Co(NH_3)_6{}^{3+}$ and $Co(sep)^{3+}$ with $Ru(CN)_6{}^{4-}$ (ref. 47). In both cases excitation in the CT band leads to the formation of the one-electron reduction product of the Co complex and the one-electron oxidation product of the Ru complex, but only for the

$$Co^{III}(NH_3)_6{}^{3+}\text{-}Ru^{II}(CN)_6{}^{4-} \text{ ------> } Co^{II}(NH_3)_6{}^{2+}\text{-}Ru^{III}(CN)_6{}^{3-} \quad (18)$$

$$Co^{III}(sep)^{3+}\text{-}Ru^{II}(CN)_6{}^{4-} \text{ -----> } Co^{II}(sep)^{2+}\text{-}Ru^{III}(CN)_6{}^{3-} \quad (19)$$

first system a net photoreaction, leading to formation of Co^{2+}, NH_3, and $Ru^{III}(CN)_6{}^{3-}$, takes place (Φ = 0.034 in DMSO (ref. 46)). This is in agreement with the known lability of the $Co^{II}(NH_3)_6{}^{2+}$ complex (ref. 55) whose ligand release apparently competes with the back electron transfer process. The $Co^{II}(sep)^{2+}$ complex is quite similar to $Co^{II}(NH_3)_6{}^{2+}$ but, because of the cage-type ligand, it cannot undergo ligand dissociation and thus it will be involved in the back electron transfer reaction with unitary efficiency. This different behavior of the otherwise similar $Co(NH_3)_6{}^{3+}$ and $Co(sep)^{3+}$ complexes can be observed in photochemical, pulse radiolysis, and electrochemical reduction processes (ref. 32). For the $Ru^{III}(NH_3)_5Cl^{2+}$-$Ru^{II}(CN)_6{}^{4-}$ ion pair (ref. 51), excitation in the CT band in aqueous solution leads to the formation of the binuclear complex $(NH_3)_5Ru^{III}NCRu^{II}(CN)_5{}^{-}$ with Φ = 2×10^{-3}. The reaction is thought to proceed via the rapid aquation (in competition with back electron transfer, Fig. 5b) of the $Ru^{II}(NH_3)_5Cl^{+}$ primary photoproduct, with successive replacement of the labile H_2O ligand by $Ru^{II}(CN)_6{}^{4-}$; the $(NH_3)_5Ru^{II}NCRu^{II}(CN)_5{}^{2-}$ product so obtained is thought to be oxidized by the other primary photoproduct, $Ru^{III}(CN)_6{}^{3-}$.

The primary (transient) photoproducts obtained upon CT excitation of ion pairs have been observed only in one case. The system studied (ref. 50) was the ion pair $Ru(NH_3)_5Py^{3+}.Fe(CN)_6{}^{4-}$ in aqueous solution where excitation with a 20-ns pulse of 1060-nm light has led to the transient formation of $Ru(NH_3)_5Py^{2+}$ and $Fe(CN)_6{}^{3-}$ with $\Phi = 0.06\pm0.03$. No permanent photoproduct was observed. The reactions involved are shown in the following scheme (ref. 38,50):

$$Ru(NH_5)Py^{3+}.Fe(CN)_6^{4-} \underset{k_r}{\overset{h\nu}{\rightleftharpoons}} Ru(NH_3)_5Py^{2+}.Fe(CN)_6^{3-} \quad (20)$$

$$^{(FC)}Ru(NH_3)Py^{2+}.Fe(CN)_6^{3-} \xrightarrow{k'_r} Ru(NH_3)_5Py^{2+}.Fe(CN)_6^{3-} \quad (21)$$

$$Ru(NH_3)_5Py^{2+}.Fe(CN)_6^{3-} \underset{k_d}{\overset{k_{-d}}{\rightleftharpoons}} Ru(NH_3)_5Py^{2+} + Fe(CN)_6^{3-} \quad (22)$$

$$Ru(NH_3)_5Py^{2+}.Fe(CN)_6^{3-} \xrightarrow{k_b} Ru(NH_3)_5Py^{3+}.Fe(CN)_6^{4-} \quad (23)$$

Reaction 20 represents the photoexcitation step with formation of the Franck-Condon charge transfer excited state. Relaxation of the Franck-Condon excited state proceeds by two competing pathways: return to the ground state of the original ion pair (k_r) or formation of the related electronic isomer (k'_r, reaction 21). The low quantum yield requires that $k'_r \ll k_r$, suggesting that the relaxation to the original ion pair occurs via relatively high-frequency (10^{12}-10^{13} s^{-1}) inner-sphere modes before the change in solvent polarization (which is determined by the dielectric relaxation frequency of water, $\sim 10^{11}$ s^{-1}) can occur. Since the reaction of $Ru(NH_3)_5Py^{2+}$ with $Fe(CN)_6^{3-}$ is not diffusion controlled, the rate of cage escape (k_{-d}, eq. 22) for the two partners of the electronic isomer is faster than the rate of back electron transfer (k_b, eq. 23) within the cage (refs. 38,50).

3 SUPRAMOLECULAR SYSTEMS

It has been known since the early years of coordination chemistry that ambidentate ligands can function as bridges between transition metal ions. This may lead to infinite-lattice type compounds (e.g., prussian blue) or to discrete supramolecular units (bi- or polymetallic complexes). The study of supramolecular systems containing transition metal complexes has been strongly stimulated by the appearance of the Hush model for optical metal-to-metal charge transfer (intervalence transfer, IT) transitions (refs. 5, 21). Recently, further interest in the synthesis and study of new inorganic supramolecular systems has arisen in connection with the problem of photoinduced charge separation (ref. 54). The energy levels and the possibility of detecting intramolecular electron transfer processes in these systems are briefly discussed in the following paragraphs.

3.1 Electronic levels

When two (or more) metallic centers M_1 M_2 (where M_1, M_2 represent the metal plus all the ligands except the bridging one) are put together in a supramolecular structure of the type

$$M_1-L_1-L_2-M_2$$

via an appropriate bridging ligand L_1-L_2, different degrees of electronic coupling between the metal centers may occur, depending on the properties of the components. The situations that arise can be classified, following the approach of Robin and Day (ref. 56), as follows.

Class I. The $M_1^+-L_1-L_2-M_2^-$ and $M_1^--L_1-L_2-M_2^+$ states are at very high energy. The electronic coupling is very small. The properties of the ensemble are essentially a superposition of those of the single components.

Class II. One of the electron-transfer states, e.g., $M_1^+-L_1-L_2-M_2^-$, is at relatively low energy. Most of the properties of the metallic centers are still consistent with their original oxidation states. However, the electronic coupling is appreciable and a number of new properties characteristic of the ensemble (e.g., $M_1-L_1-L_2-M_2$ ---> $M_1^+-L_1-L_2-M_2^-$ intervalence transfer (IT) absorption bands) also appear.

Class III. The electronic coupling is strong. The properties of the isolated components are absent and only new properties characteristic of the ensemble are observed. The new absorption bands do not involve metal-to-metal charge transfer and are better described in a delocalized MO picture.

Except for a limited number of systems involving very short and/or highly conjugated bridges (e.g., $L_1-L_2 = O^{2-}$, N_2, NCCN) most of the discrete bi- or polynuclear complexes investigated up to now exhibit class I or class II behavior (ref. 21). Thus, a localized-valence description seems to be appropriate for the discussion of the spectroscopic and photophysical behavior of most transition-metal supramolecular systems. From this point of view, these systems resemble the ion-pair systems described in section 2, and it could be tempting to discuss their energy levels along the same lines used for the ion-pair species (section 2.1). A number of peculiar features, however, must be pointed out which caution against drawing too close analogies between the supramolecular and the ion-pair case.

3.1.1. Individual levels of the subsystems. In the ion-pair systems, the two subsystems have an individual existence and the interaction between the two subsystems is purely electrostatic. Thus, it can be easily verified that the energy levels of the free and ion-paired subunits are virtually identical, except for the possibility of additional IT levels of the ensemble. In bi-nuclear (or polynuclear) complexes the identification of the "individual subsystems" is much less trivial than in the ion-pair case. In fact: (i) M_1 and M_2 as such would correspond to coordinatively unsaturated species; (ii) except

TABLE 3.

Spectroscopic and redox properties of supramolecular systems of the type M_1--L_1-L_2--M_2 (M_1 = $NCRu(bpy)_2^+$, L_1-L_2 = CN^-) in aqueous solution (refs. 54, 57-60).

M_2	λ_{max}(LMCT Ru-->bpy) (nm)	$E_{1/2}(M_1^+/M_1)$ (V, vs SCE)
---	428	+0.85
$Pt(dien)^{2+}$	416	+1.03[a]
$Ru(NH_3)_5^{2+}$	413	
$Ru(NH_3)_5^{3+}$	403	+1.07
$Ru(NH_3)_4py^{2+}$	417	+1.12
$Ru(NH_3)_4py^{3+}$	400	
$Ru(bpy)_2CN^{2+}$	400	+1.31

[a]DMF solution

for the case in which L_1 and L_2 in L_1-L_2 are separated by saturated groups, compounds resembling M_1-L_1 or M_2-L_2 are not good models; (iii) M_1-L_1-L_2 and M_2-L_2-L_1 are not always independently existent species. Moreover, even if good model compounds for the subunits are found, the covalent M_1-L_1-L_2-M_2 interaction may often perturb to a certain extent the levels of the "individual subsystems". Examples of these problems can be found in the family of binuclear complexes in which M_1 is the NC-$Ru(bpy)_2^+$ chromophore, L_1-L_2 is CN^-, and M_2 represents a variety of metal-containing moieties, (refs. 54, 57-60). In fact, while $Ru(bpy)_2(CN)_2$ is a stable molecule, the CN-M_2 complexes do not exsist as such, since they would correspond to the wrong (N-bonded) linkage isomers of otherwise stable cyanide complexes. On the other hand, the coordination of the M_2 moiety gives rise (Table 3) to sizeable shifts in the energies of the Ru->bpy MLCT states and in the oxidation potentials of the parent $Ru(bpy)_2(CN)_2$. This is due to the much better π-acceptor properties of ambidentate CN^- relative to those of the C-bonded one.

It should be noticed that in the above comparison the L_1-L_2 bridging group has been considered as part of one of the metal-containing subunits rather than part of the other. This depends on the emphasis placed on the perturbation of one subunit by the other, but is arbitrary. Moreover, it should be stressed that in many instances (as in the examples given in Table 3) the bridging group is capable of independent existence and could be usefully considered as a further subunit. This subunit may in some cases have individual energy levels that play an important role (see section 3.1.3).

3.1.2. Intervalence transfer levels. As for the ion pairs, the energies and other spectroscopic parameters of the IT transitions occurring in polynuclear metal complexes can be treated in terms of the Hush model (ref. 20) (section 1.3). According to this model, some parameters (energy, halfwidth) of the IT band depend on both the reorganizational parameter $\Delta G^{\neq}(0)$ and the driving force of the thermal electron transfer process ΔG (eqs. 8-9). While in the ion-pair systems ΔG can be straightforwardly calculated from the independently known redox potentials of the two ions (corrected, when necessary, for electrostatic work terms), the evaluation of ΔG for an electron transfer process within a supramolecular system (eq. 24) meets again with problems, as

$$M_1\text{-}L_1\text{-}L_2\text{-}M_2 \text{ -----> } M_1^{+}\text{-}L_1\text{-}L_2\text{-}M_2^{-} \qquad (24)$$

only the oxidized M_1^{+}-L_1-L_2-M_2 or the reduced M_1-L_1-L_2-M_2^{-} forms can be obtained electrochemically from the starting system. If the L_1-L_2 bridge is rigid and saturated, an appropriate electrostatic work term correction to the potentials of the oxidized and reduced species may be sufficient to obtain a correct ΔG. In several practical cases, however, through-bridge nonelectrostatic effects will operate requiring the use of suitable model systems for the estimation of ΔG (ref. 21).

For symmetrical supramolecular systems ($M_1 = M_2$, $L_1 = L_2$), the energy of the IT transition is simply given by $4\Delta G^{\neq}(0)$ (eq. 8), where the intrinsic barrier parameter can be further split into inner and outer contributions (eq. 4). It has been pointed out by several authors that a simple two-sphere model may be inadequate to calculate the outer reorganizational energy $\Delta G^{\neq}(0)_{out}$ and that more elaborate models may be required to reproduce the actual shape of the supermolecule (ref. 61). A striking example of the qualitative discrepancies that can arise when supermolecules of different structures are compared is given by the IT transitions observed in water for trans-$(NH_3)_5Ru^{II}pzRu^{II}$-$(NH_3)_4pzRu^{III}(NH_3)_5{}^{7+}$ (ref. 62) and cis-$(NH_3)_5Ru^{II}NCRu^{II}(bpy)_2CNRu^{III}(NH_3)_5{}^{5+}$ (ref. 58) at 0.59 and 0.95 μm^{-1}, respectively. The structures of the two supramolecular species are shown in Fig. 6. The IT transitions are in both cases assigned to end-to-end transfer between the terminal ruthenium pentammine groups. Although all the proposed models imply an increase in outer-sphere reorganizational energy with increasing transfer distance, the reverse ordering is experimentally observed. This discrepancy may in fact be related to a major difference between the two systems: in the cis compound the "bridge" is bent and the region between the relevant redox sites is occupied by solvent molecules, while in the trans trimer the solvent is kept away from this region by the bulky linear $pzRu(NH_3)_4pz$ "bridge". It is conceivable that the solvent

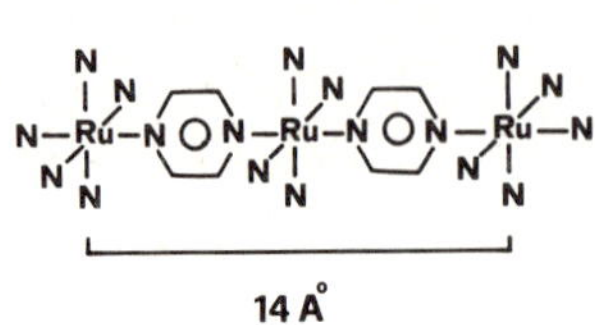

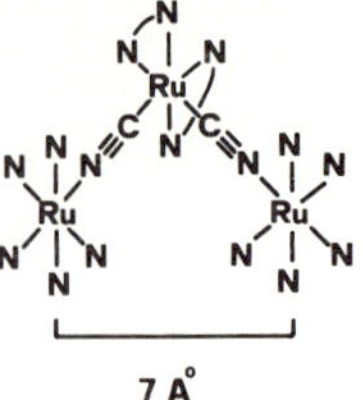

Fig. 6. Schematic representation of the structures of the trinuclear complexes trans-$(NH_3)_5RupzRu(NH_3)_4pzRu(NH_3)_5{}^{7+}$ and cis-$(NH_3)_5Ru^{II}NCRu^{II}(bpy)_2CNRu^{III}$-$(NH_3)_5{}^{5+}$ showing the transfer distance for the end-to-end IT transition.

molecules in the critical inter-site region of the cis complex will experience a substantial reorganization upon electron transfer, a contribution that is absent in the case of the trans complex.

According to the Hush model (ref. 20), the intensity of the IT transitions depends primarily on the electronic matrix element H_{if} connecting the initial and final states of the system (eq. 10), and thus on the overlap between orbitals of the M_1 and M_2 metal centers. To the extent to which the bridging group acts simply as a rigid, saturated spacer, the intensities should correlate with intermetal distance with some kind of exponential dependence (see, e.g., ref. 63). In most inorganic supramolecular systems, however, the bridging group seems to play a more subtle electronic role, with specific effects on H_{if}. The role of the bridging groups has been discussed by Day (ref. 64) in terms of a configuration interaction between the initial and final states and local charge transfer states of the bridge of the type $M_1{}^+$-$(L_1$-$L_2)^-$-M_2 or M_1-$(L_1$-$L_2)^+$-$M_2{}^-$. This mechanism is similar to that known as "superexchange" for through-bond or through-solvent long-range electron transfer (ref. 65). From this point of view, the magnitude of H_{if} should depend heavily on the energy of the local charge transfer states and thus on the redox properties of the bridging ligand. When extended L_1-L_2 bridges are used, the extent of conjugation along the bridge is also likely to play a substantial role in determining H_{if}. Values of H_{if} obtained from the IT transitions of a large number of systems involving Ru(II) and Ru(III) centers as M_1 and M_2 span a relatively wide range (ref. 21), going from a few wavenumbers to maximum typical values of ca. 0.2 μm^{-1} for cyano-bridged complexes (ref. 58). In most cases, the H_{if} values can be easily rationalized by a judicious mixture of arguments based on intermetal distance, bridge redox properties, and degree of conjugation.

An interesting point about IT transfer intensity in supramolecular structures is that concerning the possibility of symmetry or orientational effects. In the precursor complex of bimolecular electron transfer processes the two reactants have the possibility of optimizing H_{if} by exploring a number of configurations differing in mutual orientation. This is still true for ion pairs, although in some highly charged systems such orientational freedom could be somewhat restricted (refs. 29,39). In supramolecular systems with rigid bridges, on the contrary, the orientation of the M_1 and M_2 fragments is fixed and the symmetry of the orbitals involved in the electron transfer process may become relevant. In fact, it has been known for some time that in Prussian Blue analogs t_{2g}-t_{2g} IT transitions are intense while those involving locally orthogonal t_{2g}-e_g orbitals are weak (ref. 66). Also, in the adducts of $Ru(bpy)_2(CN)_2$ with Cu^{2+}_{aq} studied by Demas (ref. 67) no Ru-->Cu (t_{2g}-->e_g) IT bands were observed, in spite of the favourable redox properties. This may be compared with the very intense IT transitions exhibited by the otherwise rather similar adducts with $Ru(NH_3)_5^{3+}$ (ref. 58), for which the transition is however of the t_{2g}-t_{2g} type. It may be useful to recall here that the same symmetry selection rules applying to optical IT transitions should also hold for the corresponding thermal electron transfer processes, in spite of the difference in the nature of the inducing perturbation. In fact, in the optical electron transfer processes the transition moment is dominated by the "transfer" term (ref. 68), which polarizes the transition in the direction of the charge transfer and corresponds to a totally symmetric component of the dipole moment operator. In principle, therefore, symmetry selection rules could be used to control relative rates of intramolecular electron transfer processes within supramolecular structures.

Finally, it should again be recalled that, while originally designed to treat metal-to-metal charge transfer in mixed-valence systems, the Hush model can be conceptually extended to deal with any kind of electronic transition between localized redox centers in a supramolecular structure. Thus, for example, both the optical d-π^* metal-to-ligand charge transfer (MLCT) transition of $Ru(bpy)_3^{2+}$ and the thermal deactivation of the triplet excited state of the same orbital origin can be treated in the same terms as optical or thermal IT (ref. 25). Also, "remote" MLCT transitions from one metal center to a (nonbridging) ligand belonging to another metal center have been recently identified in a number of supramolecular inorganic systems (refs. 58, 59). Figure 7 shows the absorption spectrum of a trinuclear complex in which transitions between all of the available redox sites of the supermolecule can be resolved (ref. 59). The positions and intensities of the various transitions correlate remarkably well with qualitative expectations based on the simple Hush model.

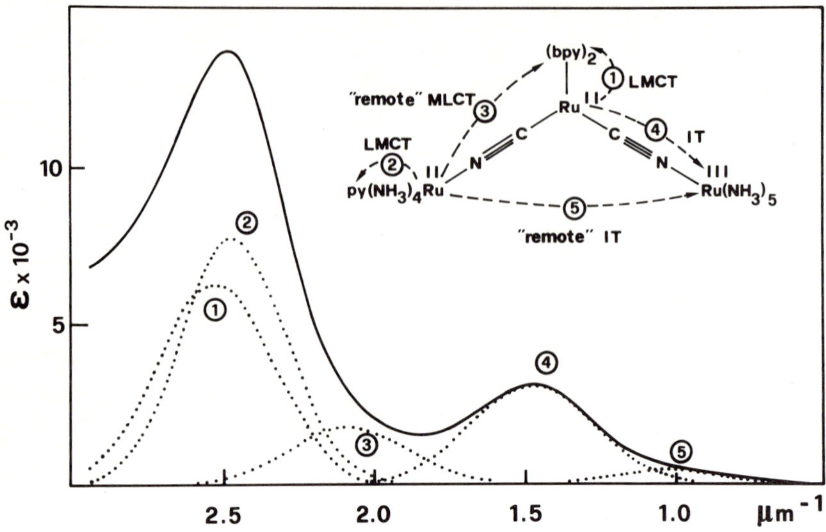

Fig. 7. Absorption spectrum of cis-py$(NH_3)_4$RuNCRu$(bpy)_2$CNRu$(NH_3)_5{}^{5+}$ resolved into component bands corresponding to the various site-to-site optical electron transfer transitions.

3.1.3. Levels of the bridging groups. Most of the common bridging groups have excited states that lie at substantially higher energies than the other relevant states of the supermolecule, and are thus of little consequence. More important are the MLCT or LMCT states that involve one of the metal centers and the bridging ligand. Even when these states lie at high energies, they may provide (see paragraph 3.1.2) a superexchange path for propagation of electronic interaction between the metal centers. In several cases, however, (e.g., when L_1-L_2 is 4,4'-bipyridine) these levels have energies comparable to those of the metal-containing subunits or to the IT states. In these cases, the bridging ligand should be better considered just as one of the several redox sites of the supermolecule that may be connected by optical or thermal electron transfer processes. Very interesting photophysical and photochemical behavior in binuclear complexes initiated by metal-to-bridging ligand excitation has been described by Sutin (ref. 50) and Meyer (ref. 69).

3.2. Monitoring the electron transfer process

With ion-pair systems a very practical, though non-specific, way to monitor a photoinduced electron transfer process is to look at the lifetime or emission intensity quenching of the local excited state of one of the two partners (photosensitizer) by the other (quencher). Depending on a number of kinetic conditions, the electron transfer process in these systems may result in dynamic or static quenching effects (Section 2.2). For supramolecular systems this possibility is precluded in a strict sense.

In a rather general sense, one can think of performing a "static quenching" experiment by comparing the excited state lifetime or emission intensity of the excited "free" subunit with those of the same subunit in the supramolecular system. Such comparisons may be quite safe in the field of weakly interacting organic supramolecular systems (ref. 70). As discussed in Section 3.1.1, for inorganic systems there may be serious difficulties in the definition of the individual subunits, in the possibility to have them as free molecules, or in the identification of suitable model systems as substitutes. Moreover, several processes besides electron transfer may lead to quenching of a local excited state in the supermolecule (e.g., perturbation of the original energy levels of the subsystem, paramagnetic or spin-orbit interactions, energy transfer). Thus, although such comparisons may be applied in some instances (refs. 58, 59, 69), care should be taken to analyze all the possible deactivation pathways before drawing conclusions about the occurrence of specific processes.

Another indirect, though more specific, way to monitor photoinduced electron transfer processes is offered by those supramolecular systems in which one of the subunits undergoes a fast, irreversible, typical reaction when oxidized or (more frequently) reduced. Practical cases are binuclear complexes containing Co(III) or Cr(III) subunits which become labile upon reduction leading to disruption of the supramolecular structure. Several of these systems have been investigated in detail by Haim (ref. 71) and Vogler (ref. 47).

When the photoinduced electron transfer within a supramolecular system is fully reversible, the only direct way to monitor the process is via transient absorbance changes in laser photolysis experiments. Few direct measurements of this type have been performed on inorganic supramolecular systems. A noticeable example is the direct observation by Sutin and coworkers (ref. 50) of the back electron transfer process shown in eq 27, following picosecond laser excitation (eq 25) and subsequent electron transfer (eq 26).

$$(NH_3)_5Ru^{II}pzRu^{III}(EDTA)^+ \text{------>} (NH_3)_5Ru^{III}pz^-Ru^{III}(EDTA)^+ \quad (25)$$

$$(NH_3)_5Ru^{III}pz^-Ru^{III}(EDTA)^+ \text{------>} (NH_3)_5Ru^{III}pzRu^{II}(EDTA)^+ \quad (26)$$

$$(NH_3)_5Ru^{III}pzRu^{II}(EDTA)^+ \text{------>} (NH_3)_5Ru^{II}pzRu^{III}(EDTA)^+ \quad (27)$$

It seems likely that direct observations of this type will become more common as the use of picosecond techniques extends throughout the field.

Acknowledgments. We would like to thank Mr. G. Gubellini for the drawings. Financial support from the Ministero della Pubblica Istruzione and Consiglio Nazionale delle Ricerche is gratefully acknowledged.

LIST OF ABBREVIATIONS

biq = 2,2'-biquinoline
bpy = 2,2'-bipyridine
CT = charge transfer
dien = diethylenetriamine
EDTA = ethylenediaminetetraacetate ion
en = ethylenediamine
IPCT = ion pair charge transfer
IP = ion pair
IT = intervalence transfer
LLCT = ligand-to-ligand charge transfer
MC = metal centered
MLCT = metal-to-ligand charge transfer
MMCT = metal-to-metal charge transfer
mnt = 1,2-dicyano-1,2-ethylenedithiolate
OSCT = outer sphere charge transfer
phen = 1,10-phenanthroline
pn = propylenediamine
py = pyridine
pz = pyrazine
sep = sepulchrate
SSCT = second sphere charge transfer
$\subset$ = symbol of inclusion

REFERENCES

1 R.D. Cannon, Electron Transfer Reactions, Butterworths, London, 1980.
2 N. Sutin, Progr. Inorg. Chem., 30 (1983) 441.
3 M. Graetzel (Ed.), Energy Resources through Photochemistry and Catalysis, Academic Press, London, 1983.
4 R.A. Marcus, Annu. Rev. Phys. Chem., 15 (1964) 155.
5 N.S. Hush, Electrochim. Acta, 13 (1968) 1005.
6 N. Sutin, in G.L. Eichorn (Ed.), Inorganic Biochemistry, Elsevier, New York, 1983, p. 611.
7 V. Balzani and F. Scandola, in M. Graetzel (Ed.), Energy Resources through Photochemistry and Catalysis, Academic Press, London, 1983, p. 1.
8 S. Efrima and M. Bixon, Chem. Phys., 63 (1976) 4358.
9 N.R. Kestner, J. Logan, and J. Jortner, J. Phys. Chem., 78 (1974) 2148.
10 J. Ulstrup, Charge Transfer in Condensed Media, Springer-Verlag, Berlin, 1979.
11 R. Engleman and J. Jortner, J. Mol. Phys., 18 (1970) 145.
12 R.A. Marcus and P. Siders, J. Phys. Chem., 86 (1982) 662.
13 M.T. Indelli, R. Ballardini, and F. Scandola, J. Phys. Chem., 88 (1984) 2547.
14 J.R. Miller, J.V. Beitz, and R.K. Huddleston, J. Am. Chem. Soc., 106 (1984) 5057.
15 J.R. Miller, L.T. Calcaterra, and G.L. Closs, J. Am. Chem. Soc., 106 (1984) 3047.
16 M.R. Wasielewski, M.P. Niemczyk, W.A. Svec, and E.B. Pewitt, J. Am. Chem. Soc., 107 (1985) 1080.
17 T.J. Meyer and J.V. Caspar, Inorg. Chem., 22 (1983) 2444.
18 T.J. Meyer, Pure Appl. Chem., 58 (1986) 1193.
19 M. McGuire and G. McLendon, J. Phys. Chem., 90 (1986) 2549.

20 N.S. Hush, Prog. Inorg. Chem., 8 (1967) 391.
21 C. Creutz, Progr. Inorg. Chem., 30 (1983) 1.
22 D. Brown (Ed.), Mixed Valence Compounds, Reidel, Dortrecht, 1980.
23 T.J. Meyer, Acc. Chem. Res., 11 (1978) 94.
24 R.A. Marcus and N. Sutin, Comments Inorg. Chem., 5 (1986) 119.
25 T.J. Meyer, Progr. Inorg. Chem., 30 (1983) 389.
26 A. Vogler and H. Kunkely, J. Am. Chem. Soc., 103 (1981) 1559.
27 A. Vogler and H. Kunkely, Angew. Chem. Int. Ed. Engl., 21 (1982) 77.
28 G.A. Crosby, R.G. Highland, K.A. Truesdell, Coord. Chem. Rev., 64 (1985) 41.
29 J.C. Curtis and T.J. Meyer, Inorg. Chem., 21 (1982) 1562.
30 A. Vogler and H. Kunkely, private communication.
31 A. Vogler and H. Kunkely, J. Chem. Soc. Chem. Commun., (1986) 1616.
32 V. Balzani, N. Sabbatini, and F. Scandola, Chem. Rev., 86 (1986) 319.
33 W. Rybak, A. Haim, T.J. Netzel, and N. Sutin, J. Phys. Chem., 85 (1981) 2856.
34 R. Frank and H. Rau, J. Phys. Chem., 87 (1983) 5181.
35 R. Ballardini, M.T. Gandolfi, and V. Balzani, Chem. Phys. Lett., 119 (1985) 459.
36 D. Rehm and A. Weller, Ber. Bunsenges. Phys. Chem., 73 (1969) 834.
37 D. Mauzerall, Ann. Rev. Phys. Chem., 33 (1982) 377.
38 A. Haim, Comments Inorg. Chem., 4 (1985) 113.
39 V. Balzani, A. Juris, and F. Scandola, in E. Pelizzetti and N. Serpone (Eds.), Homogeneous and Heterogeneous Photocatalysis, Reidel, Dordrecht, 1986, p. 1.
40 R. Ballardini, M.T. Gandolfi, and V. Balzani, Inorg. Chem., in press.
41 V. Balzani, L. Moggi, M.F. Manfrin, F. Bolletta, and G.S. Laurence, Coord. Chem. Rev., 15 (1975) 321.
42 F. Bolletta, M. Maestri, L. Moggi, and V. Balzani, J. Phys. Chem., 78 (1974) 1374.
43 J.N. Demas and J.W. Addington, J. Am. Chem. Soc., 98 (1976) 5800.
44 G. Weber, in E.C. Slater (Ed.), Flavins and Flavoproteins, Elsevier, Amsterdam, 1966, p. 15.
45 R. Larson, Acta Chem. Scand., 21 (1967) 257.
46 A. Vogler and J. Kisslinger, Angew. Chem. Int. Ed. Engl., 21 (1982) 77.
47 A. Vogler, A.H. Osman, and H. Kunkely, Coord. Chem. Rev., 64 (1985) 159.
48 J.C. Curtis and T.J. Meyer, J. Am. Chem. Soc., 100 (1978) 6284.
49 H.E. Toma, J.C.S. Dalton, (1980) 471.
50 C. Creutz, P. Kroger, T. Matsubara, T.L. Netzel, and N. Sutin, J. Am. Chem. Soc., 101 (1979) 5442.
51 A. Vogler and J. Kisslinger, J. Am. Chem. Soc., 104 (1982) 2311.
52 N. Sabbatini, A. Bonazzi, M. Ciano, and V. Balzani, J. Am. Chem. Soc., 106 (1984) 4055.
53 N. Sabbatini and V. Balzani, J. Less-Common Metals, 112 (1985) 381.
54 F. Scandola, C.A. Bignozzi, and V. Balzani, in E. Pelizzetti and N. Serpone (Eds.), Homogeneous and Heterogeneous Photocatalysis, Reidel, Dordrecht, 1986, p. 29.
55 J. Lilie, N. Shinohara, and M.G. Simic, J. Am. Chem. Soc., 98 (1976) 6516.
56 M.B. Robin and P. Day, Adv. Inorg. Chem. Radiochem., 10 (1967) 247.
57 C.A. Bignozzi and F. Scandola, Inorg. Chem., 23 (1984) 1540.
58 C.A. Bignozzi, S. Roffia, and F. Scandola, J. Am. Chem. Soc., 107 (1985) 1644.
59 C.A. Bignozzi, C. Paradisi, S. Roffia, and F. Scandola, submitted for publication.
60 C.A. Bignozzi, C. Chiorboli, S. Roffia, and F. Scandola, XI IUPAC Symposium on Photochemistry, Lisboa, 1986, p. 404.
61 B.S. Brunschwig, S. Ehrenson, and N. Sutin, J. Phys. Chem., 90 (1986) 3657.
62 A. von Kameke, G.M. Tom, and H. Taube, Inorg. Chem., 17 (1978) 1790.
63 See, e.g., G.L. Closs, L.T. Calcaterra, N.J. Green, K.W. Penfield, and J.R. Miller, J. Phys. Chem., 90 (1986) 3673.

64 B. Mayoh and P. Day, J. Chem. Soc. Dalton Trans., (1973) 846.
65 J.R. Miller and J.V. Beitz, J. Chem. Phys., 74 (1981) 6747.
66 P.S. Braterman, J. Chem. Soc. (A), (1966) 1471.
67 J.N. Demas, J.W. Addington, S.H. Peterson, and E.W. Harris, J. Phys. Chem., 81 (1977) 1039.
68 P. Day and N. Sanders, J. Chem. Soc. (A), (1967) 1536.
69 J.C. Curtis, J.S. Bernstein, and T.J. Meyer, Inorg. Chem., 24 (1985) 385.
70 See, e.g., J.A. Schmidt, A. Siemiarczuk, A.C. Weedon, and J.R. Bolton, J. Am. Chem. Soc., 107 (1985) 6112.
71 A. Haim, Progr. Inorg. Chem., 30 (1983) 273.

Chapter 5.5

Photochemical Reactions of Transition Metal Complexes Induced by Intramolecular Electron Transfer between Weakly Coupled Redox Centers

A. Vogler

1 INTRODUCTION

Light-induced electron transfer may take place as an inter- or intramolecular process. There is certainly no fundamental difference between both possibilities. However, intramolecular electron transfer is generally much better defined with regard to the geometrical arrangement of electron donor and acceptor. Unfortunately, the investigation of intramolecular light-induced electron transfer is hampered by other complications.

In most transition metal complexes the electronic interaction between metal and ligands cannot be neglected and may be rather strong. Consequently, intramolecular electron transfer involving metal and ligands takes place between coupled redox centers. In these cases it is often not quite clear what fraction of charge is transferred from the metal to the ligand and vice versa since donor and acceptor orbitals are delocalized to a certain degree. The majority of photoredox processes of metal complexes which have been reported (refs. 1-3) take place upon direct optical charge transfer (CT) excitation involving redox centers with substantial coupling. Although this review deals essentially only with systems consisting of weakly coupled donors and acceptors a short general survey of intramolecular optical CT excitation of transition metal complexes regardless of the extent of coupling is given here. CT transitions are classified according to the redox sites (ref. 4).

Ligand to Metal Charge Transfer (LMCT)

LMCT is a classical optical transition of metal complexes. Corresponding absorption bands are observed at low energies if the metal is oxidizing and the ligand reducing. Co(III) and Fe(III) complexes are well-documented examples. The colors of d^0 oxometallates such as the yellow CrO_4^{2-} and the violet MnO_4^- are caused by LMCT bands. In many cases LMCT excitation is associated with the reduction of the metal and oxidation of the ligand. Co(III) ammines undergo such photoredox reactions (refs. 1,2).

Metal to Ligand Charge Transfer (MLCT)

MLCT is another classical optical transition of metal complexes. MLCT absorption bands appear at long wavelength if the metal is reducing and a ligand provides low-energy empty orbitals. Complexes such as $[Fe(CN)_6]^{4-}$ and $[Ru(bipy)_3]^{2+}$ (bipy = 2,2'-bipyridyl) are typical cases (ref. 4). In addition, organometallics which contain a metal in a low oxidation state and π-accepting ligands such as an olefin or an aromatic molecule are characterized by low-energy MLCT bands (ref. 3). Since metal ligand bonding is not much affected intramolecular photoreactions do generally not occur on MLCT excitation. Ligands are electron rich and cannot be easily reduced to stable species. MLCT excitation is therefore often associated with photooxidation of the metal while an electron is transferred from the ligand to another molecule by an intermolecular process. In the case of $[Fe(CN)_6]^{4-}$ the solvent may act as electron acceptor (ref. 2). In the MLCT state the ligand can be also susceptible to an electrophilic attack. The addition of protons to coordinated NO (ref. 5) or carbynes (ref. 6) illustrates this type of excited state reactivity.

Ligand to Ligand Charge Transfer (LLCT)

LLCT absorptions were observed only recently. These bands appear if one ligand is reducing and another oxidizing. LLCT bands were detected in the spectra of complexes of the type (1,2-diimine)M^{II}(1,2-ethylenedithiolate) with M = Ni, Pd, Pt (refs. 7, 8) and related complexes (refs. 9, 10). The diimine with its empty π orbitals is here the electron acceptor while the dithiolate acts as donor. LLCT transitions can be also identified in the spectra of complexes containing the same ligand in an oxidized and reduced form. $[Ru^{II}(bipy)_3]^+$ is apparently composed of two bipy ligands and one in its reduced state ($bipy^-$). The electronic spectrum of this complex displays a $bipy^-$ to bipy LLCT band in the near IR (ref. 11). A photoreaction originating from a LLCT state was reported recently (ref. 7).

Intra Ligand Charge Transfer (ILCT)

A ligand itself may consist of a reducing and an oxidizing part. The spectrum of the metal complex as well as that of the free ligand should show low-energy ILCT bands. An example of an optical ILCT transition is indeed known (ref. 12).

Metal to Metal Charge Transfer (MMCT)

Ligand-bridged binuclear (or polynuclear) complexes which contain a reducing and an oxidizing metal are characterized by optical MMCT transitions at low energies (refs. 4, 13-19). There are several well-documented examples of photochemical reactions induced by intramolecular MMCT excitation (ref. 19). In the first part of this review they are discussed in some detail as typical cases of photoredox processes initiated by optical CT transitions between weakly interacting redox sites.

A light-induced electron transfer involving weakly coupled redox centers cannot only take place by direct optical CT excitation (resonance transfer). As a second possibility an intramolecular excited state electron transfer may occur (non-resonance transfer). This process does not require the presence of CT absorptions. The initial internal excitation of an electron donor or acceptor can be followed by electron transfer. Examples of this type are elaborated in the second part of this review.

The further discussion is essentially restricted to systems which are subject to a permanent photochemical change. However, reference is also made to complexes which undergo a reversible intramolecular electron transfer. In these cases the occurence of electron transfer was detected by emission or time-resolved absorption spectroscopy.

2 RESONANCE ELECTRON TRANSFER BY OPTICAL MMCT EXCITATION

The electronic coupling of metallic redox centers in binuclear complexes has been studied extensively for many years (refs. 13-18). The theory advanced by N. S. Hush provides the basis for a discussion of this interaction (refs. 15, 16, 18, 20, 21). A binuclear ligand-bridged complex of the type M_α^{red}-L-M_β^{ox} is composed of the two mononuclear components M_α^{red}-L and L-M_β^{ox}. Besides the bridging ligand both metals are coordinated to other terminal ligands. The electronic spectrum of the binuclear complex consists of the superimposed spectra of the mononuclear components if the interaction between the reducing and oxidizing metal is weak. In addition, a new absorption band appears which belongs to the optical MMCT transition from the reducing to the oxidizing metal (refs. 13-19).

With increasing metal-metal interaction the individual components loose their identity. Finally, the valence orbitals of both metals are completely delocalized and the metals do not any more exist in well-defined ("trapped") oxidation states. As a consequence the spectral features of the mononuclear components are not any more apparent and MMCT bands do not occur. They are now replaced by electronic transitions involving orbitals which are delocalized between both metals. An analysis of this absorption band provides the information on the extent of delocalization α^2.

$$\alpha^2 = \frac{4.24 \cdot 10^{-4} \cdot \varepsilon_{max} \cdot \Delta\bar{\nu}_{1/2}}{\bar{\nu}_{max} \cdot d^2}$$

The parameters are the energy of the MMCT band at its maximum $\bar{\nu}_{max}$, the molar extinction coefficient ε_{max}, the half-width of the MMCT band $\Delta\bar{\nu}_{1/2}$ and the metal-metal distance d. At complete delocalization α^2 is unity while the valencies are trapped if α^2 is much smaller than unity. Only in the latter case MMCT transitions are associated with a complete electron transfer from one metal to another. The following discussion is restricted to this type of CT interaction.

The energy of the optical MMCT transition depends on the potential difference ΔE between the redox couples $M_{\alpha}^{red/ox}$ and $M_{\beta}^{red/ox}$ and on the reorganizational energy χ.

$$\bar{\nu}_{max} = \Delta E + \chi$$

The parameter χ consists of an outer- and an inner-sphere contribution.

$$\chi = \chi_o + \chi_i$$

While χ_i is an intrinsic property of the binuclear complex χ_o depends on the reorganization of solvent environment according to

$$\chi_o = e^2\left(\frac{1}{2a_1} + \frac{1}{2a_2} - \frac{1}{d}\right)\left(\frac{1}{n^2} - \frac{1}{D}\right)$$

The parameters a_1 and a_2 are the radii of the coordination spheres of both metals, d is the metal-metal distance, n and D are the refractive index and the static dielectric constant of the solvent. This equation means simply that χ_o and hence the energy of the MMCT transition varies with the polarity of the solvent. Since the specific complexes discussed below were investigated only in aqueous solution details on the solvent dependence are not elaborated here.

The inner-sphere contribution to the reorganizational energy is a fraction of the Franck-Condon MMCT transition as shown in Fig. 1. It depends on the structural distortion which accompanies electron transfer. This reorganization which may be represented by changes of the metal-ligand bond length of one metal center varies with the oxidation state of this metal. Frequently reduction is associated with an extension of the metal-ligand distance.

The optical MMCT is a Franck-Condon transition which terminates in a vibrationally excited state of the redox isomer $M_{\alpha}^{ox}/M_{\beta}^{red}$ before it relaxes. The electron transfer cannot only be achieved by light absorption but also as a thermal process which requires the activation energy E_{th} to reach the crossing point of both potential curves. When the redox isomer exists in its vibrational ground state it may undergo a thermal back electron transfer by overcoming the activation barrier E'_{th}.

The majority of compounds which were investigated with regard to optical MMCT transitions are homobinuclear complexes containing one metal in two different oxidation states. Particularly Ru^{II}/Ru^{III} systems found much attention (refs. 13-17). These complexes are called mixed-valence (MV) compounds. In this case the term MMCT is also known as intervalence transfer (IT). The potential curves (Fig. 1) of both redox isomers of such a symmetric MV complex are located at the same energy. They are only displaced

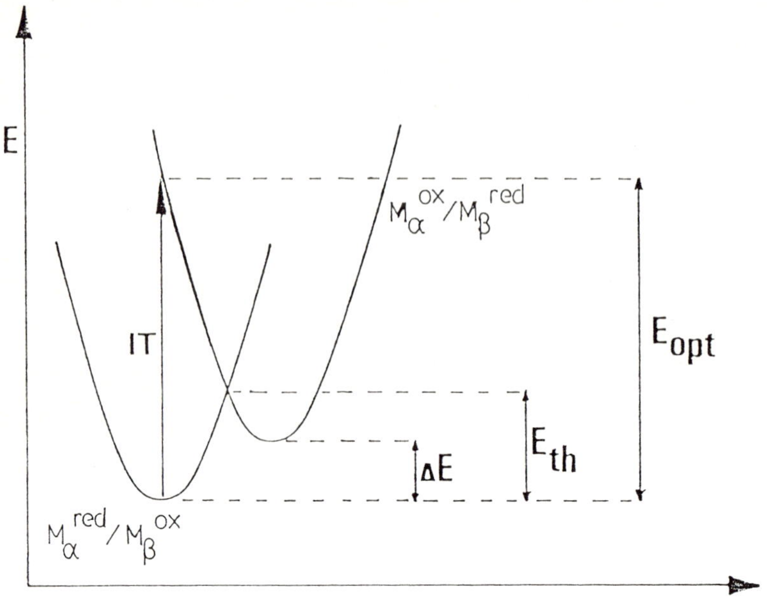

Fig. 1. Potential energy versus metal-ligand bond length for binuclear complexes containing a reducing and an oxidizing metal

horizontally due to the structural reorganization which accompanies the electron transfer. Unfortunately, for a variety of reasons these compounds are not well suited to observe a photochemical reaction leading to a permanent chemical change. First of all, in a symmetric MV complex an electron exchange does not cause a real chemical change, although the individual metal atoms have exchanged their oxidation state and hence their environment. But even in most homobinuclear complexes which are slightly asymmetric due to different ligands a rapid thermal electron exchange occurs. This situation interferes with the observation of light-induced electron transfer. Finally, the MMCT bands of the symmetric or nearly symmetric MV systems appear in the near IR which is not easily accessible by conventional irradiation sources and light detection devices. Consequently, photoactive systems should be designed according to these considerations.

Light-sensitivity will be most easily observed for strongly asymmetric binuclear complexes. They may be stable towards thermal electron exchange which requires the activation energy E_{th} and are expected to display their MMCT bands in the visible or UV region (see Fig. 1). An asymmetric system may be constructed in two ways. In homobinuclear complexes different ligands at both metals can be employed. For example, in a Ru^{II}, Ru^{III} complex a large redox asymmetry will be achieved if Ru^{II} is stabilized by π-acceptor ligands and Ru^{III} by π-donors. Much larger energy separations are possible in heteronuclear systems. The individual components are selected according to their redox potentials.

Another very important criterion for a proper choice is the anticipated reactivity of the redox isomer generated by MMCT excitation. It will not be stable but return rapidly to the starting point since the activation barrier E'_{th} for back electron transfer is rather low (Fig. 1). An irreversible formation of stable photoproducts can only be achieved if the redox isomer is able to undergo some further geometrical rearrangements. These secondary processes must be fast enough to compete with back electron transfer. For example, photoactivity is expected if $[Co(NH_3)_6]^{3+}$ is the oxidizing component of a binuclear complex. Upon reduction $[Co(NH_3)_6]^{2+}$ is formed. It is kinetically very labile and undergoes a rapid decomposition in aqueous solution. According to these considerations in 1975 we started to explore photochemical reactions induced by MMCT excitation (refs. 19, 22, 23).

$[(NC)_5Ru^{II}CNCo^{III}(NH_3)_5]^-$

The binuclear cyanide-bridged complex $[(NC)_5RuCNCo(NH_3)_5]^-$ (ref. 22) may be viewed as being composed of $[Co^{III}(NH_3)_6]^{3+}$ and $[Ru^{II}(CN)_6]^{4-}$ if the coupling between Ru and Co is weak. The assumption that $[Co(NH_3)_6]^{3+}$ can be considered as one of the components is supported by the observation that bridging cyanide which coordinates via nitrogen is similar to ammonia with regard to its ligand field strength. Weak coupling is then indicated by the absorption spectrum of $[(NC)_5RuCNCo(NH_3)_5]^-$ which is indeed composed of the spectra of $[Ru(CN)_6]^{4-}$ and $[Co(NH_3)_6]^{3+}$ (Fig. 2). The binuclear complex displays its longest-wavelength absorption at λ_{max} = 475 nm. This is the first ligand field (LF) band of $[Co(NH_3)_6]^{3+}$. The other component $[Ru(CN)_6]^{4-}$ does not absorb above 300 nm. Independent support for a weak interaction of both metals comes from the IR spectrum. Beside the bridging cyanide which absorbs at 2135 cm^{-1} the stretching vibration of the terminal cyanides appears at 2050 cm^{-1}. This is the same value as that of $[Ru(CN)_6]^{4-}$. It follows that the oxidation state of ruthenium is two in the binuclear complex since the frequency of this stretching vibration is a sensitive function of the valency of the metal.

The presence of the reducing Ru^{II} and oxidizing Co^{III} leads to the appearance of a new absorption (Fig. 2) at λ_{max} = 375 nm (ε = 690) which is

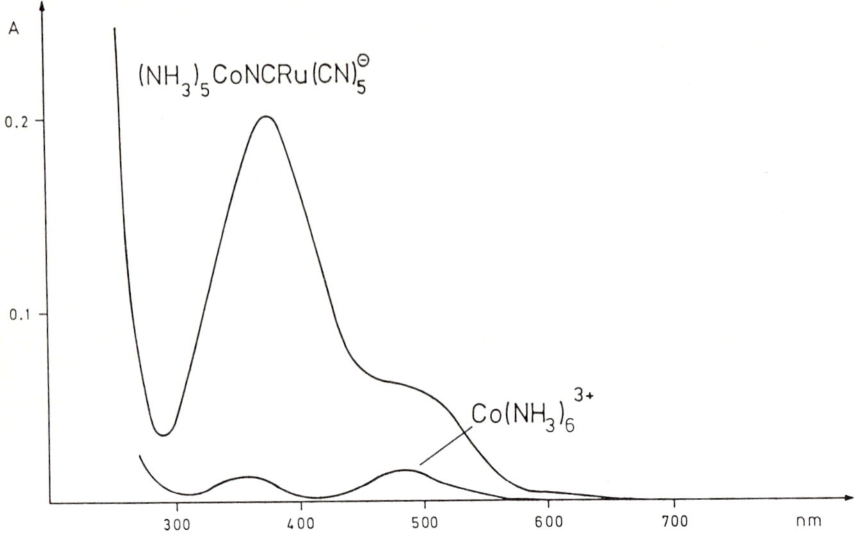

Fig. 2. Absorption spectra of aqueous $3x10^{-4}$ M $[(NH_3)_5CoNCRu(CN)_5]^-$ and $3x10^{-4}$ M $[Co(NH_3)_6]^{3+}$ at room temperature, 1-cm cell

assigned to a MMCT transition (ref. 22). An analysis of the band with regard to the Hush theory would require the knowledge of the potentials of the redox couples $Co^{III/II}$ and $Ru^{II/III}$ within the binuclear complex. These potentials are not known. However, for a very rough estimate the potentials of the mononuclear components $[Co(NH_3)_6]^{3+/2+}$ (E° = 0.11 V) (ref. 24) and $[Ru(CN)_6]^{4-/3-}$ (E° = 0.86 V) (ref. 24) may be taken. This gives a reorganizational energy of χ = 59 kcal/mol. Such a large value is certainly associated with the fact that the MMCT transition leads to the population of an antibonding e_g orbital of Co^{III} (in O_h symmetry) which causes a large distortion at Co^{II} in the redox isomer $[(NC)_5Ru^{III}CNCo^{II}(NH_3)_5]^-$. The activation barrier for thermal electron transfer from Ru^{II} to Co^{III} is then calculated to be E_{th} = 25 kcal/mol. Aqueous solutions of the binuclear complex are stable at room temperature but undergo a redox decomposition upon light absorption by the MMCT band with the quantum yield Φ = 0.46 at λ_{irr} = 366 nm.

$$[(NC)_5Ru^{II}CNCo^{III}(NH_3)_5]^- \xrightarrow{h\nu} [(NC)_5Ru^{III}CNCo^{II}(NH_3)_5]^-$$

$$[(NC)_5Ru^{III}CNCo^{II}(NH_3)_5]^- \rightarrow [Ru^{III}(CN)_6]^{3-} + Co^{2+} + 5\ NH_3$$

Because of the strong distortion of the vibrationally relaxed MMCT state, thermal back electron transfer is now associated with a substantial activation barrier of E'_{th} = 5 kcal. The decomposition of the kinetically labile Co^{II} is apparently fast enough to compete with charge recombination.

An attempt to prepare the corresponding binuclear iron complex $[(NH_3)_5Co^{III}NCFe^{II}(CN)_6]^-$ failed. Upon mixing solutions of $[Co(NH_3)_5H_2O]^{3+}$ and $[Fe(CN)_6]^{4-}$ a rapid thermal outer-sphere electron transfer from Fe(II) to Co(III) takes place (ref. 25) before the formation of the binuclear complex occurs. $[Fe(CN)_6]^{4-}$ (E° = 0.36 V vs SCE) is more reducing than $[Ru(CN)_6]^{4-}$ by 0.5 V (ref. 24). Assuming the same reorganizational energy of χ = 58 kcal/mol for both binuclear complexes $[(NH_3)_5Co^{III}NCM^{II}(CN)_5]^-$ with M = Fe and Ru the activation barrier would now be only E_{th} = 17 kcal/mol for thermal electron transfer within the hypothetical complex $[(NH_3)_5Co^{III}NCFe^{II}(CN)_5]^-$ in an aqueous solution.

$\underline{[(NC)_5Fe^{II}CNCr^{III}(NH_3)_5]^-}$

The binuclear ion $[(NC)_5Fe^{II}CNCr(NH_3)_5]^-$ consists of the components $[Fe^{II}(CN)_6]^{4-}$ and $[Cr(NH_3)_6]^{3+}$ which are weakly coupled (ref.26). The MLCT band of $[Fe(CN)_6]^{4-}$ at λ_{max} = 210 nm can be recognized in the spectrum of the binuclear complex. The typical LF bands of $[Cr(NH_3)_6]^{3+}$ are apparently obscured by the much more intense absorption at λ_{max} = 376 nm (ε = 2400) which is assigned to the MMCT transition from the reducing Fe^{II} to the oxidizing Cr^{III}.

The potential difference between the redox couples $[Fe(CN)_6]^{4-/3-}$ and $[Cr(NH_3)_6]^{3+/2+}$ is ΔE = 1.19 V. Taking this as a rough estimate for ΔE of the binuclear complex the electron transfer is associated with a reorganizational energy of χ = 49 kcal. This large value supports the assumption that the antibonding e_g orbitals at Cr(III) (in O_h symmetry) are the acceptor orbitals of the MMCT transition. The binuclear complex is quite stable in aqueous solution with regard to thermal electron transfer (E_{th} = 29 kcal) while an efficient photolysis takes place upon MMCT excitation (Φ = 0.1 nm at λ_{irr} = 366 (ref. 26):

$$[(NC)_5Fe^{II}CNCr^{III}(NH_3)_5]^- \xrightarrow{h\nu} [(NC)_5Fe^{III}CNCr^{II}(NH_3)_5]^-$$

$$[(NC)_5Fe^{III}CNCr^{II}(NH_3)_5]^- \rightarrow [Fe^{III}(CN)_6]^{3-} + Cr^{2+} + 5\ NH_3$$

The Franck-Condon state reached by the optical MMCT transition undergoes a rapid vibrational relaxation to the thermally equilibrated redox isomer Fe^{III}/Cr^{II}. The large distortion at Cr^{II} slows down back electron transfer ($E'_{th} \approx$ 2 kcal). The kinetically labile Cr^{II} undergoes a ligand displacement before charge recombination can occur. Finally, Cr^{2+} is oxidized by oxygen.

$[(NC)_5M^{II}CNCo^{III}(CN)_5]^{6-}$ with M = Fe, Ru, and Os

The IR spectra of the anions $[(NC)_5M^{II}CNCo^{III}(CN)_5]^{6-}$ (refs. 23, 26) with M = Fe, Ru, and Os are indicative of weak coupling between M^{II} and Co^{III}. The absorptions of the terminal cyanides appear at nearly the same wavenumbers as those of mononuclear cyano complexes of M^{II} and Co^{III}. The electronic spectra of the binuclear complexes are less instructive since well-separated bands do not appear. However, there is much evidence that the absorption features at λ_{max} 385 nm (ε = 630) for M = Fe, 312 nm (460) for Ru, and 360 nm (734) for Os can be assigned to MMCT transitions from M^{II} to Co^{III} (refs. 23, 26). The energy of the MMCT band varies with the reducing strength of $[M(CN)_6]^{4-}$ (M = Fe, E° = +0.36 V; Ru 0.86 V and Os 0.56 V) (refs. 24, 27).

The MMCT transitions of the binuclear complexes terminate in e_g orbitals of Co^{III}. The thermally equilibrated redox isomers $[(NC)_5M^{III}CNCo^{II}(CN)_5]^{6-}$ are expected to undergo large distortions at Co^{II}. Reorganizational energies of more than 45 kcal are estimated.

All three anions $[(NC)_5M^{II}CNCo^{III}(CN)_5]^{6-}$ are stable in solution (refs. 23, 26). The activation energies for thermal electron transfer may exceed 30 kcal. However, in all cases the aqueous complexes undergo photoredox reactions upon MMCT excitation. The redox isomers $[(NC)_5M^{III}CNCo^{II}(CN)_5]^{6-}$ dissociate in the primary photochemical step:

$$[(NC)_5M^{II}CNCo^{III}(CN)_5]^{6-} \xrightarrow{h\nu} [(NC)_5M^{III}CNCo^{II}(CN)_5]^{6-}$$

$$[(NC)_5M^{III}CNCo^{II}(CN)_5]^{6-} \rightarrow [M^{III}(CN)_6]^{3-} + [Co^{II}(CN)_5]^{3-}$$

In the absence of oxygen a complete regeneration of the binuclear complexes occurs by a thermal inner-sphere electron transfer which is simply a reversal of the photoreaction. The activation barrier for this back electron transfer was estimated to be around E'_{th} = 2 kcal. The regeneration of the binuclear complexes by this inner-sphere redox process is not surprising since all three cyanide-bridged anions are synthesized by this reaction. The iron complex was prepared by Haim and Wilmarth in 1961 according to this procedure (ref. 28). In distinction to ammine complexes of Co^{II} which decay irreversibly in aqueous solution the complex $[Co^{II}(CN)_5]^{3-}$ does not decompose.

The photolysis of all three complexes induced by MMCT excitation leads to a permanent chemical change only in the presence of air. The complex $[Co(CN)_5]^{3-}$ can be intercepted by O_2:

$$2[Co(CN)_5]^{3-} + O_2 \rightarrow [(NC)_5Co^{III}(O_2^{2-})Co^{III}(CN)_5]^{6-}$$

In acidic solution the peroxo complex decomposes to yield H_2O_2 and $2[Co^{III}(CN)_5(H_2O)]^{2-}$ while in basic solution the peroxo complex is further oxidized to the superoxo complex $[(NC)_5Co^{III}(O_2^-)Co^{III}(CN)_5]^{5-}$. The quantum yields for the formation of $[M^{III}(CN)_6]^{3-}$ are slightly wavelength-dependent due to the overlap of the MMCT bands with absorptions of other origin. The quantum yields may exceed unity because $[M(CN)_6]^{3-}$ is not only produced in the primary photochemical reaction but also by the oxidation of $[M(CN)_6]^{4-}$ by H_2O_2. The quantum yields are fairly large: Φ = 1.6 at λ_{irr} = 405 nm for M = Fe; Φ = 0.39 at λ_{irr} = 313 nm for M = Ru, and Φ = 0.32 at λ_{irr} = 366 nm for M = Os (refs. 23, 26).

3 EXCITED STATE (NON-RESONANCE) ELECTRON TRANSFER

3.1 Introduction

A photoinduced electron transfer does not only occur by direct optical excitation (resonance transfer). As an alternative an electronically excited molecule may undergo an electron transfer to or from another molecule. Intermolecular excited state electron transfer of this type has been studied intensively (refs. 14, 29, 30). Particularly the complex $[Ru(bipy)_3]^{2+}$ was used as excited state electron donor or acceptor (ref. 31). In solution intermolecular electron transfer can take place if the excited molecule has a diffusional encounter with a suitable redox partner before it returns to the ground state.

An excited state electron transfer may take place also as intramolecular process. An excited chromophoric group can undergo an electron transfer to or from another part of the same molecule. Since donor and acceptor are already in contact prior to electronic excitation a long lifetime of the excited state is not required. Even higher excited states which can not participate in bimolecular reactions due to their short lifetimes may undergo intramolecular electron transfer. While in bimolecular redox processes the structural arrangement of donor and acceptor in the encounter complex is not known intramolecular electron transfer occurs in a better defined environment. Although these features make it attractive to study intramolecular excited state electron transfer this subject has been largely neglected until a few years ago.

The recent interest in intramolecular excited state electron transfer is associated with attempts to understand the primary events of photosynthesis and to design model systems for the photosynthesis (ref. 32). In the first step an excited state electron transfer occurs which must be uphill with regard to the ground state in order to convert light into chemical energy. In simple systems this first step is followed by a rapid downhill charge recombination. In the photosynthesis a charge separation is achieved by introducing a barrier for back electron transfer. Recently model compounds have been designed to study the charge separation in detail. A system which found much attention consists of a porphyrin as excited state electron donor which is linked covalently to a quinone as electron acceptor. In addition, a carotene may be attached as a donor to accomplish charge separation over large distances (ref. 32). However, in this review the discussion is restricted to typical transition metal complexes.

Two different cases will be distinguished. First, the primary electron transfer is followed by rapid secondary processes which compete successfully with back electron transfer. As a result, the light absorption leads to a permanent chemical change. Secondly, charge recombination is rapid and a net photolysis is not observed. Luminescence and time-resolved absorption spectroscopy were applied to gain insight into the charge separation and recombination processes.

The compounds discussed below do not display CT absorption bands involving the electron donor and acceptor. In some cases this is a good indication that electronic coupling is weak. In other cases the extent of coupling is more difficult to assess since CT bands of interest may be obscured by absorptions of different origin.

3.2 IRREVERSIBLE PHOTOREDUCTION OF Co(III)

3.2.1 Aromatic Molecules as Electron Donors

In 1969 Adamson et al. observed a photoreduction of Co^{III} upon IL excitation of aqueous $[Co^{III}(NH_3)_5TSC]^{2+}$ with TSC^- = trans-4-stilbene-carboxylate (ref. 33). In a later study a detailed analysis of the photoredox products of this complex was carried out (ref. 34).

The stilbene moiety is an isolated chromophore of the complex since its absorption spectrum did not change upon coordination via the carboxylic group. Any bands which could be assigned to a CT transition from the stilbene group to Co^{III} do not appear.

The photolysis takes place according to the following reaction scheme (* excited state):

$$[Co^{III}(NH_3)_5(TSC^-)]^{2+} \xrightarrow{h\nu} [Co^{III}(NH_3)_5(TSC^-)^*]^{2+}$$

$$[Co^{III}(NH_3)_5(TSC^-)^*]^{2+} \rightarrow [Co^{II}(NH_3)_5(TSC^o)]^{2+}$$

$$[Co^{II}(NH_3)_5(TSC^o)]^{2+} \rightarrow Co^{2+} + 5NH_3 + TSC^o$$

$$[Co^{III}(NH_3)_5(TSC^-)]^{2+} + TSC^o \rightarrow Co^{2+} + 5NH_3 + TSC^- + \text{oxid. TSC}$$

$$TSC^o + O_2 \rightarrow \text{benzaldehyde + other products}$$

Upon light absorption at 313 nm the first excited $\pi\pi^*$ singlet of the TSC^- ligand is populated. This IL state is strongly reducing ($E_{1/2} \approx -2$ V vs SCE) while the Co^{III} center is a weak oxidant ($E_{1/2} = -0.03$ V) (ref. 35). Excited state electron transfer from the IL state of the stilbene group to Co^{III} has obviously a large driving force and is apparently very rapid. While the free ligand shows a strong fluorescence and undergoes a trans/cis photoisomerization these processes are not observed in the coordinated state. The electron transfer is thus much faster than other deactivation processes of the $\pi\pi^*$ singlet. In the original study (ref. 33) it was suggested that the excited IL state undergoes an energy transfer to a non-spectroscopic LMCT excited state of the complex. This explanation seems to be equivalent to an excited state electron transfer if the LMCT state can be described as a stilbene radical cation coordinated to Co^{II} by a carboxylic group (ref. 34).

The Co^{II} complex generated by excited state electron transfer is kinetically labile and decomposes before an efficient charge recombination takes place. A TSC radical is released and undergoes further reactions according to the scheme (ref. 34).

Interestingly, the lowest $\pi\pi^*$ triplet of the TSC^- ligand which can be populated by intermolecular energy transfer from biacetyl and other sensitizers induces only the trans/cis isomerization of the ligand but not the reduction of Co^{III} (ref. 34). The redox potential of the IL triplet is only $E_{1/2} \approx -0.3$ V vs SCE. Electron transfer is now apparently not fast enough to compete with other deactivation processes such as the photoisomerization.

A variety of other complexes of the type $[Co^{III}(NH_3)_5OOCR]^{2+}$ with R = aromatic group such as 1- and 2-naphthalene, 9-anthracene, 4-biphenyl shows qualitatively the same behavior as the TSC complex (refs. 37, 38). The quantum yields of Co^{2+} production was dependent very much on the nature of R. A simple correlation was not apparent.

The complexes $[\text{2-naphthyl-CONH-}(CH_2)_n\text{-COOCo}^{III}(NH_3)_5]^{2+}$ with n = 1-5 were studied in order to learn more about the structural requirements for excited state electron transfer (ref. 39). The naphthyl group as excited state donor and Co^{III} as acceptor are connected by a peptide linkage. In distinction to the complexes discussed above Co^{III} and the aromatic group are now not only separated by a carboxylic group but also by saturated and hence electronically insulating methylene groups. The basic observations were the same as those for the related complexes without intervening CH_2 groups. However, additional data were obtained since fluorescence quenching of the naphthyl moiety by Co^{III} was efficient but not complete. The observations can be described by the following reaction scheme (N = 2-naphthyl, B = peptide bridge):

$$[\text{N-B-Co}^{III}(NH_3)_5]^{2+} + h\nu \rightarrow [\text{N*-B-Co}^{III}(NH_3)_5]^{2+}$$

$$[\text{N*-B-Co}^{III}(NH_3)_5]^{2+} \xrightarrow{k_1} [\text{N-B-Co}^{III}(NH_3)_5]^{2+} + h\nu$$

$$[\text{N*-B-Co}^{III}(NH_3)_5]^{2+} \xrightarrow{k_2} [\text{N-B-Co}^{III}(NH_3)_5]^{2+} + \text{heat}$$

$$[N^*\text{-B-Co}^{III}(NH_3)_5]^{2+} \xrightarrow{k_3} [N^+\text{-B-Co}^{II}(NH_3)_5]^{2+}$$

$$[N^+\text{-B-Co}^{II}(NH_3)_5]^{2+} \xrightarrow{k_4} [N\text{-B-Co}^{III}(NH_3)_5]^{2+}$$

$$[N^+\text{-B-Co}^{II}(NH_3)_5]^{2+} \xrightarrow{k_5} Co^{2+} + 5\ NH_3 + \text{oxid. N}$$

The $\pi\pi^*$ singlet of the naphthyl group is populated by irradiation with λ = 313 nm. The excited singlet of the free ligand has a lifetime of approximately 10^{-8} s and undergoes an efficient fluorescence, which is strongly quenched in the complex due to electron transfer to Co^{III}. From the relative quantum yields of fluorescence the rate constants k_3 and quantum yields Φ_{ET} of excited state electron transfer were obtained. An increase from $k_3 = 4.9\times10^9\ s^{-1}$ and $\Phi_{ET} = 0.98$ for n = 1 to $9.2\times10^9\ s^{-1}$ and 0.99 for n = 4 was observed. At n = 5 k_3 and Φ_{ET} dropped to $6.0\times10^9\ s^{-1}$ and 0.98.

From Φ_{ET} and the experimental quantum yields of Co^{2+} formation relative rate constants for back electron transfer k_4' were calculated assuming that the rate of Co^{2+} formation k_5 is independent of n. It was found that the rate of back electron transfer reached also a maximum at n = 4.

On the basis of these results it is concluded that the actual distance of excited state and back electron transfer decreases with increasing chain length of the peptide from n = 1 to 4.

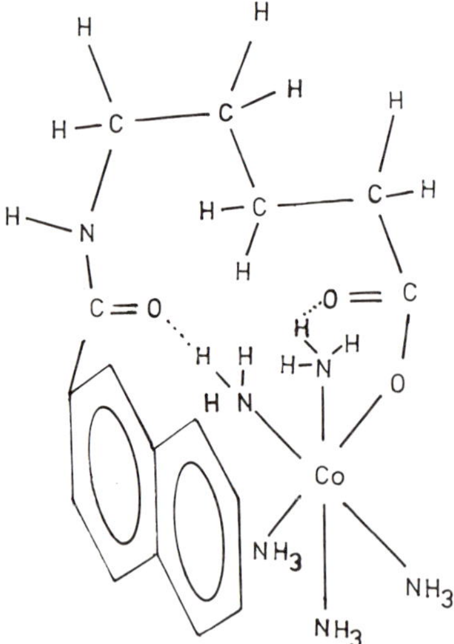

Fig. 3. Suggested structure of aqueous $[\text{2-naphthyl-CONH-}(CH_2)_4\text{-COOCo}^{III}\text{-}(NH_3)_5]^{2+}$

It is assumed that donor and acceptor come to a closer approach by an appropriate bending of the flexible peptide linkage. This back folding may be favored by hydrogen bonding between coordinated ammonia and carbonyl groups of the peptide (Fig. 3). At n = 5 electron transfer becomes less efficient. The donor-acceptor distance may now increase by an extension of the peptide.

Finally, it should be mentioned here that excited state electron transfer from an aromatic molecule to Co(III) ammines takes place also as an intermolecular reaction (refs. 38, 40). First observations were explained by the assumption that an energy transfer occurs to reactive LMCT states of the complex (ref. 40). However, more recent investigations have shown that the aromatic molecules are indeed oxidized and all observations can be explained best by an excited state electron transfer mechanism (ref. 38).

3.2.2 Metal Complexes as Electron Donors

Intramolecular excited state electron transfer between weakly coupled redox centers takes also place in complexes of the type

$[(NH_3)_5Co^{III}NH_2\text{-}(CH_2)_n\text{-}CH{=}CH_2Cu^{I}]^{4+}$

The Cu^{I} olefin π-complex is now an isolated chromophore which is characterized by an optical Cu^{I} to π*(olefin) MLCT transition (ref. 41). The interaction of this chromophore with Co is certainly weak due to the intervening saturated methylene groups. Light absorption by the Cu^{I} olefin chromophore is followed by the reduction of Co^{III} and oxidation of Cu^{I}. In aqueous solution stable redox products are formed according to the equation:

$$[(NH_3)_5Co^{III}NH_2\text{-}(CH_2)_n\text{-}CH{=}CH_2Cu^{I}]^{4+} \rightarrow Co^{2+} + Cu^{2+} + 5NH_3 + NH_2\text{-}(CH_2)_n\text{-}CH{=}CH_2$$

The MLCT state of the copper complex acts here as excited state electron donor. In the excited state it is the reduced olefin which undergoes electron transfer to Co^{III}:

$$\{(NH_3)_5Co^{III}\text{-}NH_2\text{-}(CH_2)_n\text{-}[(CH{=}CH_2^-)Cu^{II}]^*\}^{4+} \rightarrow [(NH_3)_5Co^{II}\text{-}NH_2\text{-}(CH_2)_n\text{-}CH{=}CH_2Cu^{II}]^{4+}$$

The electron transfer to Co^{III} is fast enough to compete with internal deactivation in the copper complex. The rate constant for electron transfer was estimated by time-resolved absorption spectroscopy to be $k > 10^7\text{-}10^8\ s^{-1}$. Finally, the kinetically labile Co^{II} complex undergoes a ligand displacement. The quantum yield of Co^{2+} formation decreases with an increasing number n of CH_2 groups. Beyond four to five methylene groups electron transfer is no longer competitive with internal deactivation within the Cu chromophore. The actual distance between donor and acceptor seems here to increase with the length of the connecting CH_2 chain. At a fully extended chain a maximum distance of about 7 Å for electron transfer was estimated.

Similar observations were made with binuclear Co^{III}/Cu^{I} complexes which contain a pyridine or carboxylic group instead of an amine coordinated to Co^{III} (ref. 41).

Intramolecular excited state electron transfer from a metal complex as donor to Co^{III} as acceptor occurs also in the neutral complex:

$[(NC)_5Fe^{II}pyrazineCo^{III}(NH_3)_5]$

This binuclear compound is composed of the chromophore $[Fe^{II}(CN)_5pyrazine]^{3-}$ and $[Co^{III}(NH_3)_5pyrazine]^{3+}$ (ref. 42). The metal centers do not seem to be strongly coupled. The iron containing chromophore is characterized by an optical Fe^{II} to π*(pyrazine) MLCT transition (λ_{max} = 630 nm).

Upon light absorption into this CT band an efficient reduction of Co^{III} took place according to the equation:

$$[(NC)_5Fe^{II}pyrazineCo^{III}(NH_3)_5] \xrightarrow{h\nu} [Fe^{III}(CN)_5pyrazine]^{2-} + Co^{2+} + 5NH_3$$

The processes following light absorption should be quite analogous to those of the Co^{III}/Cu^{I} complex described above. The MLCT excited state of the iron chromophore transfers an electron from the reduced pyrazine bridge to Co^{III} before it undergoes an internal deactivation.

Interestingly, the complex $[(NC)_5Fe^{III}pyrazineCo^{II}(NH_3)_5]$ which is the primary product of the electron transfer is a Fe^{II} to Co^{III} MMCT state or redox isomer of the starting complex. This MMCT state lies certainly well below the Fe to pyrazine MLCT state. The excited state electron transfer can then be also considered as an energy transfer from a MLCT to a MMCT excited state. In the absorption spectrum of the binuclear Fe^{II}/Co^{III} complex a MMCT band was not identified. However, it may be obscured by the MLCT absorption. It is also possible that a MMCT band appears in the near IR region which may have not been investigated. The expectation that a MMCT state occurs at rather low energies is supported by another observation. This MMCT state can be populated even thermally. The redox decomposition of the binuclear complex does not take place only as a photoreaction but also as a thermal process (ref. 42). Analogous results were obtained with binuclear Fe^{II}/Co^{III} complexes which contain bridging ligands similar to pyrazine (ref. 42).

3.3 Reversible Excited State Electron Transfer

3.3.1 Introduction

Intramolecular excited state electron transfer between weakly coupled redox centers described in the previous section results in a permanent chemical change. In addition, there are many systems which undergo an efficient back electron transfer regenerating the starting compound. These materials appear to be not light-sensitive. However, by means of time-resolved absorption or emission spectroscopy it is possible to demonstrate the existence of short-lived intermediates. Many of such studies were carried out in order to get more information on the charge separation process. It was assumed that back electron transfer could be retarded if the electron/hole pair can be separated over increased distances. However, this may not always be true. In suitable cases through-bond interaction seems to provide some coupling of donor and acceptor which may facilitate an efficient charge recombination over larger distances. This assumption is supported by recent observations that optical CT transitions occur also between remote redox centers. The spectrum of $[(NH_3)_5Ru^{II}pyrazineRu^{II}(NH_3)_4pyrazineRu^{III}(NH_3)_5]^{7+}$ displays a near-IR absorption which was assigned to an end-to-end MMCT transition (ref. 43). An absorption band of the ion $[(NH_3)_5Ru^{II}NCRu^{II}$-$(bipy)_2CNRu^{II}(NH_3)_4]^{4+}$ was identified as MLCT transition from Ru^{II} coordinated to ammonia to bipy of the adjacent ruthenium atom (ref. 44). As a further example M^{II} to bipy MLCT bands were detected in the spectra of $[(bipy)(H_2O)Pt^{II}NCM^{II}(CN)_4CNPt^{II}(H_2O)(bipy)]$ with M = Fe, Ru, and Os (ref. 45).

In the following discussion of reversible intramolecular excited state electron transfer individual complexes are presented. They are classified according to the donor and acceptor site of excited state electron transfer following directly optical excitation of one chromophore. Evidence of this primary excited electron transfer was obtained by the identification of subsequent slower processes.

3.3.2 Ligand to Metal Electron Transfer

In the binuclear ion $[(NH_3)_5Ru^{II}pyrazineCu^{II}(aq)]^{4+}$ the optical MLCT transition generates the excited state $[(NH_3)_5Ru^{III}(pyrazine^-)Cu^{II}]^{4+*}$ (ref. 46). This is followed by a rapid electron transfer from the reduced bridging ligand to Cu^{II} producing $[(NH_3)_5Ru^{III}pyrazineCu^{I}]^{4+}$ which undergoes regeneration of the starting complex. The metal-to-metal back electron transfer takes place with a first order rate constant of $k = 7.8x10^3\ s^{-1}$.

The mixed-valence compound $[(NH_3)_5Ru^{II}pyrazineRu^{III}(edta)]^+$ can be excited to $[(NH_3)_5Ru^{III}(pyrazine^-)Ru^{III}(edta)]^{+*}$ by light absorption into the corresponding MLCT band (ref. 47). In this excited state electron transfer takes place from the reduced pyrazine ligand to Ru^{III} coordinated to edta with $k > 10^{11}\ s^{-1}$. The complex $[(NH_3)_5Ru^{III}pyrazineRu^{II}(edta)]^+$ which is a high-energy redox isomer of the ground state or a MMCT excited state regenerates the starting complex by metal-to-metal electron transfer ($k = 0.8x10^{10}$ s^{-1}). Similar results were obtained with related mixed-valence compounds (ref. 47).

The complex $[(bipy)Re^{I}(CO)_3(py\text{-}PTZ)]^+$ contains a coordinated pyridine (py) which is linked at its 4 position to the reducing phenothiazine (PTZ) molecule via an insulating methylene group (ref. 48). An optical MLCT transition leads to the excited state $[(bipy^-)Re^{II}(CO)_3(py\text{-}PTZ)]^{+*}$ which undergoes rapid excited state electron transfer from PTZ to the oxidized Re center. In contrast to the previous examples this electron transfer is not an inner-sphere process since the donor site is isolated by the intervening CH_2 group. The result of this excited state electron transfer is a LLCT excited state with the electron distribution $[(bipy^-)Re^{I}(CO)_3(py\text{-}PTZ^+)]^{+*}$. It reverts to the ground state with $k = 4x10^8$ s^{-1}.

3.3.3 **Metal to Metal Electron Transfer**

The binuclear ion $[(bipy)_2(CO)Os^{II}LOs^{II}(o\text{-}phen)(dppe)Cl]^{3+}$ with L = 4,4'-bipy and dppe = cis-$Ph_2PCH{=}CHPPh_2$ undergoes optical MLCT excitation to form $[(bipy)(bipy^-)(CO)Os^{III}LOs^{II}(o\text{-}phen)(dppe)Cl]^{3+*}$ which is a mixed-valence compound existing only in the excited state (ref. 49). The optical transition is followed by an excited state metal-to-metal electron transfer from Os^{II} to Os^{III} with $k \approx 10^7$ s^{-1} yielding $[(bipy)(bipy^-)(CO)\text{-}Os^{II}LOs^{III}(o\text{-}phen)(dppe)Cl]^{3+}$. Subsequent electron transfer from $bipy^-$ to the remote Os^{III} regenerates the stable ground state.

In the heterobimetallic cation $[(bipy)_2Ru^{II}bipymRe^{I}(CO)_3Cl]^{2+}$ with bipym = 2.2'-bipyrimidine an optical MLCT transition terminates in the excited state $[(bipy)_2Ru^{II}(bipym^-)Re^{II}(CO)_3Cl]^{2+*}$ (ref. 50). This MLCT state undergoes metal-to-metal electron transfer from Ru^{II} to the oxidized rhenium to form $[(bipy)_2Ru^{III}(bipym^-)Re^{I}(CO)_3Cl]^{2+*}$. The production of this MLCT excited state was detected by its emission to the ground state.

3.3.4 **Ligand to Ligand Electron Transfer**

The complexes $[(bipy)_2Ru^{II}(MeQ^+)_2]^{4+}$ (ref. 51) and $[(bipy)Re^{I}\text{-}(CO)_3(MeQ^+)]^{2+}$ (ref. 52) contain the oxidizing ligand NCH_3-4.4'-bipyridinium cation (MeQ^+) which is weakly coupled to the metals. The optical transition to the MLCT state involving the promotion of an electron from the metal to the bipy ligand is followed by the reverse process as emission but only at low temperatures in a rigid matrix. In fluid solution at room temperature the MLCT states $[(bipy)(bipy^-)Ru^{III}(MeQ^+)_2]^{4+*}$ and $[(bipy^-)Re^{II}(CO)_3(MeQ^+)]^{2+*}$ undergo excited state electron transfer from the reduced $bipy^-$ to the oxidizing MeQ^+ ligand. This process generates the MLCT states $[(bipy)_2Ru^{III}(MeQ^0)(MeQ^+)]^{4+*}$ and $[(bipy)Re^{II}(CO)_3\text{-}(MeQ^\circ)]^{2+*}$ which were identified by their emission spectra. The excited state ligand-to-ligand electron transfer requires apparently an orientational mobility of the MeQ^+ ligand which is possible only in fluid solution. It was suggested that electron transfer can occur if both aromatic rings of MeQ^+ assume a coplanar arrangement. This assumption is supported by the observation that the complex $[(bipy)_2Os^{II}(CO)(N,3,3'\text{-trimethyl-4,4'-bipyridinium}^+)]^{3+}$ does not undergo this excited state ligand-to-ligand electron transfer (ref. 52). Due to a steric repulsion imposed by the methyl substituents in the 3 and 3' position of the MeQ^+ ligand a coplanar position of both rings cannot be adopted.

An excited state ligand-to-ligand electron transfer was also observed for the mixed-valence compound $[(dpte)_2ClRu^{II}LRu^{III}(bipy)_2Cl]^{3+}$ with dpte = 1,2-diphenylthioethane and L = 4,4'-bipyridine (refs. 53, 54). The optical transition from Ru^{II} to the bridging ligand L yields the MLCT state $[(dpte)_2ClRu^{III}(L^-)Ru^{III}(bipy)_2Cl]^{3+*}$. Excited state electron transfer generates another MLCT state $[(dpte)_2Ru^{III}LRu^{III}(bipy^-)(bipy)Cl]^{3+*}$. This state which cannot be reached by an optical transition from the ground state was detected by its luminescence. The emission leads to a species with the electron distribution $[(dpte)_2ClRu^{III}LRu^{II}(bipy)_2Cl]^{3+}$. This is a MMCT excited state which is rapidly deactivated to the ground state. Similar results were obtained with analogous complexes containing other bridging ligands related to 4,4'-bipy.

CONCLUSION

A light-induced intramolecular electron transfer between weakly coupled redox centers involving transition metal complexes can take place by direct optical charge transfer excitation (resonance mechanism) as well as by excited state electron transfer (non-resonance mechanism). The primary charge-separated state undergoes competing back electron transfer and secondary transformations. The efficiency of the formation of stable photoproducts is determined by this competition. It depends on various factors such as the reorganizational energy and the extent of electronic coupling between electron donor and acceptor. More studies are needed to obtain a detailed picture. Structural requirements such as the distance and the angle between donor and acceptor but also the nature of the connecting bridge play an important role which is not yet completely understood. As a long-term goal an efficient charge separation which can be utilized in an artificial photosynthesis and other applications seems to be feasible.

REFERENCES

1 A. W. Adamson and P. D. Fleischauer (Eds.), Concepts of Inorganic Photochemistry, Wiley, New York, 1975.
2 V. Balzani and V. Carassiti, Photochemistry of Coordination Compounds, Academic Press, New York, 1970.
3 G. L. Geoffroy and M. S. Wrighton, Organometallic Photochemistry, Academic Press, New York, 1979.
4 A. B. P. Lever, Inorganic Electronic Spectroscopy, Elsevier, Amsterdam, 1984.
5 W. Evans and J. I. Zink, J. Am. Chem. Soc., 103 (1981) 2635.
6 A. Vogler, J. Kisslinger and W. R. Roper, Z. Naturforsch., 38b (1983) 1506.
7 A. Vogler and H. Kunkely, J. Am. Chem. Soc., 103 (1981) 1559.
8 A. Vogler, H. Kunkely, J. Hlavatsch and A. Merz, Inorg. Chem., 23 (1984) 506 and ref. cited therein.
9 G. A. Crosby, R. G. Highland and K. A. Truesdell, Coord. Chem. Rev., 64 (1985) 41.
10 M. Haga, E. S. Dodsworth and A. B. P. Lever, Inorg. Chem., 25 (1986) 447.
11 G. A. Heath, L. J. Yellowlees and P. S. Braterman, Chem. Phys. Lett., 92 (1982) 646.
12 A. Vogler and H. Kunkely, Inorg. Chim. Acta, 54 (1981) L273.
13 H. Taube, Ann. N. Y. Acad. Sci., 313 (1978) 483.
14 T. J. Meyer, Acc. Chem. Res., 11 (1978) 94.
15 T. J. Meyer, Ann. N. Y. Acad. Sci., 313 (1978) 496.
16 D. Brown (Ed.), Mixed-Valence Compounds, Reidel, Dordrecht, 1980.
17 C. Creutz, Prog. Inorg. Chem., 30 (1980) 1.
18 N. S. Hush, Progr. Inorg. Chem., 8 (1967) 391.
19 A. Vogler, A. H. Osman and H. Kunkely, Coord. Chem. Rev., 64 (1985) 159.
20 N. S. Hush, Trans. Faraday Soc., 57 (1961) 557.
21 N. S. Hush, Electrochim. Acta, 13 (1968) 1005.
22 A. Vogler and H. Kunkely, Ber. Bunsenges. Phys. Chem., 79 (1975) 83.
23 A. Vogler and H. Kunkely, Ber. Bunsenges. Phys. Chem., 79 (1975) 301.
24 G. Milazzo and S. Caroli, Tables of Standard Electrode Potentials, Wiley, New York, 1978.
25 D. Gaswick and A. Haim, J. Am. Chem. Soc., 93 (1971) 7347.
26 A. Vogler, A. H. Osman and H. Kunkely, Inorg. Chem., 26 (1987) 2337.
27 J. C. Curtis and T. J. Meyer, Inorg. Chem., 21 (1982) 1562.
28 A. Haim and W. K. Wilmarth, J. Am. Chem. Soc., 83 (1961) 509.
29 V. Balzani, F. Bolletta, M. T. Gandolfi and M. Maestri, Top. Curr. Chem., 75 (1978) 1.
30 N. Sutin and C. Creutz, Pure Appl. Chem., 52 (1980) 2717.
31 K. Kalyanasundaram, Coord. Chem. Rev., 46 (1982) 159.
32 D. Gust, T. A. Moore, P. A. Liddell, G. A. Nemeth, L. R. Makings, A. L. Moore, D. Barrett, P. J. Pessiki, R. V. Bensasson, M. Rougée, C. Chachaty, F. C. De Schryver, M. Van der Auweraer, A. R. Holzwarth and J. S. Connolly, J. Am. Chem. Soc., 109 (1987) 846 and ref. cited therein.
33 A. W. Adamson, A. Vogler and I. Lantzke, J. Phys. Chem., 73 (1969) 4183.
34 A. Vogler and A. Kern, Z. Naturforsch., 34b (1979) 271.
35 The redox potential of Co^{III} in $[Co(NH_3)_5TSC]^{2+}$ is not known but the potential of the couple $[Co(NH_3)_5H_2O]^{3+/2+}$ (ref. 36) is taken as an approximate value.
36 R. C. Henney, H. F. Holtzclaw and R. C. Larson, J. Electroanal. Chem. Interfac. Chem., 14 (1967) 435.
37 A. Kern, Dissertation, Universität Regensburg, 1978.
38 S. Schäffl, Diplomarbeit, Universität Regensburg, 1984.
39 A. H. Osman, Dissertation, Universität Regensburg, 1987.
40 M. A. Scandola, F. Scandola and V. Carassiti, Mol. Photochem., 1 (1969) 403.
41 K. A. Norton and J. K. Hurst, J. Am. Chem. Soc., 104 (1982) 5960.
42 J. M. Malin, D. A. Ryan and T. V. O'Hallotan, J. Am. Chem. Soc., 100

(1978) 2097.
43 A. von Kameke, G. M. Tom and H. Taube, Inorg. Chem., 17 (1978) 1790.
44 C. A. Bignozzi, S. Roffia and F. Scandola, J. Am. Chem. Soc., 107 (1985) 1644.
45 A. Vogler and H. Kunkely, unpublished results.
46 V. A. Durante and P. C. Ford, J. Am. Chem. Soc., 97 (1975) 6898.
47 C. Creutz, P. Kroger, T. Matsubara, T. L. Netzel and N. Sutin, J. Am. Chem. Soc., 101 (1979) 5442.
48 T. D. Westmoreland, K. S. Schanze, P. E. Neveux, Jr., E. Danielson, B. P. Sullivan, P. Chen and T. J. Meyer, Inorg. Chem., 24 (1985) 2596.
49 K. S. Schanze and T. J. Meyer, Inorg. Chem., 24 (1985) 2121.
50 A. Vogler and J. Kisslinger, Inorg. Chim. Acta, 115 (1986) 193.
51 B. P. Sullivan, H. Abruna, H. O. Finklea, D. J. Salmon, J. K. Nagle, T. J. Meyer and H. Sprintschnik, Chem. Phys. Lett., 58 (1978) 389.
52 T. D. Westmoreland, H. Le Bozec, R. W. Murray and T. J. Meyer, J. Am. Chem. Soc., 105 (1983) 5952.
53 J. C. Curtis, J. S. Bernstein, R. H. Schmehl and T. J. Meyer, Chem. Phys. Lett., 81 (1981) 48.
54 J. C. Curtis, J. S. Bernstein and T. J. Meyer, Inorg. Chem., 24 (1985) 385.

Chapter 5.6

Photoinduced Electron Transfer in Organometallic Transition Metal Complexes

C. Giannotti, S. Gaspard and P. Krausz

1 - INTRODUCTION

In this chapter we focus our attention on complexes having at least one metal-carbon bond (M-C) or single metal-metal bond (M-M), where M is a transition metal with partly filled d or f shells.

When, after light excitation of a complex, eqn (1), a photoinduced electron transfer (ET) reaction occurs with either reductive, eqn (3), or oxidative, eqn (4), quenching of the excited state, three expected mechanisms can be schematized as follows : (A is the excited compound, B the quencher) energy transfer, eqn (2) ; reductive ET, eqn (3) ; and oxidative ET, eqn (4).

$$A + B \xrightarrow{h\nu} A^{*} + B \text{ excitation} \tag{1}$$

$$A^{*} + B \longrightarrow A + B^{*} \text{ energy transfer} \tag{2}$$

$$A^{*} + B \xrightarrow{k_3} A^{-} + B^{+} \text{ reductive ET} \tag{3}$$

$$A^{*} + B \xrightarrow{k_4} A^{+} + B^{-} \text{ oxidative ET} \tag{4}$$

While these processes can occur with very high rates and efficiencies, they are generally followed by an energy wasting back reaction, eqns (5) and (6).

$$A^{-} + B^{+} \xrightarrow{k_5} A + B \tag{5}$$

$$A^{+} + B^{-} \xrightarrow{k_6} A + B \tag{6}$$

Reductive or oxidative quenching of the excited state follows the Taube classification (ref 1) into outer or inner sphere mechanisms. We discuss in this chapter several sections covering the solvated electron, electron transfer to or from solvents, electron transfer between an organometallic complex and a quencher and between a coordination transition metal complex and an organometallic one, and in ion pair systems.

2 - Electron transfer from an organometallic complex to a scavenger with the formation of a solvated electron

Irradiation of aqueous solutions containing simple anions can lead to formation of a hydrated electron (refs 2-5). In such reactions, the photolysis of organometallic complex produces electrons which are trapped in the medium by various traps, such as N_2O, so that the quantum yield of trapped electron $\Phi e_{aq} = \Phi_{N2}$ can be obtained.

The earliest reports have been made by Shirom and Stein (refs 6,7) and by Dainton et al. (refs 8,9). Nitrous oxide has been used as a specific scavenger of the aquated electron produced by photodetachement from $[Fe(CN)_6]^{4-}$ using light of wavelength 253.7 nm, eqns (7,9). The quantum yield of reaction, eqn (9), is 0.48.

$$[Fe(CN)_6]^{4-} + N_2O \xrightarrow{h\nu} [Fe(CN)_6]^{3-} + N_2 + O^- \; (^{\bullet}OH + OH^-) \tag{7}$$

$$N_2O + e^-_{aq} \longrightarrow N_2 + O^- \tag{8}$$

$$^{\bullet}OH + [Fe(CN)_6]^{4-} \longrightarrow OH^- + [Fe(CN)_6]^{3-} \tag{9}$$

Φ_{N2} increases to a limiting value (0.66) as concentration of N_2O is increased. If isopropanol is used instead of N_2O, no gas is detected indicating that no H atoms are formed in the primary act. The following reaction, corresponding to aquation and ejection of one cyanide ligand, is also occurring, eqn (10).

$$[Fe(CN)_6]^{4-} + H_2O \xrightarrow{h\nu} [Fe(CN)_5 H_2O]^{3-} + CN^- \tag{10}$$

If $[Fe(CN)_6]^{4-}$ and N_2O are irradiated in presence of KCN, Φ_{N2} is not affected but the quantum yield of formation of $[Fe(CN)_6]^{3-}$ decreases considerably. Nitrogen gas can also be evolved, if acidified solutions of the $La^{3+}[Fe(CN)_6]^{4-}$ ion pair are irradiated, the first step is shown in eqn (11).

$$La^{3+}[Fe(CN)_6]^{4-} \xrightarrow{h\nu} La^{3+}[Fe(CN)_6]^{3-} + e^-_{aq} \qquad (11)$$

Ohno (refs 10-14) found that irradiation of hexacyanoferrate $[Fe(CN)_6]^{4-}$ with 253.7nm wavelength light in the presence of specific scavengers such as N_2O, H_3O^+, NO_3^- or acetone for the hydrated electron or with CH_3OH and C_2H_5OH for the hydrogen atom gave a photo-oxidation reaction leading to hexacyanoferrate $[Fe(CN)_6]^{3-}$, eqn (12).

$$[Fe(CN)_6]^{4-} \xrightarrow{h\nu} [Fe(CN)_6]^{3-} + e^-\,aq. \qquad (12)$$

The effect of temperature, the addition of salts, the use of D_2O as solvent, and the size of the cavity in which the reaction occurs have been investigated.

The primary act of light absorption at 253.7 nm which is responsible for the formation of a hydrated electron is supposed to be an internal excitation attributed to a charge transfer of electrons from the metal (d_{t2g}) to the ligand (CN-π antibonding t_{1u} orbital) (ref 15). This excited state could also eject an electron to the solvent, leading to the formation of a hydrated electron. The efficiency of this process may be determined by the size of the cavity occupied by the ion from which the electron is transferred to the solvent.

Photoaquation of $[Fe(CN)_6]^{4-}$ is known to occur, (refs.16-18) as reported previously, (refs 6-8), eqns (10-12), but it is established that this reaction and photoelectron production do proceed from different excited states. Irradiation in the d-d absorption bands leads primarily to aquation and irradiation in the region of the first charge transfer band gives photoelectron ejection (refs 15,17).

Flashing deaerated aqueous solutions of $[Fe(CN)_6]^{4-}$, $[Mo(CN)_8]^{4-}$, $[W(CN)_8]^{4-}$ and $[Ru(CN)_6]^{4-}$ (refs 19,20) gives a transient which lives for about 40 μs and was identified as the hydrated electron, following the general eqn (13).

$$MLn^{X} \xrightarrow{h\nu} MLn^{X+1} + e^-_{aq.} \qquad (13)$$

Steady state illumination experiments have been made, using also H_2O as electron scavenger, eqn (14).

$$2MLn^{X} + H_2O + N_2O \longrightarrow 2MLn^{X+1} + 2OH^{\cdot} + N_2 \qquad (14)$$

The OH˙ radicals recombine to give H_2O_2 which could be reoxidized by MLn^{x+1} to give oxygen, as in the following equation:

$$2MLn^{x+1} + H_2O_2 \longrightarrow 2H^+ + O_2 + 2MLn^x$$

The quantum yields of e^-_{aq} formation for $[Fe(CN)_6]^{4-}$, $[Mo(CN)_8]^{4-}$, $[W(CN)_8]^{4-}$ and $[Ru(CN)_6]^{4-}$ are, respectively 0.66, 0.28, 0.34 and 0.36, using 253.7 nm. incident radiation (Table 1). In the case of $[Mo(CN)_8]^{4-}$, $\Phi\, e^-_{aq}$ drops to nearly zero when the irradiation is done at wavelengths greater than 300 nm. Qualitative product analysis suggests that in addition to the oxidized complex, which exhibits the same stoichiometry as the starting material, there is also production of a partly aquated species in the same or next higher oxidation state as that of the starting material.

In 1977 Kalisky and Shiron (ref 21) have studied again these reactions. The limiting quantum yield $\Phi\, e^-_{aq}$ reaches a value of 0.50 at 228.8 nm for $[Ru(CN)_6]^{4-}$ and a value of 0.46 at 228.8 nm for $[W(CN)_8]^{4-}$ ion . The limiting quantum yields decrease with wavelength : 0.076 at 313nm for $[Ru(CN)_6]^{4-}$ and 0.093 at 313 nm for $[W(CN)_8]^{4-}$. Above 313 nm, the formation of the hydrated electron is significant. At 365 nm, the quantum yield is around zero for both ions (Table 1).

Finally, in the illumination of the four cyano complexes $[Fe(CN)_6]^{4-}$, $[Mo(CN)_8]^{4-}$, $[Ru(CN)_6]^{4-}$ and $[W(CN)_8]^{4-}$, electron formation occurs after irradiation in the charge transfer -to- solvent (CTTS) transition or into the charge transfer band (CT).

The charge transfer -to- solvent (CTTS) state can be reached either directly or by internal conversion from the CT excited state but not through excitation into the lower d-d band. In some cases, CT excited states may produce an electron, eqn (15),

$$[M(CN)_x]^{4-} \xrightarrow{h\nu} [M(CN)_x]^{4*-} \xrightarrow[\text{CT/or CTTS}]{} [M(CN)_x]^{3-} + e^-_{aq} \qquad (15)$$

$$[M(CN)_x]^{4-} \xrightarrow[\text{LF}]{h\nu} [M(CN)_x]^{*4-} \longrightarrow \text{photoaquation products} \qquad (16)$$

while excitation into the ligand field band (LF) gives photoaquation, eqn (16).

In 1971 Shirom and Stein, (refs 22-24) using laser pulses at 337 nm from a N_2 gas laser, reinvestigated the photolysis of $[Fe(CN)_6]^{4-}$. These authors suggest the following mechanism : excitation of $[Fe(CN)_6]^{4-}$ at wavelengths below 313 nm gives two excited states ($^1A_{1g} \rightarrow {}^1T_{2g}$ at 270 nm and $^1A_{1g} \rightarrow {}^1T_{1u}$ at

218 nm) which permit passage into the CTTS state and compete with decay into the lowest state level ($^1T_{2g}$).

$$Fe(CN)_6^{4-*}\ (^1T_{1u}\ \text{or}\ ^1T_{2g}) \begin{cases} \nearrow [Fe(CN)_6]^{4*-}\ (^1T_{1g}) \xrightarrow{H_2O} [Fe(CN)_5\ H_2O]^{3-} + CN^- & (17) \\ \searrow [Fe(CN)_6]^{4-*}\ (CTTS) \longrightarrow [Fe(CN)_6]^{3-} + e^-_{aq} & (18) \end{cases}$$

Shirom and Tomkiewicz (ref 25) report the quantum yield of photoejected electrons by using 253.7 nm UV light and N_2O as electron scavenger in different water-alcohol mixtures. The results are reported in Table 1. The escape probability of the ejected electrons from geminate recombination in various water-alcohol mixtures cannot be explained on the basis of dielectric constant, relaxation time or diffusion coefficient alone. The Mozunder theory (ref 26) provides a reasonable approach if sufficient care is taken in choosing the relaxation times. The thermalization length of the ejected electron was found to be 30 A° and seems to be independent of the solvent used and dependent upon the solute and the excitation wavelength.

Illumination of ferrocyanide in 10 M NaOH glasses at 77°K has been carried out at 228.8, 253.7 and 313 nm (ref 27). Acrylamide has been used for total scavenging of the electrons. The relative quantum yield of electron formation at the various wavelengths has been determined from the intensities of the ESR signals of the trapped electrons and the acrylamide radical formed.

The dependence of the quantum yield for electron formation on the excitation wavelength appears to be much steeper in a solid matrix than in aqueous solutions (Table 1) . The absence of translational and rotational motion and the hindrance of electron diffusion in alkaline ices at 77°K could explain the steeper dependence of the electron quantum yield on wavelength in alkaline ices as compared to aqueous solutions.

More recently, Horvath and Papp (ref 28) also reported their results concerning the formation of solvated electrons after photolysis of $[Fe(CN)_6]^{4-}$ ion in aqueous solution at 253.7 nm. They compare their results to the one they get for $[Fe(CN)_5\ PBu_3]^{3-}$. Steady illumination experiments have been made using NO_3^- and IO_3^- ions as electron scavengers to obtain quantum yields: for pentacyano complex, $\Phi\ e^-_{aq}$ is lower (0.4) than for a ferrocyanide $\Phi\ e^-_{aq}$ = 0.66. These quantum yields are dependent on the concentration of NO_3^- and IO_3^- and also the on concentration of added acetonitrile (Table 1).

To conclude, these numerous studies on the photooxidation of hexacyanoferrate, have shown that the quantum yield for the aquated electron

TABLE 1

Quantum yields $\Phi_{e^-_{aq}}$ for formation of the hydrated electron under different conditions, upon excitation of cyanide complexes

Complex	Solvent	Wavelength/nm	Scavenger	$\Phi_{e^-_{aq}}$	Ref.
$[Fe(CN)_6]^{4-}$	H_2O	253.7	N_2O	0.66	8,28
	"	"	N_2O	0.35	11
	"	313		0.1	22
	"	265		0.4	6
	"	254		0.65	22,6
	"	228		0.89	22
	"	214		0.89	22
	H_2O pH 3.01	253.7	N_2O	0.069	8
	2.52	"	N_2O	0.071	8
	2.02	"	N_2O	0.077	8
	H_2O/KCN	254	NO_3^- a	0.66-0.56	28
	H_2O/KCN	"	IO_3^- a	0.68-0.58	28
	H_2O	"	NO_3^- a	0.66-0.55	28
	H_2O/MeOH[a]	253.7	N_2O	0.66-0.29	25
	H_2O/C_2H_5OH[a]		N_2O	0.52-0.26	25
	H_2O/Ethylene glycol a		N_2O	0.47-0.14	27
	NaOH/glasses 77°K	228.8-253.7	acrylamide	6.13	27
	" "	253.7-313	"	38	27
	" "	228.8-313	"	239	27
$La^{+3}[Fe(CN)_6]^{4-}$	H_2O	253.7	N_2O	0.48	8
$[Fe(CN)_4(PBu_3)]^{3-}$	H_2O/KCN	253.7	No_3^- a	0.44-0.34	28
	H_2O/KCN	"	IO_3^- a	0.40-0.35	28
	H_2O/KCN	"	NO_3^- a	0.38-0.30	28
	H_2O/KCN	253.7	IO_3^-	0.36-0.025	28
$[W(CN)_8]^{4-}$	H_2O	228,8	N_2O	0.46	19
	"	253,7	"	0.34	19,21
	"	265	"	0.38	21
	"	313	"	0.093	21
	"	365	"	0.011	21
$[Ru(CN)_6]^{4-}$	H_2O	228.8	N_2O	0.50	21
	"	253.7	"	0.36	19,21
	"	265	"	0.37	21
	"	313	"	0.076	21
	"	365	"	0.013	21
$[Mo(CN)_8]^{4-}$	H_2O	254	N_2O	0.28	19,20
	"	>300	"	0.01	19,20

(a: for different concentration of the concerned compound)

has been determined in the presence of various electron scavengers by continuous irradiation or by flash techniques. Ainey and Dainton (ref 8), Waltz and Adamson (ref 19), and Shirom and Stein (ref 22,23) measured Φe^-_{aq} = 0.66 as limiting value using N_2O and 253.7 nm as excitation source, although, Ohno et al. (refs 10-14) obtain a different value. Φe^-_{aq} has been found to depend on the wavelength and the nature of the irradiation source (refs 22,23).

There is agreement among authors on the necessity of charge transfer (CT) excitation for photoredox processes, but there is desagreement on the nature of these CT excitations. The two active states considered are: the $^1T_{1u}$ state, excited by charge transfer from the metal to the ligand (CTTL) or the level exited by charge transfer from the central atom to the solvent (CTTS). This latter state could be formed directly from the $^1A_{1g}$ ground state or by internal conversion from the $^1T_{1u}$ and $^1T_{2g}$ states of higher energies.

It has been found probable that geminate recombination according to the modified radical pair model (refs 16, 29-32) is negligible (ref 33) but could not be excluded, (refs 12,13). Good agreement between the quantum yields calculated from the model and from experimental data determined in various solvents has been obtained (ref 25).

In halogenated solvents such as CCl_4 or $CHCl_3$, ferrocene shows an additional band at 307 nm assigned by Brand and Snedden (ref 34) to a dissociative intermolecular charge transfer. Traverso and Scandola (ref 35) have studied the photochemistry of solutions of ferrocene in a mixture CCl_4-ethanol and found that at wavelengths of 313 to 404 nm the efficiency of formation of ferricenium ion corresponds to a quantum yield of about one. When ferrocene is irradiated in ethanol or cyclohexane with 253.7 nm radiation in presence of N_2O, N_2 is obtained at a rate which is a continually increasing function of N_2O concentration. At wavelengths above 313 nm, no N_2 is formed (ref 36). A study of the influence of alkyl substituents on the quantum yield of electron transfer on H_2O (Table 2) has been made : acetyl (0.0036), phenyl (0.0050), méthyl (0.027), 1,1'-dimethyl (0.0137), 1,2-dimethyl (0.0137) and 1,1-di-t-butyl-ferrocene (0.0160). The best quantum yield is obtained for very bulky substituents, i.e 1,1-di-t-butylferrocene (Table 2). The quantum yield of this photoexcited species is found to be very much less than unity : it is deduced that it is not formed from the primary photoexcited state.

Under UV excitation at 240 nm in aqueous solutions, in the presence of N_2O as electron scavenger, ferrocenylacetate $(\pi^5C_5H_5Fe)_2$ = [Fe], [Fe] - $(CH_2)_nCO_2^-$ (n=2-4) and (4-([13] ferrocenophenyl)) butanoic acid derivatives (refs 37,38) are excited and undergo photooxidation to zwitterions _1_, _2_.

TABLE 2

Quantum yields $\Phi_{e^-_{aq}}$ of hydrated electron formed by excitation of several ferrocene derivatives (ref 36), in EtOH at 253.7 nm in presence of N_2O.

COMPLEX	Φ_{e^-}
Acetylferrocene	0.0036
Phenylferrocene	0.0050
ferrocene	0.0102
Methylferrocene	0.0127
1,1' Dimethyl-ferrocene	0.0137
1,2 Dimethyl-ferrocene	0.0137
1,1'-Di-t-butyl ferrocene	0.048

$[Fe]^+[CH_2]_nCO_2^-$

1

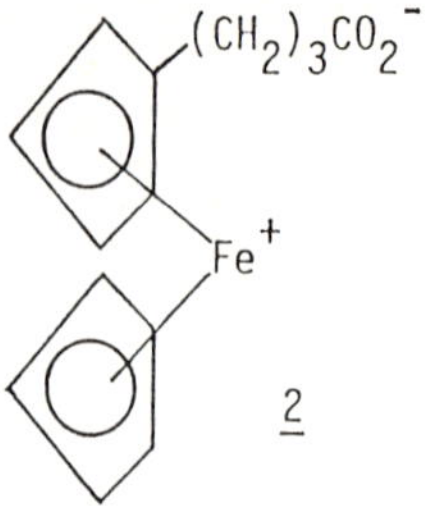

$(CH_2)_3CO_2^-$ … Fe^+

2

Addition of alcohol decreases the yield of the zwitterions by three competitive processes : a) first order decay of the ground electronic state; b) reaction with nitrous oxide to give the zwitterion ; and quenching by ROH. The lifetimes of these excited states are found to be 2μs for ferrocenyl acetate ion and 0.074 μs for ferrocenoate ion. The dipolar ion obtained by electron transfer from ferrocenylacetate is unstable, especially at high pH.

3 - Electron transfer from or to the solvent

3.1 - Electron transfer from the solvent to the organometallic complex

Hockins and Bernal (ref 39) report that photoirradiation of disodium or tetrabutylammonium pentacyanonitrosylferrate (nitroprusside) $[Fe(CN)_5NO]^{2-}$ in a variety of donor solvents such as N-N dimethylformamide, acetylacetone, or pyridine yield the ESR detectable $[Fe(CN)_5NO]^{3-}$ radical derived from electron transfer from the solvent, the same type of spectrum having been obtained by electrochemical techniques. Electrolytically and photochemically generated samples of labelled $[Fe(^{13}CN)_5\ NO]^{3-}$ were studied, and the hyperfine interactions are discussed with reference to a particular energy level scheme.

Wilkinson et al. (refs. 40,41) also report that photolysis at 254, 300 or 350 nm with several reducing agents or electrolysis gives a paramagnetic ion with a well-defined ESR spectrum corresponding to $[(CN)_5\ Fe\ (NO)]^{3-}$. The captured electron is tightly confined to the iron d_{z^2} orbital. Photolysis at 77°K in glassy methanol solution also gives $[(CN)_5Fe\ NO]^{3-}$, along with $\cdot CH_2OH$ and HCO radicals derived from the solvent. These results are explained in terms of charge transfer from the ligands to the metal (CTTM), followed by a redox reaction involving electron transfer from methanol to the ligand (CTFS) to give $(MeOH)^+$ which deprotonates to $\cdot CH_2OH$, eqns (19,20). In the process one ligand could be ejected eqn (21)

$$[Fe(CN)_5NO]^{2-} \xrightarrow[CTTM]{h\nu} {}^{*}[Fe(CN)_5(NO)]^{2-} \xrightarrow[CTFS]{solvent} [Fe(CN)_5NO]^{3-} + solvent^{+} \quad (19)$$

$$[Fe(CN)_5NO]^{2-} \xrightarrow[CTTS]{} [Fe^{III}(CN)_5(NO)]^{-} + e^{-} \longrightarrow [Fe(CN)_5(\dot{N}O)]^{3-} \quad (20)$$

$$[Fe(CN)_5(NO)]^{3-} \longrightarrow [Fe(CN)_4(NO)]^{2-} + CN^{-} \quad (21)$$

Using ESR and spin trapping techniques, it has been shown (ref 42) that photolysis of several other cyanide complexes gives outersphere oxidation of water, eqns (22,23).

$$[M(CN)_x]^{n-} + H_2O \xrightarrow{h\nu} [M(CN)_x]^{(n+1)-} + H^{+} + OH^{\bullet} \quad M = Mo,W \quad (22)$$

$$[M(CN)_x]^{n-} + H_2O \longrightarrow [M(CN)_{x-1}]^{n-} + CN^{-} \quad M=Ru \quad (23)$$

Rěhorek et al. (ref 43) studied the photolysis of octacyanomolybdate (V) complexes and give some ESR evidence of paramagnetic intermediates. The mechanisms suggested in aqueous solution are those of eqns (22,23). Several free radical species have been detected. Both outer sphere and inner sphere redox reactions, as well as substitution reactions could be considered as primary photoreactions of $[Mo(CN)_8]^{3-}$ in methanol.
Unfortunately, the spin density of such radicals are not reported and it is difficult to determine which is the primary step.

Irradiation of frozen ethanol solutions of $Mo(CO)_6$ results in bands at 465nm ascribed to $[Mo(CO)_6]^{-}$ in the excited state (ref 44).

The photolysis of alkylcobaloximes in vacuo or under inert atmosphere (refs 45,46) have been studied and it is generally assumed that homolytic cleavage of the Co-C bond occurs.

By ESR spectroscopy it has been shown that the photolysis results in the abstraction of an electron from the solvent, producing a Co^{II} complex containing all the original ligands. For methyl and benzyl (pyridinato) cobaloximes (3, 4), the reaction occurs in fluid solution, but for other complexes (5-9), the reaction occurs only in a solid matrix. (refs 45,46)

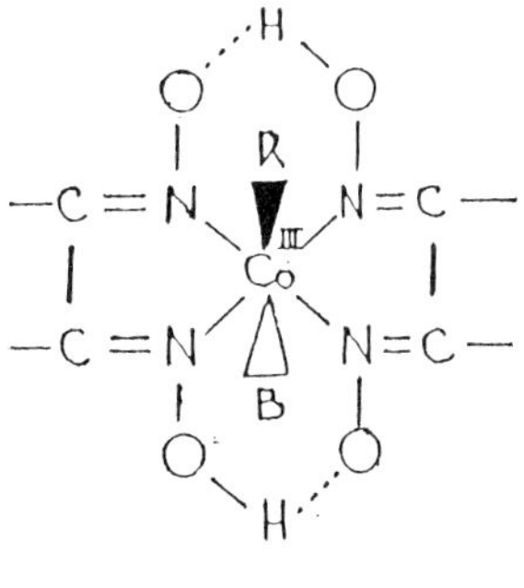

3 R = CH_3 B = Py
4 R = CH_2-Φ B = Py
5 R = isopropyl B = Py
6 R = isobutyl B = Py
7 R = n.pentyl B = Py
8 R = cyclohexyl B = Py
9 R = isopropyl B = morpholine

3.2 - Electron transfer from an organometallic complex to the solvent

Photochemical decomposition of several ferrocene derivatives has been reported by Nesmeyanov et al. (refs 47-49) and Havinga et al. (ref. 50). They described the elimination of an acid in the photolysis of ferroceneboric acid in an aqueous alkaline solution.

In carbon tetrachloride, Brand and Snedden (ref 51) observed a dissociative charge transfer absorption by ferrocene and Korner Von Gustorf et al. (ref 52) found that UV-irradiation of ferrocene in carbon tetrachloride gave ferrocenium tetrachloroferrate. This compound is also found by Spilners (ref 53) in hexachlorocyclopentadiene. Traverso and Scandola (ref 54) reported on the photochemical oxidation of ferrocene to ferrocenium ion in carbon tetrachloride-ethanol solution. Photochemical substitution of ferrocene (ref 55) in several organic halide-ethanol solutions and the general photochemistry of ferrocene has been reported (ref 56)

The UV irradiation of ferrocene in several halogenated hydrocarbon-ethanol solutions gives photochemical substitution (ref 57). By this method, ethoxycarbonyl, formyl, ethoxy, methyl, benzyl and allyl groups were photochemically introduced into ferrocene using carbon tetrachloride, chloroform, dichloromethane, benzyl chloride and allyl bromide in ethanol as solvent (ref 58). The reactions proceeded by the excitation of the charge transfer complex between ferrocene and the halogenated hydrocarbon (eqn 24).

$$Fe(C_5H_5)_2 + R\text{-}Cl \xrightarrow[EtOH]{h\nu} R'C_5H_4FeC_5H_5 \qquad \begin{array}{ll} R = CCl_3 & R' = CO_2Et \\ R = CHCl_2 & R' = CHO \\ R = CH_2Cl & R' = CH_2OEt \end{array} \qquad (24)$$

These reactions are formally substitutions at the cyclopentadienyl ring of ferrocene by halomethyl radicals which are formed by the fission of R-Cl bond of halomethanes. This product then reacts by ethanolysis of the halomethyl group giving R'. Brand and Snedden (ref 51) reported that the charge transfer absorption of ferrocene at 300-340 nm in such halogenated solvents can be attributed to electron transfer from ferrocene to the solvent.

The mechanism of photosubstitution of ferrocene in haloalkane-ethanol solutions was investigated by flash photolysis. The photoreaction is initiated by electron transfer from ferrocene to carbon tetrachloride. The absorption band of ferrocene at 440 nm is assigned to a d-d transition (ref 59). That at 325 nm is assigned to a 3d-MO*, a ring MO-MO*, or a symmetry-forbidden π-π^* transition. An absorption at shorter wavelength is due to an π-M* or n-π^* transition (ref 51)

The effect of substituents on ferrocene in the photoethoxycarbonylation of monosubstitued ferrocene in carbon tetrachloride-ethanol solution has been studied (ref 60). Electron withdrawing substituents such as chloro, bromo, iodo, cyano and acetyl allow the photoethoxycarbonylation but electron-releasing substituents such as ethyl accelerate the reaction. A mechanism could be written as follows, (ref 60) eqns (25,26).

$$(RC_5H_4)Fe(C_5H_5) + CCl_4 \rightleftharpoons [(RC_5H_4)Fe \cdots CCl_4(C_5H_5)] \xrightarrow{h\nu} (RC_5H_4)Fe^+(C_5H_5) + Cl^- + {}^{\bullet}CCl_3 \longrightarrow [(RC_5H_4)Fe^+(C_5H_5) + {}^{\bullet}CCl_3] \quad (25)$$

$$[(RC_5H_4)Fe^+(C_5H_5) + {}^{\bullet}CCl_3] \longrightarrow \text{H, CCl}_3\text{-adducts (+)} \xrightarrow{+EtOH} CO_2Et\text{-substituted ferrocenes} \quad (26)$$

UV light in halocarbon solvent induces the oxidation of ferrocene to ferricenium. No effect of visible light was observed but in aqueous acetyltrimethyl-ammonium, cetylmethylammonium bromide and sodium dodecyl sulfate micelles, the same reaction occurs under visible light (ref 61). The quantum yield is very low (4×10^{-5}) and it is not clear if electron transfer process occurs in the reaction (refs 62,63). Part of the micelle effect may be attributed to the micelle's ability to organize the donor and acceptor in close proximity (ref 64), although this cannot explain by itself a reactivity greater than that in neat CCl_4. It appears likely that the major role of the micelle is to solubilize ferrocene in a highly polar medium, providing conditions favoring formation of ionic products and perhaps decreasing back electron transfer.

Borell and Henderson (ref 65) have studied the photolysis and naphthalene-photosensitized reaction of the charge-transfer complex between ruthenocene and carbon tetrachloride, eqn (27)

$$[Ru(Cp)_2] + CCl_4 \rightleftharpoons [Ru(Cp)_2 : CCl_4] \qquad (27)$$

$$[(Cp)_2Ru : CCl_4] \longrightarrow [Ru(Cp)_2]^+ + CCl_3 + Cl^- \qquad (28)$$

The reaction forms the ruthenocenium cation, eqn (28), in the primary process. The overall quantum yield at 313 nm is 0.72 and at 366 nm is 0.52. The partial inhibition of the reaction with $SmCl_3$, oxygen or acrylamide has been studied and it has been found that at 313 nm the quantum yield for the singlet state decomposition is 0.24 and that occuring from the triplet state is 0.40. The naphthalene-photosensitized reaction proceeds through a triplet state and the reaction is inhibited at higher concentrations by competitive quenching of the naphthalene singlet state by ruthenocene itself.

The production of solvated electrons is an important photochemical reaction of various transition metal complexes (ref 66) and the reactive excited state is rather the metal-to-ligand charge transfer (CT) or the (CT) metal- to-solvent (CT) type. Some cyanide complexes undergo a clean one electron photo-oxidation without side reactions when $CHCl_3$ is used as solvent (ref 67). $CHCl_3$ is an efficient electron scavenger (ref 67). Charge transfer excitation ($\lambda_{irrad.}$ = 228 nm) of complexes such as $[Ru(CN)_6]^{4-}$, $[Mo(CN)_8]^{4-}$ and $[W(CN)_8]^{4-}$ dissolved in $CHCl_3$ are respectively oxidized to $[Ru(CN)_6]^{3-}$, $[Mo(CN)_8]^{4-}$ and $[W(CN)_8]^{3-}$ with quantum yields, respectively, Φ =0.49, 0.40 and 0.37 (Table 3). These values are in good agreement with those obtained by N_2O trapping in aqueous solutions (ref 21). When $CHCl_3$ solutions were not degassed, the quantum yields of photooxidation increased considerably (Φ=2 for $[Ru(CN)_6]^{4-}$). The photoelectrons are apparently scavenged very efficiently by oxygen which is reduced to superoxide (ref 68) which further could oxidize the unreacted complex. Longer wavelength irradiations into ligand field (LF) bands did not cause any change in $CHCl_3$ solutions of the complexes.

Low valent metal complexes containing arylisocyanides ligands $M(CNIPh)_6$ (IPh=2,6 diisopropylphenyl) exhibit intense visible absorption bands

TABLE 3

Quantum yields for photolysis of different cyanide and cyclopentadienyl organometallic complexes under different conditions

COMPLEX	wavelength	solvent	Φ	Reference
$(\pi^5C_5H_5)_2Fe$ in CTAB or SDS	485	H_2O/CCl_4	4.10^{-5}	61
$Cp_2Ru:CCl_4/N_2$	313	CCl_4/CH_3OH 1/1	0.76	65
	366	CCl_4/CH_3OH 3/1	0.49	65
$Cp_2Ru:CCl_4/N^2$ + acrylamide	313	CCl_4/CH_3OH 1/1	0.63	65
+ air	313	"	0.51	65
+ air	366	CCl_4/CH_3OH 3/1	0.33	65
+ oxygen	313	CCl_4/CH_3OH 1/1	0.38	65
+ $SmCl_3$	313	"	0.28	65
+stilbene	366	CCl_4/CH_3OH 3/1	0.50	65
+ biacetyl	313	CCl_4/CH_3OH 1/1	0.76	65
+ glyoxal	313	"	0.4	65
$[Ru(CN)_6]^{3-}$	228	$CHCl_3$	0.49	67
$[Mo(CN)_8]^{3-}$	228	$CHCl_3$	0.40	67
$[W(CN)_8]^{3-}$	228	$CHCl_3$	0.37	67
$[Ru(CN)_6]^{3-}$	228	$CHCl_3/O_2$	2	67
$Cr(CNIPh)_6$	436	$CHCl_3$	0.19	68
$Mo(CNIPh)_6$	436	$CHCl_3$	0.19	68
$W(CNIPh)_6$	436	$CHCl_3$	0.19	68

attributable to metal-to-ligand charge transfer (MLCT) transitions. Irradiation of $M(CNIPh)_6$(M=C, Mo, W) in $CHCl_3$ (ref 68) gives one electron oxidation product : $M(CNIPh)_6^+$. The following speculative mechanistic scheme proposed by H.B. Gray et al. (ref 68) can explain the results :

$$ML_6 \xrightarrow{436\ nm} ML_6^* \quad (29)$$

$$ML_6^* + CHCl_3 \longrightarrow [ML_6^+ \quad HCCl_3^{\bar{\cdot}}] \quad (30)$$

$$[ML_6^+ \ldots.HCCl_3^{\bar{\cdot}}] \longrightarrow ML_6^+ + HCCl_3^{\bar{\cdot}} \quad (31)$$

$$[ML_6^+ \ldots HCCl_3^{\bar{\cdot}}] \longrightarrow [ML_6\ Cl^+ \ldots.HCCl_2^{\bar{\cdot}}] \quad (32)$$

$$[ML_6Cl^+ \ldots HCCl_2^{\bar{\cdot}}] \longrightarrow [ML_6Cl]^+ + HCCl_2^{\bar{\cdot}} \quad (33)$$

The excited state formed via eqn (29) is quenched by chloroform, eqn (30), via electron transfer, forming the radical pair. $[ML_6^+ \ldots HCCl_3^{\bar{\cdot}}]$ which can revert back to starting material or undergo reaction, forming $HCl_3^{\cdot}$, eqns (32, 33), before eventually releasing Cl^-. This step is rate controlling and gives a quantum yield of 0.19 for the three metals (Table 3). With the complex containing a CNPh group, the chloroform radical anion $HCCl_2^-$ transfers a chlorine atom to ML_6^+, eqns 32 and 33 giving the seven-coordinate product observed (ref 68).

When a solution of methyl (triphenylphosphine) gold in $CDCl_3$ (eqns 34,35) is irradiated, a radical reaction detected by CIDNIP takes place (ref 69). The emission of CH_3 Au (PPh_3) is caused by polarized escaping methyl radicals, but the main reaction is one between $CH_3Au(PPh_3)$ and $CDCl_3$, eqn (35). CH_3Au (PPh_3) is excited either to a singlet state which decays to an excited triplet, eqn (34), or is excited to a triplet. The latter reacts with the solvent to give $Au(PPh_3)$ Cl and the triplet radical pair $\overline{CH_3^{\cdot} + CDCl_2}^T$ responsible for some of the polarization observed, eqn (35). The methyl radical mainly reacts with $CDCl_3$ to give CH_3D, eqn (37), or recombines, eqn (38), with intersystem crossing in the triplet radical pair, eqn (36).

$$CH_3\ Au\ (PPh_3) \xrightarrow{h\nu} (CH_3Au(PPh_3))^S \longrightarrow CH_3Au\ (PPh_3)^T \quad (34)$$

$$CH_3Au\ (PPh_3)^T + CDCl_3 \longrightarrow Au\ (PPh_3)Cl + \overline{CH_3^{\cdot} + {}^{\cdot}CDCl_2}^T \quad (35)$$

$$\overline{CH_3^{\cdot} + {}^{\cdot}CDCl_2}^T \longrightarrow CH_3\ CDCl_2 \quad (36)$$

$$CH_3^{\bullet} + CDCl_3 \longrightarrow CH_3D + {}^{\bullet}CCl_3 \qquad (37)$$

$$2CH_3^{\bullet} \longrightarrow C_2H_6 \qquad (38)$$

The first two reactions, eqns (34 and 35), can be written as involving electron transfer, eqn (39).

$$CH_3Au\ (PPh_3) + CDCl_3 \xrightarrow{h\nu} [CH_3\ Au^{II}(PPh_3)]^+\ [CDCl_3^{\bullet}]^- \qquad (39)$$

4 - Electron transfer from an organometallic complex to an acceptor

Traverso et al. (ref 70) report that Cp_2Fe ($Cp= \Pi^5\ C_5H_5$) and Ru Cp_2 give an association with mercury halides HgX_2 (X=Cl,Br), eqn (40). These complexes exhibit $Cp_2M \rightarrow HgH_2$, charge-transfer absorptions, respectively, at 360 nm and 280 nm. The irradiation in the charge transfer-to-metal transition band causes the oxidation of metallocenes to the metallocenium and the formation of mercurous ion. The intramolecular excitation with 280 nm light showed that conversions from intramolecular excited states of ferrocene to the charge-transfer complex $Cp_2Fe.HgCl_2$ are inefficient.

The mechanism seems to be an inner sphere electron transfer. Some excited states must decay via an electron transfer which results in oxidation of Cp_2M and formation of the HgX_2^-, eqn (40)

$$2\ Cp_2M + HgX_2 \longrightarrow 2Cp_2\ M \rightarrow HgX_2 \qquad (40)$$

$$2\ Cp_2M \rightarrow Hg\ X_2 \longrightarrow 2Cp_2\ M^+ + 2Hg\ X_2^- \qquad (41)$$

$$2\ Cp_2\ M^+ + 2HgX_2^- \longrightarrow [Cp_2\ M\ ]_2\ Hg_2\ X_4 \qquad (42)$$

This primary photoreaction could be followed by rapid dimerization of HgX_2^- to $Hg_2\ X_4^{2-}$, which could precipitate as $[Cp_4M_2Hg_2X_4]$ eqn (42). The formation of these compounds is a thermal process. The quantum yield for the formation of ferricenium and ruthenicemium is Φ=1.Demas et al. (ref. 71) have studied luminescence quenching at room temperature of $Ru(bpy)_2(CN)_2$ and Ru $(phen)_2(CN)_2$ (bpy = 2,2' bipyridine and phen= 1,10-phenanthroline) in the solvent series (water, alcohols, DMF) by several transition metal complexes such as $K_2[Ni(CN)_4]$, $K_3\ [Co(CN)_6]$, $K_3\ [(Cr(CN)_6]$, $K_3\ [Fe(CN)_6]$ and $K_4\ [Fe(CN)_6]$.

Either energy transfer and electron transfer could occur. It has been shown that the only quenching mechanism possible for Co^{2+}, Ni^{2+}, $Co(acac)_2$ (acac = acetylacetonate),$PtCl_4^{2-}$ or Cr^{III} complexes is energy transfer, while for Cu $(acac)_2$, $[Ru(bpy)_3]^{2+}$, $Ru(bpy)_2(CN)_2$, $Ru(phen)_2(CN)_2$ and Cu^{2+} (to yield Cu^+), at least in part electron transfer can occurs. Thermodynamically $[Fe(CN)_6]^{3-}$ can quench $[Ru(bpy)_3]^{2+}$ and $[Ru(phen)_3]^{2+}$ by either energy or oxidative electron transfer, as is true for the Ru-cyanide. Here electron transfer is the dominant quenching pathway.

The Re ligand charge transfer (ReLCT) excited state of fac-$[XRe(CO)_3L_2)]$ should be an oxidant powerful enough to effect the oxidation of Et_3N (ref 72). This conclusion is based on the cyclic voltammetry of the Re complexes which shows a reversible one electron reduction wave at -1.2 eV vs SCE in CH_3CN / 0.1M $[n\text{-}Bu_4N]ClO_4$. The onset of the emission is at $\simeq$ 2.5 eV, giving an excited state potential for Re^*/Re^- of $\simeq$ 1.3 V vs SCE. In the same solvent/ electrolyte system, the peak potential for the oxidation of Et_3N is + 1.0 V vs SCE. Because Et_3N has no low-lying excited states, it would appear that quenching of the photoexcited ReLCT state occurs according to eqn (43).

$$\text{fac-}[XRe(CO)_3L_2]^* + Et_3N \xrightarrow{k=10^9 M^{-1}s^{-1}} \text{fac-}[XRe(CO)_3L_2]^{\bar{\cdot}} + Et_3N^{+\cdot} \quad (43)$$

The quenching of the ReLCT excited state is followed by a permament chemical change eqns (44 and 45).

$$\text{fac-}[XRe(CO)_3L_2] \xrightarrow[Et_3N/CH_2Cl_2]{436\ nm} \text{fac-}[XRe(CO)_3LL'] \quad (44)$$

$$\text{fac-}[XRe(CO)_3LL'] \xrightarrow[Et_3N/CH_2Cl_2]{436\ nm} \text{fac-}[XRe(CO)_3L'_2] \quad (45)$$

with X=Cl, L=4-BzPyr L' = Ph-4-PyrCHOH Φ=0.20

X=I,L = 4-AcPyr, L'=Me-4-Pyr-CHOH $\Phi=0.18 \pm 0.02$

Irradiation (436 nm) of fac-$[XRe(CO)_3L_2]$ in the presence of L gives L' in good yield. A large number of L' substituted complexes have been prepared; in 1981, Wrighton et al. (refs 72-73) reported that irradiation of $[(CH_3CN)Re(CO)_3Phen]^+$ in CH_3CN containing PPh_3 and $n\text{-}Bu_4NPF_6$ yields clean substitution and gives $[(Ph_3P)Re(CO)_3Phen]^+$. PPh_3 quenches the emission of $[(CH_3CN)Re(CO)_3]Phen^+$ with linear Stern-Volmer kinetics; the associated quenching constant k_{46} is 8.2 x 10^{-9} $M^{-1}s^{-1}$. The quenching is logically associated with electron transfer, eqn (46).

$$[(CH_3CN)Re(CO)_3Phen]^{*+} + PPh_3 \xrightarrow{k_{46}} [(CH_3CN)Re(CO)_3Phen]^{\circ} + PPh_3^{+} \quad (46)$$

The oxidizing power of the excited complex is $\simeq$ + 1.5 V vs SCE, which exceeds the potential required to oxidize PPh_3. The lowest excited state of PPh_3 is too high in energy for PPh_3 to quench the excited Re complex by energy transfer at a diffusion controlled rate. Substitution follows that step, eqn (47).

$$[(CH_3CN)\ Re\ (CO)_3\ Phen]^{\circ} + L \longrightarrow [LRe(CO)_3Phen]^{\circ} + CH_3CN \qquad (47)$$

The quantum yield exceeds 1 and the explanation is based on a succession of the initiation step, eqn (46), and propagation steps, eqns (48,49).

$$[LRe(CO)_3Phen]^{\circ} + Q^{+} \longrightarrow [LRe(CO)_3Phen]^{+} + Q \qquad (48)$$

$$[LRe(CO)_3Phen]^{\circ} + [(CH_3CN)\ Re\ (CO)_3Phen]^{+} \longrightarrow [LRe(CO)_3Phen]^{+} + [(CH_3CN\ Re\ (CO)_3Phen]^{\circ} \qquad (49)$$

$$[(CH_3CNRe\ (CO)_3Phen]^{\circ} + Q^{+} \longrightarrow [(CH_3CN\ Re\ (CO)_3\ Phen]^{+} + Q = PPh_3 \qquad (50)$$

The last two equations could be representative of chain termination, eqns (49,50).

Electron transfer is energetically possible since all the $[LRe(CO)_3$ Phen]+ species are reducible electrochemically at nearly the same potential in CH_3CN containing 0.1 M $[nBu_4N]ClO_4$ = -1.2 ± 0.1 V vs SCE).

Two sets of experiments confirm the electron transfer mechanism for the substitution of CH_3CN in $[(CH_3CN)Re\ (CO)_3\ Phen]^{+}$:

1. The reduction of the $[(CH_3CN)Re\ (CO)_3Phen]^{+}$ is only quasi-reversible at 1.2 V vs SCE. Reduction of $[(CH_3CN)\ Re\ (CO)_3\ Phen]^{+}$ at controlled potentials at 1.1 V vs SCE in CH_3CN/0.1 M [n-Bu4N]ClO_4 containing L=pyridine or PPh_3 gives rapid formation of the substituted product $[LRe(CO)_3Phen]^{+}$. Many molecules of the substitution product are obtained with only small coulombic consumption.

2. In the light-induced substitution of CH_3CN by pyridine and employing Q=N,N'-dimethyl-p-toluidine as the electron transfer quencher, Q is as an efficient quencher as for Q=PPh_3, confirming that the process is likely an electron transfer. Since no substitution by pyridine occurs without an electron transfer quencher, a clean, quantum efficient substitution to yield

[(pyridine)Re(CO)$_3$Phen]$^+$ occurs when the solution contains 2 M pyridine and an electron donating quencher.

In 1978, Nadjo, Wrighton, et al. (ref 74) reported their finding on electron transfer quenching in the ground and excited states of fac ClRe(CO)$_3$L (L=1,10 phenanthroline or 4,7-diphenyl-1,10 phenanthroline) at 298 K in acetonitrile solution containing (n-Bu$_4$N)ClO$_4$. The quenchers are N,N'-dimethyl-4,4'-bipyridinium (MV^{2+}) and N,N'-dibenzyl-4,4'-bipyridinium (BV^{2+}) dications. A linear Stern-Volmer plot is obtained and from the slope the bimolecular quenching is determined to be $k_{51} \simeq 2.3 \times 10^9\ M^{-1}\ s^{-1}$. The quenching by MV^{2+} and BV^{2+} is approximately diffusion-controlled. The quenching corresponds to the following equations:

$$\text{fac-[Cl Re (CO)}_3\text{ L]}^* + Q^{2+} \xrightarrow{k_{51}} Q^{+\cdot} + \text{fac-[ClRe(CO)}_3\text{L]}^{+\cdot} \qquad (51)$$

$$Q^{2+} = MV^{2+},\ BV^{2+}$$

Flash photolysis has been carried out, allowing the detection of $MV^{+\cdot}$ and $BV^{+\cdot}$ via electron transfer. These compounds are intensively colored, with absorption maxima at about 600 nm. Such species must be transient because in the continuous photolysis at 436 nm of fac-[Cl Re(CO)$_3$L], where the emission is totally quenched by Q^{2+}, no net chemical change is observed (436 nm << 10^{-4}). In these Re-systems it has not been possible to observe directly any transient absorption attributed to fac-[ClRe (CO)$_3$L]$^{+\cdot}$. Flashes were at 500 J and the flash time was $\simeq$ 20 μs, allowing observation of transients of longer life time than $\simeq$ 75 μs.

The transient aborptions around 600 nm is bleached within a few milliseconds. Very carefully degassed solutions can be flash excited a number of times without deterioration of the sample. These facts suggest the following reaction :

$$Q^{+\cdot} + \text{fac[ClRe(CO)}_3\text{L]}^{+\cdot} \xrightarrow{k_{52}} Q^{2+} + \text{fac-[ClRe(CO)}_3\text{L]} \qquad (52)$$

which corresponds to a thermal back electron transfer reaction. Such a reaction occurs very efficiently. It has been established that k_{52} is essentially diffusion controlled in the [Ru(2,2'-bpy)$_3$]$^{3+/2+}$ couple .

Several compounds in Table 4 used as electron donors and acceptors show correlated $E_{1/2}$ values and excited state potentials with fac-[Cl Re(CO)$_3$L]. The lowest lying excited state is quenchable at fast rates by donors having oxidation potentials less positive than +1 V vs SCE, or by acceptors having reduction potentials less negative than -1.1 V vs SCE. It can be established

TABLE 4

Redox potentials E 1/2 (V/vs SCE) measured in degassed CH_3CN with 0.1 M $(nBu_4N)ClO_4$ from (refs 74,85).

	quencher	E1/2 V.v.s. SCE
1	p-dimethoxybenzene	+ 1.34
2	N,N-dimethylbenzylamine	+ 1.01
3	p-bromoaniline	+ 0.97
4	aniline	+ 0.98
5	triphenylamine	+ 0.86
6	diphenylamine	+ 0.83
7	10-methylphenothiazine	+ 0.83
8	N,N-dimethylaniline	+ 0.78
9	N,N' dimethyl-p-toluidine	+ 0.65
10	N,N'-diphenyl-p-phenylenediamine	+ 0.35
11	N,N,N'N'-tetramethyl-p-phenylenediamine	+ 0.24
12	Tetracyanoethylene	+ 0.24
13	1,1'-propylene-1,10 phenanthrolinium hexafluorophosphate	- 0.27
14	1,1'-ethylene-2,2'-bipyridinium hexafluorophosphate	- 0.36
15	N,N'- dibenzyl-4,4'-bipyridinium hexafluorophosphate	- 0.36
16	N,N'-dimethyl-4,4'-bipyridinium hexafluorophosphate	- 0.45
17	N methyl-4 cyanopyridinium hexafluorophosphate	- 0.66
18	N-methyl methylisonicotinate hexafluorophosphate	- 0.78
19	N-methyl-4-cyanopyridinium hexafluorophosphate	- 0.79
20	p-nitrobenzaldehyde	- 0.86
21	N-methyl-4 carbomethoxypyridinium hexafluorophosphate	- 0.93
22	4,4'-dinitrobiphenyl	- 1.00
23	m-nitrobenzaldehyde	- 1.02
24	4-chloronitrobenzene	- 1.06
25	4-methylnitrobenzene	- 1.20
26	4-aminonitrobenzene	- 1.34

by cyclic voltammetry that ground state potential for oxidation and reduction are respectively ≃ + 1.3 and ≃ - 1.3V vs SCE. The excited state species is found to be 2.3 V more oxidizing and reducing than the ground state.

The photochemistry of metal-metal bonds has been a subject of intense interest (ref 75). Photochemical metal-metal bond homolysis (refs 76-80), disproportionation (ref 81), and metal-metal bond dissociation (refs 77,78,80,82) are well known. It has been established that photogenerated ˙MLn radicals are both potentially stronger oxidants and reductants than their metal-metal bonded LnM-MLn complexes (ref 81).

An interesting electron transfer study has been reported by Hepp and Wrighton (ref 83). They have photolysed $M_2(CO)_{10}$ M=Mn or Re and $(\Pi^5 C_5H_5)_2 W_2(CO)_6$ in CH_2Cl_2 or CH_3CN in presence of CCl_4 and several electron oxidants A^+ (E° A^+/A vs SCE,V) : ferricenium $[Fe(\Pi^5\text{-}C_5H_5)]^+$ (+0.45) ; 1,1' dimethyl-ferricenium $[Fe(\Pi^5\text{-}C_5H_4Me)_2]^+$ (+ 0.36) ; decamethylferricenium $[Fe(\Pi^5\text{-}C_5Me_5)_2]^+$ (-0,11) ; tropylium $[C_7H_7]^+$ (0.18); or N,N' dimethyl-4,4'-bipyridinium (MV^{2+}) (-0,45). The analysis of the products shows that there is a competition between one electron oxidation and a capture by CCl_4 to give $MnCO)_5$ Cl via chlorine atom abstraction and $M(CO)_5\ ClO_4$ from oxidation by A^+. Perchlorate scavenges the 16-valence-electron $M(CO)_5$. They calculated at room temperature rate constants for the following reactions, eqns (53-59).

$$(Cp)_2W_2(CO)_6 \xrightarrow{h\nu} 2CpW(CO)_3^{\bullet} \tag{53}$$

$$Cp\ W(CO)_3^{\bullet} + CH_3CN + Fe(Cp)_2^+ \xrightarrow{k_{54}} [CpW(CO)_3\ (CH_3\ CN)]^+ + Fe(Cp)_2 \tag{54}$$

$$k_{54} = 1.43 \times 10^6\ M^{-1}s^{-1}$$

$$Mn_2\ (CO)_{10} \xrightarrow{h\nu} 2\ Mn\ (CO)_5^{\bullet} \tag{55}$$

(55)

$$Mn(CO)^{\bullet}_5 + CH_3CN + Fe(Cp)_2^+ \xrightarrow{k_{56}} Mn\ (CO)_5(CH_3CN)^+ + FeCp_2 \tag{56}$$

$$k_{56} - 7x10^6\ M^{-1}s^{-1}$$

The general mechanism could be written as in eqns (57-59).

$$\rangle M{-}M\langle \xrightarrow{h\nu} 2{\cdot}M\langle \tag{57}$$

$$\rangle M^{\cdot} + CCl_4 \xrightarrow{k_{ox}} \rangle M\text{-}Cl + {}^{\cdot}CCl_3 \qquad (58)$$

$$\rangle M^{\cdot} + A \xrightarrow[k_{ox}]{+L} \rangle M^{+}\text{-}L + A \qquad (59)$$

Several cluster complexes like the cyclopentadienylmolybdenum tricarbonyl dimer 10 and corresponding iron 11 and nickel 12 derivatives, the cobalt carbonyl dimer 13, the manganese carbonyl dimer 14, the tetracobalt dodecacarbonyl 15 and the tris iron dodecacarbonyl 16 give photoinduced electron transfer to methyl-viologen MV^{2+} 17 and quinones 18 - 20 (ref 84).

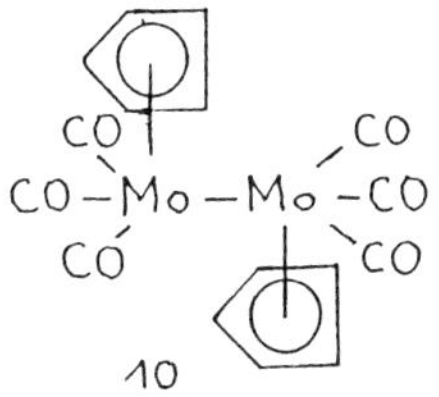

10

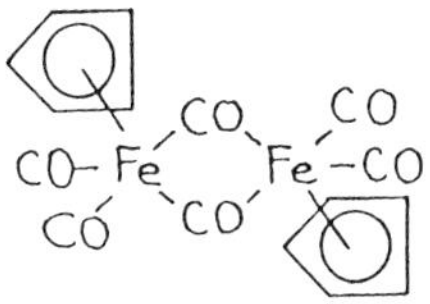

11

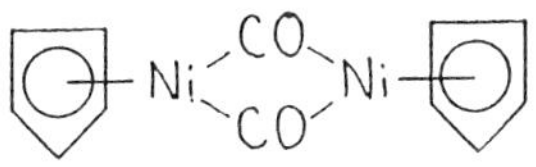

12

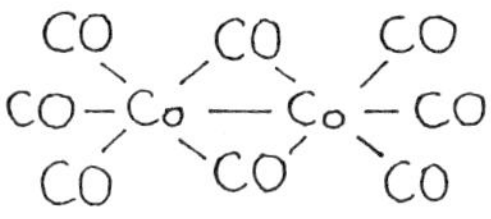

13

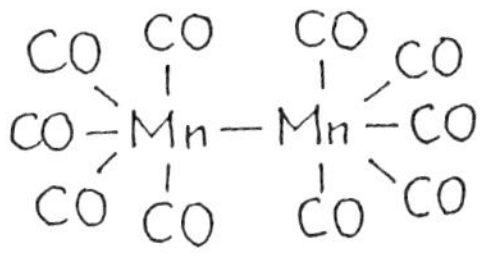

14

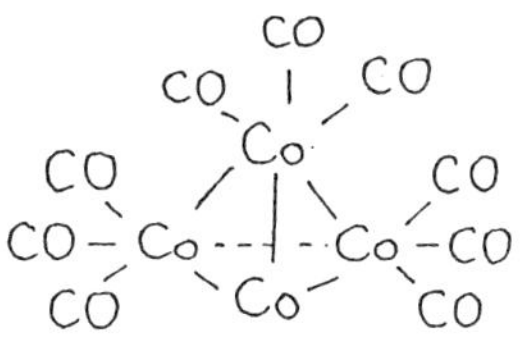

15

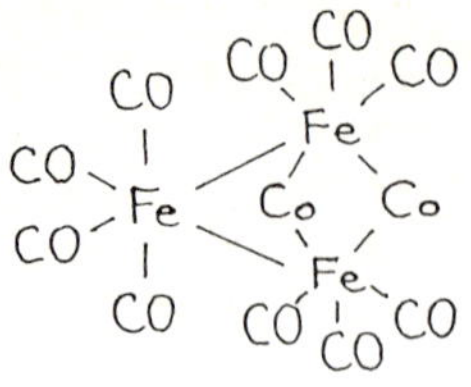

16

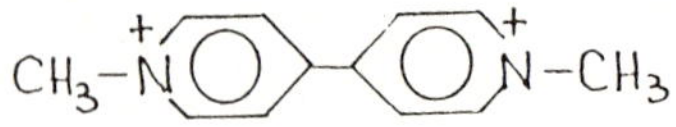

17

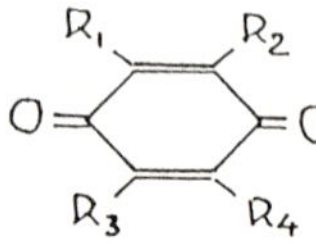

18 $(R_1=R_2=R_3=R_4=H)$
19 $R=CH_3$, $R_2=R_3=R_4$ = H
20 $(R_1=R_2=R_3=R_4=Cl)$

Any of these clusters complexes 10 - 17 give under visible irradiation (λ max =420nm) in the presence of PQ^{2+} the formation of PQ^{+}. For the quinones, compound 10 was used as a donor with chloranil 20 as the acceptor. With 20 a stable anion radical is formed; but with the other quinones 18, 19 the radical anion is not so stable. The reaction behaves differently if a small amount of water is present in the medium. Without any water in the medium, one can detect the formation of the $^{\bullet}CH_2CN$ radical derived from solvent. With a small amount of water present, one can detect an $OH^{\bullet}$ radical by spin trapping with nitrosodurene or 5,5'- dimethyl-1-pyrroline oxide.

The mechanisms are the following, (eqns 60-62) :

$$[Cp\ Mo\ (Co)_3]_2 \xrightarrow{h\nu} [Cp\ Mo\ (CO)_3]_2^{*} \longrightarrow 2Cp\ \ Mo\ (CO)_3^{\bullet} \tag{60}$$

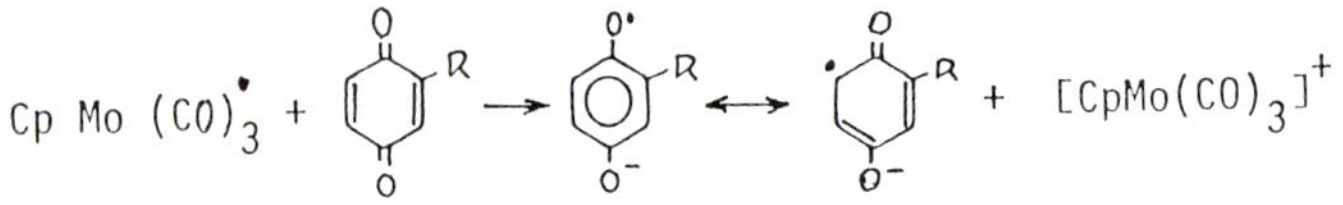

$$Cp\ Mo\ (CO)_3^{\bullet} + \text{quinone} \rightarrow \text{semiquinone radical anion} \leftrightarrow \text{semiquinone radical anion} + [CpMo(CO)_3]^+ \tag{61}$$

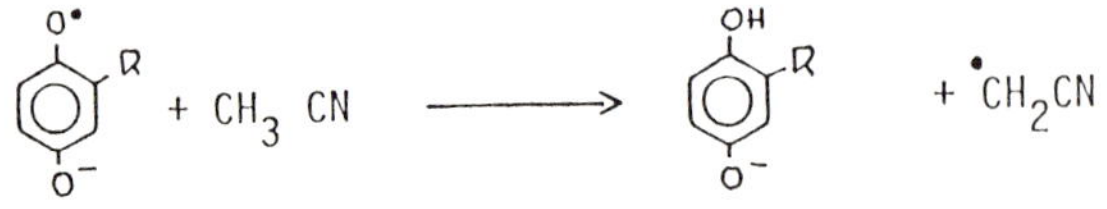

$$\text{semiquinone radical anion} + CH_3\ CN \longrightarrow \text{hydroquinone anion} + {}^{\bullet}CH_2CN \tag{62}$$

When the water is present in the medium, the radical anion can trap an electron of OH^- formed by the dissociation of the water eqn (63)

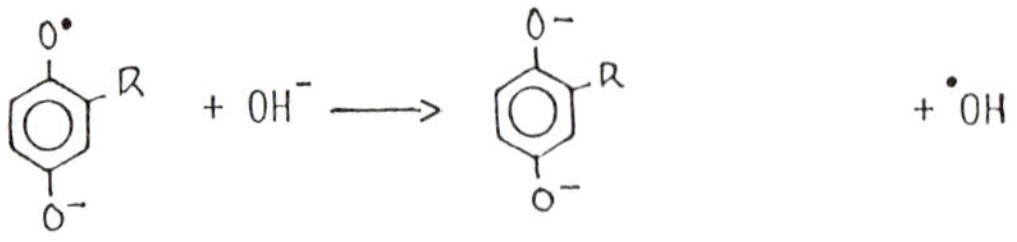

$$\text{semiquinone radical anion} + OH^- \longrightarrow \text{quinone dianion} + {}^{\bullet}OH \tag{63}$$

In 1980 Wrighton et al. (ref 85) report new studies concerning the photochemistry of R_3 $EM(CO)_3L$ (R=Ph or Me, E=Ge or Sn, M=Mn or Re, L= 1,10 phenanthroline, 2,2'-bipyridine or 2,2'-biquinoline) which have relatively long-lived excited states in solution, where bimolecular processes are possible. Specifically they have examined the excited-state electron transfer processes given in the following reactions, eqns (64,65).

$$[Ph_3\ E^+ - Re\ (CO)_3(Phen)^-]^* + Q \xrightarrow{k} [Ph_3\ E\text{-}Re\ (CO)_3\ (Phen)]^{\bar{\cdot}} + Q^{+\cdot} \quad (64)$$

$$[Ph_3E^+ -Re(CO)_3(Phen)^-]^* + P^+ \xrightarrow{k} [Ph_3\ E\ Re\ (CO)_3\ (Phen)]^+ + P^{\cdot} \quad (65)$$

They found that several of the electron donors Q and electron-acceptors P^+ quench the excited state emission of the Re complexes in CH_3CN/0.1 M $(n\text{-}Bu_4N)PF_6$ at 298°K. A large number of quenchers have been used (Table 4).

When excited state electron transfer is significantly energetically downhill, the quenching is fast, but when the energetics are less favorable the quenching constants decline. The plots are similar to the ones already reported for fac-$[Re(CO)_3\ (Phen)]^+$ (ref 74), for numerous other excited organic molecules (ref 86), for excited $Ru(bpy)^{2+}$ and related inorganic complexes (refs 87-90). But detailed quenching studies of $Ph_3SnRe(CO)_3(Phen)$ (phen = 1,10-phenanthroline) have been carried out and the quenching obeys Stern-Volmer kinetics (ref 85). Electron donors, Q, for which E° (Q^+/Q) is more negative than + 0.2 V vs SCE quench at nearly a diffusion-controlled rate. Electron acceptors, P^+, for which E° (P^+/P) is less negative than - 1V vs SCE also quench at nearly a diffusion-controlled rate. Cyclic voltammetry of the complexes in CH_3CN/0.1M $[n\text{-}Bu_4N]ClO_4$ typically shows a one-electron, reversible reduction in the -1.1 to -1.7 V vs SCE range associated with the population of the lowest available π^* orbital principaly localized on L. An irreversible oxidation current peak is observed in the range +0.5 to 0.8 V vs SCE. The M-containing oxidation product is fac-$[(CH_3CN)M(CO)_3L]^+$. Consistent with the ground state electrochemistry, quenching by reversible electron-donor quenchers (e.g., N,N,N',N'-tetramethyl-p-phenylenediamine) results in no net photoredox reaction ($\Phi < 10^{-3}$), whereas quenching by reversible electron-acceptor quenchers (e.g. N,N'-diméthyl-4,4'-bipyridinium) results in net redox chemistry to reduce the quencher and to form fac-$[(CH_3CN)M(CO)_3L]^+$ from the complex. The data are consistent with primary formation of the $[R_3EM(CO)_3L]^+$ formed by excited state electron transfer. Rate of $[R_3EM(CO_3L]^+$ cleavage is similar to the dissociative E-M bond cleavage induced by the (E-M) $\sigma b \rightarrow \pi^* L$ optical excitation.

The electronic absorption spectrum in MeCN of a mixture of $[Rh_2-(dicp)_4]^{2+}$ (dicp = 1,3-diisocyanopropane) and n-BuI peaks at 300, 320 and 553 nm. After short irradiation (1 min) with 553 nm light (refs 91,92), there is a important spectral charge and an intense absorption appears at 407 nm, while the violet color λ_{max}= 553 nm fades. Since neither the shape nor the intensity of the lowest energy band at 553 nm, assigned as an $^1A_{1g}$ or $^1A_{2u}$ (refs 93-95) transition of $[Rh_2-(dicp)_4]^{2+}$ has been changed by the addition of n-BuI (which has no absorption band in the visible) there seems to be no ground state complex formed. The resulting yellow solution contains the oxidative adduct : $[Rh_2-(dicp)_4]$ $(n-BuI)^{2+}$ as a major product. This assigment is based on a band which absorb at 407 nm assigned to a σ-σ^* transition in the Rh^{II}-Rh^{II} single bond.

The $[Rh_2(dicp)_4]^{2+}$ cation reacts with iPrI as well under irradiation at 553 nm, yielding $[Rh_2(dicp)_4\ (iPr)(I)]^{2+}$ with band at λ_{max} = 414 nm . The photochemical reaction with MeI has not been examined, since the thermal oxidative addition of MeI to $[Rh_2\ (dicp)_4]^{2+}$ occurs.

The quantum yields of such reactions are quite high, suggesting that the photoinduced oxidative addition of $[Rh_2\ -(dicp)_4]^{2+}$ with n-BuI and iPrI is a chain process. In order to understand this chain process, the effect of electron acceptors on the quantum yields was examined. The quantum yields are decreased drastically by addition of p-dinitrobenzene which is a powerful electron acceptor; radical scavengers also result in a reduction of the quantum yield.

The excited state $^*[Rh_2\ (dicp)_4]^{2+}$ acts as an excellent electron donor towards various acceptors, eqn (66).

$$^*[Rh_2(dicp)_4]^{2+}\ nBuI \longrightarrow [Rh_2\ (dicp)_4]^{3+} \cdot n\ BuI^{\bar{\cdot}} \quad (66)$$

$$[Rh_2\ (dicp)_4]^{3+} \cdot n\text{-}BuI^{\bar{\cdot}} \xrightarrow{fast} [Rh_2\ (dicp)_4(I)]^{2+\cdot} + n\text{-}Bu^{\cdot} \quad (67)$$

Electron transfer quenching is thus followed by a subsequent fast dark reaction, eqn (67), producing a radical in competition with back electron transfer to the ground state. A reaction scheme to account for all the kinetic results has been proposed (refs 91,92) which contains initiation, eqns (68, 69), propagation, eqns (70,71), and termination steps, eqs (72,73).

$$\text{Initiation :}\quad [Rh_2\ (dicp)_4]^{2+} \xrightarrow{h\nu} [Rh_2\ (dicp)_4]^{2+*} \quad (68)$$

$$[Rh_2\ (dicp)]_4^{2+*} + nBuI \longrightarrow [Rh_2\ (dicp)_4(I)]^{2+\cdot} + n\text{-}Bu^{\cdot} \quad (69)$$

Propagation : n-Bu + $[Rh_2\ (dicp)_4]^{2+} \longrightarrow [Rh_2\ (dicp)_4\ n\text{-}Bu)]^{2+}$ (70)

$[Rh_2\ (dicp)_4(n\text{-}Bu)]^{2+} + n\text{-}BuI \longrightarrow [Rh_2\ (dicp)_4(n\text{-}Bu)(I)]^{2+} + n\text{-}Bu^{\bullet}$ (71)

Termination : 2 $[Rh_2\ (dicp)_4\ (n\text{-}Bu)]^{2+} \longrightarrow$ Bu-Bu +2 $[Rh_2\ (dicp)_4]^{2+}$ (72)

2 $[Rh_2\ (dicp)_4\ (I)]\ \xrightarrow{2+} [Rh_4\ (dicp)_8\ (I)_2\ ]4^+$ (73)

Metcalf and P.Kubiak (ref 96) reported that flash photolysis of acetonitrile solutions of 21 gives rise to an intense transient absorbance

$$\left[\begin{array}{ccc} \text{Me-NC} & & \text{CN-Me} \\ | & & | \\ \text{Me-NC}-\text{Pd} & - & \text{Pd}-\text{CN-Me} \\ | & & | \\ \text{Me-NC} & & \text{CN-Me} \end{array}\right]^{2+}$$

21

with λ_{max}=405 nm, while acetonitrile solution of 21 exhibit no significant absorbance in the 380-520 nm region. This transient decayed by a second-order process assigned to recombination of photogenerated $[Pd(CNMe)_3]^+$ radicals, eqn (74).

$$[Pd_2\ (CNMe)_6]^{2+} \underset{k'_{74}}{\overset{h\nu}{\rightleftharpoons}} 2\ \ [Pd\ (CNMe)_3]^{\cdot +} \qquad (74)$$

The value of the rate expression for second order recombination

$2k'_{74}/\varepsilon_{405}$ is $4x10^4\ cm^{-1}s^{-1}$. The extinction coefficient at 405 nm ε_{405} and the rate of recombination of photogenerated $[Pd(CNMe)_3]^+$ have been determined by examining their quantitative reduction of benzylviologen (BV^{2+}) to $(BV)^+$ (E° A/A$^\circ$ vs SCE = 0.36 V), (eqn 75).

$$[Pd^I\ (CNMe)_3]^+ + BV^{2+} \xrightarrow[\text{solvent}]{k_{75}} [Pd^{II}\ (CNMe)_3\ \text{solvent}]^{2+} + BV^{+\cdot} \qquad (75)$$

$$k_{75} = 2x\ 10^8\ M^{-1}\ s^{-1}$$

The recombination of Pd $(CNMe)_3^+$ radicals to give 21 occurs with a rate constant $k_{74} = 1x10^9\ M^{-1}\ s^{-1}$ which is near the diffusion controlled limit . In the presence of other electron acceptors such as (E° (A/A$^\circ$, vs SCE, V) methylviologen (-0.45), N,N'-propylenephenanthrolinium (-0.13) dichlorobenzoquinone (-0,18) of electron transfer rate constants k_e- of respectively $3x10^7$, $2x10^8$ and $3x10^7\ M^{-1}\ s^{-1}$ are obtained. The appearance of Pd $(CNMe)_3^+$ and $BV^{+\cdot}$ is first order. The rate constant of the three substituted viologens decrease from $2x10^8$ to $3x10^7$ $M^{-1}s^{-1}$ as the driving force for electron transfer decreases from - 0.13 V to -0,45 V.

5 - Electron transfer from a transition metal complex to a transition organometallic

Balzani et al. (ref 97) describe the results obtained from the quenching of Ru $(bpy)_3^{2+}$ (bpy=2,2'- bipyridine) by several cyanide complexes which were chosen because they span a wide range of excited state energies and redox potentials, eqns (76-78).

$$^*[Ru\ (bpy)_3]^{2+} + Q \longrightarrow [Ru(bpy)_3]^{2+} + Q^* \qquad (76)$$

$$^*[Ru\ (bpy)_3]^{2+} + Q \longrightarrow [Ru(bpy)_3]^{+} + Q^+ \qquad (77)$$

$$^*[Ru\ (bpy)_3]^{2+} + Q \longrightarrow [Ru(bpy)_3]^{3+} + Q^- \qquad (78)$$

This quenching occurred by energy transfer (eqn. 76) for $[Ni(CN)_4]^{2-}$ or $[Cr(CN)_6]^{3-}$ and by reduction, eqn (77), in $[Mo(CN)_8]^{4-}$ (E_0+0.74 V, kq 3.4 10^8) and $[Fe\ (CN)_6]^{4-}$ E_0=0.36V, k_q=7.5x10^8).

With $[Co(CN)_6]^{3-}$, $[Pd(CN)_4]^{2-}$ and $[Pt\ (CN)_4]^{2-}$ quenching of the ruthenium excited state does not occur. This lack of quenching can be explained (ref 97) by the following :

1 - energy transfer is not allowed thermodynamically,

2 - oxidative quenching, even if thermodynamically not favorable, is expected to involve a high activation barrier.

At the same time, Sutin and Creutz (ref 98) reported their results on quenching $^*[Ru(bpy)_3]^{2+}$ excited state by $[Fe(CN)_6]^{4-}$, $[Fe(CN)_6]^{3-}$ and for $Mo(CN)_8^{4-}$ and proposed that the dominant quenching mechanism for at least some of these quenchers involves electron transfer, which yields oxidized quencher and $[Ru(bpy)_3]^+$. No net reaction is likely to be observed, since the back reaction should be very fast. The quenching of $^*[Ru(bpy)]^{2+}$ and also $^*[Ru(Phen)_3]^{2+}$ by $[Fe(CN)_6]^{4-}$, $[Fe(CN)_6]^{3-}$, $[Cr(CN)_6]^{3-}$, $[Co(CN)_6]^{3-}$ has also been interpreted as occurring by electron transfer (ref 99). It is suggested that the lowest excited state of $[Fe(CN)_6]^{4-}$ could be at very low energy and so the corresponding absorption could not be detected, a phenomenon which is quite common for spin forbidden transitions. The argument is that the absence of detectable visible or ir emission from $[Fe(CN)_6]^{4-}$ at 77°K either in glass or solid state, is correlated by the formation of an apparently non luminescent double salt with $[Cr(CN)_3)]^{3+}$. The lowest state of $[Fe(CN)_3]^{4-}$ may be below the energy of the emitting doublet of $[Cr(CN)_3]^{3+}$. The sensitized photochemistry of the Co^{III} complexes indicates at least some electron transfer quenching to yield Co^{II}, (ref 98).

In 1977 Serpone et al. (ref 100) reported a systematic study of bimolecular quenching of the luminescent excited state of $[Cr(bpy)_3]^{3+}$, $[Ru(bpy)_3]^{2+}$ and $[Os(bpy)_3]^{2+}$ by cyanide complexes such as : $[Mo(CN)_8]^{4-}$, $[Co(CN)_6]^{3-}$, $[Fe(CN)_6]^{4-}$, $[Fe(CN)_6]^{3-}$, $[Ru(CN)_6]^{4-}$, $[Co(CN)_6]^{3-}$ and $[Ni(CN)_4]^{2-}$

With $[Mo(CN)_8]^{4-}$, both energy transfer and oxidative electron transfer are thermodynamically favorable for $^*[Cr(bpy)_3]^{2+}$ and $^*[Ru(bpy)_3]^{2+}$. They are slightly disfavored for $^*[Os(bpy)_3]^{2+}$. With $[Cr(CN)_6]^{3-}$, energy transfer is thermodynamically favored and spin-allowed for all three donors whereas electron transfer is thermodynamically disfavored with all three bpy complexes. With $[Fe(CN)_6]^{4-}$, the most favored quenching mechanism is clearly the reductive one. Support of this assertion comes from a study of $[Ru(bpy)_3]^{2+}$ -$[Fe(CN)_6]^{4-}$ in which Creutz and Sutin (ref 98) have obtained evidence for the formation of $[Ru(bpy)_3]^+$. From the excited state energy of $[Fe(CN)_6]^{3-}$, energy transfer appears to be unlikely but with $^*[Ru(bpy)_3]^{2+}$ and $^*[Os(bpy)_3]^{2+}$, reductive quenching eqn (3) seems to be allowed.
In the system $^*[Cr(bpy)_3]^{3+}$ - $[Fe(CN)_6]^{3-}$, oxidative quenching occurs. For $[Ru(CN)_6]^{4-}$ reductive electron transfer quenching can occur since it is thermodynamically allowed . The data reported (ref 100) and the lack of knowledge of the oxidation potential of $[Co(CN)_6]^{3-}$ preclude any direct inference on the occurrence of reductive electron transfer. However, this process looks quite improbable because $[Co(CN)_6]^{3-}$ is very difficult to oxidize and $^*[Cr(bpy)_3]^{3+}$ is practically unaffected by $[Co(CN)_6]^{3-}$. The quenching of $^*[Ru(bpy)_3]^{2+}$ and $[Os(bpy)_3]^{2+}$ by $[Co(CN)_6]^{3-}$ may occur via an oxidative mechanism. This reaction is thermodynamically allowed for the $Os^{(II)}$ complex. The less disfavored process seems to be oxidative electron transfer quenching for the $Os(bpy)_3^{2-}$ -$[Ni(CN)_4]^{2-}$ system, whereas for the $[Cr(bpy)_3]^{3+}$-$[Ni(CN)_4]^{2-}$ system, a reductive electron transfer quenching is more probable.

It has also been reported by Atik and Thomas (ref 101), using pulsed laser techniques, that photoinduced electron transfer occurs from

$^{*}[Ru(bpy)_3]^{2+}$ to $[Fe(CN)_6]^{3-}$ in disodium diisooctylsulfosuccinate (AOT)/ alkane water system or potassium oleate/hexanol or cetyltrimethylammonium bromide (CTAB)/hexanol, eqn (79).

$$^{*}[Ru^{II}(bpy)_3]^{2+} + [Fe(CN)_6]^{3-} \xrightarrow[\text{micelle}]{} [Ru^{III}(bpy)_3]^{3+} + [Fe(CN)_6]^{4-} \qquad (79)$$

These ionic reactants are associated only with the water pool of the reversed micelle. Braun, Grätzel et al. (ref 102), have studied electron transfer involving $[Fe(CN)_6]^{3-}$. They use an excited organic donor D^{*} to reduce a transition metal M^{n+} . Most of the experiments were carried out using N-methylphenothiazine (MPTH) <u>22</u> or N,N'-dimethyl-5,11-dihydroindolo [3,2-b] carbazole <u>23</u> as the photoactive species.

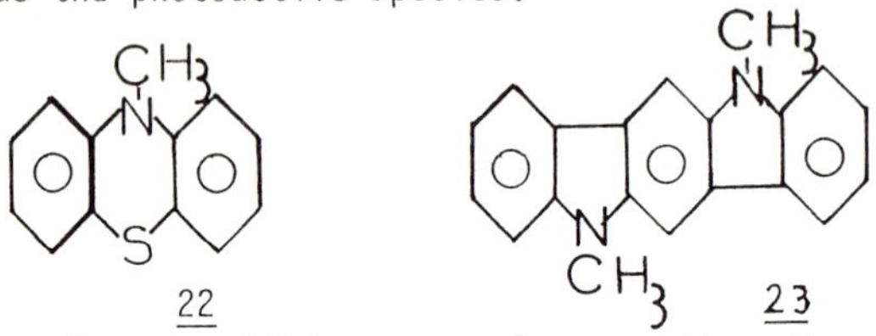

<u>22</u> <u>23</u>

These donors exhibits several properties that are suitable for mechanistic studies in surfactant aggregates. The reactive triplet excited state $(MPTH)^{T}$ is a powerful reductant. The triplet energy derived from energy transfer experiments is 2.64 eV. is close to the value determined for unsubstituted phenothiazine (ref. 103). When micelles of sodium laurylsulfate are replaced by Cu^{2+}, formation of $MPTH^{+}$ cation radical occurs whose absorption has been reported (ref 104). Oxidative quenching of $MPTH^{*}$ by Cu^{2+} ions is completed within 15 μs it occurs at an extremely rapid rate. The extinction coefficient ε_{516} $(MPTH^{+})$ is determined via quantitative oxidation of MPTH to $MPTH^{+}$ by ferricyanide. The standard redox potential E_0 $(MPTH^{+}/MPTH)$ is 0.86 V, while E_0 $[Fe(CN)_6]^{3-}$ / $[Fe(CN)_6]^{4-}$ is only 0.36 V, so that the above electron transfer is endoergic by 0.5 V and hence cannot occur spontaneously. The reaction is initiated in $Cu(LS)_2$ micellar solution via a triplet induced electron transfer to Cu^{2+}. The sequence of events are as follows : the Cu^{2+} ion receives an electron from the excited $(MPTH)^{T}$. The monovalent ion subsequently escapes into the bulk micellar solution where, in turn, it reduces $[Fe(CN)_6]^{3-}$, eqn (80). The approach of $[Fe(CN)_6]^{4-}$ to $MPTH^{+}$, which remains associated with the micelle is impaired by the negative micellar surface potential. The $MPTH^{+}$ produced in this way is relatively stable. The absorption band at 516 nm diminishes only by about 20% in a 24 h period.

$$MPTH + [Fe(CN)_6]^{3-} \longrightarrow MPTH^{+} + [Fe(CN)_6]^{4-} \qquad (80)$$

Photoinduced electron transfer involving the bis (2,9-dimethyl-1,10 phenanthroline) copper^{I} complex $[Cu(dmp)_2]^+$ and a variety of Co^{III} complexes such as trans $[Co(NH_3)_4(CN)_2]^+$ and $[Co(NH)_5(CN)]^{2+}$ have been reported (refs 105,106). With mM concentrations of Co^{III}, the quantum yields for the loss of $Cu^{(I)}$ are typically on the order of 10^{-3}. The quantum yield is sensitive to the nature of the Co^{III} used and also to the solvent. The results are interpreted in terms of a model in which the reactive species is a charge transfer excited state of $[Cu(dmp)_2]^{2+}$, eqns (81-84).

$$[Cu(dmp)_2]^+ + Co^{III} \underset{}{\overset{k}{\rightleftarrows}} [(Cu(dmp)_2]^+ . Co^{III} \quad (81)$$
$$[Cu(dmp)_2]^+ . Co^{III} \xrightarrow{h\nu} {}^*[Cu(dmp)_2]^+ . Co^{III} \quad (82)$$
$${}^*[Cu(dmp)_2]^+ . Co^{III} \xrightarrow{k'} [Cu(dmp)_2]^+ . Co^{III} + \text{energy} \quad (83)$$
$${}^*[Cu(dmp)_2]^+ . Co^{III} \longrightarrow [Cu(dmp)_2]^{2+} . Co^{II} \quad (84)$$

Electron transfer quenching may occur by a dynamic mechanism involving the diffusional encounter of the CT excited state $^*[Cu(dmp)_2]^+$ and the cobalt acceptor, eqns (85-87).

$$[Cu(dmp)_2]^+ \xrightarrow{h\nu} {}^*[Cu(dmp)_2]^+ \quad (85)$$
$${}^*[Cu(dmp)_2]^+ \longrightarrow [Cu(dmp)_2]^+ \quad (86)$$
$${}^*[Cu(dmp)_2]^+ + Co^{III} \longrightarrow [Cu(dmp)_2]^{2+} . Co^{II} \quad (87)$$

Such a mechanistic scheme can be completed in the following way, eqns (88,89).

$$[Cu(dmp)_2]^{2+} . Co^{II} \longrightarrow [Cu(dmp)_2]^+ + Co^{III} \quad (88)$$
$$[Cu(dmp)_2]^{2+} . Co^{II} \longrightarrow [Cu(dmp)_2]^{2+} + Co^{II} \quad (89)$$

Studies of the effect of pressure effect on the electron transfer quenching of a photoexcited metal complex may provide a means for investigating the chemical behavior of the excited state. Kirk and Porter (ref 107) reported the first example of a small pressure effect (ΔV= +0.5 to -2.6cm^{-3} M^{-1}) on the quenching of photo-excited $[Ru(bpy)_3]^{2+}$ by O_2, MV^{2+} and $[Co(acac)_3]$ (acac=acetylacetonate). But Saito et al. (ref 108) have measured the effect of pressure on the electron-transfer quenching of photoexcited $[Ru(bpy)_3]^{2+}$ by various metal ions such as $[Fe(CN)_6]^{4-}$, $[Mo(CN)_8]^{4-}$, $[Os(CN)_6]^{4-}$, $[Fe(CN)_6]^{3-}$, $[(CN)_5[Co\mu\text{-}O_2)Co(CN)_5]^{5-}$ and several other complexes. The $[Mo(CN)_8]^{4-}$ gives the largest activation volume (ΔV = 24.7 cm^3 M^{-1}),whereas quenching by other quenchers reacting with quenching rate constants on the order of 10^9 M^{-1} dm^{-3} s^{-1}, have very small pressure effects regardless of the charge of the quencher, except for Eu^{2+}_{aq} (ΔV = 11 cm^3 M^{-1}).

6 - Electron transfer in ion pairs

The photoreaction of the ion pair $[Co(en)_3]^{3+}$ (en= ethylene diamine) with $[Fe(CN)_6]^{4-}$ in NaCl solution has been shown to originate in an excited state of Co^{III} and this is thus one of the first examples (refs 109,110) of efficient electron transfer to a cobalt amine complex in a ligand field excited state. Solution of $[Co(en)_3]^{3+}$ and $[Fe(CN)_6]^{4-}$ exposed to visible light give rise to cherry-red color. This color can be attributed to formation of $[Cl\text{-}(en)_2\ Co\text{-}NC\text{-}Fe(CN)_5]^{2-}$ after the following electron transfer reaction, eqn (90).

$$[Co(en)_3]^{3+} + [Fe\ (CN)_6]^{4-} \xrightarrow[(LF)]{h\nu} [Co(en)_3]^{2+} + [Fe(CN)_6]^{3-} \qquad (90)$$

Since, there is a fairly efficient relaxation at all wavelenghts between 436 to 647 nm. The reactive excited state may be a thermally equilibrated excited state. This longer lived reactive state might be the quintet Co^{III} ligand field state which has a configuration $(t_{2g}^{4}\ e_{g}^{2})$ and is appropriate for facile reduction to the Co^{II} product with the $t_{2g}^{5}e_{g}^{2}$ configuration. Alternatively, it might be a charge transfer excited state of the ion pair with $[Fe(CN)_6]^{4-}$ which could be called an exciplex. The results do not clearly demonstrate reactions of the free $[Co(en)_3]^{3+}$ ion as distinct from an ion pair.

Luminescence and lifetime measurements of the tris(5-chloro-phenanthroline) osmium (II) - hexacyanoferrate (II) $[Os(5\text{-}Cl\text{-}phen)_3]^{2+}\text{-}\ [Fe(CN)_6]^{4-}$ with tris (bipyridine), and ruthenium (II) $[Ru(bpy)_3]^{2+}\text{-}\ [Fe(CN)_6]^{4-}$, systems have been performed (ref 111). The Stern-Volmer plots exhibit the upward curvature which are characteristic of static quenching. The plots yield values for second-order quenching rate constants : 1.0×10^{-3} and 0.10 M ionic strenght at respectively 1.7×10^{10} and $3.2\times10^{9}\ M^{-1}s^{-1}$ for the osmium and 5.1×10^{10} and $7.2 \times 10^{9} M^{-1}\ s^{-1}$ for the ruthenium system. These show that electron transfer in $^*[Ru(bpy)_3]^{2+}$ - $[Fe(CN)_6]^{4-}$ and in $^*[Os(5\text{-}Cl\text{-}phen)_3]^{2+}\text{-}$ $[Fe(CN)_6]^{4-}$ are respectively combination diffusion and activation controlled. Appropriate combination of the ruthenium and osmium data yields the rate constant k_{e^-} for electron transfer in the excited ion pair $^*[Os(5\text{-}Cl\text{-}phen)_3]^{2+}\text{-}\ [Fe(CN)_6]^{4-}$. At an ionic strength of 1.0×10^{-3} and at 0.10 M concentration, it has values of 8.7×10^{7} and $1.6\times10^{8}\ s^{-1}$, respectively. The mechanism can be written as follows eqns (91-95):

$$[Os(5\text{-}Cl\text{-}phen)_3]^{2+} + [Fe(CN)_6]^{4-} \xrightarrow{h\nu} {}^*[Os(5\text{-}Cl\text{-}phen)_3]^{2+} + [Fe(CN)_6]^{4-} \quad (91)$$

$${}^*[Os(5\text{-}Cl\text{-}phen)_3]^{2+} + [Fe(CN)_6]^{4-} \longrightarrow [Os(5\text{-}Cl\text{-}phen)_3]^{2+}\ [Fe(CN)_6]^{4-} \quad (92)$$

$${}^*[Os(5\text{-}Cl\text{-}phen)_3]^{2+}\ [Fe(CN)_6]^{4-} \longrightarrow [Os(5\text{-}Cl\text{-}phen)_3]^{+}\ [Fe(CN)_6]^{3-} \quad (93)$$

$$[Os(5\text{-}Cl\text{-}phen)_3]^{+}\ [Fe(CN)_6]^{3-} \longrightarrow [Os(5\text{-}Cl\text{-}phen)_3]^{2+}\ [Fe(CN)_6]^{4-} \quad (94)$$

$$[Os(5\text{-}Cl\text{-}phen)_3]^{2+}\ [Fe(CN)_6]^{4-} \longrightarrow [Os(5\text{-}Cl\text{-}phen)_3]^{2+} + [Fe(CN)_6]^{4-} \quad (95)$$

In the step (93), an inner sphere electron transfer occurs to produce cationic states from the negatively charged quencher. The chief difficulty found (ref 111) is the low ion pair formation constants at the high ionic strength, making it difficult to produce substantial quantities of ion pairs.

In the presence of excess added ligand, photolysis (ref 112) of the ionic complex $[Co(CO)_3L_2]^+$ $[Co(CO)_4]^-$ results in the formation of $[Co(CO)_2L_3][Co(CO)_4]$ but the irradiation of the ionic complex $[Co(CO)_3L_2][Co(CO)_4]$ itself with L=Bu_3P leads to dissociative loss of either CO or L to produce coordinatively unsaturated species, eqns (96-102).

$$2[Co(CO)_3L_2]\ [Co(CO)_4] \xrightarrow{h\nu} Co_2(CO)_6L_2 + Co_2(CO)_7 + CO \quad (96)$$

$$Co_2(CO)_6L_2 \longrightarrow 2\,{}^{\cdot}Co(CO)_3L \quad (97)$$

$${}^{\cdot}Co(CO)_3L + L \rightleftharpoons {}^{\cdot}Co(CO)_3\,L_2 \quad (98)$$

$${}^{\cdot}Co(CO)_3L_2\quad {}^{\cdot}Co_2(CO)_6L_2 \longrightarrow Co(CO)_3L_2^{\ +} + Co_2(CO)_6\,L_2^{\ \cdot-} \quad (99)$$

$$Co_2(CO)_6L_2^{\ \cdot-} \longrightarrow {}^{\cdot}Co(CO)_2L_2 + Co(CO)_4^{\ -} \quad (100)$$

$${}^{\cdot}Co(CO)_2L_2 + L \rightleftharpoons {}^{\cdot}Co(CO)_2\,L_3 \quad (101)$$

$${}^{\cdot}Co(CO)_2L_3 + Co_2(CO)_6L_2 \longrightarrow [Co(CO)_2L_3] + [Co_2(CO)_6L_2]^{\cdot-} \quad (102)$$

Following electron transfer in the step (99), this product may then react with $[Co(CO)_4]^-$ to give $[Co_2(CO)_6L_2]$ and $Co_2(CO)_7L$. The above scheme is similar to the one proposed by T.L. Brown et al. (ref 113) for the thermal reaction of $Co_2(CO)_8$ with tributylphosphine. For the ionic complex $[Co(CO)_3L_2]$-$Co(CO)_4$, the key step is electron transfer from the 19 electron ${}^{\cdot}Co(CO)_3L_2$

species to the neutral complex $Co_2(CO)_6L_2$. The quantum yield is 0.2.

The electronic spectra of ion pairs $[(NH_3)_5\ Co^{III}\ NC\ M^{II}\ (CN)_5]^-$ with M=Ru,Fe exhibit a new absorption band assigned (refs 114,115) to an intervalence transfer (IT) transition from $[Co^{III}$ to $M^{II}]$. Vogler and Kunkely (refs. 114,115) report very efficient formation of Co^{II} induced by irradiation into this IT region. The formed ion pair underwent a redox reaction between of $[M^{III}\ (CN)_6]^{3-}$ and Co^{2+} with a quantum yield $\Phi = 0.46$.

$$[(NH_3)_3\ Co^{III} - NC- M^{II}(CN)_5]^- \xrightarrow[IT]{hv} [(NH_3)_3\ Co^{II}\ NC\ M^{III}\ (CN)_5] \longrightarrow \quad (103)$$

$$Co^{II}aq + 5NH_3 + \ldots \qquad M = Ru,\ Fe$$

These results can be rationalized by the well-known scavenging of the kinetically labile $cobalt^{II}$ by fast aquation reactions. The aquation of the $cobalt^{II}$ prevents back electron transfer.

Hennig, Rehorek, and Coll (refs 116-119) studied aqueous solutions of the following cyanometallates: octacyanomolybdate (IV) $[Mo(CN)_8]^{4-}$, octacyanotungstate (IV) $[W(CN)_8]^{4-}$, hexacyanoferrate $[Fe(CN)_6]^{4-}$ and hexacyanoruthenate (II) $[Ru(CN)_6]^{4-}$. These complexes exhibit remarkable color changes after mixing with Fe^{II}, Cu^{II}, UO_2^{2+}, VO^{2+} and Ce^{IV} or upon addition of Co^{III} amine complexes. Addition of Cr^{III}, Co^{II}, Ni^{II}, Zn^{II}, Hg^{II} and Tl^{I} leads to no change in their electronic spectra.

In heteronuclear mixed-valence compounds between octacyanomolybdate (IV) and an appropriate counterion, light-induced electron transfer caused by IT excitation is observed . This electron transfer leads to the formation of a vibrationally-excited valence isomeric from octacyanomolybdate and a reduced metal center. $[Mo(CN)_8]^{3-}$ has a well known kinetic lability and fast cyanide splitting is expected to compete efficiently with back electron transfer.

Hennig et al. have (refs 116-119)investigated in detail the photoreactivity of mixed-valence systems based on $[Mo(CN)_8]^{4-}$. The primary step following the irradiation of $Cu^{II}/[Mo(CN)_8]^{4}$ is achieved by IT excitation of the $Cu^{II}/\ [Mo(CN)_8]^{4-}$ pair. This step is followed by consecutive thermal reactions leading to free cyanide, eqn (104).

$$Cu^{II}/\ [Mo(CN)_8]^{4-} \xrightarrow{hv} Cu^{I}\ /\ [Mo(CN)_8]^{3-} \longrightarrow Cu^{+}S;\ CN^{-}\ \text{etc}\ S = \text{scavenger} \quad (104)$$

Unfortunately the other cyanometallates under investigation do not show

comparable changes of their kinetic behavior upon changing their oxidation number and therefore they have not been investigated photochemically.
Vogler and Kisslinger (refs 120, 121) report photochemical reaction induced by an optical IT transition in Ru^{II} and Ru^{III}. On mixing aqueous solutions of $[Ru^{II}(NH_3)_5Cl]Cl_2$ and $K_4[Ru^{II}(CN)_6]$, a color develops immediately and a new absorption band with a maximum at 510 nm appears. The new absorption band is assigned to an outer-sphere Ru^{II} to Ru^{III} IT transition within the ion pair $[Ru(NH_3)_5Cl]^{2+}/[Ru(CN)_6]^{4-}$. Upon irradiation into the IT band λ <490 nm) the solution turns blue via formation of the complex $[(NH_3)_5Ru^{III}(CN)Ru^{II}(CN)_5]^-$ with a quantum yield $\Phi = 0.002$. The mechanism suggested (refs 120,121) is as follows, eqns (105-109).

$$[Ru^{III}(NH_3)_5Cl]^{2+} + [Ru^{II}(CN)_6]^{4-}) \longrightarrow [Ru^{III}(NH_3)_5\text{-}Cl]^{2+}[Ru^{II}(CN)_6]^{4-} \quad (105)$$

$$[Ru^{III}(NH_3)_5Cl]^{2+}[Ru^{II}(CN)_6]^{4-} \xrightarrow{h\nu} [Ru^{II}(NH_3)_5Cl]^{+}/[Ru^{III}(CN)_6]^{3-} \quad (106)$$

$$[Ru^{II}(NH_3)_5Cl]^{+}/[Ru^{III}(CN)_6]^{3-} \longrightarrow [Ru^{II}(NH_3)_5Cl]^{+} + [Ru^{III}(CN)_6]^{3-} \quad (107)$$

$$[Ru^{II}(NH_3)_5Cl]^{+} + H_2O \longrightarrow [Ru^{II}(NH_3)_5]H_2O^{2+} + Cl^- \quad (108)$$

$$[Ru^{II}(NH_3)_5H_2O]^{2+} + [Ru^{II}(CN)_6]^{4-} \longrightarrow [(NH_3)_5Ru^{II}(NC)Ru^{II}(CN)_5]^{2-} \quad (109)$$

$[Ru^{III}(NH_3)_5Cl]^{2+}$ and $[Ru^{II}(CN)_6]^{4-}$ give the ion pair $[Ru^{III}(NH_3)_5Cl]^{2+}/$

$[Ru^{II}(CN)_6]^{4-}$. The IT excited ion pair, eqn (106), could diffuse apart, eqn (107), with $k=5\ s^{-1}$. $[Ru^{II}(NH_3)_5Cl]^+$ aquates rapidly, eqn (108), and reacts with $[Ru^{II}(CN)_6]^{4-}$ eqn (109).Then electron transfer restores ruthenium to its stable oxidation state, yielding a blue product. The overall reaction has a very low quantum yield ($\Phi = 0.002$) because equation 106 is easily reversed thermally. This competes with the separation of the primary electron transfer product ions, eqn (107).

7 - Miscellaneous

We do not want to end this chapter without reporting interesting works on photochemical acceleration of thermal oxidative additive process and on mercury ($5d^{10}6s^2$) organometallic complexes, which are a little out of the scope of the subject but merit attention.

Osborn et al. (refs 122-125) have observed that during the photolysis of trans-$Ir(CO)(PR_3)_2$, PR_3=PMe_3 23, PMe Ph_2 24 with a variety of alkyl halides leads to marked acceleration of the thermal oxidative addition process involving a radical chain pathway, eqn(110).

$$\text{trans- Ir }(PMe_3)COCl + CH_3CH_2Br \xrightarrow[h\nu]{C_6H_6} IrClBr(CH_3CH_2)CO(PMe_3)_2 \qquad (110)$$

The oxidative addition could also occur photochemically, whereas the thermal reaction has not been observed. It must be noticed that in many of these reactions irradiation after completion of addition leads to further products. This is well illustrated by reaction of 24 with C_2H_5I in benzene; good yields of the expected adduct are obtained after 20 min., but it disapears if the irradiation is continued. An attractive mechanism for such reaction can be based upon the known reaction of Co^{II} complexes with alkyl halides (ref 126). The propagation steps consist of addition of alkyl radical to Ir (I), generating RIr^{II} which, like Co^{II}, can then abstract a halogen atom from RX, regenerating $R^{\cdot}$, eqns (111-113).

$$R^{\cdot} + Ir^{I} \longrightarrow RIr^{II} \qquad (111)$$

$$RIr^{II} + Rx \longrightarrow RIr\ x + R^{\cdot} \qquad (112)$$

$$\text{overall } Rx + Ir^{I} \longrightarrow RIrX \qquad (113)$$

Such an abstraction of halide from RX may proceed via an electron transfer step similar to that reported by Kochi et al. (ref 127) for reaction of Ni(0) eqn (114).

$$[R\text{-}Ir^{II}..\ X\text{-}R] \rightarrow [R\text{-}Ir^{III}...\ X - R^{\bar{\cdot}}] \longrightarrow [R\text{-}Ir^{III}...X^{-}...R^{-}] \longrightarrow$$

$$R - Ir^{III} - X + R^{\cdot} \qquad (114)$$

Kochi et al. (ref 128) report electron transfer from alkylmetals RM to I_2, where M = tin, lead, mercury, and suggest that for alkylmetals and iodine, the electronic excitation occurs essentially from the ground state structure of RMI_2 to a charge separated to one ($RM^{+}\ Ir^{-}$), i.e., it corresponds to an intramolecular transition $h\nu_{CT}$ within the complex involving electron transfer from RM to I_2, eqn. 115).

$$[RMI_2] \xrightarrow[CT]{h\nu} [RM^{+}I_2^{-}]^{*} \qquad (115)$$

Shulpin et al. (ref 129) report that light irradiation of frozen solution of $PtCl_6^{2-}$ and Ph_2Hg in CH_3COOH exhibited ESR spectra containing signal due to platinium III complex and the organic radical (or cation radical) at g ≃ 2, eqns (116,117).

$$Ar_2Hg + [Pt^{IV}Cl_5]^- \rightarrow [Ar_2Hg]^+ [(Pt^{III}Cl_5)^{2-}] \quad (116)$$

$$[Ar_2Hg]^+ [(Pt^{III}Cl_5)^{2-}] \quad [\sigma\text{-}ArPt^{IV}Cl_5]^{2-} + ArHg^+ \quad (117)$$

Finally we notice that Russell et al. (ref 130), report that primary or secondary alkylmercury chlorides or bromides will participate in the $S_{RN}1$ process with nitronate anions in Me_2SO or DMF, eqn. (118).

$$RHgX + R'_2C = NO_2 \xrightarrow{h\nu} RC(R')_2\ NO_2 + X^- + Hg^o \quad (118)$$

8 - Concluding remarks

Therefore, in this chapter we report ET of transition metal compounds with partially filled d of f shells which contain σ M-C or M-M bonds. Several reviews of general interest for photochemical electron transfer processes have appeared recently, most noteworthy are those of Creed and Cadwell (ref 131), Chanon et al. (ref 132-136), Meyer et al. (refs 137,138) Endicott et al. (ref 139), Balzani et al. (ref 140) and Sutin and Creutz (ref 141), where one can find specific aspects of ET of excited states of metal complexes.

A lot of work needs to be done in the future, especially in the case of ET between a metal complex with large ligands and an organic compound, in ion pair systems and in organized media, in order to obtain better understanding of photocatalytic processes and biological ET involving metal complexes.

1 - H. Taube, Electron Transfer Reactions of Complex ions in Solution, Academic Press, New York (1970).

2 - M.S. Matheson, W.A. Malac and J. Rabani, J. Phys. Chem. 67 (1963) 2613.

3 - J. Jortner, M. Ottolenghi and G. Stein, J. Phys Chem. 68 (1964) 247.

4 - J.R. Huber and E. Hayon, J. Phys. Chem. 72 (1968) 3820.

5 - R.Devonshire and J.J. Weiss, J. Phys. Chem. 72 (1968) 3815.

6 - M. Shiron and G. Stein, Nature 204 (1964) 778.

7 - G. Stein, Adv. Chem. Ser. 50 (1965) 238.

8 - P.L Airey and F.S. Dainton , Proc. Roy. Soc. A. 291 (1966) 340.

9 - P.I. Airey and F.S. Dainton, Proc.Roy.Soc. A 291 (1966) 478.

10- S.I. Ohno and G. Tsuchihashi, Bull. Chem. Soc Japan 38(1965) 1052.

11- S.I. Ohno, Bull. Chem. Soc. Japan 40 (1967) 1765.

12- S.I. Ohno, ibid 40 (1967) 1776.

13- S.I. Ohno, ibid 40 (1967) 1770.

14- S.I. Ohno, ibid 40 (1967) 1779.

15- H.B. Gray and N.A. Beach, J. Amer. Chem. Soc. 85 (1963) 2922.

16- A.W. Adamson and A.H. Sporer, J. Amer. Chem. Soc. 80 (1958) 3865.

17- A.W. Adamson, W.L. Waltz , E. Zinato, D.W. Watts, P.D Fleischauer and R.D. Lindholm, Chem. Rev. 68 (1568) 541.

18- G. Emschwiller and J. Legros, C.R. Acad. Sci. 261 (1965) 1535.

19- W.L. Waltz and A.W. Adamson, J. Phys. Chem. 73 (1969) 4250.

20- W.L. Waltz, A.W. Adamson and P.D. Fleischauer, J. Amer. Chem. Soc. 89 (1967) 3923.

21- O. Kalisky and M. Shirom J. Photochem 7 (1977) 215.

22- M. Shirom and G. Stein, J. Chem. Phys. 55 (1971) 3372.

23- M. Shirom and G. Stein, J. Chem. Phys. 55 (1971) 3379.

24- U. Lachish , A. Shafferman and G. Stein, J. Chem. Phys. 64 (1976) 4205.

25- M. Shirom and M. Tomkiewicz , J. Chem. Phys. 56 (1972) 2731.

26- A. Mozumder, J. Chem. Phys. 48 (1968) 1659.

27- M. Shirom and M. Weiss, J. Chem. Phys. 56 (1972) 3170.

28- A. Horvath and S. Papp, Acta Chimica Hungaria 115 (1984) 415.

29- A.W. Adamson and F. Basolo, Acta Chem. Scand. 9 (1955) 1261.

30- A.W. Adamson, Disc. Faraday Soc. 29 (1960) 163.

31- V. Balzani, R. Ballardini, N. Sabbatini, and L. Moggi Inorg. Chem. 7 (1968) 1398.

32- F. Scandola, C. Bartocci, and M.A. Scandola, J. Phys. Chem 78 (1974) 572.

33- R.D. Revonshire and J.J. Weiss, J. Phys. Chem. 72 (1968) 3815.

34- J.C.D. Brand and W. Snedden, Trans. Faraday Soc. 53 (1957) 894.

35- O. Traverso and F. Scandola , Inorg. Chim. Acta 4 (1970) 493.

36- J.A. Powell and S.R. Logan, J. Photochem. 3 (1974) 189.

37- E.K. Heaney and S.R. Logan, J. Chem. Soc. Perkin II (1978) 590.

38- E.K.Heaney, S.R.Logan,and W.E. Watts, J. Organometal. Chem. 150, 1978, 309.

39 - E.F. Hockings and I. Bernal, J. Chem. Soc. (1964) 5029.

40 - M.C.R. Symons, D.X. West and J.G. Wilkinson, J.Chem. Soc.(1973) 917

41 - M.C.R. Symons, J.G. Wilkinson and D.X. West, J. Chem. Soc. Dalton Trans (1982) 2041 .

42 - H. Henning, A. Rehorek, D. Rehorek and P. Thomas, Z. Chem. 22 (1982) 418.

43 - D. Rehorek, J. Salvetter, A. Hantschmann and H. Hennig, Inorg. Chim. Acta , 37 (1979) L 471.

44 - J.O. Dziegielewski, Polyhedron (1984) 1131.

45 - C. Giannotti and J.R. Bolton, J. Organomet. Chemistry 80, (1974) 379.

46 - C. Giannotti and J.R. Bolton, J. Organomet. Chemistry 110 (1976) 383.

47 - A.N. Nesmeyanov, V.A. Sazonova,U.I. Romanenko, N.A. Rodionova and G.P. Zol'nikova, Dokl. Akad. Nauk. SSSR 155 (1964) 1130.

48 - A.N. Nesmeyanov, Pure Appl. Chem. 17 (1968) 220.

49 - A.N. Nesmeyanov, V.A. Sazonova, V.I.Romanenko, V.N. Postov, G.P. Zol'nikova, V.A. Blonova and R.M. Kalyanova, Dokl. Akad. Nauk. SSSR 173 (1967) 589.

50 - H.C.H.A. Van Riel, F.C. Fischer, J. Lugtenburg and E. Havinga, Tetrahedron Lett. (1969) 3085.

51 - J.C. Brand and S. Snedden, Trans. Faraday Soc. 53 (1957) 894.

52 - E. Körner von Gustorf, H. Köller, M.J. Jun and G.O. Schenck, Chem. Eng. Tech. 35 (1963) 591.

53 - I.J. Spilners, J. Organomet. Chem. 11 (1968) 381.

54 - O. Traverso and F. Scandola, Inorg. Chim. Acta 4 (1970) 493.

55 - Y. Hoshi, T. Akiyama and A. Sugimori, Tetrahedron Lett. (1970) 1485.

56 - G.L. Geoffroy and M.S. Wrighton, Organometallic Photochemistry, Academic Press, New York (1979).

57 - T. Akiyama, Y. Hoshi, S. Goto and A. Sugimori, Bull Chem. Soc. Jap. 46 (1973) 1851.

58 - T. Akiyama, A. Sugimori and H. Hermann, Bull. Chem. Soc. Jap. 46 (1973) 1855.

59 - D.R. Scott and R.S. Becker, J. Chem. Phys. 35 (1961) 516.

60 - T. Akiyama, T. Kitamura, T. Kato, H. Watanabe, T. Serizawa and A. Sugimori, Bull. Chem. Soc Jap 50 (1977) 1137.

61 - D.M. Papsun, J.K. Thomas and J.A. Labinger, J. Organomet. Chem. 208 (1981) C36.

62 - W.G. Herkstroetter, J. Amer. Chem. Soc. 97 (1975) 4161.

63 - K. Kikichi, H. Kokubun and Kikuchi, Bull. Chem. Soc. Japan 48 (1975) 1378.

64 - M. Almgren and J.K. Thomas in K.L. Mittal, Ed. Solution Chemistry of

surfactants, Vol 2 , Plenum Press, New-York (1979), P 559.

65 - P. Borrell and E. Henderson, J. Chem. Soc (1975) 432.

66 - V. Balzani, F. Boletta, M.T. Gandolfi and M. Maestri, Topics Current Chem. 75 (1978) 1.

67 - A. Vogler, W. Losse and H. Kunkely, J. Chem. Soc., Chem. Comm. (1979) 187.

68 - K.R. Mann, H.B. Gray and G.S. Hammond, J. Amer. Chem. Soc. 99 (1977) 306.

69 - P.W.N.M. Van Leluwen, R. Kaptein, R. Huis and C.F.Roobeek, J. Organomet. Chem. 104 (1976) C44.

70 - O. Traverso, C. Chiorboli, U. Mazzi and G.L. Zucchini. Gazz. Chim. Italiana 107 (1977), 181.

71 - J.N. Demas, J.W Addington, S.H. Peterson and F.W. Harris, J. Phys. Chem. 81 (1977) 1039.

72 - S.M. Fredericks and M.S. Wrighton, J. Amer. Chem. Soc. 102 (1980) 6166.

73 - D.P. Summers, J.C. Luong and M.S. Wrighton, J. Amer. Chem. Soc. 103 (1981) 5238.

74 - J.C. Luong, L. Nadjo and M.S. Wrighton, J. Amer. Chem. Soc. 100 (1978) 5790.

75 - T.J. Meyer and J.V. Cappar, Chem. Rev. 85 (1985) 187.

76 - H. Yesaka, T. Kobayashi, K. Yasufuku, and S. Nakamura, J. Amer. Chem. Soc. 105 (1983) 6249.

77 - H.W. Walker, R.S. Herrick, R.J. Olsen, and T.L. Brown, Inorg. Chem. 23 (1984) 4550.

78 - J.L. Highey, C.R. Bock and T.J. Meyer, J. Amer. Chem. Soc. 97 (1975) 4440.

79 - M.K. Reinking, M.L. Kullberg, M.L. Cutler and C.P. Rubiak, J. Amer. Chem. Soc. 107 (1985) 3517.

80 - C. Giannotti and G. Merle, J. Organometal. Chem. 105 (1976) 97.

81 - A.E. Steigman and P.R. Tyler, Inorg. Chem 23 (1984) 527 and references therein.

82 - A.F. Hepp and M.S. Wrighton, J. Amer. Chem. Soc. 105 (1983) 5035 .

83 - A.F. Hepp and M.S. Wrighton, J. Amer. Chem. Soc. 103 (1981) 1258.

84 - C. Giannotti and G. Mousset, Tetrahedron Lett. (1980) 2155

85 - J.C. Luong, R.A. Faltynek and M.S. Wrighton, J. Amer. Chem. Soc. 102 (1980) 7893.

86 - D. Rehm and A. Weller, Ber. Bunsenges Phys. Chem. 73 (1969) 834

87 - C.R. Bock, I.J. Meyer and D.G. Whitten, J. Amer. Chem. Soc. 97 (1975) 2909.

88 - N. Sutin and C. Creutz, Adv. Chem. Ser 168 (1978) 1.

89 - D.G. Whitten, Acc. Chem. Res. 13 (1980) 83.

90 - V. Balzani, F. Bolleta, F. Scandola and R. Ballardini, Pure Appl. Chem. 51 (1979) 299.

91 - F. Fukuzumi, M. Nishigawa and T. Tanaka, Chem. Lett. (1982) 719.
92 - F. Fukuzumi, N. Nishizawa and T. Tanaka, Bull. Chem. Soc. Japan 56 (1983) 709.
93 - N.S. Lewis, K.R. Mann, J.G. Gordon II and H.B. Gray, J. Amer. Chem. Soc. 98 (1976) 7461.
94 - S.J. Milder , R.A. Golberk, D.S. Kliger and H.B. Gray, J. Amer. Chem. Soc. 102 (1980) 6761.
95 - V.M. Miskowski, G.L. Mobinger, D.S. Kliger, G.S. Hammond, N.S. Lewis, K.R. Mann and H.B. Gray, J. Amer. Chem. Soc. 100 (1978) 485.
96 - P.A. Metcalf and C.P. Kubiak J. Amer. Chem. Soc. 108 (1986) 4682.
97 - A. Juris, M.T. Gandolfi, M.F. Manfrin and V. Balzani J. Amer. Chem. Soc 98 (1976) 1047.
98 - C. Creutz, and N. Sutin, Inorg. Chem 15 (1976) 496.
99 - J.M. Delmas and J.W. Addington, J. Amer. Chem. Soc. 98 (1976) 5801.
100 - A. Juris, M.F. Manfrin, M. Maestri and N. Serpone, Inorg. Chem 17 (1977) 8.
101 - S.S. Atik and J.K. Thomas J. Amer. Chem. Soc. 103 (1981) 3543.
102 - Y. Morsi, A. Braun and M. Grätzel, J. Amer. Chem. Soc 101 (1979) 567.
103 - S.A. Alkaitis , M. Grätzel and A. Henglein , Ber. Bunsenges. Phys Chem. 79 (1975) 541.
104 - E.R. Biehl, H. Chiou, J. Keepers, S. Kennard and P.C. Reeves, J. Heterocyl. Chem. 12 (1975) 397.
105 - B. Tae Ahn and D.R. McMillin, Inorg Chem. 17 (1978) 2253.
106 - B. Tae Ahn and D.R. McMillin, Inorg. Chem 20 (1981) 1427.
107 - A.D. Kirk and G.B. Porter, J. Phys. Chem. 84 (1980) 2998.
108 - F.B. Ueno, Y. Sasaki, T. Ito and K. Saito, J. Chem. Soc. Chem. Comm (1982) 328.
109 - N.A.P. Kane-Maguire and C.H. Langford , J. Chem. Soc. Chem. Comm. (1973) 351.
110 - C.H. Langford and R. Sasseville, Can. J. Chem. 59 (1981) 647.
111 - W. Rybak, A. Haim, T.L. Netzel and N. Sutin, J. Phys Chem. 85 (1981) 2856.
112 - M.F. Mirbach , M.J. Mirbach and R.W. Wegman, Organomet. 3 (1984) 900.
113 - M. Absi-Halabi, J.D. Atwood, N.P. Forbus and T.L. Brown, J. Amer. Chem. Soc. 102 (1980) 6248.
114 - A. Vogler and H. Kunkely , Ber. Bunsenges. Phys. Chem. 79 (1975) 89.
115 - A. Vogler and H. Kunkely, Ber. Bunsenges. Phys. Chem. 79 (1975) 301.
116 - D. Rehorek, A. Rehorek, P. Thomas and H. Hennig, Inorg. Chim. Acta 64 (1982) L225.
117 - H. Hennig, A. Rehorek, D. Rehorek and Ph. Thomas, Z. Chem. 22 (1982) 388.

118 - H. Hennig, A. Rehorek, D. Rehorek, P. Thomas and D. Bäzold, Inorg. Chim. Acta 77 (1983) L11.

119 - H. Hennig, A. Rehorek, D. Rehorek and Ph. Thomas Inorg. Chim. Acta. 86 (1984) 41.

120 - A. Vogler and J. Kisslinger, Angew. Chem. Int. Ed. Engl. 21 (1982) 77.

121 - A. Vogler and J. Kisslinger, J. Amer. Chem. Soc. 104 (1982) 2311.

122 - J.A. Labinger, R.J. Braus, D. Dolphin and J.A. Osborn, J. Chem. Soc., Chem. Comm. (1970). 612.

123 - J.S. Bradley, D.E. Connor, D.Dolphin, J.A. Labinger and J.A. Osborn , J. Amer.Chem. Soc. 94 (1972) 4043.

124 - J.A. Labinger and J.A. Osborn, Inorg. Chem. 19 (1980) 3230.

125 - J.A. Labinger, J.A. Osborn and N.J. Coville, Inorg. Chem. 19 (1980) 3326.

126 - J. Halpern, Acc, Chem. Res. 3(1970) 386.

127 - T.T. Tsou and J.K. Kochi, J.Amer. Chem.Soc. 101 (1979) 6319.

128 - S. Fukuzumi and J.K. Kochi, J. Amer. Chem. Soc. 102 (1980) 2141.

129 - G.B. Shulpin, G.V. Nizova and A.T. Nikitaev, J. Organomet. Chem. 276 (1984) 115.

130 - G.A. Russell, J. Hershberger, K. Owens, J. Amer.Chem. Soc. 101 (1979) 1312.

131 - D. Creed and R.A. Caldwell, Photochem. Photobiol. 41 (1985) 715.

132 - M. Juliard and M. Chanon, Chem. Rev. 83 (1983) 425.

133 - M. Juliard and M. Chanon, Chem. Ser. 24 (1985) 11.

134 - M. Chanon, Bull. Soc.Chim. France (1982) 197.

135 - M. Chanon, Bull. Soc. Chim. France (1985) 209.

136 - M. Chanon, Acc. Chem. Res. 20 (1987) 214.

137 - T.J. Meyer, Prog. Inorg. Chem. 30 (1983) 389.

138 - T.J. Meyer and J.V. Caspar, Chem. Rev. 85 (1985) 187.

139 - J.F. Endicott, K. Kumar, T. Ramasami and F.P. Rotzinger, Prog. Inorg. Chem. 30 (1983) 141.

140 - V. Balzani, F. Bolleta, M. Ciano and M. Maestri, J. Chem. Ed. 60 (1983) 447.

141 - N. Sutin and C. Creutz, J. Chem. Ed. 60 (1983) 809.

6. APPLICATIONS

Chapter 6.1

Photocatalysis on Semiconductors

P. Pichat and M.A. Fox

1. INTRODUCTION

1.1. Scope of the Article

A topic of increasing importance for both fundamental investigations and for practical applications in solar energy conversion and in imaging technology involves photoinduced electron transfer across the gas-solid or liquid-solid interface at the surface of an excited semiconductor in contact with an appropriate redox couple. Of greatest interest are those chemical reactions induced by photochemical excitation of the surface from which the solid semiconductor emerges intact. Such photocatalysts provide a surface useful both for initiating redox half reactions and for providing a well-defined environment to control the subsequent chemical reactivity of the photogenerated intermediates.

The principles operative in this photoelectrochemical catalysis have been reviewed elsewhere (refs. 1-12) and will be discussed here only briefly to orient the reader to the conceptual basis for photocatalysis. This article will then provide a brief overview of recent advances in understanding the often complex chemistry which ensues upon irradiation of this heterogeneous system. We shall briefly discuss the bulk and surface properties which control interfacial electron transfer as well as the available methods for surface modification which can control more effectively the critical interfacial electron transfer. We will analyze the techniques for establishing the photocatalytic character and efficiency of a given surface-mediated photoinduced electron transfer and will provide representative inorganic and organic examples which illustrate the wide applicability of these photodriven events.

To date, many investigations have focused on inducing redox reactivity in simple inorganic molecules (e.g., in water splitting or carbon dioxide or nitrogen reduction), but the principles uncovered by these studies are equally important in investigating organic transformations. This area is inherently interdisciplinary, and a comprehensive compilation of all work relevant to photoinduced reactions on semiconductor surfaces would exceed the scope of this article. The examples cited here, however, should allow the reader to understand the mechanistic features of catalytic photoelectrochemical transformations and to design practical experiments using these principles.

1.2. Principles of Photocatalysis

Photoelectrochemical catalysis involves redox reactions of organic or inorganic molecules (refs. 1,13,14) initiated by the absorption of a band gap photon at or near the surface of a semiconductor particle. The adsorbed substrate of interest is not converted to an excited state, the photon having instead been used to excite the solid semiconductor.

Metals, by definition, comprise a near continuum of states between which electrons can freely move. Electronic excitation energy of a metal will be dissipated therefore by nonradiative decay much more efficiently than it can initiate electron exchange with an adsorbed substrate.

Semiconductors are characterized by energetically non-overlapping bands. Photochemical excitation of a semiconductor thus accomplishes a band-to-band electronic transition, moving an electron from the filled valence band to the vacant conduction band, Figure 1. Rapid relaxation of the high energy

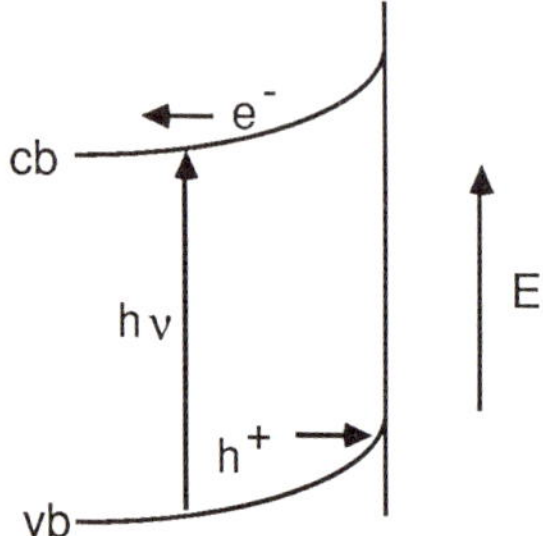

Fig. 1. Photoexcitation on n-Type Semiconductor (cb = conduction band; vb = valence band; vertical line represents surface).

electron is inhibited by the absence of intra-gap states so that an excited semiconductor particle can be envisioned as an electron-hole pair poised at the band edge potentials of the conduction and valence bands respectively. The lifetime of this charge-separated pair is sufficiently long to allow the conduction band electron to be captured by interfacial electron transfer to an appropriate adsorbed acceptor and for the valence band hole to be filled by dark interfacial electron transfer from an adsorbed donor.

When a semiconductor is exposed to a redox couple, interfacial electron transfer will occur to equilibrate the redox pair with the Fermi level of the bulk semiconductor. In an intrinsic (undoped) semiconductor, the Fermi level will lie at a potential about halfway between the bands, whereas with an n-type (negatively doped) material the Fermi level will lie just below the conduction band edge. In a p-type (positively doped) material the Fermi level will be poised just above the valence band edge. Upon equilibration with the redox couple, the bulk Fermi level moves to its equilibrium position while the surface band edge positions remain fixed. Bending of the bands is thus encountered upon moving inward from the semiconductor-electrolyte interface. Interfacial electron transfer at the surface of a particle possessing bent bands will impel electrons to move in the opposite direction from holes: in a n-doped material, electrons will move toward the bulk and holes will move toward the surface. Even if the particle dimensions are too small for full development of a space-charge layer, interfacial electron transfer is often observed, thus initiating photocatalytic redox reactivity.

2. CONTROL OF PHOTOINITIATED INTERFACIAL ELECTRON TRANSFER

2.1. Bulk Effects

2.1.1. Thermodynamic Considerations. The electrochemical potentials of the semiconductor bands, which can be established independently by any of a number of physical or electrochemical methods, (ref. 15) can be used to evaluate the thermicity of interfacial electron transfer. Since relaxation to the band edge potentials occurs within a few picoseconds after excitation, the band edge positions define the limit of attainable photopotential on an irradiated semiconductor surface. As long as the donor's oxidation potential lies less positive than the valence band edge and the acceptor's reduction potential lies less negative than the conduction band edge, electron transfer at the illuminated interface will be thermodynamically permissible, Figure 2.

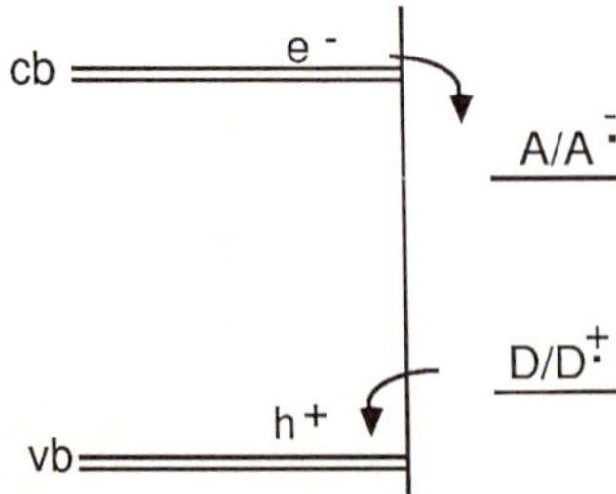

Fig. 2. Thermodynamic Requirements for Redox Potentials of Donors and Acceptors Compared with Flat Band Potentials. (cb = conduction band; vb = valence band; vertical line represents interface between the solid semiconductor and the electrolyte containing the adsorbed donor and acceptor.)

Since typical redox potentials for common organic and inorganic redox couples often lie between the band edges of easily accessible semiconductors, many electron transfer reactions will be feasible, from thermodynamic criteria. (refs. 16-17)

When donors and acceptors are both adsorbed on a photocatalyst, oxidative and reductive exchanges occur on the same surface. Provided that these adsorbed intermediates move together or diffuse away from the surface before back electron transfer occurs, novel chemistry can ensue. The generation of both oxidized and reduced species on the same surface differentiates photoelectrochemical redox reactions from conventional electrochemical transformations on inert electrodes where the electrode surface is rendered either oxidizing or reducing (but not both) by applying an external potential.

2.1.2. Stability. Photocatalytic charge separation on irradiated semiconductor surfaces is sometimes complicated by the instability of the irradiated semiconductor, especially in aqueous solutions. Metal oxides can withstand decomposition under photochemical irradiation, but narrow band gap semiconductors, which are thus responsive to visible light, often degrade rapidly upon exposure to light. In these cases, deactivation is often caused by the reactivity of the semiconductor itself, so that light induces formation of an insulating (blocking) layer or corrosion of the particle. Although surface modification by attaching catalysts or inert coverages, by chemically altering the surface composition of the semiconductor (e.g., by silicide formation), by changing the composition of the electrolyte, or by imbedding the semiconductor in a protective layer or film can at least partially control this undesired decomposition, many photocatalytic studies employ robust, highly stable metal oxides. The wide band gap materials are unfortunately less useful for solar energy conversion, since they are responsive only to ultraviolet excitation.

2.1.3. Morphology. Analogous photoelectrochemical events occur whether the semiconductor is formulated as a single crystal, a polycrystalline electrode, as a powder, or as a finely divided colloid. Although there is not a uniformly accepted definition of a colloid, we can think of a colloidal suspension of semiconductor particles as being composed of particles too small for microscopic resolution, yet which diffract light and settle out from suspension, especially upon aggregation. In practice, photoactivity is maintained as the size of the particle decreases to the colloidal range, (refs. 18-19) although the absorption characteristics, the quantum efficiency of charge separation, and the kinetics of interfacial electron transfer may be influenced by the particle size.

Effective photoelectrochemical conversion on non-metallized semiconductor surfaces requires minimal dark overpotentials for either the oxidation or the reduction half reaction so that scavenging of the photogenerated charge carriers is efficient. For some photoinduced redox reactions, metallization of the semiconductor photocatalyst will be essential, whereas for others metallization will have nearly no effect. Secondary effects of metal deposition will be covered in more detail in the following section.

2.1.4. Formation of radical ions. Three adjustable parameters (photopotential, surface adsorption and photocurrent) significantly control photocatalysis. Selectivity in activating a specific functional group in a multifunctional molecule or in activating one species from a mixture of adsorbates can, in principle, be attained by judicious choice of the semiconductor. (refs. 12,20-21) The band positions of common semiconductors are known to shift with pH and with solvent. The bands of TiO_2, for example, shift negative by about 0.7 V upon shifting the solvent from water (pH=1) to acetonitrile. (ref. 22) This shift allows for the reduction of oxygen in acetonitrile, since its potential (-0.78 V, ref. 23) is nearly isoenergetic with the conduction band position of TiO_2. Thus, oxygen can act as an electron acceptor, generating superoxide, eqn 1, and other activated oxygen species on illuminated TiO_2 particles.

$$TiO_2 \longrightarrow h^+ + e^- \xrightarrow{O_2} O_2^{\overline{\cdot}} \qquad (1)$$

Other acceptors which have reduction potentials below the conduction band edge of the illuminated semiconductor can also function in this role. The single electron reduction product derived from methyl viologen, for example, can be detected spectroscopically when a colloidal suspension of titanium dioxide is flashed in the presence of this electron acceptor. (ref. 24) The activity of an adsorbed donor requires a substrate whose oxidation potential lies less positive than the valence band edge of the semiconductor and which can therefore form a radical cation upon photoexcitation. Many photocatalytic oxidations can be rationalized on this basis. (ref. 16) In sections V and VI we cite some recent illustrative examples.

Spectroscopic evidence for the transient formation of radical cations can be obtained by many techniques. For example, when colloidal TiO_2 suspended in an acetonitrile solution containing trans-stilbene (a species which should also be exothermically oxidized by a TiO_2 valence band hole) was excited with a laser pulse, (ref. 24) a transient identical in spectroscopic features and in lifetime to an authentic sample of the stilbene cation radical (generated in the same medium via pulse radiolytic techniques) was observed. With analogous methods, time resolution for interfacial electron transfer has been studied by flash photolysis (refs. 25-29) and the dynamics of subsequent reactions of the photogenerated intermediate have been established. Electron spin resonance methods can be employed to characterize trapped electrons and intermediates. (ref. 30-31) Conductivity, (ref. 32-35) polarized infrared modulation, (ref. 36) Raman spectroscopy, (ref. 37) and rotating ring disc electrochemical techniques (ref. 38) have been successfully utilized in recent attempts to characterize redox intermediates formed photocatalytically.

Chemoselectivity can be attained, in principle, if the oxidation potential of a desired donor adsorbate lies between the valence band edges of two possible semiconductor photocatalysts. Since TiO_2 has a more positive valence band edge than does CdS, it should be the more active photocatalyst. Consistent with this idea, decarboxylation of organic acids is much more efficient on irradiated suspensions of rutile than of CdS. (ref. 39) Whether the recently reported regioselective oxidation of lactic acid (oxidation on platinized TiO_2 led to decarboxylation, while that on platinized CdS led to pyruvic acid, ref. 40) can be attributed to band position control is unclear.

Thus, the bulk properties of the irradiated semiconductor are greatly influential in controlling the observed redox chemistry. The production of radical ions via photocatalyzed single electron transfer across the semiconductor-electrolyte interface will often be a primary mechanistic step in these reactions. The chemical consequences which follow the initial electron transfer seem to be dictated, however, by surface effects.

2.2 Surface Effects

2.2.1. Adsorption sites. As in thermally activated catalytic reactions surface features play an important role in the activity and selectivity of photocatalytic processes. Therefore, they are as relevant as are energy levels and redox potentials for heterogeneous photocatalysis. Surface sites intervene in (i) the amounts and types of adsorption of the reactants, in the dissociation or activation of particular bonds, and in the control of stereochemistry, (ii) the mobility of the reactants, intermediates and products at the surface and in the kinetics of reverse reactions, and (iii) the competitive desorption of products. Their nature depends on the pretreatment and on the reaction conditions. In aqueous solutions, pH also influences the electrostatic adsorption of ions. Probe molecules, attuned to a particular spectroscopic method are used to obtain information on surface sites, e.g., acid and base sites. (ref. 41) In fact, since photocatalysis is a relatively recent discipline, methods for characterizing surface features have not yet been developed as fully as might be desired.

Kinetics experiments performed at various reactant concentrations (or pressures) can indicate whether dissociated or undissociated reactants intervene and, in some cases, can give information on the adsorbed species, on the adsorption sites (e.g., by the use of poisons, refs. 42-43), and on the adsorption competition between the reactants and the products.

For instance, the dependence of the reaction rate r on the concentration C (or pressure) of one reactant can be accounted for by the Langmuir expression $\Theta = KC/(1 + KC)$ where Θ is the surface coverage by this reactant. (refs. 44-50) By contrast, a linear relationship between $1/r$ and $C^{-1/2}$ indicates that r depends on a species arising from the dissociative adsorption of the reactant. (ref. 51) In the photocatalytic oxidation of NH_3 over TiO_2, for example, a plot of the reciprocal of the formation rate of N_2 against the reciprocal of NH_3 pressure yielded non-parallel straight lines at various O_2 pressures. This allowed one to conclude that the oxidation occurs via a Langmuir-Hinshelwood mechanism, oxygen and ammonia being adsorbed at different sites. (ref. 52)

Kinetic equations can also indicate the degree of inhibition by products. They allow one to suggest hypotheses concerning the type of adsorption and the nature (acidic or basic) of the adsorption sites. For instance, trichloroethylene would be adsorbed on Ti sites through its pi-electrons, whereas dichloroacetaldehyde (an intermediate in the degradation of trichloroethylene) would be dissociatively chemisorbed by attachment of the α-hydrogen to a surface O^{2-} ion and attraction of the resulting anion to a Ti site. (ref. 53)

Kinetic experiments offer the advantage of being simple and of being performed with inexpensive apparati. However, physical methods are necessary to obtain a more detailed picture of the phenomena occurring in the adsorbed phase.

ESR spectroscopy has been intensely used to determine the nature of adsorbed oxygen species. (ref. 54-57) It has also been successfully employed to show the appearance of methyl or triphenylmethyl radicals during the decarboxylation of acetic acid and triphenylacetic acid, respectively, over TiO_2 or Pt/TiO_2. (ref. 58)

Laser-flash excitation coupled with fast optical analysis has allowed considerable progress in understanding charge transfer between an excited semiconductor colloid and adsorbed reactants in aqueous or non-aqueous solutions. (refs. 24,59-66 The absorption bands observed have been attributed either to trapped charges or to ions derived from interfacial charge-transfer and are affected both by the surface coverage of the reactant and by its ease of oxidation or reduction.

The use of time-resolved Raman spectroscopy yields more detailed structural information on the species involved in charge transfer or resulting from it. (refs. 67-68)

2.2.2. Hydrogen spillover. Because group VIII metal deposits are necessary for efficient photocatalytic hydrogen evolution, it has been assumed a priori that hydrogen atoms (or protons) can migrate from the metal to the semiconductor and vice-versa. Several experiments have confirmed this hydrogen spillover. (ref. 69) IR spectroscopy has shown, for example, that titania surface OH groups are not isotopically exchanged by gaseous deuterium below 373 K in the absence of platinum. (ref. 70) This observation was corroborated by conductivity (refs. 71-72) and photoconductance measurements (refs. 72-73) comparing naked TiO_2 to this semiconductor supporting Pt, Rh, or Ni. The effect of treatments inducing so-called "strong metal-support interactions", i.e., decreasing hydrogen chemisorption, also supports the occurrence of hydrogen spillover. (ref. 74) This phenomenon can also account for temperature effects on this reaction, (ref. 74) and on the photocatalytic dehydrogenation of alcohols. (ref. 75) Finally, isotope effects observed in the photoassisted formation of hydrogen from deuterated or tritiated water mixtures over metal/titania also require hydrogen migration from the semiconductor to the metal. (refs. 76-77)

2.2.3. Photoadsorption. The photons which are used to activate the catalyst can modify dark adsorption equilibrium and, furthermore, can increase the lability of surface ions. One of the authors has recently published a review on this topic (ref. 78) and we only summarize its main conclusions here.

Band-gap illumination increases the number of charge carriers and can change the adsorption equilibrium of certain compounds. (In particular, the relative increase in the number of minority charge carriers can be considerable.) In theory, the species involved are those which carry a charge. With large adsorbed species, configurations with highly localized partial charge can also be affected.

Most studies of photoadsorption relate to oxygen on semiconductor oxides. The photoadsorbed amounts are low. For instance, maxima of the order of 0.3 molecules nm^{-2} have been reported for the O_2-TiO_2 system at pressures of a few hundreds Pa. (refs. 79-82) In addition, surface adsorption decreases with a decreasing degree of surface hydroxylation. (refs. 79-80,83-84) This has been attributed to the coupling of electron trapping by oxygen adsorbed species, such as eqns 2 and 3

$$O_2(ads) + e^- \longrightarrow O_2^-(ads) \tag{2}$$

and

$$O(ads) + e^- \longrightarrow O^-(ads), \tag{3}$$

and of hole trapping by OH^- groups, eqn 4, (refs. 55,79-80,83-85)

$$OH_s^- + h^+ \longrightarrow OH_s^\bullet. \tag{4}$$

As expected, gases capable of inducing surface changes affect the photoadsorption of oxygen. (ref. 86)

For fully hydrated TiO_2 samples, water photodesorption is reported to accompany O_2 photoadsorption; (ref. 84) this desorption was assumed to originate from a weakening of the bond between adsorbed water and surface-accessible titanium cations when their charge is decreased from 4+ to 3+ because of photo-induced electron trapping. This is an example of a light-induced desorption of species which were not ionosorbed.

Various kinetic laws have been reported for O_2 photoadsorption. (refs. 87-89) Significant differences arise from distinct samples, pretreatments, and ranges of oxygen pressure. Characterization of the adsorbed oxygen species has been carried out by ESR (refs. 54-57) (unfortunately, most often at 77 K) or by indirect electrical measurements (photoconductivity measurements (refs. 90-91) or variations of the work function by the vibrating capacitor method (ref. 92)) The existence of O_2^-, O^-, O_3^-, O_3^{3-} and O_2^{2-} species have been detected and kinetic expressions have been proposed for their formation. Only the first two species were stable at room temperature. Since free electrons in the semiconductor can be also captured by other electrophilic gases, e.g. NO, (ref. 91) photoadsorption is not restricted to oxygen.

The investigation of oxygen adsorption on a semiconductor suspended in a liquid is not as easy as that on a dry semiconductor. Nonetheless, the formation of O_2^-, $OH^•$ and $O_2H^•$ species over TiO_2 pigments in contact with water has been studied in connection with paint chalking. (refs. 93-94)

Numerous works dealing with the photodissociation of water over semiconductors (see M. Graetzel's chapter in this book) have finally defined the fate of oxygen, since the amounts of gaseous oxygen obtained in water splitting, if any, were far less than that expected from stoichiometry. From chemical redox titrations, the formation of titanium peroxo complexes, Figure 3,

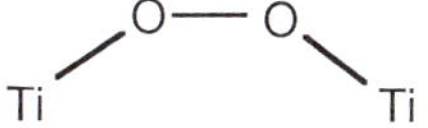

Fig. 3. A Surface-bound Titanium Peroxo Complex

was inferred. (refs. 95-97) The quantities of oxygen included within these complexes are greater than a monolayer of O_2^- because sub-surface titanium ions can be involved. (ref. 98) While illumination is required to dissociate water, the formation of peroxo complexes can probably occur in the dark and should not be considered, strictly speaking, as photoadsorption of oxygen.

2.2.4. Photodesorption. For a n-type semiconductor, intrinsic photodesorption results from the interaction of holes, produced by band-gap illumination, with negatively charged adsorbed species, generating neutral molecules which are thermally released. A clear distinction between real photodesorption and thermodesorption arising from sample heating by the illumination is not easy. Local temperature elevation can be caused by the recombination of charges produced by band-gap irradiation. The dependence of the desorption rate upon the radiant power should be definitive: linear for photodesorption at sufficiently low radiant powers (when second-order recombination of the charge carriers is not dominant), non-linear for thermodesorption (intervention of Boltzmann-type factors, exp $(-E_a/RT)$). (refs. 99-101)

Photodesorption should also be distinguished from photolysis. (refs. 99, 102,103) For unstable semiconductor oxides, oxygen evolution, eqn 5, can stem

$$O_s^{2-} + 2\,h^+ \longrightarrow 1/2\ O_2 \qquad (5)$$

from oxidation associated with the corresponding electron-consuming reaction. For instance, in the case of ZnO (refs. 99,102,104), reduced interstitial Zn^+ is produced, eqn 6:

$$Zn^{2+} + e^- \longrightarrow Zn^+ \ \text{(interstitial)} \qquad (6)$$

Since the cross-section for hole capture of oxygen ionosorbed species is much greater than that of O_s^{2-}, (ref. 102) however, photodesorption in the presence of gaseous oxygen should predominate over photolysis.

Upon illumination of several semiconductors, evolution of CO_2 has been detected. (ref. 105) This likely derives from a reaction between adsorbed oxygen and carbon-containing adsorbed species or impurities whose complete removal is arduous.

The observation of oxygen photodesorption by mass spectrometry is difficult since very low amounts are expected and since simultaneous photoadsorption takes place. However, experiments monitoring the mass peaks of oxygen isotopes indicate that oxygen photodesorption from TiO_2 and from other semiconductor oxides occurs. (refs. 106-107) Quite recently, by an indirect technique (the fluorescence of dissolved perylene as a probe for oxygen concentration in a suspension of TiO_2 in ethanol), photodesorption was found to predominate from ca. 435 to ca. 390 nm, whereas photoadsorption was more important at higher energies. Furthermore, the quantum yield was greater for photodesorption, in agreement with the oxidizing properties of titania. However, details of the experiment were not given in this brief report. (ref. 108)

2.2.5. Evidence for oxygen lability. Isotopic Exchange of gaseous $^{18}O_2$ (refs. 82,109-110) and $N^{18}O$ (ref. 111) with surface and adsorbed oxygen atoms can be used to show the lability of these oxygen atoms at room temperature under band-gap illumination. By contrast, in the dark, temperatures substantially above the ambient are generally required to allow such isotopic heteroexchanges to occur with equivalent rates. The effects of various pretreatments show that light-induced isotopic exchange involves not only adsorbed oxygen atoms, but also surface oxygen atoms which are not removed by treatment in a H_2 atmosphere at 723 K. (ref. 111)

Isotopic exchange of oxygen is a very sensitive probe in photoinduced gas-solid reactions to assess the photocatalytic oxidizing properties of semiconductor oxides. For instance, when comparing a TiO_2 sample homogeneously doped with 0.85 at % Cr^{3+} cations with undoped titania of nearly equal surface area, the rates of various photocatalytic oxidations of organic compounds decreased from ~25 to ~85 times, whereas the initial rate of oxygen isotope exchange under illumination was ~1000 times lower. (ref. 112) Similarly, whereas the photocatalytic oxidation of isobutane in the gas phase over a series of anatase samples varied by a factor of about 3, the photoinduced heteroexchange of oxygen over the same specimens was altered by a factor of about 45. (ref. 109)

On the other hand, the same type of labile adsorbed/surface atomic oxygen species which take part in $^{18}O_2$ or $N^{18}O$ isotopic exchange are also involved in photocatalytic oxidations over semiconductor oxides since the same orders of activities were found in both types of reactions for several TiO_2 specimens, (ref.109) as well as for SnO_2, ZnO, ZrO_2 samples. V_2O_5 specimen was inactive (ref. 110) and these exchanges were inhibited in the presence of an oxidizable compound. (refs. 109,111) The role of O_2 (or NO) in photocatalytic oxidation reactions is to replenish the coverage of TiO_2 in removable atomic oxygen species.

3. SURFACE MODIFICATIONS TO CONTROL INTERFACIAL ELECTRON TRANSFER

3.1. Metals

3.1.1. Choice of Metals. Depositing a metal on an n-type semiconductor to produce a photoassisted chemical reaction has two purposes: to introduce catalytic centers to accelerate some step of the reaction and to increase the separation of electron-hole pairs. The first goal can also be understood in terms of electrochemistry. For instance, the metal deposit aims at decreasing the overpotential for H_2 evolution in reactions which involve this step. The second goal derives from analogy with photoelectrochemical cells: while the photoproduced holes on the n-type semiconductor powder play the role of a photoanode, the electrons are drained to the metal particles which are equivalent to a cathode. (ref. 3)

The role of the metal deposit in diminishing charge recombination depends on several factors. The respective work functions of the semiconductor and of the metal control electron transfer between the materials. That of a semiconductor oxide such as TiO_2 is markedly affected by oxidizing or reducing pretreatments. (ref. 113) That of a group VIII metal (even metals of the 2nd or 3rd rows) depends on whether the sample has been exposed to oxygen or hydrogen. (ref. 114) This latter effect has been demonstrated with current-potentials curves obtained with electrodes formed from crystals of TiO_2, $SrTiO_3$, CdS and InP and metallic films (Pt, Rh, Ru). (refs. 115-117) In the case of powders, exposure to H_2 increases the conductivity (ref. 71) and photoconductivity (ref. 73) of Pt(or Ni)/TiO_2 samples with respect to the vacuum, whereas it has no effect in the absence of deposited metal. Furthermore, the observed half-order dependence of the conductivity or photoconductivity on H_2 pressure can be interpreted as spillover of hydrogen atoms from surface platinum atoms to the oxide, according to eqns 7 and 8.

$$H_2 + 2\ Pt_s \rightleftharpoons 2\ Pt_s\text{-}H \quad (7)$$

$$Pt_s\text{-}H + O^{2-} \rightleftharpoons Pt_s + OH^- + e^- \quad (8)$$

Consequently, group VIII metal particles can behave as charge recombination centers, either if the samples have been preoxidized (because of an increase in the work function of the semiconductor oxide) or in the presence of H_2 (because of a decrease in the work function of the metal). This behavior can counterbalance the beneficial catalytic effect of the metal and can explain the existence of optimum metal content (refs. 74,75,118,119) even in the absence of particle size effects.

All group VIII metals have been tried, (refs. 120,121) with Pt being the most frequently used, because of their well-known catalytic properties for hydrogenation and dehydrogenation reactions, as well as their utility as cathodes for H_2 evolution.

3.1.2. Preparation. Although mixtures of metals and semiconductor powders apparently yield positive results, (ref. 122) more intimate contact between the materials has been sought. Perhaps the most commonly used method of metal deposition, because it can be performed under experimental conditions similar to those of photocatalytic reactions, has been the *in situ* photoreduction of a metallic salt or complex added to the suspension of the semiconductor powder, with or without the presence of a hole scavenger, as discussed in another section of this chapter. The other method involves the reduction, in flowing H_2 at temperatures in the range 573-773 K, of a metallic salt or complex previously impregnated on the semiconductor. A preoxidation step can produce a better dispersion, i.e., a smaller particle size (refs. 123-124), of the metal. In both methods numerous parameters can affect the dispersion and the distribution of the metal particles on the semiconductor grains. (refs. 124-125)

3.1.3. Photocatalytic reactions requiring metal deposits. Metal/n-type semiconductor bifunctional photocatalysts can induce photocatalytic reactions, even if they do not catalyze the photodecomposition of water in the absence of sacrificial compounds, i.e., electron donors. (refs. 51,126-128) Several reactions which are not catalytic without group VIII metal deposits include the endergonic dehydrogenation of alcohols, some amine conversions, hydrazine decomposition, and the isotopic exchanges between deuterium and saturated hydrocarbons. These reactions will be described later.

Decarboxylation of organic acids occurs on the n-type semiconductor (TiO_2, WO_3, Fe_2O_3) even in the absence of deposited Pt. (refs. 58,129-134) However, the metal deposit was found to be beneficial, particularly in the absence of O_2, and it was suggested that it favors reduction steps, for instance, in acetic acid decarboxylation, in the formation of methane from methyl radicals. (ref. 58) By contrast, no improvement was observed for the oxidation of liquid 4-tertbutyltoluene by oxygen when replacing TiO_2 by 0.5 wt % Pt/TiO_2, probably because oxygen reduction is not the rate-limiting step. (ref. 135) From a comparison of current-potential curves obtained on a Pt electrode and on a given semiconductor electrode either in the dark or under band-gap illumination in the presence of a given redox couple, it can in principle be predicted whether the reaction rate would be higher with the platinized semiconductor powder. (ref. 3) However, the detailed analogy between electrodes and particles of small size may not always be valid. In addition, pretreatment can play a considerable role, as is shown, for instance, by the distinct behavior of Pt/TiO_2 powder samples in the isotopic heteroexchange of $^{18}O_2$ depending on how they have been pretreated. (ref. 82) Furthermore, exactly equivalent pretreatments cannot sometimes be accomplished for electrodes and powders.

Comparisons between various metals deposited on the same semiconductor have occasionally been made. They are of limited value if the metal particle size varies both with the metal and the loading and if the relationships between the loading and the reaction rate have not been determined for each metal.

3.2. Substitutional Dopants

Substitutional doping has been widely used, especially in attempts at photodissociating water, to extend the photosensitivity of large band gap semiconductors to the visible range and to control the position of the Fermi level. (For instance, it can be displaced toward more negative potentials to allow proton reduction.) In addition, it was expected and, in some cases, it has been found (ref. 136) that the voluntary introduction of impurities stabilizes the samples.

Substitution of Ti^{4+} in TiO_2 by cations of higher valency, such as Nb^{5+} ions, introduces in the lattice extra valence electrons (one per substituted cation), i.e., Nb^{5+} ions behave as complexation sites for electron donors. The opposite occurs when substituting Ti^{4+} by cations of lower valency, such as Cr^{3+} ions. Thus, Cr^{3+} ions act as complexation sites for electron acceptors. Isoelectronic substitution, such as Zr^{4+} for Ti^{4+}, (refs. 137,138) changes the energetic position of the conduction band edge.

Experimentally, semiconductors can be doped by sintering of the required salts at high temperature. Materials of low surface area are thus obtained. This is one of the reasons why doping has been attempted most frequently with electrodes, usually with the intent to photosplit water. Powders of high surface area, homogeneously doped, can be prepared in a flame reactor.

Attempts to render TiO_2 or $SrTiO_3$ anodes photosensitive to visible wavelengths by doping with various cations of lower valency than Ti^{4+} have not been successful. (refs. 137-140) High biases were required to increase hole mobility which was greatly diminished as a result of the presence of the foreign cations in the same sublattice. In contrast, microcrystalline

$La_xSr_{1-x}TiO_3$ photoanodes (donor doping) were more efficient than undoped $SrTiO_3$ without external bias, the spectral response to longer wavelengths being improved as well. (ref. 141) The use of structures containing two separate cation sublattices led to adequate hole mobility, but, because of small-polaron formation at the surface, the rate of the expected charge transfer was too slow. (refs. 137,138)

A homogeneously Cr-doped titania powder (0.85 wt. % Cr, 60 m^2g^{-1}) (ref. 112,142) was markedly less active under UV-illumination than undoped titania for oxidations (of propene, 2-propanol, aqueous solutions of oxalic acid or 4-chlorophenol (ref. 143)), for oxygen isotope heteroexchange or for hydrogen peroxide decomposition. (ref. 144) Also, the photocurrent measured in a titania slurry-electrode cell containing acetate ions, was substantially decreased when using Cr-doped TiO_2. (ref. 145) No photocatalytic activity was found at wavelengths $\geq$ 400 nm, despite the existence of an absorption in the visible region caused by dissolved Cr^{3+} ions. In agreement with these activity tests, the photoconductance under vacuum of doped titania was drastically decreased with respect to that of the undoped material and its action spectrum showed no sensitivity to visible photons. These phenomena were attributed to electron-hole recombination at the Cr^{3+} sites whose concentration (~2.5 x 10^{20} cm^{-3}) was greater than that of the photogenerated electrons. (ref. 112) These data have been corroborated recently. (ref. 146)

These results illustrate the difficulties encountered when attempting at modifying a semiconductor by doping. As in the case of metal deposits, detrimental effects can appear.

3.3. Supports

The dispersal of the semiconductor on various supports can also influence photoinduced interfacial electron transfer. While the construction and investigation of fully integrated semiconductor-support systems is an active area of materials research, its full description here is beyond the scope of this article. We do cite, however, several recent examples which demonstrate how the support can affect photoelectrochemical efficiency.

Dispersal of semiconductor particles into macroscopic gel arrays produces a porous material with high surface area. (refs. 147-149) On such amorphous porous glasses, Ti(III) is formed upon photoexcitation in the presence of oxidizable substrates, and at appropriate pH, hydrogen gas evolves. (ref. 148) With multiple layers formed by this sol-gel method, sensitizers can be strongly adsorbed for extended wavelength response. (ref. 149)

Semiconductors can also be dispersed into cationic (ref. 150) or anionic (ref. 151) vesicles, either on the inside or outside of the heterogeneous aggregate. These materials can be characterized by atomic, optical, or emission spectroscopy or by light scattering measurements, and photochemical properties can be differentiated on the inner and outer surfaces. (ref. 151) The blue shift observed for the onset of absorption by the semiconductor dispersed inside the vesicle demonstrates that the vesicle serves to limit the size of the growing particle as well as to protect it from aggregation and to reduce the effect of aging by dissolution or precipitation. When the included particles are brought into contact with a metal catalyst, e.g., Rh, hydrogen evolution is observed. (ref. 150) Differential electron transfer is observed on oppositely charged vesicles, particularly if the vesicle is stabilized by polymerization. (ref. 152) High turnover numbers and 10% quantum efficiency for hydrogen evolution can be attained on such systems. (ref. 152)

Polar or charged polymers can also act as supports for active semiconductor dispersions. Polyvinyl alcohol, for example, stabilizes colloidal TiO_2 suspensions from precipitation without quenching photoactivity. (ref. 153) Polymeric films can similarly be employed as supports, (refs. 154-157) particularly if the semiconductor can be introduced by cation exchange. Spectroscopic methods can be used to characterize the particle size and surface composition of these photoactive species. (ref. 158) Cellophane has also been used as an analogous support for CdS. (ref. 159)

Dispersal of photocatalysts into more rigid inorganic matrices is also known. For example, porous Vycor has been so employed (refs. 160-161), as have zeolites (refs. 162-163) and clays. (refs. 158,164-166) When CdS is held within porous Vycor, the cavity dimensions preclude one dimension from exceeding 40 A, and yet optical properties of particles resembling bulk are observed. (ref. 160)

Particles can also be held tightly on the surface of Vycor (ref. 161) or of colloidal silica. (ref. 167) In the latter case, the surface charge retards energy dissipative back electron transfer. (ref. 168) As on clays, the effect on photoactivity can be rationalized as deriving from the high electrostatic fields of the support. (ref. 166) In comparison with unsupported powders, clay-supported photocatalysts face mass transfer problems linked to the spacing between sheets which decrease the overall electron transfer efficiency.

Thin films of metal oxides can also be supported on the metal itself. Thus, spark anodization of titanium forms a photosensitive layer of TiO_2 on which vectorial electron transfer can be observed. (ref. 169) Such materials have also been formulated into stacked arrays for unassisted water splitting.

3.4. Coupling of Semiconductors

To reduce recombination of photoproduced charges and accordingly to improve the rates of photocatalytic reactions, coupling of appropriate semiconductors has recently been proposed. (refs. 170-171)

For instance, the generation of H_2 under visible illumination from basic Na_2S aqueous solutions was increased by a factor of 1.2 to 1.4 when using CdS mixed with 0.5 wt % RuO_2/TiO_2 instead of 1 wt % RuO_2/CdS. (ref. 170) For dehydrogenation of methanol, with visible excitation, an improvement by a factor of 1.5 to 3.3, depending on the content, was found when comparing the catalytic activity of a mixture of CdS and Pt/TiO_2 with Pt/CdS. (ref. 171)

These variations of photocatalytic activity were ascribed to electron transfer from CdS to TiO_2, eqn 9. (refs. 170-171)

$$e_{cb}^{-}(CdS) \longrightarrow e_{cb}^{-}(TiO_2) \qquad (9)$$

through a coupled aggregate, although electron transfer to a metal island on the separate particle might also suffice. (ref. 172) This transfer diminishes electron-hole recombination in CdS and facilitates the charge transfer required to generate the products, eqns 10-12

$$2\ h_{vb}^{+}(CdS) + S^{2-} \rightarrow S_n \qquad (10)$$

$$nS + S^{2-} \rightarrow S_n^{\ 2-} \qquad (11)$$

$$2\ e^{-}(RuO_2/TiO_2) + 2\ H_2O \rightarrow H_2 + 2OH^{-} \qquad (12)$$

or eqns 13-15.

$$CH_3OH \rightarrow CH_3O^{-} + H^{+} \qquad (13)$$

$$h_{Vb}^{+}(CdS) + CH_3O^{-} \rightarrow HCHO + H^{\bullet} \qquad (14)$$

$$e^{-}(Pt/TiO_2) + H^{+} + H^{\bullet} \rightarrow H_2 \qquad (15)$$

This latter transfer implies the migration of H atoms or protons from CdS to the Pt particles deposited on TiO_2 as a result of agglomeration and inter-particle collisions in the stirred suspension. The RuO_2 or Pt catalysts deposited on TiO_2 are needed to efficiently produce H_2. Aggregation of particles is certainly needed, although whether conduction band-to-conduction band transfer is required in the presence of deposited metal particles is unclear. (ref. 172)

A photoconductivity study was undertaken to verify by a direct means the existence of electron transfer between powdered semiconductors. When the samples were illuminated under vacuum, the results were interpreted in terms of electron transfer as for suspensions, but from TiO_2 to CdS, eqn 16.

$$e_{cb}^{-}(TiO_2) \rightarrow e_{cb}^{-}(CdS) \qquad (16)$$

In the presence of O_2, this transfer was limited by oxygen adsorption on TiO_2, which depletes this material of its free electrons. (ref. 173)

The photoconductivity measurements and the photocatalytic results are not inconsistent in view of the different pretreatments and different experimental conditions used. The electron density of titania was, in the latter case, increased by prolonged illumination under vacuum, whereas it was decreased by the Pt particles in methanol dehydrogenation. The presence of an electrolyte or of an alcohol should also play a role. Indeed, photoelectrochemical studies in aqueous solutions containing Na_2S and NaOH showed that CdS and TiO_2 have nearly equal flatband potentials under these conditions. (ref. 174) Consequently, the direction of the electron transfer can easily change upon varying experimental conditions.

4. DETERMINATION OF THE PHOTOCATALYTIC CHARACTER AND EFFICIENCY OF A CHEMICAL REACTION

Several criteria should be used to verify that a reaction which seems to be influenced by illumination at the surface of a semiconductor really results from the creation of electron-hole pairs and that the products are catalytically generated. These are indicated as follows.

4.1. Light Effects

Although wavelength effects are generally probed with monochromatic irradiation, the use of a monochromator is generally impossible for studying photocatalysis because of low reaction rates. However, approximate action spectra can be determined by employing a series of cut-off filters. A filter which does not transmit photons of energies equal to or greater than the semiconductor band gap permits a quick check to prove that the creation of electron-hole pairs is a prerequisite for the reaction studied.

For a given lamp and optical setting, the effect of radiant power or flux can be determined by using grids or neutral density filters whose absorption does not vary with frequency in the spectral range investigated. For several reactions involving gases, organic liquids, or aqueous solutions, reaction rates proportional to flux have been observed. (refs. 44,131,135,175- 186) Generally the maximum radiant power was of the order of 10^{16} photons $cm^{-2}s^{-1}$.

In the photocatalytic oxidation of liquid 2-propanol over rutile, a change in the dependence of the reaction rate from first to half power dependence on flux was discovered above the critical value of 2.5×10^{15} photons $cm^{-2}s^{-1}$. The latter proportionality was observed at a flux of 10^{18} photons $cm^{-2}s^{-1}$. (ref. 181) The 1/2 power flux dependence was also found for the same reaction in the gas phase over ZnO, (ref. 184) for H_2O_2 decomposition over TiO_2, (ref. 185) and for the reduction of MV^{2+} ions over colloidal CdS. (ref. 186) This square root dependence was thought to arise from the predominance of second-order charge recombination over the consumption of holes to initiate the chemical reaction. (ref. 181) Thus these measurements are very useful to assess the influence of charge recombination.

4.2. Discrimination Between Photocatalytic and Photochemical Processes

Control experiments must be conducted to evaluate any direct photochemical transformations which could occur and filters can be selected to cancel or minimize these transformations, if any. If the photochemical reactions cannot be suppressed because they are induced by photons which are also required to excite the photocatalyst employed, the possibility can still exist of discriminating between the photocatalytic reaction and the photochemical background. For that purpose, a solid which has about the same light-scattering and absorption properties as the photocatalyst under study, and which is as inactive as possible, is needed and will be used as a reference. Photons will be absorbed similarly by this solid and by the photocatalyst. Any conversion obtained in the presence of this inactive solid can be considered as a photochemical contribution to the photocatalyst conversion.

Doped semiconductors provide one such class of additives. For instance, as already mentioned, a 0.85 wt % Cr^{3+} doped-TiO_2 sample has been found about 25 to 85 times less active than an equivalent undoped sample in various photocatalytic oxidation reactions. (ref. 112) This method has been used to distinguish the photocatalytic decomposition of H_2O_2 over TiO_2 from its photochemical decomposition. (ref. 144)

The rate of a truly photocatalytic reaction should increase with the amount of catalyst up to a value corresponding to the maximum absorption of photons. Above this value, it will level off and, for very high amounts, it can even decrease, if the non-illuminated catalyst grains play a detrimental role.

4.3. Discrimination Between Thermal and Photochemical Activation

Temperatures of the catalyst layer or suspension can be compared in the dark and under illumination. However, local temperature spikes can be masked by heat dissipation by the gaseous stream or the stirred liquid. Consequently, for reactions which can be thermally catalyzed at temperatures only slightly above room temperature, the use of low radiant flux is recommended. To determine whether a given radiant flux increases the temperature of a M/TiO_2 fixed-bed catalyst, the cyclopentane-deuterium exchange in the gas phase can be used, since at 263 K only the photocatalytic reaction takes place, producing monodeuterocyclopentane at short illumination periods, whereas if local heating raises the temperature even 10 K higher, the thermally catalyzed reaction intervenes and gives rise to multiply exchanged molecules. (ref. 187)

4.4. Discrimination Between Photocatalysis and Photoassistance

The most powerful laboratory UV lamps yield only 10^{-4} einstein h^{-1} cm^{-2}. Consequently, the reactions should be carried out over relatively long periods to ensure conversions many fold greater than those which could correspond to stoichiometric reactions involving preadsorbed or preexisting non-renewable species (assuming a maximum coverage in these species of the order of 10 nm^{-2}, if their exact number is unknown).

4.5. Quantum Yields and Chemical Yields

The quantum yield of formation of a product is the number n_p of molecules formed for each quantum q of radiation absorbed. It has a value at every time (differential) or it can relate to a given period of time (mean or integral). Obviously, it applies only if the reaction is really photocatalytic. For a catalytic bed or suspension, q cannot be measured accurately because of light scattering and reflection by the catalyst grains. This may explain why quantum yields are not reported in many photocatalysis investigations. However quantum yield evaluations are indispensable to compare catalysts, reactions and, more generally, results from various laboratories. The use of a monochromator will increase the accuracy of determination of q but will decrease that of n_p, so that its advantage in this context is not obvious.

In view of the difficulties in determining accurate quantum yields, an agreement on standard photocatalysts (for example, one with a metal deposit, another bare) has been suggested. (ref. 188) In other words, the activity of a given material in a given reaction would be related to that of a standard photocatalyst under the same conditions, so that effects of different illumination settings and reactors will be minimized.

The chemical yield represents the ratio of the number of molecules of a desired product which are formed to the number of reactant molecules consumed in the photoreactor. It is of primary importance for synthesis. If laboratory lamps which have low energy output are employed, it will be necessary, even in the best cases, to dilute the reactant(s) in an inert solvent to allow for complete conversions.

5. PHOTOCATALYTIC INORGANIC TRANSFORMATIONS

5.1. Oxidations

Many inorganic compounds have redox potentials which are less positive (more energetic) than the valence band of wide band-gap semiconductor oxides. Consequently, they can be easily oxidized in the presence of oxygen over these solids. Table 1 summarizes some of the photocatalyzed oxidations which have been reported.

Many of these studies have already been reviewed. Therefore, in the following paragraphs, we will only consider examples which illustrate the photocatalytic oxidation of inorganic compounds.

TABLE 1.

Representative Inorganic Photoelectrochemical Oxidations

CO	$\xrightarrow[O_2]{\text{n-type Sc*}}$	CO_2
NH_3	$\xrightarrow[O_2]{\text{n-type Sc*}}$	N_2, N_2O
$SO_2(SO_3^{2-})$	$\xrightarrow[O_2]{\text{n-type Sc*}}$	$SO_3(SO_4^{2-})$
CN^-	$\xrightarrow[O_2]{\text{n-type Sc*}}$	CNO^-
X^-	$\xrightarrow[O_2]{\text{n-type Sc*}}$	$X_2(XO^-)$
NO_2^-	$\xrightarrow[O_2]{\text{n-type Sc*}}$	NO_3^-(ref.189)

5.1.1. CO oxidation. The oxidation of CO has been a favorite reaction, possibly because most groups engaged in this field were simultaneously working in catalysis where it is a well-documented test reaction. It presents the advantage of dealing only with gaseous reactants and products over a large temperature range. Conversely, because of its lack of selectivity, it is not very informative so that the reaction is primarily of academic interest.

Studies conducted prior to 1981 have been reviewed. (refs. 190-191) Most of them attempted, via kinetics, conductivity, ESR, and isotopic exchange, to determine the nature of the active species and the mechanism of reduction on a number of semiconductor oxides, TiO_2 and ZnO being the most frequently examined. No general picture emerges; in particular, there is no general agreement about the route for formation of the active dissociated oxygen species. Not surprisingly, the nature of the semiconductor and the conditions of the reaction intervene. The Mars and van Vrevelen mechanism (surface oxido-reduction) and the Eley-Rideal mechanism (CO reacting from the gas phase) have been proposed, depending on the initial CO pressure range.

To our knowledge, since 1981, only two reports have mentioned the use of CO. (refs. 192-193) In one case, its oxidation was employed to determine activity at wavelengths $\geq$400 nm of a TiO_2 sample doped with some undefined nitrogen-containing impurity. (ref. 192) In the other case, its photoreduction was used as a supplementary evidence for light-induced electron transfer, eqn 17,

$$M^{n+}-O^{2-} \longrightarrow M^{(n-1)+}-O^{-} \qquad (17)$$

for V_2O_5, MoO_3 and CrO_3 deposited on Vycor glass (silica), since in the excited state the length of the M-O bond is increased, which facilitates the removal of the oxygen atom by CO. (refs. 193-194) This is an example of the use of a molecule as a probe for electron transfer.

5.1.2. The water-gas shift reaction. The water-gas shift reaction, eqn 18,

$$H_2O + CO \longrightarrow CO_2 + H_2 \qquad (18)$$

or eqn 19,

$$2OH^- + CO \longrightarrow CO_3^{2-} + H_2 \text{ (basic aqueous solutions)} \qquad (19)$$

is slightly exergonic at room temperature and has been investigated in connection with the numerous studies dealing with water dissociation (see M. Graetzel's chapter in this book). It has been reported to occur over SiC, (ref. 195) CdS (ref. 195) and ZnS (refs. 195-196) powders. In the two former cases, loading with Rh (0.5 wt %) was examined. It considerably increased the yield as might be expected since H_2 is a product in agreement with previous studies using 2 wt % Pt/TiO_2 (ref. 178) powder or a Pt/TiO_2 (100) single crystal. (ref. 197) Quantum yields in the range 5×10^{-3} to 8×10^{-3} have been indicated. (refs. 178,196) Carbon monoxide does not intervene directly in electron transfer associated with the photosplitting of water but does allow the elimination of oxygen from the semiconductor surface. In addition, in its presence, the photocorrosion of ZnS (ref. 196) (and probably of CdS) is much weaker.

5.1.3. SO_2 (or SO_3^{2-}) oxidation. The gas-phase photocatalytic oxidation of SO_2 to SO_3 over TiO_2 was reported many years ago. (ref. 190) It also occurs in aqueous solutions over UV-illuminated TiO_2, ZnO (unstable), Fe_2O_3, WO_3 and CdS. (ref. 198) A quantum yield of ca. 0.16 was reported on TiO_2. Quite recently, the formation of sulfate ion was employed as a test for various TiO_2 samples. (ref. 199) The activation energy was low (25 to 28 kJ mol^{-1}), as is expected for a reaction whose fundamental step involves photons, and did not vary when TiO_2 was loaded by photodeposited Pd, Pt or Ru. Thus, the rate determining step of the oxidation occurred on the semiconductor and not on the metal. However, an increase in the reaction rate for metal loaded TiO_2 was found and was attributed to a better separation of charge. Optimum loadings were not determined, but, in our opinion, as the metal particle size increased with the content, comparisons between various loadings are not straightforward. The suggested mechanism is quite general and simply assumes the reaction of holes with sulfite ions and hydroxide groups and the reaction of electrons with oxygen. (ref. 198)

5.1.4. Halide ion oxidation. The oxidation of halide ions by photoproduced holes has drawn the early attention of the photoelectrochemist. Prior to 1979 it was assumed that I^- ions and, to a lesser extent, Br^- ions and Cl^- ions, can be oxidized over UV-illuminated semiconductor anodes, such as TiO_2, ZnO, or WO_3. (ref. 200) However, the products were not titrated. Later on, rotating ring-disk electrodes of TiO_2, (refs. 38,201-203) WO_3, or Fe_2O_3 (ref. 203) were used to compare various reducing species, among which were halide ions (especially I^-). The observed pH dependence was explained by the redox potential-pH relationship.

The first study examining photoinduced halide oxidation on powdered titania established that Cl^- ions withstand oxidation, whereas the conversion of Br^- ions was ca. 80 times slower than that of I^- ions. For these latter ions, the initial quantum yield was of the order of 0.02, after correcting the results for the homogeneous reaction in acidic medium and for an oxygen flow which should hamper the reverse reaction to some extent. Halogens or hypohalite ions were obtained in acidic or basic medium, respectively. (ref. 44) By contrast, a latter study (ref. 204) claimed the oxidation of Cl^- ions over TiO_2 at acidic pH, provided the back reaction is avoided by continuously removing Cl_2 from the photoreactor. In addition, platinization of TiO_2 (10 wt %) by photodeposition improved the rates because of a better separation of charges. However, the reaction of Br_2, for instance, on Pt particles forms $PtBr_6^{2-}$ and casts some doubt on the use of these deposits. (ref. 205) Furthermore, the method of titration might be responsible for the high quantum yields reported, even for Cl_2.

Concerning the mechanism, the use of cyclohexene as a probe (ref. 205), reacting with Br^-, Br_2 or BrO^-, allowed one to conclude that the initial oxidation step in anhydrous acetonitrile produces Br atoms and to reject the oxidation of Br^- by active dissociated oxygen species forming BrO^- as had been previously suggested. (ref. 44) The associated reduction reaction involves oxygen since this gas or another electron-acceptor is required. However, the oxidation would be limited in the absence of a step which impedes the accumulation of the reduced oxygen species. In acidic medium, the overall reaction, eqn 20

$$1/2\ O_2 + 2\ e^- + 2H^+ \longrightarrow H_2O \qquad (20)$$

has been proposed in agreement with the effect of pH. (ref. 204) However, the oxidation of I^- and Br^- still takes place in basic medium. (ref. 44)

5.2. Reductions

Metal cations and several simple inorganic molecules have reduction potentials which lie positive of the conduction band edges of common semiconductors, and can then, at least in principle, be reduced by photogenerated electrons. The extensive work on proton reduction (for hydrogen gas generation) is discussed in another chapter of this book. Here, we deal with three areas: (a) photoassisted metal deposition; (b) CO_2 reduction; and (c) N_2 reduction.

5.2.1. Photoassisted metal deposition. Electron transfer from band-gap illuminated semiconductors to metallic cations has been known for a long time. Originally, interest in this process lies in imaging systems. More recently, this reduction method has commonly been used for *in situ* preparation of photocatalysts by research groups which were generally not familiar with procedures employed in catalysis, i.e., impregnation of the powder semiconductor with a metal salt, followed by reduction in H_2. Although it yields, under certain conditions, well-dispersed metal particles (vide infra), (ref. 20) it adds an additional element to the difficulty in comparing results of different laboratories, since the photocatalysts were prepared in small amounts and therefore were not frequently characterized. Obviously, the same materials can be utilized as catalysts in reactions which do not occur under illumination and this new method of catalyst elaboration has lately drawn attention. Finally, electron transfer between illuminated semiconductors and metallic cations can be used to recover or eliminate them from dilute solutions, which has obvious interest if these cations are either precious or toxic.

From a thermodynamic viewpoint, a metallic cation can be reduced by free electrons of a n-type semiconductor provided that its reduction potential is less negative than the semiconductor Fermi level, eqn 21.

$$M^{n+}(ads) + ne^- \longrightarrow M^o(ads) \qquad (21)$$

The valence band holes should be able to oxidize simultaneously either water or another more easily oxidizable species adsorbed on the semiconductor. These species should of course not be capable of reducing the M^{n+} cations.

The following overall reactions represent the oxidation steps as derived from photoelectrochemical measurements, for water, eqn 22 or for acetate, eqn 23.

$$n/2\ H_2O + n\ h^+ \longrightarrow n/4\ O_2 + n\ H^+ \quad (22)$$

$$n\ CH_3CO_2^- + n\ h^+ \longrightarrow n\ CO_2 + n/2\ C_2H_6 \quad (23)$$

In other words, electrons are transferred either from water or from some other species to the metallic cations via the excited semiconductor. Proton-to-metal atom ratios corresponding to the stoichiometry of the above equations have been found for the deposition of Pt or Ag on powdered TiO_2 from H_2PtCl_6, Na_2PtCl_6 (ref. 125) or $AgNO_3$. By contrast, apart from one case, (ref. 207) Ag/O_2 ratios for deposition on titania samples were generally greater than expected. (refs. 208-209) An inverse correlation between the density of surface hydroxyl groups and the volume of oxygen involved was attributed to the formation of peroxo titanium complexes. (ref. 209) Recent kinetics results, dealing with Pt and Ag deposition on TiO_2 from dilute solutions, (ref. 125) showed that the amount of oxygen evolved was initially several times greater than expected and that it progressively decreased and finally became nil. The initial excess of oxygen was explained by photodesorption (vide supra). Once most metallic cations have been reduced, electrons become available for oxygen ionosorption on titania. In addition, oxygen reacts with surface platinum atoms, eqn 24.

$$2\ Pt_s + O_2 \longrightarrow 2\ Pt_s\text{-}O \quad (24)$$

Finally, for TiO_2 samples which have not been made free of organic contaminants by calcination, oxygen can be consumed by the photocatalytic oxidation of these compounds.

When the cation/metal redox potential corresponds to an energy too high to allow electron transfer from the excited semiconductor, the electron transfer can occur in the other direction, i.e., the cation is oxidized to a higher valence state and can be deposited as an insoluble oxide. For instance, PbO_2 and Tl_2O_3 were identified by x-ray diffraction as being the deposits obtained on a Pt/rutile powder from Pb^{2+} and Tl^+ nitrate solutions. (ref. 210) Since a linear relationship was found between the decrease in O_2 (gaseous and dissolved) and the removal of Pb^{2+} ions, the overall reaction proposed was eqn 25.

$$Pb^{2+} + 1/2\ O_2 + H_2O \longrightarrow PbO_2 + 2\ H^+ \quad (25)$$

In the absence of O_2, the coupled redox reactions can involve a second metallic cation, as illustrated by the simultaneous deposition of PbO_2 and Pd, eqn 26, from mixed nitrate solutions on a rutile single crystal. (ref. 211)

$$\begin{aligned} &Pb^{2+} + 2\ H_2O + 2h^+ \longrightarrow PbO_2 + 4\ H^+ \\ &Pd^{2+} + 2e^- \longrightarrow Pd \end{aligned} \quad (26)$$

If applications are sought, the semiconductor should be stable in illuminated water solutions and should not be expensive. From these viewpoints, titania powders are good candidates. Besides, some TiO_2 samples are the most efficient semiconductors. The redox potentials of most noble metals allow them to be deposited on this material, whereas Ni and Cu are not. (ref. 125) This difference can be of great interest in separating precious and common metals. For instance, 93% of the gold contained in a 1M HCl solution was recovered, whereas Cu^{2+}, Ni^{2+} and Zn^{2+} ions, also present, remained dissolved. (ref. 212) By adjusting various parameters which affect the kinetics, (ref. 125) separation of precious metals from one another can also be performed. (ref. 213)

For model solutions containing only a salt of the metal studied, the removal can be achieved to the analytical detection limit. (ref. 125) Obviously this limit should depend on a number of factors and for waste solutions the procedure is likely to be not so efficient. Further studies are needed in this direction.

Quantum yields ranging from 0.08 to 0.2 and from 0.235 to 0.385 were found for $AgClO_4$ reduction over TiO_2 (ref. 214) and ZnO samples, (ref. 215) respectively, and the value reported was ca. 0.05 for that of $PtCl_6^{2-}$ over TiO_2. (ref. 125) Considering that one or four quanta of light are required for Ag^+ or Pt^{4+}, respectively, these values are close enough. Because of the high cost of the metals thus recovered, even low quantum yields might be acceptable. Furthermore, optimization of photoreactors and photocatalysts has not been made and the addition of hole scavengers might also improve efficiency. (ref. 212)

Transmission electron microscopy has shown that after illumination for several hours of a suspension of TiO_2 the deposited Pt forms agglomerates with diameters up to 100 nm. (ref. 125) In other words, most of the semiconductor surface remains free of metal and thereby accessible to photons, i.e., the deposition can be continued or the photocatalyst can be reused in another platinum solution.

In some cases, the state of the metal or cation was characterized by XPS. (refs. 216,217) For platinum on TiO_2, the higher the acetic acid concentration, the more reduced was the adsorbed Pt. (ref. 216) It was even reported that reduction to Pt^o was not achieved in the absence of acetate ions. (ref. 217) However, other XPS results show Pt and Ag deposited on TiO_2 in the presence of water with sacrificial donors are metallic. (ref. 218)

The question arises as to whether the electron donors used for metal photodeposition act only as hole scavengers or whether they form radicals capable of transferring electrons to the conduction band or of directly reducing the metallic cations. Formation of radicals was inferred from quantum yields greater than 1. (ref. 219) The effect of the nature of various electron donors on the Pt dispersion on TiO_2 was thought to be the result of electron transfer from radicals with various oxidation potentials to the Pt cations. (ref. 220) Alternatively, since the deposition rate, and accordingly the illumination time necessary to obtain catalysts with equal metal loadings, changed with the electron donor, variations in the extent of metal dispersion could also arise from these differences in illumination time.

Indeed, the illumination time is a crucial factor in obtaining small metal particles. (refs. 125,221) In the absence of an electron donor other than water, the metal is initially deposited as crystallites regularly dispersed on the semiconductor grains if the semiconductor is itself homogeneous, but, after longer illumination times, the metal particles agglomerate. Under similar conditions, the particle size depends on the nature of the metal; for instance, average diameters of 1 nm for Pt and 3 to 5 nm for Ag are observed. (ref. 221)

Photodeposition of bimetallic catalysts can be achieved. For example, Pt-Pd and Pd-Ag deposits have been prepared on TiO_2 with a probable alloy formation according to STEM analysis. The Pt-Pd/TiO_2 catalyst contained mainly ~1 nm crystallites and some with diameters in the range of 2 to 3 nm, but the Pd-Ag/TiO_2 was poorly dispersed. (ref. 221) More studies are needed to assess the possibilities of the method. Successive depositions might be of interest.

This method also allows metal deposition on a number of oxides and sulfides. (ref. 221) XPS analysis showed the absence of surface chloride when using $PtCl_6^{2-}$ ions. (ref. 218)

5.2.2. CO_2 reduction. Let us consider first the reactions which require the transfer of two electrons per CO_2 molecule, eqns 27-29.

$$2\ CO_2(g) + 2\ H^+ + 2\ e^- \longrightarrow H_2C_2O_4 \quad \Delta E^o = -0.475V \tag{27}$$

$$CO_2(g) + 2\ H^+ + 2\ e^- \longrightarrow HCOOH\ (aq) \quad \Delta E^o = -0.2V \tag{28}$$

$$CO_2(g) + 2\ H^+ + 2\ e^- \longrightarrow CO(g) + H_2O \quad \Delta E^o = -0.106V \tag{29}$$

The two latter reactions correspond to ΔG^o values which are not very high: 38.6 and 20.5 kJ mol^{-1}, respectively, i.e., of the same order of magnitude as those of the dehydrogenation reaction of simple alcohols which occur with high quantum yield. (ref. 51) Their redox potentials are only slightly negative, which thermodynamically allows the possibility of transferring electrons from several types of illuminated semiconductors. In fact, in recent studies, CO and HCOOH were the principal products, although methanal, methanol and even methane have also been detected.

A review on photochemical fixation of CO_2 appeared in 1983 and is recommended as a background. (ref. 222)

The list of pure semiconductors which have been used for photocatalyzed CO_2 reduction includes: TiO_2, (refs. 222-224) ZnO, (ref. 223) WO_3, (ref. 223) $SrTiO_3$, (refs. 222,225) $BaTiO_3$, (ref. 222) $LiNbO_3$, (ref. 222) CdS, (ref. 223) GaP, (ref. 223) and SiC. (ref. 223) Attempts were also made with $SrTiO_3$ onto which various transition metal oxides were deposited by impregnation, (ref. 225) as well as with TiO_2 doped with Nb^{5+} or Cr^{3+}, or loaded with RuO_2. (ref. 224) Several experimental conditions have been employed: aqueous suspensions or semiconductor-coated plates, solar collectors or mercury lamps. Formic acid was generally the main product, methanal and methanol being also formed. Unfortunately, the maximum yields, despite these many efforts, remained on the order of some mol h^{-1}, which corresponded to maximum energy conversion efficiencies in the range of 1×10^{-4} to 4×10^{-4}. Also, decreases in activity were observed within 24 h at the best. In basic medium, the yields were still worse. (ref. 226) In other words, no practical conversion of CO_2 has yet been achieved. (ref. 224) As in the case of N_2 reduction, even the catalytic nature of the process is questionable. (ref. 224)

The strong adsorption of CO_2 on some oxides forming very stable carbonate, carboxylate and/or bicarbonate surface species can be one of the reasons for the low activities found, since the surface is thus poisoned. Also, downhill reverse reactions are possible. For instance, oxalic acid is unstable, even in the dark, over some semiconductors. (ref. 45) In addition, the products can undergo other photocatalytic processes: oxidations (since the presence of traces of oxygen are difficult to avoid) or decarboxylation. (ref. 206)

Illumination of semiconductor cathodes allows one to accomplish CO_2 reduction at voltages less negative than those necessary in the dark. Whereas a polyaniline-coated Si electrode produced principally formic acid in aqueous solution with a current efficiency of 0.03 to 0.05, (ref. 227) the same research group indicated that, by using a p-CdTe (100) cathode, CO was formed with a current efficiency of 0.9, in a system which was stable for 24 h in a DMF solution with 5% water (the organic solvent increased the solubility of CO_2) containing tetraalkylammonium salts (which would favor CO_2 reduction (ref. 228)). No deposited catalyst was needed. By contrast, in order to avoid the formation of H_2, a competitive process, Pb or Zn were deposited on a p-GaP cathode, which led to the formation of formic acid (and CO) with a current efficiency up to ca. 0.5. (ref. 229) Undoubtedly, the photoelectrochemical reduction of CO_2, using p-type semiconductors, has reached a stage which is encouraging.

5.2.3. $\underline{N_2 \text{ reduction}}$. The six-electron reduction of N_2 to NH_3 in the presence of water over a band-gap illuminated semiconductor was reported in 1977, (ref. 230) eqn 30.

$$N_2(g) + 3\ H_2O(l) \rightarrow 2\ NH_3(g) + 3/2\ O_2(g) \qquad \Delta G_{298\ K}=766\ kJmol^{-1} \tag{30}$$

The maximum yield obtained by the authors in this and further studies was 23 mol $h^{-1}g^{-1}$ (i.e., ~46 mol $h^{-1}m^{-2}$). (ref. 231) It required very strict conditions for the solid used: anatase prepared by hydrolysis of titanyl sulfate heated at 1273 K for 1 to 3 h to produce 20 to 40% rutile which was impregnated with 0.2 wt % Fe_2O_3. (ref. 231) Such special prerequisites illustrate the difficulty of N_2 reduction and, at the same time, cast some doubt on the possibility of reproducing the preparation of the semiconductor. The catalytic nature of the reaction was questioned on the basis of the criteria indicated in this chapter. (ref. 232)

Attempts to improve the yields have been made and have been reviewed to mid-1984 inclusive. (ref. 223) In addition to TiO_2, (ref. 233-235) several semiconductors, (refs. 236-238) even some with band-gaps < 1.17 eV (energetically equivalent to the redox potential corresponding to eq. 30), (ref. 237) and a Fe^{3+}-containing 5A zeolite, (ref. 237) have been tried in various conditions, and the published studies certainly represent only a fraction of the work which has been undertaken. The yields remained desperately low: they were rarely > 1 mol h^{-1} and often smaller than that. Considering the accuracy of ammonia analysis and the possible presence of nitrogen-containing impurities in the solids and solutions employed, it is not even sure, in some cases, that the ammonia found stemmed from the reduction of N_2.

However, in studies where great care was taken, yields of nearly 6 mol h^{-1} were observed (ref. 233) with characterized solids prepared by coprecipitating Fe_2O_3 and TiO_2 and containing 0.5 to 1 wt. % Fe^{3+} in solid solution. (ref. 239) These yields corresponded up to 1.6 OH^- layers h^{-1} and were sustained for periods of time greater than 24 h. Despite these figures, the authors themselves concluded that "no firm statement about the catalytic nature of the reduction can be made". (ref. 233)

The low efficiency of N_2 reduction can be explained by adsorption considerations. The stability of the N_2 molecule is not only reflected in the ΔG value of eq. 30, but in a very small adsorption constant. Attempts to modify the titania surface in order to provide sites capable of activating the N_2 molecule (ref. 240) have not been successful because of increased electron-hole recombination. (ref. 241) By contrast, ammonia is chemisorbed on the acid sites existing on most semiconductor oxides such as TiO_2, (refs. 242-244) which facilitates further conversion of this compound. In particular, its photocatalytic oxidation by oxygen to N_2 and N_2O over TiO_2 (and probably over other photosensitive oxides) easily occurs. (ref. 52) Furthermore, the ionosorption of oxygen on the n-type semiconductor should impede electron transfer to N_2.

These limitations cause one to be pessimistic about the likelihood of substantially improving the yields achieved so far and hence of foreseeing a practical procedure: a rate of 1 mol $h^{-1}m^{-2}$ means more than a century to produce 1 mole of ammonia at the laboratory scale! However, such yields can account for the slow reduction of N_2 in nature under sunlight on sands containing titanium and iron oxides. (refs. 231,245) In this latter case, very large areas are available.

A recent paper proposed the use of a dinitrogen complex $[Ru(HEDTA)N_2]^-$ and of CdS, loaded with Pt and RuO_2 by photodeposition, to effect N_2 reduction under visible light illumination. (ref. 246) The photocatalyst would decompose water to O_2 and active hydrogen atoms (ref. 247) which would reduce coordinated N_2 to NH_3. Because of its lability, ammonia would not remain bound and would be replaced by N_2 continuously bubbled through the solution. Within 6 h, a turnover number of 200 mol NH_3 per mol of complex was found, and the production continued for about 24 h, but with a gradual decrease, although the complex was recovered unchanged. The proposed use of a dinitrogen complex is interesting since it obviates the difficulty imposed by the low affinity of N_2 for the semiconductor surface. On the other hand, CdS is certainly a poor photocatalyst for the oxidation of ammonia by the oxygen produced. However, it is quite surprising that the reduction of N_2 by hydrogen atoms completely dominates over the formation of hydrogen on Pt. (ref. 247) Besides, there is no indication of the photocorrosion of CdS.

The photoelectrochemical reduction of nitrogen in a photoelectrochemical cell has been attempted using p-GaP (poorly stable), n-TiO_2 and n-$LuRhO_3$ electrodes. (ref. 248) Current conversion efficiencies in the range 4-10% were found and ammonia was detected. However, it was concluded that most of it was formed from nitrite or nitrate ions (vide infra) contained in the electrolyte (KOH). (ref. 249)

The overall reaction, eqn 31,

$$N_2 + 6\ H^+ + 2\ Al \longrightarrow 2\ NH_3 + 2\ Al^{3+} \tag{31}$$

which is exergonic, was observed in a photoelectrochemical cell containing an illuminated p-GaP cathode and an aluminum anode. (ref. 250) Since this latter electrode was consumed, the process is in fact a photoenhanced reduction of N_2 by aluminum. Under short-circuit conditions, the formation rate of NH_3 was ~0.1 mmol h^{-1} cm^{-2} (on GaP), i.e., orders of magnitude greater than over powdered semiconductors lacking sacrificial donors. Both types of electron transfer were made easier by a special nonaqueous electrolyte containing titanium tetraisopropoxide. In particular, this complex was first reduced to a state wherein N_2 could be bound before accepting electrons. Other nonaqueous electrolytes could however be used and traces of water served as a proton source. (ref. 251) The current efficiency remained < 1%.

5.2.4. Reactions related to N_2 reduction.

5.2.4.1. Reduction of NO_2^- and NO_3^- ions. The conversion of NO_2^- ions to ammonia in alkaline aqueous sulphide solutions over $SrTiO_3$, TiO_2, CdS and a CdS-ZnS mixture was achieved under visible illumination. (ref. 252-253) S^{2-} ions probably facilitate the reduction by trapping photoproduced holes and by forming polysulphides, whereas SO_3^{2-} ions were inefficient in playing the same role despite a less positive oxidation potential. No mechanism, however, was suggested for this rather complex reduction whose maximum yield was ca. 9 mol h^{-1} over 40 h for a TiO_2 amount corresponding to ~1 m^2 surface area.

The same research group has also studied the still more difficult reduction of NO_3^- ion to hydroxylamine and, above all, to ammonium ions, according to the overall eqns 32 and 33,

$$NO_3^- + 7\ H^+ + 6\ e^- \longrightarrow NH_2OH + 2\ H_2O \qquad (32)$$

$$NO_3^- + 10\ H^+ + 8\ e^- \longrightarrow NH_4^+ + 3\ H_2O \qquad (33)$$

in aqueous phosphoric acid solutions, either over $SrTiO_3$ and TiO_2 powders or in a photoelectrochemical cell using an illuminated p-GaP cathode and a Pt anode. (ref. 254) With light flux in the range 30 to 40 mW cm^{-2}, the product concentrations were in the μM range and underwent erratic fluctuations. Consequently, the efficiency was estimated to be too low for a viable process even for pollutant removal. (ref. 254)

5.2.4.2. N_2H_4 decomposition. Hydrazine has been suggested as an intermediate in N_2 reduction to NH_3. (ref. 231) Nevertheless only insignificant amounts have been detected (ref. 230) and this is consistent with the photocatalytic decomposition of N_2H_4 in deaerated aqueous solutions over anatase, although loading with platinum metals is required for a sustained yield, eqn 34. (refs. 255-256)

$$2\ N_2H_4 \longrightarrow 2\ NH_3 + N_2 + H_2 \qquad (34)$$

The quantum yield is presumably high, since the quantity of H_2 obtained is greater than that obtained from methanol under similar conditions. The electron-consumption and rate-determining pathway was proposed to be proton reduction. From the effects of various parameters it was inferred that N_2 and NH_3 result from hole-initiated steps, eqns 35 and 36, possibly,

$$N_2H_4 + h^+ \longrightarrow N_2H_3^{\bullet} + H^+ \qquad (35)$$

$$2\ N_2H_3^{\bullet} \longrightarrow N_4H_6 \longrightarrow N_2 + 2\ NH_3 \qquad (36)$$

although other elementary steps were not excluded depending on the conditions. (ref. 256)

6. PHOTOCATALYTIC ORGANIC TRANSFORMATIONS

6.1. Oxidations

The capture of a photogenerated hole by an adsorbed organic molecule is thermodynamically feasible for many substrates. Since the valence band edges of most common metal oxide and metal chalcogenide semiconductors lie positive of the oxidation potentials of many common organic functional groups, both pure compounds or mixtures can be attained. If this broad range of reactivity is controlled, chemical selectivity can be attained. Since an extensive review on this topic is available, only a survey of recent work is presented here.

Distinction should be made between photocatalytic oxidations and dehydrogenations (or reductions/hydrogenations), both of which give oxidized (or reduced) organic products. A photocatalytic oxidation involves a primary electron transfer to the photogenerated hole, producing, at least transiently, a single electron oxidized intermediate. A photocatalytic dehydrogenation, however, may involve loss of hydrogen atoms from the organic substance, without intermediate radical ion formation. Dehydrogenations are driven by the affinity of metal cocatalysts for hydrogen atoms and are accompanied by the evolution of hydrogen gas. A clear mechanistic distinction between the two routes can be difficult. It is obvious however that hydrogen evolution, via photocatalyzed dehydrogenation, can only be accomplished in the absence of oxygen. A parallel dichotomy exists for photocatalytic organic reductions. A survey of organic photooxidations and dehydrogenations is available elsewhere (ref. 257) but representative examples and mechanistic features which afford control of photoinduced interfacial electron transfer are treated here briefly. Because of space limitations, it is far from complete.

6.1.1. Alcohols. Alcohols are among the most widely studied organic species to be investigated as photoelectrochemically oxidizable substrates. Upon excitation of semiconductor suspensions in the presence of alcohols, dehydrogenation ensued, producing an isolable carbonyl compound, eqn 37, often in nearly quantitative chemical yield.

$$RR'CHOH \xrightarrow{SC^*} RR'C{=}O + H_2 \qquad (37)$$

SC = semiconductor suspension

Anatase was more active than rutile in the liquid phase photooxidation of 2-propanol, an observation which is parallel to that observed with gaseous alcohol on platinized titania. (ref. 258) Few reports of quantum efficiencies are available.

In liquid phase photocatalytic alcohol dehydrogenation, an optimum platinum loading of TiO_2 of 0.1 to 1 wt % was observed, with platinum islands having diameters of approximately 2 nm. (refs. 75,259) Other metals can also catalyze the reductive half reaction, with activity decreasing in the order Pt > Rh > Pd > Ru > Ir. (ref. 260) (Note, however, that these comparisons are only valid if they refer to metal particles of the same size.) Transition metal oxide additives influenced photoactivity, sometimes in the reverse direction from thermal catalytic oxidations. Surface-bound vanadium and molybdenum oxides, for example, increased the rate, and decreased the selectivity, of thermal oxidation of methanol on titanium dioxide, (ref. 261) but decreased its photocatalytic activity. Steric and doping effects can be at the origin of these phenomena. The addition of alkali metal salts similarly shifted the selectivity in the photooxidation of aqueous methanol toward water, and caused further oxidation of partially oxidized products derived from methanol. (ref. 262)

Chemical selectivity was observed, in that primary or secondary aliphatic, (refs. 51,75,118,119) terpenes, (refs. 127,263) cyclic, (ref. 263) and aromatic (ref. 263) alcohols could be selectively oxidized to the corresponding aldehydes or ketones without appreciable overoxidation, over platinized TiO_2, eqn 38.

$$C_7H_{15}CH_2OH \xrightarrow[PhH,\ O_2]{TiO_2^*} C_7H_{15}CHO + H_2 \qquad (38)$$

Yields were dependent on the structure of the alcohol, with primary alcohols being oxidized in 60-98% yield and secondary alcohols being converted in ~30%.

In the photodehydrogenation of mixtures of ethanol and D_2O, D_2 was found to be the major gaseous product (88%). (ref. 264) This study, however, did not correct for isotopic exchange and is therefore ambiguous. The ratio of hydrogen to methane evolved was about 14, implying that evolution of H_2 proceeded about four times faster than does evolution of CO_2 and CH_4.

With polyols, e.g., sorbitol, the magnitude of photocurrent produced on derived TiO_2 depended on molecular structure (chain length and the number of OH groups). The selectivity, which was pH dependent, was ascribed to the high concentration of OH groups at the photoactivated surface when long chain polyols were adsorbed. (ref. 265)

In the photooxidation of phenol induced by excited titanium dioxide, photocatalytic formation of hydroxy radical was implicated as the reactive species, since the organic products had incorporated oxygen, eqn 39. (refs. 50,266)

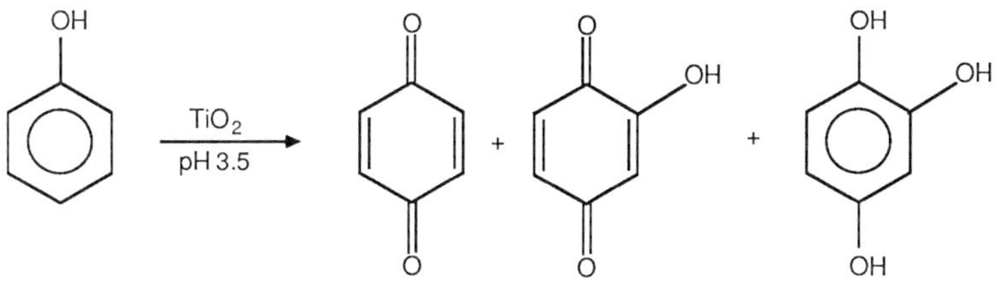

(39)

The chemical yield of hydroquinones was high (~90%) at low conversion and the quantum yield was wavelength dependent, ranging up to ~12%. Upon further photolysis, aldehydes, acids, and CO_2 could be obtained.

6.1.2. Amines. Primary amines can be similarly oxidized and dehydrogenated on irradiated semiconductor suspensions to form Schiff bases, (refs. 267-268) which undergo coupling with the starting amine. Different pathways can be induced by varying initial concentration of the amine and the identity of the photocatalyst, eqn 40. (ref. 267)

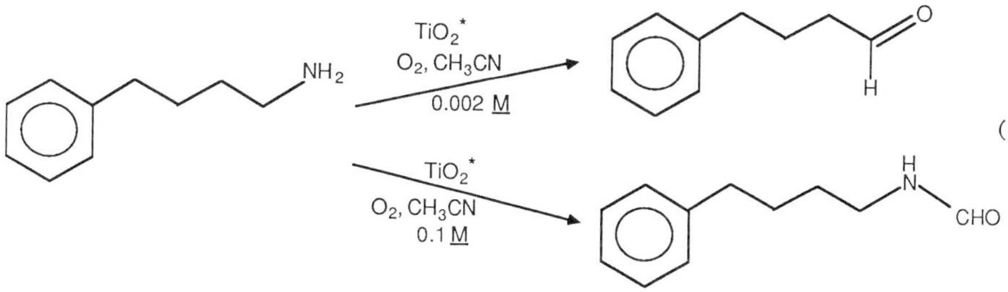

(40)

N-formylation product was obtained in 30-46% yield. Diamines can be cyclized photoelectrochemically, (ref. 269) and primary amines can be converted photocatalytically to secondary amines. The efficiency of photocatalytic production of amines from alcohols and ammonia varied with alcohol structure: ethanol > methanol > 2-propanol > t-butanol. Conducting the reaction under hydrogen improved the yield of amines, a process which was inhibited by oxygen. (ref. 270)

Carbon-carbon coupling is observed upon single electron oxidation of amines bearing alpha hydrogens. On colloidal ZnS, for example, alpha coupling, presumably through a radical produced by deprotonation of the photogenerated radical cation, can be observed, eqn 41. (ref. 271)

$$Et_3N \xrightarrow[H_2O]{ZnS^*} \text{[Et}_2\text{N-CH(CH}_3\text{)-CH(CH}_3\text{)-NEt}_2\text{]} \quad (41)$$

With toluidines, for example, azo products are formed. (ref. 272)

The catalytic oxidation of lactams to imides, eqn 42, (ref. 273)

$$\text{N-R-2-piperidone} \xrightarrow[H_2O, O_2]{TiO_2^*} \text{N-R-glutarimide} \quad (42)$$

and of N-acylamides to imides introduces oxygen into the organic substrate, but if the reaction is conducted in the presence of Cu(I), unsaturation via dehydrogenation is induced instead. Chemical yields of 50-90% have been attained.

6.1.3. Sulfides. Organosulfur compounds are sensitive to photocatalytic oxidative conditions. Thioethers can.be cleanly converted to the corresponding sulfoxides, eqn 43. (ref. 274)

$$C_6H_5(CH_2)_2\text{-S} \xrightarrow{TiO_2^*} C_6H_5(CH_2)_2\text{-SO} \quad (43)$$

A qualitative yield was observed with di-n-hexylsulfide. The photocatalyzed oxidation of dimethyl sulfide is complex, giving a product mixture which depends on the initial concentration of thioether. A Stevens rearrangement occurs on TiO_2, CdS, or ZnSe, presumably through a cation radical. (ref. 275)

Disulfides can be formed by photocatalytic oxidation of thiols, as in the CdS-photocatalyzed conversion of cysteine to cystine. (ref. 276) Since superoxide dismutase inhibited the conversion, photoinduced electron transfer was thought to be responsible for the observed transformations. Such organosulfide oxidations may be environmentally important since naturally occurring hematite suffers a photoassisted dissolution in the presence of thiols. (ref. 277)

6.1.4. Acids. Amino acids can be converted to peptides by semiconductor catalyzed routes. For example, oligopeptides have been isolated upon irradiation of semiconductor suspensions in the presence of amino acids with quantum efficiencies ranging from 6×10^{-3} to 1.1×10^{-2}. (ref. 278) The molecular weight distribution of the peptides could be partially controlled by choice of the sensitizer: diglycine was formed twice as efficiently on platinized TiO_2 as on platinized CdS, while the yield of pentaglycine was four times higher on CdS/Pt than on TiO_2/Pt.

The photocatalytic production of amino acids from simple inorganic molecules has also been reported as a possible abiotic route to these substances. Low efficiencies were observed in the conversion of mixtures of methane, ammonia and water to several amino acids on platinized TiO_2. (ref. 279) When glucose replaced methane as the carbon source, both amino acids and peptides were formed. (ref. 280) Higher quantum efficiencies (0.2-0.4) were reported in the conversion of alpha-keto acids or alpha-hydroxy acids to the corresponding alpha-amino acids. (ref. 281) Moderate levels of enantiomeric selectivity (optical yields of about 50%) were reported when chiral starting materials were employed. When alpha,beta-unsaturated acids were used as substrates for the amino acid synthesis, Michael-like products were isolated. (ref. 281)

Photo-Kolbe decarboxylation occurs when monocarboxylic acids are subjected to photocatalytic oxidative conditions. At the surface of an irradiated n-TiO_2 single crystal or polycrystalline electrode, acetic acid produced ethane, whereas methane formation became dominant on irradiated powders. (refs. 129,206,282-283) The ratio of alkyl coupling to reduction could be controlled by altering the identity of the semiconductor, the extent and identity of the deposited metal, or the pH of the solution. (ref. 131) Presumably, on a powder, a photogenerated radical will persist until back electron transfer generates an anion, protonation of which allows for surface desorption. That alkyl radicals are involved has been established in spin-trapping experiments, (ref. 284) although deuterium incorporation seems to implicate the eventual protonation of an anionic intermediate. (ref. 129)

With several transition metal oxide semiconductor powders, the decarboxylation of oxalic acid was found to be controlled by surface properties and the presence of recombination centers, which in turn depended on preparation method. (ref. 45) Analogous results have also been observed in the photodecarboxylation of pyruvic acid (ref. 285) and formic acid. (ref. 286) The primary photochemical product in these photocatalyzed reactions is thought to be a surface-bound hydroxy radical since water is required for the reaction.

6.1.5. Hydrocarbons. Radical intermediates formed via hydrogen abstraction by photogenerated hydroxy or hydroperoxy radicals can also be formed in the photoelectrochemically induced oxidation of hydrocarbons. The semiconductor-catalyzed photooxidation of toluene to cresols, eqn 44, for example, although with lower quantum efficiency (~1%)

$$X\text{-}C_6H_4\text{-}CH_3 \xrightarrow{TiO_2^*} X\text{-}C_6H_4\text{-}CH_2OH + X\text{-}C_6H_3(OH)\text{-}CH_3 \qquad (44)$$

may proceed through a photo-Fenton (radical) mechanism. (refs. 287-288) The active radical might derive from the reduction of oxygen (ref. 287) or from the oxidation of water. (ref. 289) Incorporation of ^{18}O into the photogenerated phenol from $^{18}O_2$ was taken as evidence that $OH^{\bullet}$ derives from O_2 at high pH and from water at low pH, (ref. 290) although this explanation incorrectly assumes slow ^{18}O exchange with the surface. Competitive trapping of the photogenerated conduction band electron by adsorbed protons can account for the reduced contribution of oxygen at lower pH. Water oxidation cannot be solely responsible for the observed chemistry, however, since toluene can be oxidized to benzyl alcohol as a neat liquid. (ref. 291) Here the intervention of the hydrocarbon radical cation seems possible.

Products of intermediate oxidation level may also be involved in the photocatalytic reactions of hydrocarbons and fossil fuels. (ref. 292) Isotope exchange between cyclopentane and deuterium can be observed on bifunctional platinum/titanium dioxide photocatalysts. Better control of the exchange at either the gas-solid or liquid-solid interface is afforded with photoelectrochemical than thermal catalysis.

Evidence for the involvement of radical cations in the photooxidation of olefins is based on relative reactivity, substituent effects, and the direct observation of radical cations in flash photolysis experiments. (refs. 16,24) Many oxidizable arenes (ref. 293) and dienes (ref. 294) similarly exhibit oxidative cleavage or rearrangement chemistry consistent with initial formation of a radical cation. Oxidative cleavage is attained by interception of the surface bound radical cation with superoxide or adsorbed oxygen. With alkanes or simply substituted alkenes, however, the capture of a photogenerated hole is often thermodynamically forbidden. Thus, radicals formed by hydrogen abstraction by activated oxygen species represent the critical intermediates. The relative ratio of oxygenation to complete decomposition can be controlled, at least to some extent, by judicious choice of the photocatalyst.

The secondary dark reactions of the photocatalytically generated intermediates can lead to different products from the same intermediates when on a surface. The radical cation of diphenylethylene, for example, gives completely different products upon photoelectrochemical activation than upon electrochemical oxidation at a metal electrode or by single electron transfer in homogeneous solution, eqn 45. (ref. 295) The photoelectrochemical transformation is quantitative, whereas the electrooxidation and homogeneous oxidation give complex mixtures with lower chemical efficiency.

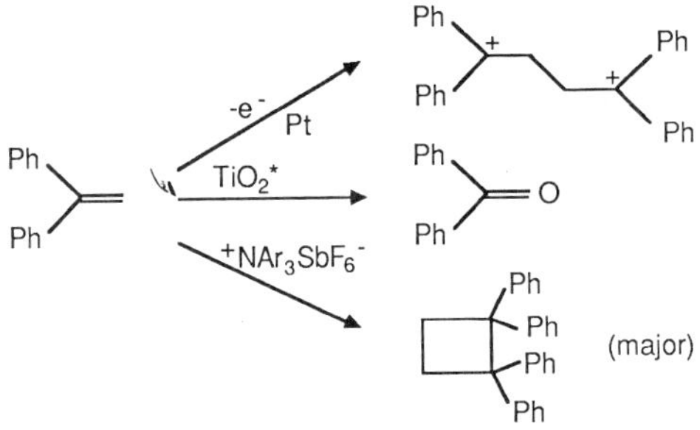

Ph
Ph
$-e^-$
Pt
TiO_2^*
$^+NAr_3SbF_6^-$
Ph
Ph
Ph
Ph
Ph
Ph
O
Ph
Ph
Ph
Ph
(major)

(45)

The divergent chemistry attained from the 1-methylnaphthalene cation radical formed on irradiated TiO_2 powders and by homogeneously dispersed single electron oxidants, eqn 46, (ref. 293)

CH_3 TiO_2^* O_2 CO_2H NC–⟨⟩–$\overset{*}{CN}$ O_2 CH_2OH (46)

can be similarly explained by surface effects. Here the semiconductor surface delivers the coadsorbed oxygen or superoxide more rapidly than C-H deprotonation can ensue, whereas radical formation via deprotonation dominates the chemistry in solution. Chemical yields from ring cleavages ranged from 50-90%.

The hydroxy radical is probably the significant intermediate in the photocatalytic mineralization of alkyl, (ref. 46) vinyl, (ref. 53) and aryl (ref. 296) halides. These reactions are of great environmental importance.

Thus, oxidative transformations of organic substrates can be readily understood as emanating from either photogenerated surface adsorbed radical cations or from radicals formed by activated oxygen radicals (surface oxides or adsorbed hydroxy, hydroperoxy or peroxy radicals). Photoelectrochemical catalysis not only generates the reactive species by interfacial electron transfer, but also controls the subsequent activity of the surface adsorbed intermediates.

6.2. Reductions

Only a small number of photocatalytic organic reductions are known. Since oxygen is such a ready acceptor in the reductive half reaction, it must be specifically removed if other reagents are to be reduced. Because of the only modestly negative potential of electrons at the conduction band edge of conveniently accessible semiconductors, few organic substrates can be directly reduced. Some cationic organic reagents do indeed fill this role, and viologens, for example, have been frequently used as photoelectrochemical relays. (ref. 18) The photoelectrochemical reduction of $NADP^+$ on irradiated CdS has also been reported. (ref. 297)

Two types of organic reductions can be discussed: those in which the organic substrate acts as an electron acceptor and is reduced by the conduction band electron and those in which sacrificial oxidation of solvent or additive reduces protons, producing hydrogen. On metallized surfaces, the photogenerated hydrogen attacks an adsorbed organic via normal, dark catalytic hydrogenation routes. Since oxygen will quickly scavenge hydrogen on metal surfaces, the latter route can be operative only under deaerated conditions.

The first route requires that the organic substrate possess a reduction potential less negative than the conduction band edge. Although band positions are pH dependent, most common semiconductors have conduction band edges positive of -1.0 V, even under the most favorable conditions. Thus, only unusual molecules or substrates multiply substituted with electron withdrawing groups can be directly reduced on irradiated semiconductor surface. Thus, most of the observed organic reductions occur via photogenerated hydrogen. Organic reductions are therefore much more relevant to the numerous studies of water splitting than are the photocatalytic oxidations. Photocatalytic hydrogenation of double and triple bonds can be similarly attained with sulfide as a sacrificial donor, eqn 47. (ref. 298)

$$\text{H-C}\equiv\text{C-H} \xrightarrow[\substack{S^{2-}\\ H_2O}]{\text{CdS*/Pt}} \text{H-CH}_2\text{CH}_2\text{-H} \tag{47}$$

With CdS metallized with platinum or rhodium as the photocatalyst, hydrogenation is about four times as efficient as hydrogen evolution, with the efficiency depending on the solution phase pH and the identity and size of the metal catalyst island. The photocatalyzed hydrogenation is quantitative, but slow.

Olefins, vinyl ethers, and the double bond of alpha,beta-unsaturated enones could be photoelectrochemically hydrogenated on TiO_2 with ethanol acting as the electron source. With 2-methyl-2-pentene, hydrogenation occurred in 63% yield without side products, eqn 48. (ref. 299)

$$\text{2-methyl-2-pentene} \xrightarrow[\text{EtOH}]{TiO_2^*/Pt} \text{2-methylpentane} \quad (48)$$

The direct electron-transfer photoelectrochemical reduction of the N=N double bond of the diaryl azo dye methyl orange can be accomplished on colloidal titanium dioxide. (refs. 300-301) The reaction was sensitive to pH and surfactants. Cationic surfactants increased the efficiency of oxidative cleavage, (ref. 302) by inhibiting charge recombination, while polyvinyl alcohol instead favored reduction.

Aldehydes and ketones can also be photocatalytically reduced: for example, ZnS solutions prepared as cold oxygen-free aqeuous suspensions of $ZnSO_4$ and Na_2S induce efficient photodisproportionation of aldehydes, i.e., a photo-Canizzaro reaction. (ref. 303)

6.3. Cleavages

The observation of formal cycloadditions and retrocycloadditions on irradiated semiconductor suspensions is certainly reasonable in view of analogous reactions recently discovered in homogeneous solution. The photocatalysis by ZnO or CdS of the ring-opening of a strained cyclobutane, which could also be opened in the dark by a single electron oxidant, ceric ammonium nitrate, (ref. 304) was the first such reaction. On TiO_2, over-oxidation of the multiple bonds was observed. The valence isomerization of hexamethyldewarbenzene to hexamethylbenzene, proceeds through a surface-bound cation radical chain mechanism since the observed quantum yield was greater than 1. (ref. 305) Different surface treatments, areas, impurities, and surface structures did not dramatically influence the rate of reaction. Only ring closure could be observed, however, in the photodimerization of phenyl vinyl ether, which proved to be irreversible. (refs. 306-307) Neither particle size, surface area, nor crystal structure appeared to significantly influence the dimerization, (ref. 308) of this reaction or the similar dimerization of N-vinylcarbazole. (ref. 309)

Intramolecular cycloadditions are also known. Norbornadiene, for example, can be cyclized upon irradiation of ZnO, ZnS, CdS, or Ge semiconductors. (ref. 310)

6.4. Rearrangements and Isomerizations

Both positional and geometric isomerizations can be induced photocatalytically. For example, substituted styrenes can achieve thermodynamic geometric equilibration via a radical ion when exposed to band-gap excited CdS. (ref. 311) The reaction can be quenched by electron donors (methoxybenzenes or pyrenes) and can be controlled by the ratio of substrate to catalyst, by light intensity, and by temperature. Oxygen inhibited the reaction. (ref. 312) A non-linear Hammett plot indicated that a change in mechanism occurred with substituents of greater electron donating ability. (ref. 313)

In an analogous study on ZnS with simple alkenes, high turnover numbers were observed at active sites where trapped holes derived from surface states (sulfur radicals from zinc vacancies or interstitial sulfur) play a decisive role. (ref. 314)

7. CONCLUSIONS

Many chemical conversions can be induced by photoelectrochemical activation of light sensitive semiconductor surfaces. This photochemical activation creates an electron-hole pair, which can initiate interfacial electron transfer. The oxidized and reduced species thus produced undergo dark secondary reactions which are greatly influenced by the surface on which they are produced. The observed chemistry is thus controlled by band edge positions, by surface features, and by adsorption effects. The surface can bring together coadsorbates and can control the local polarity of photogenerated intermediates. Oxidations, reductions of both inorganic and organic substrates, as well as, in this latter case, formal cycloadditions and retrocycloadditions, geometric and positional isomerizations can be rationally conducted photocatalytically. Establishing the factors which control selectivity will continue to offer a significant intellectual challenge.

8. ACKNOWLEDGEMENTS

Expenses incurred in writing this article were covered by a NATO travel grant. This article arose from a scientific collaboration made possible by support of the US-France NSF/CNRS Collaborative Research Program.

9. REFERENCES

1. M.A. Fox, Top. Org. Electrochem. 1 (1986) 177.
2. A.J. Bard, J. Photochem. 10 (1979) 50.
3. A.J. Bard, Science 207 (1980) 139.
4. A.J. Bard, J. Phys. Chem. 86 (1982) 172.
5. R. Memming, "Electroanalytical Chemistry", A.J. Bard, ed., Marcel Dekker, New York, 1979, p. 1.
6. M.S. Wrighton, Accts. Chem. Res. 12 (1979) 303.
7. H.O. Finklea, J. Chem. Ed. 60 (1983) 325.
8. A.J. Nozik, Faraday Disc. 70 (1980) 7.
9. T. Freund and W.P. Gomes, Catal. Rev. 3 (1969) 1.
10. J.A. Turner, J. Chem. Ed. 60 (1983) 327.
11. B. Parkinson, J. Chem. Ed. 60 (1983) 327.
12. H. Gerischer, "Photoelectrochemistry, Photocatalysis, and Photoreactors", M. Schiavello, ed., D. Reidel Publishers, Dordrecht, 1985, p. 39.
13. S.J. Teichner and M. Formenti, ibid., p. 457.
14. P. Pichat, ibid., p. 425.
15. A.J. Nozik, Ann. Rev. Phys. Chem. 29 (1978) 189.
16. M.A. Fox, Accts. Chem. Res. 16 (1983) 314.
17. For typical values for inorganic couples, see "Electrochemistry of the Elements", A.J. Bard, ed., Vols. I-XV.
18. M. Graetzel, "Energy Resources through Photochemistry and Catalysis", M. Graetzel, ed., Academic Press, New York, 1983, p. 71; N. Serpone, D.K. Sharma, J. Moser, and M. Graetzel, Chem. Phys. Lett. (1987) in press.
19. M. Graetzel, Accts. Chem. Res. 14 (1981) 376.
20. K. Kalyanasundaram, Solar Cells 15 (1985) 93.
21. H. Gerischer and F. Willig, Top. Curr. Chem. 61 (1976) 33.
22. Kabir-ud-Din, R.C. Owen, and M.A. Fox, J. Phys. Chem. 85 (1981) 1679.
23. T. Kano, T. Oguchi, H. Sakuragi, and K. Tokumaru, Tet. Lett. **21** (1980) 467.
24. M.A. Fox, B.A. Lindig, and C.C. Chen, J. Am. Chem. Soc. 104 (1982) 5828.
25. D. Bahnemann, A. Henglein, and L. Spanhel, Faraday Discuss. Chem. Soc. 78 (1984) 151.
26. M. Evenor, S. Gottesfeld, Z. Harzion, D. Huppert, and S.W. Feldberg, J. Phys. Chem. 88 (1984) 6213.

27. J. Kiwi and C. Morrison, J. Phys. Chem. 88 (1984) 6146.
28. V.M. Mekler, A.I. Kotel'nikov, G.I. Likhtenshtein, A.P. Kaplun, and V.I. Shvets, Biofizika 29 (1984) 779.
29. Y. Nosaka and M.A. Fox, J. Phys. Chem. 90 (1986) 6526.
30. A. Leaustic, F. Babonneau, and J. Livage, J. Phys. Chem. 90 (1986) 4193.
31. R.F. Howe and M. Graetzel, J. Phys. Chem. 89 (1985) 4495.
32. M. Kunst, G. Beck, and H. Tributsch, J. Electrochem. Soc. 131 (1984) 954.
33. K. Itoh, R. Baba, and A. Fujishima, Chem. Phys. Lett. 135 (1987) 521.
34. J.H. Richardson, S.B. Deutscher, A.S. Maddux, J.E. Harrar, D.C. Johnson, W.L. Schmelizinger, and S.P. Perone, J. Electroanal. Chem. 109 (1980) 95.
35. P.V. Kamat and M.A. Fox, J. Phys. Chem. 87 (1983) 59.
36. P.W. Yang and H.L. Casal, J. Phys. Chem. 90 (1986) 2422.
37. Q. Feng and T.M. Cotton, J. Phys. Chem. 90 (1986) 983.
38. J. Riefkohl, L. Rodriguez, L. Romero, G. Sampoll, and F.A. Souto, J. Phys. Chem. 90 (1986) 6068.
39. R.E. Schwerzel, Ext. Abstr. Electrochem. Soc. 83 (1983) 513.
40. H. Harada, T. Sakata, and T. Ueda, J. Am. Chem. Soc. 107 (1985) 1773.
41. K. Tanabe, "Catalysis", J.R. Anderson and M. Boudart, eds., Springer-Verlag, Vol. 2, 1981, p. 231.
42. J. Cunningham, B.K. Hodnett, M. Ilyas, E.M. Leahy, and J.P. Tobin, J. Chem. Soc., Faraday Trans. 1 78 (1982) 3297.
43. P. Pichat, M.N. Mozzanega, and H. Courbon, J. Chem. Soc., Faraday Trans. 1 83 (1987) 697.
44. J.M. Herrmann and P. Pichat, J. Chem. Soc., Faraday Trans. 1 76 (1980) 1138.
45. J.M. Herrmann, M.N. Mozzanega, and P. Pichat, J. Photochem. 22 (1983) 333.
46. C.Y. Hsiao, C.L. Lee, and D.F. Ollis, J. Catal. 82 (1983) 418.
47. D.F. Ollis, C.Y. Hsiao, L. Budiman, and C.L. Lee, J. Catal. 88 (1984) 89.
48. T. Nguyen and D.F. Ollis, J. Phys. Chem. 88 (1984) 3386.
49. K. Okamoto, Y. Yamamoto, H. Tanaka, and A. Itaya, Bull. Chem. Soc. Jpn. 58 (1985) 2023.
50. V. Augugliaro, L. Palmisano, A. Sclafani, C. Minero, and E. Pelizzetti, Toxicological and Environmental Chemistry in press.
51. P. Pichat, J.M. Herrmann, J. Disdier, H. Courbon, and M.N. Mozzanega, Nouv. J. Chim. 5 (1981) 627.
52. H. Mozzanega, J.M. Herrmann, P. Pichat, J. Phys. Chem. 83 (1979) 2251.
53. D.F. Ollis, Environ. Sci. Tech. 19 (1983) 480.
54. P. Meriaudeau, and J.C. Vedrine, J. Chem. Soc., Faraday Trans. I 72 (1976) 472.
55. A.R. Gonzalez-Elipe, G. Munuera, and J. Soria, J. Chem. Soc., Faraday Trans. 1 76 (1980) 1535.
56. M. Che and A.J. Tench, Adv. Catal. 31 (1982) 77.
57. M. Che and A.J. Tench, Adv. Catal. 32 (1983) 1.
58. B. Kraeutler, C.D. Jaeger, and A.J. Bard, J. Am. Chem. Soc. 100 (1978) 4903.
59. J. Moser and M. Graetzel, Helv. Chim. Acta 65 (1982) 1436.
60. J. Moser and M. Graetzel, J. Am. Chem. Soc. 105 (1983) 6547.
61. T. Nakahira and M. Graetzel, J. Phys. Chem. 88 (1984) 4006.
62. D. Bahnemann, A. Henglein, J. Lilie, and L. Spanhel, J. Phys. Chem. 88 (1984) 709.
63. D. Bahnemann, A. Henglein, and L. Spanhel, Faraday Discuss. Chem. Soc. 88 (1984) 151; A. Henglein, Pure Appl. Chem. 56 (1984) 1215.
64. P.V. Kamat, J. Photochem. 28 (1985) 513.
65. J. Moser, M. Graetzel, D.K. Sharma, and N. Serpone, Helv. Chim. Acta 68 (1985) 1686.
66. J.J. Ramsden and M. Graetzel, Chem. Phys. Lett. 132 (1986) 269.
67. R. Rossetti, S.M. Beck, and L.E. Brus, J. Am. Chem. Soc. 106 (1984) 980.
68. R. Rossetti and L.E. Brus, J. Am. Chem. Soc. 106 (1984) 4336.
69. P. Pichat, Nouv. J. Chem. 11 (1987) 135.
70. M. Primet, P. Pichat, and M.-V. Mathieu, J. Phys. Chem. 75 (1971) 1216.
71. J.-M. Herrmann and P. Pichat, J. Catal. 78 (1982) 425.

72. J.-M. Herrmann and P. Pichat, in "Spillover of Adsorbed Species", Studies in Surf. Sci. Catal. 17, G.M. Pajonk, S.J. Teichner, and J.E. Germain, Eds., Elsevier, 1983, p. 77.
73. J. Disdier, J.-M. Herrmann, and P. Pichat, J. Chem. Soc., Faraday Trans. I 79 (1983) 651.
74. H. Courbon, J.-M. Herrmann, and P. Pichat, J. Catal. 95 (1985) 539.
75. P. Pichat, M.N. Mozzanega, J. Disdier, and J.-M. Herrmann, Nouv. J. Chim. 6 (1982) 559.
76. R. Baba, S. Nakabayashi, A. Fujishima, and K. Honda, J. Phys. Chem. 89 (1985) 1902.
77. K. Watanabe, K. Ichimura, N. Inoue, and I. Matsuura, J. Phys. Chem. 90 (1986) 866.
78. P. Pichat, "Homogeneous and Heterogeneous Photocatalysis", E. Pelizzetti and N. Serpone, Eds., D. Reidel, Dordrecht, 1986.
79. R.I. Bickley and F.S. Stone, J. Catal. 31 (1973) 389.
80. A.H. Boonstra and C.A.H.A. Mutsaers, J. Phys. Chem. 79 (1975) 1694.
81. G. Munuera and F. Gonzalez, Rev. Chim. Minerale 4 (1967) 207.
82. H. Courbon, J.-M. Herrmann, and P. Pichat, J. Phys. Chem. 88 (1984) 5210.
83. R.I. Bickley and R.K.M. Jayanty, Disc. Faraday Soc. 58 (1974) 194.
84. G. Munuera, V. Rives-Arnau, and A. Saucedo, J. Chem. Soc., Faraday Trans. I 75 (1979) 736.
85. A.R. Gonzalez-Elipe, G. Munuera, J. Sanz, and J. Soria, J. Chem. Soc., Faraday Trans. I 76 (1980) 1535.
86. A.R. Gonzalez-Elipe, G. Munuera, and J. Soria, React. Kinet. Catal. Lett. 18 (1981) 367.
87. Yu. P. Solonitsyn, Kin. Catal. 7 (1966) 424.
88. A.M. Volodin, V.S. Zakharenko, and A.E. Cherkashin, React. Kin. Catal. Lett. 18 (1981) 321.
89. S. Bourasseau, J.-R. Martin, F. Juillet, and S.J. Teichner, J. Chem. Phys. 71 (1974) 122.
90. J.-M. Herrmann, J. Disdier, and P. Pichat, J. Chem. Soc., Faraday Trans. I 77 (1981) 2815.
91. P. Pichat, J.-M. Herrmann, H. Courbon, J. Disdier, and M.-N. Mozzanega, Can. J. Chem. Eng. 60 (1982) 27.
92. S. Bourasseau, J.-R. Martin, F. Juillet, and S.J. Teichner, J. Chim. Phys. 70 (1973) 1467, 1472; 71 (1974) 1017.
93. H.G. Volz, G. Kaempf, H.G. Fitzky, and A. Klaeren, in "Photodegradation and Photostabilization of Coatings", S.P. Pappas, and F.H. Winslow, Eds., A.C.S. Symp. Ser. 151 (1981) 163 and references therein.
94. E.M. Ceresa, L. Burlamacchi, and M. Visca, J. Mater. Sci. 18 (1983) 289.
95. E. Yesodharan and M. Graetzel, Helv. Chim. Acta 66 (1983) 2145.
96. E. Yesodharan, S. Yesodharan, and M. Graetzel, Solar Energy Mater. 10 (1984) 287.
97. J. Kiwi and M. Graetzel, J. Mol. Catal. 39 (1987) 63.
98. J.R. Harbour, J. Tromp, and M.L. Hair, Can. J. Chem. 63 (1985) 204.
99. J. Cunningham, in "Comprehensive Chemical Kinetics", C.H. Bamford, and C.F.H. Tipper, Eds., Elsevier, Vol. 19, Chap. 3, 1984, 291.
100. S. Baidyaroy, W.R. Bottoms, and P. Mark, Surf. Sci. 28 (1971) 517.
101. P. Genequand, Surf. Sci. 25 (1971) 643.
102. M. Grade, in "Current Topics in Materials Sci.;, E. Kaldis, Ed., North Holland Publ. Co., 7 (1981), 339.
103. S.R. Morrison, "The Chemical Physics of Surfaces", Chap. 9, Plenum Press (1977).
104. P. Bonasweicz and W. Hirschwald, ibid. 410.
105. D. Lichtman, and Y. Shapira, CRC Crit. Rev. Solid State Mater. Sci. 8 (1978) 93 and references therein.
106. H. Courbon, M. Formenti, F. Juillet, A.A. Lisachenko, J.-R. Martin, and S.J. Teichner, Kinetics and Catalysis 14 (1973) 84.
107. J. Cunningham, B. Doyle, D.J. Morrissey, and N. Samman, "Proc. 6th Int. Congr. Catal.", G.C. Bond, and P.B. Wells, Eds., Chem. Soc. London, Vol. 2, 1976, 1093.

108. D. Oelkrug, W. Flemming, R. Fuellemann, R. Guenther, W. Honnen, G. Krabichler, M. Schaefer, and S. Uhl, Pure Appl. Chem. 58 (1986) 1207.
109. H. Courbon, M. Formenti, and P. Pichat, J. Phys. Chem. 81 (1977) 550.
110. H. Courbon and P. Pichat, C.R. Acad. Sci. 285C (1977) 171.
111. H. Courbon and P. Pichat, J. Chem. Soc., Faraday Trans. I 80 (1984) 3175.
112. J.-M. Herrmann, J. Disdier, and P. Pichat, Chem. Phys. Lett. 108 (1984) 618.
113. H.W. Chung, W.J. Lo, and G.A. Somorjai, Surf. Sci. 64, (1977) 588.
114. N. Yamamoto, S. Tonomura, T. Matsuoka, and H. Tsubomura, Surf. Sci. 92 (1980) 400.
115. D.E. Aspnes and A. Heller, J. Phys. Chem. 87 (1983) 4919.
116. A. Heller, Catal. Rev.-Sci. Eng. 26 (1984) 655.
117. A. Heller, Science 223 (1984) 1141.
118. E. Borgarello and E. Pelizzetti, Chim. Ind. 65 (1983) 474.
119. E. Pelizzetti, E. Borgarello and N. Serpone, "Photoelectrochemistry, Photocatalysis, and Photoreactors", M. Schiavello, ed., D. Reidel Publishers, Dordrecht, 1985, p. 305.
120. J.-M. Lehn, J.-P. Sauvage and R. Ziessel, Nouv. J. Chim. 4 (1980) 355.
121. S. Teratani, C. Sungborn, K. Tanaka and Y. Takagi, China-Japan-USA Symp. Heterog. Catal., Dalian (China), 1982, A.22J.
122. T. Kawai and T. Sakata, J. Chem. Soc., Chem. Commun. (1980) 694.
123. I. Ait-Ichou, M. Formenti, B. Pommier, and S.J. Teichner, J. Catal. 90 (1985) 298.
124. R.L. Moss, in "Experimental Methods in Catalytic Research", R.B. Anderson and P.T. Dawson, Eds., Academic Press, 1976, Vol. 2, p. 43.
125. J.-M. Herrmann, J. Disdier, and P. Pichat, J. Phys. Chem. 90 (1986) 6028.
126. A. Mills and G. Porter, J. Chem. Soc., Faraday Trans. I 78 (1982) 3659.
127. P. Pichat, J. Disdier, M.-N. Mozzanega, and J.-M. Herrmann, Proc. 8th Int. Cong. Catal., Verlag Chemie-Dechema, Weinheim; Deerfield Beach, Florida; Basel, Vol. III, 1984, p. 487.
128. I. Ait-Ichou, D. Bianchi, M. Formenti, and S.J. Teichner, in "Homogeneous and Heterogeneous Photocatalysis", E. Pelizzetti and N. Serpone, Eds, D. Reidel, 1986, 433.
129. B. Kraeutler and A.J. Bard, J. Am. Chem. Soc. 100 (1978) 5985.
130. I. Izumi, F.-R.F. Fan, and A.J. Bard, J. Phys. Chem. 85 (1981) 218.
131. H. Yoneyama, Y. Takao, H. Tamura, and A.J. Bard, J. Phys. Chem. 87 (1983) 1417.
132. S. Sato, J. Phys. Chem. 87 (1983) 3531.
133. I. Izumi, W.W. Dunn, K.O. Wilbourn, F.-R.F. Fan, and A.J. Bard, J. Phys. Chem. 84 (1980) 3207.
134. H.L. Chum, M. Ratcliff, F.L. Posey, J.A. Turner, and A.J. Nozik, J. Phys. Chem. 87 (1983) 3089.
135. P. Pichat, J. Disdier, J.-M. Herrmann, and P. Vaudano, Nouv. J. Chim. 10 (1986) 545.
136. A. Sobczynski and J.M. White, J. Molec. Catal. 29 (1985) 379.
137. J.B. Goodenough, in "Study Inorg. Chem. 3 (Solid State Chem.)" 1983, p. 829 and references therein.
138. G. Bin-Daar, M.P. Dare-Edwards, J.B. Goodenough, and A. Hamnett, J. Chem. Soc., Faraday Trans. 1 79 (1983) 1199.
139. G. Campet, M.P. Dare-Edwards, A. Hamnett, and J.B. Goodenough, Nouv. J. Chim. 4 (1980) 501.
140. S. Zielinski and A. Sobczynski, Acta Chim. Hungarica 120 (1985) 229.
141. B. Odekirk and J.S. Blakemore, J. Electrochem. Soc. 130 (1983) 321.
142. P. Pichat, J.-M. Herrmann, J. Disdier, M. Mozzanega and H. Courbon, "Catalysis on the Energy Scene", S. Kaliaguine and A. Makay, eds., Elsevier, 1984, p. 319.
143. G.H. Al'Sayyed and P. Pichat, unpublished results.
144. B. Jenny and P. Pichat, "Book of Abstracts, 6th Int. Conf. Photochemical Conversion and Storage of Solar Energy", Paris, 1986, C-29.
145. B. Jenny and P. Pichat, "Proc. 10th IUPAC Symp. Photochemistry, Presses Polytechniques Romandes", 1984, 321.

146. A. Sobczynski, J. Molec. Catal. 39 (1987) 43.
147. H. Toyuki, M. Itami, K. Hotta, and Y. Kawamoto, Nippon Kagaku Kaishi 9 (1984) 1363.
148. L. Kruczynski, H.D. Gesser, C.W. Turner, and E.A. Speers, Nature 291 (1981) 399.
149. K. Kalyanasundaram, N. Vlachopoulous, V. Krishnan, A. Monnier, and M. Graetzel, J. Phys. Chem. 91 (1987) 2342.
150. R. Rafaeloff, Y.-M. Tricot, F. Nome, and J.H. Fendler, J. Phys. Chem. 89 (1985) 533.
151. Y. Degani and I. Willner, J. Chem. Soc., Perkin Trans 2 1 (1986) 37.
152. Y.-M. Tricot, A. Emeren, and J.H. Fendler, J. Phys. Chem. 89 (1985) 4721.
153. S. Nishimoto, B. Ohtani, H. Shirai, and T. Kagiya, J. Polym. Sci., Polym. Lett. Ed. 23 (1985) 141.
154. D. Meissner, R. Memming, and B. Kastening, Chem. Phys. Lett. 96 (1983) 34.
155. M. Krishnan, J.R. White, M.A. Fox, and A.J. Bard, J. Am. Chem. Soc. 105 (1983) 7002.
156. A.W.H. Mau, C.B. Huang, N. Kakuta, A.J. Bard, A. Campion, M.A. Fox, J.M. White, and S.E. Webber, J. Am. Chem. Soc. 106 (1984) 6537.
157. A. Makhmadmurodov, Yu. A. Gruzdkov, E.N. Savinov, and V.N. Parmon, Kinet. Catal. 27 (1986) 121; translated from Kinet. Katal. 27 (1986) 133.
158. F.-R.F. Fan, H.-Y. Liu, and A.J. Bard, J. Phys. Chem. 89 (1985) 4418.
159. B.H. Milosavljevic, and J.K. Thomas, J. Am. Chem. Soc. 108 (1986) 2513.
160. T. Kennelly, H.D. Gafney, and M. Braun, J. Am. Chem. Soc. 107 (1985) 4431.
161. M. Anpo, N. Aikawa, Y. Kubokawa, M. Che, C. Louis, and E. Giamello, J. Phys. Chem. 89 (1985) 5017.
162. E.A. Shishkina, I.S. Kolomnikov, T.V. Lysyak, A.V. Rudnev, E.P. Kalyazin, and Yu.Ya. Kharitonov, Dokl. Phys. Chem. 277 (1984) 658; translated from Dokl. Akad. Nauk SSSR 277 (1984) 650.
163. T.L. Pettit and M.A. Fox, J. Phys. Chem. 90 (1986) 1353.
164. R.D. Stramel, T. Nakamura, and J.K. Thomas, Chem. Phys. Lett. 130 (1986) 423.
165. T. Nakamura and J.K. Thomas, Langmuir 1 (1985) 568.
166. O. Enea and A.J. Bard, J. Phys. Chem. 90 (1986) 301.
167. D.N. Furlong, O. Johansen, A. Launikonis, J.W. Loder, A.W.-H. Mau, and W.H.F. Sasse, Aust. J. Chem. 38 (1985) 363.
168. I. Willner, J.M. Yang, C. Laane, J.W. Otvos, and M. Calvin, J. Phys. Chem. 85 (1981) 3277.
169. E. Smotkin, A.J. Bard, A. Campion, M.A. Fox, T. Mallouk, S.E. Webber, and J.M. White, J. Phys. Chem. 90 4604 (1986).
170. N. Serpone, E. Borgarello, and M. Graetzel, J. Chem. Soc., Chem. Commun. (1984) 342.
171. N. Serpone, E. Borgarello, E. Pelizzetti, and M. Barbeni, Chim. Ind. 67 (1985) 318.
172. A. Sobczynski, A.J. Bard, A. Campion, M.A. Fox, T. Mallouk, S.E. Webber, and J.M. White, J. Phys. Chem. (1987) in press.
173. P. Pichat, E. Borgarello, J. Disdier, J.-M. Herrmann, E. Pelizzetti, and N. Serpone, J. Chem. Soc., Faraday Trans. 1, in press.
174. N. Serpone, P. Pichat, J.-M. Herrmann, and E. Pelizzetti, "Photoinduced Charge Separation and Energy Migration in Supramolecular Species", D. Reidel, Dordrecht, 1987, in press.
175. M. Formenti, F. Juillet, and S.J. Teichner, Bull. Soc. Chim. Fr. (1976) 1315.
176. P. Pichat, J.-M. Herrmann, J. Disdier, and M.N. Mozzanega, J. Phys. Chem. 83 (1979) 3122.
177. M. Bideau, B. Claudel, and M. Otterbein, J. Photochem. 14 (1980) 291.
178. S.M. Fang, B.H. Chen, and J.M. White, J. Phys. Chem. 86 (1982) 3126.
179. M. Daroux, D. Klvana, M. Duran, and M. Bideau, Can. J. Chem. Eng. 63 (1985) 668.
180. H. Nakanishi, C. Sanchez, M. Hendewerk and G.A. Somozjai, Mat. Res. Bull. 21 (1986) 137.

181. P.R. Harvey, R. Rudham, and S. Ward, J. Chem. Soc., Faraday Trans. 1 79 (1983) 1381 and 2975.
182. I.M. Fraser and J.R. MacCallum, J. Chem. Soc., Faraday Trans. 1 82 (1986) 2747.
183. J. Cunningham and B.K. Hodnett, J. Chem. Soc., Faraday Trans. 1 77 (1981) 2777.
184. B. Jenny and P. Pichat, "Sixth Internat. Conf. Photochem. Conver. Storage Solar Energy", Abstr. C-31 (1986).
185. R.W. Fessenden and P.V. Kamat, Chem. Phys. Lett. 123 (1986) 233.
186. T.A. Egerton and C.J. King, J. Oil. Col. Assoc. 62 (1979) 386.
187. H. Courbon, J.M. Herrmann, and P. Pichat, J. Catal. 72 (1981) 129.
188. R.I. Bickley, "Photoelectrochemistry, Photocatalysis, and Photoreactors", M. Schiavello, Ed., D. Reidel, Dordrecht, 1985, p. 581.
189. Y. Hori, A. Nakatsu and S. Suzuki, Chem. Lett. (1985) 1429.
190. M. Formenti and S.J. Teichner, "Catalysis", C. Kemball, Ed., The Chem. Soc. of London, Vol. 2, 1978, p. 87.
191. R.I. Bickley, "Catalysis", C. Kemball, Ed., The Chem. Soc. of London, Vol. 5, 1982, p. 308.
192. S. Sato, Chem. Phys. Lett. 123 (1986) 126.
193. M. Anpo, I. Tanahashi, and Y. Kubokawa, J. Phys. Chem. 86 (1982) 1.
194. Y. Kubokawa and M. Anpo, "Adsorption and Catalysis on Oxide Surfaces", M. Che and G.C. Bond, Eds., Elsevier, Amsterdam, 1985, p. 127.
195. D.H.M.W. Thewissen, A.H.A. Tinnemans, M. Eeuwhorst-Reinten, Timmer, and A. Mackor, Nouv. J. Chim. 7 (1983) 73.
196. H. Kisch and W. Schlamann, Chem. Ber. 119 (1986) 3483.
197. S. Tsai, C. Kao, and Y. Chung, J. Catal. 79 (1983) 451.
198. S.N. Frank and A.J. Bard, J. Phys. Chem. 81 (1977) 1484.
199. K.H. Stadler and H.P. Boehm, Z. Phys. Chem. 144 (1985) S.9.
200. See for example: K. Micka and H. Gerischer, J. Electroanal. Chem. 38 (1972) 397.
201. A. Fujishima, T. Inoue, and K. Honda, J. Am. Chem. Soc. 101 (1979) 5582.
202. K. Hirano and A.J. Bard, J. Electrochem. Soc. 127 (1980) 1056.
203. T. Kobayashi, H. Yoneyama, and H. Tamura, J. Electroanal. Chem. 122 (1981) 133.
204. B. Reichmann and C.E. Byvik, J. Phys. Chem. 85 (1981) 2255.
205. M.A. Fox and T.L. Pettit, J. Org. Chem. 50 (1985) 5013.
206. B. Krauetler and A.J. Bard, J. Am. Chem. Soc. 100 (1978) 4317.
207. H. Hada, Y. Yonezawa, and M. Saikowa, Bull. Chem. Soc. Jpn. 55 (1982) 2010.
208. S. Nishimoto, B. Ohtani, H. Kajiwara, and T. Kagiya, J. Chem. Soc., Faraday Trans. 1 79 (1983) 2685.
209. Y. Oosawa and M. Graetzel, J. Chem. Soc., Chem. Commun. (1984) 1630.
210. K. Tanaka, K. Harada, and S. Murata, Solar Energy 36 (1986) 159.
211. T. Kobayashi, Y. Taniguchi, H. Yoneyama, and H. Tamura, J. Phys. Chem. 87 (1983) 768.
212. E. Borgarello, R. Harris, and N. Serpone, Nouv. J. Chim. 9 (1985) 743.
213. E. Borgarello, N. Serpone, G. Emo, R. Harris, E. Pelizzetti, and E. Minero, Inorg. Chem. 25 (1986) 4499.
214. H. Hada, Y. Yonezawa, M. Ishino, and H. Tanemura, J. Chem. Soc., Faraday Trans. 1 78 (1982) 2677.
215. H. Hada, H. Tanemura, and Y. Yonezawa, Bull. Chem. Soc. Jpn. 51 (1978) 3154.
216. W.W. Dunn and A.J. Bard, Nouv. J. Chim. 5 (1981) 651.
217. C. Sungbom, M. Kawai, and K. Tanaka, Bull. Chem. Soc. Jpn. 57 (1984) 871.
218. M. Guenin and P. Pichat, unpublished results.
219. K.H. Stadler and H.P. Boehm, "Proceedings of the 8th International Congress on Catalysis", Vol. IV, Weinheim, Deerfield Beach, Florida, Verlag Chemie-Dechema, Basel, 1984, p. 803.
220. H. Nakamatsu, T. Kawai, A. Koreeda, and S. Kawai, J. Chem. Soc., Faraday Trans. 1 82 (1985) 527.

221. J.M. Herrmann, J. Disdier, P. Pichat, and C. Leclercq, "Preparation of Catalysts IV", B. Delmon, P. Grange, P.A. Jacobs and G. Poncelet, Eds., Elsevier, Amsterdam, 1987, p. 285.
222. M. Halmann, in "Energy Resources Through Photochemistry and Catalysis", M. Graetzel, Ed., Academic Press, 1983, p. 507.
223. T. Inoue, A. Fujishima, S. Koishi, and K. Honda, Nature 277 (1979) 637.
224. M. Halmann, V. Katzir, E. Borgarello, and J. Kiwi, Solar Energy Mat. 10 (1984) 85.
225. M. Ulman, A.H.A. Tinnemans, A. Mackor, B. Aurian-Blajeni, and M. Halmann, Int. J. Solar Energy 1 (1982) 213.
226. M. Halmann and K. Zuckerman, in "Homogeneous and Heterogeneous Photocatalysis", E. Pelizzetti and N. Serpone, Eds., D. Reidel, Dordrecht, 1986, p. 521.
227. B. Aurian-Blajeni, I. Taniguchi, and J. O'M. Bockris, J. Electroanal. Chem. 149 (1983) 291.
228. I. Taniguchi, B. Aurian-Blajeni, and J. O'M. Bockris, Electrochim. Acta 29 (1984) 923.
229. S. Ikeda, M. Yoshida, and K. Ito, Bull. Chem. Soc. Japan 58 (1985) 1353.
230. G.N. Schrauzer and T.D. Guth, J. Am. Chem. Soc. 99 (1977) 7189.
231. G.N. Schrauzer, T.D. Guth, J. Salehi, N. Strampach, N.-H. Liu, and M.R. Palmer, in "Homogeneous and Heterogeneous Photocatalysis", E. Pelizzetti, and N. Serpone, Eds., D. Reidel, 1986, and references therein.
232. H. Van Damme and W.K. Hall, J. Am. Chem. Soc. 101 (1979) 4373.
233. M. Schiavello, and A. Sclafani, in "Photoelectrochemistry, Photocatalysis and Photoreactors", M. Schiavello, Ed., D. Reidel, 1985, and references therein.
234. H. Miyama, N. Fujii, and Y. Nagae, Chem. Phys. Lett. 74 (1980) 523.
235. P.P. Radford and C.G. Francis, J. Chem. Soc., Chem. Commun. (1983) 1520.
236. Q. Li, K. Domen, S. Naito, T. Onishi, and K. Tamaru, Chem. Lett. (1983) 321.
237. N.N. Lichtin and M. Vijayakumar, J. Indian Chem. Soc. 63 (1986) 29.
238. E. Endoh, J.K. Leland, and A.J. Bard, J. Phys. Chem. 90 (1986) 6223.
239. D. Cordishi, N. Burriesci, F. D'Alba, M. Petrera, G. Polizzotti, and M. Schiavello, J. Solid State Chem. 56 (1985) 182.
240. A. Ozaki and K. Aika, "Catalysis", J.R. Anderson and M. Boudart, Eds., Springer-Verlag, Vol. 1, 1981, p. 87.
241. P. Pichat and M. Formenti, unpublished results.
242. M. Primet, P. Pichat, and M.-V. Mathieu, J. Phys. Chem. 75 (1971) 1221.
243. G.D. Parfitt, J. Ramsbotham, and C.H. Rochester, Trans. Faraday Soc. 67 (1971) 841.
244. G. Busca, H. Saussey, O. Saur, J.-C. Lavalley, and V. Lorenzelli, Applied Catal. 14 (1985) 245.
245. N.R. Dhar, E.V. Sechacharynbu, and S.K. Mukerji, Ann. Arg. 11 (1941) 83.
246. M.M. Taqui Khan, R.C. Bhardwaj, and C. Bhardwaj, Indian J. Chem. 25A (1986) 1.
247. M.M. Taqui Khan, R.C. Bhardwaj, and C.M. Jadhav, J. Chem. Soc., Chem. Commun. (1985) 1690.
248. S. Grayer and M. Halmann, J. Electroanal. Chem. 170 (1984) 363.
249. M. Halmann, J. Electroanal. Chem. 181 (1984) 307.
250. C.R. Dickson and A.J. Nozik, J. Am. Chem. Soc. 100 (1978) 8007.
251. M. Koizumi, H. Yoneyama, and H. Tamura, Bull. Chem. Soc. Jpn. 54 (1981) 1682.
252. M. Halmann and K. Zuckerman, J. Chem. Soc., Chem. Commun. (1986) 455.
253. M. Halmann and K. Zuckerman, in "Homogeneous and Heterogeneous Photocatalysis", E. Pelizzetti, and N. Serpone, Eds., D. Reidel, 1986.
254. M. Halmann, J. Tobin, and K. Zuckerman, J. Electroanal. Chem. 209 (1986) 405.
255. Y. Oosawa, J. Chem. Soc., Chem. Commun. (1982) 221.
256. Y. Oosawa, J. Chem. Soc., Faraday Trans. 1 80 (1984) 1504.
257. M.A. Fox, Top. Curr. Chem. 142 (1987) 72.

258. R.B. Cundall, R. Rudham, and M. Salim, J. Chem. Soc., Faraday Trans. I 72 (1976) 1642; F.H. Hussein and R. Rudham, J. Chem. Soc., Faraday Trans. I 80 (1984) 2817.
259. I.A. Ichou, M. Formenti, and S.J. Teichner, Stud. Surf. Sci. Catal. 19 (1984) 297.
260. S. Teratani, J. Nakmichi, K. Taya, and K. Tanaki, Bull. Chem. Soc. Japan 55 (1982) 1688.
261. T. Carlson and G.L. Griffin, J. Phys. Chem. 90 (1986) 5896.
262. S. Naito, Chem. Commun. (1985) 1211.
263. F.H. Hussein, G. Pattenden, R. Rudham, and J.J. Russell, Tetrahedron Lett. (1984) 3363.
264. T. Sakata and T. Kawai, Chem. Phys. Lett. 80 (1981) 341.
265. O. Enea, Electrochim. Acta 31 (1986) 405; O. Enea and A.J. Bard, Nouv. J. Chim. 9 (1985) 691.
266. K. Okamoto, Y. Yamamoto, H. Tanaka, M. Tanaka, and A. Itaya, Bull. Chem. Soc. Japan 58 (1985) 2015.
267. M.A. Fox and M.J. Chen, J. Am. Chem. Soc. 105 (1983) 4497; M.A. Fox and J.N. Younathan, Tetrahedron 42 (1986) 6285.
268. B. Ohtani, H. Osaki, S. Nishimoto, and T. Kagiya, Chem. Lett. (1986) 1075.
269. S. Nishimoto, B. Ohtani, T. Yoshikawa, and T. Kagiya, J. Am. Chem. Soc. 105 (1983) 7180.
270. H. Miyama, Y. Nosaka, T. Fukushima, and H. Toi, J. Photochem. (1986) 1553.
271. S. Yanagida, T.A. Zuma, H. Kawakami, H. Kizumoto, and H. Sakurai, Chem. Commun. (1984) 21.
272. M.A. Hema, V. Ramakrishnan, and J.C. Kuriacose, Ind. J. Chem. 15B (1977) 947; op. cit., 16B (1978) 619; H. Kasturirangen, V. Ramakrishnan, and J.C. Kuriacose, J. Catal. 69 (1981) 216.
273. J.W. Pavlik and S. Tantayanon, J. Am. Chem. Soc. 103 (1981) 6755.
274. R.S. Davidson and J.E. Pratt, Tetrahedron Lett. 24 (1983) 5903.
275. W. Hoyer and K.D. Asmus, "Sixth Internat. Conf. Photochem. Conver. Storage Solar Energy, Abstr C-55" (1986).
276. J.D. Spikes, Photochem. Photobio. 34 (1981) 549.
277. T.D. Waite, A. Torikov, and J.D. Smith, J. Colloid Interface Sci. 112 (1986) 412.
278. J. Onoe, T. Kawai, and S. Kawai, Chem. Lett. (1985) 1667.
279. H. Reiche and A.J. Bard, J. Am. Chem. Soc. 101 (1979) 3127.
280. T. Kawai, M. Fujii, S. Kambe, and S. Kawai, Electrochem. Soc. Japan Abstr E214, (1983) p. 211.
281. T. Sakata, J. Photochem. 29 (1985) 205.
282. H. Reiche, W.W. Dunn, K. Wilbourn, F.R.F. Fan, and A.J. Bard, J. Phys. Chem. 84 (1980) 3207.
283. D.N. Furlong, D. Wells, and W.H.F. Sasse, Austr. J. Chem. 39 (1986) 757.
284. C.D. Jaeger and A.J. Bard, J. Phys. Chem. 83 (1979) 3146.
285. K. Miyachita, T. Sakata, K. Nakamura, T. Kawai, and T. Sakata, Photochem. Photobio. 39 (1984) 151.
286. I. Willner and Z. Goren, Chem. Commun. (1986) 172.
287. M. Fujihara, Y. Satoh, and T. Osa, Nature 293 (1981) 206.
288. M. Fujihara, Y. Satoh, and T. Osa, Bull. Chem. Soc. Japan 55 (1982) 666.
289. Y. Simamura, H. Misawa, T. Oguchi, T. Kanno, H. Sakuragi, and K. Tokumaru, Chem. Lett. (1983) 1691.
290. K. Takagi, T. Fujioka, Y. Sawaki, and H. Iwamura, Chem. Lett. (1985) 913.
291. M. Fujihara, Y. Satoh, and T. Osa, J. Electroanal. Chem. 126 (1981) 277.
292. K. Hashimoto, T. Kawai, and T. Sakata, J. Phys. Chem. 88 (1984) 4083.
293. M.A. Fox, C.C. Chen, and J.N. Younathan, J. Org. Chem. 49 (1984) 1969.
294. M.A. Fox, D.A. Sackett, and J.N. Younathan, Tetrahedron 43 (1987) 1643.
295. M.A. Fox, C.C. Chen, K.H. Park, and J.N. Younathan, Am. Chem. Soc. Sympos. Ser. 278 (1985) 69.
296. D.F. Ollis, J. Catal. 97 (1986) 569.
297. I.J. Fan and Y.C. Chien, Sci. Sinica 21 (1978) 663.
298. A.J. Frank, Z. Goren, and I. Willner, Chem. Commun. (1985) 1029.

299. H. Yamataka, N. Seto, J. Ichihara, T. Hanafusa, and S. Teratani, Chem. Commun. (1985) 788.
300. G.T. Brown and J.R. Darwent, J. Phys. Chem. 88 (1984) 4955.
301. G.T. Brown and J.R. Darwent, J. Chem. Soc., Faraday Trans. I 80 (1984) 1631.
302. J.R. Darwent and A. Lepre, J. Chem. Soc., Faraday Trans II 82 (1986) 1457.
303. S. Yanagida, Y. Ishimaru, and C. Pac, J. Am. Chem. Soc. (1986) in press.
304. K. Okada, K. Hisamitsu, and T. Mukai, Chem. Commun. (1980) 941; K. Okada, K. Hisamitsu, Y. Takahashi, T. Hanaoka, T. Miyashi, and T. Mukai, Tetrahedron Lett. 25 (1984) 5311.
305. H. Al-Ekabi and P. deMayo, J. Phys. Chem. 90 (1986) 4075.
306. R.A. Barber, P. deMayo, and K. Okada, Chem. Commun. (1982) 1073.
307. A.M. Draper, M. Ilyas, P. deMayo, and V. Ramamurthy, J. Am. Chem. Soc. 106 (1984) 6222.
308. M. Ilyas and P. deMayo, J. Am. Chem. Soc. 107 (1985) 5093.
309. H. Al-Ekabi and P. deMayo, Tetrahedron (1988) in press.
310. S. Lahiry and C. Haldar, Solar Energy 37 (1986) 71.
311. H. Al-Ekabi and P. deMayo, Chem. Commun. (1984) 1231.
312. H. Al-Ekabi and P. deMayo, J. Phys. Chem. 89 (1985) 5815.
313. T. Hasegawa and P. deMayo, Chem. Commun. (1985) 1534.
314. S. Yanagida, K. Mizumoto, and C. Pac, J. Am. Chem. Soc. 108 (1986) 647.

Chapter 6.2

Intramolecular Electron Transfer: History and Some Implications for Artificial Photosynthesis

John S. Connolly and James R. Bolton

1. INTRODUCTION

Natural photosynthesis is a unique photoconversion system that transforms sunlight into stored chemical energy in the form of carbohydrates and other products. Photosynthesis, of course, has had the advantage of approximately three billion years of evolutionary development that has resulted in the assembly of many complex molecules collected together in an elegant macromolecular structure. This structure, the reaction center protein, performs the dual function of harvesting photons and storing some of the sun's energy by synthesizing molecules rich in chemical energy. The reactions are highly specific and are optimized for the survival of the plant or microorganism (1). Purely synthetic systems, on the other hand, offer the prospect of much greater flexibility and higher solar efficiencies, together with the possibility of "fine-tuning" the chemistry for production of specific fuels and chemicals. The processes of interest include conversion to electricity, photochemical water splitting and photochemical reduction of molecular nitrogen and carbon dioxide (2,3).

In this review we have relied on other chapters for background on the mechanisms of excited-state and ground-state electron-transfer (ET) systems. We shall restrict our coverage to purely organic donor-acceptor (D-A) molecules and shall not cover, for example, inorganic binuclear complexes, which are discussed elsewhere in this monograph. Within these limits we shall present a brief review of the current knowledge of the structure and function of the reaction center protein in photosynthesis, set forth a brief outline of the theory and techniques used in the study of photoinduced electron transfer (PET), discuss the factors affecting the rates of PET and present a thorough review of the work on PET in covalently linked D-A molecules, particularly those containing a porphyrin as the chromophore and electron donor and various moieties (such as quinones) as the acceptor. Several earlier reviews (2-9) have covered various aspects of this topic.

1.1 The Reaction Center Protein in Natural Photosynthesis

The primary photochemistry of photosynthesis takes place within highly specialized reaction center (RC) proteins spanning specific membranes: the thylakoid membrane in the chloroplasts of green plants and algae and the chromatophore membranes of photosynthetic bacteria (10). Most of the chlorophyll and other pigments in the membrane function as an antenna system to gather photons and channel the energy to the RC (11). In essence the antenna system acts as a kind of luminescent concentrator by amplifying the flow of excitation energy into the RC by over a factor of 300 (1).

In both photosystems (PSI and PSII) of green-plant and algal systems the photochemically active component (P700 or P680) is a chlorophyll a (Chl a) species. The primary photochemical step (following light absorption and excitation transfer to the RC) then involves ET from the excited state of the donor Chl a to an acceptor species. In PSII the acceptor is plastoquinone; in PSI the acceptor is not as well characterized but probably involves an iron-sulfur center (12).

Nature has chosen perhaps the simplest of all photochemical reactions, namely electron transfer, to trap the energy of solar photons. In effect, the RCs of photosynthesis are solar cells, converting light to electrical potential, which is then used to drive the much slower biochemical reactions that ultimately lead to generation of carbohydrates and other plant materials. The RC "solar cells" are very efficient; the quantum yield is almost unity (*i.e.*, one electron delivered into the circuit for every photon absorbed), and the overall efficiency of solar energy conversion to electricity is ~16%, which is as good as, or better than, most commercial silicon solar cells (1).

The detailed structure of the RC proteins in green plants and algae is not well known; however, single crystals of the reaction centers of a photosynthetic bacterium recently have been grown and their X-ray structure determined (13-15). One view of the RC structure is shown in Chapter 1.4. Only one side is functional in the *in vivo* ET process even though the structure appears to have C_2 symmetry; the asymmetry lies in the surrounding protein matrix. The electron donor is a dimer of bacteriochlorophyll (BChl) in a staggered coplanar arrangement. Within 2.8 ps (16) an electron is transferred from the primary donor to a bacteriopheophytin (BPheo) molecule; within ~200 ps the electron is further transferred to one of two ubiquinones (see ref. 17 for a discussion of this ET step and Chapter 1.4 for a thorough discussion of the overall kinetics). There is another BChl molecule between the primary donor pair and the BPheo molecule that may mediate electron transfer; however, there is no evidence that the electron resides on this entity for any measurable length of time. From the detailed structure it is clear that several molecules, oriented in a specific way by the protein, are necessary to carry out efficient charge separation and also to maximize the conversion of photon energy into Gibbs free-energy of the products. This complexity presents an interesting, though severe, challenge to model builders.

1.2 Why try to mimic the reaction center protein?

Over the past decade there has been growing activity in the design of model systems, "synthetic reaction centers", to mimic the process of primary charge separation in natural photosynthesis (2,18). The primary objective is to understand how to induce electron transfer out of an excited singlet state of a donor (D) to a nearby acceptor (A) on a time-scale faster than other competing processes, *viz.*, fluorescence, internal conversion and intersystem crossing. It is equally important to understand how to stabilize the resulting charge-separated state with respect to back electron transfer to the ground state. In terms of the processes depicted schematically in Fig. 1, the goal is to devise systems in which the following requirements are met for ET from an excited singlet state

$$k_{et}^{S} \gg (k_f + k_{ic} + k_{isc}) \tag{1}$$

and

$$k_s \gg (k_r^S + k_{-et}^S + k_{-et}^T) \tag{2}$$

Model compounds are much easier to study than natural RCs since it is possible to carry out a systematic variation of structure and environment to provide insight into the important molecular and structural factors that can lead to achievement of the above kinetic conditions.

1.3 Approaches to artificial photosynthesis

Two basic approaches have been taken towards modeling the photosynthetic RC. The first involves attempts to model the structure and function of the (bacterio)chlorophyll dimer or "special pair"; we shall not cover this approach here (for comprehensive reviews, see refs. 2 and 4); instead we shall concen-

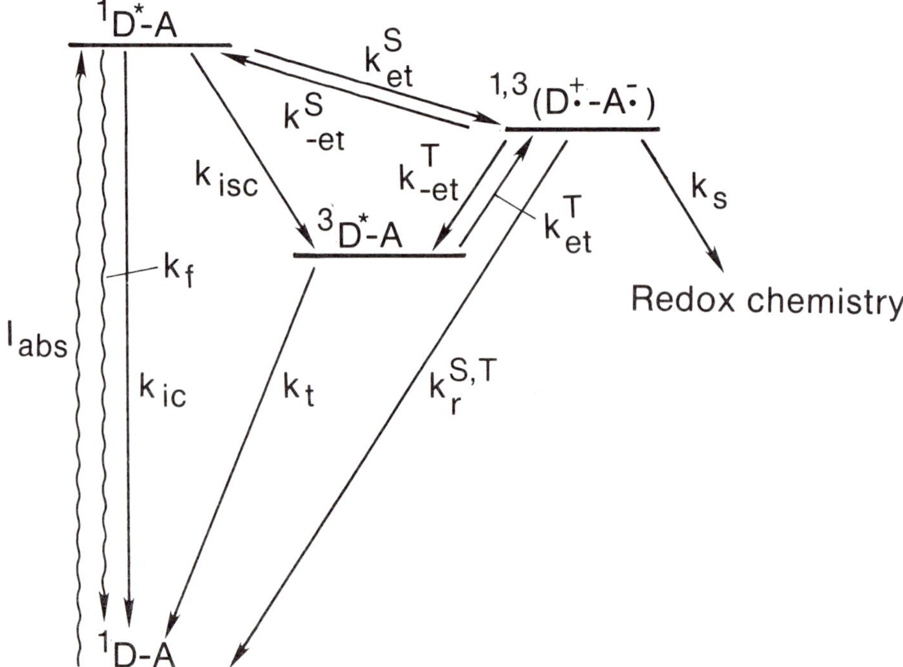

Figure 1. Energy-level diagram and rate constants for PET in a generalized D–A system. Here, the energy of the radical-ion-pair (RIP) state ($D^{+\cdot}$–$A^{-\cdot}$) is arbitrarily placed midway between the excited singlet and lowest triplet state. The superscripts S and T on the rate constants denote singlet- and triplet-state processes, respectively. (After ref. 2).

trate on the second approach, namely the synthesis and characterization of donor and acceptor entities covalently linked through a molecular bridge or bridges. These model systems are designed to study intramolecular ET in well-defined molecular structures and environments.

2. Photoinduced Electron Transfer (PET)

We shall be concerned with the general PET reaction

$$D + A \xrightarrow{h\nu} D^{\cdot+} + A^{\cdot-} \quad (3)$$

where the ET step can proceed either by reductive quenching of the excited state of A by the donor D or by oxidative quenching of the excited state of D by the acceptor A. Most of the PET systems studied are of the latter type so we shall focus on reaction 4 to discuss the general features of PET

$$D^{*} + A \longrightarrow D^{\cdot+} + A^{\cdot-} \quad (4)$$

where D^{*} is either the first excited singlet state or the lowest triplet state of D.

If the reduction potentials satisfy the requirement that

$$\varepsilon^{\circ}_{D^{\cdot+}/D} > \varepsilon^{\circ}_{A/A^{\cdot-}} > \varepsilon^{\circ}_{D^{\cdot+}/D^{*}}$$

reaction 4 will be exergonic (see Fig. 1), and some fraction of the photon energy will be stored in the redox products. However, the back reaction

$$D^{\cdot+} + A^{\cdot-} \longrightarrow D + A \quad (5)$$

is also exergonic and hence spontaneous. The design problem is to achieve a fast rate for reaction 4 and, at the same time, introduce kinetic barriers that will retard reaction 5 and hence achieve a high net quantum yield of redox products. Some strategies for circumventing the problem of reverse electron transfer are discussed by Mauzerall in Chapter 1.6 and by Rabani in Chapter 3.4.

Reaction 4, as written, is a bimolecular process. Many bimolecular PET reactions have been studied (see refs. 7 and 19 for comprehensive reviews); however, we shall restrict our attention to covalently linked donors and acceptors

$$D^{*}\text{–}A \longrightarrow D^{\cdot+}\text{–}A^{\cdot-} \quad (6)$$

in which the PET reaction is unimolecular and first-order.

2.1 Electron-transfer theory: a brief outline

The most widely used and successful theory of ET is that of Marcus (20-23), which is based on the premise that the precursor state (D^*–A) must reorganize the nuclear coordinates of both the complex and the surrounding environment to the point where the energy of the reorganized precursor state is the same as that of the successor state immediately following ET. A full development of the theory is beyond the scope of this review, but we shall state several of the key equations and define the principal factors. (See, *e.g.*, refs. 24 and 25 and Chapter 1.3 for thorough and comprehensive expositions of the theory.)

In the classical Marcus theory of ET the unimolecular rate constant k_{et} (in s^{-1}) is given by

$$k_{et} = \kappa_{el}\nu_n \exp\left[- \frac{(\lambda + \Delta G^{\circ\prime})^2}{4\lambda kT} \right] \tag{7}$$

where λ is the reorganization energy needed to distort the precursor complex and its surroundings to the equilibrium configuration of the successor complex; κ_{el} is the electronic transmission coefficient and ν_n is the frequency of passage (nuclear motion) through the transition state ($\nu_n \simeq 10^{13}\ s^{-1}$). $\Delta G^{\circ\prime}$ is the Gibbs free-energy change in the overall ET reaction, with a correction term ΔG_{corr} to account for any specific interactions (such as Coulombic attraction) between $D^{\dot{+}}$ and $A^{\dot{-}}$ in the product successor complex, *i.e.*, $\Delta G^{\circ\prime} = \Delta G^{\circ} + \Delta G_{corr}$ (usually $\Delta G_{corr} < 0$). Note that eq. 7 is a transition-state theory expression with

$$\Delta G^{\ddagger} = \frac{(\lambda + \Delta G^{\circ\prime})^2}{4\lambda} \tag{8}$$

Thus as $-\Delta G^{\circ\prime}$ increases, $\Delta G^{\ddagger}$ decreases, and k_{et} is maximal when $-\Delta G^{\circ\prime} = \lambda$. However, in the region where $-\Delta G^{\circ\prime} > \lambda$, k_{et} is predicted to *decrease*. This has been called the *inverted region* (22). The solid curve in Fig. 2 illustrates how $\log(k_{et})$ varies with $\Delta G^{\circ\prime}$ as predicted by eq. 7.

The reorganization energy λ is usually divided into two contributions

$$\lambda = \lambda_i + \lambda_o \tag{9}$$

where λ_i arises from critical internal vibrational modes in the precursor complex and is usually represented by a sum of harmonic potential energies (25), and λ_o accounts for the effect of the surrounding medium. If a two-sphere dielectric continuum model is assumed (22,23,26), then λ_o is given by

$$\lambda_o = \frac{e^2}{4\pi\epsilon_o}\left[\frac{1}{2r_D} + \frac{1}{2r_A} - \frac{1}{r_{DA}}\right]\left[\frac{1}{\epsilon_{op}} - \frac{1}{\epsilon_s}\right] \qquad (10)$$

where r_D and r_A are the donor and acceptor radii, respectively, r_{DA} is the D–A center-to-center distance, ϵ_{op} is the optical dielectric constant ($\epsilon_{op} = n^2$, where n is the refractive index) and ϵ_s is the static dielectric constant.

Brunschwig *et al.* (26) have analyzed these two contributions to λ and have shown that for a wide variety of transition-metal complexes $\lambda_i \simeq 0.2$-0.3 eV; for most of the large organic D–A systems considered here, λ_i is not likely to exceed 0.2 eV. λ_o varies from near zero for non-polar solvents where $\epsilon_{op} \simeq \epsilon_s$ to $\geqslant 1$ eV for polar solvents. Hence in most cases of interest here, where studies have been carried out in moderately polar solvents, λ_o is the dominant contribution to λ.

Equation 7 is applicable for *adiabatic* ET, *i.e.*, where $\kappa_{el} \simeq 1$. However, many (perhaps most) of the D–A systems studied fall into the class of *non-adiabatic* ET reactions where $\kappa_{el} \ll 1$. In this case the problem must be treated quantum mechanically. One widely used model (27), based primarily on earlier work of Jortner (27a), assumes a single, averaged effective internal vibrational

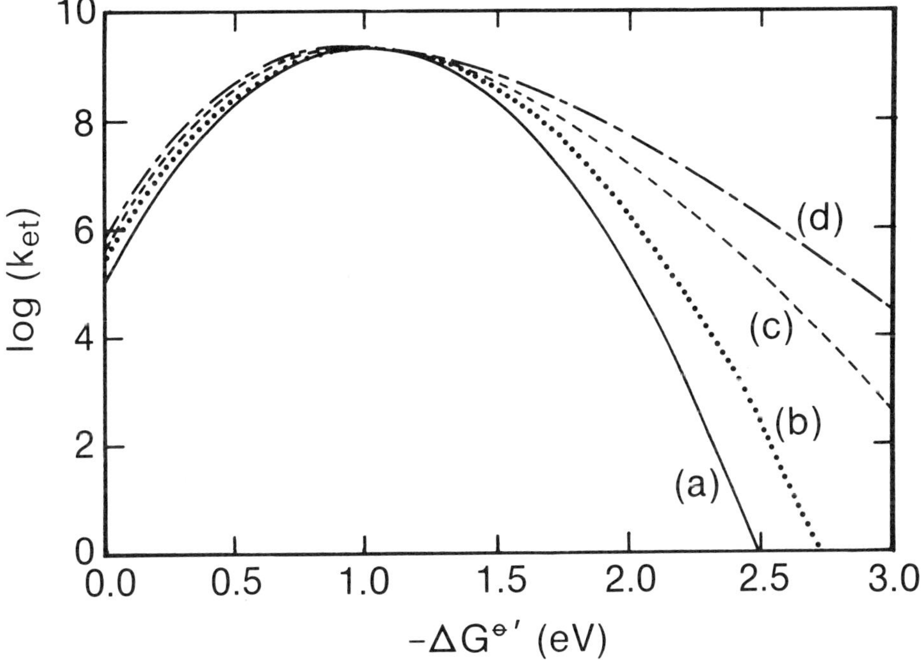

Figure 2. $\ell og(k_{et})$ *vs.* $-\Delta G^{\ominus\prime}$, where k_{et} is calculated from the classical Marcus equation (eq. 7) (curve a) or the semiclassical Marcus equation (eq. 11), with ν = 750 (curve b), 1500 (curve c) and 2250 cm^{-1} (curve d).

frequency ν in the precursor complex that contributes to λ_i; the surrounding medium is treated classically as a dielectric continuum. In this semiclassical model k_{et} is given by

$$k_{et} = \frac{2\pi}{\hbar} H_{ps}^2 (4\pi\lambda_o kT)^{-1/2} \sum_{m=0}^{\infty} (e^{-S} S^m/m!) \exp\left[- \frac{(\lambda_o + \Delta G^{\circ\prime} + mh\nu)^2}{4\lambda_o kT} \right] \quad (11)$$

where $S = \lambda_i/h\nu$ and H_{ps} is an electronic coupling energy ($H_{ps} = \langle\psi_p^o|\hat{\mathcal{H}}_{el}|\psi_s^o\rangle$), where $\hat{\mathcal{H}}_{el}$ is the electronic Hamiltonian and ψ_p^o and ψ_s^o are the electronic wave functions for the precursor and successor complexes, respectively. In the high-temperature limit (where $kT \gg h\nu$), eq. 11 reduces to

$$k_{et} = \frac{2\pi}{\hbar} H_{ps}^2 (4\pi\lambda kT)^{-1/2} \exp\left[- \frac{(\lambda + \Delta G^{o\prime})^2}{4\lambda kT} \right] \quad (12a)$$

which is the same as eq. 7 with

$$\kappa_{el}\nu_n = \frac{2\pi}{\hbar} H_{ps}^2 (4\pi\lambda kT)^{-1/2} \quad (12b)$$

The dotted lines in Fig. 2 illustrate the result of using eq. 7 for various values of ν. Note that in the normal region (*i.e.*, where $-\Delta G^{\circ\prime} < \lambda$) the classical and semiclassical curves are quite similar. When a high-frequency mode makes an important contribution to λ, eq. 12a is a good approximation in the normal region, but an expression such as eq. 11 is better in the inverted region (27b).

2.2 Techniques used in studies of photoinduced electron transfer

We should first mention the skills of the organic chemists who have synthesized the elegant, complex molecules discussed in this review. It is clear that none of the detailed photophysical studies would have been possible without their efforts.

Once a D–A compound has been synthesized and characterized, it is important to establish the energy levels (*e.g.*, Fig. 1). The spectroscopic levels can be measured by optical absorption and emission techniques, but the energy of the $D^{\cdot+}$–$A^{\cdot-}$ state requires special mention. In certain cases emission from this state has been observed from which its energy is readily obtained. However, in most cases this energy must be obtained indirectly from electrochemical measurements, usually by cyclic voltammetry or polarography.

The energy of the $D^{\cdot+}$–$A^{\cdot-}$ state with respect to the ground state is given by

$$\Delta G^{\circ\prime}(D^{\cdot+}-A^{\cdot-}) = e(\mathcal{E}^{\circ}_{D^{\cdot+}/D} - \mathcal{E}^{\circ}_{A/A^{\cdot-}}) + \Delta G_{corr} \quad (13)$$

where the $\mathcal{E}^{\ominus}$ values are reduction potentials for the indicated couples. These reduction potentials should be measured in the same solvent in which the photophysical measurements are made (2,3), although this is not always possible. Accordingly, Weller (28) proposed an equation (based on the Born equation) by which reduction potentials measured in a reference solvent (*e.g.*, acetonitrile) could be corrected for another solvent. However, Bolton and co-workers, on the basis of careful photophysical (29) and electrochemical (29a) measurements in 16 solvents, have recently questioned the applicability of this treatment, particularly for solvents of medium to low polarity (*i.e.*, $\epsilon_s < 15$).

The electrochemical studies on **1E2** by Archer *et al.* (29a) deserve special mention: This is the only case in which the solvent dependence of the reduction potentials of P and Q have been measured in the *actual* linked porphyrin-quinone (PQ) molecules rather than in model compounds of the separate entities. These measurements were essential to a quantitative understanding of the solvent dependence of PET in 1E2 carried out by Schmidt *et al.* (29) (*vide infra*).

The most important measurements, of course, are detection of the $D^{\dot{+}}$–$A^{\dot{-}}$ state and accurate determinations of rate constants for ET from the excited state D^*–A (singlet or triplet) to the charge-separated state $D^{\dot{+}}$–$A^{\dot{-}}$ and for the back reaction from $D^{\dot{+}}$–$A^{\dot{-}}$ to the ground state. Four techniques have been used.

(i) Fluorescence properties. If the rate constants for the photophysical processes by which the excited singlet state relaxes (k_f, k_{ic} and k_{isc} in Fig. 1) remain constant between the D–A system and a similarly structured reference molecule that does not undergo PET, then the forward ET rate constant from this state may be determined from the relation (see, *e.g.*, ref. 30).

$$k_{et}^{f} = \frac{1}{\tau_1} - \frac{1}{\tau_2} \tag{14}$$

where τ_1 is the fluorescence lifetime of $^1D^*$ in D–A and τ_2 is the corresponding lifetime of the reference molecule. For example, if D is a porphyrin and A is a quinone, the hydroquinone-linked porphyrin would be a suitable reference (30).

Fluorescence intensities have sometimes been used to infer ET rates in linked D–A systems, the assumption being that all excited-state interactions (at least in geometrically constrained systems) are "dynamic" and not "static" in the classical sense. However, Connolly and co-workers (31,32) found in two related, flexibly linked porphyrin-anthraquinone systems that the fluorescence intensities did not reflect the observed (ns) lifetimes, even though there was no evidence of ground-state interactions [see Sections 4.3 (iii) and (iv)]. These results demonstrate that PET kinetics inferred from fluorescence intensities alone should be viewed with a great deal of caution.

(ii) Transient absorption spectroscopy. The $D^{\dot{+}}$–$A^{\dot{-}}$ state is a radical-ion-pair (RIP); hence, the optical spectra of both the oxidized donor and the reduced acceptor should be detectable, at least in principle. Both ns and ps laser flash photolysis techniques have been used to detect transient intermediates in PET reactions. In very few cases, however, has unambiguous spectral evidence been obtained for *both* the $D^{\dagger}$ and $A^{\overline{\cdot}}$ moieties in a linked D-A system. This is particularly difficult in the case of benzoquinone-containing molecules; the absorption bands of interest are in the UV (32a), which is not an easy region in which to work on the ps time scale, and the RIP states of these compounds are generally too short-lived to be observed by ns laser techniques (2,32b,33, 34), except when such processes occur from long-lived triplet states (35). Thus, ps absorption spectroscopy has been more useful in PET studies (see Chapters 1.4, 1.7 and 2.1).

(iii) Transient microwave conductivity. Warman and co-workers (36,37) have used transient microwave conductivity to follow the rate of decay of the $D^{\dagger}$–$A^{\overline{\cdot}}$ RIP state in the ns time domain. This technique is based on the fact that a large change in dipole moment occurs on PET; this in turn causes a significant change in the conductivity of the medium, which is monitored by a change in the dielectric loss. The latter can be measured quantitatively by a change in the microwave power reflected by a cavity containing the sample (38).

(iv) Electron paramagnetic resonance (EPR) spectroscopy. If the lifetime of the RIP state is sufficiently long, EPR spectroscopy can be used to detect its presence and, in some cases, the kinetics as well. McIntosh *et al.* (39) used EPR to reveal intramolecular spin-spin interactions between two photo-induced entities in a flexibly linked porphyrin-quinone molecule. However, it is unclear whether the species they observed was indeed the $P^{\dagger}$–$Q^{\overline{\cdot}}$ state or a product (*e.g.*, containing a semiquinone radical) because an optical spectrum of the quinone radical-anion could not be detected (*vide supra*).

3. Factors Affecting Intramolecular Electron-Transfer Rates

If a successful synthetic model of the natural RC is to be assembled, it is essential to understand the various molecular, structural and environmental factors that affect the ET rates. Although this review is concerned primarily with intramolecular PET, a number of studies of thermal ET (see, *e.g.*, refs. 27,40-42) have been carried out that reveal many of the important factors affecting ET rate constants. In this Section we shall refer to these thermal ET studies as we discuss the various factors; then in Section 4, we shall show how these factors have been examined in a variety of intramolecular PET studies.

3.1 Energetics

As shown in eqs. 6, 11 and 12 and in Fig. 2, Marcus theory predicts that k_{et} should rise as $-\Delta G^{\circ\prime}$ increases, reach a maximum value when $-\Delta G^{\circ\prime} = \lambda$ and then decrease again in the inverted region where $-\Delta G^{\circ\prime} > \lambda$. There have been many attempts to test these predictions of the inverted region (see, *e.g.*, refs. 43-45), but until recently these experiments failed to reveal the predicted decrease in k_{et} at high exergonicities. The first experimental indication of the existence of the inverted region came from a study of the reactions of trapped electrons in a rigid, glassy matrix (42), but definitive observations of a significant decrease in rate in the inverted region were not made until 1984. These studies concerned intermolecular ET reactions between the biphenyl radical anion and various acceptors in a rigid low-temperature glass (27) and intramolecular ET in solution from the biphenyl radical anion to various acceptors across a rigid steroid linkage (40). Miller *et al.* (27) pointed out that the probable reason that earlier studies had failed to reveal the inverted region was that they had involved bimolecular ET reactions in solution where the observed second-order rate constant is usually diffusion-limited. It is only in a rigid matrix or in a covalently linked D-A molecule that rapid first-order ET rate constants can be measured without the complications of rate-limiting diffusion.

3.2 Geometry: distance and orientation effects

The electronic coupling energy H_{ps} (eqs. 11 and 12) and the reorganization energy λ (eqs. 9 and 10) are the only factors that depend significantly on the center-to-center distance r_{DA} between D and A. λ depends only weakly on r_{DA}, so H_{ps} should be the major factor in the dependence of k_{et} on distance. Since H_{ps} involves integration over electronic wave functions, it is usually assumed to decrease exponentially with r_{DA}

$$H_{ps}(r_{DA}) = H_{ps}(r_o)\exp[-\beta(r_{DA} - r_o)/2] \qquad (15)$$

where r_o is a reference distance; the factor of ½ in the exponent accounts for the fact that eqs. 12 are expressed in terms of H_{ps}^2. β depends on both the absolute and relative donor, acceptor and bridge energetics (see Section 3.3).

The pre-exponential factor given in eq. 12b can thus be written

$$\kappa_{el}(r_{DA})\nu_n = \kappa_{el}(r_o)\nu_n\exp[-\beta(r_{DA} - r_o)] \qquad (16)$$

Hence, in a series of D-A molecules with the same values of $\Delta G^{\circ\prime}$ and λ but differing r_{DA}, the rate constants should follow the relation

$$k_{et}(r_{DA}) = k_{et}(r_o)\exp[-\beta(r_{DA} - r_o)] \qquad (17)$$

It is common to choose the reference distance such that $\kappa_{el}(r_o) = 1$ (ref. 25); this is equivalent to saying that the reaction becomes adiabatic as $r_{DA} \rightarrow r_o$. In practice, it is found that r_o corresponds approximately to the sum of the van der Waals radii of D and A. From transition-state theory, ν_n is expected to be $\sim 10^{13}$ s^{-1}. Hence from eq. 12b, $H_{ps}(r_o) \approx 0.025$ eV for a typical case with $\lambda \approx 1$ eV. This value of H_{ps} represents an approximate dividing line between adiabatic and non-adiabatic ET processes, although the exact demarcation may depend on the details of the nuclear dynamics.

Miller and co-workers (27), in the study cited above, were the first to test eq. 17; they found $\beta = 1.2$ Å^{-1}. Closs *et al.* (41) carried out an elegant study of ET from the biphenyl radical anion to naphthalene covalently linked by a series of rigid steroid bridges with center-to-center distances of 6.2 to 17.4 Å and edge-to-edge distances from 3.8 to 10.3 Å. For the same stereoisomers, a good fit of k_{et} to eq. 17 was found with $\beta = 1.0$ Å^{-1}. They pointed out that this is 20% smaller than the value they found for intermolecular ET reactions between the same two entities (27), although this difference may be within the experimental error. They also found that intramolecular $H_{ps}(r_o)$ is about six times larger than for the intermolecular case and suggested that through-bond coupling is an important factor in the covalently linked systems. Closs, Miller and co-workers (41) interpreted their data in terms of an exponential decline of H_{ps} with the number of saturated bonds between the donor and acceptor; thus the rate decreases by about a factor of 10 for every two σ-bonds between the two moieties.

Closs *et al.* (41) also examined various stereoisomers of most of the structures studied. They found that, for a given linking group, intramolecular ET rates varied by as much as a factor of ~10, some of which could be ascribed to structural factors. However, most of this variation probably arises from changes in the relative conformation of the biphenyl and naphthalene groups; an equatorial-equatorial conformation appeared to be the most favorable.

A theoretical treatment of the effect of mutual donor-acceptor orientation on ET rates has been presented recently by Cave *et al.* (46,47). The approach uses a one-electron oblate potential-well model in which the donor electron in a filled well interacts with an acceptor represented as an empty oblate potential well. The results of this analysis indicate that ET rates should be very sensitive functions of both the orientation and the nature of the orbitals involved.

3.3 Nature of the linkage: through-bond *vs.* through-space transfer

Larsson (48-50) has developed an extended Hückel treatment which indicates that the bridge and/or solvent molecules between the donor and acceptor can influence the ET rates of non-adiabatic processes. These calculations show that the rates in a given D-A system should be sensitive to the molecular orbitals: *viz.*, energies, overlap (donor-bridge and bridge-acceptor) and symmetry. These theoretical results also indicate that transfer through σ-orbitals of the bridge can be as effective as transfer through π-orbitals, and that transfer can occur with equal facility through either filled or empty orbitals.

Miller (51) has also discussed the role of the orbitals in the molecular entities between the donor and acceptor (either bridge or solvent) in terms of a "superexchange" mechanism based on the theoretical model of Beratan *et al.* (52), which is depicted in Fig. 3. In this model, interactions through the bridging group (or solvent molecules in the case of intermolecular ET) occur *via* a virtual mechanism either by electrons coupling through the lowest unoccupied antibonding orbitals of the bridge or by holes transferring through the highest occupied orbitals. Most cases studied seem to have involved the former mechanism, for which the superexchange model predicts that β in eqs. 15-17 should vary linearly with $\ell n(B_-)$, where B_- is the ionization potential of the donor in the condensed medium (see Fig. 3). Miller (51) has presented evidence in support of this relatively weak dependence of β on B_-.

3.4 Medium effects: solvent and temperature

In Marcus theory, the solvent has two major influences on k_{et}. A change of solvent will affect λ_o (and hence λ) through changes in both ϵ_s and ϵ_{op} (eq. 10) and will also result in changes in reduction potentials, which in turn will affect $\Delta G^{\ominus\prime}$ (eq. 13). Schmidt *et al.* (29) have shown that the solvent can change λ in the range 0.5 to 1.2 eV and $\Delta G^{\ominus\prime}$ by as much as 0.3 eV. These solvent effects can alter k_{et} by over two orders of magnitude (53). Quantitative treatments of the variation of $\Delta G^{\ominus}$ with solvent are discussed in Section 4.3 in connection with specific molecular types.

Aside from the clear temperature-dependent terms in eqs. 7, 11 and 12a, λ will vary through the temperature dependence of both ϵ_s and ϵ_{op}, and $\Delta G^{\ominus\prime}$ may also be slightly temperature dependent. As yet no single study has been carried out in which all of these terms have been examined experimentally.

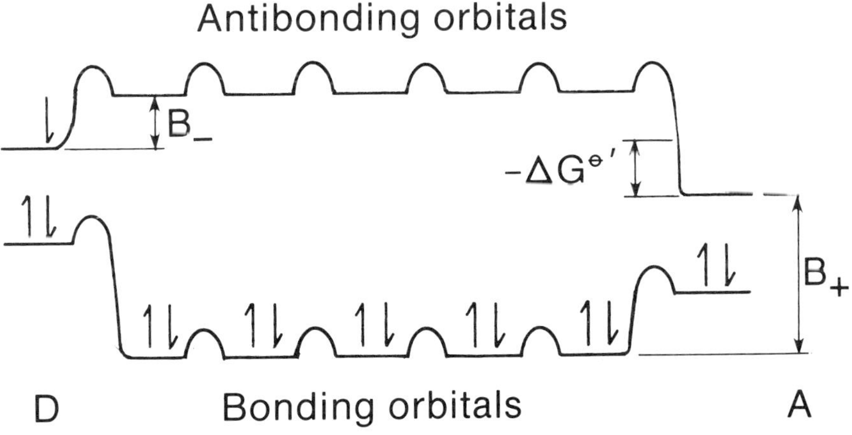

Figure 3. Energy-level diagram for the superexchange model of ET through the orbitals of a bridging group or solvent molecules. B_- and B_+ are, respectively, the ionization potential of the donor and the "hole" ionization potential of the acceptor in the condensed medium. (After ref. 51).

In addition, there is the very important question regarding the role of solvent motion. For example, under what conditions can k_{et} in a polymer matrix ($\eta \geqslant 10^{15}$ cp) approach the rates observed in fluid solvents ($\eta \leqslant 10$ cp)? The role of dielectric relaxation in limiting the formation rates of charge-transfer (CT) states has been investigated for several years by Kosower and co-workers (54-57) and has recently been addressed in the context of PET by other groups (58-62). When ET rates are very fast ($\geqslant 10^9$-10^{10} s^{-1}), solvent relaxation may become rate limiting; k_{et} is then proportional to $\tau' = (\epsilon_{op}/\epsilon_s)\tau_1$, where τ_1 is the dielectric relaxation time of the solvent. Calef and Wolynes (63) have developed a theory of relaxation in ET reactions; this topic has been reviewed by Kosower and Huppert (57) and by Calef (Chapter 1.9).

4. Survey of Intramolecular Photoinduced Electron-Transfer Reactions

Studies of intramolecular ET, in which the donor and acceptor are covalently linked via a bridge group, have provided a rich store of results to test theories of ET. In this Section we discuss specifically cases in which intramolecular ET is preceded by photon absorption. In most cases ET then occurs directly from the excited singlet state. Reverse ET rates from the charge-separated product to the ground state are often available for the same compound, and in many cases these fall in the inverted region. In a few cases, intersystem crossing to the triplet state has been shown to occur followed by ET from that state.

Since the late 1960's many examples of linked D-A molecules have been studied. We have divided these compounds into three classes: singly linked, capped and triad molecules; summaries of their structures and properties are given in Tables 1, 2 and 3, respectively. In the discussion that follows, we refer to these molecules by alphanumeric codes; the number preceding each letter symbol designates the Table in which that compound is listed (*e.g.*, 1A1 refers to Table 1, group A, molecule 1).

One hundred twenty-five papers are cited in these tables, representing nearly 300 individual molecular systems; clearly, space limitations preclude a thorough discussion of every compound. It should be noted that each of these studies has, in some way, built on the results of its predecessors, and each has made an important contribution to our understanding of PET. The individual papers highlighted in the following discussion were selected on the basis of their contributions to specific aspects that we chose to emphasize.

4.1 Early work (1969-1978)

The earliest work on intramolecular PET involved studies of CT absorption bands in intramolecular D-A complexes. Verhoeven *et al.* (64) synthesized compounds 1R1-23, which consist of a series of methyl- and methoxy-benzene derivatives as donors and the 4-cyanopyridinium cation as the acceptor, separated by one or two methylene groups. This group was the first to observe distinct, intramolecular CT bands in a linked D-A system.

In a parallel development arising from extensive studies of exciplexes (*vide infra*), Okada *et al.* (65) synthesized the series of compounds **1P1-4** in which the length of the saturated hydrocarbon bridge was varied from zero to three carbon atoms. These compounds were prepared for the purpose of defining the geometric requirements for formation of exciplexes between anthracene (A) and N,N′-dimethylaniline (D). To their surprise, this group observed a new, long-wavelength fluorescence, especially in polar solvents. This emission was found to shift markedly to longer wavelengths with increasing solvent polarity, indicating that the emitting state had a large dipole moment. From these observations, they speculated that the emission originated from a CT state ($A^{\cdot -}...D^{\cdot +}$) in which ET had occurred from D to A^*. Similarly, for the compounds **1S1,2**, Pasman *et al.* (66) also observed long-wavelength emission that shifted to longer wavelength as the solvent polarity was increased. They likewise observed these effects in the conformationally restricted compounds **1S3-5** and commented on the importance of "through-bond" as opposed to "through-space" interactions in mediating ET from the methoxybenzene chromophore to the dicyanoethylene acceptor.

4.2 Exciplexes *vs.* charge-transfer states

The definitions of exciplexes and CT states tend to vary from author to author. Förster (67) defined an exciplex as "any kind of well-defined complex which exists in electronically excited states." It is usually assumed that some degree of electronic orbital overlap is required between the two interacting components. When these components are aromatic, a "sandwich" or cofacial configuration is assumed to be optimal. The stability of an exciplex state $(DA)^*$ arises from admixtures of four resonance structures:

$$D^*A \longleftrightarrow (DA)^* \longleftrightarrow D^+A^- \longleftrightarrow D^-A^+$$

An exciplex is characterized by a broad, structureless, red-shifted emission. However, there is *no difference in the absorption spectrum compared to the sum of the spectra of the separated components.*

A CT state can be defined as a product state in which electron transfer from D to A is complete, including stabilization (reorganization) of the solvent around D^+A^-. If the concentrations are sufficiently high, or if D and A are linked in an intramolecular entity, a broad, red-shifted CT absorption band can sometimes be observed. Mataga and Ottolenghi (68) explain this as follows: In the case of a CT state, "the excited complex state is produced by the very act of light absorption within a predominantly charge-transfer band, in a process involving only limited geometrical changes." In exciplexes, on the other hand, "charge transfer occurs after population of a locally excited monomer state."

The distinction between exciplexes and CT states is blurred by the fact that a broad, red-shifted emission can often be observed in cases where CT absorption is also observed. Thus for a given D-A pair, there probably exists a continuous range of states, from distinct exciplexes in non-polar media to distinct CT states in polar solvents (69,70). This is further indicated by the fact that exciplex emissions are further red-shifted and diminished in intensity with increasing solvent polarity (67).

In summary: exciplexes emit but do not have distinct absorption bands, while for CT states the opposite is generally true. As solvent polarity is increased, the bands in both cases shift to lower energies and decrease in intensity. CT states sometimes emit, especially if they are at high energy, in which case the distinction becomes blurred; *i.e.*, the absorption band can be assigned to charge transfer but the emitting state may not be strictly D^+A^- in character. When there is no ground-state absorption and no excited-state emission, but the lifetime of the local D^* state is diminished and there is a new excited-state absorption, it is uncertain whether the new absorption is really that of the RIP state or of a precursor that does not fully evolve to the solvent-separated RIP.

Weller and co-workers (71) have referred to a continuum of states:

$$D^*A \xrightarrow{1} (DA)^* \xrightarrow{2} (D^+A^-)_C \xrightarrow{3} (D^+)_S(A^-)_S$$

where the last two entities are the contact ion pair and solvent-separated ion pair, respectively, and the latter represents the RIP state. This terminology has been used to describe intermolecular processes (69,70) but apparently has not been invoked in the case of intramolecular interactions. Such a continuum of states should also be possible in linked D-A systems, especially when the distances between the two moieties are constrained to ~ 5 Å. Indeed, solvent dielectric relaxation rates should be expected to limit step 3. Thus, PET rates inferred from fluorescence intensities or measured by fluorescence lifetimes reflect steps 1 and 2, while the kinetics of RIP formation measured by ps absorption techniques will be limited by step 3.

4.3 Singly linked molecules

Molecules 1A2 and 1E1 were the earliest and simplest linked porphyrin-quinone systems to be synthesized. 1A1 has four benzoquinone moieties attached directly at the *meso*-positions of the porphine ring (72) and is discussed at the end of this Section together with all of the 1An molecules; 1E1 has a single benzoquinone attached by a peptide link. Tabushi *et al.* (73) found that the fluorescence of 1E1 was strongly quenched relative to that of the non-linked porphyrin; they attributed this effect to intramolecular ET.

The first molecules in which intramolecular ET was demonstrated definitively were the diester-linked compounds 1H1-3 synthesized by Kong and Loach

(74,75) and studied by Loach and co-workers (76,77). These systems also showed strong fluorescence quenching as compared to appropriate controls. Ho *et al.* (78) likewise synthesized compound 1H2 and reported a light-induced EPR signal [also seen by Loach *et al.*, (77)], which was shown to arise from the RIP formed by ET from the porphyrin to the quinone.

This early work was followed by further studies on these molecules and on the related diamide-linked compounds 1I1-3 (30,32b,33,34,79,79a). However, in these compounds conformational heterogeneity, arising from the length and flexibility of the linkages, caused problems in interpreting the results with respect to specific ET rates since the D-A distances could not be well characterized. Nevertheless, it was possible to interpret most of the observations in terms of two families of conformers present in both the ground and excited states: "extended", in which ground-state interactions between the two chromophores are minimal, and "complexed" in which the two chromophores are close enough to exhibit broadening and red-shifting of the porphyrin Soret band (30). The two sets of conformers are manifested by two distinct fluorescence lifetimes (30), two triplet lifetimes (32b,33) and two triplet EPR spectra (34).

Schmidt *et al.* (35) carried out a thorough analysis of the photophysics and photochemistry of the simpler porphyrin-amide-quinone molecule 1E2 in benzonitrile in terms of the general energy-level diagram shown in Fig. 1. k_{et}^{S} was determined from the fluorescence lifetimes of 1E2 and its hydroquinone analogue using eq. 14. The formation and decay of the RIP intermediate were followed by ps absorption spectroscopy; the rise-time of the donor cation-radical absorption matched the decay of the excited singlet state, thus confirming the validity of eq. 14. The value of k_{r}^{S} was obtained from the decay kinetics of the RIP to the ground state as determined from the decay profile of the transient optical absorption. k_{et}^{T} was inferred from the enhanced decay of the triplet-state absorption following ns laser flash photolysis.

As indicated in Table 1, nearly 220 singly linked D-A molecules have now been synthesized, but only in a few cases have they been subjected to thorough photophysical and photochemical analyses. In the following discussion we concentrate on those studies that reveal the important factors that influence the rates of both forward and reverse ET.

(i) Dependence on the exergonicity of the reaction. Wasielewski and co-workers (80-83) studied the series of PQ molecules 1D1-6 in which the donor and acceptor are linked by a rigid triptycene group. They were able to measure the forward and reverse ET rate constants for each of these six molecules by using a combination of ps and ns absorption spectroscopy. The Gibbs free-energy differences for charge separation $\Delta G_{cs}^{\circ\prime}$ (*i.e.*, between the excited singlet state $^{1}P^{*}Q$ and the RIP state $P^{+}Q^{-}$) and for charge recombination $\Delta G_{cr}^{\circ\prime}$ (*i.e.*, between the $P^{+}Q^{-}$ state and the PQ ground state) were estimated from the measured reduction potentials (eq. 13). The PET rate data provide a striking demonstration of the

Marcus *inverted region*. There is no doubt that the rates become slower for very negative values of $\Delta G^{\ominus\prime}$, although the actual values must be interpreted cautiously because butyronitrile ($\lambda \simeq 1.0$ eV) and toluene ($\lambda \simeq 0.3$ eV) are shown on the same diagram. Equation 12a indicates that $\log(k_{et})$ should be a function of both λ and $\Delta G^{\ominus\prime}$. Moreover, ΔG_{corr} was calculated from the Weller equation (28), the validity of which has been questioned (29). Nevertheless, the important conclusion to be drawn from this work (81) is that *for the same molecule in the same solvent*, the rate constant for the reverse reaction (which should be in the inverted region) is smaller than that for the forward reaction. This observation was also made by Schmidt *et al.* (35) for molecule **1E2**.

In spite of our reservations regarding the quantitative interpretation, it is clear from the work of Wasielewski's group (81) as well as the earlier work of Miller *et al.* (40) on thermal ET, that the existence of the inverted region has been solidly confirmed.

(ii) Solvent effects. For the porphyrin-amide-quinone **1E2**, Schmidt and coworkers (53) found that the PET rate from the excited singlet state varies from 1.4×10^7 to 2.3×10^9 s^{-1} in a series of 22 solvents. More recently, this group (29) has shown that if the solvent dependence of *both* λ *and* $\Delta G^{\ominus\prime}$ are taken into account, a reasonable correlation with Marcus theory can be obtained. They used the logarithmic form of eq. 12a

$$\log (k_{et}\lambda^{1/2}) = \log \left[\frac{2\pi}{\hbar} \frac{H_{ps}^2}{(4\pi kT)^{1/2}} \right] - \frac{\Delta G^{\ddagger}}{2.303\ kT} \qquad (18)$$

(where $\Delta G^{\ddagger}$ is defined in eq. 8) and found an excellent correlation with a slope close to the predicted value of -1.00 in two binary solvent mixtures. This is shown in Fig. 4, where $\Delta G^{\ddagger}$ was obtained using a simple Coulombic attraction term (eq. 13). They also tested other assumptions for ΔG_{corr}, including the Weller equation (28). In the latter case the correlation was very poor, which indicates that further work is necessary to test the validity of the Born equation for estimating solvation energies.

(iii) Dependence on the distance between the donor and acceptor. Wasielewski and Niemczyk (84) examined the series of zinc porphyrin-anthraquinone molecules **1D6-8** in which the number of saturated carbon atoms between the porphyrin and the anthraquinone was varied from one to three. They measured the rate constants for the forward reaction from the porphyrin excited singlet state and for the reverse ET reaction from the RIP to the ground state for each of these compounds in butyronitrile and found that both rates decrease exponentially with the estimated *edge-to-edge* distance (Fig. 5a). The fit parameters β and $k_{et}(r_o)$ (eq. 17) are listed in Table A.

Similar studies of distance dependence were carried out on the rigidly linked dimethoxynaphthalene-dicyanoethylene molecules **1S21-25** in benzene (36,37,

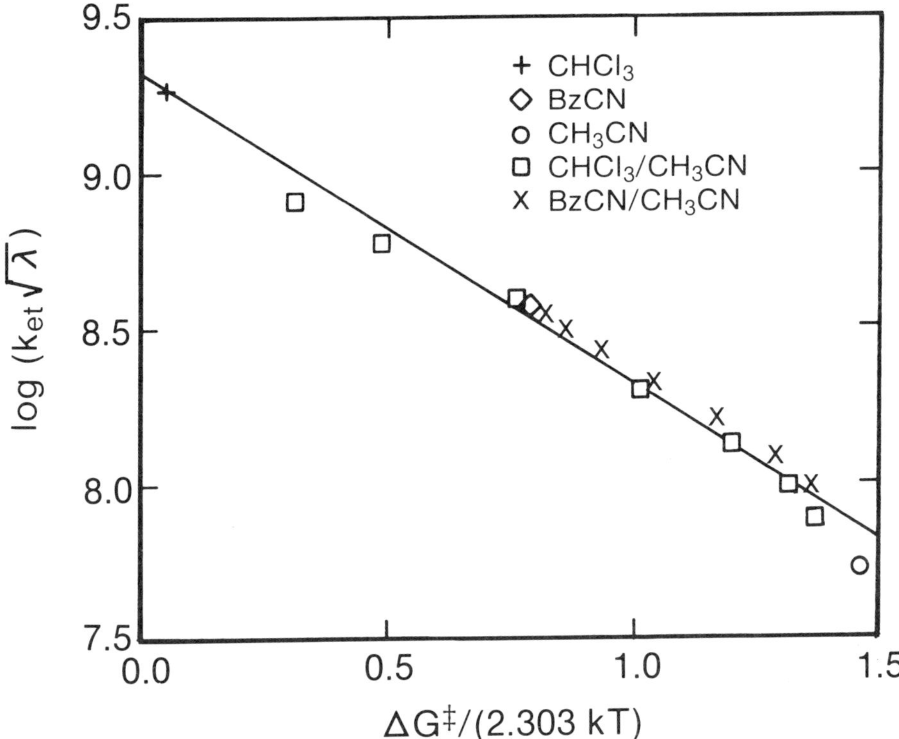

Figure 4. Plot of PET rates for molecule 1E2 in three neat solvents [$CHCl_3$, CH_3CN and benzonitrile (BzCN)] and two binary solvent mixtures according to eq. 18. (□) $CHCl_3$-CH_3CN, (x) BzCN-CH_3CN. λ (eqs. 9 and 10) was calculated from measurements of ϵ_s and ϵ_{op} for each mixture, and $\Delta G^{\ominus}$ was estimated from electrochemical measurements corrected with a simple Coulombic attraction term. The straight line represents the least-squares unit slope for all 17 data points. (From ref. 29).

85-87). The forward ET rates were obtained from fluorescence lifetimes of the photo-excited donor, except for 1S21,22, where the rates were too fast ($\geq 10^{11}$ s^{-1}) to measure. However the reverse ET rates were measured for all five systems by the powerful technique of time-resolved microwave conductivity (36-38). The distance dependence for both forward and reverse ET (see Fig. 5b) is apparently much weaker in these systems than in the PQ-type depicted in Fig. 5a.

For both classes of molecules, the intercepts for the forward and reverse steps are markedly different. This indicates that the electronic coupling between each excited-state precursor and the resulting RIP is considerably stronger than the coupling between the latter state with its respective ground state, a conclusion that is not surprising.

Wasielewski and Niemczyk (84) suggested that the reason for the stronger

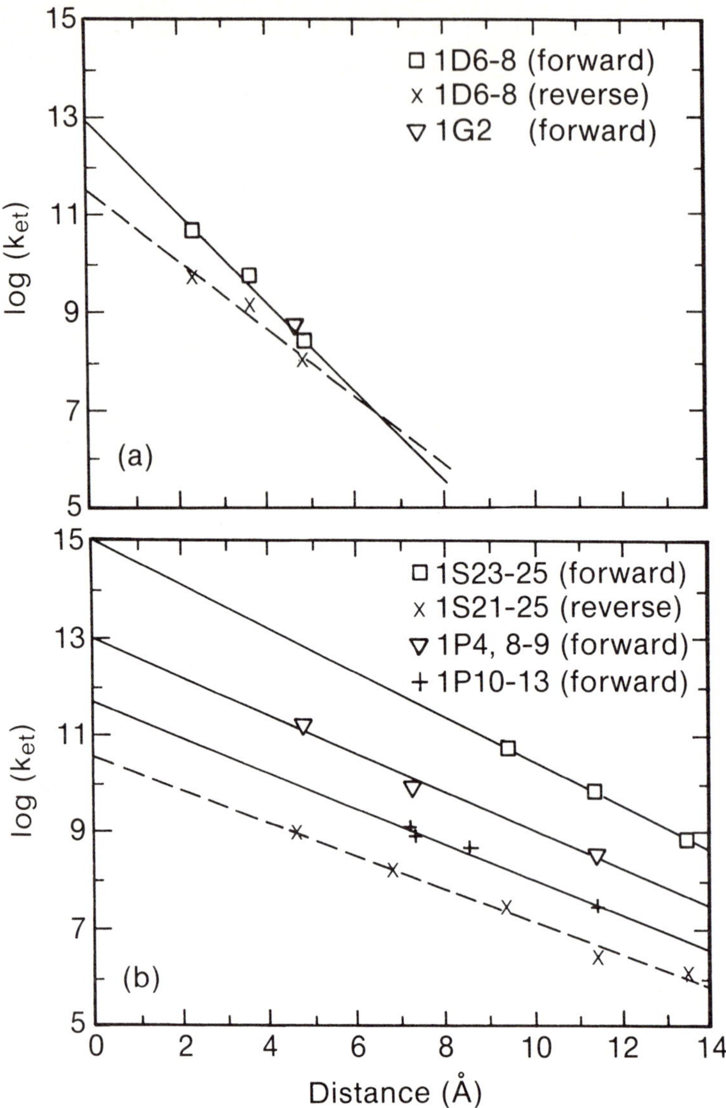

Figure 5. Plots of $\ell og(k_{et})$ *vs.* edge-to-edge distances in a variety of linked donor-acceptor systems. (a) Forward (□) and reverse (x) ET rate constants for the zinc porphyrin-anthraquinone compounds 1D6-8 (84) and the forward ET rate for 1G2 (▽) (31,32), all in butyronitrile. (b) Forward (□) and reverse (x) ET rate constants for the dimethoxynaphthalene-dicyanoethylene compounds 1S21-25 in benzene (after refs. 36,37,85-87) compared with the forward ET rates for three anthracene-dimethylaniline systems (▽) 1P4 (89) and 1P8-9 (90,91) and four analogous pyrene-dimethylaniline molecules (+) 1P10-13 (91,92), all in CH_3CN. The mutual orientations between donor and acceptor are restricted only in 1D6 (Fig. 5a) and in 1S21-25 (Fig. 5b). The reverse rate constant in 1S25 reflects some charge recombination back to the excited singlet state of the donor, resulting in delayed fluorescence (37), and is not included in the least-squares fit. The photoexcited chromophore in both types of 1Pn molecules is the electron acceptor while the reverse is true in the other systems.

distance dependence in their PQ systems as compared with the **1Sn** molecules arises from the fact that the excited state of the dimethoxynaphthalene chromophore in the latter lies about 3.5 eV above the ground state as opposed to 2.1 eV for the Zn porphyrin. The higher energy of the former allows it to couple more effectively with the antibonding orbitals of the linkage by superexchange.

An important consideration in understanding ET rates in non-rigid systems is that the mutual orientations of donor and acceptor are randomized within the range of bond rotations permitted by the linkage. The mutual orientations between donor and acceptor in the PQ molecules (Fig. 5a) are restricted only in **1D6**, whereas all of the **1S21-25** systems (Fig. 5b) are quite rigid (36,37,85-88). The close similarity of the electron-transfer rate for **1G2** (ester link) with that for the more rigid adamantane-linked **1D8** (see Fig. 5a) could indicate that torsional mobility is indeed less important in determining the quenching rates than are the donor-acceptor distance imposed by the number of saturated carbon atoms between them (31,84).

The implication that orientation may not be important stands in contrast with the results of Closs *et al.* (41) (see Section 3.1). Thus, it could be argued that the linearity in Fig. 5a is fortuitous, since the rotational freedom of the anthraquinone acceptor with respect to the Zn porphyrin donor in these molecules is also a function of the length of the linking bridge. This could explain the much stronger distance dependence in Fig. 5a as compared with the trends depicted in Fig. 5b. [See also the discussion of the **1Gn** compounds in Section 4.3 (iv).]

Table A. Fit parameters (eq. 17) for distance-dependent PET data in Figure 5.

Molecular symbol	ET step	Solvent	β $Å^{-1}$	$k_{et}(r_o)$[a] s^{-1}	Reference
1D6-8	forward[b]	C_3H_7CN	2.08	8.1×10^{12}	84
" "	reverse[b]	" "	1.59	3.1×10^{11}	84
1S23-25	forward[b]	C_6H_6[c]	1.04	9.6×10^{14}	85-87
1S21-24	reverse[b]	" "	0.83	5.2×10^{10}	36,37
1P4,8-9	forward	CH_3CN	0.89	7.9×10^{12}	89-91
1P10-13	forward	" "	0.83	4.3×10^{11}	91,92

[a]Most authors have used ν_n for this term, but from eqs. 12 and it is clear that this designation is valid only when $\lambda = -\Delta G^{\circ\prime}$.

[b]Fit parameters recalculated from data given in cited references.

[c]Oevering *et al.* (88) report $\beta = 0.85\ Å^{-1}$ and $k_{et}(r_o) = 2.5 \times 10^{15}\ s^{-1}$ from a composite plot of data obtained in four other solvents.

The data for compounds **1S21-25** are the most extensive available for any homologous series of linked D–A compounds and, because of their rigidity, the D–A distances are the most reliable. Similarly, molecules **1D6-8** are the best singly linked PQ systems available for studies of distance dependence, although there exist analogous data for some "triad" molecules (Section 4.5). A definitive test of the superexchange model must await synthesis of a series of D–A molecules with rigid links such as those in **1S21-25** but containing low-energy chromophores such as porphyrins.

There do exist, however, linked systems with geometries less well-defined than **1S21-25** but with chromophores that also absorb in the near-UV region. Molecules **1Pn**, in which ET occurs *to the photoexcited acceptor* (anthracene or pyrene), were among the earliest linked D–A systems to be synthesized. Forward ET rates have been measured for **1P4** (89) and for **1P8-13** (90-92) in CH_3CN. In all seven of these molecules, there is free rotation of the D and A moieties about the linking bridge, and additional flexibility is provided by the intervening methylene group at either end of the bridge. Accordingly, we have estimated the mean D–A distances in these molecules by molecular mechanics calculations (35). The data are plotted in Fig. 5b, and the fit parameters are listed in Table A.

The trends are remarkably linear, especially considering that **1P10** and **1P11** contain phenyl and biphenyl links (fragments **Em** and **En**, Appendix 2), respectively, whereas both **1P12** and **1P13** have naphthyl links. The close similarity of the slopes (see Table A) with those for both forward and reverse ET in **1S21-25** tends to confirm the suggestion (84) that diminished superexchange is responsible for the strong distance dependence of the ET rates in **1D6-8**.

Another interesting aspect of the results displayed in Fig. 5b is that the distance dependence (*i.e.*, β) does not appear to be strongly affected by the nature of the linking hydrocarbon bridge. One would expect that the aromatic bridges in 1P8-13 might be more effective than saturated norbornane-type spacers in facilitating long-distance PET (see below). However, comparisons between such highly dissimilar molecular systems are risky.

Leland *et al.* (93) measured the forward ET rate constants for compounds **1B5,6** which contain one and two bicyclooctane spacers in the linkage. They found that the rate decreased by at least a factor of 10^3 when the extra spacer was introduced (adding ~4 Å to the separation distance). This is consistent with a value of $\beta \geqslant 1.7$ Å^{-1}, a value similar to that found by Wasielewski and Niemczyk (84). Beratan (94) has presented a detailed theoretical analysis of this system, and has predicted that β should be $\simeq 1.8$ Å^{-1} for forward ET and $\simeq 1.0$ Å^{-1} for the reverse reaction, in good agreement with the experimental values observed for 1B5,6 but somewhat smaller than those for **1D6-8**.

The analysis of the data in Fig. 5 should be interpreted with care. In constructing these plots, it is assumed that λ and $\Delta G^{\circ\prime}$ do not change within a given series. However, as discussed earlier, both λ_o and the work-correction term in $\Delta G^{*\prime}$ are, to some extent, distance dependent.

(iv) Nature of the linkage. It appears from the preceding discussion that intramolecular ET is largely mediated through the bonds of the linkage and that "through-space" effects are relatively minor. The importance of the nature of the bridge can be seen in the comparisons shown in Table B. The bridge length for all of these molecules is ~4-5 Å, and yet the rate constants can differ by as much as ~2,000 (ref. 95). The superexchange model accounts qualitatively for the faster rate for phenyl *vs.* bicyclooctane bridges because of the lower-energy antibonding orbitals in the former. The factor of ~12 difference between the two amide isomers 1E2 and 1F1 is a striking indication that studies of intramolecular ET in closely related molecules such as these can reveal important details concerning the ability of certain groups to act as "molecular wires" (96) for ET across a bridge. These data suggest that a peptide chain is likely to be much more favorable for ET than is a saturated hydrocarbon chain. This may explain why PET is so effective in certain protein systems (see, *e.g.*, 97-101).

The effects of various types of hydrocarbon bridges on electron-tunneling rates have been analyzed by Onuchic and Beratan (102) according to a theoretical treatment that takes topological as well as energetic considerations into account. They conclude that both constructive and destructive interference effects are important in mediating the rates, and that the former give larger ET matrix elements than the latter. They predict that, other factors being equal, a spirocyclobutane-bridged system should show higher ET rates than other types

Table B. PET rate constants for linked porphyrin-quinone molecules with different bridges.

Donor-Bridge-Acceptor
(Pb) (Qa)

Molecule[a]	Bridge[b]	k_{et}^{S}/s^{-1} [c]	Reference
1B13	Ab	1.5×10^7	103
1E2	Ba	8.1×10^8	53
1C1	Aa	2.7×10^9	95
1F1[d]	Bc	9.9×10^9	104

[a]See Table 1.
[b]See Appendix 2.
[c]Measured in CH_2Cl_2.
[d]Molecule 1F1 has a *p*-amino group instead of a *p*-methyl group in the phenyl ring opposite the bridge.

of hydrocarbon-linked D–A molecules. This type of bridge has not yet been tested experimentally (see Table 1).

Before turning to the capped and triad molecules, we should mention the other singly linked compounds listed in Table 1. As a guide to this rather complex compilation, we discuss the molecules by types in the order listed.

1An: These are the simplest of the linked porphyrin-quinone systems, with the acceptors attached directly at one or more *meso*-positions of the porphine ring. Molecules **1A1-2** (72) were examined by ps laser techniques by Netzel and coworkers (105,106). These studies showed marked quenching of the porphyrin excited singlet state, accompanied by repopulation of the ground state and enhanced yields of the porphyrin triplet state, from which they claimed a rapid ($k_{et} \geqslant 3 \times 10^{10}\ s^{-1}$) ET to the quinone moiety occurred. Although this reaction scheme is consistent with the kinetic observations, it cannot be easily understood in terms of the energetics and has also been criticized on other grounds (4).

The fluorescence properties of **1A2** and the analogous **1A3-5** molecules were studied by Harriman and Hosie (107), who found various degrees of quenching by attached electron acceptors as well as donors. They also made some comparative bimolecular studies, and found in both cases a general trend of rates with the redox potentials of the quenchers. An interesting aspect of the intramolecular results, which appears to have been overlooked, is the poor correlation between the fluorescence lifetimes and yields. In general, the observed lifetimes were longer than one would anticipate from the diminished yields. (See also the discussion of the **1Gn** molecules, below.) A PQ molecule similar to **1A1**, but containing a single anthraquinone attached at a porphine *meso*-position (**1A6**), has been synthesized recently (108), but it photophysical behavior has not yet been analyzed in detail.

In all of these directly linked PQ systems, it appears that the donor-acceptor distances are too short to permit long-lived (*i.e.*, $\geqslant$ 1 ns) charge separation to be observed. Thus, when the forward ET step is fast, so also is recombination back to the ground state. This is one aspect of PET experiments on these systems that is in very good agreement with theory.

1Bn: Structurally, this is the simplest class of <u>bridged</u> PQ molecules. The **1B1-6** systems studied by the Cal Tech group (93,94,109,110) are based on octamethylporphines (both Zn and free-base) to which are attached rigid spacers consisting of a phenyl ring and either one or two bicyclo[2.2.2]octane groups. Although there is rotational freedom in these systems, the center-to-center distances between the porphyrin donor and benzoquinone acceptor are fixed at ~10.8, 14.8 and 18.8 Å, respectively. Unfortunately, no ps absorption studies have been carried out on these important molecules, but there is no reason to believe that the observed fluorescence quenching is due to anything other than electron transfer. As expected (see Fig. 5), the PET rates decrease exponentially with distance. Beratan (94) has analyzed these results theoretically in terms of the effects of linker topology on the electron-tunneling matrix elements.

An important result for compound 1B5 was evidence of PET in a rigid glass at 77 K (93). Nonexponential fluorescence decay was attributed to frozen rotational conformations. This is the only such system studied to date that does not "shut off" at low temperatures, most likely because of the large value of $-\Delta G^{\circ}$ (> 1.0 eV). Nevertheless, this result indicates that the PET rates in this system are not strongly temperature dependent, but do depend critically on "structural control of every degree of freedom along the reaction coordinate," presumably because of the non-adiabaticity of the ET process (93).

Room-temperature fluorescence studies on a similar molecule 1B13, which contains only a single phenyl-bicyclooctane spacer, have been carried out by Bolton's group (103). The important result of this work was the finding that the PET rates are optimal in medium-polarity solvents with high refractive indices rather than in solvents with high ϵ_S. This observation points once again to the difficult problem of determining the actual $\Delta G^{\circ\prime}$ values in these model systems.

Finally, Joran *et al.* (110) have synthesized the only existing set of PQ molecules that constitute a truly homologous series. These compounds are based on molecule 1B5 together with six additional systems (1B7-12) containing benzoquinone moieties carrying substituents that provide a range of reduction potentials spanning nearly 0.6 eV. PET rates were determined from fluorescence-lifetime measurements in four solvents, and the rate constants were found generally to follow the dependence on Gibbs free-energy predicted by Marcus theory. However, electrochemical measurements were not carried out in all solvents, so the uncertainty in $\Delta G^{\circ\prime}$ is rather high.

1Cn: These molecules are analogous to the 1Bn systems, but have only aromatic hydrocarbon rings as spacers that confine the donor and acceptor moieties at fixed distances (95). The ET rate constants for 1C2 increase with solvent polarity (ϵ_S), in contrast with the results for 1B13 (see above); the rates for 1C1 are either near or beyond the limit of the detection technique (~3×10^{11} s^{-1}), so the solvent dependence is not yet clear. However, it is clear that k_{et} for 1C1 is ~10 times greater than for 1C2, in agreement with the result found by Michel-Beyerle and co-workers (90-92) for the pairs 1P8,10 and 1P9,11. Evidence for a superexchange mechanism is provided by comparisons of the PET rates for 1C2 with those of three other PQ systems containing different linking groups but with comparable donor-acceptor separations (see Table B).

1Dn: Most of these elegant porphyrin-based systems, synthesized and characterized by Wasielewski and co-workers (80-84), have been discussed previously in the context of the distance and energy dependence of PET rates. Very recently, this group carried out studies on four analogous compounds containing chlorin instead of porphyrin donors, *viz.*, pyropheophorbide a (**1D9**,**10**) and pyrochlorophyllide a (**1D11**,**12**) (ref. 111). As in the case of the **1D1-6** systems, the bridge is a rigid triptycene spacer that holds the attached benzoquinone (**1D9**,**11**) or naphthoquinone (**1D10**,**12**) at a fixed distance. The only conformational freedom in these molecules is restricted rotation about the 2-vinyl group on the chlorin ring to which the spacer is bonded. Since the free-base and Mg macrocycles and the attached quinones have quite different redox potentials, the measured $-\Delta G^{\circ\prime}$ values span a range from ~0.2 eV for the least exergonic forward ET step to ~1.6 eV for the most exergonic recombination process. The rates measured in both directions show that forward ET in these molecules is in the Marcus normal region while reverse ET is in the inverted region. Thus, there can no longer be any doubt regarding the validity of this important theoretical concept.

1En: These PQ systems carry various types of single-amide bridges attached to either the *ortho*- (**1E1**, ref. 73) or *para*- (**1E2**, ref. 53) positions of a phenyl ring located at a *meso*-position of the porphyrin. The relevant features of these compounds were discussed at the beginning of this Section.

1Fn: The research group at Arizona State University (ASU) has synthesized and studied this class of PQ compounds (104). Like the **1E2** systems, the bridges consist of simple amides covalently attached at the *para*-position of a porphyrin *meso*-phenyl ring. The simplest (**1F1**) carries one methylene group immediately adjacent to the terminal benzoquinone and is an isomer of **1E2**, except for the *para*-NH_2 group on the phenyl ring opposite that attached to the bridge. The PET rates observed for these two bonding arrangements differ considerably (Table B). This may be due to differences in through-bond interactions, but changes in exergonicity and/or donor-acceptor orientation may also play a role.

Although these bridges are flexible (especially as more methylene groups are inserted into the chain), Gust *et al.* (104) have used high-field NMR to resolve the time-averaged conformations of these molecules in fluid solution. From this analysis, the mean porphyrin-quinone distances (on the NMR time scale) were derived. The dependence of $\log(k_{et})$ on edge-to-edge distance for **1F2-4** was found to be much weaker than in the case of **1D6-8** (see Fig. 5a and Table A), with β = 0.6 $Å^{-1}$ and $k_{et}(r_o) \sim 3 \times 10^{10}$ s^{-1}. Similar fit parameters were obtained for the corresponding carotenoporphyrin-quinone triads **3C1-6** (see Section 4.5 and Table 3). This weak distance dependence was tentatively ascribed to conformational heterogeneity and/or molecular motion during the lifetime of the donor excited singlet state, both of which arise from bridge flexibility.

As in the case of virtually all other PQ systems, PET in the 1Fn systems is strongly temperature dependent (104,177), so it is unlikely that studies in the usual low-temperature glassy matrices would yield any further information on orientation or distance effects. However, these molecules (as well as the related triads) may be excellent candidates for PET experiments in room-temperature matrices such as polymers (Sections 5.3 and 5.5).

Tien and co-workers (112,113) have observed a photoelectric response in bilayer lipid membrane (BLM) preparations containing PQ molecules 1F4-7, as discussed below (Section 5.2). They also synthesized a compound with two quinone moieties attached, and found a reduced photoresponse as compared to its single-quinone analogue. No explanation was offered for this curious observation.

1Gn: This class has an anthraquinone (AQ) acceptor, an ester bridge and either a free-base (1G1) or Zn porphyrin (1G2) donor. The latter compound is closely related to 1D8 in that both contain anthraquinone (AQ) as the acceptor and the mean porphyrin-AQ distances are about 4.5 Å (31,32); the PET rates in these two compounds are also quite similar (Fig. 5a), even though the geometry is considerably more constrained in 1D8. This suggests that distance rather than orientation plays the dominant role in these systems. Alternatively, the agreement between the two rates may be fortuitous; thus, more effective superexchange in the ester linkage of 1G2 relative to the saturated hydrocarbon bridge in 1D8 may compensate for slower ET rates in unfavorable conformations [see also, Section 4.3 (iii) and Table B].

In contrast, the energetics for PET in 1G1 are unfavorable, the excited singlet state of the donor and the radical-pair being essentially isoenergetic. Indeed the fluorescence lifetimes are indistinguishable from those of a related reference porphyrin. However, differences as much as ~15% were observed in the relative fluorescence intensities, following a qualitative trend with solvent polarity. It was suggested (31) that the interactions responsible for this "static" quenching are related to charge transfer, although no net charge separation could be observed (32).

Parallel studies of interactions between non-linked AQ and the free-base porphyrin showed quite efficient quenching of fluorescence lifetimes, with bimolecular rate constants approaching the diffusion-controlled limits (31). The geometry in intermolecular interactions is not constrained, and efficient fluorescence quenching (perhaps *via* exciplex formation) can occur without formation of radical ions. However, when the geometries accessible to the two moieties are restricted, even by a relatively flexible link, net electron transfer may be a requirement for diminished fluorescence lifetimes to be observed (31). These results point out how little is known about *inter*molecular porphyrin-quinone interactions compared to the behavior of the linked systems.

1Hn: The current abundance of linked PQ molecules began with the synthesis of **1H1-3** by Kong and Loach in 1978 (74). The characteristic feature of the linkage is an n-methylene diester bridge, which is sufficiently flexible that no conformational heterogeneity is manifested at room temperature (in contrast to the 1In compounds discussed below). 1H2 was later studied by Ho *et al.* (78) who observed a persistent light-induced EPR spectrum at low temperatures. The temperature-dependent decay profiles of these signals showed a distribution of recombination rates, suggesting a distribution of porphyrin-quinone distances. The other pertinent features of these systems were discussed at the beginning of Section 4.3.

1In: Like the 1Hn compounds, these early contributions to the field, synthesized and studied by Bolton and co-workers (30,32b,33,34,79,79a) have flexible n-methylene bridges, but with diamide rather than diester links. These systems were also discussed in detail at the beginning of Section 4.3.

1Jn: The photophysical and photochemical properties of these interesting linked PQ compounds (114,115) have not been studied. However, they are included here since the synthetic route itself includes a self-sensitized photochemical step that apparently involves PET from the porphyrin triplet state. The progress of the reactions was followed by the CIDNP technique, and the yields were found to be remarkably high (~32%). This approach might be explored further in the search for new porphyrin-based systems with donors, bridges and acceptors exhibiting specific, desired properties.

1Kn: The research groups at Osaka University have synthesized and studied these octa-alkylporphines linked to benzoquinone *via* n-methylene chains (116-118). These systems were among the first for which both the forward and reverse ET rate constants were measured using ps laser techniques. There is a general trend of decreasing forward k_{et} with the length of the linking hydrocarbon chain (1K1-3), but the reverse rates are difficult to correlate with donor-acceptor distances, probably because of the flexibility of the bridge (2). These studies later lead to the assembly of some analogous triad molecules (see Section 4.5).

1L1: Only one system has been synthesized in this class, which contains a sandwiched *bis*etioporphyrin donor, an aliphatic hydrocarbon bridge and an unsubstituted benzoquinone acceptor (119). It is structurally similar to 1K3, but has a longer (by one methylene group) linking bridge. The fastest of three forward PET rates was found to be ~10 times faster than in 1K3, but it is uncertain whether this arises from more favorable geometry or more favorable energetics. The former could be ascribed to the greater flexibility of the *n*-pentane bridge, which permits closer interaction of the quinone with the *bis*porphyrin. The latter effect cannot be ruled out, however, because such dimeric porphyrins are generally easier to oxidize than their monomeric counterparts (120).

1Mn: These compounds have *meso*-tetraphenylporphine (TPP) donors attached to a trinitrobenzene (TNB) acceptor by simple ether linkages at an *ortho-*, *meta-* or *para*-position of a TPP phenyl ring (121). Computer models of the structures and the NMR spectra both indicate that conformational freedom is restricted by these short linkages. Strong fluorescence quenching was found at room temperature, and a stable (*i.e.*, several min) light-induced EPR signal was observed at low temperatures (100-140 K). It is uncertain, however, whether these EPR spectra represent the $TPP^{+\cdot}$-$TNB^{-\cdot}$ RIP or, perhaps more likely, a subsequent product.

1Nn: This large class of molecules is characterized by having various types of viologens as the acceptors (122-128). The reduction potentials of these groups are generally more negative than that of benzoquinone, so these compounds might provide useful comparisons with several existing PQ systems. However, no analogous PQ molecules have yet been synthesized with the same bridges as in the 1Nn compounds.

Only a few photophysical measurements have been carried out on these molecules. Fluorescence quenching in **1N1-33** was found to depend both on the site of attachment of the linking bridge and on its length (122). No long-lived radical products were detected, in contrast to the case of the similar **1N34-36**. Reverse ET rates of the latter group appear to be relatively insensitive to the length of the methylene chain in the bridge (123). A slight trend of decreasing rate with chain length was observed (Table 1), but this cannot be readily correlated with distance because of bridge flexibility. The slow charge-recombination rate constants listed in the Table reflect the spin multiplicity of the RIP, which is formed from the triplet state of the porphyrin donor. This explains the sensitivity of these rates to an external magnetic field, being slower in the case of 1N36 by a factor of ~10 with an applied field of $\geqslant$ 0.1 T (124).

These linked porphyrin-viologen molecules are interesting also because viologens have been used extensively as relays in photochemical water-photolysis schemes. Indeed, compounds **1N37,38** have shown the ability to evolve molecular H_2 when coupled with hydrogenase and/or Pt catalysts (125-128). As in the case of **1N34-36**, the forward ET reactions appear to emanate only from the porphyrin triplet state.

Detailed photophysical studies of this class of molecules would be of further interest because of their solubility in high-polarity solvents, including water. They thus afford a means of comparing intramolecular PET mechanisms with intermolecular processes in diffusionally constrained vesicle systems, many of which have water-soluble porphyrins and viologens incorporated in them (see Section 5.2).

1On: Like the 1Nn systems, these compounds have a porphyrin donor and a methylviologen acceptor attached by either a simple methylene link (1O1) or by flexible alkyl-ester bridges (1O2,3). A general correlation of ET rates with chain length was observed for the latter two molecules, with forward/reverse ratios on the order of 150 (129). Intramolecular motion in the RIP state was invoked to explain the slow charge-recombination rates. A similar argument was made in the case of the diamide-linked PQ compounds 1I1-3 (33).

1Pn: This class is distinguished from most of the others by virtue of the fact that ET occurs through (or across) an n-methylene bridge from a ground-state donor (dimethylaniline) to an excited-state acceptor (anthracene or pyrene). Molecules 1P1-4 were among the first D-A compounds for which lifetimes of CT states were measured (65). A comprehensive study of the solvent dependence of CT emission in 1P1-7 was carried out by Masaki *et al.* (130). Also, as noted in Section 4.3 (iii), compound 1P4 was the subject of ps laser studies of intramolecular ET as early as 1974 (89,131-134).

The most extensive recent studies of this class of compounds have been carried out by Michel-Beyerle and co-workers (90-92). In addition to the distance-dependence discussed in Section 4.3 (iii), they found solvent and temperature regimes in which the PET rates are limited by relaxation of the surrounding dielectric medium (61).

1Qn: Mutai (135) studied these *p*-nitrobenzyl derivatives of aromatic amines. CT emission was observed for all of the systems, but only in the case of the simpler aromatic donors (1Q3-5) was there a correlation with the electron-donating ability of the ring. The emission was interpreted as arising from through-space interactions. However, at that time the concept of through-bond ET had not been developed.

1Rn: This broad class of linked molecules includes aryl, methoxyphenyl and dimethylaniline as donors, simple n-methylene linkages as bridges and *para*-substituted pyridinium ions, picric acid and various phthalimide and naphthalimide derivatives as acceptors (64,136-141). These compounds were the subject of an extensive, systematic study by Verhoeven's group of the effects of molecular geometry on CT absorption and emission and eventually led to the concept of through-bond ET (139). Unfortunately, no detailed photophysical measurements of ET rate constants were carried out. It would be very useful to re-examine several of these compounds with the fast laser techniques now available.

1Sn: The donors in these compounds consist of mostly methoxy-substituted aromatic rings (benzene and naphthalene) or aniline moieties, and the acceptors are cyano- or alkylester-substituted ethylenes. These molecules can be further grouped into two classes, distinguished by the type of rigid, saturated hydrocarbon spacer between the donor and acceptor. Early work concentrated on studies of intramolecular CT absorption (66,143) and emission (66,142-146) bands in systems with aromatic (1S1,2) and saturated hydrocarbon (1S3-20) spacers.

Later work (36,37,85-88) involved extensive and very important photophysical studies on systems with similar D-A pairs but linked by rigid, norbornyl-type hydrocarbon bridges. These are often referred to as "Paddon-Row" molecules, after the chemist who first synthesized them. The great advantage of these compounds lies in the precise geometries in which the donor and acceptor are held. Because of this constraint, relatively simple Hückel calculations (147,148) have yielded predicted ET rates in quite good agreement with the experimental data. The latter were discussed in Section 4.3 (iii).

1T1: This compound has been the subject of detailed analyses of very fast ET rates in the regime where solvent dynamics must be considered (see Sections 3.4 and 4.2). Kosower and co-workers (54-57) have interpreted the intramolecular ET rates in this molecule as being controlled by the solvent relaxation time. Only recently has this phenomenon been shown experimentally to be important also in cases of net electron transfer (59-62).

4.4 Capped molecules

As noted in the previous Section, a single, flexible linkage can lead to problems in interpretation of the rate data in terms of specific rate constants, largely because the conformation(s) cannot be specified. An alternative strategy has been to synthesize molecules in which the acceptor is linked to the donor *via* two or more bridges so that the acceptor is held in a relatively rigid, "capped" position over the donor. Table 2 contains a summary of the structures and properties of several such systems.

The first capped molecules to show intramolecular charge-transfer interactions (2A1-5) were synthesized by Schroff *et al.* (149,150). This work was followed by preparation of the highly similar 2A6-9 systems (151). As in the case of the analogous linear compounds (1R30-37), long-wavelength CT bands were observed in the absorption and emission spectra. These bands were shown to decrease in intensity and shift to the blue as the chain length of the bridges was increased. The intensities of both the CT absorption and emission were found to depend strongly on the relative donor-acceptor geometry, being much stronger in the asymmetrical 2A6 than in the symmetrical 2A7. The k_{et} values for 2A6 calculated from the fluorescence lifetimes, show a dependence on solvent polarity, from which it was inferred that a solvent-induced geometry change was involved. However, it is uncertain whether distance or orientation is the major factor.

Zsom *et al.* (152) examined single crystals of 2A6,7 and confirmed the charge-transfer nature of these transitions from polarized absorption spectra.

The solvent dependence of both the CT absorption and emission spectra of molecules **2A8,9** was studied by Borkent *et al.* (151). They found that, in contrast to **2A6**, the CT absorption was red-shifted and its intensity was enhanced with increasing solvent polarity; at the same time, both the yields and lifetimes of the CT emission were diminished.

Compounds **2B1-4**, the first capped molecules containing a porphyrin donor and a quinone acceptor, were synthesized by Ganesh and Sanders (153,154). The bridges consist of a combination of ester and n-methylene units. Using NMR spectroscopy, Sanders and co-workers (155,156) observed distinct conformational differences between the free-base and metalloporphyrin species. Specifically, with Mg in the center of the ring (**2B3,4**), one of the quinone oxygens is coordinated intramolecularly with the metal, whereas in the free-base species (**2B1,2**) the quinone lies roughly coplanar with the porphyrin.

Irvine *et al.* (157) examined in detail the forward and reverse ET rates for **2B2** and its Mg analogue (**2B4**) in a variety of solvents and found strong evidence for the Marcus inverted region. However, a quantitative interpretation of these data may be skewed because they did not allow for the solvent dependence of λ and they used the Weller equation (28) to calculate ΔG_{corr} in eq. 13.

Lindsey and Mauzerall (158) reported the synthesis of the quinone-capped porphyrins **2C1,2**; these are the only compounds reported to date in which the donor and acceptor are anchored by four covalent linkages. The porphyrin fluorescence is quenched in the Zn compound **2C2** but not in the free-base analogue **2C1** (159). This was ascribed to electron transfer from the energetically more favorable excited singlet state of the Zn porphyrin to the quinone. From laser flash photolysis studies, Lindsey *et al.* (159) concluded that the route for formation of the RIP was *via* the triplet state of the porphyrin, with a risetime of ~150 ns and a remarkably long lifetime of ~1.4 μs (*vide infra*).

Despite the presence of four bridging groups, the quinone in **2C2** is not held rigidly above the plane of the porphyrin. In a more detailed photophysical study of this molecule, Lindsey and co-corkers (160) found two ET rate constants for the excited singlet state, which they ascribed to slowly equilibrating "introverted" and "extroverted" conformers in which the estimated interplanar porphyrin-quinone separations are, respectively, 6.5 and 8.5 Å. Extensive data on the fluorescence lifetimes of **2C2** show that that the faster of the two ET rate constants is independent of temperature over the range of 80-300 K and is only slightly solvent dependent, varying by less than a factor of four over the range $2.5 \leqslant \epsilon_s \leqslant 40$, thus tending to confirm an electron-tunneling mechanism.

The triplet state of **2C2** shows a small activation energy (~6 kJ/mol) for ET while that of **2C1** (*i.e.*, with the free-base porphine) shows no reaction (160). These results can be explained by the proximity of the energies of the triplet state and the RIP at about 1.4 eV in **2C1**. The long lifetime (~1.4 μs) of the presumed RIP in **2C2** may involve differences in orbital symmetry between the RIP and the ground state (161,162); at these distances (6.5-8.5 Å) reverse ET, even if in the Marcus inverted region, should be much faster. Moreover, such a long lifetime suggests that, if spin multiplicity is a consideration, triplet-singlet spin rephasing is unusually slow (35). The major problem, which is not unique to these molecules, is that incontrovertible evidence has not yet been presented that this long-lived species is indeed the RIP and not a product, *e.g.*, containing a protonated semiquinone radical instead of the quinone anion radical.

Morgan and Dolphin (163) synthesized the interesting **2D1-2** molecules with two quinones joined to a single porphyrin, one above the plane and the other below. However, nothing has been reported so far concerning their photophysical behavior.

A similar *bis*quinone porphyrin cyclophane **2F1** has been synthesized by the Heidelburg group (164,165). An important aspect of this compound is that it is the only linked PQ molecule for which a structure has been determined by X-ray techniques (166). In the solid state, the two quinones are coplanar with the porphyrin and are symmetrically displaced at a distance of 3.4 Å. Nevertheless, the photophysical behavior of this molecule in fluid solution is not simple. **2F1** shows fluorescence lifetimes varying from ~300 ps in toluene to ~60 ps in dimethylacetamide with a small (~2 kJ/mol) activation energy (167). In contrast, the single-quinone analogue **2E1** shows essentially the same lifetime (~500 ps) in a variety of solvents (167). Many of these capped molecules show fluorescence-decay profiles that are best fit by a distribution of lifetimes, as expected if the quinone is able to undergo even restricted rotation.

Osuka *et al.* (168) prepared the capped PQ molecules **2G1-3** by their photochemical synthesis route about the same time as the singly linked analogues **1J1-4** (114,115). Workers in this field will look forward to photophysical characterizations of all of these ether-linked PQ systems.

The quinone has been replaced by other types of acceptors in this same general structure--methylviologen in **2H1,2** (169), the corresponding neutral bipyridyl in **2H3-10** (170), and pyromellitimide in **2I1-9** (171). They all generally behave in a manner similar to the quinone-bearing **2B1-4** molecules discussed previously. The reduction potentials of these acceptors are sufficiently different that a very wide range of ΔG° (~1.0 eV) values is encompassed.

As in the case of **2B1-4**, Leighton and Sanders (169), as well as Abraham *et al.* (172), observed "switching" of conformations in **2H1,2**; *i.e.*, in the metalloporphyrin species the acceptor is internally ligated to a central metal ion (either Mg or Zn) in the macrocycle but is roughly coplanar with the porphyrin in the free-base compound. Irvine *et al.* (157) carried out extensive ps studies of forward and reverse ET for **2H1,2** in several solvents. The results show, as now expected, that forward ET is in the normal region and charge recombination is in the inverted region. As in the case of the parallel studies on the **2Bn** compounds (157), quantitative interpretation of the rate data in terms of Marcus theory is obscured by the lack of appropriate corrections for solvent effects.

Similar ps laser studies of both forward and reverse ET rates for **2In** in several solvents were carried out by Cowan *et al.* (173). This approach was extended by Harrison *et al.* (62) to include some temperature-dependent data. They found, as in virtually every other case, that PET "shuts off" at low temperatures, an effect they ascribed to dielectric relaxation. This work was the first to invoke limitations of solvent dynamics on PET rates in a linked D-A system containing a porphyrin.

Synthesis of molecules **2J1** and **2K1**, both consisting of *bis*porphyrins linked to a pyromellitimide acceptor was reported by Cowan *et al.* (173), who also characterized some of the photophysical properties of these complex systems. For the cofacial *bis*porphyrin **2J1** they found strong quenching of fluorescence and concluded, ". . . photochemically, the two porphyrins are acting in concert as a 'special pair'. . ." For **2K1**, on the other hand, which has a side-by-side *bis*-porphyrin, they found relatively little fluorescence quenching. These conclusions were supported by ps laser studies from which ET was found to be very rapid in **2J1** but quite slow in **2K1**. The primary importance of these results is that they demonstrate that the proximity of donor and acceptor is not a sufficient condition for efficient ET; the two entities must also be in a favorable geometric configuration, or at least be flexible enough to be able to explore favorable conformations. These results are also of interest because this system models the "special pair" electron donor in photosynthesis.

These capped D-A molecules are elegant from a synthetic point of view and have yielded some very interesting photophysical results. Nevertheless, the rates for reverse ET are generally too fast to allow effective trapping of the photon energy. Also, with the possible exception of the **2J1** and **2K1** systems, these capped molecules do not represent good models of the photosynthetic RC, where none of the donor or acceptor entities are in capped configurations (13-15). It is likely that better models of the natural RC will be found among the "triad" molecules discussed in the next Section.

4.5 Triad molecules

As noted in Section 1, the primary processes of photosynthesis involve a series of sequential ET steps. Since 1983, five molecular systems have been synthesized that mimic the sequential ET feature of biological RCs. Each contains three molecular entities, hence the term "triad." Two types have been studied: $D-A_1-A_2$ where D is the absorber and A_1 and A_2 are the primary and secondary electron acceptors, respectively, and D_2-D_1-A, where D_1 is the absorber and primary donor, A is the acceptor and D_2 is a secondary donor. Table 3 summarizes the structures and properties of these systems.

3A1, which is the only $D-A_1-A_2$ system in Table 3, was synthesized and characterized by Nishitani *et al.* (174). They found evidence for sequential ET from D to A_1 to A_2, with a lifetime of the $D^{\dot{+}}-A_1-A_2^{\dot{-}}$ RIP state of ~300 ps, as compared to ~130 ps for the corresponding porphyrin linked to a single quinone in 1K2 (118). The reason that addition of a second quinone stabilized the RIP by only a factor of ~2 is probably related to the flexibility of the molecule, which allows a closer approach between the two oppositely charged moieties in the triad than in the singly linked molecule.

At about the same time, the ASU group (175-176b) reported the synthesis of the carotenoid-porphyrin-quinone (C-P-Q) compounds 3B1 and 3C1-6, which are of the type D_2-D_1-A. Formation and decay of both $P^{\dot{+}}$ and $C^{\dot{+}}$ were followed by ns and ps absorption spectroscopy (177,104); the observed kinetics are consistent with the following reaction scheme

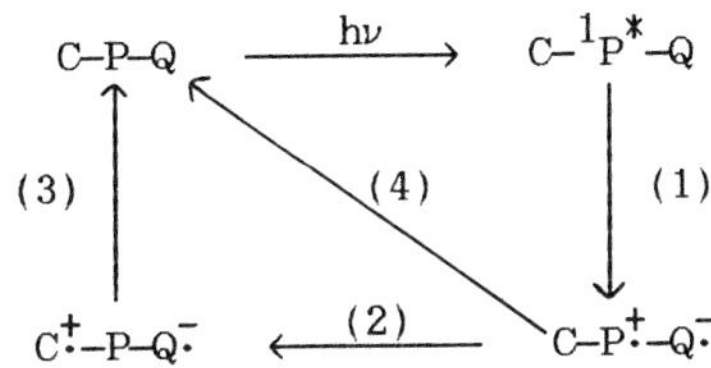

Gust *et al.* (177,177a) found that the charge-recombination reaction actually occurs in two steps and involves a $C-P^{\dot{+}}-Q^{\dot{-}}$ intermediate. Indeed, the activation energy for step 3 (11.3 kJ mol^{-1}) is in good agreement with the calculated energy difference between the $C-P^{\dot{+}}-Q^{\dot{-}}$ and $C^{\dot{+}}-P-Q^{\dot{-}}$ states.

Studies of C-P-Q triads have been extended recently to several related molecular systems. For the series 3C1-6, Gust, Moore and their many collaborators (104) found that the ET rate constants decrease exponentially with the distance between the donor and acceptor with β = 0.6 $Å^{-1}$ (eq. 17). This value, which is considerably smaller than the value of ~2.1 $Å^{-1}$ found for 1D6-8 (Fig. 5a), may arise from the amide group in the linking bridge in contrast to the hydrocarbon spacers in the latter molecules. In any case, the important feature of these triad molecules is that the lifetime of the charge-separated state is considerably lengthened (by a factor of ~100) as compared to the analogous porphyrin-quinone "diads" 1F1-4 (Section 4.3). Also, 3B1 has shown the capability of facilitating light-induced charge separation across a bilayer lipid membrane (178) (see Section 5.2).

Compounds 3D1 and 3D2 are also of the C-P-Q (D_2-D_1-A) type (179). These are the closest of the synthetic model systems to natural photosynthesis in several respects: (1) The absorber and primary electron donor (P) is pyropheophorbide a (Pyropheo a), a derivative of Chl a. (2) The initial electron acceptor is naphthoquinone, a close homologue of the menaquinone contained in bacterial RCs (10-17). (3) In addition to sequential ET, as in the above reaction scheme, these triads--like all of the other C-P-Q compounds--exhibit efficient intramolecular energy transfer: singlet excitation energy is transferred from the carotenoid to the Pyropheo a moiety, and triplet energy flows in the opposite direction. Thus these systems mimic both the spectral sensitization and photoprotection functions of biological molecular ensembles. The latter aspect is discussed in Section 5.3.

Triad **3E1**, synthesized and studied by Wasielewski *et al.* (82,83), is also of the D_2-D_1-A type. The photophysical behavior of this molecule follows the same scheme as for the C-P-Q compounds discussed previously. The rate constant for step 2 (in butyronitrile) was found to be 1.4×10^{10} s^{-1}, whereas that for step 3 was only 4.1×10^{5} s^{-1}, as compared to 5.4×10^{9} s^{-1} for the reverse ET rate constant in the simpler diad **1D6**. Thus, in this case, addition of a secondary donor (dimethylaniline) prolongs the lifetime of the charge-separated RIP state by over four orders of magnitude. Moreover, the RIP state in **3E1** lies ~1.4 eV above the ground state and thus represents considerable storage of the excited-state energy.

This is a rapidly moving field and the future will certainly see the preparation of many more triads, "tetrads" and perhaps even more complex molecular assemblies. These molecules will prove to be important in developing our understanding of mechanisms that facilitate efficient photogeneration of long-lived charge-separated states with significant energy storage.

5. Future Prospects and Problems to be Resolved

Although impressive progress has been made in the past decade towards the assembly of an efficient and stable artificial photosynthesis system, this field is still only in its early stages. Much remains to be done, not only to synthesize new molecules, but also to develop a more comprehensive theoretical framework to guide both the synthesis and the experimental design.

5.1 Synthesis of additional linked D-A systems

A rigid D-A complex almost certainly will be required in a functional, synthetic photoreaction center. However, we believe there is a strong case for further studies of flexibly linked as well as rigid molecules from which the necessary information can be inferred. Because this field is largely synthesis-limited, it is important to note that flexible molecules are generally easier to synthesize than rigid ones for the purpose of elucidating a specific factor. The major limitation of flexible molecules, of course, is that the orientation(s) of donor and acceptor cannot be specified; this is one of the major factors that remain to be understood (46,47). Also, more systems, with the same donor and acceptor but with different bridges, are needed to sort out the effects of distance, orientation and other bridge-mediated factors (104) (see also, Section 5.4).

5.2 Vectorial electron transfer in membranes, vesicles and polymers

Coupling with the "outside world" will be required in an artificial photosynthetic system just as it is in the natural organisms. Thus one test of the usefulness of these linked D-A complexes is their ability to effect *vectorial* electron transfer. Planar bilayer lipid membranes (BLMs) containing electron donors or acceptors have been studied for many years by Tien and co-workers as models of biological membranes (180), including the thylakoid membrane of chloroplasts (181-183).

These studies have been extended in the last few years to include pigmented BLMs containing the linked porphyrin-quinone complexes **1F5-7** (112,113). Enhanced photovoltage and photocurrent signals were observed with these linked molecules in BLMs as compared with preparations containing the non-linked components. The pigmented BLM thus behaves like an organic semiconductor interposed between two aqueous compartments containing a secondary electron donor on one side and a secondary acceptor on the other. Presumably, even higher photoefficiencies could be achieved in BLM preparations if the PQ molecules were oriented in the membrane. In similar experiments, Moore, Gust and co-workers (178,184) have demonstrated that triads of the type **3B1** are able to induce ET across a phospholipid bilayer and thus produce a steady-state photocurrent across the membrane.

PET experiments in BLM preparations are related to studies in vesicles (8, 185,186), at least in terms of the microheterogeneity of the pigment environment. In both cases, the membrane serves the very important function of prolonging the lifetime of useful charge separation by removing either an electron or hole from the initially formed RIP state. This approach is discussed by Baral and Fendler in Chapter 3.2. To date, these have been the major experimental techniques used to understand the role of the medium in facilitating PET and stabilizing the redox products.

In the RCs of photosynthetic bacteria, the structure of the protein is no doubt crucial with respect to maintaining the distance and also, most probably, the orientation of the various donors and acceptors. However, it is important to note that there are no covalent bonds between any of these components (13-15), at least not in the same sense that we understand such links in simple model systems. Moreover, the reorganization energy appears to be small in natural RCs (25) whereas it is quite high ($\geqslant$0.5 eV) in all of the linked D-A systems studied to date. This factor is critical in determining the fraction of photon energy that can be converted to stored Gibbs free-energy.

A functional system of artificial photosynthesis will require that we learn how to mimic not only the donors and acceptors, but also the biological environment itself. The role of protein membranes in facilitating the primary electron-transfer process between non-linked D-A pairs is only beginning to be understood (97-101). Accordingly, we need to learn more about the environment in which the molecular constituents of photosynthetic RCs are embedded, in particular, the effects of dielectric relaxation (54-63) and whether dielectric anisotropy (31) plays a role in governing the kinetics. In other words we should try to understand what properties of the natural environment facilitate electron transfer and what factors are responsible for the minimal reorganization-energy requirements.

To study such effects in biological RCs themselves would be very difficult. However, the use of polymer matrices, both natural (97-101) and synthetic (187), may lead to a better understanding of the role of the medium in light-induced electron transfer. Just as simple models of linked donors and acceptors have given us experimental validation of the Marcus theory, even fairly crude models of biological matrices should allow us to investigate medium control on photophysical and photochemical processes in a systematic way.

5.3 Quantum efficiencies, solar efficiencies and stability

In addition to electron transfer, mimicry of biological quantum harvesting and energy transfer is also an important consideration in the design of synthetic systems. Bensasson *et al.* (188) have suggested that synthetic solar energy conversion systems based on porphyrins or chlorophylls might be expected to inherit the problems of light-gathering efficiency and photodegradation shown by the *in vivo* systems. The need for very high molecular stability (*i.e.* low net quantum yield of degradation, Φ_d) under solar irradiance conditions has been stated in quantitative terms by Connolly and Turner (3): Suppose each chromophore in a molecular system absorbs one photon per second and its desired useful lifetime (1/e) is to be, say, 10 years; over this time it would absorb $\sim 10^8$ photons, and the *maximum* allowable value of Φ_d would be only 10^{-8}!

Both aspects have been addressed by Gust, Moore and their co-workers who have synthesized and characterized an elegant series of covalently linked carotenoporphyrins (188-192) and carotenopheophorbides (179,193,194), similar in structure to 3B1. These systems demonstrate both light-harvesting and photoprotective functions. Specifically, the carotenoid moiety in 3B1 can transfer energy absorbed in the 450-550 nm region to the lower-lying excited singlet state of the porphyrin or chlorin electron donor. In addition, the short-lived carotenoid triplet state acts as a trap for donor triplet-state energy, which could otherwise induce photodegradation by energy transfer to molecular O_2 to yield the highly reactive singlet-oxygen species, $O_2(^1\Delta_g)$ (see, *e.g.*, ref. 195).

Another important consideration is that symmetric porphyrins have rather weak absorptions in their lowest allowed electronic transitions (196). In contrast, chlorophylls (197,198) and bacteriochlorophylls (199) have higher oscillator strengths in these bands (200), and would therefore be preferable from a spectroscopic standpoint. Covalently linked chlorins or bacteriochlorins are of interest also because their significantly red-shifted absorption enhances the theoretical thermodynamic conversion efficiencies (201-204). These pigments also possess a wide variety of photophysical (196) and electrochemical (205) properties, which allows "fine-tuning" of the energetics for photoinduced charge separation (2,3).

5.4 Future studies and recommendations

We believe there is a need for some "standard" reference D-A molecules that research groups working in this area can use to calibrate their own methods and techniques. Our candidates are molecules **1D1**, synthesized by Wasielewski and Niemczyk (80), or **1E2**, synthesized by Schmidt *et al.* (53). Both contain a free-base porphine donor and benzoquinone acceptor (see Table 1). The former incorporates a rigid triptycene bridge and has the further advantages that it is stable, crystalline, can be synthesized in gram quantities, and the ET rates have been measured in several solvents (81,83). The latter compound contains a simple amide link, which although less rigid than the spacer in **1D1**, nevertheless affords only limited geometric heterogeneity; this molecule is, in our view, the most well-characterized PQ system with respect to its photophysical (29,35,53) and electrochemical (29a) properties. Additional studies using different techniques and approaches would enhance the overall photophysical profiles of these interesting D-A systems. In this same context, we encourage all research groups to set aside quantities of their own molecules (particularly ones that are difficult to synthesize) for further study by others.

From the discussion presented in previous Sections, it seems clear that Marcus theory can explain most of the experimental results obtained to date. However, further experimental validation is required in order to sort out factors other than the Gibbs free-energy dependence, especially effects of distance and orientation. Thus there is a continuing need for studies of homologous series of linked D-A molecules in which a single factor is varied systematically. Whenever possible, the spectroscopy, photophysics and electrochemistry for each system should be studied in the same solvent (2,3). Only in this way can the rates of the forward and reverse reactions be characterized quantitatively with respect to the energetics.

Donors: Free-base and metalloporphyrins sometimes show quite disparate behavior with respect to understanding ET rates in terms of Marcus theory (see, *e.g.*, ref. 62). This may be due in part to effects on ΔG° by ligation of the central metal ion (perhaps by adventitious solvent impurities), or to changes in ligation during the excited-state lifetime (199). Whatever the reason, even when ΔG° values are designed to be similar by using appropriate acceptors, the ET rates for free-base and metalloporphyrins in the same solvent are often quite different.

Bridges: One cannot compare ET rates with systems containing the same donor and acceptor but greatly different types of bridges. Studies designed to examine, *e.g.*, effects of distance or orientation should be done with molecules containing bridges with similar groups. For example, hydrocarbon and ester links can be expected to show quite different superexchange effects. Even so, it is clear that our current understanding of through-bond ET is rudimentary, since two PQ systems with isomers of the same amide link show markedly different ET rates (see Table B).

Acceptors: Anthraquinone, naphthaquinone and benzoquinone do not constitute a homologous series of acceptors because, *inter alia*, of differences in orbital overlap (31). To date, the only systematic study of a homologous series of acceptors with the same donor and bridge has been that of Joran *et al.* (110). Also, as noted in Section 2.2, only in a very few cases has a reduced acceptor moiety been detected spectroscopically and its formation kinetics compared with those of the oxidized donor.

Solvents: Comparisons of ET rates for a given D-A system obtained in different solvents have rarely yielded satisfactory results. The reason most likely has to do with the fact that, even though necessary, it is not always possible to make accurate determinations of ΔG° and λ in every solvent. The most careful study to date has been that of Schmidt *et al.* (29) who used measurements of $\Delta G^{\circ\prime}$, ϵ_{op} and ϵ_{s} in binary solvent mixtures to calculate λ (see Fig. 4). We encourage more such studies in a variety of binary solvent mixtures.

There is also the major question regarding accurate estimations of ΔG_{corr} (eq. 13) (28,29). That is, the true $\Delta G^{\circ\prime}$ for a given ET process is not given simply by the sum of the redox potentials. For this reason, there is also a need for studies in some "standard" solvents for reference; we suggest benzonitrile, dichloromethane and benzene (or toluene) as examples of high-, medium- and low-polarity solvents, respectively. In any case, more studies of existing D-A molecules would likely yield rich information on the solvent dependence of PET processes.

Another reason that ET rates in different solvents are not always comparable may be due to dielectric relaxation (54-63). Breakdown of Marcus theory is to be expected when the rates become very fast (*e.g.*, $k_{et} \geqslant 5 \times 10^{9}\ s^{-1}$) because of limitations imposed by the solvent dynamics. In addition to temperature effects on $\Delta G^{\circ\prime}$, slower solvent relaxation is probably the major reason that PET rates are diminished at low temperatures in virtually every D-A system studied to date. However, these are problems that the photosynthetic organisms do not seem to experience.

6. Conclusions

The results of all of the intramolecular studies mentioned in this review demonstrate that these simple model systems continue to provide considerable insight into the fundamental mechanisms governing the rates of PET. Nevertheless, the fact that we continue to learn something from virtually every new intramolecular D-A system indicates that we are a long way from having optimized the parameters that affect light-induced electron transfer.

It is clear that the rate of the initial step is governed to some extent by the nature of the linking bridge between the two moieties. Thus, at the D-A distances required to inhibit the back reaction, through-bond ET will generally be faster than through-space interactions. However, the relative contributions of the bonding arrangement in the link (*e.g.*, superexchange) and the constraint that it imposes on the geometry are only beginning to be sorted out. As yet, we still do not know enough to predict the optimum distance and orientation for any given D-A pair, how these two factors depend on each other, or how they affect other molecular properties, especially photophysics and electrochemistry.

This is a very exciting field of research, and remarkable progress has been made over the past decade. The next ten years promise to be even more productive as our understanding of linked D-A complexes increases, particularly with respect to through-bond *vs.* through-space ET and medium effects in proteins and other polymeric matrices. We think the prospects are quite encouraging that a true, synthetic photoreaction center for artificial photosynthesis will be developed before the end of the century.

Acknowledgments

We wish to thank Drs. D.N. Beratan, D. Gust, S. Larsson, R.A. Marcus, D.C. Mauzerall, T.A. Moore, H.-T. Tien, J.W. Verhoeven, J.M. Warman and M.R. Wasielewski for reviewing the manuscript and for many helpful comments. We are also grateful to Dr. Roberta Ross and the staff of the Radiation Chemistry Data Center, University of Notre Dame Radiation Laboratory, for their generous and reliable assistance during our many forays into their extensive bibliographic data base. We are happy to acknowledge the very special assistance of the Academic Computing Services staff at the University of Colorado at Boulder and the staff of the Computing and Communications Services Center at The University of Western Ontario, who facilitated the flow of our voluminous BITNET correspondence, without which this review would have been much more difficult to write. Preparation of this review was supported by the Division of Chemical Sciences, Office of Energy Research, U.S. Department of Energy (JSC) and an Operating Grant from the Natural Sciences and Engineering Research Council of Canada (JRB). Finally, we gratefully acknowledge an Associated Western Universities grant to JRB, which helped make this aspect of our long collaboration possible.

TABLE 1. SINGLY LINKED DONOR-ACCEPTOR MOLECULES

These molecules have the general structure: D–B–A, where B is a bridge linking the donor D and the acceptor A.

Molecule Symbol	Structure D^b	B^b	A^b	k_{et}^{f} [a] $/10^9\ s^{-1}$	k_{et}^{r} [a] $/10^9\ s^{-1}$	Solvent
1A1	Pa	-	Qa	~30	~0.1	CH_2Cl_2
1A2	Px	-	tetra-Qa	-	-	-
1A2	Px	-	tetra-Qa	-	-	C_6H_6
1A3	Px	-	tetra-Nc	-	-	"
1A4	Px	-	tetra-Nd	-	-	"
1A5	Px	-	tetra-Ca	-	-	"
1A6	Pa	-	Qp	-	-	CH_2Cl_2
1B1	Pm	Aa	Qa	5.8	-	C_6H_6
1B2	Pm	Ab	Qa	0.015	-	"
1B3	Pm	Ac	Qa	<0.004	-	"
1B2,3	Pm	Ab,c	Qa	-	-	-
1B4	Pp	Aa	Qa	220	-	C_6H_6
1B5	Pp	Ab	Qa	1.8	-	"
1B6	Pp	Ac	Qa	<0.02	-	"
1B5	Pp	Ab	Qa	5.0	-	CH_3CN
1B6	Pp	Ac	Qa	<0.008	-	"
1B7	Pp	Ab	Qb	0.59	-	CH_3CN
1B8	Pp	Ab	Qc	1.01	-	"
1B9	Pp	Ab	Qd	5.00	-	"
1B10	Pp	Ab	Qe	4.03	-	"
1B11	Pp	Ab	Qf	7.65	-	"
1B12	Pp	Ab	Qg	2.37	-	"
1B13	Pb	Ab	Qa	0.015	-	CH_2Cl_2

a. k_{et}^{f} and k_{et}^{r} represent k_{et}^{S} and k_{r}^{S}, respectively, unless ET occurs from the triplet state of the photoexcited chromophore. In this case the entries in these columns denote, respectively, k_{et}^{T} and k_{r}^{T} (or $k_{r}^{T} + k_{-et}^{T}$). (See Fig. 1).

Techniques	Comments	References
d,i	ET from triplet state; also studied porphyrin with four quinones at the *meso*-positions	105,106
		" "
a,f		72
a-c,h	Also studied several other *meso*-	107
"	substituted derivatives	"
"	" " " "	"
"	" " " "	"
b,c,f	No diminution of fluorescence lifetimes in two solvents	108
a,f	Also some data in butyronitrile	109
"	" " " " "	"
"	" " " " "	"
j	Theoretical analysis of ET through bridge orbitals	94
a,f	Also some data in butyronitrile	109
"	" " " " " " "	"
"	" " " " " " "	"
b,i	ET rates measured in 3 other solvents;	93
"	evidence for PET in 1B5 at 77 K	"
b,i	ET rates measured in 3 other solvents	110
"	" " " " " " " " "	"
"	" " " " " " " " "	"
"	" " " " " " " " "	"
"	" " " " " " " " "	"
"	" " " " " " " " "	"
b	ET rates measured in 7 other solvents	103

b. Structure fragments are shown in Appendices 1, 2 and 3 for Donors (D), Bridges (B) and Acceptors (A), respectively.
c. See Appendix 4 for code to Techniques.

TABLE 1. (cont'd)

Molecule Symbol	Structure D	B	A	k^f_{et} /10⁹ s⁻¹	k^r_{et} /10⁹ s⁻¹	Solvent
1C1	Pb	Aa	Qa	50	-	CH_3CN
1C2	Pb	Ad	Qa	5.7	-	"
1D1-6	Pa,d	Ae	Qs-u	-	-	CH_2Cl_2
1D1	Pa	Ae	Qs	110	3.6	C_3H_7CN
1D2	Pa	Ae	Qt	200	3.1	"
1D4	Pd	Ae	Qs	250	170	"
1D5	Pd	Ae	Qt	170	130	"
1D6	Pd	Ae	Qu	~48	~3.7	"
1D6	Pd	Ae	Qu	43	5.4	C_3H_7CN
1D7	Pd	Af	Qu	5.2	1.5	"
1D8	Pd	Ag	Qu	0.24	0.10	"
1D9	Pheo[d]	Ae	Qs	20	2.0	C_3H_7CN
1D10	" "	Ae	Qt	71	6.7	"
1D11	Chl[e]	Ae	Qs	200	110	"
1D12	"	Ae	Qt	400	130	"
1E1	Pa	Bh	Qa	-	-	CH_3Cl
1E2	Pb	Ba	Qa	0.81	-	CH_2Cl_2
1E2	Pb	Ba	Qa	-	-	various
1E2	Pb	Ba	Qa	0.41	0.16	C_6H_5CN
1F1	Pc	Bc	Qa	9.9	-	CH_2Cl_2
1F2	Pb	Bd	Qa	0.74	-	"
1F3	Pc	Be	Qa	0.56	-	"
1F4	Pc	Bf	Qa	0.20	-	"
1F5	Ps	Bm	Qa	-	-	-
1F6	Ps	Bn	Qa	-	-	-
1F7	Ps	Bo	Qa	-	-	-

d. Pyropheophorbide a (similar to donor in 3D2) but with bridge linked at the 2-vinyl group.

TABLE 1. (cont'd)

Techniques	Comments	References
b,f	ET rates measured in 5 other solvents	95
"	" " " " " " " "	"
a,d,f,i		80
a,d,e	Evidence for Marcus inverted region	81,82
"	in reverse ET rates; also measured	" "
"	rates in toluene	" "
"		" "
"		" "
a,b,d,i	ET rates decrease exponentially with	84
" "	increasing donor-acceptor distance	"
" "		"
a,c,d,	Reverse ET rates show further evidence	111
f,h,i	for Marcus inverted region as in 1D6	"
" "		"
" "		"
a,f	Quenching of fluorescence intensities	73
b,f	ET rates measured in 21 other solvents	53
b,i	ET rates correlate with Marcus theory	29,29a
b,d,e	ET also observed from the triplet state	35
	and reverse ET to the ground state	
b,f,k	Also studied triads with linked	104
b,f	carotenoid (see Table 3)	"
"	" " " " "	"
"	" " " " "	"
n	Photoelectric response in bilayer	112,113
n	lipid membranes; also studied a	" "
n	porphyrin-diquinone molecule	" "

e. Pyrochlorophyllide a linked as above.

TABLE 1. (cont'd)

Molecule Symbol	Structure D	B	A	k_{et}^{f} $/10^9$ s^{-1}	k_{et}^{r} $/10^9$ s^{-1}	Solvent
1G1	Pe	Ch	Qp	~0.3	-	C_3H_7CN
1G2	Pb	Ch	Qp	-	-	various
1H1	Pb	Ca	Qa	-	-	C_2H_5OH
1H2,3	Pb,e	Cb	Qa	-	-	CH_2Cl_2
1H2	Pb	Cb	Qa	-	-	CH_3OH
1H1,2	Pb	Ca,b	Qa	-	-	CH_2Cl_2
1H3,4	Pe	Ca,b	Qa	-	-	"
1H1-4	Pb,e	Ca,b	Qa	-	-	CH_3CN
1I1-3	Pb	Cd-f	Qa	-	-	-
1I1-3	Pb	Cd-f	Qa	-	-	CH_2Cl_2
1I2	Pb	Ce	Qa	0.41	-	"
1I2	Pb	Ce	Qa	-	-	mTHF
1I1-3	Pb	Cd-f	Qa	-	-	CH_2Cl_2
1I1-3	Pb	Cd-f	Qa	-	-	CH_2Cl_2
1J1	Pb	Da	Qa	-	-	-
1J1	Pb	Da	Qa	-	-	-
1J2	Pb	Da	Qo	-	-	-
1J3	Pb	Dc	Qa	-	-	-
1J4	Pb	Dc	Qb	-	-	-

TABLE 1. (cont'd)

Techniques	Comments	References
a,b,d	ET rates measured in 8 other solvents; fluorescence yields do not reflect lifetimes in polar solvents	31,32
a,b,h	No diminution of fluorescence lifetimes in 9 solvents; quenching of fluorescence yields is solvent dependent	31,32
f,h	First linked porphyrin-quinone molecule	74
"		75
g	EPR spectra and kinetics	78
a-c,g,h	Fluorescence lifetimes non-exponential;	76
a-c,g	also observed EPR signals in vesicles	"
a-c,g-i	Observed long-lived EPR spectrum of of PET product in lipid vesicles	77
c,f,h		79
a,b	Complexed and extended conformers	30
a-c,e,h	" " " " " " "	32b,33
e,g	EPR and optical spectra of triplets; two sets of spectra and lifetimes	34
g,h	EPR and optical spectra	39
i	Redox potentials of linked molecules	79a
f	Photochemical synthesis	114
a,f,i	" " " "	115
"	"; also Cl, Br and methyl analogs of Qo	"
"		"
"		"

TABLE 1. (cont'd)

Molecule Symbol	Structure D	B	A	k^f_{et} /10^9 s^{-1}	k^r_{et} /10^9 s^{-1}	Solvent
1K1-6	Po,r	Eb,d,f	Qa	-	-	-
1K1-3	Po	Eb,d,f	Qa	-	-	C_6H_6,CH_3CN
1K1	Po	Eb	Qa	~100	<50	C_6H_6
1K2	Po	Ed	Qa	14	7.7	"
1K3	Po	Ef	Qa	1.7	1.7	"
1K7	Pn	Ed	Qa	8.6	-	CH_3CN
1L1	Pz	Ee	Qa	17	-	CH_3CN
1M1-9	Pa,d,f	Da-c	Na	-	-	toluene
1N1-12	Pd	Dd-f (n=3,4,6,8)	Va	-	-	-
1N13-33	Pd	Dd-f (n=3-6,8,10,12)	Vc	-	-	-
1N34	Pd	Dd(n=4)	Vb	-	0.00091	CH_3CN/H_2O
1N35	Pd	Dd(n=6)	Vb	-	0.0017	" "
1N36	Pd	Dd(n=8)	Vb	-	0.0022	" "
1N34-36	Pd	Dd (n=4-8)	Vb	-	-	H_2O
1N37	Px'	Dg (n=2-5)	Va	-	-	H_2O
				-	-	H_2O
1N38-42	Pe	Dg (n=2-6)	Va	-	-	H_2O
1O1	Pt	Ec	Va	~25	~25	DMSO
1O2	Pt	Ei	Va	1.1	~0.0067	"
1O3	Pt	Ej	Va	0.7	~0.0067	"

TABLE 1. (cont'd)

Techniques	Comments	References
a,f,h,k		116
b,d	Approximate lifetimes determined	117
"	Biphasic decay of excited singlet state	118
"	" " " " " " " " "	"
d	" " " " " " " " "	"
b,d	" " " " " " " " "	"
b,f	Non-exponential fluorescence decay	119
a,c,f-h	Observed light-induced EPR spectra	121
a,b,f	Also studied molecules with four viologen groups	122
a,b,f		122
a,e	ET occurs from porphyrin triplet state	123
"	" " " " " " " " "	"
"	" " " " " " " " "	"
e	Reverse ET rates affected by magnetic fields	124
a,c,e,f	Triplet lifetimes diminished with linked Va vs. reference compound	125
e	H_2 evolution reported; ET occurs from porphyrin triplet state	126,127
e,f	H_2 evolution reported only for 1N38	128
b,d	2 acceptor groups; non-exponential	129
"	fluorescence decay	"
"	" " "	"

TABLE 1. (cont'd)

Molecule Symbol	Structure D	B	A	k^f_{et} /10^9 s^{-1}	k^r_{et} /10^9 s^{-1}	Solvent
1P1	Da	-	Aa	-	-	CH_3CN
1P2-4	Da	Ea-c	Aa	-	-	"
1P1	Da	-	Aa	-	-	various
1P2-4	Da	Ea-c	Aa	-	-	"
1P5-7	Da	Ea-c	Ab	-	-	"
1P4	Da	Ec	Aa	1.1	<0.02	*n*-hexane
1P4	Da	Ec	Aa	-	-	*i*-pentane
1P4	Da	Ec	Aa	140	-	CH_3CN
1P4	Da	Ec	Aa	-	-	gas jet
1P8	Da	Em	Aa	7.6	-	CH_3CN
1P9	Da	En	Aa	0.34	-	"
1P10	Da	Em	Ab	0.83	-	"
1P11	Da	En	Ab	0.03	-	"
1P12	Da	Eo	Ab	1.1	-	CH_3CN
1P13	Da	Ep	Ab	0.47	-	"
1P10-13	Da	Em-p	Ab	-	-	C_3H_7CN $C_3H_6(OH)_2$
1Q1	Ab	Bp	Nd	-	-	various
1Q2	Ac	Bp	Nd	-	-	"
1Q3	Ma	Bp	Nd	-	-	"
1Q4	An	Bp	Nd	-	-	"
1Q5	Aa	Bp	Nd	-	-	"
1R1-3	Aa,m,n	Ea	Pa	-	-	H_2O and
1R4-15	Ma-l	Ea	Pa	-	-	C_2H_5OH
1R16,17	Aa,Ma	Eb	Pa	-	-	"
1R18-20	Me,g,h	Eb	Pa	-	-	"
1R21-23	Mi,k,l	Eb	Pa	-	-	"
1R24	Af	Ec	Nb	-	-	CH_3Cl

TABLE 1. (cont'd)

Techniques	Comments	References
a-c,h	Measured lifetimes of anthracene fluor-	65
"	escence and CT state in 9 solvents	"
a,c,h	Solvent dependence of CT emission	130
"	in 31 solvents	"
"		"
c,d,f,h	First ps study of intramolecular ET	131
c,d	Two CT states detected	132,133
d	Extended and folded conformations	89
c,d	Formation of CT state in a cooled jet	134
b	ET rates also measured in hexane	90,91
"	" " " " " " " "	" "
"	" " " " " " " "	" "
"	" " " " " " " "	" "
b	ET rates also measured in hexane	91,92
"	" " " " " " " "	" "
b	ET rates limited by solvent dielectric	61
"	relaxation at low temperatures	"
a,c	Study of CT emission	135
"	" " " "	"
"	" " " "	"
"	" " " "	"
"	" " " "	"
c,h	CT absorption bands correlated with	64
"	with ionization potential of donor	"
"	" " " " " " "	"
"	" " " " " " "	"
"	" " " " " " "	"
h,k	Shows a discrete CT absorption	136

TABLE 1. (cont'd)

Molecule Symbol	Structure D	B	A	k^f_{et} $/10^9\ s^{-1}$	k^r_{et} $/10^9\ s^{-1}$	Solvent
1R25-27	Da	Ea,c,d	Rc	-	-	CH_2Cl_2
1R28	Ad	Ea	Rc	-	-	"
1R29	Ae	Ec	Rc	-	-	"
1R30,31	Ma	Ea-d	Ra	-	-	CH_2Cl_2
1R32,33	Da	Ec	Ra,b	-	-	"
1R34-37	Ma	Ea-d	Ra	-	-	"
1R38-41	Ma	Eq-t	Pb	-	-	C_2H_5OH
1R42-54	Ma	Fa-f,h-n	Pb	-	-	"
1R55-58	Dc	Eb-d,g	Ra	-	-	cyclo-hexane
1S1,2	Ma,e	Eb	Ec	-	-	various
1S3-5	Ma	Gh-j	Ec	-	-	"
1S6	Mp	Gg	Ec	>100	0.053	Et-O-Et
1S7	Mp	Gg	Eb	0.066	-	CH_3CN
1S8	De	Gk	Ea	-	-	*n*-hexane
1S9	Ea	Gf	Ea	-	-	" "
1S10,11	Eb	Gl	Ea,f	-	-	" "
1S12,13	Eb	Gf	Ea,g	-	-	" "
1S14,15	Db,c	Gf	Ej	-	-	2-propanol
1S16	Db	Gf	Ek	-	-	" "
1S17	Db	Gf	Eh	0.17	0.0017	dioxane
1S18	Db	Gf	Ei	0.08	0.0016	" "
1S6	Mp	Gg	Ec	>100	-	CH_3CN
1S7	Mp	Gg	Eb	0.066	-	"
1S19	Mp	Gi	Ea	>100	-	"
1S20	Mp	Gi	Ed	>100	-	"

TABLE 1. (cont'd)

Techniques	Comments	References
h	CT bands most pronounced with	137
"	Ea and Ec bridges	"
"	" " " "	"
a,h	Strong fluorescence quenching	138
"	with Ea and Eb bridges	"
"	CT absorption bands in all compounds	"
h,k	Detailed study of conformational	139,140
"	dependence for ET from CT bands	" "
a-c,e,h	Fluorescence strongly quenched	141
	except with Eg bridge	
c,h	CT absorption and emission bands observed	66
"	" " " " " " " " "	"
a,b	Weak CT emission observed; some	142
"	studies in methanol	"
c,h	CT absorption and emission bands observed	143
"	" " " " " " " " "	"
"	" " " " " " " " "	"
"	" " " " " " " " "	"
a,c,k,m	CT emission interpreted in terms of	144
" "	a trichromophoric model	"
a,b,f,h,l	CT state has a dipole moment of 25 D;	145
" "	rate data in 9 other solvents	"
a,h	ET data in 4 other solvents	146
"	" " " " " " "	"
"	" " " " " " "	"
"	" " " " " " "	"

TABLE 1. (cont'd)

Molecule Symbol	Structure D	B	A	k^{f}_{et} /10^9 s^{-1}	k^{r}_{et} /10^9 s^{-1}	Solvent
1S21	Mr	Ga	Ea	>50	0.53	Bu-O-Bu
1S22	Mr	Gb	Ea	>50	0.29	" "
1S23	Mr	Gc	Ea	15	-	" "
1S21	Mr	Ga	Ea	>100	~1	C_6H_6
1S22	Mr	Gb	Ea	>100	0.17	"
1S23	Mr	Gc	Ea	52	0.031	"
1S24	Mr	Gd	Ea	7.2	0.0028	"
1S25	Mr	Ge	Ea	0.73	0.0014	"
1S21-23	Mr	Ga-c	Ea	-	-	-
1T1	Dg	-	Ac	2.5	-	CH_3OH

TABLE 1. (cont'd)

Techniques	Comments	References
b,i	CT emission also observed	88
"	" " " " " "	"
"	" " " " " "	"
b,f,h,i,l	ET rates measured in 6 other solvents;	36,37,85-87
" "	reverse ET measured by transient	" " " "
" "	microwave conductivity; CT emission	" " " "
" "	observed; delayed donor fluorescence	" " " "
" "	also observed for **1S25**	" " " "
j	Good correlation of theory with experimental ET rates	147,148
b	ET rates correlate with solvent	54-57
	dielectric relaxation rates	" "

TABLE 2. CAPPED DONOR-ACCEPTOR MOLECULES

These molecules have the general structure:

```
      A
     / \
   B1   B2
     \ /
      D
```

where B1 and B2 are bridges between donor D and acceptor A.

Molecule Symbol	Structure D	B1	B2	A	k_{et}^{f} [a] /$10^9 s^{-1}$	k_{et}^{r} [a] /$10^9 s^{-1}$
2A1	Ms	Ee	Ee	Rd	-	-
2A2	Ms	Ef	Ef	Rd	-	-
2A3	Ms	Ed	Eh	Rd	-	-
2A1,2	Ms	Ee,f	Ee,f	Rd	-	-
2A4	Mt	Ec	Eg	Re	-	-
2A5	Mu	Ed	Ee	Rd	-	-
2A6	Ap	Ef	Ef	Rd	0.18	-
2A6	Ap	Ef	Ef	Rd	0.04	-
2A7	Ap	Ec	Eh	Rd	0.04	-
2A8	Ms	Ef	Ef	Re	-	-
2A9	Ms	Eh	Ed	Re	-	-
2B1	Pu	Cm	Cm	Qr	-	-
2B2	Pu	Cn	Cn	Qr	-	-
2B3	Pv	Cm	Cm	Qr	-	-
2B4	Pv	Cn	Cn	Qr	-	-
2B1-4	Pu,v	Cm,n	Cm,n	Qr	-	-
2B1-6	Pu-w	Cm,n	Cm,n	Qr	-	-
2B2	Pu	Cn	Cn	Qr	250	50
2B4	Pv	Cn	Cn	Qr	285	50

a. See Table 1.

Solvent	Techniques	Comments	References
CH_2Cl_2	h	CT absorption bands	149
"	"	observed	"
"	"		"
-	f	Full synthesis and	150
-	"	characterization	"
-	"	reported; plus 19	"
		other compounds of	
		the same type	
CH_2Cl_2	a,b,c,h	CT absorption and	151
cyclo-	" "	emission studied	"
hexane	" "	" " " "	"
-	h	Polarized CT absorp-	152
-	"	tion spectra of	"
		single crystals	
$CDCl_3$	f,k		153,154
"	"		" "
"	"		" "
"	"		" "
$CDCl_3$	a,k	Quinone found to be	155,156
CH_3CN	"	perpendicular to	" "
		porphyrin ring	
CH_2Cl_2	d,i	Evidence for inverted	157
"	"	region; four other	"
		solvents studied	

TABLE 2. (cont'd)

Molecule Symbol	Structure D	B1	B2	A	k^{f}_{et} $/10^{9}\,s^{-1}$	k^{r}_{et} $/10^{9}\,s^{-1}$
2C1,2	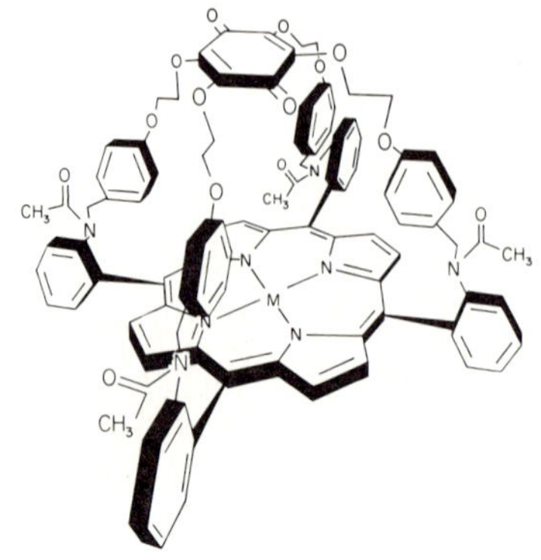 2C1: M = H_2; 2C2: M = Zn				-	-
2C2					1.5, 0.3	>0.03
2C2					0.0067	0.000
2C2					1.9, 0.6	-
2D1	Py	Ee	Ee	Qr	-	-
2D2	Py	Ef	Ef	Qr	-	-
2E1					-	-
2E1					2.0	-
2E1					2.2	-

TABLE 2. (cont'd)

Solvent	Techniques	Comments	References
CH_3CN	a,f	Fluorescence quenching in Zn compound but not in free-base	158
CH_3CN	b	Two ET processes from singlet state	159,160
CH_3CN	e	ET from triplet state; reverse ET very slow	" "
DMA:EtOH* 1:4	b	Negligible activation energy (80-300 K)	160
$CDCl_3$	f,k		163
"	"		"
	f		165
toluene	b		167
MeOH:EtOH 1:4	"		"

*DMA = dimethylacetamide

TABLE 2. (cont'd)

Molecule Symbol	Structure D	B1	B2	A	k^{f}_{et} $/10^{9}\,s^{-1}$	k^{r}_{et} $/10^{9}\,s^{-1}$
2F1		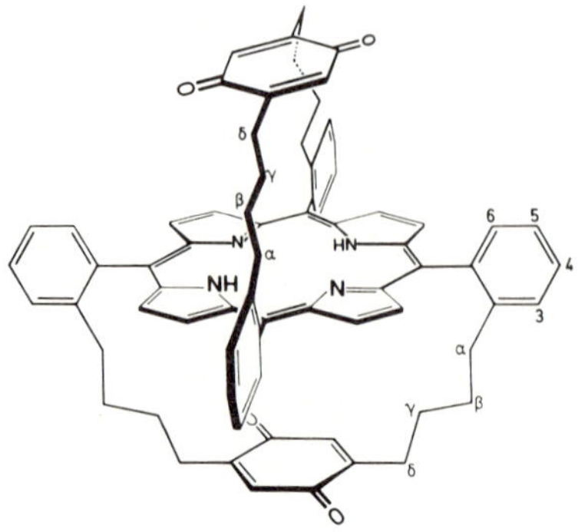			-	-
2F1					10.5	-
2F1					-	-
2G1		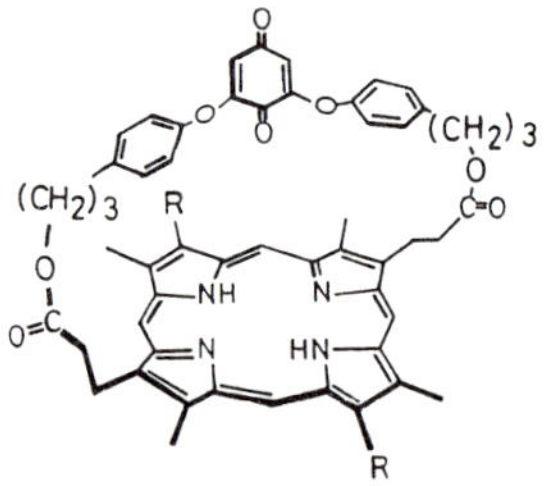			-	-
2G2		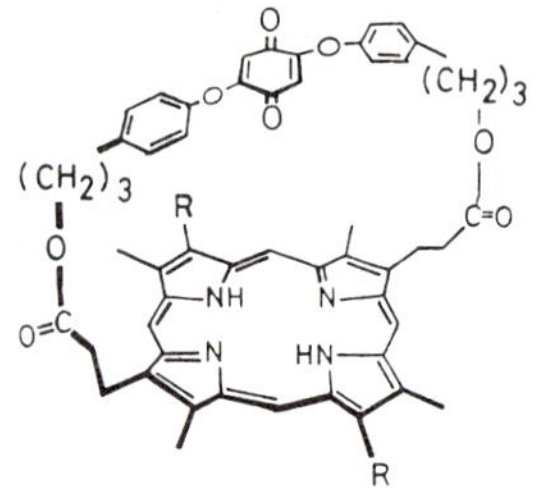			-	-
2G3		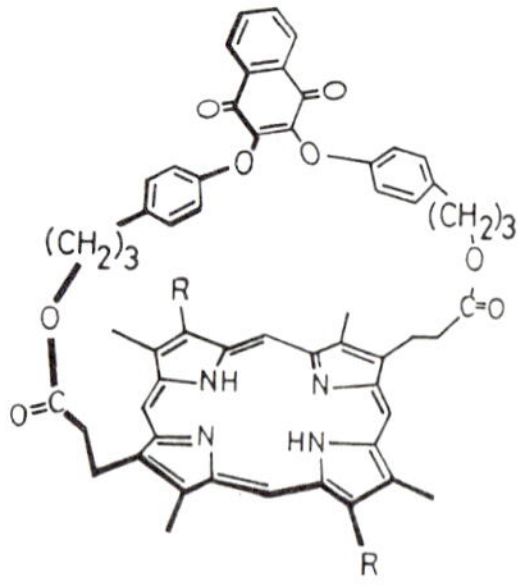			-	-

TABLE 2. (cont'd)

Solvent	Techniques	Comments	References
toluene	a,f		164,165
MeOH:EtOH 1:4	"	Activation energy = 2 kJ/mol	" "
benzene	a,c,m	X-ray structure shows that porphyrin and quinones are coplanar at a distance of 3.4 Å; fluorescence quenching also observed	166
-	f	Photochemical synthesis	168
-	f	Photochemical synthesis	168
-	f	Photochemical synthesis	168

TABLE 2. (cont'd)

Molecule Symbol	Structure D	B1	B2	A	k_{et}^{f} /10^9 s^{-1}	k_{et}^{r} /10^9 s^{-1}
2H1,2	Pu	Cn,o	Cn,o	Vd	-	-
2H3,4	Pu	Cn,o	Cn,o	Ve	-	-
2H3-8	Pu,w	Cm-o	Cm-o	Ve	-	-
2H9,10	Pv	Cn,o	Cn,o	Ve	-	-
2H1	Pu	Cn	Cn	Vd	333	82
2H2	Pu	Co	Co	Vd	143	21
2I1-3	Pu	Cm,n,p	Cm,n,p	Re	-	-
2I4-6	Pv	Cm,n,p	Cm,n,p	Re	-	-
2I7-9	Pw	Cm,n,p	Cm,n,p	Re	-	-
2I2	Pu	Cn	Cn	Re	50-800	4-100
2I1	Pu	Cm	Cm	Re	>600	42
2I2	Pu	Cn	Cn	Re	>600	67
2I3	Pu	Co	Co	Re	325	94
2I4	Pv	Cm	Cm	Re	200	30
2I5	Pv	Cn	Cn	Re	250	21
2I6	Pv	Co	Co	Re	430	24
2I7	Pw	Cm	Cm	Re	120	19
2I8	Pw	Cn	Cn	Re	175	15
2I9	Pw	Co	Co	Re	470	45

TABLE 2. (cont'd)

Solvent	Techniques	Comments	References
CH_3CN	a,f,k		169
"	"		"
$CDCl_3$	f,k	Intramolecular	170,172
"	"	coordination between bipyridyl and metal	
CH_3CN	d	Evidence for Marcus	157
CH_2Cl_2	"	inverted region	"
CH_2Cl_2	a,f,k	Fluorescence quench-	171
"	"	ing decreases with	"
"	"	increasing distance	"
various	a,d,h	Strong fluorescence quenching reported	173
CH_2Cl_2	a,b,d,i	ET rates measured in	62
"	" "	7 other solvents;	"
"	" "	temperature depend-	"
"	" "	ence reported; evi-	"
"	" "	dence for ET rates	"
"	" "	limited by solvent	"
"	" "	dielectric relaxation	"
"	" "	at low temperatures	"
"	" "	" " " "	"

TABLE 2. (cont'd)

Molecule Symbol	Structure D	B1	B2	A	k_{et}^{f} $/10^9 s^{-1}$	k_{et}^{r} $/10^9 s^{-1}$
2J1	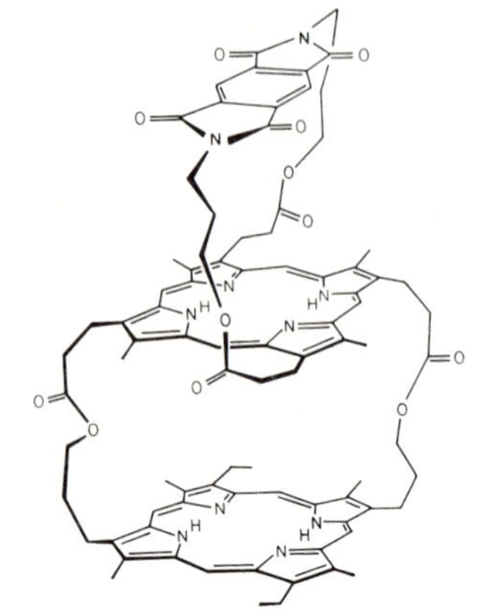				100-400	9-78
2K1	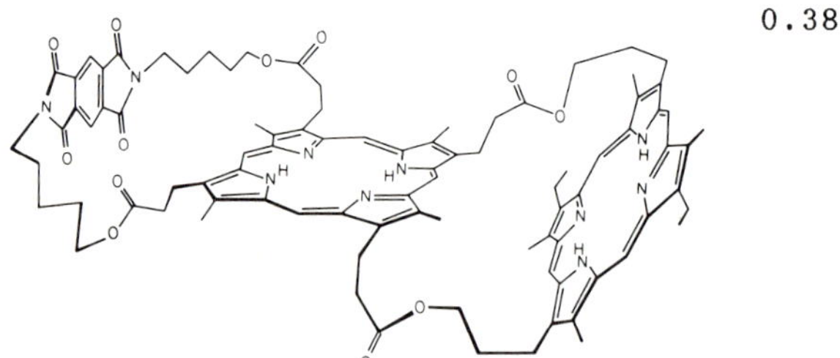				0.38-0.65	<0.01

TABLE 2. (cont'd)

Solvent	Techniques	Comments	References
various	d,f,h		173
various	d,f,h		173

TABLE 3. TRIAD DONOR-ACCEPTOR MOLECULES

Molecule Symbol	Structure	$k^f_{et}(1)^a$ $/10^9 s^{-1}$	$k^f_{et}(2)^a$ $/10^9 s^{-1}$
3A1	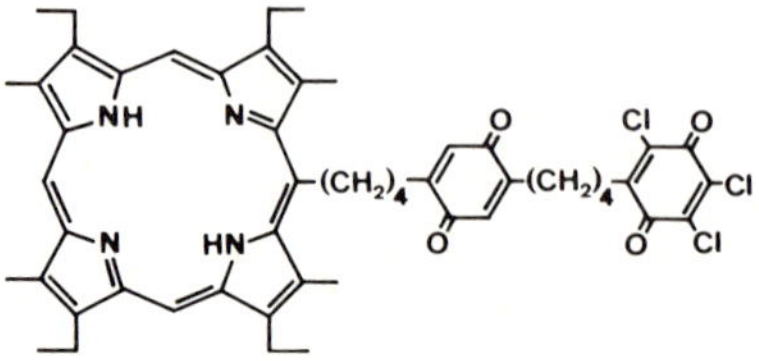 NH, N, HN, N, $(CH_2)_4$, O, O, $(CH_2)_4$, Cl, O, Cl, O, Cl	>30 - -	- - -
3B1	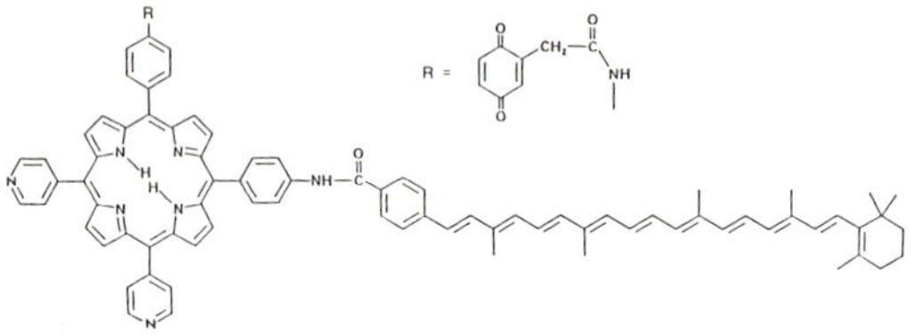 R = [quinone] CH_2—C(O)—NH; —NH—C(O)—	- -	~10 -
3Cn	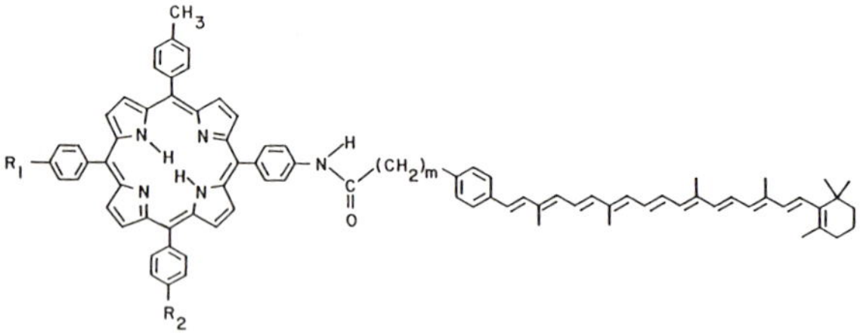 CH_3; R_1; N—H; $(CH_2)_m$; C=O; R_2		
3C1	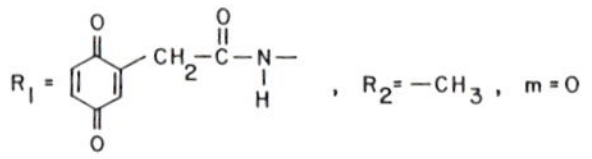 R_1 = [quinone] CH_2—C(O)—N(H)—, R_2 = —CH_3, m = 0	9.7	-
3C2	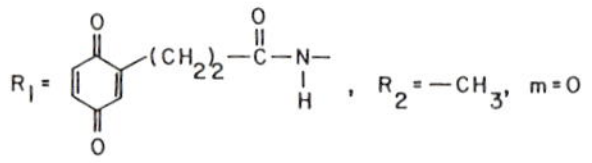 R_1 = [quinone] $(CH_2)_2$—C(O)—N(H)—, R_2 = —CH_3, m = 0	8.3	-
3C3	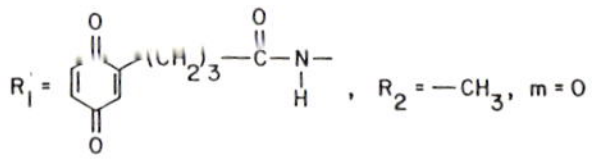 R_1 = [quinone] $(CH_2)_3$—C(O)—N(H)—, R_2 = —CH_3, m = 0	3.9	-
3C4	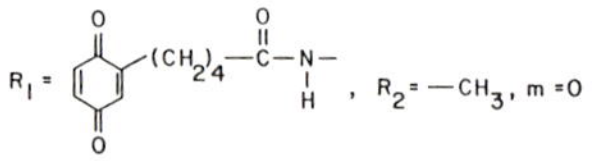 R_1 = [quinone] $(CH_2)_4$—C(O)—N(H)—, R_2 = —CH_3, m = 0	1.5	-
3C5	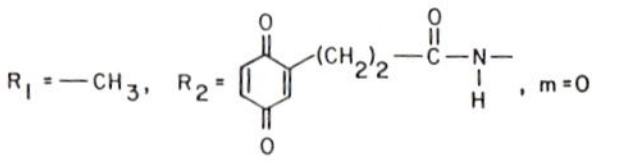 R_1 = —CH_3, R_2 = [quinone] $(CH_2)_2$—C(O)—N(H)—, m = 0	6.9	-
3C6	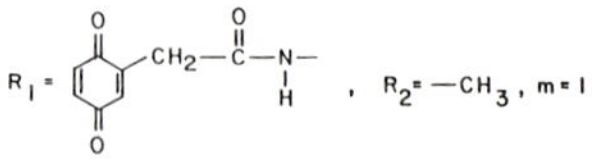 R_1 = [quinone] CH_2—C(O)—N(H)—, R_2 = —CH_3, m = 1	-	-

a. All rate constants involve singlet-state processes; $k^f_{et}(1)$ and $k^f_{et}(2)$ represent the two forward steps and k^r_{et} is for charge

k_{et}^{r} [a] /$10^9 s^{-1}$	Solvent	Techniques	Comments	References
3.3	C_6H_6	d,f		174,118
3.3	dioxane	"		" "
2.5	THF	"		" "
0.0040	CH_2Cl_2	d-f		175-176b
-	-	n	transmembrane electron transfer	178
0.0034	CH_2Cl_2	b,e,d,k	Activation energy = 11.3 kJ mol^{-1}; distance dependence of rate constants obtained from NMR structural analysis	177,103
0.0035	"	" "		" "
0.0032	"	" "		" "
0.0030	"	" "		" "
0.0035	"	" "	" "	" "
0.00099	"	" "	" "	" "

a. (cont'd) recombination of the RIP to the ground state.

TABLE 3. (cont'd)

Molecule Symbol	Structure	$k^f_{et}(1)$ /$10^9 s^{-1}$	$k^f_{et}(2)$ /$10^9 s^{-1}$
3D1	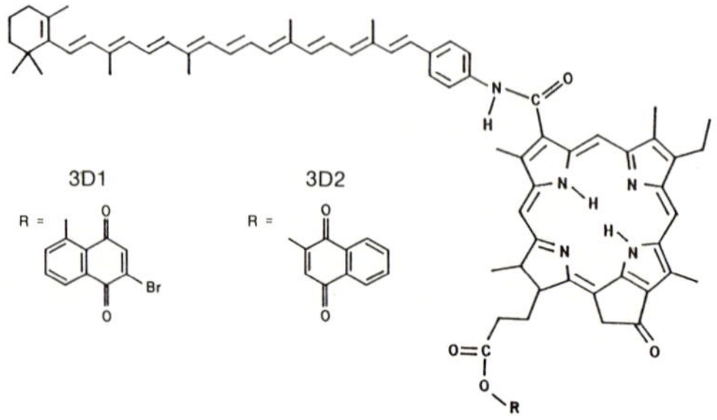	-	-
3D2		-	-
3E1	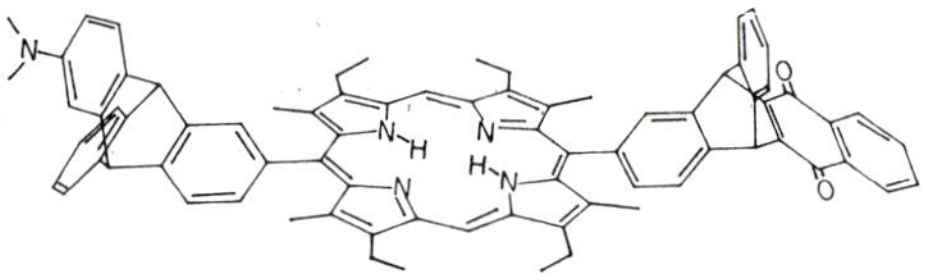	110	14

TABLE 3. (cont'd)

k^r_{et} /$10^9 s^{-1}$	Solvent	Techniques	Comments	References
0.0083	CH_2Cl_2	d,e	Also observed energy transfer	179
0.0083	"	"	from carotenoid to porphyrinic macrocycle	"
0.00041	C_3H_7CN	d,e,i		82,83

Appendix 1. Donors

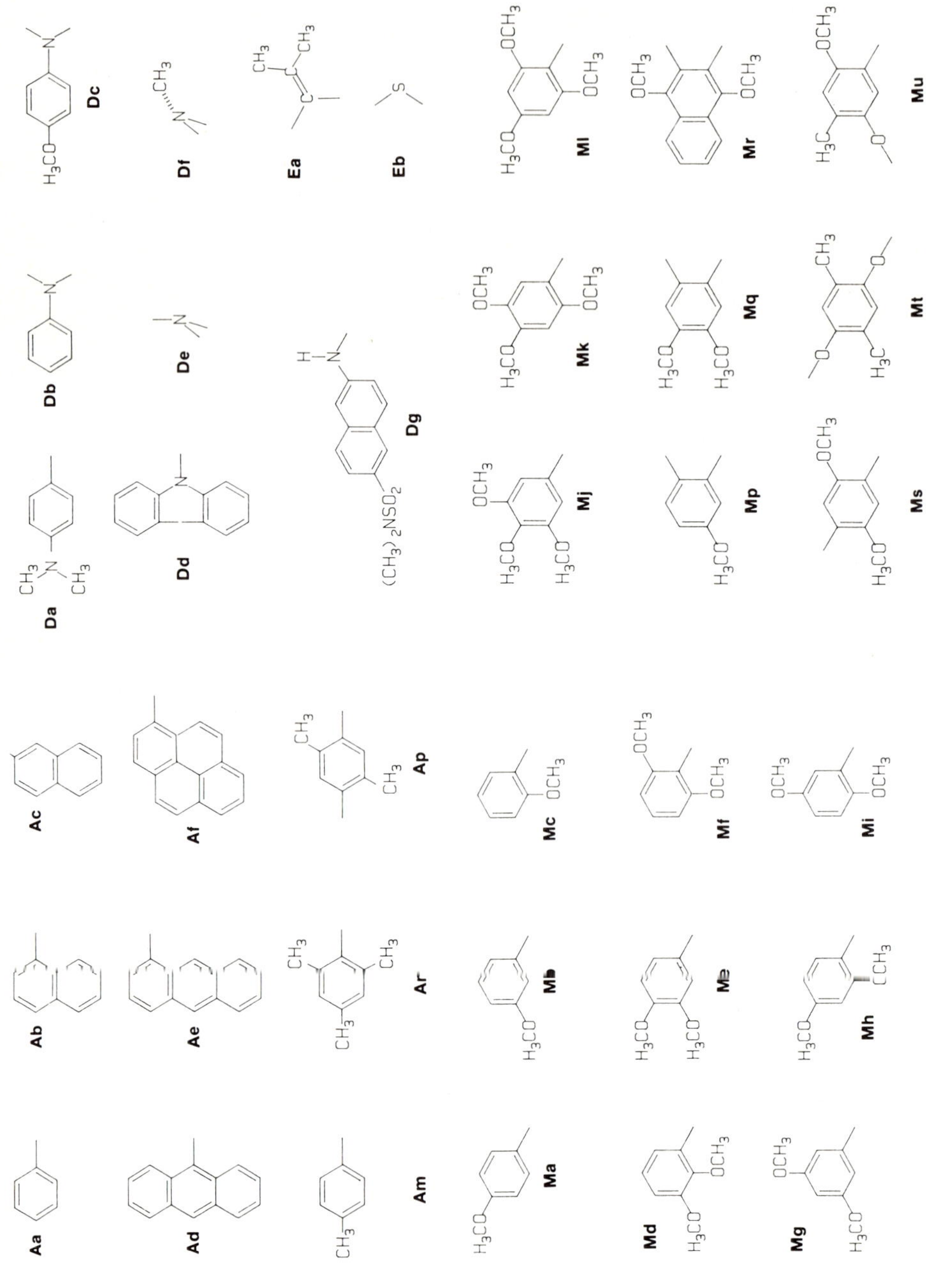

Aa Ab Ac Da Db Dc

Ad Ae Af Dd De Df

Am Ar Ap Dg Ea Eb

Ma Mb Mc Mj Mk Ml

Md Me Mf Mp Mq Mr

Mg Mh Mi Ms Mt Mu

Appendix 1. Donors (cont'd)

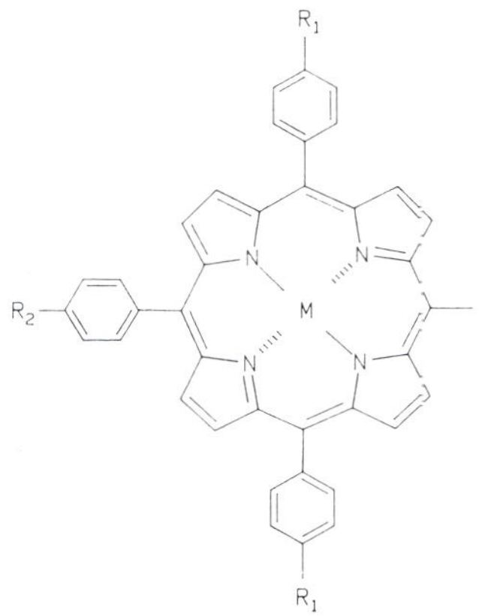

Pa: $R_1 = R_2 = H; M = H_2$
Pb: $R_1 = R_2 = CH_3; M = H_2$
Pc: $R_1 = CH_3; R_2 = NH_2; M = H_2$
Pc: $R_1 = R_2 = H; M = Zn$
Pe: $R_1 = R_2 = CH_3; M = Zn$
Pf $R_1 = R_2 = H; M = Cu$

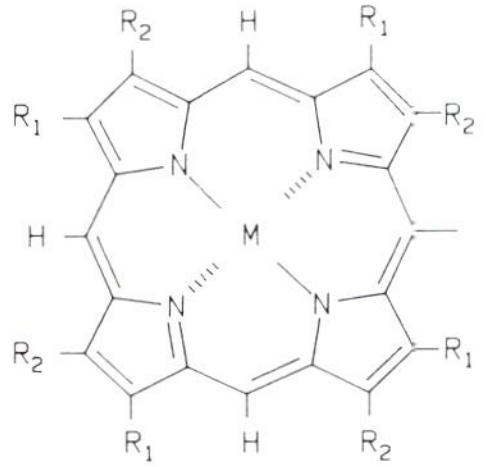

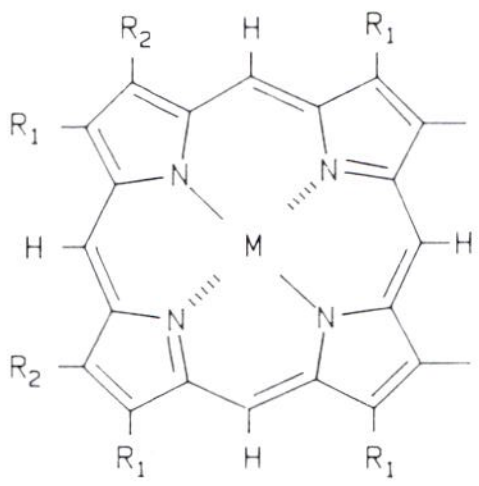

Pt: $R_1 = CH_3; R_2 = C_2H_5; M = Zn$

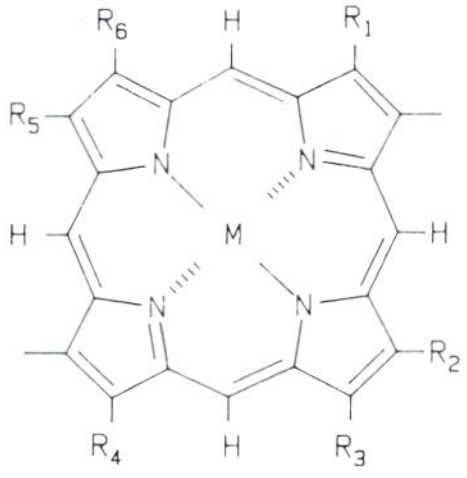

Pu: $R_1 = R_2 = R_4 = R_5 = CH_3; R_3 = R_6 = C_2H_5; M = H_2$
Pv: $R_1 = R_2 = R_4 = R_5 = CH_3; R_3 = R_6 = C_2H_5; M = Mg$
Pw: $R_1 = R_2 = R_4 = R_5 = CH_3; R_3 = R_6 = C_2H_5; M = Zn$
Py: $R_1 = R_3 = R_4 = R_6 = CH_3; R_2 = R_5 = C_2H_5; M = H_2$

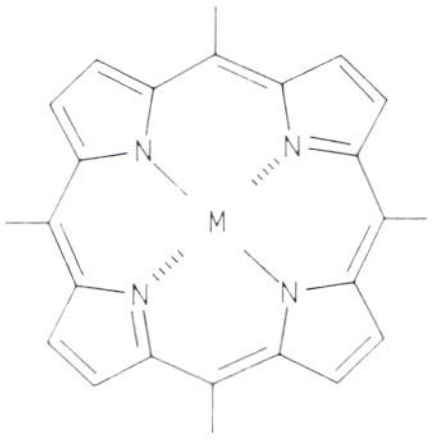

Pm: $R_1 = R_2 = CH_3; M = H_2$
Pn: $R_1 = CH_3; R_2 = C_2H_5; M = H_2$
Po: $R_1 = R_2 = CH_5; M = H_2$
Pp: $R_1 = R_2 = CH_3; M = Zn$
Pq: $R_1 = CH_3; R_2 = C_2H_5; M = Zn$
Pr: $R_1 = R_2 = C_2H_5; M = Zn$

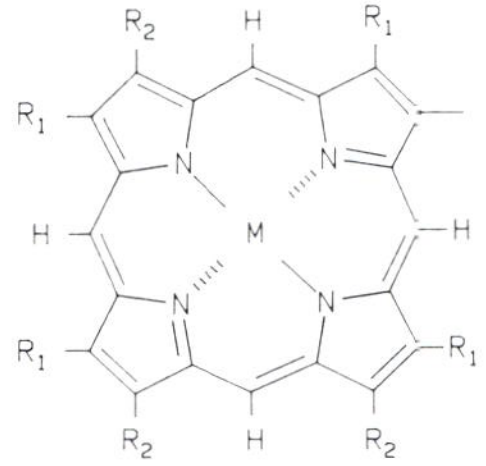

Px: $M = H_2$
Px': $M = Zn$

P3: $R1 = CH_3; R_2 = C_5H_{11}; M = H_2$

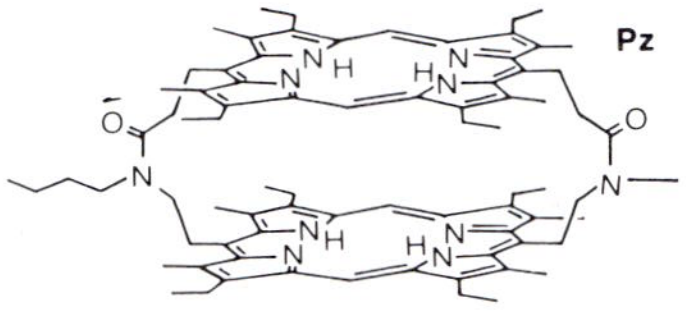

Pz

Appendix 2. Bridges

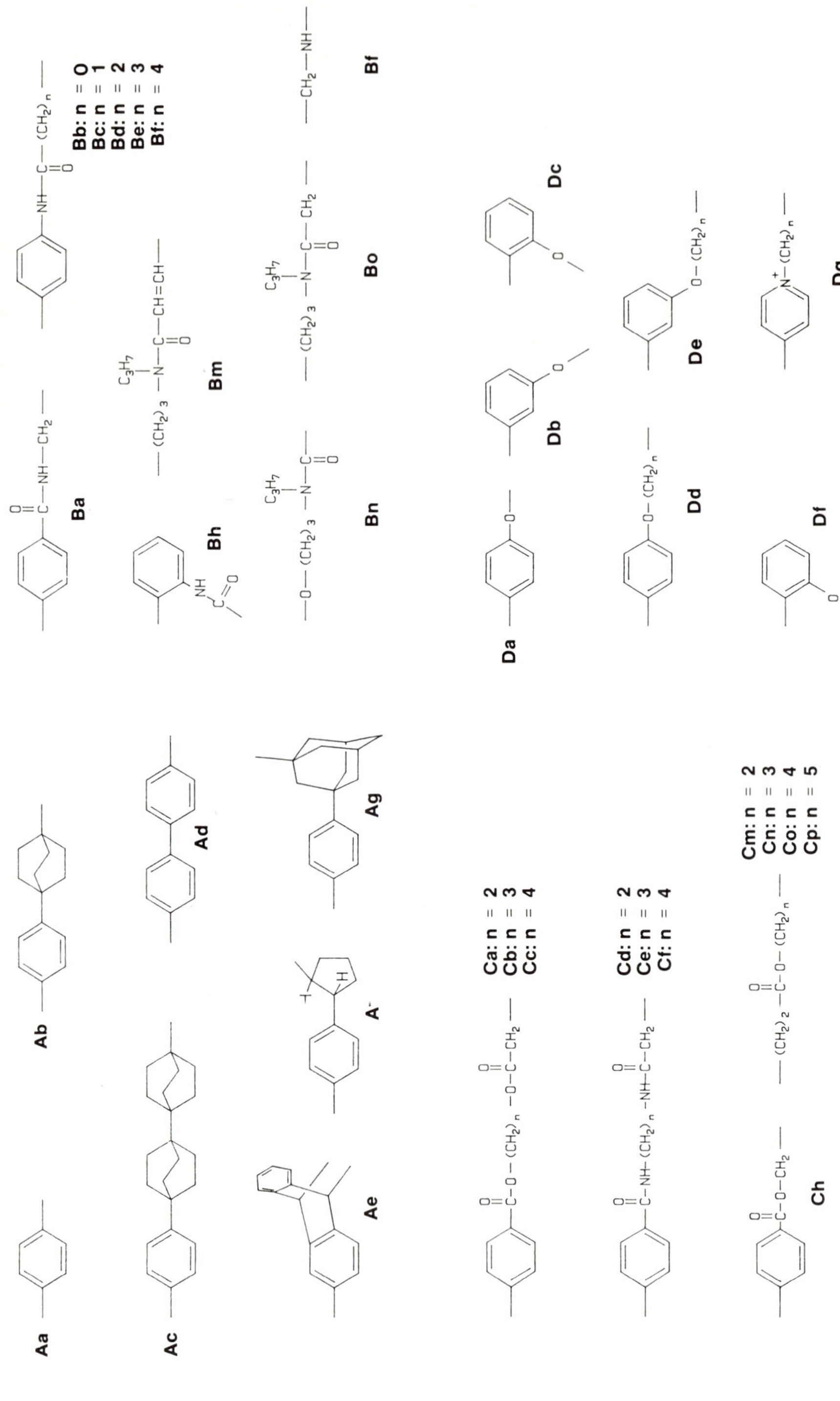

Aa

Ab

Ac

Ad

Ae

A⁻

Ag

C(=O)–NH–CH₂–

Ba

–NH–C(=O)–(CH₂)ₙ–

Bb: n = 0
Bc: n = 1
Bd: n = 2
Be: n = 3
Bf: n = 4

NH–C(=O)

Bh

–(CH₂)₃–N(C_3H_7)–C(=O)–CH=CH–

Bm

–O–(CH₂)₃–N(C_3H_7)–C(=O)–

Bn

–(CH₂)₃–N(C_3H_7)–C(=O)–CH₂–

Bo

–CH₂–NH–

Bf

–C(=O)–O–(CH₂)ₙ–O–C(=O)–CH₂–

Ca: n = 2
Cb: n = 3
Cc: n = 4

–C(=O)–NH–(CH₂)ₙ–NH–C(=O)–CH₂–

Cd: n = 2
Ce: n = 3
Cf: n = 4

–C(=O)–O–CH₂–

Ch

–(CH₂)₂–C(=O)–O–(CH₂)ₙ–

Cm: n = 2
Cn: n = 3
Co: n = 4
Cp: n = 5

Da

–O–

O

Db

O

Dc

–O–(CH₂)ₙ–

Dd

O–(CH₂)ₙ–

De

O

(CH₂)ₙ–

Df

N⁺–(CH₂)ₙ–

Dg

Appendix 2. Bridges (cont'd)

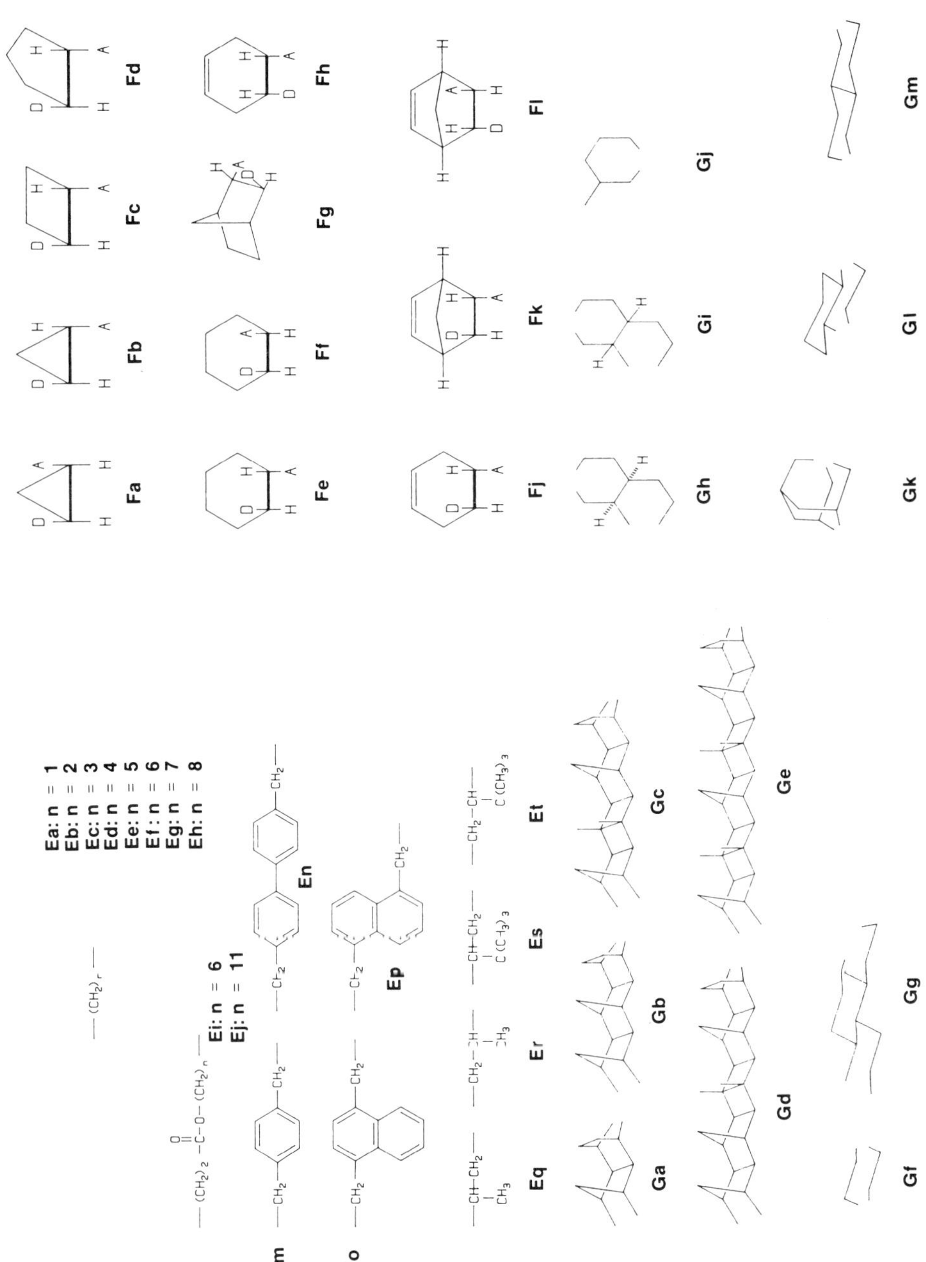

Appendix 3. Acceptors

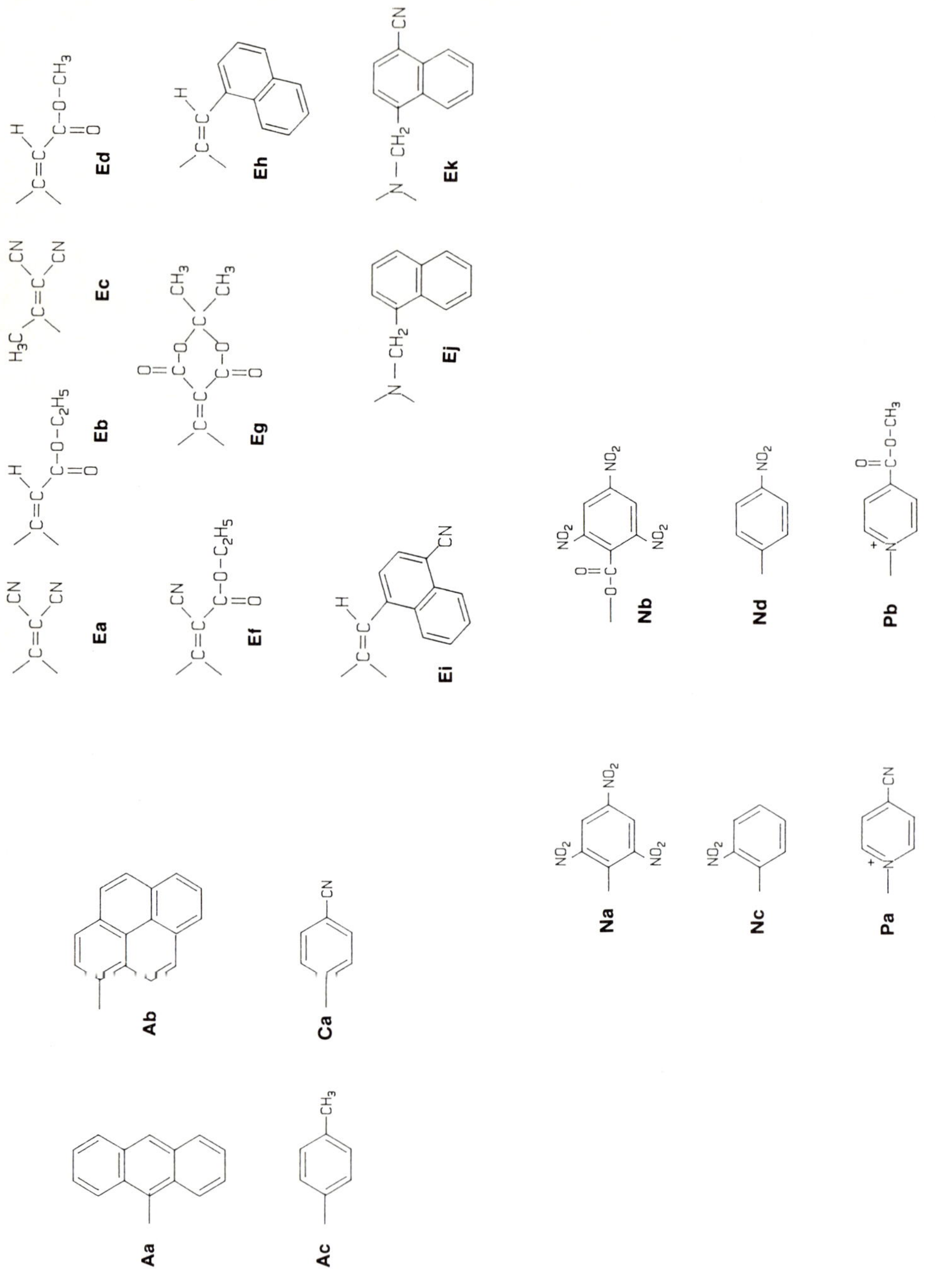

Ea Eb Ec Ed

Ef Eg Eh

Ei Ej Ek

Nb Nd Pb

Na Nc Pa

Ab Ca

Aa Ac

Appendix 3. Acceptors (cont'd)

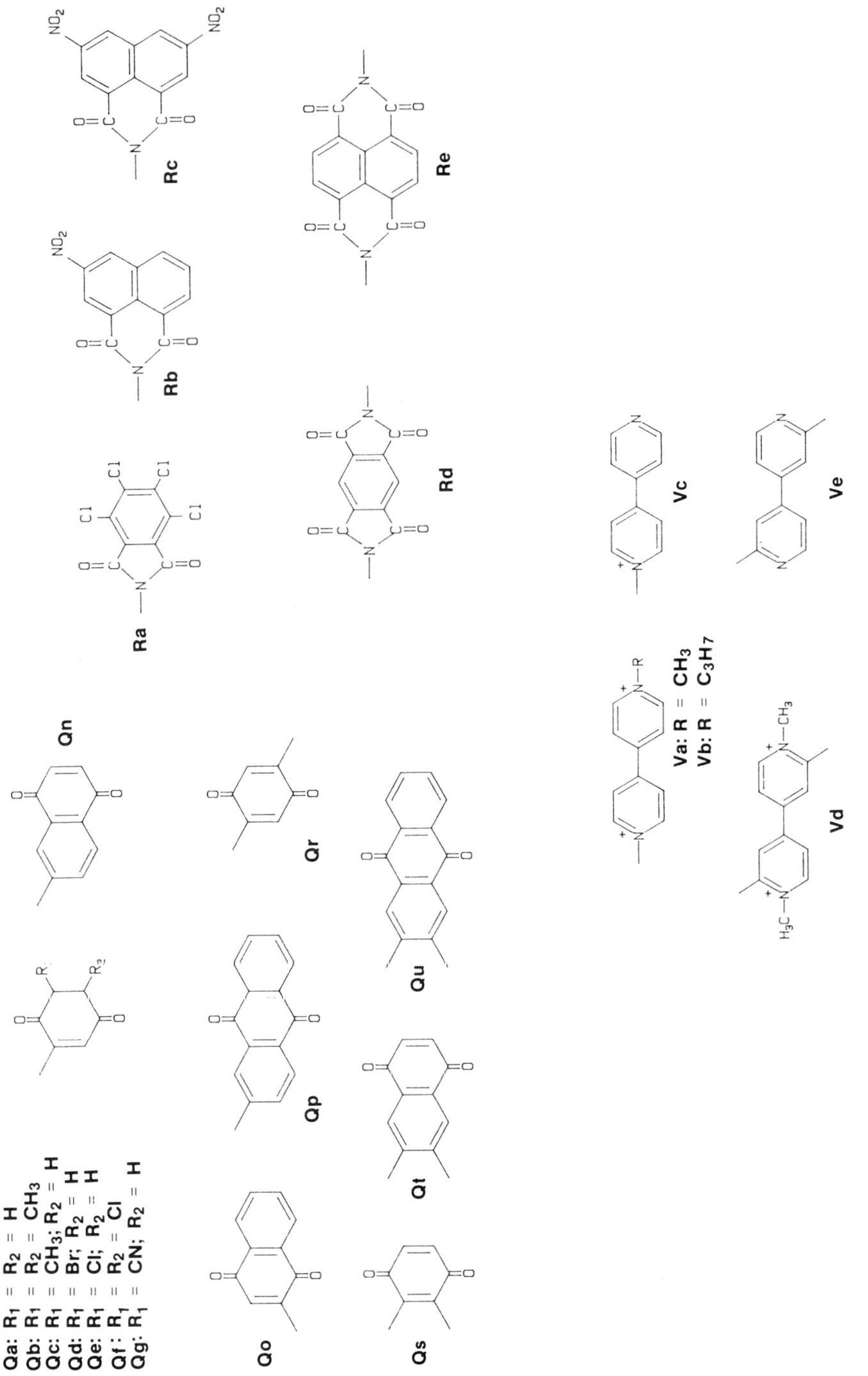

Qa: $R_1 = R_2 = H$
Qb: $R_1 = R_2 = CH_3$
Qc: $R_1 = CH_3$; $R_2 = H$
Qd: $R_1 = Br$; $R_2 = H$
Qe: $R_1 = Cl$; $R_2 = H$
Qf: $R_1 = R_2 = Cl$
Qg: $R_1 = CN$; $R_2 = H$

Qn

Qo

Qp

Qr

Qs

Qt

Qu

Ra

Rb

Rc

Rd

Re

Va: $R = CH_3$
Vb: $R = C_3H_7$

Vc

Vd

Ve

Appendix 4. Techniques

a. fluorescence yields
b. fluorescence lifetimes
c. fluorescence spectra
d. picosecond absorption spectra and kinetics
e. nanosecond absorption spectra and kinetics
f. synthesis reported
g. electron paramagnetic resonance (EPR) spectra
h. UV/visible spectra
i. redox potentials
j. theory
k. NMR structure analysis
l. transient microwave conductivity
m. X-ray structure
n. membrane photoelectric response

References

1. J.R. Bolton, Solar Electricity: Lessons Gained from Photosynthesis, in: M.H. Chisholm (Ed.), Inorganic Chemistry: Toward the 21st Century, ACS Symposium Series No. 211, American Chemical Society, Washington, D.C., 1983, pp. 1-19.

2. J.S. Connolly, The Role of Porphyrins and Chlorophylls in Artificial Photosynthesis, in: J. Rabani (Ed.), Photochemical Conversion and Storage of Solar Energy, 1982, Part A, The Weizmann Science Press of Israel, Jerusalem, 1982, pp. 175-204.

3. J.S. Connolly and J.A. Turner, Status and Prospects for Solar Photochemistry, in: A.M. Braun, M.-T. Maurette and E. Oliveros (Eds.), Photochemical Conversions, Presses Polytechnique Romandes, Lausanne, Switzerland, 1983, pp. 73-129.

4. S.G. Boxer, Model Reactions in Photosynthesis, Biochim. Biophys. Acta, 726 (1983) 265-292.

5. A. Harriman, The Role of Porphyrins in Natural and Artificial Photosynthesis, in: M. Grätzel (Ed.), Energy Resources through Photochemistry and Catalysis, Academic Press, New York, 1983, pp. 163-215.

6. M. Julliard and M. Chanon, Photoelectron-Transfer Catalysis: Its Connections with Thermal and Electrochemical Analogues, Chem. Rev., 83 (1983) 425-506.

7. D. Creed and R.A. Caldwell, Photochemical Electron Transfer Reactions and Exciplexes, Photochem. Photobiol., 41 (1985) 715-739.

8. J.H. Fendler, Photochemical Solar Energy Conversion. An Assessment of Scientific Accomplishments, J. Phys. Chem., 89 (1985) 2730-2740.

9. M.A. Fox, Photoinduced Electron Transfer in Organic Systems: Control of Back Electron Transfer, in: D.H. Volman, G.S. Hammond and K. Gollnick (Eds.), Advances in Photochemistry, Vol. 13, John Wiley & Sons, New York, 1986, pp. 237-327.

10. R.K. Clayton, Photosynthesis: Physical Mechanisms and Chemical Patterns, Cambridge University Press, Cambridge, 1980, pp. 79-87.

11. C.A. Wraight, Current Attitudes in Photosynthesis Research, in: Govindjee (Ed.), Photosynthesis: Energy Conversion by Plants and Bacteria, Academic Press, New York, 1982, pp. 17-61.

12. W.W. Parson and B. Ke, Primary Photochemical Reactions, in: Govindjee (Ed.), Photosynthesis: Energy Conversion by Plants and Bacteria, Academic Press, New York, 1982, pp. 331-385.

13. J. Deisenhofer, O. Epp, K. Miki, R. Huber and H. Michel, X-ray Structure of a Membrane Protein Complex: Electron Density Map at 3 Å Resolution and a Model of the Chromophores of the Photosynthetic Reaction Centre from *Rhodopseudomonas viridis*, J. Mol. Biol., 180 (1984) 385-398.

14. J. Deisenhofer, O. Epp, K. Miki, R. Huber and H. Michel, Structure of the Protein Subunits in the Photosynthetic Reaction Centre of *Rhodopseudomonas viridis* at 3 Å Resolution, Nature (London), 318 (1985) 618-624.

15. C.-H. Chang, D. Tiede, J. Tang, U. Smith, J. Norris and M. Schiffer, Structure of *Rhodopseudomonas sphaeroides* R-26 Reaction Center, FEBS Lett., 205 (1986) 82-86.

16. J.-L. Martin, J. Breton, A.J. Hoff, A. Migus and A. Antonetti, Femtosecond Spectroscopy of Electron Transfer in the Reaction Center of the Photosynthetic Bacterium *Rhodopseudomonas sphaeroides* R26: Direct Electron Transfer from the Dimeric Bacteriochlorophyll Primary Donor to the Bacteriopheophytin Acceptor with a Time Constant of 2.8 ± 0.2 ps, Proc. Natl. Acad. Sci. USA, 83 (1986) 957-961.

17. W.W. Parson, Photosynthetic Bacterial Reaction Centers - Interactions among the Bacteriochlorophylls and Bacteriopheophytins, Ann. Rev. Biophys. Bioeng., 11 (1982) 57-80.

18. J.J. Katz and J.C. Hindman, Current Status of Biomimetic Systems for Solar Energy Conversion, in: J.S. Connolly (Ed.), Photochemical Conversion and Storage of Solar Energy, Academic Press, New York, 1981, pp. 27-78.

19. G.R. Seely, The Energetics of Electron Transfer Reactions of Chlorophyll and Other Compounds, Photochem. Photobiol., 27 (1978) 639-654.

20. R.A. Marcus, On the Theory of Oxidation-Reduction Reactions Involving Electron Transfer. I, J. Chem. Phys., 24 (1956) 966-978.

21. R.A. Marcus, Exchange Reactions and Electron Transfer Reactions including Isotopic Exchange: Theory of Oxidation-Reduction Reactions involving Electron Transfer. 4. A Statistical-Mechanical Basis for Treating Contributions from Solvent, Ligands and Inert Salt, Faraday Discuss. Chem. Soc., 29 (1960) 21-31.

22. R.A. Marcus, On the Theory of Electron-Transfer Reactions. VI. Unified Treatment for Homogeneous and Electrode Reactions, J. Chem. Phys., 43 (1965) 679-701.

23. R.A. Marcus, Electron, Proton and Related Transfers, Faraday Discuss. Chem. Soc., 74 (1982) 7-15.

24. M.D. Newton and N. Sutin, Electron Transfer Reactions in Condensed Phases, Ann. Rev. Phys. Chem., 35 (1984) 437-480.

25. R.A. Marcus and N. Sutin, Electron Transfers in Chemistry and Biology, Biochim. Biophys. Acta, 811 (1985) 265-322.

26. B.S. Brunschwig, S. Ehrenson and N. Sutin, Solvent Reorganization in Optical and Thermal Electron-Transfer Processes, J. Phys. Chem., 90 (1986) 3657-3668.

27. J.R. Miller, J.V. Beitz and R.K. Huddleston, Effect of Free Energy on Rates of Electron Transfer between Molecules, J. Am. Chem. Soc., 106 (1984) 5057-5068.

27a. J. Jortner, Temperature Dependent Activation Energy for Electron Transfer between Biological Molecules, J. Chem. Phys., 64 (1976) 4860-4867.

27b. P. Siders and R.A. Marcus, Quantum Effects for Electron-Transfer Reactions in the "Inverted Region", J. Am. Chem. Soc., 103 (1981) 748-752.

28. A. Weller, Photoinduced Electron Transfer in Solution: Exciplex and Radical-Ion-Pair Formation Free Enthalpies and Their Solvent Dependence, Z. Phys. Chem. (Wiesbaden), 133 (1982) 93-98.

29. J.A. Schmidt, J.-Y. Liu, J.R. Bolton, M.D. Archer and V.P.Y Gadzekpo, Intramolecular Photochemical Electron Transfer. 5. Solvent Dependence of Electron Transfer in a Porphyrin-Amide-Quinone Molecule, J. Am. Chem. Soc., (submitted, 1988).

29a. M.D. Archer, V.P.Y. Gadzekpo, J.R. Bolton, J.A. Schmidt and A.C. Weedon, Solvent Dependence of Photochemical Electron-Transfer Rates in a Covalently Linked Porphyrin-Quinone Molecule, J. Chem. Soc., Faraday Trans. 2, 82 (1986) 2305-2313.

30. A. Siemiarczuk, A.R. McIntosh, T.-F. Ho, M.J. Stillman, K.J. Roach, A.C. Weedon, J.R. Bolton and J.S. Connolly, Intramolecular Photochemical Electron Transfer. 2. Fluorescence Studies of Linked Porphyrin-Quinone Compounds, J. Am. Chem. Soc., 105 (1983) 7224-7230.

31. J.S. Connolly, J.K. Hurley, W.L. Bell and K.L. Marsh, Inter- and Intramolecular Quenching of Porphyrin Excited States by Quinones, in: V. Balzani (Ed.), Supramolecular Photochemistry, D. Reidel Publishers, Dordrecht, The Netherlands, NATO ASI Series C: Mathematical and Physical Sciences, Vol. 214, 1987, pp. 299-318.

32. J.K. Hurley, W.L. Bell, K.L. Marsh, M.R. Wasielewski and J.S. Connolly, Light-Induced Electron Transfer in Ester-Linked Porphyrin-Anthraquinone Molecules (in preparation, 1988).

32a. T. Shida (personal communication, 1983). (See ref. 46 in ref. 39.)

32b. J.S. Connolly, K.L. Marsh, D.R. Cook, J.R. Bolton, A.R. McIntosh and A. Siemiarczuk, Effects of Conformation and Molecular Motion in Light-Induced Intramolecular Electron Transfer (to be submitted, 1988).

33. J.S. Connolly, K.L. Marsh, D.R. Cook, J.R. Bolton, A.R. McIntosh, A. Siemiarczuk, A.C. Weedon and T.-F. Ho, Mechanisms of Light-Induced Intramolecular Electron Transfer: Effects of Conformation and Molecular Motion, Sci. Pap. Inst. Phys. Chem. Res. (Jpn.), 78 (1984) 118-128.

34. A.R. McIntosh, J.R. Bolton, J.S. Connolly, K.L. Marsh, D.R. Cook, T.-F. Ho and A.C. Weedon, Electron Paramagnetic Resonance and Optical Evidence for Two Distinct Porphyrin Triplet States in Linked Porphyrin-Quinone Molecules, J. Phys. Chem., 90 (1986) 5640-5646.

35. J.A. Schmidt, A.R. McIntosh, A.C. Weedon, J.R. Bolton, J.S. Connolly, J.K. Hurley and M.R. Wasielewski, Intramolecular Photochemical Electron Transfer. 4. Singlet and Triplet Mechanisms of Electron Transfer in a Covalently Linked Porphyrin-Amide-Quinone Molecule, J. Am. Chem. Soc., 110 (1988) in press.

36. J.M. Warman, M.P. de Haas, M.N. Paddon-Row, E. Cotsaris, N.S. Hush, H. Oevering and J.W. Verhoeven, Light-Induced Giant Dipoles in Simple Model Compounds for Photosynthesis, Nature (London), 320 (1986) 615-616.

37. J.M. Warman, M.P. de Haas, H. Oevering, J.W. Verhoeven, M.N. Paddon-Row, A.M. Oliver and N.S. Hush, Donor, Acceptor, and Self-Quenching of the Giant-Dipole State of a Rigid, σ-Bond Separated, Donor-Acceptor Molecular Assembly, Chem. Phys. Lett., 128 (1986) 95-99.

38. M.P. de Haas and J.M. Warman, Photon-Induced Molecular Charge Separation Studied by Nanosecond Time-Resolved Microwave Conductivity, Chem. Phys. 73 (1982) 35-53.

39. A.R. McIntosh, A. Siemiarczuk, J.R. Bolton, M.J. Stillman, T.-F. Ho and A.C. Weedon, Intramolecular Photochemical Electron Transfer. 1. EPR and Optical Absorption Evidence for Stabilized Charge Separation in Linked Porphyrin-Quinone Molecules, J. Am. Chem. Soc., 105 (1983) 7215-7223.

40. J.R. Miller, L.T. Calcaterra and G.L. Closs, Intramolecular Long-Distance Electron Transfer in Radical Anions. The Effects of Free Energy and Solvent on the Reaction Rates, J. Am. Chem. Soc., 106 (1984) 3047-3049.

41. G.L. Closs, L.T. Calcaterra, N.J. Green, K.W. Penfield and J.R. Miller, Distance, Stereoelectronic Effects and the Marcus Inverted Region in Intramolecular Electron Transfer in Organic Radical Anions, J. Phys. Chem., 90 (1986) 3673-3683.

42. J.V. Beitz and J.R. Miller, Exothermic Rate Restrictions on Electron Transfer in a Rigid Medium, J. Chem. Phys., 71 (1979) 4579-4595.

43. D. Rehm and A. Weller, Kinetics of Fluorescence Quenching by Electron and H-Atom Transfer, Isr. J. Chem., 8 (1970) 269-271.

44. V. Balzani, F. Scandola, G. Orlandi, N. Sabbatini and M.T. Indelli, The Non-Adiabaticity Problem of Outer-Sphere Electron-Transfer Reactions. Reduction and Oxidation of Europium Ions, J. Am. Chem. Soc., 103 (1981) 3370-3378.

45. V. Balzani and F. Scandola, Photochemical Electron Transfer Reactions in Homogeneous Solutions, in: J.S. Connolly (Ed.), Photochemical Conversion and Storage of Solar Energy, Academic Press, New York, 1981, pp. 97-125.

46. R.J. Cave, S.J. Klippenstein and R.A. Marcus, A Semiclassical Model for Orientation Effects in Electron Transfer Reactions, J. Chem. Phys., 84 (1986) 3089-3098.

47. R.J. Cave, P. Siders and R.A. Marcus, Mutual Orientation Effects on Electron Transfer between Porphyrins, J. Phys. Chem., 90 (1986) 1436-1444.

48. S. Larsson, Electron Transfer in Chemical and Biological Systems. Orbital Rules for Nonadiabatic Transfer, J. Am. Chem. Soc., 103 (1981) 4034-4040.

49. S. Larsson, Electron Transfer in Proteins, J. Chem. Soc., Faraday Trans. 2, 79, (1983) 1375-1388.

50. S. Larsson, Electron-Exchange Reaction in Aqueous Solution, J. Phys. Chem., 88 (1984) 1321-1323.

51. J.R. Miller, Controlling Charge Separation through Effects of Energy, Distance and Molecular Structure on Electron Transfer Rates, Nouv. J. Chim., 11 (1987) 83-89.

52. D.N. Beratan, J.N. Onuchic and J.J. Hopfield, Limiting Forms of the Tunneling Matrix Element in the Long-Distance Bridge-Mediated Electron-Transfer Problem, J. Chem. Phys., 83 (1985) 5325-5329.

53. J.A. Schmidt, A. Siemiarczuk, A.C. Weedon and J.R. Bolton, Intramolecular Electron Transfer. 3. Solvent Dependence of Fluorescence Quenching and Electron-Transfer Rates in a Porphyrin-Amide-Quinone Molecule, J. Am. Chem. Soc., 107 (1985) 6112-6114.

54. E.M. Kosower and D. Huppert, Solvent Motion Controls the Rate of Intramolecular Electron Transfer in Solution, Chem. Phys. Lett., 96 (1983) 433-436.

55. E.M. Kosower, H. Kanety, H. Dodiuk, G. Striker, T. Jovin, H. Boni and D. Huppert, Intramolecular Donor-Acceptor Systems. 7. Solvent Dielectric Relaxation Effects on the Photophysics of 6-(Phenylamino)-N,N-Dimethyl-2-Naphthalene-Sulfonamides, J. Phys. Chem., 87 (1983) 2479-2484.

56. E.M. Kosower, Mechanism of Fast Intramolecular Electron-Transfer Reactions, J. Am. Chem. Soc., 107 (1985) 1114-1118.

57. E.M. Kosower and D. Huppert, Excited State Electron and Proton Transfers, Ann. Rev. Phys. Chem., 37 (1986) 127-156.

58. H. Sumi and R.A. Marcus, Dielectric Relaxation and Intramolecular Electron Transfers, J. Chem. Phys., 84 (1986) 4272-4276.

59. M. McGuire and G. McLendon, How Do Solvent Relaxation Dynamics Affect Electron-Transfer Rates? A Study in Rigid Solution, J. Phys. Chem., 90 (1986) 2549-2551.

60. J.N. Onuchic, D.N. Beratan and J.J. Hopfield, Some Aspects of Electron-Transfer Dynamics, J. Phys. Chem., 90 (1986) 3707-3721.

61. H. Heitele, M.E. Michel-Beyerle and P. Finckh, The Influence of Dielectric Relaxation on Intramolecular Electron Transfer, Chem. Phys. Lett., 137 (1987) 237-243.

62. R.J. Harrison, B. Pearce, G.S. Beddard, J.A. Cowan and J.K.M. Sanders, Photoinduced Electron Transfer in Pyromellitimide-Bridged Porphyrins, Chem. Phys., 116 (1987) 429-448.

63. D.F. Calef and P.G. Wolynes, Classical Solvent Dynamics and Electron Transfer. 1. Continuum Theory, J. Phys. Chem., 87 (1983) 3387-3340.

64. J.W. Verhoeven, I.P. Dirkx and Th.J. de Boer, Studies of Inter- and Intra-molecular Donor-Acceptor Interactions - IV. Intramolecular Charge Transfer Phenomena in Substituted N-Aralkyl-Pyridinium Ions, Tetrahed., 25 (1969) 4037-4055.

65. T. Okada, T. Fujita, M. Kubota, S. Masaki, N. Mataga, R. Ide, Y. Sakata and S. Misumi, Intramolecular Electron Donor-Acceptor Interactions in the Excited State of (Anthracene)-$(CH_2)_n$-(N,N-Dimethylaniline) Systems, Chem. Phys. Lett., 14 (1972) 563-568.

66. P. Pasman, J.W. Verhoeven and Th.J. de Boer, Fluorescence of Intramolecular Electron Donor-Acceptor Systems: The Importance of Through-Bond Interaction, Chem. Phys. Lett., 59 (1978) 381-385.

67. Th. Förster, Excimers and Exciplexes, in: M. Gordon and W.R. Ware (Eds.), The Exciplex, Academic Press, New York, 1975, pp. 1-21.

68. N. Mataga and M. Ottolenghi, Photophysical Aspects of Exciplexes, in: R. Foster (Ed.), Molecular Association, Including Molecular Complexes, Academic Press, 1979, Vol. 2, pp. 1-78.

69. J.K. Roy and D.G. Whitten, Medium Effects on the Lifetime and Reactivity of Triplet Exciplexes, J. Am. Chem. Soc., 94 (1972) 7162-7164.

70. J.K. Roy, F.A. Carroll and D.G. Whitten, Spectroscopic Studies of Formation and Decay of Triplet Exciplexes. Evidence for a Limited Role of Charge-Transfer Interactions in a Nonpolar Solvent, J. Am. Chem. Soc., 96 (1974) 6349-6355.

71. H. Knibbe, K. Röllig, F.P. Schäfer and A. Weller, Charge-Transfer Complex and Solvent-Shared Ion Pair in Fluorescence Quenching, J. Chem. Phys., 47 (1967) 1184-1185.

72. J. Dalton and L.R. Milgrom, A Novel Porphyrin with Weak Fluorescence Due to Intramolecular Electron Transfer Quenching, J. Chem. Soc., Chem. Commun. (1979) 609-610.

73. I. Tabushi, N. Koga and M. Yanagita, Efficient Intramolecular Quenching and Electron Transfer in Tetraphenylporphyrin Attached with Benzoquinone or Hydroquinone as a Photosystem Model, Tetrahed. Lett. (1979) 257-260.

74. J.L.Y. Kong and P.A. Loach, Covalently Linked Porphyrin Quinone Complexes as RC Models, in: P.L. Dutton, J.S. Leigh and A. Scarpa (Eds.), Frontiers of Biological Energetics, Academic Press, New York, 1978, Vol. 1, pp. 73-82.

75. J.L.Y. Kong and P.A. Loach, Syntheses of Covalently Linked Porphyrin-Quinone Complexes (1), J. Heterocyclic Chem., 17 (1980) 737-744.

76. J.L.Y. Kong, K.G. Spears and P.A. Loach, Photochemical Characterization of Covalently Linked Porphyrin-Quinone Complexes, Photochem. Photobiol., 35 (1982) 545-553.

77. P.A. Loach, J.A. Runquist, J.L.Y. Kong, T.J. Dannhauser and K.G. Spears, Model Systems for the Primary Photochemical Events of Photosynthesis and Electron Transfer in Bioenergetic Membranes, in: K.M. Kadish (Ed.), Electrochemical and Spectrochemical Studies of Biological Redox Components, Advances in Chemistry Series No. 201, American Chemical Society, Washington, D.C., 1982, pp. 515-561.

78. T.-F. Ho, A.R. McIntosh and J.R. Bolton, Intramolecular Photochemical Electron Transfer in a Linked Porphyrin-Quinone Molecule as a Model for the Primary Step of Photosynthesis, Nature (London), 286 (1980) 254-256.

79. T.-F. Ho, A.R. McIntosh and A.C. Weedon, Synthesis and Properties of Linked Porphyrin-Quinone Molecules Designed as Models of the Reaction Center in Photosynthesis, Can. J. Chem., 62 (1984) 967-973.

79a. J.H. Wilford, M.D. Archer, J.R. Bolton, T.-F. Ho, J.A. Schmidt and A.C. Weedon, Redox Potentials of Some Covalently Linked Porphyrin-Quinone and Related Molecules, J. Phys. Chem., 89 (1985) 5395-5398.

80. M.R. Wasielewski and M.P. Niemczyk, Photoinduced Electron Transfer in *meso*-Triphenyltriptycenylporphyrin-Quinones. Restricting Donor-Acceptor Distances and Orientations, J. Am. Chem. Soc., 106 (1984) 5043-5045.

81. M.R. Wasielewski, M.P. Niemczyk, W.A. Svec and E.B. Pewitt, Dependence of Rate Constants for Photoinduced Charge Separation and Dark Charge Recombination on the Free Energy of Reaction in Restricted-Distance Porphyrin-Quinone Molecules, J. Am. Chem. Soc., 107 (1985) 1080-1082.

82. M.R. Wasielewski, M.P. Niemczyk, W.A. Svec and E.B. Pewitt, High-Quantum-Yield Long-Lived Charge Separation in a Photosynthetic Reaction Center Model, J. Am. Chem. Soc., 107 (1985) 5562-5563.

83. M.R. Wasielewski, M.P. Niemczyk, W.A. Svec and E.B. Pewitt, Ultrafast Electron Transfer in Biomimetic Models of Photosynthetic Reaction Centers, in: Antennas and Reaction Centers of Photosynthetic Bacteria, M.E. Michel-Beyerle (Ed.), Springer-Verlag, Berlin, 1985, 242-249.

84. M.R. Wasielewski and M.P. Niemczyk, Distance-Dependent Rates of Photoinduced Charge Separation and Dark Charge Recombination in Fixed-Distance Porphyrin-Quinone Molecules, in: M. Gouterman, P.M. Rentzepis and K.D. Straub (Eds.), Porphyrins - Excited States and Dynamics, ACS Symposium Series No. 321, American Chemical Society, Washington, D.C., 1986, pp. 154-165.

85. J.W. Verhoeven, M.N. Paddon-Row, N.S. Hush, H. Oevering and M. Heppener, Through-Bond Charge Transfer Interaction and Photoinduced Charge Separation, Pure Appl. Chem., 58 (1986) 1285-1290.

86. H. Oevering, M.N. Paddon-Row, M. Heppener, A.M. Oliver, E. Cotsaris, J.W. Verhoeven and N.S. Hush, Long-Range Photoinduced Through-Bond Electron Transfer and Radiative Recombination *via* Rigid Nonconjugated Bridges: Distance and Solvent Dependence, J. Am. Chem. Soc., 109 (1987) 3258-3269.

87. M.N. Paddon-Row, A.M. Oliver, J.M. Warman, K.J. Smit, M.P. de Haas, H. Oevering and J.W. Verhoeven, Factors Affecting Charge Separation and Recombination in Photo-excited Rigid Donor-Insulator-Acceptor Compounds, J. Phys. Chem. (submitted, 1987).

88. N.S. Hush, M.N. Paddon-Row, E. Cotsaris, H. Oevering, J.W. Verhoeven and M. Heppener, Distance Dependence of Photoinduced Electron Transfer through Nonconjugated Bridges, Chem. Phys. Lett., 117 (1985) 8-11.

89. M.K. Crawford, Y. Wang and K.B. Eisenthal, Effects of Conformation and Solvent Polarity on Intramolecular Charge Transfer: A Picosecond Laser Study, Chem. Phys. Lett., 79 (1981) 529-533.

90. H. Heitele and M.E. Michel-Beyerle, Electron Transfer through Aromatic Spacers in Bridged Electron-Donor-Acceptor Molecules, J. Am. Chem. Soc., 107 (1985) 8286-8288.

91. H. Heitele, M.E. Michel-Beyerle, P. Finckh and W. Rettig, Role of Intramolecular Bridges on Electron Transfer in Donor/Acceptor Systems, in: J. Jortner and B. Pullman (Eds.), Tunneling, 1986, pp. 333-344.

92. H. Heitele, M.E. Michel-Beyerle and P. Finckh, Electron Transfer through Intramolecular Bridges, Chem. Phys. Lett., 134 (1987) 273-278.

93. B.A. Leland, A.D. Joran, P.M. Felker, J.J. Hopfield, A.H. Zewail and P.B. Dervan, Picosecond Fluorescence Studies on Intramolecular Photochemical Electron Transfer in Porphyrins Linked to Quinones at Two Different Fixed Distances, J. Phys. Chem., 89 (1985) 5571-5573.

94. D.N. Beratan, Electron Tunneling through Rigid Molecular Bridges: Bicyclo-[2.2.2]octane, J. Am. Chem. Soc., 108 (1986) 4321-4326.

95. T.-F. Ho and J.R. Bolton, Intramolecular Photochemical Electron-Transfer. 6. Porphyrin-(Phenyl)$_n$-Quinone Molecules and the Role of the Bridging Group, (in preparation, 1988).

96. A.F. Janzen and J.R. Bolton, Photochemical Electron Transfer in Monolayer Assemblies. 2. Photoelectric Behavior in Chlorophyll a/Acceptor Systems, J. Am. Chem. Soc., 101 (1979) 6342-6348.

97. S.S. Isied, Long-Range Electron Transfer in Peptides and Proteins, in: S.J. Lippard (Ed.), Prog. Inorg. Chem., 32 (1984) 443-517.

98. P.S. Ho, C. Sutoris, N. Liang, E. Margoliash and B.M. Hoffman, Species Specificity of Long-Range Electron Transfer within the Complex between Zinc-Substituted Cytochrome c Peroxidase and Cytochrome c, J. Am. Chem. Soc., 107 (1985) 1070-1071.

99. H.B. Gray, Long-Range Electron Transfer in Blue Copper Proteins, Chem. Soc. Rev., 15 (1986) 17-30.

100. S.L. Mayo, W.R. Ellis, Jr., R.J. Crutchley and H.B. Gray, Long-Range Electron Transfer in Proteins, Science, 233 (1986) 948-952.

101. G. McLendon, J.R. Miller, K. Simolo, K. Taylor, A.G. Mauk and A.M. English, Thermal and Photoinduced Long Distance Electron Transfer in Proteins and in Model Systems, in: A.B.P. Lever (Ed.), Excited States and Reactive Intermediates, ACS Symposium Series No. 307, American Chemical Society, Washington, D.C., 1986, pp. 150-165.

102. J.N. Onuchic and D.N. Beratan, Molecular Bridge Effects on Distant Charge Tunneling, J. Am. Chem. Soc., 109 (1987) 6771-6778.

103. J.R. Bolton, T.-F. Ho, S. Liauw, A. Siemiarczuk, C.S.K. Wan and A.C. Weedon, Light-Induced Intramolecular Electron Transfer from a Porphyrin Linked to a *p*-Benzoquinone by a Rigid Spacer Group, J. Chem. Soc., Chem. Commun. (1985) 559-560.

104. D. Gust, T.A. Moore, P.A. Liddell, G.A. Nemeth, L.R. Makings, A.L. Moore, D. Barrett, P.J. Pessiki, R.V. Bensasson, M. Rougeé, C. Chachaty, F.C. De Schryver, M. Van der Auweraer, A.R. Holzwarth and J.S. Connolly, Charge Recombination in Carotenoporphyrin-Quinone Triads: Synthetic, Conformational, and Fluorescence Lifetime Studies, J. Am. Chem. Soc., 109 (1987) 846-856.

105. T.L. Netzel, M.A. Bergkamp, C.-K. Chang and J. Dalton, A Comparison of Ultrafast Electron Transfers in Porphyrin-Quinone and Magnesium-Free-Base Diporphyrin Molecules: Mimicking Photosynthetic Charge Separations, J. Photochem., 17 (1981) 451-460.

106. M.A. Bergkamp, J. Dalton and T.L. Netzel, Quenching of the Singlet Excited States of Meso-Substituted Porphines by *p*-Benzoquinone under Unimolecular and Bimolecular Conditions: Evidence for Electron Transfer in Competition with Vibrational Relaxation, J. Am. Chem. Soc., 104 (1982) 253-259.

107. A. Harriman and R.J. Hosie, Fluorescence Quenching Effect of Substituted Tetraphenylporphyrins, J. Photochem., 15 (1981) 163-167.

108. W.L. Bell and J.S. Connolly, Synthesis and Photophysical Studies of a Directly Linked Porphyrin-Anthraquinone Molecule (in preparation, 1988).

109. A.D. Joran, B.A. Leland, G.G. Geller, J.J. Hopfield and P.B. Dervan, Models for Photochemical Electron Transfer at Fixed Distances. Porphyrin-Bicyclo[2.2.2]octane-Quinone and Porphyrin-Bibicyclo[2.2.2]octane-Quinone, J. Am. Chem. Soc., 106 (1984) 6090-6092.

110. A.D. Joran, B.A. Leland, P.K. Felker, A.H. Zewail, J.J. Hopfield and P.B. Dervan, Effect of Exothermicity on Electron-Transfer Rates in Photosynthetic Molecular Models, Nature (London), 327 (1987) 508-511.

111. M.R. Wasielewski, D.G. Johnson and W.A. Svec, Photoinduced Electron Transfer in Fixed Distance Chlorophyll-Quinone Donor-Acceptor Molecules, in: V. Balzani (Ed.), Supramolecular Photochemistry, D. Reidel Publishers, Dordrecht, The Netherlands, NATO ASI Series C: Mathematical and Physical Sciences, Vol. 214, 1987, pp. 255-266.

112. N.B. Joshi, J.R. Lopez, H.-T. Tien, C.-B. Wang and Q.-Y. Liu, Photoelectric Effects in Bilayer Lipid Membranes Containing Covalently Linked Porphyrin Complexes, J. Photochem., 20 (1982) 139-151.

113. C.-B. Wang, H.T. Tien, J.R. Lopez, Q.-Y. Liu, N.B. Joshi and Q.-Y. Hu, Photoelectrochemical Properties of Bilayer Lipid Membranes Containing Covalently Linked Porphyrin-Quinone and Other Complexes, Photobiochem. Photobiophys., 4 (1982) 177-184.

114. K. Maruyama, H. Furuta and A. Osuka, Photoinduced Cross-Coupling Reaction between Porphyrin and Quinone, Chem. Lett. (1986) 475-478.

115. A. Osuka, S. Morikawa, K. Maruyama, S. Hirayama and T. Minami, An Efficient Photochemical Synthesis of Conformationally Restricted Quinone-Substituted Porphyrins, J. Chem. Soc., Chem. Commun. (1987) 359-361.

116. S. Nishitani, N. Kurata, Y. Sakata, S. Misumi, M. Migita, T. Okada, N. Mataga, Syntheses of a Series of Octaethylporphyrin-Benzoquinone Linked Molecules as a Model for the Primary Process of Photosynthesis, Tetrahed. Lett., 22 (1981) 2099-2102.

117. M. Migita, T. Okada, N. Mataga, S. Nishitani, N. Kurata, Y. Sakata, S. Misumi, Picosecond Time-Resolved Observation of Photoinduced Charge Separation from the Singlet Excited State of Porphyrin-Quinone Model Systems, Chem. Phys. Lett., 84 (1981) 263-266.

118. N. Mataga, A. Karen, T. Okada, S. Nishitani, N. Kurata, Y. Sakata and S. Misumi, Picosecond Dynamics of Photochemical Electron Transfer in Porphyrin-Quinone Intramolecular Exciplex Systems, J. Phys. Chem., 88 (1984) 5138-5141.

119. Y. Sakata, S. Nishitani, N. Nishimizu, S. Misumi, A.R. McIntosh, J.R. Bolton, Y. Kanda, A. Karen, T. Okada and N. Mataga, Synthesis of a Model Compound for the Photosynthetic Electron Transfer, Tetrahed. Lett., 26 (1985) 5207-5210.

120. M.R. Wasielewski, W.A. Svec and B.T. Cope, *Bis*(chlorophyll)cyclophanes. New Models of Special Pair Chlorophyll, J. Am. Chem. Soc., 100 (1978) 1961-1962.

121. G.B. Maiya and V. Krishnan, Intramolecular Electron Transfer in Donor-Acceptor Systems. Porphyrins Bearing a Trinitroaryl Group, J. Phys. Chem., 89 (1985) 5225-5235.

122. G. Blondeel, D. De Keukeleire, A. Harriman and L.R. Milgrom, Fluorescence of Covalently Bound Zinc Porphyrin-Viologen Complexes, Chem. Phys. Lett., 118 (1985) 77-82.

123. H. Nakamura, A. Motonaga, T. Ogata, S. Nakao, T. Nagamura and T. Matsuo, Microenvironmental Control of Photoinduced Electron Transfer and the Reverse Reactions in Porphyrin-Viologen Linked System, Chem. Lett. (1986) 1615-1618.

124. H. Nakamura, A. Uehata, A. Motonaga, T. Ogata and T. Matsuo, Reaction Control of Photoinduced Electron Transfer in Porphyrin-Viologen Linked System by the Use of External Magnetic Fields, Chem. Lett. (1987) 543-546.

125. I. Okura, N. Kaji, S. Aono, and T. Nishisaka, Photoredox Properties of Viologen-Linked Porphyrins, Bull. Chem. Soc. Jpn., 60 (1987) 1243-1247.

126. I. Okura, N. Kaji, S. Aono and T. Nishisaka, Photoinduced Hydrogen Evolution with Viologen-Linked Porphyrin in a Micellar System, Bull. Chem. Soc. Jpn., 59 (1986) 3967-3968.

127. N. Kaji, S. Aono and I. Okura, Photoinduced Hydrogen Evolution with Viologen-Linked Water-Soluble Zinc Porphyrins, J. Mol. Catal., 36 (1986) 201-203.

128. S. Aono, N. Kaji and I. Okura, Photoinduced Hydrogen Evolution with a Viologen-Linked Porphyrin, J. Chem. Soc., Chem. Commun. (1986) 170-171.

129. Y. Kanda, H. Sato, T. Okada and N. Mataga, Formation of a Long-Lived Intramolecular Radical Ion Pair from the Singlet Excited State of Porphyrin-Methylviologens Combined with Flexible Chains, Chem. Phys. Lett., 129 (1986) 306-309.

130. S. Masaki, T. Okada, N. Mataga, Y. Sakata and S. Misumi, On the Solvent-Induced Changes of Electronic Structures of Intramolecular Exciplexes, Bull. Chem. Soc. Jpn., 49 (1976) 1277-1283.

131. T.J. Chuang, R.J. Cox and K.B. Eisenthal, Picosecond Studies of the Excited Charge-Transfer Interactions in Anthracene-(CH_2)-N,N,-Dimethylaniline Systems, J. Am. Chem. Soc., 96 (1974) 6828-6831.

132. Y. Wang, M.K. Crawford and K.B. Eisenthal, Intramolecular Excited-State Charge-Transfer Interactions and the Role of Ground-State Conformations, J. Phys. Chem., 84 (1980) 2696-2698.

133. Y. Wang, M.K. Crawford and K.B. Eisenthal, Picosecond Laser Studies of Intramolecular Excited-State Charge-Transfer Dynamics and Small-Chain Relaxation, J. Am. Chem. Soc., 104 (1982) 5874-5878.

134. P.M. Felker, J.A. Syage, W.R. Lambert and A.H. Zewail, Direct Observation of Intramolecular Energy Transfer by Selective Picosecond Laser Excitation of a Single Chromophore in Jet-Cooled Molecules, Chem. Phys. Lett., 92 (1982) 1-3.

135. K. Mutai, The Study of Across-Space Intramolecular Charge-Transfer Interaction by Fluorescence Spectroscopy, Chem. Commun. (1970) 1209-1210.

136. H.A.H. Craenen, J.W. Verhoeven and Th.J. de Boer, Studies of Inter- and Intra-molecular Donor-Acceptor Interactions - IX. Intramolecular Charge Transfer Interactions in Aralkyl 2,4,6-Trinitrobenzoates, Tetrahed., 27 (1971) 2561-2566.

137. H.A.H. Craenen, J.W. Verhoeven and Th.J. de Boer, Studies of Inter- and Intra-molecular Donor-Acceptor Interactions - X. Intramolecular Charge Transfer Interactions in Substituted N-Aralkyl-Phthalimides and 1,8-Naphthalimides, Rec. Trav. Chim. Pays-Bas, 91 (1972) 405-416.

138. J.H. Borkent, J.W. Verhoeven and Th.J. de Boer, Fluorescence from Inter- and Intramolecular Charge-Transfer Complexes, Tetrahed. Lett. (1972) 3363-3366.

139. A.J. de Gee, J.W. Verhoeven, W.J. Sep and Th.J. de Boer, Through-Bond Charge-Transfer Interaction in N-(*p*-Methoxyphenylalkyl)-Pyridinium Ions, J. Chem. Soc., Perkin Trans. II (1975) 579-583.

140. A.J. de Gee, W.J. Sep, J.W. Verhoeven and Th.J. de Boer, Optical Activity of Intramolecular Charge-Transfer Transitions, J. Chem. Soc., Perkin Trans. II (1975) 670-675.

141. J.H. Borkent, A.W.J. de Jong, J.W. Verhoeven and Th.J. de Boer, Photophysics of Intramolecular Electron Donor-Acceptor Systems; N-Carbazolyl-$(CH_2)_n$-Tetrachlorophthalimide, Chem. Phys. Lett., 57 (1978) 530-534.

142. P. Pasman, N.W. Koper and J.W. Verhoeven, Photoinduced Long-Range Electron Transfer in Rigid Bichromophoric Molecules, Recl.: J. Roy. Neth. Chem. Soc., 101 (1982) 363-364.

143. P. Pasman, F. Rob and J.W. Verhoeven, Intramolecular Charge-Transfer Absorption and Emission Resulting from Through-Bond Interaction in Bichromophoric Molecules, J. Am. Chem. Soc., 104 (1982) 5127-5133.

144. G.F. Mes, H.J. van Ramesdonk, J.W. Verhoeven, Photoinduced Electron Transfer in Polychromophoric Systems. 2. Protonation-Directed Switching between Tri- and Bichromophoric Interaction, J. Am. Chem. Soc., 106 (1984) 1335-1340.

145. G.F. Mes, B. de Jong, H.J. van Ramesdonk, J.W. Verhoeven, J.M. Warman, M.P. de Haas and L.E.W. Horsman-van den Dool, Excited-State Dipole Moment and Solvatochromism of Highly Fluorescent Rod-Shaped Bichromophoric Molecules, J. Am. Chem. Soc., 106 (1984) 6524-6528.

146. P. Pasman, G.F. Mes, N.W. Koper and J.W. Verhoeven, Solvent Effects on Photoinduced Electron Transfer in Rigid, Bichromophoric Systems, J. Am. Chem. Soc., 107 (1985) 5839-5843.

147. S. Larsson and A. Volosov, Distance Dependence in Photoinduced Intramolecular Electron Transfer, J. Chem. Phys., 85 (1986) 2548-2554; (see erratum, J. Chem. Phys., 86 (1987) 5223).

148. S. Larsson and A. Volosov, Distance Dependence in Photoinduced Intramolecular Electron Transfer. Additional Remarks and Calculations, J. Phys. Chem., 91 (1987) (in press).

149. L.G. Schroff, A.J.A. van der Weerdt, D.J.H. Staalman, J.W. Verhoeven and Th.J. de Boer, Electron Donor-Acceptor Cyclophanes - I, Tetrahed. Lett. (1973) 1649-1652.

150. L.G. Schroff, R.L.J. Zsom, A.J.A. van der Weerdt, P.I. Schrier, N.M.M. Nibbering, J.W. Verhoeven and Th.J. de Boer, Electron Donor-Acceptor Cyclophanes II. Synthesis and Electron-Induced Fragmentation of Cyclophanes Derived from 1,2:4,5-Benzene- or 1,8:4,5-Naphthalene-Tetracarboxylic Diimides, Recl.: J. Roy. Neth. Chem. Soc., 95 (1976) 89-93.

151. J.H. Borkent, J.W. Verhoeven and Th.J. de Boer, Charge-Transfer Fluorescence from Donor-Acceptor Cyclophanes; Influence of Geometry and Solvent Polarity, Chem. Phys. Lett., 42 (1976) 50-53.

152. R.L.J. Zsom, L.G. Schroff, C.J. Bakker, J.W. Verhoeven, Th.J. de Boer, J.D. Wright and H. Kuroda, Polarized Absorption Spectra of Some Electron-Donor-Acceptor Cyclophanes, Tetrahed., 34 (1978) 3225-3232.

153. K.N. Ganesh and J.K.M. Sanders, Quinone-Capped Metalloporphyrins: Synthesis and Coordination Chemistry, J. Chem. Soc., Chem. Commun. (1980) 1129-1131.

154. K.N. Ganesh and J.K.M. Sanders, Quinone-Capped Porphyrins: Synthesis and Some Chemical Properties, J. Chem. Soc., Perkin Trans. I (1982) 1611-1615.

155. K.N. Ganesh, J.K.M. Sanders and J.C. Waterton, Quinone-Capped Porphyrins: NMR Studies of Static and Dynamic Stereochemistry and Coordination Properties, J. Chem. Soc., Perkin Trans. I (1982) 1617-1624.

156. P. Leighton and J.K.M. Sanders, Quinone-Capped Porphyrins as Model Photosynthetic Systems: Use of Metal Coordination to Control Chromophore Orientation and Interaction, J. Chem. Soc., Chem. Commun. (1985) 24-25.

157. M.P. Irvine, R.J. Harrison, G.S. Beddard, P. Leighton and J.K.M. Sanders, Detection of the Inverted Region in the Photoinduced Intramolecular Electron Transfer of Capped Porphyrins, Chem. Phys., 104 (1986) 315-324.

158. J.S. Lindsey and D.C. Mauzerall, Synthesis of a Cofacial Porphyrin-Quinone via Entropically Favored Macropolycyclization, J. Am. Chem. Soc., 104 (1982) 4498-4500.

159. J.S. Lindsey, D.C. Mauzerall and H. Linschitz, Excited-State Porphyrin-Quinone Interactions at 10 Å Separation, J. Am. Chem. Soc., 105 (1983) 6528-6529.

160. J.S. Lindsey, J.K. Delaney, D.C. Mauzerall and H. Linschitz, Photophysics of a Cofacial Porphyrin-Quinone Cage Molecule and Related Compounds: Fluorescence Properties, Flash Transients and Electron-Transfer Reactions, J. Am. Chem. Soc., 110 (1988) in press.

161. H. Linschitz, Panel Report on the Role of Porphyrins and Chlorins in Artificial Photosynthesis: in: J. Rabani (Ed.), Photochemical Conversion and Storage of Solar Energy - 1984, Part B, The Weizmann Science Press of Israel, Jerusalem, 1982, pp. 67-76

162. B. Stevens, Solute Reencounter Effects. 7. Generalized Orbital and State Correlations for the Photochemical Production and Geminate Charge Neutralization of Radical-Ion Pairs, J. Phys. Chem., 88 (1984) 702-706.

163. B. Morgan and D. Dolphin, The Synthesis of Porphyrins Doubly Linked to Quinones by Hydrocarbon Chains, Angew. Chem. Intl. Ed. Eng., 24 (1985) 1003-1004.

164. J. Weiser and H.A. Staab, Synthesis of a Porphyrin Sandwiched between Two Parallel *p*-Benzoquinone Units, Angew. Chem. Int. Ed. Engl., 23 (1984) 623-625.

165. J. Weiser and H.A. Staab, A New Benzoquinone-Bridged Porphyrin, Tetrahed. Lett., 26 (1985) 6059-6062.

166. C. Krieger, J. Weiser and H.A. Staab, Molecular Structure and Fluorescence Behaviour of a Benzoquinone/Porphyrin/Benzoquinone Sandwich Molecule, Tetrahed. Lett., 26 (1985) 6055-6058.

167. J. Weiser, D.C. Mauzerall and H.A. Staab, Electron Transfer in Cyclophane Porphyrin-Quinones (in preparation, 1988).

168. A. Osuka, H. Furuta and K. Maruyama, Synthesis of Quinone-Capped Porphyrins by Porphyrin Self-Photosensitized Reaction, Chem. Lett. (1986) 479-482.

169. P. Leighton and J.K.M. Sanders, Viologen-Capped Porphyrins as Model Photosynthetic Systems: Intra- and Intermolecular π-π Interactions, J. Chem. Soc., Chem. Commun. (1984) 856-857.

170. P. Leighton and J.K.M. Sanders, A Molecular Switch for Control of Conformation: Strained Intramolecular Coordination in 4,4'-Bipyridyl-Capped Zinc Porphyrins, J. Chem. Soc., Chem. Commun. (1984) 854-856.

171. J.A. Cowan and J.K.M. Sanders, Pyromellitimide-Bridged Porphyrins as Model Photosynthetic Systems. 1. Synthesis and Steady-State Fluorescence Properties. J. Chem. Soc., Perkin Trans. I (1985) 2435-2437.

172. R.J. Abraham, P. Leighton and J.K.M. Sanders, Coordination Chemistry and Geometries of Some 4,4'-Bipyridyl-Capped Porphyrins. Proton- and Ligand-Induced Switching of Conformations, J. Am. Chem. Soc., 107 (1985) 3472-3478.

173. J.A. Cowan, J.K.M. Sanders, G.S. Beddard and R.J. Harrison, Modeling the Photosynthetic Reaction Centre: Photoinduced Electron Transfer in a Pyromellitimide-Bridged 'Special Pair' Porphyrin Dimer, J. Chem. Soc., Chem. Commun. (1987) 55-58.

174. S. Nishitani, N. Kurata, Y. Sakata, S. Misumi, A. Karen, T. Okada and N. Mataga, A New Model for the Study of Multistep Electron Transfer in Photosynthesis, J. Am. Chem. Soc., 105 (1983) 7771-7772.

175. D. Gust, P. Mathis, A.L. Moore, P.A. Liddell, G.A. Nemeth, W.R. Lehman, T.A. Moore, R.V. Bensasson, E.J. Land and C. Chachaty, Energy Transfer and Charge Separation in Carotenoporphyrins, Photochem. Photobiol., 37S (1983) S46.

176. T.A. Moore, D. Gust, P. Mathis, J.-C. Mialocq, C. Chachaty, R.V. Bensasson, E.J. Land, D. Doizi, P.A. Liddell, W.R. Lehman, G.A. Nemeth and A.L. Moore, Photodriven Charge Separation in a Carotenoporphyrin-Quinone Triad, Nature (London), 307 (1984) 630-632.

176a. T.A. Moore, P. Mathis, D. Gust, A.L. Moore, P.A. Liddell, G.A. Nemeth, W.R. Lehman, R.V. Bensasson, E.J. Land and C. Chachaty, Energy Transfer and Photoinduced Charge Separation in a Carotenoporphyrinquinone Triad Molecule, in: C. Sybesma (Ed.) Adv. Photosynth. Res., Proc. 6th Int. Congr. Photosynth., 1983 (pub. 1984), Vol. 1, Nijhoff, The Hague, The Netherlands, pp. 729-732.

176b. D. Gust and T.A. Moore, A Synthetic System Mimicking the Energy Transfer and Charge Separation of Natural Photosynthesis, J. Photochem., 29 (1985) 173-184.

177. D. Gust, T.A. Moore, L.R. Makings, P.A. Liddell, G.A. Nemeth and A.L. Moore, Photodriven Electron Transfer in Triad Molecules: A Two-Step Charge Recombination Reaction, J. Am. Chem. Soc., 108 (1986) 8028-8031.

177a. D. Gust and T.A. Moore, Electron Transfer in Model Systems for Photosynthesis, in: V. Balzani (Ed.), Supramolecular Photochemistry, D. Reidel Publishers, Dordrecht, The Netherlands, NATO ASI Series C: Mathematical and Physical Sciences, Vol. 214, 1987, pp. 267-282.

178. P. Seta, E. Bienvenue, A.L. Moore, P. Mathis, R.V. Bensasson, P. Liddell, P.J. Pessiki, A. Joy, T.A. Moore and D. Gust, Photodriven Transmembrane Charge Separation and Electron Transfer by a Carotenoporphyrin-Quinone Triad, Nature (London), 316 (1985) 653-655.

179. P.A. Liddell, D. Barrett, L.K. Makings, P.J. Pessiki, D. Gust and T.A. Moore, Charge Separation and Energy Transfer in Carotenopyropheophorbide-Quinone Triads, J. Am. Chem. Soc., 108 (1986) 5350-5352.

180. H.T. Tien and S.P. Verma, Electronic Processes in Bilayer Lipid Membranes, Nature (London), 277 (1970) 1232-1234.

181. H.T. Tien, Photoelectric Bilayer Lipid Membrane: A Model for the Thylakoid Membrane, in: J.M. Olson and G. Hind (Eds.), Chlorophyll-Proteins, Reaction Centers and Photosynthetic Membranes, Brookhaven Symposia in Biology, 28 (1976) 103-131.

182. H.T. Tien, Photoeffects in Pigmented Bilayer Lipid Membranes, in: J. Barber (Ed.), Photosynthesis in Relation to Model Systems, Topics in Photosynthesis, 3, Elsevier/North Holland, New York, 1979, pp. 445-456.

183. H.T. Tien, Chlorophyll-Containing Bilayer Lipid Membranes: A Model for the Thylakoid Membrane, in: G. Akoyunoglou (Ed.), Photosynthesis I. Photophysical Processes - Membrane Energization, Balaban International Science Services, Philadelphia, 1981, pp. 253-262.

184. T.A. Moore, D. Gust, A.L. Moore, R.V. Bensasson, P. Seta and E. Bienvenue, Transmembrane Charge Transfer in Model Systems for Photosynthesis, in: V. Balzani (Ed.), Supramolecular Photochemistry, D. Reidel Publishers, Dordrecht, The Netherlands, NATO ASI Series C: Mathematical and Physical Sciences, Vol. 214, 1987, pp. 283-297.

185. J.K. Hurst and D.H.P Thompson, Mechanisms of Oxidation-Reduction across Vesicle Bilayer Membranes: An Overview, J. Membr. Sci., 28 (1986) 3-29.

186. J.K. Hurst, D.H.P. Thompson and J.S. Connolly, Photooxidation of Tetra-anionic Sensitizer Ions by Dihexadecyl Phosphate Vesicle-Bound Viologens, J. Am. Chem. Soc., 109 (1987) 507-515.

187. J.E. Guillet, Y. Takahashi, A.R. McIntosh and J.R. Bolton, A New Polymeric Model for the Active Site in Artificial Photosynthesis, Macromolecules, 18, (1985) 1788-1790.

188. R.V. Bensasson, E.J. Land, A.L. Moore, R.L. Crouch, G. Dirks, T.A. Moore and D. Gust, Mimicry of Antenna and Photoprotective Carotenoid Functions by a Synthetic Carotenoporphyrin, Nature (London), 290 (1981) 329-332.

189. G. Dirks, A.L. Moore, T.A. Moore and D. Gust, Light Absorption and Energy Transfer in Polyene-Porphyrin Esters, Photochem. Photobiol., 32 (1980) 277-280.

190. A.L. Moore, G. Dirks, D. Gust and T.A. Moore, Energy Transfer from Carotenoid Polyenes to Porphyrins: A Light-Harvesting Antenna, Photochem. Photobiol., 32 (1980) 691-695.

191. A.L. Moore, A.M. Joy, R. Tom, D. Gust, T.A. Moore, R.V. Bensasson and E.J. Land, Photoprotection by Carotenoids during Photosynthesis: Motional Dependence of Intramolecular Energy Transfer, Science, 216 (1982) 982-984.

192. D. Gust, T.A. Moore, R.V. Bensasson, P. Mathis, E.J. Land, C. Chachaty, A.L. Moore, P.A. Liddell and G.A. Nemeth, Stereodynamics of Intramolecular Triplet Energy Transfer in Carotenoporphyrins, J. Am. Chem. Soc., 107 (1985) 3631-3640.

193. P.A. Liddell, G.A. Nemeth, W.R. Lehman, A.M. Joy, A.L. Moore, R.V. Bensasson, T.A. Moore and D. Gust, Mimicry of Carotenoid Function in Photosynthesis: Synthesis and Photophysical Properties of a Carotenopyropheophorbide, Photochem. Photobiol., 36 (1982) 641-645.

194. M.R. Wasielewski, P.A. Liddell, D. Barrett, T.A. Moore and D. Gust, Ultrafast Carotenoid to Pheophorbide Energy Transfer in a Biomimetic Model for Antenna Function in Photosynthesis, Nature (London), 322 (1986) 570-572.

195. K.L. Marsh and J.S. Connolly, Solvent Effects on the Rate of Bacteriochlorophyll a Photooxidation, J. Photochem., 25 (1984) 183-195.

196. M. Gouterman, Optical Spectra and Electronic Structure of Porphyrins and Related Rings, in: D. Dolphin (Ed.), The Porphyrins, Vol. 3, Part A, Academic Press, New York, 1978, pp. 1-165.

197. L.L. Shipman, Oscillator and Dipole Strengths for Chlorophyll and Related Molecules, Photochem. Photobiol., 26 (1977) 287-92.

198. J.S. Connolly, A.F. Janzen and E.B. Samuel, Fluorescence Lifetimes of Chlorophyll a: Solvent, Concentration and Oxygen Dependence, Photochem. Photobiol., 36 (1982) 559-563.

199. J.S. Connolly, E.B. Samuel and A.F. Janzen, Effects of Solvent on the Fluorescence Properties of Bacteriochlorophyll a, Photochem. Photobiol., 36 (1982) 565-574.

200. G.R. Seely and J.S. Connolly, Fluorescence of Photosynthetic Pigments *in Vitro*, in: Govindjee, J. Amesz and D.C. Fork (Eds.), Light Emission by Plants and Bacteria, Academic Press, New York, 1986, pp. 99-133.

201. J.R. Bolton, Solar Fuels, Science, 202 (1978) 705-710.

202. J.R. Bolton, A.F. Haught and R.T. Ross, Photochemical Energy Storage - An Analysis of Limits, in: J.S. Connolly (Ed.), Photochemical Conversion and Storage of Solar Energy, Academic Press, New York, 1981, pp. 297-330.

203. M.L. Buhl, Jr., R.E. Bird, R.V. Bilchak, J.S. Connolly and J.R. Bolton, Thermodynamic Limits on Conversion of Solar Energy to Work or Stored Energy - Effects of Temperature, Intensity and Atmospheric Conditions, Sol. Energy, 32 (1984) 75-84.

204. J.R. Bolton, S.J. Strickler and J.S. Connolly, An Analysis of Limiting and Realizable Efficiencies in the Solar Photolysis of Water to Hydrogen and Oxygen, Nature (London), 316 (1985) 495-500.

205. R.H. Felton, Primary Redox Reactions of Metalloporphyrins, in: D. Dolphin (Ed.), The Porphyrins, Vol. 5, Part C, Academic Press, New York, 1978, pp. 53-125.

Chapter 6.3

Solar Energy Harvesting

Michael Grätzel

1. LIGHT ENERGY HARVESTING SYSTEMS

Fourteen years ago the world was facing an acute energy crisis. A sharp increase in the price of oil and dwindling supplies of fossile resources rendered the scientific community aware of the necessity to explore possibilities for new energy supplies. The use of sunlight to produce alternative and environmentally clean fuels, such as hydrogen appeared as an attractive strategy and a large interdisciplinary effort was soon on its way.

Scientists knew of course that in their strive to achieve this goal a lot could be learned from photosynthesis. Nature has built up a fascinating device to make use of sunlight in order to drive a thermodynamically uphill reaction, i.e. the reduction of carbon dioxide to carbohydrates by water. The photosynthetic units assembled in the thylakoide membranes comprise antenna pigments for light energy harvesting and a reaction center consisting of two photosystems in series. Several key features as to how photosynthetic energy conversion operates are known. Light induced charge separation is achieved through judicious spatial arrangements of the chlorophyll and elements of the electron transport chain across the membrane. Cooperative interaction between these components allows the electron transfer to proceed in a vectorial fashion: positive charges are accumulated at the inside of the vesicles while the negative counter charges are transferred to the outer surface. Enzymes play the role of catalysts that couple the charge separation events to fuel-generating reactions, i.e. the oxidation of water to oxygen and reduction of NAD^+ to NADH.

While artificial photoconversion devices should not attempt to imitate all the intricacies of natural photosynthesis, it is inconceivable that the challenging task of driving endergonic chemical reactions such as the cleavage of water into hydrogen and oxygen could be accomplished without suitable engineering on the molecular level. Microheterogeneous solution systems have therefore become a topic of primary interest. The most simple equivalent of a biological membrane is a surfactant micelle. These are aggregates of approximately spherical structure which form spontaneously in aqueous solutions of ionic or nonionic detergents. The hydrocarbon tails of the latter form the interior and the polar headgroups the surface of these assemblies. Typically, about 60-100 detergent molecules are associated in one micelle. In the case of ionic micelles there exists an electrostatic potential gradient extending from the surface into the surrounding aqueous solution. Thus, some of the most important features of biological membranes such as the presence of microscopically small hydrophobic regions as well as the charged lipid water interface are mimicked by micellar assemblies.

One of the features of the natural photosynthetic system which is not represented by the micelles is the presence of an occluded water case separated from the bulk aqueous phase by the thylakoid membrane.

Vesicle aggregates are appropriate model systems to mimick this particular molecular organisation. Lipid vesicles, also known as liposomes, are closed bilayer membranes separating an occluded inner aqueous phase from the bulk solution, Fig. 1. As in the case of micellar assemblies, they are constituted by amphiphilic compounds containing a hydrophilic and hydrophobic moiety. The preference of an amphiphile to aggregate into a liposome rather than a micelle is in general linked to the presence of two long alkyl chains in the molecule. For example, a very important class of natural liposome forming agents are the phospholipids or phosphoglycerides which are found in cellular membranes. Among the phosphoglycerides present in higher plants and animals, phosphatidyl choline (lecithin) derivatives such as 1-palmitoyl-2-oleoyl-phosphatidyl choline (PC):

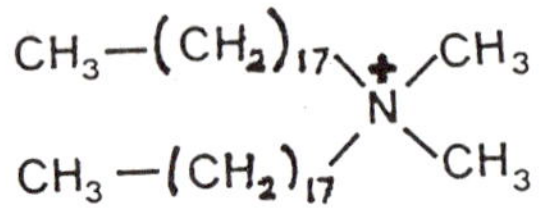

$$[CH_3-(CH_2)_{17}]_2N^+(CH_3)_2 \qquad \text{DODAC}$$

$$[CH_3-(CH_2)_{15}-O]_2P(=O)O^- \qquad \text{DHF}$$

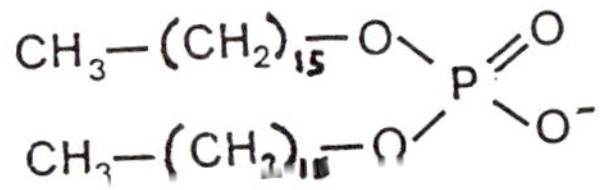

are most abundant. Its overall shape is roughly rectangular, the two fatty acid chains being approximately parallel to each other, whereas the phosphoryl choline moiety points in the opposite direction.

During the past years a large synthetic effort has been undertaken to provide artificial analogues to these natural liposome forming agents (1,2). Prominent examples are the cationic surfactant dioctadecyl dimethyl ammonium bromide (DODAC) and the anionic dihexadecyl phosphate (DHP).

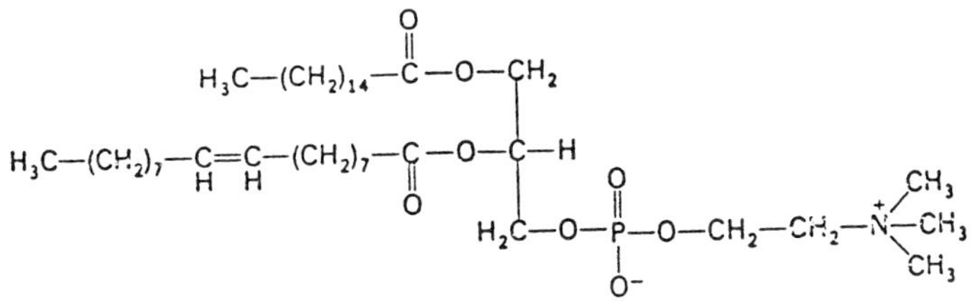

$$H_3C-(CH_2)_{14}-\overset{O}{\overset{\|}{C}}-O-CH_2$$

$$H_3C-(CH_2)_7-\underset{H}{C}=\underset{H}{C}-(CH_2)_7-\underset{O}{\underset{\|}{C}}-O-C-H$$

$$H_2C-O-\overset{O}{\overset{\|}{P}}(O^-)-O-CH_2-CH_2-\overset{+}{N}(CH_3)_3$$

The formation of bilayer vesicles from these surfactants is a spontaneous process in aqueous solution. As in the case of micelles, hydrophobic interactions are the major driving force for aggregation. Water molecules are released from the hydrocarbon tails of the amphiphiles as these tails become sequestered in the nonpolar interior of the bilayer resulting in a large entropy increase. Moreover, there are attractive van der Waals forces between the hydrocarbon tails which are closely packed in the membrane. Finally, there are favorable electrostatic and hydrogen bonding interactions between the polar head groups and the surrounding water molecules. Altogether, these factors afford a large gain in the free energy which for vesicles is much higher than in the case of micelle formation. As a result, liposomes are formed at concentrations which are several orders of magnitude below the CMC of the single chain surfactant analogues. Practically all the amphiphiles are in the aggregated state the concentration of monomers being neglibibly small.

In practice, liposomes are produced by suspending a suitable amphiphile, such as phosphatidyl choline in an aqueous medium which thereafter is subjected to sonication to yield a dispersion of vesicles that are quite uniform in size. Alternatively, vesicles can be prepared by rapidly mixing a solution of lipid in ethanol with water. This can be accomplished by injecting the lipid through a needle. Vesicles formed by these methods are nearly spherical and have a diameter of 300 to 1000 Å depending on the lipid used and other conditions, such as ionic strength and temperature.

In the electron transfer experiments discussed below the solution composition in the interior aqueous compartment of the liposomes is frequently different from that of the exterior bulk phase. Such a condition can be readily achieved by forming the vesicles in the presence of the agent which is to be entrapped in the inside of the vesicles. The suspension is subsequently dialysed or passed over a gel filtration column to remove the agent from the outside phase to which other substances can be subsequently added. An important question is how long such an unsymmetric distribution of solutes can be sustained across the membrane. The rate of migration through a bilayer membrane of the phosphatidyl choline type depends greatly on the nature of the diffusing molecule. Thus, the permeability coefficient for sodium ions is 10^{-12} cm/s while that for glucose is between 10^{-6} and 10^{-7} cm/s. Surprisingly the permeability of bilayer membranes for water is in general very high, the coefficient being 10^{-2} to 10^{-3} cm/s. Since the width of the membrane is of the order of 40 Å, this implies that the transit time of a water molecule is at most a few hundred microseconds, while that of a sodium ion is about a hundred hours. Therefore, one can neglect on the time scale the experiments dealing with photo-driven electron transfer and in the absence of ionophores, complications due to transmembrane diffusion of ionic species. It should be noted, however, that the permeability of bilayer vesicles with regard to certain ions may be dramatically increased under illumination with lights in the presence of membrane bound sensitizer (3).

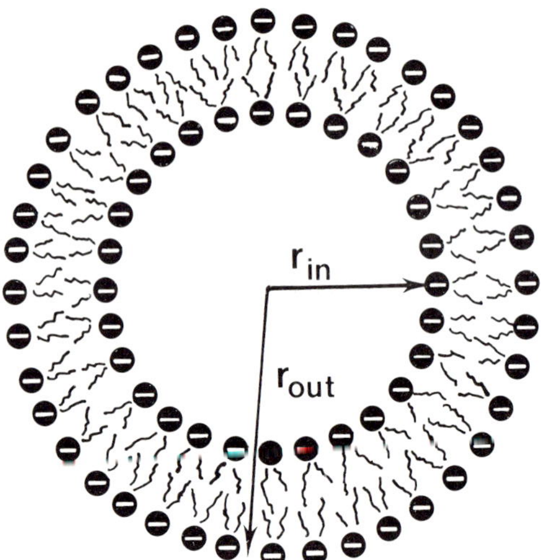

Fig. 1. Schematic illustration of the structure of liposomes.

Another type of molecular organisation which is of great interest to light energy harvesting and conversion systems are semiconductors. With regard to light induced charge separation, they have several advantages over the molecular assemblies dealt with in the previous chapters. For example, the diffusion of mobile charge carriers in semiconductors is very fast. Even for a material such as TiO_2, which is characterized by a heavy effective electron mass, the diffusion constant of the electron is at least 10^4 times larger than that of a molecular charge carrier in a micelle or vesicle. This presents an important advantage for heterogeneous photoreactions, where the achievement of high efficiency requires rapid displacement of photogenerated species from the interior of the light harvesting unit to the surface. Furthermore, since the chemical transformations usually involve redox reactions at the surface they can be mediated by derivatizing the semiconductor with suitable functional groups or by deposition of catalysts.

The configurations that are of interest in this chapter are semiconductor particles as well as their polycrystalline films. Particulate systems are commonly distinguished by their size and in Fig. 2 we have applied this classification to semiconductors. The two major categories to be discerned are colloids and macroparticles. The latter have a size exceeding approximately

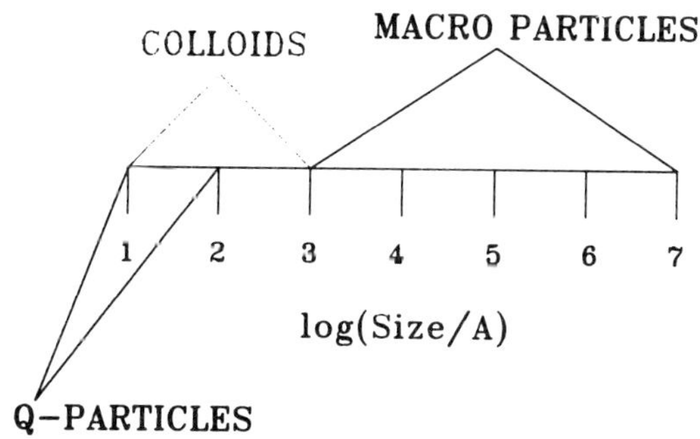

Fig. 2. Classification of Semiconductor Particulate Systems.

10^3 Å and form turbid suspensions. Colloids are smaller particles and give clear solutions. Amongst the colloids we distinguish semiconductors with normal optical and electronic behavior from those that display quantum size effects ("Q" particles). Quantum size effects occur when the Bohr radius of the first exciton in the semiconductor becomes commensurate with or larger than that of the particle. They manifest themselves in a blue shift of the fundamental absorption edge and of the luminescence maximum with decreasing particle size (4). Since the Bohr radius of the charge carriers is related to their respective effective mass:

$$r_B = \frac{h^2 \varepsilon \varepsilon_o}{c^2 \Pi m^*} \qquad (1)$$

it depends on the semiconductor material. For example, in the case of CdS, $m^*_{e-} = 0.2\ m_{e-}$, $\varepsilon = 8.9$ and $r_B = 24$ Å, while for TiO_2 $m^*_{e-} = 30m_{e-}$ and $\varepsilon = 170$, yielding $r_B = 3$ Å. Thus, quantum size effects are expected for particles with radii below 25 Å and 3 Å for CdS and TiO_2, respectively.

The shape and crystalline structure of semiconductor particles is strongly influenced by the preparation methods. If so desired, materials with uniform size and high crystallinity can be prepared. Chemical precipitation or controlled hydrolysis and subsequent polymerisation are the most frequently applied methods. Depending on preparation conditions, one obtains in this way semiconductor particles of spherical shape with amorphous or crystalline structure. As an example, we report in Fig. 3 electron microscopy and quasi-elastic light scattering results obtained from a hydrosol of TiO_2. The latter

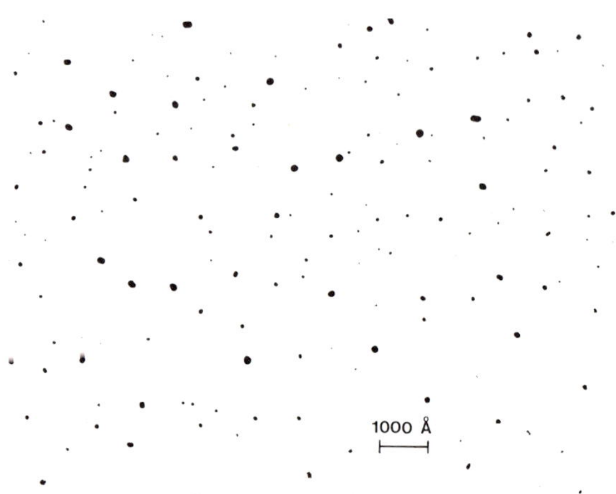

Fig. 3. Electron microscopy analysis of the size distribution and shape of colloidal TiO_2 particles.

was prepared by hydrolysis of $TiCl_4$ at low temperature (5). These data reveal that there is a distribution of particle sizes, the average diameter being 12 nm.

A particularly interesting development during recent years is the synthesis of monodisperse inorganic particles. Earlier work in this field concerned insulators, such as SiO_2 (6) or noble metals, such as Pt or Au (7). More recently, Matijevic et.al. (8) have elaborated methods for the preparation of monodispersed sols for a large number of compounds, including semiconducting oxides and chalcogenides. Apart from their application in the fabrication of special ceramics, such aggregates are also attractive from the viewpoint of heterogeneous catalysis and electron transfer reactions. An important advantage is that these particles can be produced with extremely narrow size distribution and well defined geometry. For example, it is possible to obtain spherically shaped aggregates with a smooth surface. The monodisperse nature of these sols facilitates the kinetic analysis of energy and electron transfer processes, whose rate is particle size dependent. Moreover, applications in heterogeneous catalysis and light-energy conversion can also be envisaged. These particles could constitute suitable building blocks or supports for more complex units. Their preceise geometry should allow for the engineering of systems with suitable functionality, optimising their performance in catalysis and heterogeneous electron transfer.

Apart from colloidal semiconductors there is currently great interest in thin film semiconductors as light harvesting units. The advantage of films with regard to dispersions of particles is that the electron transfer occurs in a vectorial fashion: the depletion layer field that develops at the semiconductor/solution interface allows for the local separation of positive and negative charge carriers produced by light. These charge carriers migrate to opposite sides of the membrane where they can induce thermodynamically uphill chemical reactions or are used for electrical power generation.

In order to be economically viable these films must consist of polycrystalline semiconductors. However, with such materials poor conversion yields are usually obtained due to the charge carrier recombination at grain boundaries and other defects reducing the minority carrier diffusion length. A strategy employed by us to overcome this problem is to use wide band gap oxide films in conjunction with a sensitizer.

The operation of a sensitized photoelectrochemical device containing a n-type semiconductor is outlined in Fig. 4. The dye must be selected such that its ground state redox potential lies within the band gap of the semiconductor while that of the excited state lies above the conduction band edge. It is clear from the diagram that a photon energy less than the band gap of the semiconductor may still excite the dye molecule to this state, from which a possible

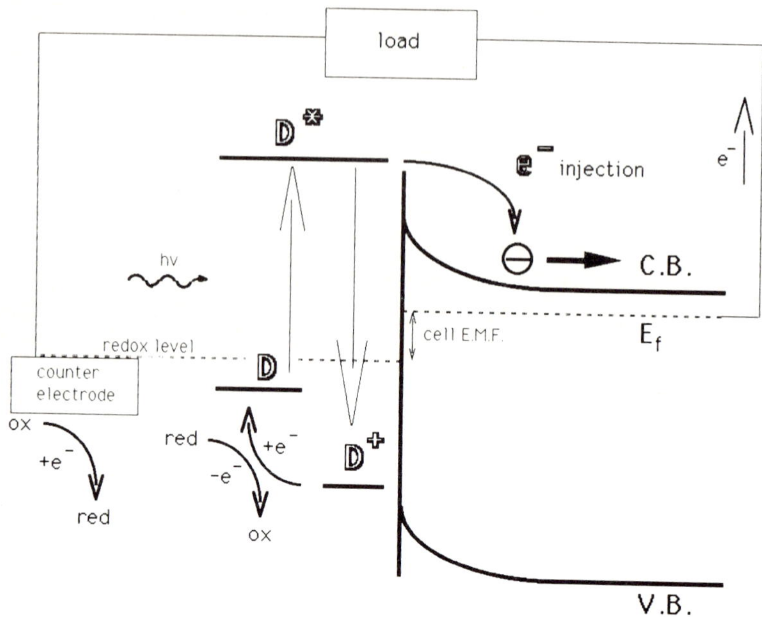

Fig. 4. Mechanism of dye sensitization and charge transfer at the surface of polycrystalline n-type semiconductor films.

decay mode is by electron loss to the n-type semiconductor. The dye molecule, now in an oxidized state, returns to the initial state by reaction with an electron donor (reducing) agent present in the electrolyte. At the cathode, a reduction reaction takes place. If the anode reaction is the reverse of the cathodic one, then the cell is a current producing system, analogous to a solid state photovoltaic cell. If this is not the case, for example if the two reactions are respectively the oxidation and reduction of water, then the cell is photosynthetic and in this example produces oxygen and hydrogen gases.

A system which meets these energy level conditions may nonetheless remain inefficient on mechanistic grounds. For example, if the dye is merely dissolved in the electrolyte, the excited state is so rapidly quenched, typically within 10^{-8} sec., that diffusion to the electrode surface and charge transfer may be ruled out. As a rule, surface-attached species alone can contribute to the sub-band gap photoresponse of the device. On such a modified semiconductor surface, only the first adsorbed monolayer can transfer charge, thicker dye layers tending to be insulating. However, on a flat surface optical absorption by monolayers is weak, at most a few percent, so that the sensitization effect is feeble. We have developed a system where the semiconductor (TiO_2) surface is highly porous (roughness factor ca. 200) and the dye is chemically modified to enhance adsorption. Below, we show representative dye structures employed in these studies:

A

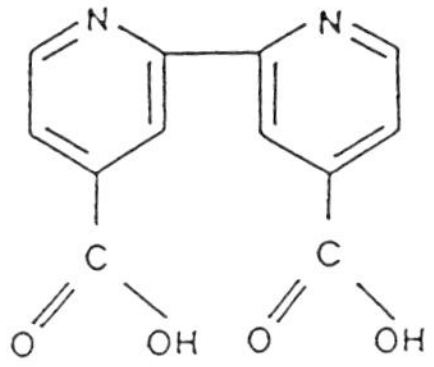

L

RuL_3

B

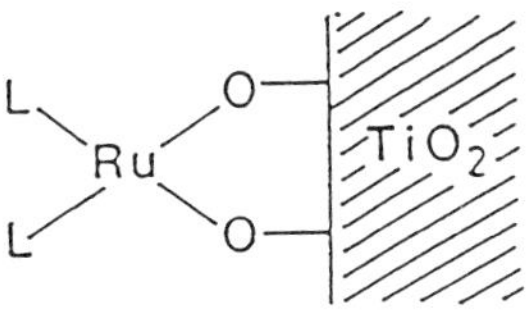

C

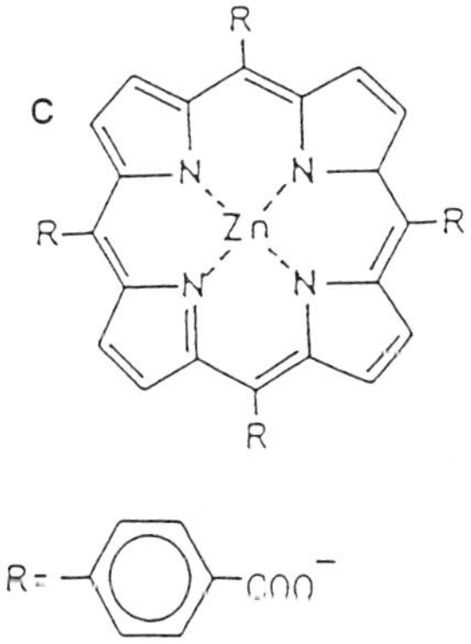

R=

D

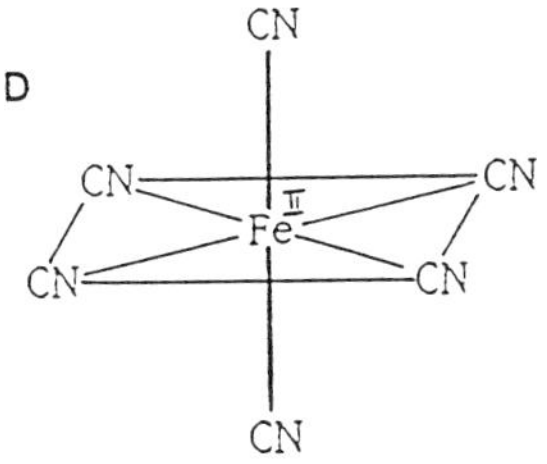

In chromophores A and C, the adhesion to the TiO_2 surface is greatly enhanced by the carboxylate group. In B the Ru complex is chemically attached via oxygen bridges. Ferrocyanide (D) forms a charge transfer complex with Ti^{IV} ions at the TiO_2 surface which greatly enhances the visible light response of this semiconductor (9).

On the porous TiO_2 layer (the fractal dimension of the layer is approximately 2.7), there is in consequence sufficient dye present for effective absorption of visible light, while relating intimate contact with the semiconductor to facilitate charge transfer. Transport losses in the semiconductor are minimal. Though functionally similar to a conventional n-type photoanode, the dye-sensitized semiconductor operates by electron injection and is therefore a majority carrier device. The high recombination losses due to disorder in a semiconductor structure where the photoexcited electroactive charge carriers are holes, are not encountered in the present case. Similar effects were also observed by Matsumura et.al. (10) and Alonso et.al. (11) on sintered ZnO electrodes sensitized by rose bengal dye although lower overall efficiencies were obtained. The TiO_2 layers used by us display an unprecendented incident photon to current conversion efficiency of up to 70% (12) in the visible. The progress in the development of these devices has recently been so rapid that practical applications in solar conversion devices can be envisaged in the near future.

2. ELECTRONIC EXCITATION ENERGY CONDUCTION IN MICELLAR ANTENNA SYSTEMS

Micellar systems have been widely used to study fundamental aspects of energy and electron transfer reactions in organized systems. These processes are distinguished by their unique kinetic features which differ from those observed in homogeneous solutions. We have recently reviewed the characteristics of light energy harvesting and light-induced charge separation processes in simple and functional surfactant aggregates (13). Therefore, we restrict ourselves in the following to the discussion of excitation energy conduction in a micellar antenna system. This is an intriguing model system for the antenna pigments in chloroplasts and has received surprisingly little attention so far.

The case will have been considered where electronic excitation energy is conducted along the micellar surface to a reactive trap. This process also occurs in the light energy harvesting units in photosynthesis. Recall that plants absorb sunlight via the antenna pigments consisting of chlorophyll and carotenoides. The excitation energy is subsequently transferred via a hopping process along the thylakoide vesicle surface to the reaction centers where light induced charge separation takes place. There are about 700 antenna molecules for each reaction center. A micellar assembly imitating the biological unit can be conceived in the following way: a functional surfactant is used to form the micelles whose head group is constituted by the sensitizer S. Incorporated in the micellar surface is a host molecule (T) which functions as an energy trap, Fig. 5. The study of the quenching of the S^* luminescence by T should yield information on the energy transfer rate within the micellar surface.

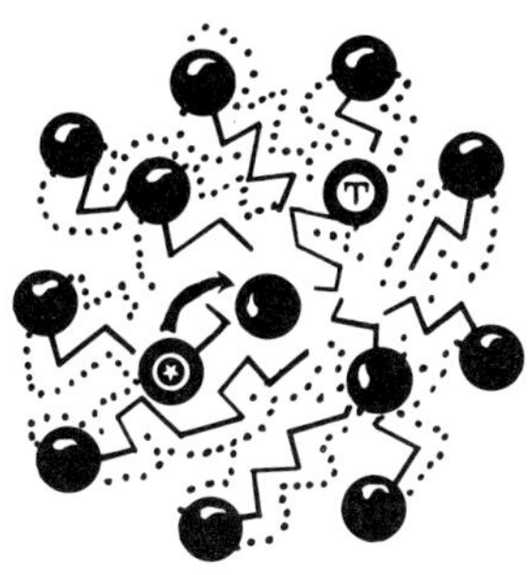

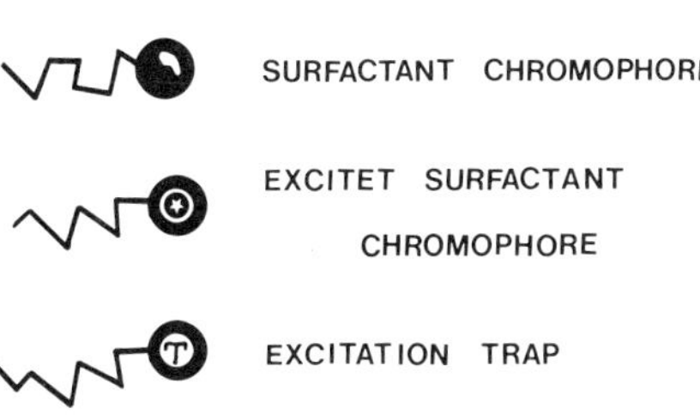

Fig. 5. Electronic excitation energy transfer within an array of micellized chromophores.

We model the kinetics of such an energy transfer reaction by considering a sequence of elementary steps:

$$\begin{array}{ccccccc} S^* & \xrightarrow{P_{tr}} & S & \xrightarrow{P_{tr}} & S & \xrightarrow{P_{tr}} & \ldots B \\ \downarrow P_d & & \downarrow P_d & & \downarrow P_d & & \end{array} \quad (2)$$

which involve hopping of excitation energy from a chromophore to an adjacent one. The excitation is quenched when it reaches a B site, i.e. a chromophore located next to the quencher. If a quencher has n neighbours, the number of A molecules which are available for energy conduction is:

$$A = A_o - n\,Q \quad (3)$$

where A_o and Q are the average number of A_o and quencher molecules per micelle respectively.

In Eq. (2) P_{tr} and P_d are the probabilities for energy transfer and deactivation of the excited chromophore respectively, i.e.

$$P_{tr} = k_{tr}/\Sigma k \quad \text{and} \quad P_d = (k_r + k_{nr})/\Sigma k \tag{4}$$

with $\Sigma k = k_r + k_{nr} + k_{tr}$

The rate constants k_r, k_{nr} and k_{tr} refer to radiative and non-radiative deactivation and energy transfer, respectively. Immediately after excitation of a chromophore within the micellar assembly, the probability for luminescence emission is given by:

$$P_f^o = P_A \cdot P_f \tag{5}$$

where $P_A = A/A^o$ is the fraction of choromophores that are not located in the immediate surrounding of the quencher where instantaneous quenching would occur and $P_f = k_f/\Sigma k$ is the probability for luminescence emission. Let $P_f(m)$ be the probability that luminescence occurs after m steps:

$$P_f^{(m)} = P_{tr}^{m+1} \cdot P_F^m \tag{6}$$

The total probability of fluorescence is:

$$P_f^{total} = \phi/\phi^o = \sum_{m=o}^{m=\infty} P_A^{m+1} P_{tr}^m P_f = P_A \cdot P_F \cdot \frac{1}{1 - P_A P_{tr}} \tag{7}$$

which can be written in the form:

$$\frac{\phi^o}{\phi} = \frac{A^o}{A} (1 + \frac{k_{tr}}{\Sigma k} \cdot n \frac{Q}{A^o} \tag{8}$$

A similar equation was employed by Shinitzky (14) to analyse the rate of energy transfer in micellar assemblies of sodium 2 hexadecylamino 6-naphthalene sulfonate (HNS^-) and N-palmitoyl L-tryptophan (PLT). In this case the quencher employed was N-hexadecyl pyridinium chloride. The rate constants for energy transfer between adjacent fluorophores were determined as 7×10^8 s^{-1} and 5.5×10^8 s^{-1} for HNS^- and PLT micelles, respectively.

One problem with employing Eq. (8) for this type of analysis is that it does not take into account the statistical nature of the distribution of quencher molecules over the host aggregates. Thus, instead of using the average occupation of a micelle by a quencher, i.e. the quantity of Q in Eq. (3), the probability of finding chromophores that are not in contact with quencher molecules is formulated more correctly by:

$$P_A(i) = \frac{A^o - i \cdot n}{A^o} \tag{9}$$

Here i indicates the number of quencher per micelle which is given by the Poisson distribution. The probability that a micelle with i quencher associations will fluoresce is:

$$P_f(i) = P_A(i)\, P_F \cdot \frac{1}{1 - P_A(i)\, P_{tr}} \tag{10}$$

The ratio of luminescence quantum yields in the presence and absence of quencher is therefore given by (15):

$$\Theta/\Theta^o = e^{-Q} \sum_{i=o}^{\infty} \frac{(Q)^i}{i!} \; \frac{P_A(i)}{1 - P_A(i)\, P_{tr}} \tag{11}$$

It should be noted that in the derivation of Eq. (11) we have not considered a dipolar (Förster) type energy transfer mechanism. Therefore, it is strictly applicable only for short range interactions such as triplet energy transfer between identical chromophores. In many cases, in particular when excited singlet states are involved, excitation energy transfer can occur over a long range and does not require collisional contact of the reactants. This so-called non-radiative or Förster-type mechanism will become prominent when the emission spectrum of the donor and the absorption spectrum of the acceptor show good overlap and, moreover, the electronic transitions accompanying the energy transfer are allowed. The latter condition is fulfilled, for example, in the case of singlet energy transfer. The first-order rate constant for Förster-type excitation energy transfer is given by the expression:

$$k_t = \frac{1}{\tau} \left(\frac{R_o}{R} \right)^6 \tag{12}$$

where τ is the lifetime of the excited donor singlet in the absence of acceptor and R_o the critical distance for which $k_{tr} = 1/\tau$, i.e. the energy-transfer rate and that for all other processes of deactivation are equal. The parameter R_o can be calculated from:

$$R_0^6 = \frac{9\ \ln 10 \cdot K^2 \cdot 10^{23}}{128\pi^5 n^4 N_A} \int_0^\infty F_D(\lambda)\varepsilon_A(\lambda)\lambda^4 d\lambda \tag{13}$$

Here κ^2 is a geometric factor depending on the relative position of the transition moments for donor emission and acceptor absorption, respectively. (The value of κ^2 is 2/3 if the molecular rotation is fast compared to the rate of energy transfer and $\kappa^2 = 0.476$ if during the transfer of energy there is no molecular rotation. In both cases the two transition moment vectors are assumed to be in a random position with respect to each other.) In Eq. (13) n is the refractive index of the solvent and N_A Avogadros's number. $F_D(\lambda)$ is the spectral distribution of the donor luminescence which has to be normalized such that:

$$\phi_o = \int_o^\infty F_D(\lambda) d\lambda \tag{14}$$

The absorption spectrum of the acceptor ^1A is expressed by the decadic molar extinction coefficient $\varepsilon_A(\lambda)$.

Recently Edniger et.al. have analyzed the kinetics of singlet excitation energy transport along a spherical micellar surface. A statistical mechanical theory was developed and the predictions from this model were compared to experimental results obtained with mixed micelles of triton X-100 and octadecyl rhodamine B. The latter was excited by a picosecond laser pulse and the time course of intramicellar energy transfer between different rhodamine B molecules was followed by monitoring the temporal decay of the fluorescence polarization. In the analysis of the data the assumption was made that the distribution of the chromophore over the micelles follows the Poisson law:

$$\frac{[M_i]}{[M]} = \bar{n}_s^{\,i} \exp(-\bar{n}_s)/i! \tag{15}$$

when $[M_i]$ represents the concentration of micelles containing i fluorophores, $[M]$ is the total micelle concentration and $\bar{n}_s$ is the average occupancy of the micelle by the florophore. The orientation averaged Förster radius for rhodamine B corresponds to 51.5 Å. The surfactant derivative has a slightly lower R_o value; i.e. 48 Å, since it is oriented at the micellar surface. Using this Förster radius and a size of 74 Å for the triton X-100 micelles, very good agreement between theory and experimental results was obtained. The assembly of the fluorophore on the micellar surface was found to render resonance energy transfer particularly effective. Thus, at an average occupance of 2.2 octadecyl rhodamine B molecules per micelle, the probability that the excitation is on the molecule which originally absorbed the light decreases to 50% within a time delay of only 5 ns.

3. LIGHT INDUCED CHARGE SEPARATION IN SEMICONDUCTOR SYSTEMS

The optimal use of semiconductors in solar light harvesting devices requires the knowledge of their solid state characteristics. Some of the electronic properties and their relation to light induced charge separation will now be discussed.

3.1. Band Edge Positions

In Fig. 6 we list the band gap and band edge position for a number of ionic and covalent materials. The data refer to conditions where the semiconductor is in contact with aqueous electrolyte of pH 1 and have been derived from flat band potential determinations via capacity measurements. The knowledge of the band edge position is particularly useful in photocatalysis. For example, in Fig. 6 we have indicated the standard potentials for several redox couples which indicate the thermodynamic limitations for the photoreactions that can be carried out with the charge carriers. For example, if a reduction of the species in the electrolyte is to be performed the conduction band position of the semiconductor has to be positioned above the relevant redox level.

Note that the ordinate in Fig. 6 presents internal energy and not free energy. The free energy of an electron hole pair is smaller than the energy of the band gap. The reason for this behavior is that the electron-hole pairs have a significant configurational entropy arising from the large number of translational states accessible to the mobile carriers in the conduction and valence band.

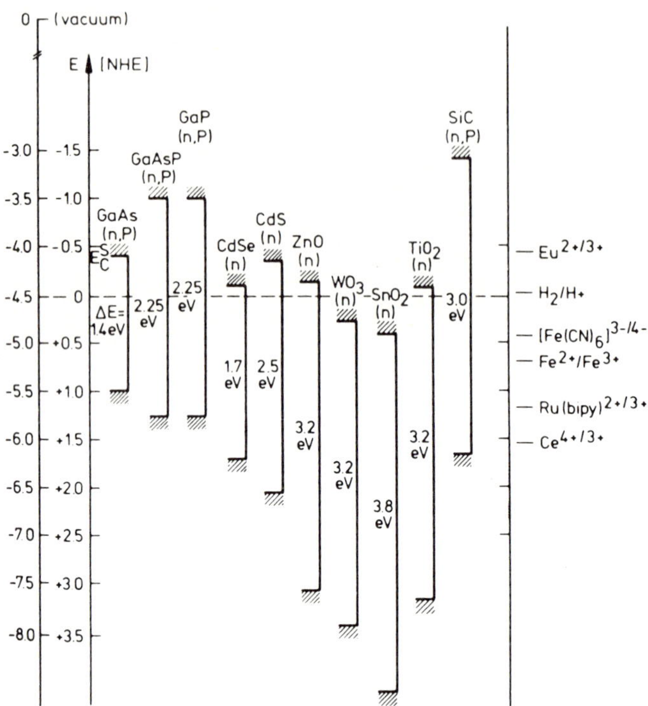

Fig. 6. Band edge position of several semiconductors in contact with aqueous electrolyte at pH 1.

The free energy of the charge carriers generated by photoexcitation of semiconductors is directly related to their chemical potential.

In the dark, under thermal equilibrium, the chemical potential of the electron is equal to that of the hole and corresponds to the Fermi level of the solid. The position of the Fermi level with respect to the band edges is given by Eq. (16) and (17):

$$E_f = E_{cb} + kT \ln (c_{e-} / {}^{1}n_{cb}) = \mu_{e-} \tag{16}$$

$$E_f = E_{vb} - kT \ln (c_{h+} / {}^{1}n_{vb}) = \mu_{h+} \quad h^+ \tag{17}$$

Under illumination the system departs from equilibrium and the chemical potential of the electron is different from that of the hole. As a result, the Fermi level splits into two quasi-Fermi levels, one for the electron and one for the hole. This is illustrated schematically for an n-type semiconductor in Fig. 7.

The maximum work that can be done by an electron-hole pair is given by the equation:

$$w = \mu_{e-} - \mu_{h+} = \text{photovoltage x e} \tag{18}$$

where e is the elementary charge. Using Eq. (16) and (17) to express the chemical potential one obtains:

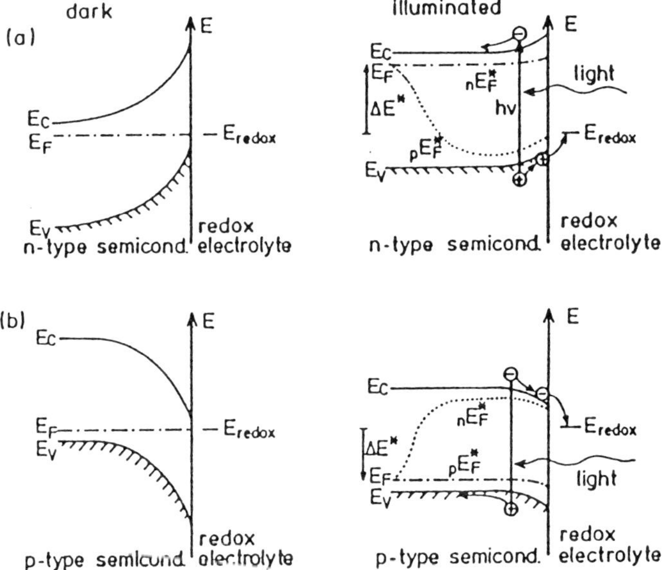

Fig. 7. Quasi-Fermi levels and photovoltages at an illuminated semiconductor/electrolyte contact with depletion layers.

$$\mu_{e-} = E_{cb} + kT \ln (c^{*}_{e-}/{}^{1}n_{cb}) \tag{19}$$

and

$$\mu_{h+} = E_{vb} - kT \ln(c^{*}_{h+}/{}^{1}n_{vb}) \tag{20}$$

where c^{*}_{e-} and c^{*}_{h+} are the nonequilibrium concentrations of electrons and holes respectively which depend on light intensity. Since at low light level, when the optical transition is far away from saturation, the concentration of minority carriers is proportional to the light intensity I, w and the open circuit photo-voltage increase with lnI.

3.2. Space Charge Layers and Band Bending

The generation of a space charge layer requires the transfer of mobile charge carriers between the semiconductor and the electrolyte. When an electroactive species is present in the electrolyte, the charge transfer can take place directly across the semiconductor solution interface. Alternatively, in the absence of a suitable redox couple in solution, the semiconductor can be polarized by applying an external bias voltage across the junction via an ohmic contact mounted at the back of the electrode. Within the space charge layer the valence and conduction bands are bent. Here, four different situations may be envisaged. These are illustrated in Fig. 8 for an n-type semiconductor in contact with an electrolyte. If there is no space charge layer, the electrode is at the flat band potential. If charges are accumulated at the semiconductor side which have the same sign as the majority charge carriers, one obtains an accumulation layer. If, on the other hand, majority charge carriers deplete into the solution, a depletion layer is formed. The excess space charge within this layer is given by immobile ionized donor states.

The depletion of majority carriers can go so far that their concentration at the surface decreases below the intrinsic level. If the electronic equilibrium is maintained, the concentration of holes in this region of the space charge layer exceeds that of electrons. As a consequence, the Fermi level is closer to the valence than the conduction band and the semiconductor is p-type at the surface and n-type in the bulk. This is called an inversion layer.

The illustration in Fig. 8 refers to n-type semiconductors. For p-type materials, analogous considerations apply, holes being the mobile charge carriers and immobile, negatively charged acceptor states forming the excess space charge within the depletion layer. The bands bend downwards in this case.

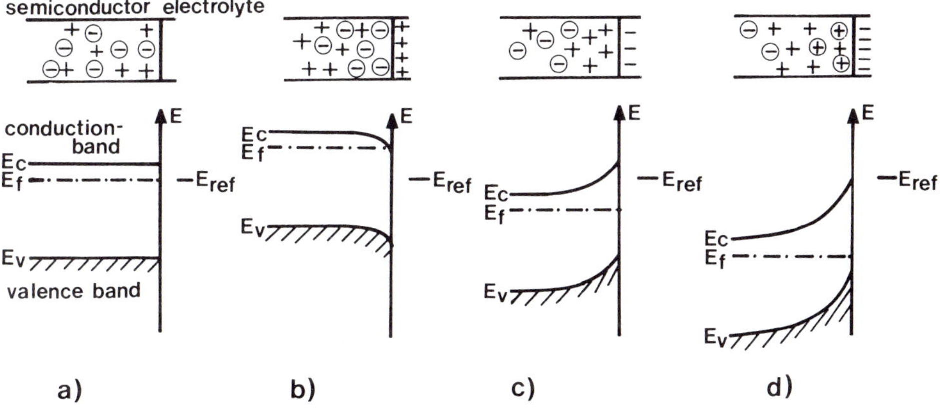

Fig. 8. Space charge layer formation at the n-type semiconductor-solution interface:
a) Flat band situation;
b) Accumulation layer;
c) Depletion layer;
d) Inversion layer.

In order to derive an expression for the width of the depletion layer, we use the definition $dq = C_{sc}d\phi$, where C_{sc} is the capacity per unit area of the space charge layer and q is the excess charge per unit area, and expresses it in the form:

$$q^2 = 2\varepsilon\varepsilon_o \int_{\phi_s}^{\phi_b} \rho\,(\phi)\,d\,\phi \tag{21}$$

Here ϕ_s and ϕ_b are the electrostatic potentials at the semiconductor surface and the bulk, respectively and ρ is the charge density. Using the Poisson equation to express ρ and making the assumption that the density of ionized donors and acceptors is constant within the depletion layer one obtains:

$$\phi_s - \phi_b = (kT/4e)\,(W/L_D)^2 = \Delta\phi_{sc} \tag{22}$$

where W is the width of the depletion layer and L_D is the Debye length:

$$L_D = (\varepsilon_o \varepsilon T/2e^2\,{}^1N)^{1/2} \tag{23}$$

${}^1N$ is the concentration of ionized dopants, expressed in number of ions per cm^3. From Eq. (22) it is apparent that for a given potential drop within the semiconductor, the width of the space charge layer is proportional to the Debye length, i.e. it decreases with the square root of dopant concentration.

In Fig. 7 we presented the position of the band edges of a variety of semiconductors. These results were obtained from measuring the differential capacity of the semiconductor/electrolyte junction. The semiconductor is subjected to reverse bias and the differential capacity is determined as a function of the applied potential. The space charge capacity of the semiconductor (C_{sc}) is in series with that of the Helmholtz layer (C_H) present at the electrolyte side of the interface. Thus, the total capacity is given by:

$$1/C = 1/C_{sc} + 1/C_H \tag{24}$$

Applying a parallel plate capacitor model to the space charge region one can express C_{sc} in terms of the depletion layer width:

$$C_{sc} = \varepsilon \cdot \varepsilon_o/W \tag{25}$$

Expressing W by Eq. (22) one obtains:

$$1/C^2 = \frac{2}{\varepsilon\,\varepsilon_o\,e\,{}^1N}\,(\Delta\phi_{sc}) \tag{26}$$

A similar equation was derived by Schottky:

$$1/C_{sc}^2 = \frac{2}{\varepsilon \varepsilon_o e^1 N} \left(|\Delta\phi_{sc}| - \frac{kT}{e} \right) \quad (27)$$

and this is frequently used in the evaluation of the capacity measurements. It is assumed that the condition $C_H >> C_{sc}$ applies and, therefore, that the measured capacity is that of the space charge layer. Since the capacity of the Helmholtz layer is relatively large ($\sim 10 \mu F/cm^2$) and does not vary significantly with the applied potential, this assumption is reasonable. Note, however, that it breaks down for semiconductors with large dielectric constants and high doping levels where C_H and C_{sc} are of comparable magnitude (16). Also, this condition is not fulfilled for semiconductors having a high density of surface states. A plot of the reciprocal capacity against the applied voltage (Mott-Schottky plot) gives a straight line in the depletion region and this is extrapolated to $1/C_{sc} = 0$ in order to evaluate the flat band potential. From the slope of the straight line, the concentration of ionized dopant can be derived.

Flat band potentials have meanwhile been determined for a large number of semiconductors under different experimental conditions (17). Apart from the semiconductor material these depend on the nature and composition of the electrolyte. Thus, specific adsorption of ions can shift the flat band position significantly. Of particular importance in this respect are potential determining ions. Potential determining ions play the same role in semiconductors as do electrons in metal electrodes. Changing their concentration changes the potential drop across the electrolytic double layer. For example, H^+ and OH^- are potential determining ions for oxide materials. With semiconducting oxides, Nerustian behavior is observed as a rule implying that the potential difference between the semiconductor and solution changes at room temperature by 59 mV for every pH unit.

$$(\phi_s^{ox} - \phi^{sol}) = \text{const} - 0.059 \text{ ph} \quad (28)$$

So far we have discussed the space charge layer in semiconducting electrodes. In spherical particles the potential distribution is obtained from the solution of the linearized Poisson Boltzmann equation (18).

$$\Delta\phi = \left(\frac{kT}{6e}\right) \left(\frac{r - (r_o - W)}{L_D}\right)^2 \left(1 + \frac{2(r_o - W)}{r}\right) \quad (29)$$

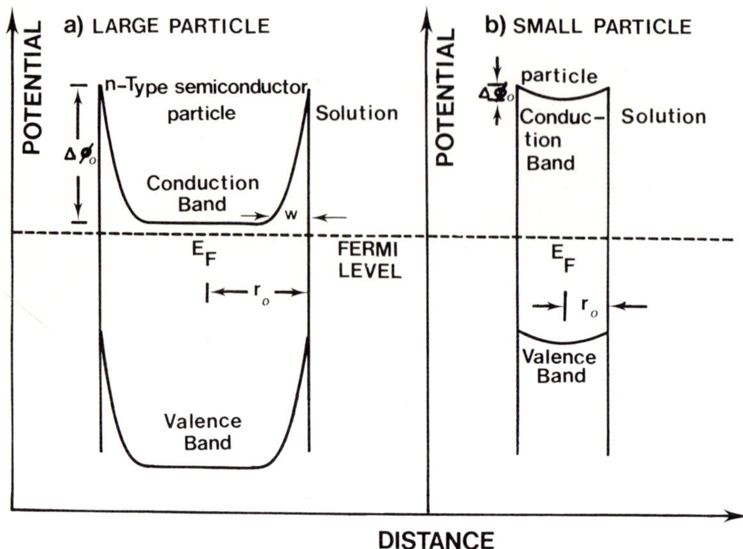

Fig. 9. Space charge layer formation in a large and small semiconductor particle in equilibrium with a solution redox system for which the Fermi level is E_F. The small particle depletes almost completely of charge carriers. Hence, its Fermi potential is located approximately in the middle of the band gap and band bending is negligibly small.

A graphical illustration of the potential distribution for an n-type semiconductor particle, which is equilibrium with a solution containing a redox system for which the Fermi level is E_F, is shown in Fig. 9. We discuss here two limiting cases of Eq. (19) which are particularly important for light induced electron transfer for semiconductor dispersions. If the size of the particles is much larger than the depletion layer width ($r_o \gg W$), the condition $r_o \simeq r$ holds within the depletion layer and Eq. (19) simplifies to:

$$\Delta\phi = \frac{kT}{2e}\left(\frac{r - (r_o - W)}{L_D}\right)^2 \quad (30)$$

For $r = r_o$ one obtains:

$$\Delta\phi_o = \frac{kT}{2e}\left(\frac{W}{L_D}\right)^2 \quad (31)$$

where $\Delta\phi_o$ corresponds to the total potential drop within the semiconductor particle. Note that Eq. (21) is identical to Eq. (22) which was derived for semiconductor electrodes. Apparently, for large particles, the depletion layer is the same as that for planar electrodes.

For very small (colloidal) semiconductor particles, the condition $W = r_o$ holds and Eq. (19) reduces to:

$$\Delta\phi = \frac{kT}{6e}\left(\frac{r}{L_D}\right)^2 \tag{32}$$

For the total potential drop within the particle one obtains:

$$\Delta\phi_o = \frac{kT}{6e}\left(\frac{r_o}{L_D}\right)^2 \tag{33}$$

From this equation and Fig. 9 it is apparent that the electrical field in colloidal semiconductors is usually small and that high dopant levels are required to produce a significant potential difference between the surface and the center of the particle. For example, in order to obtain a 50 mV potential drop in a colloidal TiO_2 particle with $r_o = 6$ nm, a concentration of 5×10^{19} cm^{-3} of ionized donor impurities is necessary. Undoped TiO_2 colloids have a much smaller carrier concentration and the band bending within the particles is therefore negligibly small. If majority carriers deplete from a colloidal semiconductor in solution and the particles are too small to develop a space charge layer the electrical potential difference resulting from the transfer of charge from the semiconductor to the solution has to drop in the Helmholtz layer, neglecting diffuse layer contributions. As a consequence, the position of the band edges of the semiconductor particle will shift. The same considerations hold for the transfer of minority carriers formed by photoexcitation of the semiconductor particle. For example if after photoexcitation of a colloidal n-type particle holes are transferred rapidly to an acceptor in solution while electrons remain in the particle a negative shift in the conduction band edge at the surface will take place.

The depletion layer at the semiconductor-solution interface, sometimes referred to as Schottky barrier, plays an important role in light induced charge separation. The local electrostatic field present in the space charge layer serves to separate the electron pairs generated by illumination of the semiconductor, Fig. (10). For n-type materials, the direction of the field is such that holes migrate to the surface where they undergo a chemical reaction, while the electrons drift through the bulk to the back contact of the semiconductor and subsequently through the external circuit to the counter electrode. Charge carriers which are photo-generated in the field-free space of the semiconductor can also contribute to the photocurrent. In solids with low defect concentration the lifetime of the electron hole pairs is long enough to allow for some of the minority carriers to diffuse to the depletion layer before they undergo recombination.

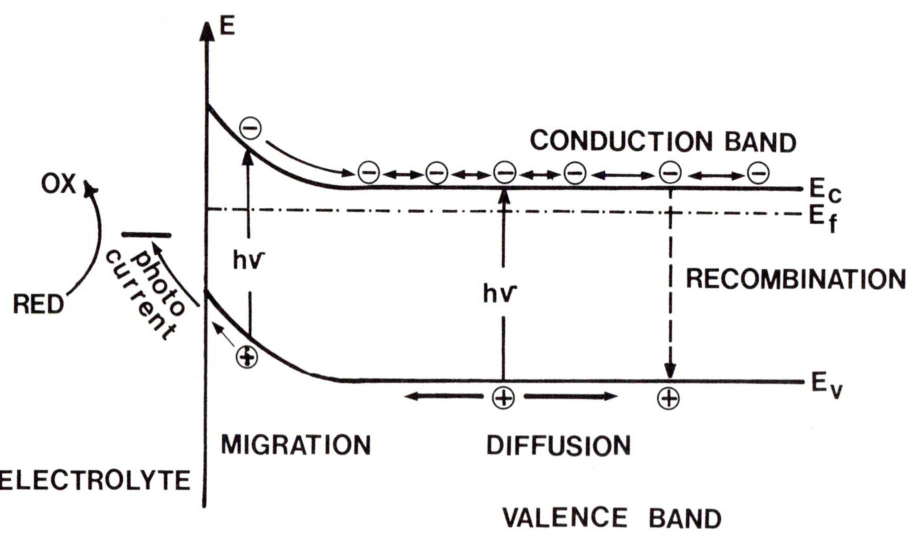

Fig. 10. Light induced charge separation assisted by the local electrostatic field present in the depletion layer of a semiconductor electrode in contact with an electrolyte.

Neglecting recombination within the space charge layer Gärtner (19) obtained for the photo-current density the expression:

$$i_{ph} = eI_o \; (1 - \exp(-\alpha W)/(1 + \alpha L) \quad (34)$$

where αI_o is the incident light intensity, l is the reciprocal absorption length and L is the minority carrier diffusion length. The latter is equal to the square root of the product of the diffusion coefficient and the mean carrier diffusion length. The latter is equal to the square root of the product of the diffusion coefficient and the mean carrier lifetime ($L = (D \times \tau)^{1/2}$. The quantum efficiency obtained from Eq. (30) by linearizing the exponential is:

$$\eta = (L + W)/(1/\alpha + L) \tag{35}$$

This equation predicts that the quantum yield of photocurrent generation should increase steeply as soon as the potential of the semiconductor electrode departs from the flat band value and a depletion layer is formed.

The experimental data usually deviate from this behavior in that the rise in the photocurrent is less steep than predicted by the Gärtner model. In addition, the photocurrent onset is frequently shifted from V_{fb} to a potential where the band bending is already several hundred millivolts. This behavior can be accounted for by recombination of charge carriers (Schockley-Read mechanism) which takes place mainly at the surface. Also, the Gärtner model assumes that the reaction of the minority carrier with solutes is so fast that it has no effect on the photocurrent. However, there is evidence that this interfacial electron or hole transfer can become an important factor, controlling the efficiency of photocurrent generation under near flat band conditions.

In the case of colloidal semiconductors, the band bending is small and charge separation occurs via diffusion. The absorption of light leads to the generation of electron-hole pairs in the particle which are oriented in a spatially random fashion along the optical path. These charge carriers subsequently recombine or diffuse to the surface where they undergo chemical reactions with suitable solutes or catalysts deposited onto the surface of the particle. Applying the random walk model in order to describe the motion of the charge carriers one obtains for the average transit time from the interior of the particle to the surface the expression:

$$T_d = r_o^2/\pi^2 D \tag{36}$$

For colloidal semiconductors τ is at most a few picoseconds. Thus for colloidal TiO_2 ($D_{e-} = 2 \times 10^2 cm/s$) with a radius of 6 nm the average transit time of the electron is 3 ps.

In the presence of a depletion layer, the transit time of the minority carriers is further reduced since, in this case, the Debye length has to be used in Eq. (32) instead of the particle radius (20). However, as pointed out by Curran and Lamouche (21), the potential difference between particle surface and interior has to be at least 50 mV in order for migration to dominate over diffusions.

It should be noted that the random walk model breaks down for particles exhibiting quantum size effects. Here the wave function of the charge carrier spreads over the whole semiconductor cluster and they do not have to undergo diffusional displacement to accomplish reactions with species present at the surface.

Since in colloidal semiconductors the diffusion of charge carriers from the interior to the particle surface can occur more rapidly than their recombination, it is feasible to obtain quantum yields for photoredox processes approaching unity. Whether such high efficiencies can really be achieved, depends on the rapid removal of at least one type of charge carrier, i.e. either electrons or holes, upon their arrival at the interface. This underlines the important role played by the interfacial charge transfer kinetics which we shall treat in further detail below.

3.3. Experimental Methods for the Study of Light Induced Charge Separation in Semiconductor Particles and Electrodes

We distinguish stationary from pulsed methods. For semiconductor electrodes, the stationary method consists in the measurement of the photocurrent as a function of applied bias. It is difficult to derive from such measurements the rate constant for the interfacial charge transfer step. It was shown in the previous chapter that the photocurrent is influenced by charge carrier recombination in the bulk and at the surface. Moreover, the heterogeneous charge transfer is often preceded by the trapping of the minority carriers in surface states. Thus, the overall photocurrent contains contributions from several processes which are difficult to discern. Although the individual rate constants for the interfacial charge transfer cannot be derived from such measurements, relative rates for minority carrier reactions with different scavengers present in the electrolyte can readily be determined this way (22). In addition, the efficiency of various redox species in suppressing the photocorrossion of semiconductor electrodes can be determined from such photocurrent measurements.

Stationary methods have also been applied to suspensions of semiconductor particles (23,24) and colloids (25). These experiments employ a conventional electrochemical cell, the semiconductor particles being dispersed in the compartment of the working electrode. Excitation by band gap irradiation produces electron-hole pairs in the particles. The electrons or holes are subsequently transferred to a suitable relay molecule dissolved in the electrolyte. This in turn transports the charge to the collector electrode producing a current which is measured as a function of the applied potential. These investigations yield valuable information on the position of the Fermi level within the semiconductor particles and on the kinetic features of heterogeneous electron transfer reactions involving colloidal semiconductors on electrodes.

With colloidal semiconductors the diffusion of the particles is rapid enough to allow for the experiments to be performed in the absence of relay compounds. In this case, the exchange of mobile charge carriers between the semiconductor colloid and the collector electrode can be directly measured yielding information on the rate of the interfacial redox reaction as well as the energetic position of the electronic energy bands within the particle (25).

Perone et.al. (26) have developed pulsed laser techniques for the study of light induced processes on semiconductor electrodes. Transient coulostatic photopotentials could be studied with a ca. 10 ns time resolution. It was suggested that the photopotential decay arises from space charge relaxation. A similar technique was developed by Gottesfeld et.al. (27,28). Following their work, several authors investigated pulsed laser induced photopotentials or photocurrents at different semiconductors such as TiO_2 (29), WSe_3 (30), Fe_2O_3 (31), PtS_2 (32) and CdSe (33). Laser induced transients at the rhodamine B/SnO_2 interface resulting from spectral sensitization were also studied (34). More recently, Willig et.al. (35) have succeeded in obtaining very high time revolution by using picosecond laser pulse excitation to study recombination and interfacial charge transfer with semiconductor electrodes.

A direct access to the analysis of heterogeneous charge transfer reaction is provided by laser excitation of solutions of semiconductor particles, in particular colloids. The latter are very attractive for time resolved studies since they yield transparent solutions. This allows for the ready analysis of the extremely fast interfacial electron transfer events occurring in these systems. Examples will be presented in the following chapters which illustrate that the primary events of light induced charge separation, carrier recombination and trapping as well as the heterogeneous redox events at the semiconductor surface leading to chemical change, are frequently within the picosecond time domain. Only be exploiting the transparent nature of semiconductor sols can such processes be directly monitored.

Among the time resolved experimental methods which are currently applied in the analysis of heterogeneous electron transfer events in semiconductor dispersions picosecond laser excitation coupled with kinetic spectroscopy, e.g. UV/visible or Raman, gives the highest resolution which is in the 10^{-12} s time domain. Apart from colloids, the analysis of fast electron transfer reactions on suspensions of larger particle or on dry powders has made good progress. Time resolved reflexion and microwave loss spectroscopy have so far given the most promising results (36-39).

3.4. Dynamics of Charge Carrier Trapping and Recombination in Small Semiconductoı Particles

We shall illustrate the principle of these investigations by reporting studies with colloidal titanium dioxide (anatase) particles having a diameter of 120 Å. Irradiation of such colloidal solutions in the presence of a hole scavenger such as polyvinyl alcohol or formate ions results in the accumulation of electrons in the particles. As a result, the solution assumes a beautiful blue color under illumination. It was found that up to 300 electrons can be stored in one TiO_2 particle. (A TiO_2 particle of 120 Å size has about 3600 conduction band states. Therefore, at most 10% of the available states are occupied by electrons.) The absorption spectrum of these stored electrons is shown in Fig. 11.

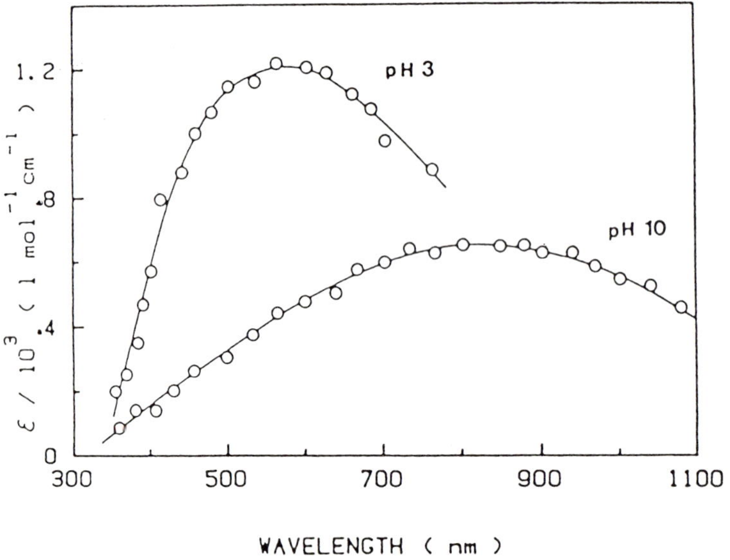

Fig. 11. Absorption spectrum of conduction band electrons in colloidal TiO_2 particles at pH 3 and pH 10.

The possibility of accumulating charges in a reaction space of minute dimensions is a very important feature of colloidal semiconductor particles. It should be recalled that light energy conversion processes such as the cleavage of water in hydrogen and oxygen or the reduction of carbon dioxide involve without exception multi-electron transfer processes. Hence, there is a need for structural organization that provides the possibility of charge storage. In natural photosynthesis, this function is assumed by the plastoquinone assembly in the thylakoid membrane. The colloidal TiO_2 particles investigated here provide an inorganic equivalent for such electron reservoirs.

One might expect that the storage of 300 electrons in a 50 Å-sized particle should produce a large negative shift in its potential with respect to the surrounding aqueous solution. However, the TiO_2/water interface is distinguished by a large double layer capacity, which is 85 $\mu F/cm^2$ at the point of zero charge and increases to 200-300 $\mu F/cm^2$ in alkaline solution (40). The negative shift in the potential of the particle during accumulation of 300 electrons is therefore only 25 to 90 mV. This explains why such highly charged particles can be preserved in anaerobic aqueous environment for at least several weeks without giving rise to significant hydrogen generation. The situation appears to be different with colloidal CdS particles where laser excitation has been reported to lead to huge negative shifts in the particle potential resulting in ejection of negative charge carriers to produce hydrated electrons (41).

The electron spectrum was found to be sensitive to the pH of the solution. Under alkaline conditions the electron absorption is very broad and has a maximum around 800 nm. Lowering the pH to 3 produced a pronounced blue shift in the spectrum which under these conditions shows a peak at 620 nm. The sensitivity of the electron absorption to the solution pH would indicate that they are located in the surface region of the particles. This has been confirmed by recent ESR experiments which show that under acidic conditions the electrons are trapped at the TiO_2 surface in the form of Ti^{3+} ions (42). Using redox titration, we have recently been able to determine the extinction coefficient of the trapped electrons (43). For the colloidal solutions of pH 3 the extinction coefficient at 600 nm is 1200 M^{-1} cm^{-1}.

Taking advantage of the characteristic optical absorption of trapped electrons in the colloidal TiO_2 particles, we have recorded their recombination with free and trapped holes in the picosecond to microsecond domain (44). Fig. 12 shows the temporal evolution of the transient spectrum after excitation of TiO_2 with a frequency tripled (353 nm) Nd laser pulse of ca. 40 ps duration.

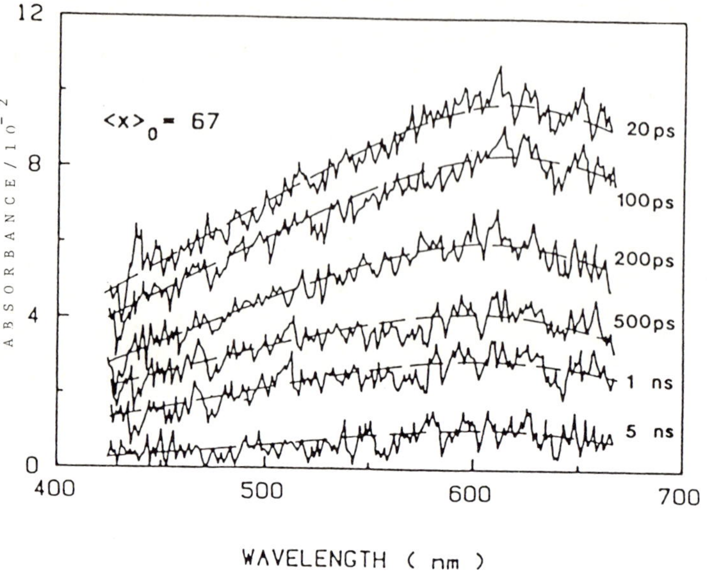

Fig. 12 Transient spectrum observed at various time intervals after picosecond excitation of colloidal TiO_2. Conditions: (TiO_2) = 17 g/l, pH 2.7, Ar saturated solution, optical pathlength 0.2 cm. Average number of electron-hole pairs present initially in one TiO_2 particle is 67.

In Fig. 12 the spectrum of the trapped electron develops within the leading edge of the laser pulse indication that the trapping time of the electron is less than 40 ps. This is in accordance with the predictions of Eq. (36). The electron absorption decay is due to recombination with valence band holes.

$$TiO_2\ (e^-_{tr} + h^+) \xrightarrow{k_r} TiO_2 \quad (37)$$

We have conceived a stochastic model to analyze the kinetics of this reaction. Since the recombination takes place between a restricted number of charge carriers restricted to the minute reaction space of a 120 Å-sized colloidal TiO_2 particle, it cannot be treated by conventional homogeneous solution kinetics. The time differential of the probability that a particle contains x electron-hole pairs at time t is given by:

$$dP_x(t)/dt = k(x+1)^2\, P_{x+1}(t) - kx^2\, P_x(t) \quad (38)$$

where x = 0, 1, 2, ...

This system of differential equations is to be solved subject to the condition that the initial distribution of electron-hole pairs over the particles follows Poisson statistics. The average number of pairs present at time t, $\langle x\rangle(t)$, can be calculated by means of the generating function technique (44) yielding:

$$\langle x\rangle (t) \quad + \quad \sum_{n=1}^{\infty} c_n \exp(-n^2 kt) \tag{39}$$

where

$$c_n = 2\exp(-\langle x\rangle_o)\,(-1)^n \sum_{i=n}^{\infty} \frac{\langle x\rangle_o^{\,i}}{(n+i)\,!} \prod_{j=1}^{n} (n-i-j) \tag{40}$$

The paramter $\langle x\rangle_o$ is the average number of pairs present at $t = 0$.

Two limiting cases of Eq. (39) are particularly relevant: when $\langle x\rangle_o$ is very small, Eq. (39) becomes a simple exponential and the electron-hole recombination follows a first order rate law. Conversely, at high average initial occupancy of the semiconductor particles by electron-hole pairs, i.e. $\langle x\rangle_o > 30$, Eq. (39) approximates to a second order rate equation:

$$\langle x\rangle (t) = \frac{\langle x\rangle_o}{1+\langle x\rangle_o kt} \tag{41}$$

In Fig. 12 the initial concentration of electron-hole pairs was sufficiently high to allow for evaluation of the recombination process by the second order rate equation, Eq. (41). This analysis gives for the recombination rate coefficient the value 3.2×10^{-11} $cm^3 s^{-1}$ corresponding to a lifetime of 30 ns for an electron-hole pair in a colloidal TiO_2 particle with a size of 120 Å. From this value one derives a diffusion length of 2.2×10^{-5} cm for the electron.

Experiments carried out at low laser fluence showed that hole trapping at the surface can compete with the recombination. In aqueous dispersions of oxide semiconductors such as TiO_2, hydroxide ions are present at the surface which can trap positive holes. Note that there are two types of OH^- groups, basic and acid ones (45). The former are attached to one Ti^{4+} ion while the latter bridge two Ti^{4+} sites. Since the electron density is higher on the basic OH^-, it is expected that these act as the hole traps. The laser photolysis experiments with colloidal TiO_2 yield a rate constant of 4×10^5 s^{-1} for this reaction. The recombination rate of electron-hole pairs has recently been determined for other semiconductor colloids than TiO_2. For example, Nosaka and Fox (46) have derived a value of $(9 \pm 4) \times 10^{-11}$ cm^3 s^{-1} for the recombination coefficient for Cds particles.

A simple and effective way to influence the charge carrier recombination dynamics is by introducing suitable dopants in the semiconductor. For example, substitutional doping of TiO_2 with Fe^{III} leads to a drastic retardation of the electron-hole recombination rate. This is the origin of the well known photochromism exhibited by Fe^{III} containing TiO_2 particles (47,48). Under illumination in an inert gas atmosphere, a black-green color is produced which fades very slowly after the light has been turned off. Recent combined EPR and laser photolysis studies with Fe^{III}-doped TiO_2 colloids in aqueous solution have helped to elucidate the mechanism of this photochromic effect (49). Incorporation of Fe^{III} ions in the colloidal particles leads to the formation of substitutional Fe^{III} on Ti^{IV} lattice sites and to charge compensating oxygen vacancies:

$$Fe_2O_3 \rightarrow 2|Fe^{III}|'_{Ti} + 3O_o + V_o^{\bullet\bullet} \tag{42}$$

The former act as traps for valence band holes:

$$|Fe^{III}|'_{Ti} + h^+ \rightarrow |Fe^{IV}|_{Ti}, \ \Delta H^o = -0.12 \text{ eV} \tag{43}$$

while the conduction band electrons are scavenged by Ti^{4+} ions at the particle surface:

$$V_o^{\bullet\bullet} + e^-_{cb} \rightarrow V_o^{\bullet}, \ \Delta H^o = -0.08 \text{ eV} \tag{44}$$

Fig. 13 illustrates the mechanism of light induced charge separation in Fe^{III} doped TiO_2 particles. Through the trapping reaction the electrons and holes are immobilized on sites that are locally separated. The recombination probably involves tunnelling of the charge carriers through the solid. In such a case, the rate constant for recombination decreases exponentially with the distance R separating the trapped electron from the trapped hole:

$$R = v \cdot \exp(-2R/a_o) \tag{45}$$

where v is a frequency factor which depends on the Franck-Condon overlap integral and attaining a maximum value of 10^{14} s^{-1} for a burnerless reaction and a_o is the radius of the hydrogenic wave function of the trapped carrier. As a consequence, charge carriers that are relatively far apart ($R > 30$ Å) recombine slowly, their survival time being in the minutes to hours time domain.

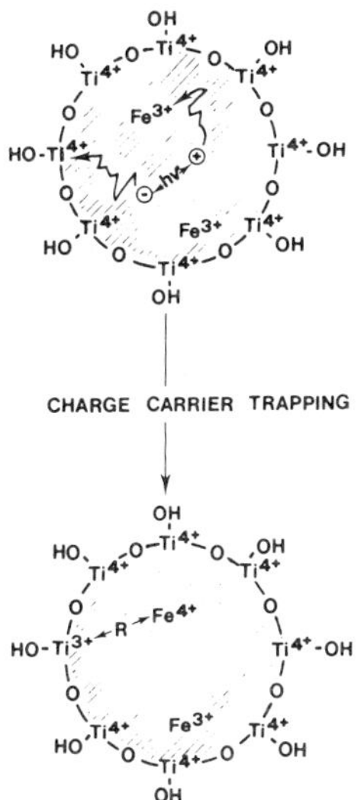

Fig. 13. Inhibition of electron-hole recombination in Fe^{III} doped TiO_2 particles. Light induced charge separation through local separation of trapped electron-hole pairs.

4. APPLICATIONS

In the following we shall illustrate applications of semiconductor particles and films in photocatalysis as well as the production of electricity from light.

4.1. Photocleavage of hydrogen sulfide

The cleavage of hydrogen sulfide by visible light

$$H_2S \rightarrow H_2 + S \qquad \Delta G^o = 8 \text{ kcal/mol} \tag{46}$$

is of industrial importance since H_2S occurs widely in natural gas fields and is produced in large quantities as an undesirably by-product in the coal and petroleum industry (50). It is a thermodynamically uphill reaction and we illustrate in the following how this process can be driven by visible light using a semiconductor particle as a light harvesting unit. The H_2S cleavage process might be used in industrial procedures where H_2S or sulfides are formed as a waste product whose rapid removal and conversion into a fuel, i.e. hydrogen, are desired. Also, in an intriguing fashion, these systems mimic the function of photosynthetic bacteria that frequently use sulfide as electron donors for the reduction of water to hydrogen.

The operation of the semiconductor particle is illustrated schematically in Fig. 14. Band gap excitation produces electron-hole pairs which diffuse to the surface of the particle where two redox reactions occur simultaneously. The valence band holes oxidize sulfide to sulfur while the conduction band electrons reduce protons to hydrogen. In order to accelerate the latter reaction, one deposits a catalyst such as Pt, Rh or RuO_2 on the particle surface. When sulfides such as CdS are used as semiconductors, the hole reaction with S^{2-} is very fast and the overall process occurs with a very high quantum yield approaching 50% (51-53).

PARTICLE

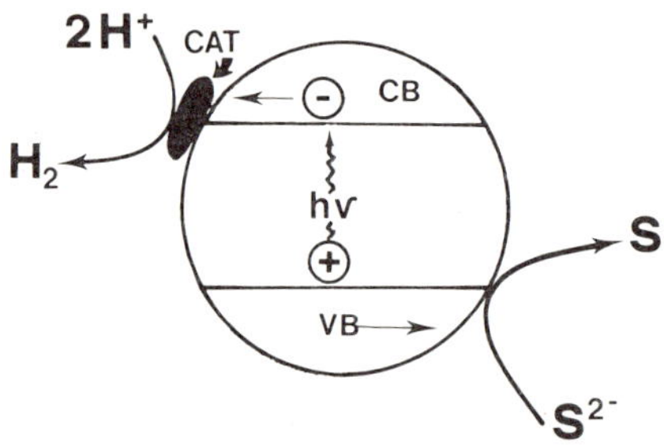

ELECTRODE

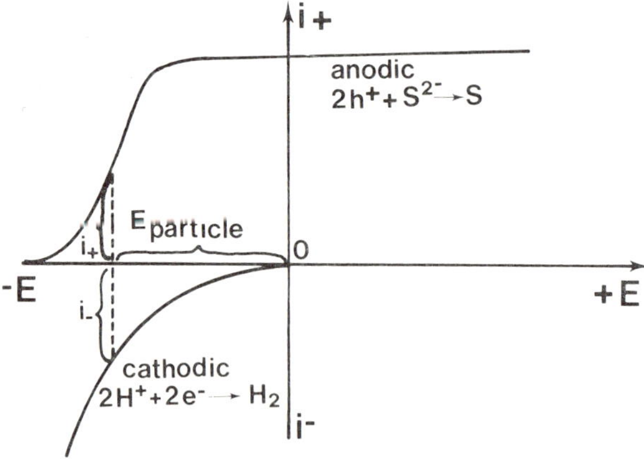

Fig. 14. Light Induced Cleavage of H_2S

In Fig. 14. we have separated the overall reaction (46) in the two individual redox processes which take place simultaneously at the semiconductor surface. These can be analyzed individually be illuminating a n-type semiconductor electrode, made from the same material as the particle, in the sulfide solution. A photo current is observed at a voltage positive of the flat band potential, V_{fb} when the electrode is polarized anodically. The cathodic dark current is determined with the semiconductor electrode onto which the same catalyst is deposited as on the particle. Since the overall current in the particle suspension is zero, we can use the cathodic and anodic current potential curves in the diagram to evaluate the particle potential under illumination. It corresponds to the potential for which $i = 0$ or $i_+ = i_-$. Furthermore, the reaction rate v can be derived from these current values via:

$$v = -\frac{dn(H_2S)}{dt} = 2FS\,i_+ = 2FS\,i_- \qquad (47)$$

where F is Faraday's constant and S is the total surface of the particles present in solution.

4.2. Photodecomposition of Water, Solar Energy Conversion

A process which is of particular importance for future energy supplies is the photodecomposition of water into hydrogen and oxygen.

$$H_2O \xrightarrow{h\nu} H_2 + 0.5\,O_2 \qquad \Delta G^o_{298} = 237 \text{ kJ/mol} \qquad (48)$$

Hydrogen is an energy vector with very desirable properties. First, it is a powerful fuel: the energy storage capacity is 120000 J/g as compared to 40000 for oil and 30000 for coal. Furthermore, hydrogen has the advantage over conventional (fossil) or nuclear energy sources that its combustion in a thermal engine or a fuel cell does not result in pollution of the environment. Reaction (48) requires an input of 237 kJ of free enthalpy for each mol of hydrogen produced under standard conditions.

The following economic analysis is instructive: consider a solar collector containing a catalytic system that performs reaction (48) with a 10% overall efficiency. At an average solar irradiance of 200 W/m^2, 66 m^3 of hydrogen can be obtained per year from each square meter of collector surface. At the current price for hydrogen of $0.2/$m^3$, this would represent a value of $13.-. If a ten percent return on the initial investment per year is considered to be reasonable, the expenses for the installation of the water splitting unit should not be higher than $130/$m^2$. The fulfilment of this condition could well be within reach by employing systems that are based on colloidal semiconductor dispersions or polycrystalline electrodes.

The efficiency of a solar energy converter is given by the ratio of work produced to incident power. The incident power is obtained by integrating over the solar emission spectrum. The efficiency of conversion of light into chemical energy is obtained by calculating the flux of solar photons that is absorbed by the converter. The majority of endergonic photochemical reactions have a threshold wavelength, λ_g; photons of energy less than this value cannot be absorbed and hence cannot contribute to the energy conversion. If a semiconductor is used as a light harvesting unit, the relation between band gap energy E_g, expressed in eV, and threshold wavelength is λ_g (nm) = $1240/E_g$. Let $N(\lambda)$ be the incident solar flux in the wavelength region between λ and $\lambda + d\lambda$, and $\alpha(\lambda)$ the extinction coefficient of the absorber. The absorbed flux of photos is then given by:

$$J_{abs} = \int_0^{\lambda_g} N(\lambda)\alpha(\lambda)d\lambda \tag{49}$$

The available solar power E (in W/m^2) at the band gap wavelength λ_g is given by:

$$E = J_{abs}\frac{hc}{\lambda_g} \tag{50}$$

where h is Planck's constant and c the velocity of light. The fraction η_E of incident solar power available to initiate photochemistry is then:

$$\eta_E = \frac{E}{\int_0^{\infty} E(\lambda)d\lambda} \tag{51}$$

where the denominator in Eq. (51) represents the total solar power. Curve I in Fig. 15 is a plot of this η_E (for AM 1.2 radiation), as a function of λ_g, for the ideal case where $\alpha(\lambda) = 1.0$ for $\lambda \leqslant \lambda_g$. The η_E has a maximum value of $\sim$ 47% at 1110 nm, but it is rather broad (> 45% over the wavelength range of 800 to 1300 nm). Curve I represents an ideal limit; in real life, there are further thermodynamic and kinetic limitations on the conversion of light energy into chemical energy. These have been examined in detail by Ross (54), Bolton (55), Almgren (56) and others. Taking into account the thermodynamic limitations alone, curve II in Fig. 15, the maximum drops to about 32% at a threshhold wavelength of 840 nm.

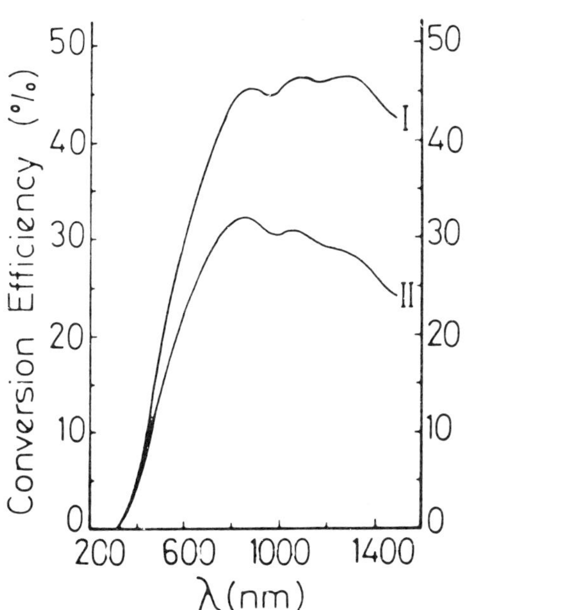

Fig. 15

For conversion of sunlight to electricity, the thermodynamic limit may well be approached. For example, a gallium arsenide solar cell has been reported with an efficiency (AM 1.4) of 23% (λ_g = 920 nm). However, if instead of electricity production one proposes to drive an endoergic chemical reaction such as the cleavage of water by light, the requirement of energy storage imposes additional constraints. As a result, the efficiency of solar energy conversion will be further reduced. These "kinetic" or "overvoltage" losses depend on the reaction rate as well as on the activity of the catalysts required to convert redox equivalents into fuels. For example, in the water cleavage process, electrons are used to produce hydrogen from water while the positive charge carriers perform the oxidation of water to oxygen. These reactions require the presence of a catalyst, and the overvoltage losses in these catalytic reactions are expected to lie between 0.4 and 0.8 V depending on the rate of water decomposition, i.e. on light intensity. If 0.8 V per absorbed photon have to be sacrificed for light energy storage, the threshold wavelength for carrying out the water photolysis is 611 nm and the maximum overall efficiency is ca. 17%. Considerably higher efficiencies can be achieved with devices where two photosystems operate in series, eg. natural photosynthesis or tandem photovoltaic cells (55).

a/ SENSITIZER / RELAY PAIR

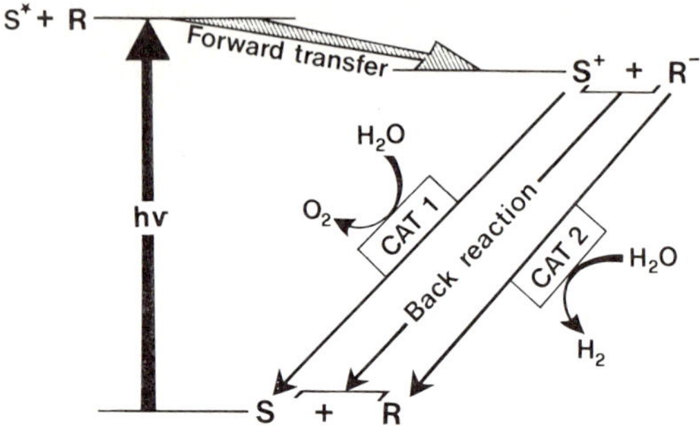

b/ SENSITIZER / COLLOIDAL SEMICONDUCTOR

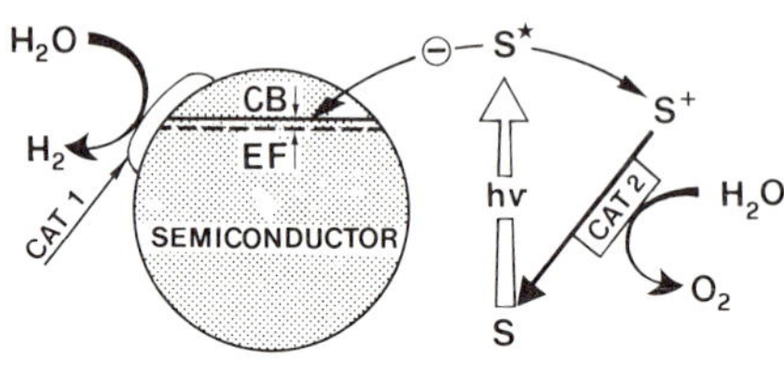

c/ COLLOIDAL SEMICONDUCTOR

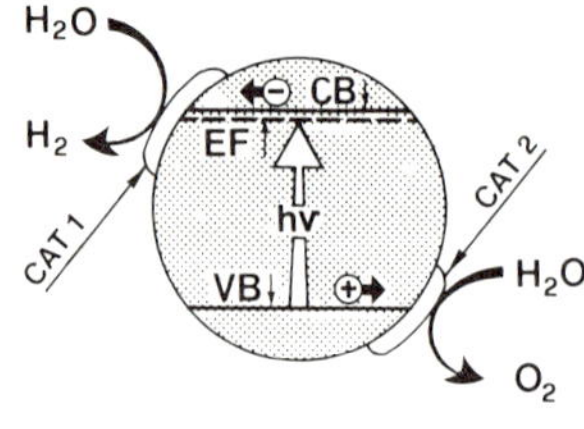

Fig. 16

Three suitable microheterogeneous light harvesting units capturing photons, converting their energy into chemical potential and using it to decompose water, are shown in Fig. 16. Case (a) represents a two-component system containing a sensitizer (S) and an electron relay (R). Light promotes electron transfer from S to R whereby the energy-rich radical ions S^+ and R^- are produced. Thus, light functions here as an electron pump operating against a gradient of chemical potential. A similar process may be conducted in a system where the sensitizer acts as electron acceptor in the excited state and the electron relay is oxidized. Such photoinduced redox reactions have been studied in great detail both from the experimental and the theoretical point of view. The rate of electron transfer between excited sensitizer and relay is expected to approach the diffusion-controlled limit as soon as the driving force for the reaction exceeds a few hundred millivolts. Conversely, the backward electron transfer between S^+ and R^-, which is thermodynamically strongly favored, is almost always diffusion-controlled. This poses a severe problem for the use of such systems in energy conversion devices as it limits the lifetime of the radical ions to, at most, several milliseconds under solar light intensity. In the first chapter we have given examples of how suitable molecular assemblies may be developed that allow the retardation of this undesired back reaction.

With regard to the choice of the sensitizer and the relay, compounds have to be used that are suitable from the viewpoints of both light absorption and redox potentials and that undergo no chemical side reactions in the oxidation states of interest. The sensitizer should have good absorption features with respect to the solar spectrum. Also, its excited state should be formed with high quantum yield and have a reasonably long lifetime, and the electron transfer reaction must occur with high efficiency, viz. good solvent cage escape yield of the redox products. The redox properties of the donor-acceptor relay must obviously be tuned to the fuel-producing transformation envisaged. If, for example, water cleavage by light is to be achieved, then the thermodynamic requirements are such that $E^o(S^+/S) > 1.23$ V (NHE) and $E^o(R/R^-) < 0$ V (NHE) under standard conditions. In the design of sensitizer/relay couples suitable for photoinduced water decomposition, excellent progress has been made over the last few years (57). A number of systems converting more than 90% of the threshold light energy required for excitation of the sensitizer into chemical potential have been explored. Also, in many cases the reduced relay and oxidized sensitizer are thermodynamically capable of generating H_2 and O_2 from water:

$$R^- + H_2O \rightarrow 0.5\ H_2 + OH^- + R \tag{52}$$

$$2S^+ + H_2O \rightarrow 0.5\ O_2 + 2H^+ + 2S \tag{53}$$

Noteworthy examples are sensitizers such as $Ru(bipy)_3^{2+}$ and derivatives, while viologens, metal ions or metal ion complexes have been used as relays. The catalysts employed are usually ultra-fine particles: RuO_2 for water oxidation and, typically, Pt for water reduction. There are significant kinetic problems with system (a) since the intervention of these catalysts has to be rapid and specific at the same time. Furthermore, the cross reaction of the reduced relay with O_2 will lead to a photostationary state where the water photolysis is short-circuited. Nevertheless, these problems can be overcome by judicious design of the microheterogeneous reaction system (57). For example, a recent successful embodiment of such a system involves RuO_2-loaded sepiolite as support, Eu^{3+}-substituted Al_2O_3 particles as relay and Pt as hydrogen evolution catalyst. Another recently discovered water cleavage system is comprised of colloidal TiO_2 loaded onto SiO_2 as a carrier for the $Ru(bipy)_3^{2+}$ sensitizer, a viologen derivative as a relay and colloidal Pt as a hydrogen evolution catalyst (59).

A second type of light harvesting unit suitable for water cleavage by visible light is illustrated in Fig. 16 (b). The sensitizer is adsorbed onto a colloidal semiconductor particle, and no electron relay is required. The excited state of the sensitizer injects an electron into the conduction band of the semiconductor where it is channelled to a catalytic site for hydrogen evolution. A second catalyst mediates oxygen generation from S^+ and H_2O whereby the original form of the sensitizer is regenerated. Several successful attempts to split water by visible light in this way have recently been described (60) and earlier work has been reviewed (57). These systems offer promising prospects, in particular in view of recent developments (61) which include the use of a dimeric ruthenium-EDTA complex as a water oxidation catalyst (62).

In the third configuration shown in Fig. 16 (c), the semiconductor particle itself is excited by band gap illumination forming electron-hole pairs. The former afford hydrogen formation while the latter give rise to O_2 generation. There are many examples for the photodecomposition of water with this type of system (57). In most cases TiO_2 or $SrTiO_2$ were used as a semiconductor material and this restricts the photoactivity to the UV region of the spectrum. However, lower band gap materials such as Cds (63) and V_2O_5 (64) have recently been employed and show activity in the visible. In this case a highly active water oxidation catalyst is required to promote water oxidation at the expense of corrosion.

For example at the illuminated n-CdS/water interface photocorrosion occurs under formation of sulfur:

$$2h^{+} + CdS \rightarrow Cd^{2+} + S \qquad (54)$$

or, in the presence of oxygen to the generation of $CdSO_4$. However, by depositing RuO_2, Ru_2O_3 (57) or Pt (65) particles on the CdS surface the photocorrosion can be prevented and the hole reaction directed towards water oxidation.

$$4h^{+} + 2H_2O \rightarrow O_2 + 4H^{+} \qquad (55)$$

Apart from semiconductor particles, it is possible to cleave water by sunlight using semiconductor electrodes or membranes. The principle of such a device was shown in Fig. 4, where we had chosen the case of a n-type semiconductor electrode as an example. When such an electrode is brought in contact with aqueous electrolyte, negative charge carriers are transferred spontaneously from the semiconductor to the solution phase resulting in the formation of a Schottky type potential barrier. As a consequence, an electrical space charge, i.e. a depletion layer, is built up underneath the surface of the semiconductor. Light is absorbed mostly in this region and the presence of an electrical field separates the electrons from the holes (the situation is analogous to that of the p-n junction in a photovoltaic cell). Holes are driven to the surface where oxidation of water to oxygen occurs, while the electrons migrate through the back contact of the electrode and the external circuit to the counter electrode, where hydrogen is evolved. The energetic position of conduction and valence bands must be such that they straddle the water reduction, $E_f(H_2O/H_2)$ and oxidation $E_f(H_2O/O_2)$ potentials. Only a few semiconducting materials, such as TiO_2 (anatase), CdS and $SrTiO_3$, satisfy this condition. An additional requirement is that the semiconductor does not decompose under illumination. While this is the case for TiO_2 (65), and $SrTiO_3$, the band gap of these materials is too large to allow for efficient solar energy conversion.

A photoelectric device containing an n-type semiconductor was outlined in Fig. 4. As an example we show in Fig. 17 the action spectrum for the photocurrent observed during visible light irradiation of RuL_3 (L = 2,2'-bipyridyl, 4,4'-dicarboxyl acid) coated electrodes. The solution contained I^- as electron donor action spectrum and is characterized by a maximum at 470. The incident monochromatic photon to current conversion efficiency, defined as the number of

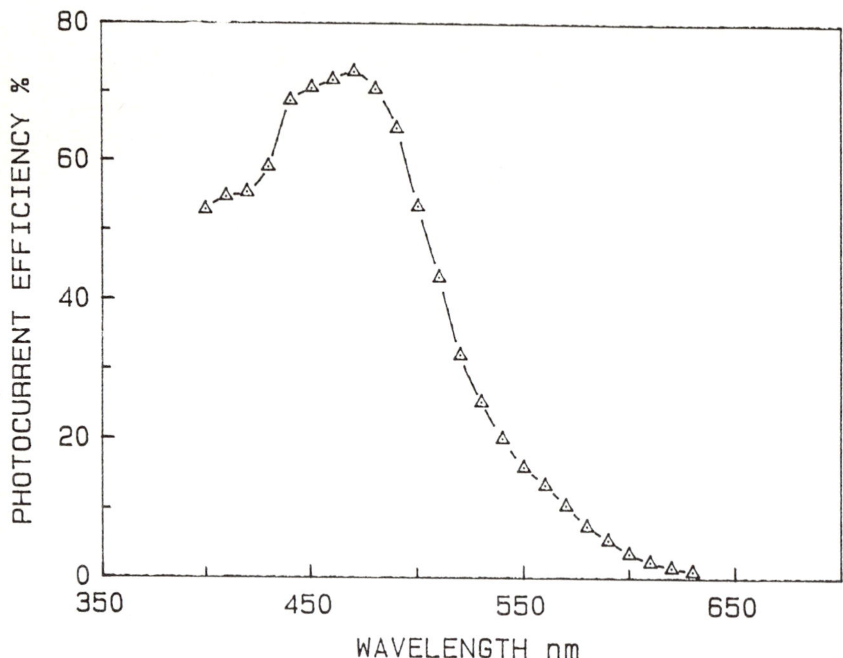

Fig. 17. Spectral characteristics of incident photon to current conversion efficiencies for RuL_3 coated fractal TiO_2 electrodes immersed in 1M Kl solution, electrode potential 0.2 V (SCE).

incident photons, was calculated from the equation:

$$\eta(\%) = \frac{1.24 \times 10^3 \times \text{photocurrent density } (\mu A/cm^2)}{\text{wavelength (nm)} \times \text{photon flux } (W/m^2)} \quad (56)$$

It exceeds 70% in the region close to the spectral maximum of the sensitizer. Such high efficiencies are unprecedented and constitute a veritable breakthrough in the field of spectral sensitization.

Fig. 18 presents the photocurrent potential curve measured during illumination (470 nm light) of the $Fe(CN)_6^{4-}$ - coated electrode under similar conditions as the photocurrent action spectrum. The steady state photocurrent has the onset at -0.3 V (SCE) which is close to the flat band potential of TiO_2 at this pH. It rises steeply with increasing potential reaching a plateau at -0.1 V. This indicates that high fill factors can be obtained with this type of electrode in the conversion of light to electricity. Preliminary tests with such a cell configuration indicate power conversion efficiencies exceeding 6% in the visible.

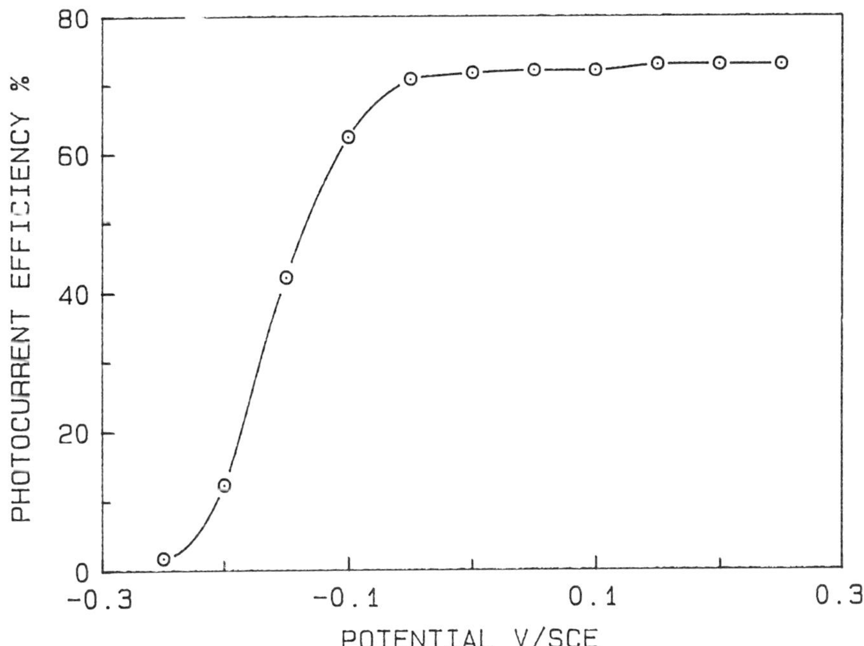

Fig. 18. Incident photon to current conversion efficiency at 470 nm for RuL_3 coated electrodes as a function of applied potential. Conditions as in Fig. 17.

5. CONCLUDING REMARKS

The investigation of photochemical reactions in semiconductor dispersions has a long history which dates back to the invention of silver halide photography and the work on oxides by Eibner (66) and Tamman (67) at the beginning of this century. The discoveries of Emil Baur in Switzerland (68) provided an important step towards the understanding of these processes. Thus, Baur and his coworkers were the first to show that aqueous ZnO dispersions produced oxygen under illumination in the presence of electron acceptors such as Ag^+ ions. The results were correctly interpreted in terms of a "molecular electrolysis" model, i.e. the photo-generation of oxidizing and reducing equivalents and their subsequent reaction with water and Ag^+ ions, respectively.

A wealth of knowledge has been acquired in the domain of photo-electrochemistry during the last 10 years (57, 69). Research in this area continues to advance at a very rapid pace, a recent highlight being the incorporation of colloidal semiconductors in molecular assemblies such as vesicles and inverted micelles (70, 71) and the efficient light energy conversion on fractal TiO_2 layers. The exploration of organized molecular assemblies as minute reaction media to control kinetically the dynamics of photo-initiated redox events is another focal point in the field of artificial photosynthesis.

The discovery of highly active redox catalyst allowed for the coupling of light induced charge separation to fuel generating steps. Following the initial work on ultrafine Pt and RuO_2 particles as mediators for the reduction and oxidation of water, respectively, this field has traversed a very active phase. The essential factors controlling the activity of the noble metal particles are established and a significant effort has gone into optimizing these systems, apart perhaps from the oxygen generation catalyst which still needs to be improved.

The advent of colloidal semiconductor particles has rendered feasible the direct observation of light induced charge carrier generation, their recombination and reaction with catalysts or molecular acceptors. These investigations will continue to thrive and more sophisticated assemblies, such as surface derivatized particles, will soon be scrutinized. Water cleavage by visible light remains the primary target of such studies although other processes, for example the photochemical splitting of hydrogen sulfide, are also of great interest. Much remains to be done, and the task is perhaps one of the most challenging one of the most challenging ones facing scientists.

REFERENCES

1 T. Kunitake and Y. Okahata, J.Am.Chem.Soc., Vol. 97, 1977, p.3860.

2 J.H. Fendler, Acc.Chem.Res., Vol. 13, 1980, p. 7.

3 L.Y.C. Lee, J.K. Hurst, M. Politi, K. Kurihara and J.H. Fendler, J.Am. Chem., Vol. 105, 1983, p. 370.

4 G.C. Papavassiliou, Prog. Solid State Chem., Vol. 12, 1979, p. 185.

5 J. Moser and M. Grätzel, Helv.Chim.Acta, Vol. 65, 1982, p. 1436.

6 W. Stäber, A. Fink and E. Bohn, J.Coll.Interf.Sci., Vol. 26, 1982, p. 62.

7 J. Turkevich, K. Aika, L.L. Ban, I. Okura and S. Namba, J.Res.Inst.Catal., Hokkaido Univ., Vol. 24, 1976, p. 54.

8 E. Matijevic, Langmuir, Vol. 2, 1986, p. 12.

9 E. Vrachnou, N. Vlachopoulos and M. Grätzel, J.Chem.Soc.Chem.Comm. in press.

10 M. Matsumura, S. Matsudaira, H. Tsubomura, M. Takata and H. Yanagida, Ind. Eng.Chem.Prod.Res.Dev., Vol. 19, 1980, p. 415.

11 N. Alonso, V.M. Beley, P. Chartier and V. Ern, Revue Phys.App., Vol. 16, 1981, p. 5.

12 N. Vlachopoulos, P. Liska, A.J. McEvoy and M. Grätzel, European Conf. on Surface Science, Lucerne, Switzerland, 1987.

13 M. Grätzel, "Heterogeneous Photochemical Electron Transfer Reactions", CRDC Press, Baton Rouge, Florida, USA, 1987.

14 M. Shinitzky, Chem.Phys.Lett., Vol. 18, 1973, p. 247.

15 M.D. Ediger and M.D. Fayer, J.Chem.Phys., Vol. 78, 1963, p. 2518.

16 H. Gerischer, "Photoelectrochemistry, Photocatalysis and Photoreactors", M. Sciavello,(Ed.),NATO ASI Series C, Vol. 146, D. Reidel, Dordrecht, 1985.

17 R. Memming, "Comprehensive Treatise in Electrochemistry", B.E. Conway et. al., (Ed.), Vol. 7, Plenum Press, New York, 1983, p. 529.

18 W.J. Albery and P.N. Bartlett, J.Electrochem.Soc., Vol. 131, 1984, p. 315.

19 W. Gärtner, Phys.Rev., Vol. 116, 1959, p. 84.

20 W.J. Albery and P.N. Bartlett, J.Electrochem.Soc., Vol. 131, 1985, p. 315.

21 J.S. Curran and D. Lamouche, J.Phys.Chem., Vol. 87, 1983, p. 5405.

22 A. Fujishima, T. Inoue and K. Honda, J.Am.Chem.Soc., Vol. 101, 1979, p. 5582.

23 M.D. Ward and A.J. Bard, J.Phys.Chem., Vol. 86, 1982, p. 3599.

24 M. Fuji, T. Kawai and S. Kawai, Chem.Phys.Lett., Vol. 106, 1984, p. 517.

25 W.J. Albery, P.N. Bartlett and J.D. Porter, J.Electrochem. Soc., Vol. 131, 1984, p. 2892.

26 S.P. Perone, J.H. Richardson, S.B. Deutscher, J. Rosenthal and J. Ziemer, J.Electrochem.Soc., Vol. 127, 1980, p. 2580; J.Phys.Chem., Vol. 85, 1981, p. 341.

27 Z. Herzion, N. Croitoru and S. Gottesfeld, J.Electrochem.Soc., Vol. 128, 1981, p. 551.

28 S. Gottesfeld and S.W. Feldberg, J.Electroan.Chem.Interf.Electrochem., Vol. 146, 1963, p. 47.

29 P.V. Kamat and M.A. Fox., J.Phys.Chem., Vol. 87, 1983, p. 59.

30 S. Prybyla, W.S. Struve and B.A. Parkinson, J.Electrochem.Soc., Vol. 131, 1984, p. 1587.

31 K. Itoh, M. Nakao and K. Honda, J.Appl.Phys., Vol. 57, 1985, p. 5493.

32 W. Jaegermann, T. Sakata, E. Janata and H. Tributsch, J.Electroanal.Chem. Interfac.Electrochem., Vol. 189, 1985, p. 65.

33 R.H. Wilson, T. Sakata, T. Kawai and K. Hashimoto, J.Electrochem.Soc., Vol. 132, 1985, p. 1082.

34 A. Fripiat and A. Kisch-De Mesmaeker, J.Electrochem.Soc., Vol. 134, 1987, p. 66.

35 K. Bitterling and F. Willig, J.Electroana.Chem., Vol. 204, 1986, p. 211.
36 F. Wilkingson, Trans.Fraday Soc.
37 J.M. Warman, M.P. De Hass, M. Grätzel and P.P. Infelta, Nature, Vol. 310, 1984, p. 306.
38 M. Kunst, G. Beck and H. Tributsch, J.Electrochem.Soc., Vol. 131, 1984, p. 954.
39 M.R.V. Sahyun, Chem.Phys.Lett, Vol. 112, 1984, p. 571.
40 T.W. Healy and L.R. White, Adv.Colloid, Interface Sci., Vol. 9, 1978, p. 303.
41 Z. Alfassi, D. Bahnemann and A. Henglein, J.Phys.Chem., Vol. 86, 1982, p. 4656.
42 R. Howe and M. Grätzel, J.Phys.Chem., Vol. 89, 1985, p. 4495.
43 U. Kölle, J. Moser and M. Grätzel, Inorg.Chem., Vol. 24, 1985, p. 2253.
44 G. Rothenberger, J. Moser, M. Grätzel, N. Serpone and D.K. Sharma, J.Am. Chem.Soc., Vol. 107, 1985, p. 8054.
45 H.P. Boehm, Disc.Faraday Soc., Vol. 52, 1971, p. 264.
46 Y. Nosaka and M.A. Fox, J.Phys.Chem., Vol. 90, 1986, p. 6521.
47 W.A. Weyl and T. Förland, Ind.Eng.Chem., Vol. 42, 1950, p. 257.
48 F.K. McTaggart and J. Bear, J.Appl.Chem., Vol. 5, 1955, p. 643.
49 J. Moser and M. Grätzel, Helv.Chim.Acta, in press.
50 J.B. Hyne, "Sulfur: New Sources and Uses", M.E.D. Raymont, (Ed.), ACS Symposium Series 183, Amer.Chem.Soc., Washington, D.C., 1982; Chem.Tech., 628, 1982.
51 E. Borgarello, K. Kalyanasundaram, M. Grätzel and E. Pelizzetti, Helv. Chim.Acta, Vol. 65, 1982, p. 243.
52 N. Bühler, K. Meier and J.-F. Reber, J.Phys.Chem., Vol. 88, 1984, p. 3261.
53 D.H.M.W. Thewissen, K. Timmer, M. Eenhorst-Reinten, A.H.A. Tinnemans and A. Mackor, Nouv.J.Chim., Vol. 7, 1983, p. 191.
54 R.T. Ross, J.Chem.Phys., Vol. 45, 1966, p. 1; R.T. Ross and M. Calvin, Biophys.J., Vol. 7, 1967, p. 595.
55 J.R. Bolton, Science, Vol. 202, 1978, p. 705.
56 M. Almgren, Photochem.Photobiol., Vol. 27, 1978, p. 603.
57 K. Kalyanasundaram, M. Grätzel and E. Pelizzetti, Coord.Chem.Rev., Vol. 69, 1986, p. 57.
58 F. Borgaya, D. Challal, J.J. Fripiat and H. Van Damme, Nouv.J.Chimie, Vol. 9, 1985, p. 721.
59 A.J. Frank, I. Willner, Z. Gorch and Y. Degani, J.Am.Chem.Soc., Vol. 109, 1987, p. 3568.
60 N. Getoff et.al., Proc.Int.Conf.Hydrogen Energy, Vienna, 1986.
61 D. Duonghong, N. Serpone and M. Grätzel, Helv.Chim.Acta, Vol. 67, 1984, p. 1012.
62 M.M T. Khan et.al., Proc. Int. Conf. Hydrogen Energy, Vienna, 1986.
63 M.M.T. Khan, R.C. Bhandwaj and C.M. Jadhar, J.Chem.Soc.Chem.Comm., 1985, p. 1690.
64 S.A. Naman, S.M. Aliwi and K. Al-Emaru, Nouv.J.Chim., Vol. 9, 1985, p. 687.
65 A. Fujishima and K. Honda, Nature, London, Vol., 237, 1972, p. 37.
66 A. Eibner, Chem.Zeit., Vol. 35, 1911, p. 735.
67 G. Tamman, Z.Anorg.Chem., Vol. 114, 1920, p. 15.
68 E. Baur and A. Perret, Helv.Chim.Acta, Vol. 7, 1924, p. 910.
69 K. Kalyanasundaram, Solar Cells, Vol. 15, 1985, p. 93.
70 Y.M. Tricot and J.H. Fendler, in "Homogeneous and Heterogeneous Photocatalysis", E. Pelizzetti and N. Serpone, (Eds.), NATO ASI Series C, Vol. 174, D. Reidel, Dordrecht, 1986, p. 241 ff.
71 H.J. Watzke and J.H. Fendler, J.Phys.Chem., Vol. 91, 1987, p. 854.

Chapter 6.4

Polymerizations Induced by Photochemical Electron Transfer

Y. Shirota

1 INTRODUCTION

Charge-transfer phenomena involving complete one-electron transfer in the electronically excited state have been widely recognized in numerous photochemical reactions of electron donor - electron acceptor systems (refs. 1-5). There is growing evidence that the exciplex or ion radicals generated by electron transfer are intermediates in a number of bimolecular photochemical reactions. Photochemical reactions of electron donor - electron acceptor systems that involve monomers as one or both components often lead to polymerizations (ref. 6). Such polymerizations as initiated by means of charge-transfer interactions of monomers in the electronically excited state have been conventionally termed photoinduced charge-transfer polymerizations.

In addition to this type of photopolymerization, there are other types of photopolymerization which include polymerizations by direct irradiation of monomers, polymerizations in the presence of photoinitiators, and solid-state photopolymerizations of diolefins or diacetylenes. Among these, photopolymerizations initiated in the presence of photoinitiators have been in practical use for the production of printing plates and for uv curing of coatings (ref. 7). A variety of photoinitiators are known which include initiators for both free radical and cationic polymerizations, some of which involve electron transfer in generating the initiating species.

This article discusses both photoinduced charge-transfer polymerizations, in which monomers are involved in charge-transfer interactions, and photopolymerizations using photoinitiator systems that involve charge-transfer phenomena in the generation of the species which initiates polymerization.

2 PHOTOCHEMICAL REACTIONS OF ARYL VINYL MONOMERS

First, photochemical reactions of vinyl monomers alone are described and then photoinduced electron-transfer reactions are discussed in comparison with excited-state reactions.

The photochemical reactions of aryl vinyl monomers alone under deaerated conditions generally lead to either polymerization or cyclodimerization. The reaction course depends on the structure of the specific aryl vinyl monomer.

2.1 Vinyl monomers containing aromatic hydrocarbons

The photochemical reactions of styrene, p-methoxystyrene or 2-vinylnaphthalene alone produce mainly head-to-head cyclodimers, namely, cis- and trans-1,2-disubstituted cyclobutanes. Reactions by direct irradiation and triplet-sensitized reactions give different cis/trans ratios.

Whereas the direct photolysis of styrene in benzene or dichloromethane favors the cis-isomer by a factor of 6, photochemical reactions induced by triplet sensitizers favor the trans-isomer by a factor of 3.4 (ref. 8). In the case of p-methoxystyrene, the cis/trans ratio in the cyclobutane cyclodimers produced by direct irradiation in acetonitrile has been reported to be 87/13 (ref. 9).

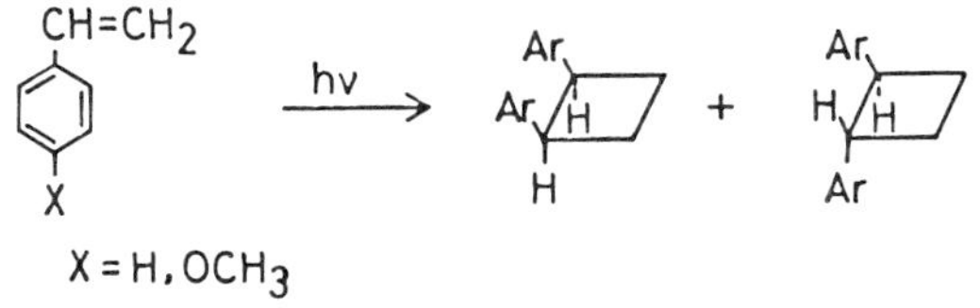

$X = H, OCH_3$

The photochemical reactivities of 2-vinylnaphthalene and 1-vinylnaphthalene are quite different. Direct irradiation of a benzene solution of 2-vinylnaphthalene (VN) with 334 nm light from a high-pressure mercury lamp produces predominantly cis-1,2-di(2-naphthyl)cyclobutane, and the trans-isomer, in the ratio of ca. 10:1, together with a small amount of VN oligomers. On the other hand, the benzophenone-sensitized reaction of VN by the irradiation with 365 nm light yields the cis- and trans-1,2-di(2-naphthyl)cyclobutanes in the ratio of ca. 1:4 via the VN triplet state. These results show that whereas the photochemical reaction of VN by direct irradiation proceeds predominantly via the VN excited singlet state to give the cis-cyclodimer probably via a cis-pairwise excimer, the reaction via the excited triplet state of VN proceeds in a stepwise process, probably involving a dimeric biradical intermediate to yield cyclodimers rich in the thermodynamically more stable trans-isomer (ref. 10). Preliminary results show that 1-vinylnaphthalene is more reactive than 2-vinylnaphthalene in the triplet state (ref. 11).

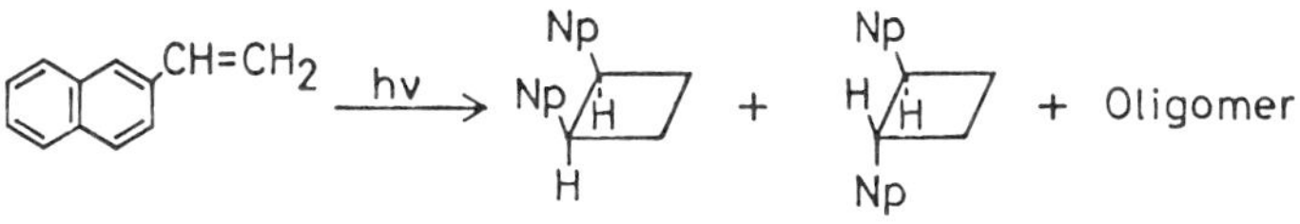

Np: 2-naphthyl

2.2 N-Vinylcarbazole

The photochemical reactivity of N-vinylcarbazole (VCZ) differs from that of vinyl monomers containing aromatic hydrocarbons such as styrene or vinylnaphthalene. The photochemical reaction of VCZ by direct irradiation at 365 nm under deaerated conditions leads to inefficient free-radical polymerization in certain solvents such as THF, acetone, and DMSO (refs. 12,13). Although the initiation mechanism for the photoradical polymerization of VCZ has not been fully clarified, it is suggested that it involves the excited triplet state of VCZ. It should be noted that oxygen or certain solvents such as dichloromethane, nitrobenzene, etc. act as electron acceptors for the excited-state VCZ, changing the excited-state reaction of VCZ, i.e., photoradical polymerization, into an electron-transfer reaction (refs. 12,13). As described in section 3.2, cyclodimerization of VCZ takes place via the VCZ cation radical. Care should be taken, since VCZ is very sensitive to trace amounts of acidic impurities present as contaminants in the solvent or electron acceptor, which initiate cationic polymerization of VCZ in the dark (ref. 14).

$$\text{CH=CH}_2\text{(N-carbazolyl)} \xrightarrow[\text{THF, acetone}]{h\nu,\ 10^{-6}\,\text{Torr}(\sim 0.13\,\text{mPa})} -\!\!(\text{CH-CH}_2)_n\!\!-\text{(N-carbazolyl)}$$

A unique example of photopolymerization of a related monomer has been reported (ref. 15). Crystals of 9-ethyl-3-vinylcarbazole are polymerized under photoirradiation or mechanical shear stress without an initiator. The polymerization proceeds by a cationic mechanism. An intramolecular zwitterion has been suggested to be responsible for the initiation of the polymerization.

3 PHOTOPOLYMERIZATIONS OF THE ELECTRON DONOR - ELECTRON ACCEPTOR SYSTEMS INVOLVING MONOMERS AS COMPONENTS

3.1 General aspects

Photochemical reactions of weak electron donor - electron acceptor systems involving vinyl monomers as components lead to polymerizations and/or small-molecule reactions via charge-transfer interactions or electron transfer in the electronically excited state. Electron-transfer reactions will be distinguished from excited-state reactions.

Photoinduced charge-transfer polymerizations have been observed in three systems: electron donor monomer - electron acceptor, electron acceptor monomer - electron donor, and electron donor monomer - electron acceptor monomer. In the first two systems, homopolymerization of an electron donor or electron acceptor monomer takes place. The photochemical reaction of the electron donor monomer - electron acceptor monomer system may result in ionic homopolymerization of each monomer and/or free-radical copolymerization of a donor monomer with an acceptor monomer. Small-molecule reactions often accompany or sometimes predominate over polymerizations.

Several features of photoinduced charge-transfer polymerizations have been noted. Photoinduced charge-transfer polymerizations generally proceed via monomer ion radicals generated by photochemical electron transfer. Photoinduced charge-transfer polymerizations involve ionic and free-radical polymerizations. Photoinduced charge-transfer polymerizations are frequently accompanied by interrelated small-molecule reactions, typically cyclodimerizations of aryl vinyl monomers. They both occur via a common intermediate, monomer ion radicals. The predominance of either polymerization or small-molecule reactions depends strongly on reaction systems and reaction conditions. Reactivities of aryl vinyl monomers differ in N-vinylcarbazole and in vinyl monomers containing aromatic hydrocarbons, e.g., styrene and vinylnaphthalene. Oxygen acts as an electron acceptor for some strong electron donor monomers in the electronically excited state to induce electron-transfer reactions. In particular, the effect of oxygen on the photochemical reaction of N-vinylcarbazole is striking. Certain solvents may also act as electron acceptors or electron donors for electron donor or electron acceptor monomers.

3.2 N-Vinylcarbazole - electron acceptor systems

Photochemical reactions of aryl vinyl monomers in the presence of an electron acceptor or electron donor differ from those of vinyl monomers alone. Among vinyl monomers, N-vinylcarbazole (VCZ) has been studied most intensively with regard to both thermally and photochemically induced charge-transfer polymerizations.

Historically, it was found in 1960s that N-vinylcarbazole is spontaneously polymerized by a cationic mechanism in solution in the presence of catalytic amounts of an electron acceptor via the formation of a charge-transfer complex (refs. 16,17). In the late 1960s, the occurrence of photoinduced cationic polymerization of VCZ was reported for several systems, e.g., VCZ - nitrobenzene, VCZ - $NaAuCl_4 \cdot 2H_2O$ in nitrobenzene (ref. 18), VCZ - chloranil or bromanil in benzene (ref. 19), VCZ - 2,4,7-trinitrofluorenone in nitrobenzene (ref. 20), and VCZ - CBr_4 in benzene (ref. 21). Photoinduced cationic polymerization attracted attention since photopolymerizations were usually induced by free-radical initiators. The polymerization rate in some of the above photocationic polymerization systems was shown to be proportional to the square root of the light intensity (refs. 21,22). It had also been observed earlier that when the VCZ - chloranil system is irradiated in methanol, a cyclodimer of VCZ, trans-1,2-dicarbazol-9-yl-cyclobutane, is produced (ref. 17); however, neither the interrelation between the polymerization and cyclodimerization nor the nature of the photoinduced charge-transfer polymerization was resolved at that time. Later studies have shown a complete picture including primary processes and follow-up reaction pathways of the photochemical reaction of the VCZ - electron acceptor system.

(i) Primary processes. Charge-transfer and electron-transfer processes in photoinduced polymerization systems have been demonstrated by the measurement of emission spectra and by means of flash or laser photolysis. Dynamic quenching of the VCZ fluorescence accompanied by the appearance of exciplex fluorescence has been observed for some systems, e.g., VCZ - dimethyl terephthalate (ref. 23), VCZ - fumaronitrile (FN) (ref. 24) and VCZ - diethyl fumarate (DEF) (ref. 24) in nonpolar solvents. In polar solvents only quenching of the VCZ fluorescence occurs, probably because of the predominance of the electron-transfer process. The quenching of the VCZ fluorescence by FN or DEF is a diffusion-controlled process.

The photochemical formation of a transient VCZ cation radical and an electron acceptor anion radical in charge-transfer polymerization systems was first demonstrated for the VCZ - chloranil (CA) system in a few solvents, e.g., 1,2-dichloroethane, acetone and acetonitrile by means of flash photolysis (ref. 25), and later for some other systems, e.g., VCZ - tetracyanobenzene (ref. 26), VCZ - pyromellitic dianhydride (PMDA) (ref. 27) and VCZ - dimethylterephthalate (ref. 28), by means of flash or laser photolysis. The anion radicals of electron acceptors, e.g., CA (absorption maxima: 446 nm with a shoulder band at 421 nm) and PMDA (absorption maxima: 615 and 665 nm), have been identified by comparison with the spectra of the chemically prepared anion radicals. The VCZ cation radical, which shows absorption maxima at 619 and 780 nm, has been identified by comparison with the spectrum of the authentic VCZ cation radical generated by γ-ray irradiation in a sec-butyl chloride glass at 77 K (ref. 25), which is known to produce the solute cation radical exclusively. Photochemically generated ion radicals exist as an ion pair or free ions as suggested from the decay process of the transient species. For example, the decay of the PMDA anion radical produced in the VCZ - PMDA system in 1,2-dichloroethane at room temperature is of first order at an initial stage ($t < 100$ μsec), and the second-order kinetics are then followed ($t > 100$ μsec) (ref. 27). This can be rationalized as an initial fast decay of the ion pair, followed by a slower decay of free ions. In acetonitrile the decay follows second-order kinetics over the whole time regime measured ($t > 20$ μsec); this can be understood as predominant formation of free ions in acetonitrile.

The VCZ cation radical, as demonstrated by flash or laser photolysis, acts as an initiator for polymerizations and interrelated reactions. A question arises as to whether the exciplexes or excited-state charge-transfer complexes act as direct initiating species for polymerization. The evidence for the intermediacy of the exciplex in bimolecular photochemical reactions, e.g., (2+2) photocycloadditions, has been obtained by observing comparable quenching of both exciplex fluorescence and reaction by an added quencher (refs. 29,30). Pyridine, which does not quench VCZ monomer fluorescence, selectively dose quench the exciplex fluorescence of the VCZ - FN system. The follow-up 1:1 alternating copolymerization of VCZ with FN is also quenched by pyridine (ref. 31). Therefore, emitting exciplexes that are detected for several polymerization systems in nonpolar solvents are suggested to be precursors to the initiating ion radicals.

The photoabsorption in polymerization systems involves two processes, i.e., selective excitation of a ground-state charge-transfer complex and local excitation of either a donor or an acceptor, followed by electron transfer. Although it is obvious that only the latter mechanism is operative for the system where no charge-transfer interaction exists in the ground state, the above two processes are operative for the system involving the formation of a stable ground-state charge-transfer complex (refs. 19,32). Primary processes of the photoinduced reaction of VCZ in the presence of an electron acceptor are described in Scheme 1.

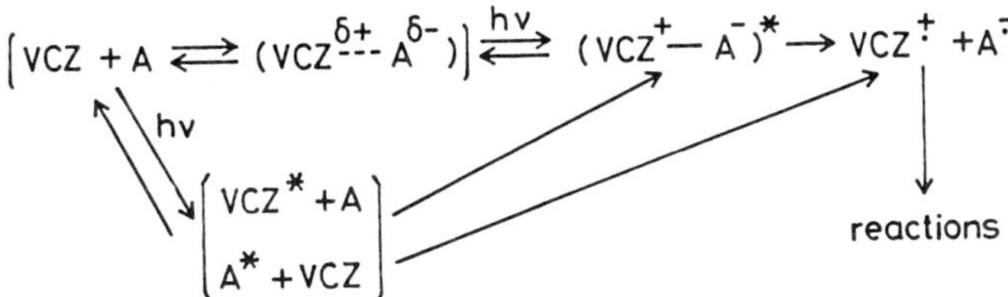

Scheme 1. Primary processes of the photoinduced reaction of VCZ in the presence of an electron acceptor.

(ii) Follow-up reactions. The photochemical reaction of VCZ in the presence of an electron acceptor takes several courses: that is, cationic homopolymerization of VCZ, cyclodimerization of VCZ to give trans-1,2-dicarbazol-9-yl-cyclobutane and free-radical homopolymerization of VCZ or free-radical copolymerization of VCZ with an electron acceptor monomer. These reaction paths are general with various organic electron acceptors expect for those that undergo dissociative electron capture, e.g., CCl_4 and CBr_4, and the reaction course is primarily controlled by the basicity, i.e., cation-solvating ability, of the solvent (refs. 33-35,13) (Scheme 2). That is, in less basic solvents such as benzene, nitrobenzene, dichloromethane, and 1,2-dichloroethane, cationic polymerization of VCZ proceeds, whereas the cyclodimerization of VCZ takes place exclusively in moderately basic solvents such as acetone, methyl ethyl ketone, acetonitrile, and methanol. In stronger basic solvents such as N-methyl-2-pyrrolidinone, dimethylformamide (DMF) and dimethyl sulfoxide (DMSO), both cyclodimerization and radical polymerization of VCZ take place simultaneously; radical polymerization is increasingly favored over cyclodimerization with an increase in the solvent basicity. In the most basic solvent examined, hexamethylphosphoric triamide (HMPA), cyclodimerization does not occur; only radical polymerization of VCZ takes place. Solvent control of the reaction course has been further demonstrated by using a binary solvent system, acetone-DMSO (ref. 35), and as described later, by using electron acceptor monomers, e.g., maleic anhydride, fumaronitrile, and diethyl fumarate.

A distinction between cationic and radical polymerizations has been made from additive effects and copolymerization studies with added monomers. When an electron acceptor monomer capable of undergoing copolymerization with VCZ is used as an electron acceptor, the radical homopolymerization of VCZ, which proceeds in strongly basic solvents, is replaced by radical copolymerization of VCZ with the electron acceptor monomer. The fluorescence spectra of resulting polymers also provide information on a distinction between cationic and radical propagation mechanisms since the tacticity of resulting polymers differs in the two mechanisms (ref. 36).

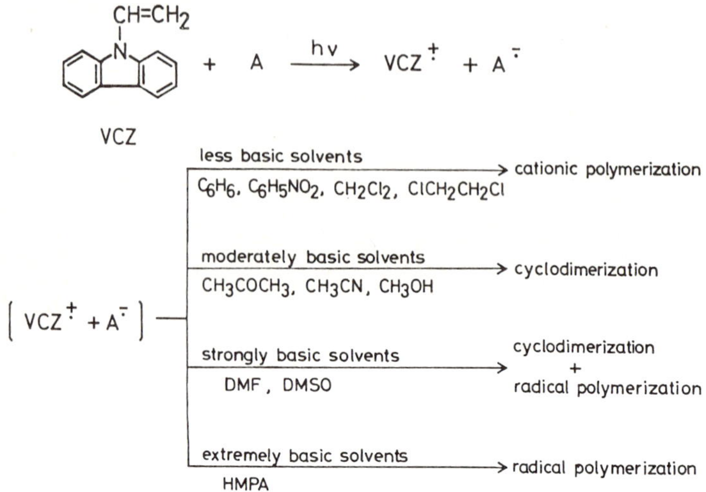

Scheme 2. Multiple reaction courses in the photochemical reaction of VCZ in the presence of an electron acceptor.

The multiplicity of the reactions and solvent control of the reaction course have been explained in terms of the common intermediates, i.e., the VCZ cation radical generated by photochemical electron transfer and the VCZ dimer cation radical formed by the reaction of the VCZ cation radical with the VCZ monomer. The VCZ monomer is known to be basic and readily undergoes cationic polymerization. Thus in less basic solvents, cationic reactivity of the VCZ cation radical and the VCZ dimer cation radical predominates over radical reactivity, and cationic polymerization proceeds. In basic solvents, however, cationic reactivity of the VCZ cation radical and the VCZ dimer cation radical is suppressed by cation solvation and either cyclodimerization or radical polymerization proceeds.

The cyclodimerization of VCZ has been shown to proceed via a novel cation radical chain reaction (refs. 37,38). The mechanism for the VCZ cyclodimerization involves back electron transfer from the VCZ monomer or from the electron acceptor anion radical (in the termination step) to the VCZ dimer cation radical. The open-chain dimer cation radical and cyclobutane cation radical have been suggested to be interconvertible (refs. 38,39).

For the cyclodimerization of VCZ to take place, cation solvation by basic solvents or stabilization of the cationic center by an additive seems to be necessary in order to suppress competing cationic polymerization of VCZ. Also, the electron acceptor anion radical as a counterpart for the VCZ cation radical should be stable (refs. 12,33,34). In more basic solvents, back electron transfer to the VCZ dimer cation radical is impeded by strong solvation, and hence, radical addition of the VCZ monomer may occur, leading to polymers (Scheme 3).

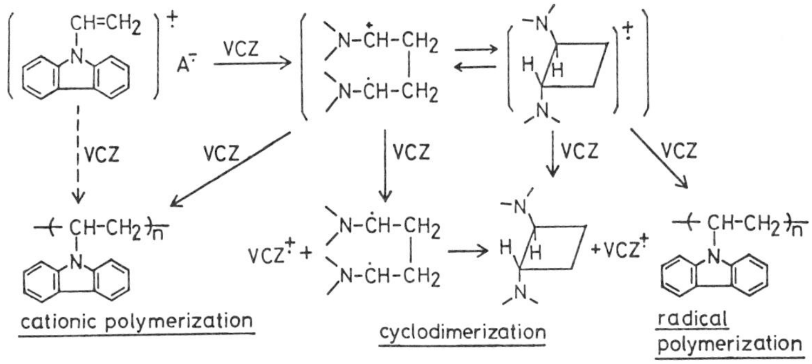

Scheme 3. Photochemical reactions of VCZ in the presence of an electron acceptor.

Factors that determine the reaction course include the polarity and basicity of the solvent; the stability of the anion radical of the electron acceptor; the radical trapping ability of the electron acceptor or of the solvent; and the polymerizability of the electron acceptor (refs. 33-35). Among these, solvent basicity is decisive in determining the reaction course via the intermediacy of the VCZ cation radical. When photochemical electron transfer does not occur in a system using a very weak electron acceptor in combination with a nonpolar solvent, neither cationic polymerization nor cyclodimerization of VCZ takes place. When the anion radical of the electron acceptor generated as a counterpart for the VCZ cation radical is unstable, the multiple reaction courses described above are not observed. The VCZ - CCl_4 and VCZ - CBr_4 systems, where CCl_4 and CBr_4 undergo dissociative electron capture, do not yield cyclodimer even in polar, basic solvents but do produce polymers in the high vacuum system. The solvent, nitrobenzene, in which cationic homopolymerization of VCZ takes place, is relatively weak in cation-solvating ability, but at the same time it acts as an efficient radical trap to inhibit radical polymerization. Even when the reaction involves the photoinduced electron-transfer process, the directly initiating species for the polymerization is not necessarily the monomer or the dimer cation radical. In VCZ - CCl_4 and VCZ - CBr_4, the electron acceptor undergoes dissociative electron capture to give X^- and a reactive $\cdot CX_3$ (X = Cl, Br) radical. The cationic photopolymerization of VCZ by $(n\text{-}C_4H_9)_4N^+\ AuX_4^-$ (X = Cl, Br) has been suggested to be initiated by the hydrogen halide produced by the photochemical decomposition of the metal salt (ref. 40).

(iii) Oxygen and solvents as electron acceptors. Oxygen acts as an electron acceptor and plays the role of co-oxidant, exerting striking influences on photochemical reactions of VCZ. It has been shown that the photochemical reaction of VCZ alone in polar solvents in the presence of dissolved oxygen differs completely from that in a deaerated system using high vacuum techniques (0.13 mPa or 10^{-6} Torr); the features of the photochemical reaction of the VCZ - oxygen system in polar solvents are essentially the same as those of VCZ - organic electron acceptor systems under high vacuum (refs. 12,13). That is, cyclodimerization of VCZ occurs preferentially in polar, basic solvents such as acetone, ethyl methyl ketone, or methanol, and simultaneous cyclodimerization and free-radical polymerization proceeds in stronger basic solvents such as DMF and DMSO. On the other hand, cationic polymerization of VCZ proceeds in polar, nonbasic solvents such as dichloromethane or 1,2-dichloroethane. Thus the photochemical reactions of the VCZ - oxygen - polar solvent system are most satisfactorily explained in terms of the intermediacy of the VCZ cation radical and superoxide generated by electron transfer from the excited-state VCZ to the ground-state oxygen. Free-radical polymerization in stronger basic solvents may include partial contribution of the polymerization via the excited triplet state. Only catalytic amounts of oxygen are necessary for the cyclodimerization of VCZ to occur in polar, basic solvents. In fact, the cyclodimerization of VCZ proceeds in acetone even under 10^{-1} Torr evacuation because a chain mechanism is initiated by the VCZ cation radical, whereas only inefficient photoradical polymerization of VCZ occurs in acetone under high vacuum (10^{-6} Torr).

VCZ
hν
in CH_3COCH_3
10^{-6} Torr (~0.13 mPa)
free-radical polymerization
O_2
cyclodimerization
N: carbazolyl

$$\text{VCZ} + O_2 \xrightarrow{h\nu} [\text{VCZ}^{+\cdot} + O_2^{-\cdot}] \longrightarrow \text{cyclodimer}$$

In a basic, but less polar solvent, tetrahydrofuran (THF), no cyclodimerization of VCZ takes place but photoradical polymerization of VCZ is significantly accelerated in the presence of oxygen. It is understood that no electron transfer occurs in a less polar solvent, THF, but that the VCZ excited triplet state is populated via charge-transfer interactions of the excited singlet-state VCZ with the ground-state oxygen. The possibility cannot be excluded, however, that peroxides formed by the reaction of THF with oxygen may participate in the initiation of the photoradical polymerization (ref. 13).

Cyclodimerization of VCZ occurs not only by direct irradiation but also by sensitization with acetophenone, benzophenone, or Rhodamine 6G in methanol only in the presence of oxygen (refs. 41,42). In this case, oxygen acts as a co-oxidant. It is suggested that stable superoxide is formed by successive electron transfer from the sensitizer anion radical to oxygen as a counterpart for the VCZ cation radical.

Solvent control of the reaction course and the effect of oxygen on the electron-transfer reaction of VCZ described above have also been shown in the studies of radiation-induced reaction of VCZ (ref. 43) and photochemical reaction of VCZ in a binary solvent system, nitrobenzene-benzonitrile (ref. 44). At a high concentration of benzonitrile in the aerated system, cyclodimerization is favored, whereas at a high concentration of nitrobenzene, only cationic polymerization of VCZ takes place. The VCZ cation radical has been generated not only by direct photochemical electron transfer from VCZ to the electron acceptor but also by means of electron-transfer sensitization, i.e., electron transfer from VCZ to a photochemically formed cation-radical species (ref. 45).

Photocationic polymerization of VCZ proceeds in dichloromethane even in the absence of oxygen or an added electron acceptor (ref. 13). The polymerization of VCZ in the absence of an electron acceptor in dichloromethane is accompanied by the formation of carbazole and other unidentified products (ref. 46), although in the presence of catalytic amounts of an electron acceptor, cationic polymerization of VCZ proceeds almost exclusively. The solvent, dichloromethane, as well as nitrobenzene may act as an electron acceptor. The VCZ fluorescence is not quenched by dichloromethane. It is suggested that photoinduced electron transfer from VCZ in the excited triplet state to dichloromethane takes place to give the VCZ cation radical or photochemical reaction of the excited triplet-state VCZ with dichloromethane via charge-transfer interactions occur to form hydrogen chloride; these species may initiate cationic polymerization of VCZ. The photopolymerization of VCZ in dichloromethane in the presence of a free-radical initiator, AIBN, has also been studied (refs. 47, 48); the polymerization is complex, being propagated through noninteracting cationic and radical mechanisms.

3.3 N-Vinylcarbazole - electron acceptor monomer systems

The photochemical reaction of the system of N-vinylcarbazole (VCZ) - electron acceptor monomer (1:1 mol ratio), such as maleic anhydride (MAn), fumaronitrile (FN), and diethyl fumarate (DEF), involves cationic homopolymerization of VCZ, cyclodimerization to yield trans-head-to-head cyclobutane cyclodimer, and free-radical copolymerization of VCZ with the electron acceptor monomer to give an 1:1 alternating copolymer (refs. 24,46). It is known that VCZ undergoes free-radical copolymerization with MAn (ref. 49), FN (refs. 50,52,53) and DEF (refs. 51-53) in the presence of 2,2'-azobisisobutyronitrile as an initiator to give 1:1 alternating copolymers regardless of the monomer feed composition.

As observed for the VCZ - electron acceptor system, the reaction course of the VCZ - electron acceptor monomer system also depends strongly on the basicity of solvents (refs. 24,46) (Scheme 4). In this system, 1:1 alternating radical copolymerization competes with cationic homopolymerization of VCZ in a less basic solvent (benzene). A significant difference in the product distribution between the VCZ - MAn and VCZ - FN systems has been observed. The results of polymerization rates and molecular weights of resulting polymers in benzene show that 1:1 alternating radical copolymerization predominates over cationic homopolymerization of VCZ for the VCZ - FN system, whereas cationic homopolymerization of VCZ is favored for the VCZ - MAn system (ref. 46).

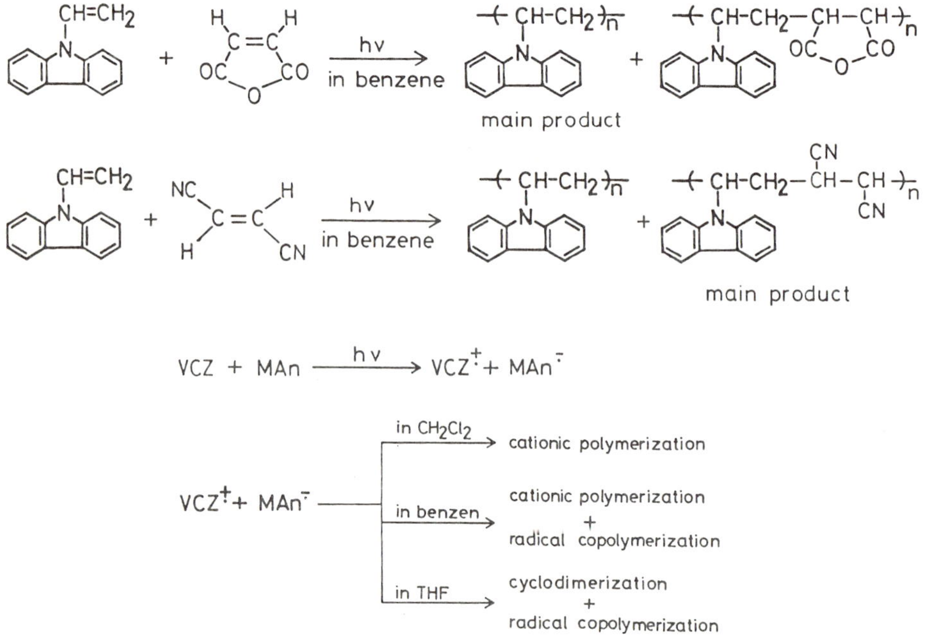

VCZ + MAn $\xrightarrow[\text{in benzene}]{h\nu}$ $\text{-(CH-CH}_2\text{)}_n\text{-}$ (main product) + $\text{-(CH-CH}_2\text{-CH-CH)}_n\text{-}$

VCZ + FN $\xrightarrow[\text{in benzene}]{h\nu}$ $\text{-(CH-CH}_2\text{)}_n\text{-}$ + $\text{-(CH-CH}_2\text{-CH(CN)-CH(CN))}_n\text{-}$ (main product)

$$\text{VCZ} + \text{MAn} \xrightarrow{h\nu} \text{VCZ}^{+\cdot} + \text{MAn}^{-\cdot}$$

$\text{VCZ}^{+\cdot} + \text{MAn}^{-\cdot}$

in CH_2Cl_2 → cationic polymerization

in benzen → cationic polymerization + radical copolymerization

in THF → cyclodimerization + radical copolymerization

Scheme 4. Solvent effects on the photochemical reaction of the VCZ - MAn system.

The photochemical reactions of the VCZ - electron acceptor monomer (MAn, FN) systems are best explained in terms of the intermediacy of ion radicals generated by electron transfer from the electronically excited singlet-state VCZ to the electron acceptor monomer and by the ionization of the excited-state charge-transfer complex. The occurrence of electron transfer is supported from highly exothermic free-energy change for electron transfer as estimated from the Rehm-Weller equation (ref. 54); the ΔG values are approximately -153 and -112 kJ mol^{-1} for the VCZ - MAn and VCZ - FN systems, respectively.

3.4 N-Vinylcarbazole - electron donor systems

N-Vinylcarbazole (VCZ) usually acts as an electron donor monomer; however, the VCZ - amine system provides an example of the system where VCZ may act as an electron acceptor in the electronically excited state. The photochemical reaction of the VCZ - amine system is characterized by three features (ref. 55): (1) Photoradical polymerization of VCZ is induced in benzene and greatly accelerated in acetone in the presence of amines such as diethylamine (DEA), triethylamine (TEA), tri-n-propylamine (TPrA), and N,N-diethylaniline (DEAn). (2) Triphenylamine (TPA) inhibits photoradical polymerization of VCZ. Benzylamine (BzA) and diphenylamine (DPA) are also ineffective for the photoradical polymerization of VCZ; reactions in the presence of these amines produce oligomers of VCZ in acetone. (3) The cyclodimerization of VCZ that occurs in acetone in the presence of dissolved oxygen is quenched in the presence of amines; free radical polymerization of VCZ takes place instead. However, when oxygen is bubbled into the system, polymerization is inhibited.

The addition of TPA, TEA, TPrA or DEAn does not affect significantly the rate of free-radical polymerization of VCZ initiated by 2,2'-azobisisobutyronitrile. Therefore, efficient initiation processes are operative for the photoradical polymerization of VCZ in the presence of certain amines. No charge-transfer interaction between VCZ and amines in the ground state is observed. Whereas the weak electron-donating amines, such as DEA, TEA, TPrA, and BzA, cause no appreciable quenching of the VCZ fluorescence, the stronger electron-donating amines, such as DEAn and TPA, quench significantly the VCZ fluorescence. Related to the VCZ - amine system, the fluorescence quenching process of the carbazole - amine system has been studied by means of laser photolysis (ref. 56). The results show that proton transfer and hydrogen atom transfer occur in the carbazole - TEA - acetonitrile system, whereas electron transfer is the main process for the fluorescence quenching in the carbazole - N,N-dimethylaniline - acetonitrile system.

It is suggested that the system of VCZ - strong electron-donating amine, e.g., DEAn and TPA, in acetone involves electron transfer from the amine to the electronically excited singlet state of VCZ to generate the VCZ anion radical and the amine cation radical. In the case of DEAn, proton transfer from the amine cation radical to the VCZ anion radical follows, leading to a neutral radical, which initiates radical polymerization of VCZ (Scheme 5). In the case of TPA which has no hydrogen available for transfer, electron transfer occurs, but the resulting VCZ anion radical and TPA cation radical do not act as initiating species for the polymerization of VCZ.

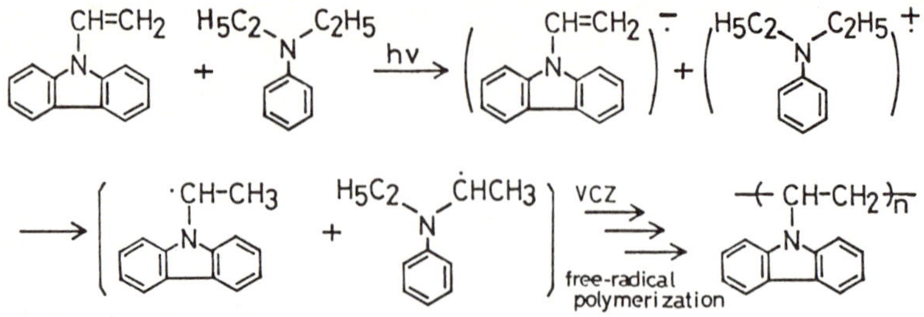

Scheme 5. Proposed mechanism for photoradical polymerization of VCZ in the presence of N,N-diethylaniline.

The initiation mechanism for the photoradical polymerization of VCZ in the presence of weak electron-donating amines, such as DEA, TEA, and TPrA, is not clear. It is suggested that the neutral radical which initiates polymerization is produced via charge-transfer interaction in the electronically excited state in view of the fact that n-hexane or 2-propanol is ineffective as a hydrogen source for the initiation of the polymerization. Benzylamine and DPA act as degradative chain transfer agents. Quenching of the cyclodimerization of VCZ by amines is attributed to electron transfer from amines to the VCZ cation radical formed in the reaction system to give neutral VCZ and cation radicals of amines. Photoradical copolymerization proceeds by irradiation of the VCZ - N,N-diethylaminoethyl methacrylate system in benzene with 365 nm light (ref. 57).

3.5 Aryl vinyl monomer - electron acceptor systems

(i) Photocyclodimerizations via monomer cation radicals. In general, photochemical reactions of aryl vinyl monomers in the presence of an electron acceptor, e.g., tetracyanobenzene (TCNB), pyromellitic dianhydride (PMDA), m- and p-dicyanobenzene (DCB), etc., in acetonitrile lead to the formation of cyclodimers via the intermediacy of monomer cation radicals generated by photochemical electron transfer. The structures of cyclodimers formed as the major product in photoinduced electron-transfer reactions depend on aryl vinyl monomers. That is, N-vinylcarbazole (VCZ) (refs. 12,17,38), p-methoxystyrene (refs. 9,58) and p-N,N-dimethylaminostyrene (ref. 59) produce trans-1,2-disubstituted cyclobutanes. As in the case of VCZ (ref. 12), the photocyclodimerization of p-methoxystyrene proceeds with oxygen as an electron acceptor in acetonitrile by excitation of a charge-transfer complex in the ground state (ref. 9). Phenyl vinyl ether also undergoes photocyclodimerization in the presence of an electron acceptor in acetonitrile to yield both trans- and cis-1,2-diphenoxycyclobutanes (ref. 60). As reported for VCZ (ref. 38), the photocyclodimerization of phenyl vinyl ether also proceeds through a cation radical chain reaction (ref. 61). The stereochemistry of the cyclobutane cyclodimers formed from phenyl vinyl ether is influenced by the monomer concentration and the reaction temperature (ref. 62). While a constant ratio of 6:4 (trans : cis) is obtained in the monomer concentration range from 0.1 to 2.0 mol dm^{-3}, the trans/cis ratio markedly decreases with a decrease in the monomer concentration in the concentration range from 0.1 to 0.002 mol dm^{-3}. The trans/cis ratio decreases with increasing reaction temperature, even in the higher monomer concentration range. Excitation of the charge-transfer complexes of indene or 1,1-dimethylindene with dimethylmaleic anhydride in acetonitrile produces anti-head-to-head cyclobutane dimers as the major product via cation radicals (ref. 63).

CH=CH2 + A →(hν, in CH_3CN) [CH=CH2 / OCH3]$^{+\cdot}$ + $A^{-\cdot}$ → Ar / Ar (cyclobutane)

Ar: —OCH3

On the other hand, photoinduced electron-transfer reactions of 1,1-diphenylethylene (ref. 64), styrene and its derivatives, e.g., p-methylstyrene and α-methylstyrene (ref. 65), in acetonitrile give rise to 1,2,3,4-tetrahydronaphthalene derivatives as the major product via the monomer cation radical as an intermediate. The α-methylstyrene cation radical is also formed by cation radical transfer from phenanthrene cation radical to α-methylstyrene (ref. 66).

CH_3–C=CH_2 + A →(hν, in CH_3CN) [CH_3–C=CH_2]$^{+\cdot}$ + $A^{-\cdot}$ → H_3C, Ph / CH_3 (tetrahydronaphthalene)

Both types of photocyclodimerization involve common intermediates, monomer cation radicals. Head-to-head cyclobutane cyclodimers are produced in the manner as proposed for the photocyclodimerization of VCZ, whereas 1,2,3,4-tetrahydronaphthalene derivatives result from intramolecular electrophilic attack on the aromatic ring by the cationic center in the open-chain dimer cation radical intermediate, followed by back electron transfer (ref. 64). The reactivity of the dimer cation radical depends on the structure of vinyl monomers. There may be a correlation between structures of vinyl monomers and reactivities of dimer cation radicals of vinyl monomers. When the cationic center is stabilized by conjugation with heteroatoms, or when the intramolecular electrophilic attack on the aromatic ring is structurally difficult, cyclobutane cyclodimers are obtained as major products. On the other hand, the photoinduced electron-transfer reaction of 1,1-diphenylethylene involves competing (2+2) cyclodimerization to give the cyclobutane cyclodimer and (2+4) cyclodimerization to yield the 1,2,3,4-tetrahydronaphthalene derivatives. The product distribution of this reaction is very dependent on the concentration of

1,1-diphenylethylene (ref. 67). The mechanism outlined in Scheme 6 has been presented to account for the data. That is, the interception of the geminate ion radical pair leads to the biradical, which cyclizes exclusively to the cyclobutane, whereas the dimer cation radical, formed from the separated monomer cation radical, undergoes both 1,4- and 1,6-cyclization to give the cyclobutane and the tetrahydronaphthalene derivatives (ref. 67).

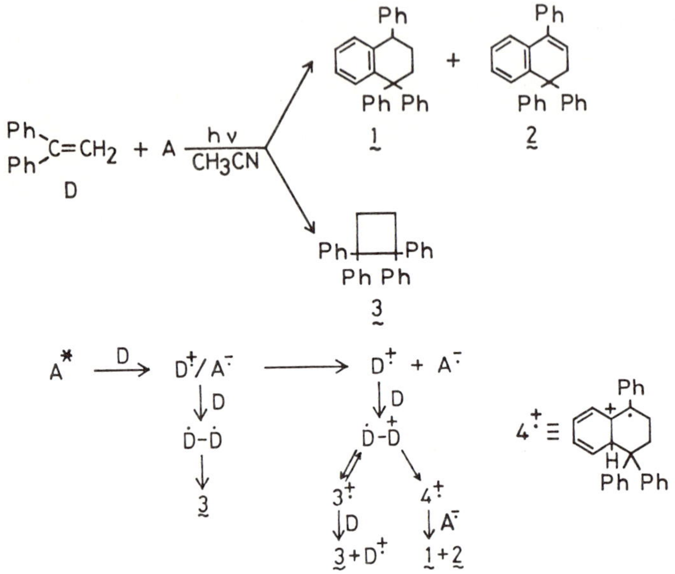

Scheme 6. Competing photocyclodimerizations of 1,1-diphenylethylene via ion radicals (ref. 67).

(ii) Photopolymerization. Photochemical reactions of styrene and α-methylstyrene in the presence of an electron acceptor such as TCNB or PMDA in dichloromethane or 1,2-dichloroethane involve both cationic polymerization and cyclodimerization to give 1,2,3,4-tetrahydronaphthalene derivatives as a major product via the common intermediates, the monomer cation radical and the dimer cation radical (refs. 68,69). The solvent effect on the reaction course observed for the styrene - electron acceptor and α-methylstyrene - electron acceptor systems, namely, exclusive cyclodimerization in a polar, basic solvent, acetonitrile, and competing cationic polymerization and cyclodimerization in polar, nonbasic solvents such as dichloromethane or 1,2-dichloroethane, is more or less in accord with that observed for the VCZ - electron acceptor system. The relative contributions of cationic polymerization and cyclodimerization in the photochemical electron-transfer reactions of styrene and α-methylstyrene in dichloromethane or 1,2-dichloroethane vary greatly with initial monomer concentration and reaction temperature; however, when the molecular weights of resulting polymers are considered, cyclodimerization to yield 1,2,3,4-tetrahydronaphthalene derivatives as a major product is predominant even in nonbasic solvents (such as dichloromethane and 1,2-dichloroethane) which ordinarily favor cationic polymerization (ref. 68).

$$\alpha\text{-MeSt} + A \xrightarrow[\text{in } CH_2Cl_2]{h\nu} [\alpha\text{-MeSt}^{+\cdot} + A^{-\cdot}]$$

α-MeSt

cationic polymerization → $-\!\!(C(CH_3)(Ph)-CH_2)\!\!-_n$

cyclodimerization → 1,2,3,4-tetrahydronaphthalene dimer (H_3C, Ph, CH_3)

The reactivities of monomer cation radicals, dimer cation radicals, and monomers themselves greatly differ in VCZ and in vinyl monomers containing aromatic hydrocarbons (such as styrene and α-methylstyrene) in the ease of cationic polymerization as well as in the structure of the resulting cyclodimers. The VCZ monomer is characteristic in that it is highly susceptible to cationic polymerization relative to styrene and α-methylstyrene.

The results of the effects of temperature and monomer concentration on the photochemical reactions of the α-methylstyrene - PMDA or TCNB and styrene - TCNB systems in dichloromethane show that the polymer yield increases with decreasing temperature and with decreasing monomer concentration, although the results in the literature regarding the dimer yield are not consistent with each other (refs. 68,69). The photoinduced cationic polymerization is initiated by the monomer cation radical produced by photochemical electron transfer from styrene or α-methylstyrene to the electron acceptor. Ion pairs tend to dissociate into free ions at low monomer concentration and at low temperature because of the greater polarity of the solution. Cationic polymerization is initiated by both the ion pair and free ions and that the 1,2,3,4-tetrahydronaphthalene cyclodimer is produced from the ion pair but not from the free ion since the dimer yield decreases with decreasing temperature and decreasing monomer concentration (ref. 69).

Photocationic polymerization of α-methylstyrene with TCNB as an electron acceptor in dichloromethane or 1,2-dichloroethane takes place both by the selective excitation of a ground-state charge-transfer complex and the excitation of TCNB (ref. 70), as in the case of the VCZ - electron acceptor system. The cationic polymerization at -30°C is propagated by both free ions and ion pairs, as suggested by the molecular weight distribution of the resulting polymer determined by gel-permeation chromatography, which is bimodal. At temperatures higher than -30°C, the polymerization is propagated by ion pairs and the molecular weight distribution becomes monomodal (ref. 71). It has been reported that photochemical electron transfer from α-methylstyrene to TCNB occurs from both the excited singlet and triplet states of the charge-transfer complex, and that ion radicals formed from the triplet state react with the monomer to induce cationic polymerization, whereas those formed from the singlet state do not initiate polymerization (ref. 72).

Styrene also undergoes photopolymerization by selective excitation of a ground-state charge-transfer complex with PMDA in dichloromethane or 1,2-dichloroethane. The polymerization proceeds by independent cationic and radical mechanisms (ref. 73). The radiation-induced reaction of styrene in dichloromethane has also been studied; the reaction involves not only cationic and radical polymerizations but also dimerizations to give 1-phenyl-1,2,3,4-tetrahydronaphthalene as a major product and cyclobutane cyclodimers via a common intermediate, the monomer cation radical (ref. 74).

Photopolymerization of 2-vinylnaphthalene (VN) is initiated in sulfolane only in the presence of zinc chloride to produce poly(vinylnaphthalene) (ref. 75). Increasing the concentration of metal salt increases the rate of polymerization until a limiting ratio of salt to monomer is achieved. The formation of a complex, $(VN)_2$-$ZnCl_2$, has been suggested to be involved in the initiation process.

3.6 Aryl vinyl monomer - electron acceptor monomer systems

As described in section 3.5, photochemical reactions of aryl vinyl monomers in the presence of an electron acceptor in acetonitrile generally lead to cyclodimerization via monomer cation radicals generated by electron transfer, whereas those in dichloromethane or 1,2-dichloroethane involve competing polymerization and cyclodimerization. On the other hand, the photochemical reaction of aryl vinyl monomers in the presence of an electron acceptor monomer that is capable of undergoing radical copolymerization tends to give copolymers by a free-radical mechanism, probably via the intermediacy of ion radicals.

Styrene, vinyltoluene, and t-butylstyrene are copolymerized by a free-radical mechanism with maleic anhydride (MAn) in dichloromethane or N,N-dimethylformamide to yield copolymers with approximately 1:1 compositions by irradiation with an argon ion laser (ref. 76). The rates of copolymerization are enhanced with increasing electron-donating property of the donor monomer, namely, in the order of t-butylstyrene - MAn > vinyltoluene - MAn > styrene - MAn.

$$C_6H_5CH=CH_2 + \text{HC=CH (OC-O-CO)} \xrightarrow{h\nu} \text{-(CH(C}_6\text{H}_5\text{)-CH}_2\text{-CH(CO)-CH(CO)-O)}_n\text{-}$$

2-Vinylnaphthalene (VN) and 9-vinylanthracene have also been reported to be copolymerized with fumaronitrile (FN) in dichloromethane or sulfolane by irradiation with a pulsed nitrogen laser at 337 nm (ref. 77). The copolymer resulting from the VN – FN system in sulfolane contains a high percentage of sulfolane. A mechanism for the incorporation of sulfolane is suggested to involve removal of a hydrogen atom from the solvent molecule by the polymer chain radical, followed by the reaction of the solvent radical with a monomer to grow a new polymer chain. Little information is available regarding photochemical aspects of the reaction of these systems.

The photochemical reaction of the aryl vinyl monomer – electron acceptor monomer system has been studied in more detail with regard to VN – MAn (ref. 78), VN – FN (ref. 79), and VN – diethyl fumarate (DEF) (ref. 80) systems in order to gain insight into the correlation between the multiplicity of an excited-state molecule and reaction pathways including the occurrence of electron transfer. Since the photochemical reaction of VN alone by direct irradiation proceeds predominantly via the excited singlet state and sensitization with benzophenone leads to the triplet-state reaction of VN (ref. 10), the VN – electron acceptor monomer system seems to be suitable for studying the problem described above. The results show that the photochemical reaction of the VN – electron acceptor monomer system by direct irradiation contrasts with the triplet-sensitized reaction.

The VN – electron acceptor monomer (MAn, FN, and DEF) systems show charge-transfer interactions both in the ground and electronically excited states. The fluorescence of VN in a dilute solution is sharply quenched by the addition of the electron acceptor monomer. In the case of FN and DEF, a weak, broad exciplex fluorescence is observed in a nonpolar solvent, benzene, but in a polar solvent, acetonitrile, only strong quenching of the VN fluorescence occurs. No exciplex fluorescence is observed with MAn. The bimolecular rate constants for the quenching of the VN fluorescence by the electron acceptor monomers in benzene and acetonitrile are diffusion controlled. The electronic absorption spectra of the reaction solution of the VN – electron acceptor monomer system are distinctly red-shifted compared with those of VN or the electron acceptor monomer alone of the same concentration; this is due to the formation of a ground-state charge-transfer complex.

The VN – MAn, VN – FN, and VN – DEF systems take similar reaction courses, although the product distribution differs depending on the reaction system. Direct irradiation of the VN – electron acceptor monomer systems with light of the wavelength longer than 334 nm, both in benzene and in acetonitrile, produces a copolymer of VN with the electron acceptor monomer as a main product and two kinds of cyclodimer of VN, i.e., 1,2,3,4-tetrahydro-4-(2-naphthyl) phenanthrene and 1,2-di(2-naphthyl)cyclobutane. Selective excitation in the charge-transfer bands of the VN – MAn and VN – FN systems in benzene with light of the wavelength longer than 365 nm resulted in the formation of the same products with decreased yields. The copolymers of VN with the electron acceptor monomers obtained by the photochemical reaction are similar to those obtained by the 2,2'-azobisisobutyronitrile-initiated copolymerizations in their ir, uv, and ^{1}H nmr spectra and copolymer compositions. By contrast, the benzophenone (BP)-sensitized reactions of the VN – electron acceptor monomer systems in both benzene and acetonitrile by irradiation with light of the wavelength longer than 365 nm yield cycloadducts of VN with the electron acceptor monomer, i.e., 3-(2-naphthyl)cyclobutane-1,2-dicarboxylic anhydride, 1-naphthyl-2,3-dicyanocyclobutane, and 1-naphthyl-2,3-dicarboethoxycyclobutane for the VN – MAn, VN – FN, and VN – DEF systems, respectively, as main products together with copolymers and VN cyclodimers. Four stereoisomeric cycloadducts are formed in the VN – FN and VN – DEF systems, and one of the two isomeric cycloadducts is predominantly formed in the VN – MAn system. As compared with the VN – FN and VN – DEF systems, the VN – MAn system is characteristic in that the yields of the products other than the copolymer and the cycloadduct are low (refs. 78,79).

$$Np{-}CH{=}CH_2 + NC{-}CH{=}CH{-}CN \xrightarrow{h\nu} {-}\!\!(CH(Np){-}CH_2{-}CH(CN){-}CH(CN))\!\!{-}_n$$

Np: 2-naphthyl

The results indicate that the reaction mechanisms greatly differ in the reaction by direct irradiation and in the triplet-sensitized reaction using BP. The free-energy change for electron transfer from the excited singlet-state VN to the ground-state electron acceptor monomer is highly exothermic; the values of ΔG are –115, –75, and –55 kJ mol^{-1} as estimated from the Rehm-Weller equation (ref. 54). It is concluded that the photochemical reaction of the VN – electron acceptor monomer system by direct irradiation proceeds via the ion radicals generated by electron transfer from the excited singlet-state VN to the ground-state electron acceptor monomer and by the ionization of the charge-transfer complex in the electronically excited singlet state. By contrast, the electron-transfer process does not contribute dominantly in the BP-sensitized reaction. Electron transfer in the triplet state may be energetically unfavorable because of the smaller excitation energy in the case of the systems of weaker electron acceptor monomers, VN – DEF and VN – FN systems.

The copolymerization of VN with the electron acceptor monomer is most reasonably explained in terms of the initiation via ion radicals generated in the excited singlet state, followed by free-radical propagation, whereas the cycloadduct results from the reaction in the excited triplet state. In the BP-sensitized reaction of the VN – MAn system, triplet-energy transfer from BP to MAn does not occur, and hence the cycloadduct results from the reaction of the triplet-state VN with the ground-state MAn, possibly via a triplet exciplex (Scheme 7). In the VN – FN and VN – DEF systems, however, triplet-energy transfer occurs from BP to both VN and the electron acceptor monomer. The cycloadducts of VN with FN or DEF are derived from the reaction of the triplet-state VN with the ground-state VN or DEF and possibly from the reaction of the triplet-state FN or DEF with the ground-state VN, possibly via the triplet exciplex, as reported for the photocycloaddition of phenanthrene to dimethyl fumarate (ref. 81). Photoisomerization of the electron acceptor monomer also takes place.

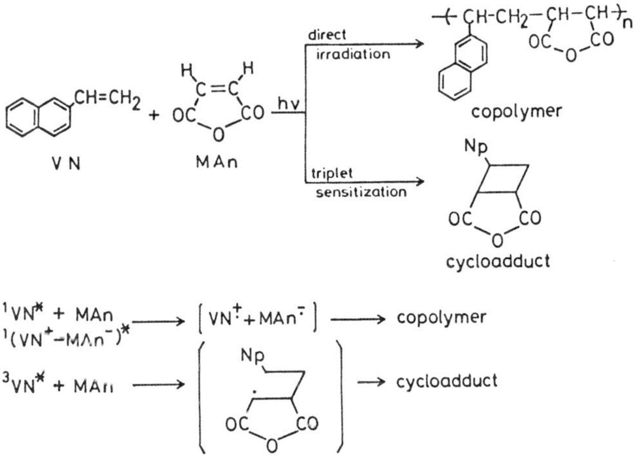

Scheme 7. Photochemical reaction of the VN – MAn system (main reaction pathway).

The presence of Lewis acids or metal salts in free-radical copolymerization systems composed of electron donor and electron acceptor monomers often leads to alternating regulation of resulting copolymers (refs. 82,83). It has been reported that free-radical copolymerization of styrene with DEF in the presence of zinc bromide in methanol proceeds by irradiation with 254 nm light; 1:1 copolymers are formed initially but changes occur after the early stages of the reaction (ref. 84). Other photoinduced free-radical copolymerizations of aryl vinyl monomers as electron donor monomers with electron acceptor monomers in the presence of metal salts include styrene – methyl methacrylate (MMA) – alkylaluminum chloride system (ref. 85), MMA – MAn – zinc chloride system (ref. 86), styrene - MMA - boron trichloride system (ref. 87), and styrene - acrylonitrile - MAn - zinc chloride terpolymerization system (ref. 86). Alternating regulation of the copolymer of styrene with MMA has been accomplished by using boron trichloride as a metal salt. In general, electron acceptor monomers form complexes with metal salts, becoming more powerful electron acceptors for electron donor monomers. The electron acceptor monomers complexed with metal salts form ternary molecular complexes with electron donor monomers, which have been suggested to be involved in both initiation and propagation processes of the copolymerization (refs. 84-87).

3.7 Aryl vinyl monomer - electron donor systems

The effect of charge-transfer interactions on the photochemical reaction of aryl vinyl monomers has been studied (ref. 10). The fluorescence of 2-vinylnaphthalene (VN) is sharply quenched by the addition of amines such as triethylamine (TEA), tri-n-propylamine (TPrA), tri-n-butylamine (TBuA) and 1,4-diazobicyclo[2,2,2]octane (DABCO) in benzene with rate constants of 1.4×10^9, 2.1×10^9, 2.8×10^9, and 6.5×10^9 dm^3 mol^{-1} s^{-1}, respectively. The VN - TEA, VN - TPrA, VN - TBuA systems exhibit exciplex fluorescence in benzene, while no exciplex fluorescence is observed for the VN - DABCO system. The results show that charge-transfer interactions between VN in the excited singlet state and amines in the ground state significantly alter the product distribution in the photochemical reaction of VN in benzene, increasing the trans/cis ratio of 1,2-di(2-naphthyl)cyclobutane from ca. 0.1 in the absence of amines to ca. 2.0 in the presence of DABCO (1:1 mol ratio). The amines in the ground state quench the reaction via the VN excited singlet state that gives the cis-head-to-head cyclobutane cyclodimer, enhancing deactivation to the ground state. The population of the VN triplet state, which favors the formation of the trans-cyclodimer, is enhanced via charge-transfer interactions. Related to this, the effect of charge-transfer interactions on the photochemical reaction of anthracene in the presence of dimethylaniline has been reported (ref. 88).

On the other hand, photochemical reactions of aromatic hydrocarbons (ref. 89-92) and aryl olefins, e.g., trans-stilbene (ref. 93), styrene (ref. 94) and VN (ref. 95), with amines in polar solvents are known to proceed via ion radicals generated by electron transfer, yielding reduction products and aminated adducts of aromatic compounds. The aromatic hydrocarbon - amine photoreduction system reportedly serves as an photoinitiator of added monomers (see section 4.5).

Photocycloaddition reaction of 1-vinylpyrene or styrene as an electron acceptor monomer with 4-dimethylaminostyrene as an electron donor monomer takes place in nonpolar solvents to yield trans- and cis-1-(4-dimethylaminophenyl)-2-pyren-1-ylcyclobutane, or cis-1-(4-dimethylaminophenyl)-2-phenylcyclobutane via an exciplex intermediate. In polar solvents, however, copolymerization mainly proceeds via ion radical intermediates (ref. 96).

3.8 Other systems

Photocationic polymerization of isobutylene is induced by VCl_4, $TiCl_4$, or $TiBr_4$ in bulk or in a heptane solution under irradiation with uv or visible light (ref. 97). The photocationic polymerization proceeds also by selective excitation of a ground-state charge-transfer complex in bulk at 589 nm using a monochromatic sodium light. An isobutylene cation radical is formed by the excitation of the isobutylene - VCl_4 complex, which acts as an intermediate for the initiation of the photocationic polymerization of isobutylene (ref. 98). Photocationic copolymerization of isobutylene with isoprene in the presence of VCl_4 has also been carried out (ref. 99).

Methyl methacrylate (MMA) undergoes photoradical polymerization in the presence of iodine monobromide (ref. 100), sulfur dioxide (ref. 101), or organic antimonies (ref. 102). It has been suggested that charge-transfer complexes of MMA with these electron acceptors are involved in the formation of initiating free radicals.

It is known that photoirradiation of the naphthalene - acrylonitrile system in an alcohol solution affords cyclobutane adducts and substitution products via charge-transfer interactions in the excited singlet state of naphthalene (refs. 103,104). On the other hand, it has been reported that acrylonitrile (AN) polymerizes in the presence of a catalytic amount of naphthalene in DMF or alcohol under irradiation with 313 nm light (ref. 105). The polymerization proceeds by a free-radical mechanism. The polymerization of AN takes place also by photosensitization with benzophenone in the presence of naphthalene (ref. 106). Thus the polymerization of AN is initiated via both the excited singlet and triplet states of naphthalene. Radical intermediates capable of initiating free-radical polymerization are produced in the system. The photoinitiating efficiencies of several aromatic hydrocarbons in AN polymerization correlate with the rate constants for the quenching of the fluorescences of aromatic hydrocarbons by AN and that the initiation rate of the polymerization increases with an increase in solvent polarity, although the ratio $k_p/k_t^{\frac{1}{2}}$ of the rate constants for propagation and termination reactions is not influenced by a change in the polarity of the reaction medium (ref. 107). These results suggest that the polymerization by direct irradiation takes place via electron transfer. The polymerization of AN photoinitiated by naphthalene, anthracene, phenanthrene and pyrene is accelerated by a mixture of zinc chloride, acetate, or nitrate (ref. 108).

Other photoinduced charge-transfer polymerization systems that have been reported include free-radical polymerization of AN in the presence of isobutyl vinyl ether (ref. 109), anionic polymerization of nitroethylene in THF (ref. 110), simultaneous cationic polymerization of cyclohexene oxide and anionic polymerization of nitroethylene (ref. 111), and cationic polymerization of cyclohexene oxide in the presence of MAn (ref. 112).

4 POLYMERIZATIONS USING PHOTOINITIATORS

Photopolymerization using photoinitiators has been widely used in the production of printing plates and in uv curing of coatings. Usually free-radical polymerizations of multifunctional acrylate monomers with photoinitiators are employed. Photoinitiators for free-radical polymerization include azo and peroxy compounds, carbonyl compounds such as benzophenone, benzoin, benzoin alkyl ethers, and benzil dimethyl ketal, halogenated compounds, sulfur-containing compounds such as mono- and di-sulfides, acyl phosphine oxides, ylides, electron donor - acceptor system, etc. Photoinitiators for cationic polymerization are also available. Photoinitiators absorb photoenergy and produce a reactive intermediate, which reacts with a monomer to initiate polymerization, by a mechanism such as fragmentation, hydrogen abstraction, electron transfer, etc. Photochemistry of photoinitiators, e.g., aromatic carbonyl compounds and acyl phosphine oxides, has been studied by means of flash photolysis, absorption and emission spectroscopy, and measurement of photopolymerization rates (refs. 113-117).

4.1 Charge-transfer complexes as photoinitiators

Charge-transfer complexes act as photoinitiators for polymerization of added monomers. Combinations of maleic anhydride (MAn) with an ether, aldehyde, or a ketone reportedly cause cationic photopolymerization of isobutyl vinyl ether (refs. 118,119). Photolysis of a THF - MAn charge-transfer complex generates a free radical, which initiates radical polymerization of methyl methacrylate (MMA) (refs. 120,121). Copolymerization of trans-stilbene with MAn photoinitiated by the THF - MAn system has also been reported (ref. 122).

It has been reported that photopolymerization of MMA is initiated by various charge-transfer complexes such as THF - bromine (Br_2) (ref. 123), THF - sulfur dioxide (ref. 124), quinoline - Br_2 (ref. 125), quinaldine and lutidine - Br_2 (ref. 126), isoquinoline - chlorine (ref. 127), poly(N-vinylcarbazole) - Br_2 (ref. 128), etc. Kinetics and other results suggest a free-radical mechanism for the propagation of the polymerization. Charge-transfer complexes photodecompose to generate a halogen atom as initiator for the polymerization. An alternative mechanism involves initial complexation between the photoinitiator charge-transfer complex and the monomer MMA, followed by the photochemical formation of a free radical (refs. 121,122,127).

4.2 Aromatic carbonyl compound – amine systems as photoinitiators

Aromatic carbonyl compound – amine systems using benzophenone, fluorenone, thioxanthone or benzil in conjunction with a tertiary amine serve as photo-initiating systems for the free-radical polymerization of methyl methacrylate (MMA), methyl acrylate, and other monomers (refs. 129-131). Ketyl and alkyl-amine radicals as the initiators for polymerization are produced either by hydrogen-abstraction reaction via a triplet exciplex or via electron transfer, followed by proton transfer (Scheme 8). Which mechanism is operative depends on the specific reaction system, including the solvent. Generally photoradical polymerizations are retarded or inhibited by oxygen; however, photo-polymerizations using the aromatic ketone – amine system as a photoinitiator are reportedly insensitive to oxygen inhibition (refs. 132,133). The alkylamine radical readily reacts with oxygen in a radical chain manner, consuming oxygen present in the system (ref. 134).

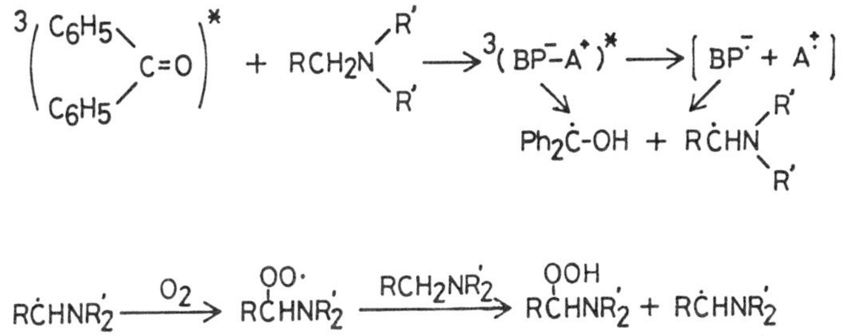

$$^3\left(\begin{matrix}C_6H_5\\ \ \end{matrix}\!\!>C{=}O\right)^* + RCH_2NR'_2 \longrightarrow {}^3(BP^{\bar{}}\text{-}A^{\dot{+}})^* \longrightarrow [BP^{\dot{-}} + A^{\dot{+}}] \longrightarrow Ph_2\dot{C}\text{-}OH + R\dot{C}HNR'_2$$

$$R\dot{C}HNR'_2 \xrightarrow{O_2} R\overset{OO\cdot}{C}HNR'_2 \xrightarrow{RCH_2NR'_2} R\overset{OOH}{C}HNR'_2 + R\dot{C}HNR'_2$$

Scheme 8. Photochemical reaction of benzophenone with a tertiary amine.

N,N-Dialkylaminoethyl methacrylate, a monomer containing an alkylamino group, undergoes photoradical polymerization in the presence of benzophenone as a photoinitiator (ref. 135). This system is characteristic in that N,N-dialkylaminoethyl methacrylate functions both as a co-initiator and as a monomer and that the polymerization rate is accelerated by oxygen.

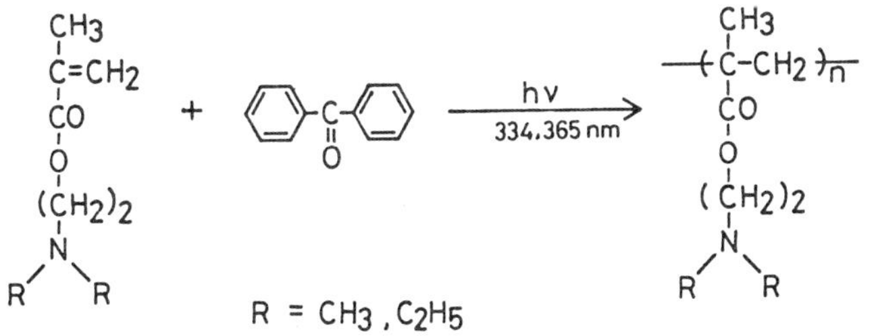

$$CH_2{=}C(CH_3)CO\text{-}O\text{-}(CH_2)_2\text{-}NR_2 + PhC(=O)Ph \xrightarrow[334,365\ nm]{h\nu} \text{-(}C(CH_3)(CO\text{-}O\text{-}(CH_2)_2\text{-}NR_2)\text{-}CH_2\text{)}_n\text{-}$$

$$R = CH_3,\ C_2H_5$$

4.3 Onium salts as photoinitiators

Photoradical polymerizations are subject to oxygen retardation or inhibition; this is unfavorable in practical application of photopolymerization. In order to overcome the problem of oxygen inhibition, new classes of photoinitiators for cationic polymerization have been developed which include aryl diazonium salts (ref. 136) and onium salts such as diaryliodonium and triarylsulfonium salts with non-nucleophilic complex metal halide anions (refs. 137,138). Photocationic polymerizations of epoxides, cyclic ethers, lactones, vinyl ethers, etc., have been carried out using these photoinitiators (refs. 136-141).

Aryl diazonium salts and onium salts decompose by photoirradiation to generate Lewis acids and proton, respectively, which initiate cationic polymerization (refs. 136-138) (Scheme 9). Initiation of free-radical polymerization by onium salts has also been reported (ref. 142). Modified mechanisms for photodecomposition of onium salts have been proposed in order to account for product analysis data (refs. 140,142,143).

$$ArN_2^+BF_4^- \xrightarrow{h\nu} ArF + BF_3 + N_2$$

$$Ar_3S^+X^- \xrightarrow{h\nu} Ar_2S^{+\cdot}X^- + Ar\cdot$$

$$Ar_2S^{+\cdot}X^- \xrightarrow{RH} Ar_2S^+HX^- + R\cdot$$

$$Ar_2S^+HX^- \longrightarrow Ar_2S + HX$$

Scheme 9. Proposed mechanisms for photodecomposition of aryldiazonium salts and triarylsulfonium salts.

Diaryliodonium and triarylsulfonium salts are thermally stable photoinitiators; however, the spectral response of these compounds is limited to the short wavelength uv region (below 300 nm). In order to extend the spectral response to a longer wavelength region, a novel series of triarylsulfonium salt photoinitiators has been prepared (refs. 144,145). The spectral response can be extended into 300-400 nm region by the utilization of photosensitizers such as anthracene, perylene, and phenothiazine (ref. 146). The proposed mechanism for photosensitization involves energetically favorable electron transfer from the excited-state photosensitizer to the ground-state onium salt. The resulting photosensitizer cation radical is thought to initiate polymerization of the

$$S \xrightarrow{h\nu} S^*$$

$$S^* + \underset{(Ph_2I^+X^-)}{Ph_3S^+X^-} \longrightarrow S^{+\cdot}X^- + \underset{(Ph_2I\cdot)}{Ph_3S\cdot}$$

$$\underset{(Ph_2I\cdot)}{Ph_3S\cdot} \longrightarrow \underset{(PhI)}{Ph_2S} + Ph\cdot$$

$$S^{+\cdot} + \text{monomer} \longrightarrow \text{polymer}$$

Scheme 10. Mechanism for photosensitization of onium salts.

reactive monomer either directly or indirectly (Scheme 10).

4.4 Semiconductors as photocatalysts

Photosensitized reactions of heterogeneous systems using solid-phase semiconductors have been the subject of recent extensive studies (ref. 147). The photosensitized reaction process involves electron transfer between the semiconductor and a substrate. The semiconductor absorbs a photon, and the resulting hole-electron pair is separated by virtue of band bending at the surface of the semiconductor immersed in an electrolyte solution containing a redox couple. When a n-type semiconductor is used, photogenerated holes in the valence band migrate to the semiconductor/solution interface, where a substrate is oxidized by reaction with holes, undergoing further reaction.

There have been a few reports which deal with the application of such heterogeneous photocatalytic process to the initiation of polymerization. Free-radical polymerization of methyl methacrylate has been shown to take place by uv irradiation of partially platinized n-type TiO_2 semiconductor powders suspended in an acetonitrile solution containig glacial acetic acid or sodium salt of γ-hydroxybutyric acid (ref. 148). Free radicals produced at the TiO_2 surface by reaction of carboxylate anions with photogenerated holes serve as initiating species of polymerization. It has been shown that direct photopolymerization competes with the polymerization initiated by a photo-electrochemical process (refs. 148,149). Styrene has been reported to undergo photoolectrochemical polymerization at a TiO_2 single-crystal electrode as a working electrode with a platinum counter electrode in dichloromethane containing tetra-n-butylammonium perchlorate as a supporting electrolyte (ref. 149). 1-Vinylpyrene is polymerized photoelectrochemically at a shorter band-gap n-GaAs single-crystal electrode under visible light irradiation (hv>460 nm) to yield poly(1-vinylpyrene) with a weight-average molecular weight of 9600 (ref. 150). It has been suggested that 1-vinylpyrene cation radical generated by reaction with photogenerated holes is responsible for the initiation of the polymerization. Semiconductor powders of CdS, CdSe, and Fe_2O_3 are also used for the photoelectrochemical polymerization of 1-vinylpyrene (ref. 151).

4.5 Dyes as photosensitizers and miscellaneous systems

Spectral sensitization of polymerization in the visible wavelength region is performed by the use of dyes as photosensitizers (ref. 152). The initiation process of the dye-sensitized polymerization is mostly based on electron transfer. Dyes in the electronically excited state function as either electron acceptors or electron donors. Usually photoreducible dyes that act as electron acceptors have been used. The sensitization by photoreducible dyes such as thiazine and xanthene dyes include two processes involving mostly the triplet state of dyes (Scheme 11). One is the Oster process (ref. 153), where reduced dyes react with oxygen to form the hydroxyl radical which initiates polymerization. Ascorbic acid or phenylhydrazine is used as the reducing agent. In another process, the excited-state dye oxidizes a catalyst to generate free radicals which serve as initiators. Sulfinates, tertiary amines, or organometallics are used as a catalyst.

A $D \xrightarrow{h\nu} {}^1D^* \longrightarrow {}^3D^*$

$${}^3D^* \xrightarrow{R} DH_2$$

$$DH_2 + \frac{1}{2}O_2 \longrightarrow DH\cdot + \cdot OH \xrightarrow{\text{monomer}} \text{polymerization}$$

B $D^+ \xrightarrow{h\nu} D^{+*}$

$$D^{+*} + CH_3ArSO_2^- \longrightarrow D\cdot + CH_3ArSO_2\cdot \xrightarrow{\text{monomer}} \text{polymerization}$$

D : dyes R : reducing agents

Scheme 11. Dye-sensitized photopolymerization.

A photosensitive system, composed of barium acrylate (monomer), methylene blue (dye), sodium salt of p-toluensulfinic acid (catalyst), sodium salt of p-nitrophenylacetic acid (fixing agent), and water (solvent), is potentially used for display of optically recorded information. In this system, light-scattering colloidal polymer particles are formed upon exposure to visible light, producing high resolution, continuous images. Optical fixation is carried out upon exposure to uv light by which aci-anion is formed from sodium salt of p-nitrophenylacetic acid to desensitize the polymerization (ref.154). The excited singlet state of the dye is quenched by the sulfinate ion without reaction and that the triplet state of the dye undergoes a redox reaction with the sulfinate, although the product rapidly reverts back to starting material when the monomer is absent. The sulfonyl radical is suggested to be the polymerization initiator (ref. 155). Topochemical polymerization of diacetylenes is photosensitized by visible light by co-crystallization of a diacetylene monomer with phenazine as a complex crystal (ref. 156). After completion of the photopolymerization, phenazine can be removed from the crystals by extraction with the usual organic solvents. Electron transfer from the diacetylene to the excited-state phenazine has been suggested to be responsible for the initiation of polymerization (ref. 157).

The involvment of electron transfer leading to the formation of free radicals as initiators has been suggested for polymerization of methyl methacrylate using photoinitiator systems such as ion (III) - amine - carbon tetrachloride (refs. 158,159) and aromatic hydrocarbon - amine in N,N-dimethylformamide (ref. 160). Free radicals are formed by fragmentation or proton transfer following electron transfer. Cationic photopolymerization of tetrahydrofuran (THF) is initiated by copper and silver salts upon irradiation with 254 nm light. Ligand-to-metal charge-transfer excitation is responsible for the initiation of the polymerization (ref. 161). Cationic polymerizations of THF and alkyl vinyl ethers are induced in the presence of oxidants such as $AgPF_6$ and onium salts and photochemically active sources of free radicals such as benzoin alkyl ethers (ref. 162). Electron donor free radicals generated photochemically undergo one-electron oxidation to the corresponding carbocations, which act as initiators for the cationic polymerization (ref. 162).

5 CONCLUDING REMARKS

Photoinduced polymerization has been the subject of extensive kinetic studies since the control of the initiation rate can be achieved by modification of the light intensity. Absolute values of propagation and termination rate constants have been obtained for some systems. In this article, attention has been focussed on electron-transfer process involved in photopolymerization from the standpoint of organic chemistry. Photoinduced electron-transfer polymerizations of vinyl and related monomers have been discussed in relation to their low-molecular photochemical reactions. Polymerizations involve homo- and co-polymerizations, and both ionic and free-radical propagations. A variety of reaction systems involving electron transfer leading to free radicals may serve as photoinitiators for free-radical polymerizations of added monomers. Conversely, information obtained from polymerizations of added monomers can be utilized to gain insight into the intermediates generated in the reaction system since polymerization is induced by a very low concentration of intermediates due to the chain nature of polymerization.

6 REFERENCES

1 S.L. Mattes and S. Farid, Organic Photochemistry, ed. by A. Padwa, Vol. 6, Marcel Dekker Inc., New York, 1983, pp. 233-326.
2 R.S. Davidson, Adv. Phys. Org. Chem., 19, 1983, pp. 1-130.
3 F.D. Lewis, Adv. Photochem., 13, 1986, pp. 165-235.
4 M. Chanon and M.L. Tobe, Angew. Chem. Int. Ed. Engl., 21, 1 (1981).
5 M. Julliard and M. Chanon, Chem. Rev., 83, 425 (1983).
6 Y. Shirota and H. Mikawa, J. Macromol. Sci., Rev. Macromol. Chem., 16(2), 129 (1977/78); Mol. Cryst. Liq. Cryst., 126, 43 (1985).
7 V.D. McGinniss, Photogr. Sci. Eng., 23, 124 (1979).
8 W.G. Brown, J. Am. Chem. Soc., 90, 1919 (1968).
9 M. Kojima, H. Sakuragi, and K. Tokumaru, Tetrahedron Lett., 22, 2889 (1981).
10 Y. Shirota, A. Nishikata, T. Aoyama, J. Saimatsu, S.-C. Oh, and H. Mikawa, J. Chem. Soc., Chem. Commun., 1984, 64.
11 Y. Hatanaka and Y. Shirota, unpublished results.

12 Y. Shirota, K. Tada, M. Shimizu, S. Kusabayashi, and H. Mikawa, J. Chem. Soc., Chem. Commun., 1970, 1110.
13 K. Tada, Y. Shirota, and H. Mikawa, J. Polym. Sci., Polym. Chem. Ed., 11, 2961 (1973).
14 T. Natsuume, Y. Shirota, H. Hirata, S. Kusabayashi, and H. Mikawa, J. Chem. Soc., Chem. Commun., 1969, 289; ibid., 1969, 189; Polym. J., 1, 181 (1970).
15 S. Tazuke, O. Supakron and T. Inoue, J. Polym. Sci., Polym. Chem. Ed., 20, 2239 (1982).
16 H. Scott, G.A. Miller, and M.M. Labes, Tetrahedron Lett., 1963, 1073.
17 L.P. Ellinger, Polymer, 5, 559 (1964); 6, 549 (1965).
18 S. Tazuke, M. Asai, S. Ikeda, and S. Okamura, J. Polym. Sci., Part B, 5, 453 (1967).
19 M. Shimizu, K. Tanabe, K. Tada, Y. Shirota, S. Kusabayashi, and H. Mikawa, J. Chem. Soc., Chem. Commun., 1970, 1628.
20 M. Yamamoto, S. Nishimoto, M. Ohoka, and Y. Nishijima, Macromolecules, 3, 706 (1970).
21 O.F. Olaj, J.W. Breitenbach, and M.F. Kaufman, J. Polym. Sci., Part B, 9, 877 (1971).
22 M. Asai, K. Kameoka, Y. Takeda, and S. Tazuke, J. Polym. Sci., Part B, 9, 247 (1971).
23 M. Yamamoto, T. Ohmichi, M. Ohoka, K Tanaka, and Y. Nishijima, Progr. Polym. Phys. Jpn., 12, 457 (1969).
24 K. Tada, Y. Shirota, and H. Mikawa, J. Polym. Sci., Part B, 10, 691 (1972).
25 Y. Shirota, K. Kawai, N. Yamamoto, K. Tada, H. Mikawa, and H. Tsubomura, Chem. Lett., 1972, 145; Y. Shirota, K. Kawai, N. Yamamoto, K. Tada, T. Shida, H. Mikawa, and H. Tsubomura, Bull. Chem. Soc. Jpn., 45, 2683 (1972).
26 Y. Taniguchi, Y. Nishina, and N. Mataga, Chem. Lett., 1972, 221; Bull. Chem. Soc. Jpn., 46, 1646 (1973).
27 Y. Shirota, T. Tomikawa, T. Nogami, N. Yamamoto, H. Tsubomura, and H. Mikawa, Bull. Chem. Soc. Jpn., 47, 2099 (1974).
28 M. Yamamoto, M. Ohoka, K. Kitagawa, S. Nishimoto, and Y. Nishijima, Chem. Lett., 1973, 745.
29 R.A. Caldwell and L. Smith, J. Am. Chem. Soc., 96, 2994 (1974).
30 D. Creed and R.A. Caldwell, J. Am. Chem. Soc., 96, 7369 (1974).
31 H. Fujioka and Y. Shirota, unpublished results.
32 M. Shimizu, K. Tada, Y. Shirota, S. Kusabayashi, and H. Mikawa, Makromol. Chem., 176, 1953 (1975).
33 K. Tada, Y. Shirota, S. Kusabayashi, and H. Mikawa, J. Chem. Soc., Chem. Commun., 1971, 1169.
34 K. Tada, Y. Shirota, and H. Mikawa, Macromolecules, 6, 9 (1973).
35 K. Tada, Y. Shirota, and H. Mikawa, Macromolecules, 7, 549 (1974).
36 K. Murai and Y. Shirota, unpublished results.
37 R.A. Crellin, M.C. Lambert, and A. Ledwith, J. Chem. Soc., Chem. Commun., 1970, 682.
38 A. Ledwith, Acc. Chem. Res., 5, 133 (1972).
39 P. Beresford, M.C. Lambert, and A. Ledwith, J. Chem. Soc., C, 2508 (1970).
40 M. Asai and S. Tazuke, Macromolecules, 6, 818 (1974).
41 R.A. Carruthers, R.A. Crellin, and A. Ledwith, J. Chem. Soc., Chem. Commun., 1969, 252.
42 R.A. Crellin and A. Ledwith, Macromolecules, 8, 93 (1975).
43 S. Tagawa, Y. Tabata, S. Arai, and M. Imamura, J. Polym. Sci., Polym. Lett. Ed., 12, 545 (1974).
44 K. Hamanoue, H. Teranishi, M. Okamoto, Y. Furukawa, S. Tagawa, and Y. Tabata, J. Polym. Sci., Polym. Chem. Ed., 18, 91 (1980).
45 S. Tazuke and N. Kitamura, J. Chem. Soc., Chem. Commun., 1977, 515; N. Kitamura and S. Tazuke, Bull. Chem. Soc. Jpn., 53, 2594 (1980).
46 Y. Yoshida and Y. Shirota, unpublished results.
47 D.R. Terrell, Polymer, 23, 1045 (1982).

48 R.G. Jones and N. Khalid, Eur. Polym. J., 18, 285 (1982).
49 Y. Shirota, K. Takemura, H. Mikawa, T. Kawamura, and K. Matsuzaki, Makromol. Chem., Rapid Commun., 3, 913 (1982).
50 Y. Shirota, A. Matsumoto, and H. Mikawa, Polym. J., 3, 643 (1972).
51 M. Yoshimura, Y. Shirota, and H. Mikawa, J. Polym. Sci., Part B, 11, 457 (1973).
52 Y. Shirota, M. Yoshimura, A. Matsumoto, and H. Mikawa, Macromolecules, 7, 4 (1974).
53 M. Yoshimura, H. Mikawa, and Y. Shirota, Macromolecules, 11, 1085 (1978).
54 D. Rehm and A. Weller, Isr. J. Chem., 8, 259 (1970).
55 Y. Shirota, H. Fujioka, and H. Mikawa, J. Polym. Sci., Polym. Lett. Ed., 16, 425 (1978).
56 H. Masuhara, Y. Tohgo, and N. Mataga, Chem. Lett., 1975, 59.
57 G.-J. Jiang and Y. Shirota, unpublished results.
58 M. Yamamoto, T. Asanuma, and Y. Nishijima, J. Chem. Soc., Chem. Commun., 1975, 53.
59 T. Asanuma, M. Yamamoto, and Y. Nishijima, J. Chem. Soc., Chem. Commun., 1975, 56.
60 S. Kuwata, Y. Shigemitsu, and Y. Odaira, J. Chem. Soc., Chem. Commun., 1972, 2; J. Org. Chem., 38, 3803 (1973).
61 T.R. Evans, R.W. Wake, and O. Jaenicke, The Exciplex ed. by M. Gordon and W.R. Ware, Academic Press, Inc., New York, 1975, pp. 345.
62 K. Mizuno, H. Kagano, T. Kasuga, and Y. Otsuji, Chem. Lett., 1983, 133.
63 S. Farid and S.E. Shealer, J. Chem. Soc., Chem. Commun., 1973, 677.
64 R.A. Neunteufel and D.R. Arnold, J. Am. Chem. Soc., 95, 4080 (1973).
65 T. Asanuma, M. Yamamoto, and Y. Nishijima, J. Chem. Soc., Chem. Commun., 1975, 608.
66 T. Asanuma, Y. Gotoh, A. Tsuchida, M. Yamamoto, and Y. Nishijima, J. Chem. Soc., Chem. Commun., 1977, 485.
67 S.L. Mattes and S. Farid, J. Am. Chem. Soc., 105, 1386 (1983).
68 T. Gotoh, M. Yamamoto, and Y. Nishijima, J. Polym. Sci., Polym. Lett. Ed., 17, 143 (1979).
69 Y. Yamamoto, M. Irie, Y. Yamamoto, and K. Hayashi, J. Chem. Soc., Perkin Trans., II, 11, 1517 (1979).
70 M. Irie, S. Tomimoto, and K. Hayashi, J. Polym. Sci., Part B, 8, 585 (1970); J. Polym. Sci., Polym. Chem. Ed., 10, 3235 (1972).
71 Y. Yamamoto, M. Irie, and K. Hayashi, Polym. J., 8, 437 (1976).
72 M. Irie, H. Msuhara, K. Hayashi, and N. Mataga, J. Phys. Chem., 78, 341 (1974).
73 M. Shimizu, Y. Yamamoto, M. Irie, and K. Hayashi, J. Macromol. Sci., Chem., 10, 1607 (1976).
74 T. Gotoh, M. Yamamoto, and Y. Nishijima, J. Polym. Sci., Polym. Chem. Ed., 19, 1047 (1981).
75 D.H. Davies, D.C. Phillips, and J.D.B. Smith, J. Polym. Sci., Polym. Chem. Ed., 15, 2673 (1977).
76 R.K. Sadhir, J.D.B. Smith, and P.M. Castle, J. Polym. Sci., Polym. Chem. Ed., 21, 1315 (1983).
77 M.A. Williamson, J.D.B. Smith, P.M. Castle, and R.N. Kauffman, J. Polym. Sci., Polym. Chem. Ed., 20, 1875 (1982).
78 S.-C. Oh, K. Yamaguchi, and Y. Shirota, Polym. Bull., 18, 99 (1987).
79 K. Yamaguchi, S.-C. Oh, and Y. Shirota, Chem. Lett., 1986, 1445.
80 K. Yamaguchi, S. Masumi, and Y. Shirota, unpublished results.
81 D. Creed, R.A. Caldwell, M. McKenney Ulrich, J. Am. Chem. Soc., 100, 5831 (1978).
82 J.M.G. Cowie (ed.), Alternating Copolymers, Plenum Press, New York, 1985.
83 H. Hirai, J. Polym. Sci., Macromol. Rev., 11, 47 (1976).
84 D.H. Davies, D.C. Phillips, and J.D.B. Smith, J. Polym. Sci., Polym. Chem. Ed., 10, 3253 (1972).
85 N.G. Gaylord, S.S. Dixit, and B.K. Patnaik, J. Polym. Sci., B, 9, 927 (1971).

86 I. Capek and J. Barton, Makromol. Chem., 181, 241 (1980); J. Barton and I. Capek, Ibid., 182, 3513 (1981).
87 H. Hirai, K. Takeuchi, and M. Komiyama, J. Polym. Sci., Part C, Polym. Lett., 25, 181 (1987).
88 N.C. Yang, D.M. Shold, and B. Kim, J. Am. Chem. Soc., 98, 6587 (1976).
89 R.S. Davidson, J. Chem. Soc., Chem. Commun., 1969, 1450.
90 J.A. Barltrop, Pure Appl. Chem., 33, 179 (1973).
91 S.G. Cohen, A. Parola, and G.H. Parsons, Jr., Chem. Rev., 73, 141 (1973).
92 S.-C. Oh, Y. Shirota, H. Mikawa, and S. Kusabayashi, Chem. Lett., 1986, 2121.
93 F.D. Lewis, Acc. Chem. Res., 12, 152 (1979).
94 S. Toki, S. Hida, S. Takamuku, and H. Sakurai, Nippon Kagaku Kaishi, 1984, 152.
95 S.-C. Oh and Y. Shirota, unpublished results.
96 A. Tsuchida, M. Yamamoto, and Y. Nishijima, J. Chem. Soc., Perkin Trans., II, 1986, 239.
97 M. Marek and L. Toman, J. Polym. Sci., Polym. Symp., 42, 339 (1973); M. Marek, ibid., 56, 149 (1977).
98 M. Marek and L. Toman, Makromol. Chem., Rapid Commun., 1, 161 (1980).
99 L. Toman, J. Pilar, J. Spevacek, and M. Marek, J. Polym. Sci., Polym. Chem. Ed., 16, 2759 (1978).
100 P. Ghosh and H. Banerjee, J. Polym. Sci., Polym. Chem. Ed., 16, 633 (1978).
101 P. Ghosh and S. Chakraborty, Eur. Polym. J., 15, 137 (1979).
102 H. Matsuda, T. Isaka, and N. Iwamoto, Makromol. Chem., 179, 539 (1978).
103 R.M. Bowman and J.J. McCullough, J. Chem. Soc., Chem. Commun., 1970, 948.
104 R.M. Bowman, T.R. Chamberlain, C-W. Huang, and J.J. McCullough, J. Am. Chem. Soc., 96, 692 (1974).
105 J. Barton, I. Capek, and P. Hrdlovic, J. Polym. Sci., Polym. Chem. Ed., 13, 2671 (1975).
106 I. Capek and J. Barton, J. Polym. Sci., Polym. Chem. Ed., 13, 2691 (1975).
107 I. Capek, J. Barton, and J. Danciger, J. Poly. Sci., Polym. Chem. Ed., 17, 943 (1979).
108 I. Capek and J. Barton, J. Polym. Sci., Polym. Chem. Ed., 17, 937 (1979).
109 S. Tazuke and S. Okamura, J. Polym. Sci., A-1, 7, 715 (1969).
110 M. Irie, S. Tomimoto, and K. Hayashi, J. Polym. Sci., Polym. Lett. Ed., 10, 699 (1972).
111 M. Irie, S. Tomimoto, and K. Hayashi, J. Polym. Sci., Polym. Chem. Ed., 11, 1859 (1973).
112 R.K. Sadhir, J.D.B. Smith, and P.M. Castle, J. Polym. Sci., Polym. Chem. Ed., 23, 411 (1985).
113 N.S. Allen, F. Catalina, P.N. Green, and W.A. Green, Eur. Polym. J., 22, 49 (1986); Ibid., 22, 871 (1986).
114 R. Kuhlmann and W. Schnabel, Polymer, 17, 419 (1976).
115 S.P. Pappas and R.A. Asmus, J. Polym. Sci., Polym. Chem. Ed., 20, 2643 (1982).
116 A. Merlin and J.-P. Fouassier, Makromol. Chem., 181, 1307 (1980).
117 T. Sumiyoshi, W. Schnabel, A. Henne, and P. Lechtken, Polymer, 26, 141, (1985).
118 K. Takakura, K. Hayashi, and S. Okamura, J. Polym. Sci., Part B, 3, 565 (1965).
119 H. Yamaoka, K. Takakura, K. Hayashi, and S. Okamura, J. Polym. Sci., Part B, 4, 509 (1966).
120 C.E.H. Bawn, A. Ledwith, and A. Parry, J. Chem. Soc., Chem. Commun., 1965, 490 (1965).
121 J. Barger and M. Lazar, J. Polym. Sci., Part A-1, 6, 3109 (1968).
122 M.L. Hallensleben, Eur. Polym. J., 9, 227 (1973).
123 P. Ghosh and A.N. Banerjee, J. Polym. Sci., Polym. Chem. Ed., 15, 203 (1977).
124 P. Ghosh and S. Jana, and S. Biswas, Eur. Polym. J., 16, 89 (1980).
125 P. Ghosh and P.S. Mitra, J. Polym. Sci., Polym. Chem. Ed., 13, 921 (1975).

126 M.K. Mishra, S. Lenka, and P.L. Nayak, J. Polym. Sci., Polym. Chem. Ed., 19, 2457 (1981).
127 S. Lenka, P.L. Nayak, and S. Nayak, J. Polym. Sci., Polym. Chem. Ed., 22, 429 (1984).
128 P. Ghosh and T.K. Ghosh, J. Polym. Sci., Polym. Chem. Ed., 22, 2295 (1984).
129 M.R. Sandner, C.L. Osborn, and D.J. Trecher, J. Polym. Sci., Polym. Chem. Ed., 10, 3173 (1972).
130 A. Ledwith and M.D. Purbrick, Polymer, 14, 521 (1973).
131 P.K. Sengupta and S.K. Modak, J. Macromol. Sci.-Chem., A20, 798 (1983).
132 E. Wang, M. Li, H. Kong, and S. Voong, Chem. Abstr., 96, 163262h (1982); 97, 110433d (1982).
133 G. Smets, S.N.E. Hamouly, and T.J. Oh, Pure Appl. Chem., 56, 439 (1984).
134 R.F. Bartholomew and R.S. Davidson, J. Chem. Soc., C, 2342, 2347 (1971).
135 G.-J. Jiang, Y. Shirota, and H. Mikawa, Polym. Photochem., 7, 311 (1986).
136 S.I. Schlesinger, Photogr. Sci. Eng., 18, 387 (1974).
137 J.V. Crivello and J.H.W. Lam, Macromolecules, 10, 1307 (1977).
138 J.V. Crivello and J.H.W. Lam, J. Polym. Sci., Polym. Chem. Ed., 17 , 977, 1047, 1059 (1979).
139 S.P. Pappas and J.H. Jilek, Photogr. Sci. Eng., 23, 140 (1979).
140 R.S. Davidson and J.W. Goodin, Eur. Polym. J., 18, 589 (1982).
141 M. Tsunooka, T. Ueda, and M. Tanaka, J. Polym. Sci., Polym. Chem. Ed., 22, 2217 (1984).
142 S. Kondo, M. Muramatsu, and K. Tsuda, J. Macromol. Sci.-Chem., A19, 999 (1983).
143 S.P. Pappas, B.C. Pappas, and L.R. Gatechair, J. Polym. Sci., Polym. Chem. Ed., 22, 69 (1984).
144 J.V. Crivello, D.A. Conlon, and J.L. Lee, Polym. Bull., 14, 279 (1985).
145 W.R. Watt, H.T. Hoffman, Jr., H. Pobiner, L.J. Schkolnick, and L.S. Yang, J. Polym. Sci., Polym. Chem. Ed., 22, 1789 (1984).
146 S.P. Pappas, L.R. Gatechair, and J.H. Jilek, J. Polym. Sci., Polym. Chem. Ed., 22, 77 (1984).
147 M.A. Fox, Acc. Chem. Res., 16, 314 (1983).
148 B. Kraeutler, H. Reiche, A.J. Bard, and R.G. Hocker, J. Polym. Sci., Polym. Lett. Ed., 17, 535 (1979).
149 B.L. Funt and S.-R. Tan, J. Polym. Sci., Polym. Chem. Ed., 22, 605 (1984).
150 P.V. Kamat, R. Basheer, and M.A. Fox, Macromolecules, 18, 1366 (1985).
151 P.V. Kamat and R.V. Todesko, J. Polym. Sci., Part A, Polym. Chem., 25, 1035 (1987).
152 K.I. Jacobson and R.E. Jacobson, Imaging Systems, Focal Press Ltd, London, 1976, pp. 181-198.
153 G. Oster, Nature, 173, 300 (1954).
154 J.D. Margerum, L.J. Miller, and J.B. Rust, Photogr. Sci. Eng., 12, 177 (1968).
155 J.D. Margerum, A.M. Lackner, M.J. Little, and C.T. Petrusis, J. Phys. Chem., 75, 3066 (1971).
156 B. Tieke and G. Wegner, Makromol. Chem., 179, 2573 (1978); 182, 133 (1981).
157 C. Bubeck, T.H. Nguyen Xuan, H. Sixl, B. Tieke, and D. Bloor, Ber. Bunsenges. Phys. Chem., 87, 1149 (1983).
158 T. Okimoto, M. Takahashi, Y. Inaki, and K. Takemoto, Angew. Chem. Makromol. Chem., 38, 81 (1974).
159 Y. Inaki, M. Takahashi, and K. Takemoto, J. Macromol. Sci.-Chem., A9, 1133 (1975).
160 H. Kubota and Y. Ogiwara, J. Appl. Polym. Sci., 28, 2425 (1983).
161 M.E. Woodhouse, F.D. Lewis, and T.J. Marks, J. Am. Chem. Soc., 100, 996 (1978).
162 F.A.M. Abdul-Rasoul, A. Ledwith, and Y. Yağci, Polymer, 19, 1219 (1978); Polym. Bull., 1, 1 (1978).

Chapter 6.5

Information Management and Storage

M.V. Alfimov and V.A. Sazhnikov

1 INTRODUCTION

In order to store information some changes must be produced in the recording medium during the recording process. Evidently, the scope of photography includes all means of producing of the visible images by the action of light. In any photographic system absorbed photons can induce an irreversible photochemical or photophysical reaction in a receptor layer.

It is known that in many photoprocesses no charge separation is necessary for response to the light stimulus. However, as one can see from the proceding chapters the absorption of a quantum of light often leads directly or indirectly to the separation of oppositely charged particles. This may be considered as a triggering action which initiates a sequence of events resulting in permanent chemical or physical alterations.

The purpose of this text is to illustrate that the light-induced electron transfer is the most important and fundamental primary act used presently in various conventional and unconventional photographic systems.

Really, the formation of latent image in the most widely used silver halide systems is well-known example of permanent change initiated by a photoionizing trigger event (ref. 1). On the other hand, it was recently shown (refs. 2-11) that the photoinduced electron transfer is also widely used in non-silver photochemical information recording processes. Furthermore, through the efforts of many investigators it has been established that the charge transfer from a sensitizing dye to the substrate is a dominant mechanism for photosensitization in many imaging systems of practical interest.

It is apparent that in general case the occurence of electron transfer following light absorption may be understood as an internal phenomenon in some electron-donor-acceptor system (refs. 12-14). The corresponding equations for the direct or indirect mechanism of electron transfer may be written in the following forms (ref. 14):

$$D + A \xrightarrow{h\nu} D^{+} + A^{-} \tag{1}$$

$$D + Z + A \xrightarrow{h\nu} D^{+} + Z^{-} + A \longrightarrow D^{+} + Z + A^{-} \tag{2}$$

where D, A, and Z are donor, acceptor, and intermediate compound (ion, molecule, semiconductor, etc.), respectively.

We shall attempt to show which evidences exist at the present time for the interpretation of the primary photochemical act in different photographic systems as a photoinduced electron transfer (direct or indirect) from some donor to an appropriate acceptor.

This text is not a full review of all cases where the reactions of photochemical oxidation or reduction are explored for photographic data registration. It aims only at a brief review of main groups of light-sensitive materials that are of interest, in our opinion, from this point of view.

2 PHOTOINDUCED CHARGE-TRANSFER PROCESSES IN IMAGING SYSTEMS BASED ON SEMICONDUCTORS

According to literature data (refs. 6,10), the silver halides and electrophotographic systems currently are the most significant parts of imaging technology (approximately 60 and 30% of the total, respectively, in terms of materials, equipment and supplies). Let us first focus our attention on the AgHal photographic process.

As is well known, the conventional silver halide photographic material is one of the most efficient systems used for recording light signals. The AgHal system is extremely versatile and can be applied in almost any imaging application. In spite of many attempts to replace it, the silver halide crystal remains presently the most widely used photoimaging entity.

2.1 Silver halide photographic systems

The various silver halide image sytems are based on the photolysis and development of silver halide microcrystals (refs. 1,8). In the simplest form the light-sensitive coating consist of an array of microcrystals (grains) dispersed in a gelatin. A typical silver grains are usually about 0.5 μm in diameter.

After exposure the grains in exposed areas of photographic layer may be reduced (developed) to metallic silver by a specially formulated reducing agent solution. The main advantage of the development reaction is that each microcrystal of silver halide which has absorbed anywhere few light quanta is fully reduced during the development. Since a typical silver grain contains approximately 10^9 silver ions and the entire grain is converted to metallic silver after development, this means that the overall amplification of approximately 10^8 is achieved.

The developability is attributed to the formation in the microcrystals of small (containing ⩾ 4 Ag atoms) silver aggregates, the latent image specks, which catalyze the reduction of the grains. Hence, the visual photographic image is obtained as a final result of two independent processes - the building of the latent image itself and its intensification by the development.

After development the image is fixed by dissolving the unexposed emulsion grains in sodium or ammonium thiosulfate and rendered permanent by washing and drying.

The elementary photographic process in silver halides which upon exposure leads to the formation of a developable latent image undoubtedly represents a complicated network of reactions. Unfortunately, several fundamental questions in the physics of the primary processes occurring in exposed AgHal microcrystals are still open. Our present knowledge of energetic and kinetic data of such events as formation and migration of charge carriers, trapping and detrapping processes, ionic reactions, etc., is far from being complete.

In extremely general terms the latent image formation in silver halides is basically described in the well-known model of Gurney and Mott, which emphasizes the occurence of electronic and ionic steps in the concentration of silver atoms. Due to excellent agreement with the many experimental observations, this theory proposed by Gerney and Mott (ref. 15) in 1938 is still generally accepted now.

2.1.1 The Gerney-Mott theory

From the viewpoint of our discussion, it should be noted that the model proposed by Gerney and Mott is actually based on the qualitative quantum-mechanical description of the one-electron transfer from the higher to lower energetic levels, while the absorption of a photon leads to an initial excitation of electron to high level and thus starts this process. In the light of present-day knowledge of photochemical and photophysical processes in molecules, complexes and semiconductors is obvious that this idea was of greatest importance.

The Gerney-Mott theory was based on following facts which were known at the time the theory was formulated. Firstly, the illumination of large crystals of silver halides (direct photolysis) leads to the formation of colloidal particles of metallic silver; in a partially printed out emulsion also the silver is not found in atomic form, but, in each halide grain, in a relatively small number of specks of colloidal metal. Secondly, the absorption of light by silver halides is accompanied by photoconductivity; this photoconductivity is due to the motion of photoelectrons, generated on the absorption of light quanta. Thirdly, silver halide crystals display the phenomenon of ionic dark conductivity which ceases at a sufficiently low temperature; this current is carried by silver ions which must therefore be assumed to be capable of moving through the lattice.

It was suggested by Gerney and Mott that the absorption of a photon by a grain transfers an electron from a halide ion to the conduction band levels in which it can diffuse around the grain. After some period of diffusion the electron will be immobilized at some singularity which provides an energy level below the conduction levels, usually reffered to as trap. The theory was not concerned with the exact nature of the electron traps. Gerney and Mott assumed only that in the crystal there exist metastable positions at cracks or impurities at which the electron may temporarily be trapped and from which it will be released later by thermal motion.

On the other hand, the ionic conductivity was explained as follows: in thermal equilibrium a few of the positive ions will have left their normal positions and will be found in the "interlattice" positions; both the displaced ion, and the "hole", will be able to move through the crystal. Thus, the silver atoms are capable of moving through the crystal in the form of separate electrons and interstitial ions, Ag_i^+.

It was further assumed that the latent image is a submicroscopic speck of metallic silver. If an electron comes in contact with a speck of silver it will be captured; then the silver ions in interlattice positions may be attracted to the negatively charged particles of silver. Thus, in the presence of metallic silver specks on or in the crystal the mobile silver atoms would collect on these traps by a mechanism which involves the trapping of the electrons with the subsequent attraction of the mobile silver ions by the electrostatic field.

Of course, other works, both on conventional emulsions and model systems, raised more questions as to the details of electronic and ionic processes in latent image formation and provided proposals for modifications and extensions of the Gerney-Mott mechanism (see, for example, ref. 16). However, the Gerney-Mott principle that both electronic and ionic stages are important in latent image formation seems to be consistent with all known facts.

Finally, it was suggested by Gerney and Mott that the crucial act in phenomenon of spectral sensitization of silver halides by dyes is the transfer of an electron from sensitizer to substrate. They noted that if energy level of the excited dye lies above the conduction levels of AgHal, the electron can freely move from the dye molecule into the crystal, when latent image formation will take place as before.

In this contex it should be emphasized that now is firmly established that other semiconductors behave in the similar way as has been observed in the silver halides, thus allowing to discuss the phenomenon of spectral sensitization in all cases from a common point of view.

Let us now consider some experimental results concerning the primary processes in silver halides.

2.1.2 Experimental results

If it is assumed that the primary process in silver halide photography is the generation of electrons and holes by exposure, then the conductivity of silver halide crystals should be increased by exposure through the libration and displacement of electronic charge carriers. Accordingly, the studies of the electrical and photoelectrical properties of silver halides have been prominent for many years in the development of photographic theory.

The use of pulsed electric fields to study the motion and trapping of photoelectrons was described by Hamilton and Brady (refs. 17-22) in their experiments with flash exposures on emulsions in high electric fields. If a strong electric field exists within the grains, the photoelectrons can drift toward the positive electrode and combine with silver ions preferentially on the sides of the grains nearest that electrode.

It was found that exposures made with no applied field showed a completely random distribution of photolytic silver in grain; when, however, the polarization pulses were used the masses of photolytic silver was assymmetricaly distributed toward the side of the grain to which the electrons should have been displaced.

This result is consistent with the Gerney-Mott theory and thus affords evidencee for the view that free electrons are produced and subsequently trapped, and that the photolytic silver forms when these trapped electrons are neutralized by mobile silver ions. The same effect was observed (after development) for latent-image specks by under the very lowest levels of exposure. This show that the latent-image specks appear to follow the same mechanism of formation as the print-out silver. Thus, the primary role of electrons was demonstrated and it was shown that they may be separated from the positive holes by an electric field.

In this connection, it may also be of interest to point out that the positive hole (electron deficiency on the halide ion) can be displaced by acceptance of an electron from a neighbouring halide ion (this is equivalent to the movement of a positive charge in the opposite direction). Gerney and Mott have assumed that in some way the positive holes are bound in the crystal and are without significance for the formation of the latent image. However, it was clear shown by Hamilton and Brady that the immobile holes are not an inherent property of pure silver halide. It was observed that on the side of the grains opposite the silver deposit occurs a diffuse bromo-gelatin cloud. Hence, the holes can diffuse to the surface where they will convert surface halide ions into halogens atoms. These atoms may diffuse over the grain surface and react with suitable compounds or with latent image silver. The mobilty and lifetime of holes in the silver halide crystals were directly measured (10^{-2}-10^{-3} cm^2/V sec and 15 μsec, respectively).

The technique of flash exposures in pulsed electric fields was applied also to study quantitatively the motion of interstitial silver ions in silver halide emulsion grains. The light flashes were delayed after the beginning of the external electric field pulses, and the latent-image displacement was used to determine the decay rate of the internal field, from which ionic conductancee could be evaluated. On the other hand, if the light flash occurs first, with an interval before application of the field, only those electrons which survive the interval will be displaced. The controlled delays between the light flash and the electric field may be therefore used to determine the length of time that the electrons remain free before final trapping.

It was found that the room-temperature ionic current in the silver bromine grains of the emulsion is two powers of ten higher than that reported for large silver bromide crystals of high purity. This means that ionic conductance in such grains is extrinsic, due to surface effects (with an activation energy about 0.42 eV). Strong evidence for the importance of the surface was provided by the fact that ionic conductance was markedly reduced by silver ion complexing agents (such as phenylmercaptotetrazole) adsorbed to the grain surface.

On the other hand, this reduction of the conductance was accompanied by an increase in the high intensity reciprocity law failure of the emulsion, showing that these mobile silver ions are also involved in the ionic step during the growth of latent image centers. (The silver halide materials obey the so-called "reciprocity law" $D = f(It) = const$ only over a limited range of intensity, I, and time, t, values. For high intensity-short time or low intensity-long time exposures photographic emulsions undergo a sizeable sensitivity loss because a lower probability of formation of latent image centers. This phenomenon is called reciprocity failure).

Finally, it was shown that a striking increase in electron lifetime is caused by silver-complexing agents (from about 3 μsec to about 100 μsec). It was therefore concluded that these silver ions , which are responsible for the dark conductance of the grain, and which combine with photoelectrons during the growth of the latent image specks, also are instrumental in the initial trapping of electrons, as the first stage in latent image formation.

The analogous experiments have been made on emulsion grains containing cadmium. When cadmium divalent cation enters the lattice substitutionally, it replaces two silver ions and decreases the concentration of interstitials. It was established that this method of reduction of the concentration of intersitials also decreases the ionic conductivity, correspondingly increases the electron lifetimes in the grains, and introduces a high-intensity reciprocity-low failure in the emulsion. These findings confirm the above conclusion that the silver ions play a major role in the initial trapping of photoelectrons during latent-image formation.

In accordance with these observations Hamilton and Brady have concluded that in a grains of primitive emulsion there are no permanent electron traps as such, but that physical imperfactions provide a great many shallow traps. An electron trapped at one of these sites normally escapes again within a very short time. If, however, during the short interval of its localization in a shallow trap, it is neutralized by a mobile silver ion, the energy with which it is bound to the center should be appreciably increased, and configuration is stabilized. Shematic diagram of the proposed trapping process is as follows:

$$[\text{free e}^-] \dashrightarrow [\text{e}^- \text{ in shallow trap}] \dashrightarrow [\text{silver atom}] \quad (3)$$

Later the measurements of photoelectron lifetimes was carried out by Kellog (ref. 23) and by Hasegawa and Sakaguchi (ref. 24) through the use of microwave techniques (9.3 GHz). These experiments are based on the principle that forced oscillatory motion of the charged carriers dissipates microwave energy. Since their lifetimes are very long compared with the repetition period of the microwave field (e.g., 0.1 nsec at 10 GHz), the carriers in the microcrystals will be moved back and forth until they disappear by recombination or become immobile by trapping. Because the mobility of free holes is much less than that of the free electrons, the microwave technique measures the photoelectron lifetime.

It was found by Kellog that after light pulse (with durations of 0.5-3 μsec) the decay of the measured photoresponce is first order and the lifetime of the photoexcited electrons in a 0.5-μm cubic AgBr emulsion is 1.7 μsec at room temperature and 4.8 msec at 77 K. In the experiments of Hasegawa and Sakaguchi the observed decrease of the number of mobile carriers occured in the time interval about 8 μsec. Broadly speaking, these results are in good agreement with observations of Hamilton and Brady.

However, at present there exists some discrepancy in the literature values for photoelectrons lifetimes. More recently, Deri and Spoonhower (ref. 25) have studied the microwave (36 GHz) photoconductivity in silver halide emulsions on the 5-100-nsec time scale. Nanosecond light pulses (460 nm, fwhm 3-4 nsec) were used for excitation of microcrystas.

According to Deri and Spoonhower, the microwave photosignal observed for primitive octahedral emulsions may be described by simple exponential decays having time constants of 10-40 nsec. Attempts to find the effect of multiple flashes (up to 60 flashes) on the observed photokinetics have failed. This means that the traps must be reset on a time scale faster than ≈100 msec (the time between laser pulses). Formation of latent image centers did not decrease the electron lifetime. It may be assumed therefore that the observed decay is more efficient than that associated with electron trapping at silver centers.

The photoelectron lifetimes show no dependence on intensity of the light pulses. Since recombination of holes and electrons is thought to be a major reason for reciprocity failure, especially at high intensity, this finding indicates that the decay process is not associated with electron-hole recombination.

In agreement with the results of Hamilton and co-workers it was found a correlation between electron kinetics and ionic properties: (i) photoelectron lifetime increases with increasing grain size when ionic conductivity decreases; (ii) addition of compound which forms complex with interstitial ions reduces the ionic conductivity and increases photoelectron lifetime.

In general, the photoelectron kinetics observed by Deri and Spoonhower occur in the time interval 10-100 nsec after exposure. These decays are significantly shorter than those previously reported for primitive AgBr emulsions of comparable grain size and of an external applied field. For measurements of the Dember signal the charge carriers must be generated inhomogeneously; due to unequal rates of diffusion of the photoelectrons and photoholes there is a net electron diffusion current moving from the region of higher light absorption to that of lower absorption.

A high intensity flash exposure can create in an emulsion grain a large number of photoelectrons which will finally deposit several latent image centers in a grain. The overall process needs some finite time which may be estimated by measuring the Dember effect with another (low intensity) light flash delayed for various time intervals from the first high intensity flash exposure. The relation of the ratio of this standart Dember signal of previously unexposed grains to the similar signal obtained after a high intensity light flash to Δt was investigated. The time at which this ratio achieves its stationary value may be regarded as the time where the aggregation process of silver atoms is completed.

According to the obtained results, the processes of latent image formation continue up to 100 μsec at room temperature and is gradually prolonged at lower temperatures (e.g., up to 300 μsec at -10°C). It is assumed that after light flash (3 μsec) the photoelectrons are rapidly distributed over a number of shallow electron traps; each trapped electron can combine with an interstitial silver ion to form a photolytic silver atom. Thus, the photoelectrons produce in a relatively short period dispersed photolytic silver atoms which then aggregate to form stable silver clusters in the grain. Since the aggregation naturally requires the thermal diffusion or the thermal ionization of the photolytic silver atoms as a kinetic driving force, it is regarded as rather slow process. The completion of the aggregation needs about 100 μsec at room temperature.

Summarizing we could say that the primary processes in photoexcited silver halide crystal may be considered as the indirect (via traps) electron transfer from the halide anion (donor) to the interstitial silver cation (acceptor) followed by the irreversible process of aggregation of the photolytic silver atoms. The reactions can be schematically summarized as:

$$Br^- \xrightarrow{h\nu} Br + e^- \quad (4)$$

$$e^- + Ag_i^+ \longrightarrow Ag \quad (5)$$

$$Ag + Ag_n \longrightarrow Ag_{n+1} \quad (6)$$

2.1.3 The problem of spectral sensitization

As is well known, spectral sensitization of the photographic system is an increase in its sensitivity to a wavelengths longer than those strongly absorbed by the photosensitive compound.

Let us now review very briefly some observations which indicate on the electron-transfer mechanism of this widely studied phenomenon.

In the Gerney-Mott electron-transfer model of sensitization process, excitation of the dye molecule must be followed by an electron transfer to the conduction band of the silver halide; on the other hand, an adsorbed electron-deficient free radical of the dye with only one electron in its highest occupied orbital can constitute the primary positive hole. For example, the one-electron oxidation of a cationic cyanine dye is expected to initially generate the corresponding radical dication. In principle, the dye molecule may be regenerated via the acceptance of an electron from some donor molecule or from donor state in the forbidden gap of a semiconductor (this should lead to the appearance of a trapped hole at the corresponding level). In the energy-transfer mechanism, on the other hand, the sensitization would result in an increase of photoconductivity without influencing the dye which is found to be again in its ground state after donating the energy.

Hence, in the case of electron-transfer mechanism of spectral sesitization, the occurence of free electrons in the grains, the quenching of fluorescence of the dyes, and the formation of new absorption bands and ESR signals of dye radicals could be observed. Furthermore, the presence in the photosensitive layer of other molecules having some electron-donor or electron-acceptor properties could supposedly alter the quantum yield of sensitization.

First of all, exposure to light absorbed by dye, but not appreciably absorbed by silver halide, can cause the formation of a latent image having essentially the same properties as the latent image formed by radiation absorbed by the silver halide. In agreement with this suggestion it was shown by Saunders, Tyler and West (ref. 28) that such exposures produce free electrons, the displacement of which per unit electric field strength was the same as that of photoelectrons produced by exposure of dyed or undyed crystals to radiation absorbed only by the silver bromide (the dye used in this case was thiacarbocyanine bromide). This similarity in the production of latent image in the two processes can reflect a similarity in the mechanism of formation: whether excitation is caused by radiation absorbed by the dye or by the silver halide crystal, free photoelectrons appear in the crystal and participate in the formation of latent image in the same way.

In addition, for the readily reducible dyes, methylene blue and phenosafranine, electron trapping by dyes was demonstrated. For example, a definite reduction in the number of field-displaceable electrons, excited by radiation of wavelength 254 nm, was caused by an adsorbed layer of methylene blue. It is consistent with a corresponding model of desentization in which these well-known desensitizers can act as alternative electron traps to the sensitivity centers. Consequently, the electrons are prevented for forming the latent image.

Levy, Lindsey and Dickson (ref. 29) have used a technique for the measurement of Dember effect in order to study the direction of photocurrents in silver halides with adsorbed sensitizing dyes. In accordance with above observations, they have shown that in the presence of sensitizing dyes adsorbed on AgHal exposure in the spectral region where only the dyes absorb light results in electron flow from the dye into the silver halide. With a desensitizing dye, e.g., phenosafranine, there is no photocurrent with exposure to light absorbed only by the dye, and exposure in the inherent region of the AgHal initiates photoelectron currents that move from the silver halide into the dye.

It is evident that in the Gerney-Mott model the spectral-sensitizing action of dye adsorbed on semiconductor must depend on the specific disposition of the ground and excited levels of the dye relative to the bands of semiconductor. If the level of the excited state of the dye molecule matches more or less the bottom of the conduction band of silver bromide, an electron can be injected there and the dye acts as a typical sensitizer for the photographic process. In a similar manner, if the ground level of the excited dye molecule is near the top of the valence band of semiconductor, this would enhance the injection of a free hole in the semiconductor. On the contrary, in the case of a primary energy transfer by dipole-dipole or exchange interaction, the suitable energy acceptors have only to be present within the semiconductor surface but the correlation of energy levels of the dye and the semiconductor is not needed.

The well-known fundamental works of Gilman and co-workers (refs. 30-32) have strongly supported the relationships between the photographic activities of dyes and the position of their energy levels. The relative location of the energy levels can be estimated from spectral data, from half-wave potentials, and from quantum-chemical calculations. It was shown, for instance, that dyes having reduction potentials more negative than E_{red}= - 1.1 V are good sensitizers; on the other hand, the dyes with reduction potentials between -0.72 and -0.12 merely act as desensitizers. Analogously, ground state dyes act as strong reducing agents for silver halides if their oxidation potentiall E_{ox} is less than +0.08 V (such substances are typical foggants).

As a general rule, electron transfer from excited molecules in solutions has usually been studied through the quenching of their fluorescence. Since the lifetime of excited singlet state is defined as reciprocal of the rates of deactivation, the fluorescence quenching is an exceedingly specific and convenient technique for measuring the reactivity of S_1 toward various primary processes. Therefore, one approach to the determination of the rate constant for the sensitization step is measurement of the fluorescence lifetimes for dyes adsorbed to silver halid[illegible] and comparison of these lifetimes to those obtained for the same dyes in photographically inert medium such as gelatin. For example, Muenter (ref. 33) has found that fluorescence lifetimes and quantum yields of two carbocyanine dyes sufficiently decreased (e.g., from 2.06 ns to 0.30 ns in the case of 3,3'-diethylthiacarbocyanine) when they were displaced from a gelatin onvironmevt to the surface of silver bromide. It may be concluded from these data that sensitization rate constants are in the range of 10^{9}-10^{10} s^{-1} for these dyes.

Generally speaking, the direct observation of transient absorption spectra of dye radicals in the picosecond and nanosecond time range would be the most strong evidence of the electron transfer mechanism (in this connection see ref. 34). In the case of energy transfer, only the excited singlet state of a sensitizer should occur in the primary photochemical process and therefore only singlet-singlet transient absorption (as a rule, in a different spectral region) should be observed. Until now, however, the picosecond technique have not been applied to problems of spectral sensitization. Let us therefore discuss the results of some microsecond flash experiments.

Exposure of a dye to an intense flash must partially deplete the ground state of the dye. This may be shown by temporary bleaching of the characteristic dye absorption band and by the formation of a new absorption bands for the transient species. Such flash spectrophotometry (with half-width of light pulse about of 3 μsec) was used by Ehrlich (ref. 35-36) to detect transient species formed during irradiation of cyanine dyes dispersed in gelatin films and in photographic silver halide emulsion systems. Since the flash photolysis of undyed grains also induces transient absorptions in the 500 and 650 nm spectral regions, the change in the maximum ground-state absorption of the dye was monitored rather than the formation of the new long-wavelength transient absorption. The change in the ground-state absorption at time zero relative to the absorption maximum reflects the maximum possible number of electrons that may be transferred to the grain if it is assumed that the bleaching is the result of electron transfer.

A series of spectral sensitizing dyes, varying systematically in reduction and oxidation potential, was examined for the ability to sensitize the AgBrI grains. It was found that the shapes of the curves the initial optical transient conversion efficiency and relative photographic quantum efficiency vs. reduction potential are strikingly similar, both in ordering of the dyes and the threshold responces. The threshold dye energy levels, in terms of the oxidation and reduction potentials, that limit the spectral sensitization of photographic processes also limit the initial transient optical absorption of the bleached ground state of the dye. This dependence suggests that the mechanism of spectral sensitization involves the injection of electrons into emulsion grains.

The light-induced ESR signal of some sensitizing dyes in emulsions was observed by Tani (refs. 37-38). He showed that upon irradiating with UV and visible light the AgBr emulsion grains (≈ 0.7 μm) which adsorbed dye (e.g., 5,5'-dichloro-3,3'-9-triethyl-thiacarbocyanine p-toluensulfonate) gave charp structureless ESR signals (at g = 2.005) at room temperature (as well as at 77 K) and gave no signals in the dark. The undyed grains neither in the dark nor under exposure to light gave ESR signal. The excitation spectrum of the ESR signal was essentially in accord with dye absorption spectrum.

The ESR signals were only observed for the dyes whose highest occupied levels were by more than 0.3 eV higher than the top of the valence band of AgBr. The highest occupied energy levels of the dyes which exhibited the ESR signals were higher than those of the dyes which did not exhibit signals. Since significant desensitization was observed only in the presence of dyes which exhibited the ESR signal, it was suggested that the dye positive holes were responsible for the desensitization.

$$\text{Dye} + \text{Br} \dashrightarrow \text{Dye}^{+} + \text{Br}^{-} \qquad (7)$$

$$\text{Dye}^{+} + e^{-} \dashrightarrow \text{Dye} \qquad (8)$$

It was obtained that the light-induced ESR signal of dyes has decayed in several minutes, with decay time depending on the kind of dye; by varying the interval between exposure and development it was established that latent images formed during exposure also faded in several minutes in the presence of such dyes.

It should be noted that recently Siegel et al. (ref. 39) have found that during photolysis of the sensitizing polymethine dyes adsorbed on silver halide crystals the dyes are reduced and neutral radicals are formed (it was shown that significant differences exist in UV-VIS and EPR-spectra between the neutral and bicationic radicals produced by electrochemical oxidation and reduction). It is obvious, that a rapid-response ESR spectrometer with a nanosecond time resolution is needed for study of such radical intermediates (in this connection see ref. 40).

Finally, it is important to note that although there are much indirect evidences for the electron-transfer mechanism of spectral sensitization, some experimental results may be nevertheless interpreted more suitably on the basis of an energy-transfer model. In this context it is to be emphasized that in the past few years Steiger et al. (refs. 41-43) have clearly shown that the energy transfer mechanism is possible for silver halides, but its relative contribution in the dye sensitization of the photographic process is, however, very small. This conclusion is based on the following observations.

In order to discriminate between these two mechanisms, the sensitizers were positioned at well-defined distance from the AgBr surface by using fatty acid monolayers as insulating spacers. The possibility of sensitization by traces of the sensitizer that might be at the surface due to imperfection of the layer system was ruled out by deposition of an efficient electron acceptor (dication paraquat) layer. It was found that the fluorescence of sensitizing dyes is completely quenched at the surface of silver halide monocrystal and the residual sensitization of AgBr with the dye monolayer at a distance of 54 A (1% of the sensitization observed in the case of contact) is due to energy transfer. The sensitization by monolayers of dye which is not capable in its excited state to inject the electrons into the AgBr is possible only due to energy transfer. In such case the sensitization in contact was only 6 times stronger than at d = 54 A. Since the energy transfer efficiency should be of similar magnitude for all dyes, the sensitization in the first case, which is at direct contact about 100 times larger than at d = 54 A can be obviously essentially due to electron injection. Furthermore, for the case of conventional photographic emulsion containing microcrystals of AgHal embedded in gelatin the decreasing of the electronic interaction between AgHal surface and the sensitizer was achieved by using as spectral sensitizer a cyanine dye chemically bonded to gelatin. It was obtained that only a small fraction of all bonded dye molecules that are able to approach the crystal surface closely enough to enable direct electron tranfer between the excited dye and silver halide may act as sensitizer. It was therefore concluded that the contribution of energy transfer must be negligible with respect to electron transfer in conventional photographic emulsions.

In addition, let us consider the phenomenon of supersensitization (SS). It is known that in some cases the efficiency of an sensitizing dye may be greatly enhanced by the addition of a small quantity of an organic compound which of itself may or may not be a sensitizer. It is likely that SS has no single cause or mechanism, but the most important mechanism may be connected with the hole-trapping action of supersensitizer. Gillman (ref. 44) investigated the efficiency of sensitization of 1,1'-diethyl-2,2'-cyanine chloride in the presence of supersensitizers with different oxidation potentials and has found that the efficiency of supersensitizers progressively increased as their oxidation potentials decreased.

This conclusion was confirmed by Iwasaki et al. (ref. 45) in their photoelectrochemical study of SS in silver halides. In this technique the semiconductors are used as an electrode of an electrochemical cell with sensitizing dyes adsorbed at the semiconductor surface from the electrolyte. In their work Iwasaki et al. used as the electrode the crystal plate of AgCl. A spectral sensitizers, 1,1'-diethyl-2,2'-cyanine chloride, was dissolved in water-methanol solution; such supersensitizers as 2-(p-diethylaminostyryl)-benzothiazole and hydroquinone (half-wave oxidation potentials versus SCE are about 0.77, 0.52, and 0.76 V, respectively) were studied. It was observed that the photocurrents considerably increased with the addition of supersensitizers (amounts of hydroquinone required for effective SS were more than a hundred times those of the first supersensitizer in accordance with its high oxidation potential).

Recently, electron-transfer reactions involving oxidation and reduction of a photoexcited cyanine dye in model monolayer assemblies have been investigated by Penner and Mobius (ref. 46). N-stearyl substituted cyanine dye was used as a photosensitizer and the N,N'-dioctadecyl derivative of the viologen molecule as a strong electron acceptor. The quenching of the cyanine fluorescence as a function of the average distance between viologen molecules in the layer adjacent to the dye and the formation of viologen radical produced by electron transfer was observed. This radical was stable for many minutes, and its absorption spectrum could be measured around 400 nm. Its formation was followed as a function of illumination time and reached a maximum yield; the amount of radical at steady state was directly proportional to the concentration of viologen molecules in the layer. Since cyanine fluorescence (the dye was incorporated into the monolayer as J aggregates) was about 90% quenched, the electron transfer must occur on a time scale $\leqslant 10^{-10}$ sec. These results indicate that the fluorescence is quenched in a fast electron-transfer process that is followed by a slow reaction leading to the observed stable radical in competition with efficient recombination.

When both acceptor and donor (the reducing agent such as 4-heptadecyl derivate of 7,8-dihydroxycoumarin which may also quench the dye-aggregate fluorescence) are incorporated into the assembly, the yield of radical from accptor is enhanced (by a factor of 1.5-3.5). Consequently, it may be concluded that the electron transfer from excited dye to acceptor is supersensitized by the electron donor. Because the donor is itself a quencher of dye fluorescence, the mechanism of SS could involve either reaction

$$\text{Dye}^{*} + \text{Donor} \dashrightarrow \text{Dye}^{-} + \text{Donor}^{+} \quad (9)$$

or

$$\text{Dye}^{+} + \text{Donor} \dashrightarrow \text{Dye} + \text{Donor}^{+} \quad (10)$$

or both.

It is obvious, on the other hand, that the electron loss to oxygen and moisture can be a desensitization pathway for an emulsion. Vacuum outgassing substantially reduces the oxygen and moisture content of emulsion coatings and can thereby improve the efficiency of spectral sensitization (ref. 47).

Summarizing we could say that in first approximation the simplest direct-electron-transfer model of spectral sensitization of silver halides may be reasonable assumed at present. This means that the light-induced electron transfer is the primary photochemical act in spectrally sensitised silver halides.

2.2 Electrophotographic systems

Electrophotography (refs. 48,49) seems to be presently the most convenient methods of reproducing printed matter.

This process uses a conducting plate coated with an inorganic or organic photoconductor such as amorphous selenium, polycrystalline zinc oxide, poly-N-vinylcarbasole (PVCz), etc. The surface of the photoconductor is charged, usually by a corona discharge. The formation of homocharge take place by the adsorption to the surface of the layer of negative oxygen ions or positive nitrogen ions from a corona discharge in air.

Then the plate is exposed to the optical image which is to be recorded. The absorbed photons create electron-hole pairs in the photoconductor, which migrate and more or less neutralize the charge in each local area depending upon the intensity of the illumination. Hence, the exposure produces on the surface of the photosemiconductor a latent image which results from the non-uniform distribution of the surface charge.

The plate is covered with a charged pigment, called toner, which clings to the surface in a pattern corresponding to the distribution of the remaining charge. The pigment is transferred to paper without disturbing its distribution, and the paper is heated to melt the pigment, forming a permanent record.

Electrophotographic layers can be prepared by using various semiconducting powders and various binding materials. Now we shall briefly discuss basic properties of ZnO and PVCz layers and models describing their behaviour.

2.2.1 Electrophotographic zinc oxide layers

Zinc oxide (refs. 48-51) is a semiconductor with a wide band gap of about 3 eV. Due to native point defects like interstitial zinc or oxygen vacancies also crystals without intentional additions always show n-type conductivity.

In electrophotography zinc oxide is used in the form of a powder. In layers with a normal amount of a binder there are direct contacts between semiconductor microcrystals. Electrostatic charging is achieved by negative oxygen molecules from a corona, giving rise to a space charge layer in the near surface region of the ZnO coating. The corresponding electrical field effects the separation of electron-hole pairs, generated by UV irradiation and prevents their recombination. The holes are attracted to the negatively charged surface and neutralize adsorbed O^{2-}, thus discharging the exposed regions. Consequently, the discharge in ZnO-binder layers may be regarded as a transfer of electrons from the negatively charged surface of the layer to the earthed support.

It is now well known that in electrophotographic systems similarly to the case of usual silver halide materials the spectral sensitization of photoconductivity for the visible spectral region may be achieved by adsorption of organic dyes (refs. 51-55). The first report of this phenomenon was published by Putzeiko and Terenin (ref. 51, ch. 3). They found that the Dember effect in the pressed powder is sensitized to visible light by adsorbed eosin, erythrosin, rhodamine B, or several cyanine dyes. The known data indicate that zinc oxide can be sensitized with any common dye whereas desensitization is nearly impossible on account of its low conduction and valence bands (-4.25 and -7.45 eV, respectively, in comparison, for example, with -3.23 and -5.76 for the AgBrI (ref. 55)).

Let us consider the following experimental evidences which were obtained, among others, in support of a mechanism of spectral sensitization in ZnO by electron transfer.

Daltrozzo and Tributsch (ref. 52) have used the electrochemical technique in order to study the mechanism of spectral sensitization of ZnO single crystal by rhodamine B. They established that dye sensitized photocurrent across the ZnO-electrolyte interface is strongly dependent on the pH value of the electrolyte (with increasing pH value, the photocurrent decreases between pH 2 and 10 by two orders of magnitude). On the other hand, it was found that the absorption and fluorescence spectra of dye is practically independent of pH. This means that the observed pH dependence of sensitized photocurrents is connected with a pH-dependent double layer which is produced by an exchange of H^+ and OH^- ions between the ZnO surface and the electrolyte. Since the potential jump in the double layer cannot influence energy transfer reactions, it can be concluded that the excited rhodamine B at a ZnO-electrolyte interface sensitizes this semiconductor by electron transfer into the conduction band of ZnO.

The electrophotographic discharge method was used by Eckenbach (ref. 56) in order to test the regeneration of sensitizing dye (Rose Bengal) in ZnO binder layers. In the case of electron transfer the dye is left as a positive radical and it must first return to its ground state in order to react once more. It was observed that the normalized dye sensitivity shows a continual strong decrease with increase of the number of discharge cycles. This decay occurs exclusively in the combination of light absorption by the dye and immediate removal of the resulting photoelectrons in an external electric field. Exposure with visible light but without charge transport has practically no deleterious effect. Parallel to the decrease of dye sensitivity was observed a change in its optical absorption spectrum (bleaching of the dye). It was established that by carefully eliminating the sourses of external electrons (in analogous experiments with silver halides the photolytic silver can behave as an autocatalict) each dye molecule sensitize only once. With favourable conditions, on the other hand, it is able to react at least 10 times (for example, in the cases of prolonged dark intervals between the discharge cycles).

2.2.2 Organic photoconductors in electrophotography

Poly-N-vinylcarbazole, the most sensitive photoconducting polymer ref. 51, ch. 2,4), is a p-type conductor. This was shown, for example, by Meier et al. (ref. 49, p. 163-177) with the help of the photo-flash method. The charge carriers, formed by a short duration flash (of about 1 μsec) in a thin surface layer of the surface cell, can migrate to the dark electrode under the influence of an electric field of suitable sign, thus indicating a short current pulse. This pulse can be caused by holes (I_p) if the illuminated electrode is positive, or electrons (I_n) if the front-electrode is negative. PVCz proved to be a p-type conductor, since $I_p/I_n > 1$.

Analogously to the sensitization of ZnO, it was observed that on the addition of dyes (i.e. methylene blue, crystal violet or rhodamine B) the photoconductivity occurred in a region where PVCz itself exhibits no absorption. As an example, it was observed by Meier that in layers that are sensitized with dyes the ratio I_p/I_n decreases with an increasing dye concentration, thus showing that by means of the dye the electron concentration formed on illumination increases in the polymer semiconductor, whereas the hole concentration decreases.

Of course, the polymer semiconductor materials containing the sensitizing dye differ essentially from the inorganic semiconductor: whereas on silver halide grains or zinc oxide microcrystals the dye molecules are adsorbed as a monomolecular layer or as aggregates, in the polymer semiconductors the dye molecules are uniformly dissolved (analogously to the systems utilized in dye-sensitized photopolymerization). In this connection, it is to be of interest the results of recent work of Fischer and Bronstein-Bonte (ref. 57) in which the quenching of the rhodamine B fluorescence has been used to examine the efficincy of photoinduced electron transfer in a rigid amorphous polymer matrix. The data obtained show that the electron transfer reaction of RhB with several donor and acceptor quenchers in solution-cast films of poly(N-vinylpyrrolidone depends on the redox potential of the quencher and that electron transfer quenching can take place over very long distances. For the samples where the free energy for the electron transfer event was about -1 eV the distance over which the electron could be transferred approached 15 A.

In the case of PVCz it is also interesting to discuss the sensitization technique with charge transfer complex (refs. 58-61). It may be noted that the discovery of this effect in systems containing a 1:1 complex of PVCz with 2,4,7-trinitrofluorenone (TNF) provided a means for the practical utilization of organic materials in the field of electrophotographic technology.

The equilibrium constant for complex formation in PVCz-TNF system is approximately 0.5 liter $mole^{-1}$ thus indicating the formation of weak complex with little electron transfer in the ground state (ref. 58). The complex exhibits absorption spectrum in the range 450-600 nm. It was found by Gill (ref. 59) that the hole and electron drift mobilities in thin films of PVCz-TNF complexes are extremely low ($<10^{-6}$ cm^2/V sec) and are strongly field and temperature dependent. As the TNF concentration is increased, the hole mobility decreases rapidly, while the electron mobility increases. It was suggested by Gill that charge carriers are highly localized with transport taking place by a thermally assisted intermolecular hopping process (hole transport is due to hopping via uncomplexed carbazole states, while electron transport is due to hopping between TNF sites).

Recently, Yokoyama and Mikawa (ref. 60,61) have observed electric field-induced exiplex and charge-transfer fluorescence quenching in PVCz films doped with weak electron acceptors such as dimethyl terephthalate (DMTP), p-dicyanobenzene (2CNB), 1,2,4,5-tetracyanobenzene (4CNB). By adding a weak electron acceptor , e.g., DMTP, a broad exciplex fluorescence appears at around 460 nm, while an undoped PVCz film exhibits only eximer fluorescence at 410 nm. By applying an electric field ($\simeq 10^5$ V/cm), only exciplex fluorescence was found to be quenched. The authors assigned this quenching to the field-assisted thermal dissociation of an ion pair into free carriers, competing with a fluorescing process. When a relatively stronger acceptor, 4CNB, is doped in a PVCz film, charge-transfer absorption (in the visible region around 475 nm) and charge-transfer fluorescence (with a maximum at 620 nm) can be observed. This charge-transfer fluorescence also quenched by an electric field. As in the case of TNF, the sensitized photocurrent was obtained in the visible region corresponding to the charge transfer band. 4CNB may therefore provide a favorable model system for study of the carrier photogeneration mechanism in the PVCz-TNF system.

In summary, it can be seen from the above consideration that in the first approximation the primary process of charge carrier generation in electrophotographic layers can be described in terms of photoinduced electron transfer between some donors and acceptors followed by irreversible charge separation in the presence of external electric field.

3 PHOTOINDUCED ELECTRON TRANSFER IN SOME UNCONVENTIONAL IMAGING SYSTEMS BASED ON PHOTOSENSITIVE MOLECULES AND COMPLEXES

Non-silver photochemical imaging systems currently are a small but significant part of imaging technology (approximately 10% of the total according to the refs. 6,10). They have been used in those applications where sensitivity is not of prime importance. The reasons for the replacement of silver halides include the need for grainless images, dry processing and more direct access to recorded information in specific applications which modern technology demands (refs. 8,10).

3.1 Intramolecular charge transfer in aromatic diazonium salts

Reproduction materials, depending on the coupling reaction of diazonium salts with phenols, napthols, or other suitable compounds, to give images composed of azodyes have been known for a long time (refs. 4-8,10, 62-70). Now diazo-type photosensitive materials are extensively used in photographic copying of engineering plans and for making copies of silver-halide microfilm originals.

The diazos employed are nearly all benzenediazonium salts having a substituted amino group or a nitrogen containing saturated heterocyclic radical in the para position. To prevent premature coupling in coated materials, the layers are stabilized with acids. When diazo compounds are exposed to light the yellow diazo colour disappear, nitrogen is evolved and colourless decomposition products are formed.

$$R\text{-}N{=}N^{+}Cl^{-} + H_2O \xrightarrow{h\nu} R\text{-}OH + N_2\uparrow + HCl \qquad (11)$$

Subsequent treatment with ammonia raises the pH sufficiently to allow coupling to take place to form a dye in the image area.

$$R\text{-}N{=}N^{+}Cl^{-} + R_1\text{-}OH \longrightarrow R\text{-}N{=}N\text{-}R_1OH + \text{Adduct} \qquad (12)$$

Another method for the photographic applications of diazo compounds is the vesicular or bubble process. Vesicular processes utilize the nitrogen evolved during diazo decomposition. After exposure the material is briefly heated causing the coating to soften and the nitrogen to expand. The image is formed by this expansion of the librated N_2 within the binder which results in the formation of light-scattering vesicles. When the material cools the plastic becomes rigid once more, but now contains bubbles in the exposed areas. A second overall exposure decomposes the remaining diazo and the nitrogen gas diffuses through the layer to the atmosphere without forming bubbles.

Lee and Voisin (ref. 63) have recently desribed the method for improving the photosensitivity of diazotype materials in which azo dye and vesicular processes are combined in the same film to create a high-density recorded image.

From theoretical and practical standpoints, the most interesting diazo compounds are the aryl diazonium salts para-substituted with strongly electron-donating substituents, e.g., dialkylamino groups, which have been the staples of the diazotype industry during the long time (ref. 64-67). Such molecules contain the strongest electron withdrawing substituent, the diazonium group, in conjugation with a strong electron donor. It seems reasonable to assume that intramolecular charge transfer can occur in such aryl diazonium compounds under excitation with light.

Several theoretical works on the electronic absorption spectra of the aromatic diazonium salt have been early reported and the remarkable change of the electronic spectra of these molecules have been interpreted as the result of the configurational mixing of local excitations and intramolecular charge-transfers.

For example, Sukigara et al. (ref. 65) have studied the electronic structure and electronic absorption spectra of benzenediazonium cation and several mono-substituted benzenediazonium cations. In the calculation they took into consideration two perpendicular π-electron systems. The mono-substituted cations were separated into the two componenets, the electron donor and the electron acceptor. It was obtained that the α-nitrogen atom of benzenediazonium cation bears almost all the positive charge of the cation.

Recently, Mustroph, Epperllein and Pietsch (ref. 66) have given a very full analysis of the factors contributing to the origin of the absorption bands in diazonium compounds. The absorption behaviour of the various diazonium ions was explained on the basis of simple perturbation-theoretical MO considerations.

It was established that in this approximation the highest occupied MO ψ_1 of the benzene diazonium ion is localized on the phenyl ring at 100% and the second highest occupied MO ψ_2 at 98%, while the lowest unoccupied MO ψ_{-1} is localized on the diazonium group at 84%. An analysis of the orbital configurations involved in the electron transitions shows that the longest wavelength band corresponds at 92% to an electron transition from ψ_1 to ψ_{-1}, and the second band corresponds at 95% to an electron transition from ψ_2 to ψ_{-1}. Thus to a good approximation these bands correspond to intramolecular electron transfer transitions from the phenyl ring into the diazonium group.

This statement is supported by the good correlation of the known ionization potentials of the substituted benzenes with the known excitation energies of the corresponding diazonium ions. According to the general charge-transfer theory, in the MO model the intramolecular charge-transfer excitation energies can be approximately expressed as

$$\Delta E_{CT} = IP_D - EA_A - C \qquad (13)$$

where IP_D is the ionization potential of donor (here the various substituted phenyl rings), EA_A is the electron affinity of the acceptor (here the diazonium group), and C is Coulomb term. The authors have obtained two straight lines for the substituents in 4-position and in 2- and 3-position, correspondingly.

The initial stage of photochemical decomposition reaction of benzenediazonium cation and p-diazobenzoquinone was discussed by Tsuda and Oikawa (ref. 67). Using molecular orbital calculations with taking into the account of all valence electrons and configuration interactions these authors investigated the adiabatic potential curves following the photochemical decomposition. It was concluded theoretically that benzenediazonium cation is a stable and planar compound in the ground state. When it absorbs light, the lenght of the C-N bond between the benzene ring and diazo group is lengthened to 1.4 A following the bending of the terminal nitrogen and the middle nitrogen out of the molecular plane. The stabilization energy obtained by the bending is used as activation energy of C-N bond cleavage; the photochemical decomposition reaction proceeds therefore without activation energy and with high quantum efficiency.

The key intermediate in photochemical decomposition of arenediazonium salts is the aryl cation, Ar+. Cox et al. (refs. 68,69) have shown that UV photolysis at 77 K of solution in polymer films of arenediazonium salts substituted at the 4 position by dialkylamino groups yields aryl cation, which appears to be stabilized in its triplet configuration by the NH_2 group.

The possibility of spectral sensitization of of diazo compounds was extensively investigated (ref. 70). For instance, Yamase et al. (ref. 71) have found that the photolysis of various diazonium compounds (e.g., 4-diazidodiphenylamine sulfate, 4DS) in aqueous solutions may be spectrally sensitized by dyes (D) in the presence of certain activators, such as 1,4-diazabicyclo[2,2,2]octane (DABCO), sodium p-toluene sulfinate (STS), or thiourea. Methylene blue (MB), phenosafranine, rose bengal were used as the sensitizing dyes. During the photolysis of 4DS in the presence of MB and activator, a little bleaching of MB was observed. The quantum yields for 4DS decomposition was about 0.15-0.30 for an aerobic systems, the decomposition rate was proportional to the first order of light intensity. The quantum yield was pH sensitive and oxygen was necessary for dye sensitization in the STS system at pH of 5 to 7 .

From these experiments it was concluded that the decomposition occur through a stage involving an electron transfer to diazonium cations from unprotonated semiquinone dye formed by a photoredox reaction with activators. It was suggested that the dye-activator complex is formed by interaction of the excited triplet dye with activators. This photoinduced complex between unprotonated semimethylene blue and sulfonyl radical may react with oxygen or acrylamide monomer to yield the isolated semimethylene blue which is effective for dye sensitization. The rapid decrease of the quantum yields of the dye-sensitized photolysis in the aerobicsystem in alkaline conditions was explained by the assumption that oxygen quenching of the complex became markedly dominant with an alkaline solution. The reaction related to the sensitization may be written as follows:

$$D \xrightarrow{h\nu} D^*$$

$$D^* + STS \longrightarrow (D \cdots STS)^* \longrightarrow D^- + STS^+$$

$$D^- + 4DS \longrightarrow D + N_2 + \text{diphenylamine radical}$$

3.2 Electron transfer in systems containing charge-transfer complexes between haloalkanes and amines

The free-radical systems (refs. 4-8,10,11, 72-74) are based on a photoinitiated reaction between an organic amines (such as diphenylamine, indole, merocyanine dye base, vinylidene, leuco crystal violet, etc.) and a haloalkanes (such as carbon tetrabromide,iodoform, etc.). The organic dye formed in this photochemical reaction can be used as image. The initial latent image can be intensified by a factor of 10^3 or more in some materials of this type. The amplification by optical development is effected by exposing the material to long wavelength radiation, that is absorbed by the photoproduct dye. Fixing of the image may be carried out either by heat which volatilises carbon tetrabromide from the coated layer or by dissolving the haloalkanes and amines in a suitable solvents.

In these systems amine is used as the color former and carbon tetrabromide or iodoform as the activator. The image may be formed by condensation reactions between tribromomethyl radicals and aromatic amines to form triphenylmethane dyes or by conversion the dye bases to its coloured forms in the presence of the hydrobromic acid.

The diphenylamine-carbon tetrabromide (DPA-CBr_4) system which gives rapid color formation was studied by Sprague et al. (ref. 72). The dye formed may be isolated by extraction with methanol. It was found that it contains the bromide ion and is therefore a salt; the elemental analysis and absorption spectrum correspond to a triphenylmethane dye, so-called Opal Blue SS. Consequently, the function of the halogen compound is to provide the bridge carbon atom linking the three phenyl groups. It was recently shown (ref. 73) that in this system is also formed 9-diphenylaminoacridine.

In the case of cyanine-base free-radical photosystem the image dye was identified as the hydrobromide of the corresponding base (ref. 74). Thus, the effect of exposure to light is to convert the nonpolar starting base to an ionized salt.

The formation of tribromomethyl radical and bromine radical which abstracts hydrogen atom from a hydrogen donor to form hydrobromic acid may occur via the photolysis of CBr_4:

$$CBr_4 \xrightarrow{h\nu} \dot{C}Br_3 + Br^{\bullet} \qquad (14)$$

However, it seems reasonable to assume that the observed high sensitivity of free-radical systems at relative long wavelenths (for example, peak sensitivity of DPA-CBr_4 system is in the range 370 - 410 nm) is connected with the formation of corresponding charge-transfer complex.

Really, the aromatic amines have been known to be typical electron donors and therefore the amine-polyhalomethane system can be regarded as a typical electron donor-acceptor system which involves the electron transfer via charge-transfer complex or exciplex. Thus, earlier work of Matsuda et al. (ref. 75) has clearly demonstrated charge transfer in the excited state for the initial photoinduced processes in the solutions of N,N-dimethylaniline and leuco malachite green in carbon tetrachloride. Upon ultraviolet irradiation (320-400 nm) the solutions showed a remarkable photoconductivity. Since the solutions of these amines in inert solvents such as n-hexane or benzene are not photoconductive, the photocurrent can be caused by charge-transfer interaction between excited amines and CCl_4.

It should be noted that complexes in the ground state between aromatic amines and halomethanes may be so loose as to approach the nature of contact pairs. For example, Iwasaki et al. (ref. 76) have shown that the charge-transfer complex in the ground state between DPA and CCl_4 plays only minor role in the photochemical reaction in solutions. The fluorescence of DPA is considerably quenched by the addition of small amounts of CCl_4 to the solution of DPA in n-hexane or methanol, whereas no obvious effect of CCl_4 is observed on the absorption spectrum of DPA. Further, flash photolysis was carried out with a duration half-time of 30 μsec. The transient species in solutions were formed only when CCl_4 was present. It was obtained that transient absorption in methanol with maximum at 675 nm is quite similar to that observed by photolysis and γ-radiolysis and can be therefore attributed to the diphenylamine cation radical, DPA^{+}. The transient spectra in n-hexane show a band with maxima at 720 and 780 nm and may be assigned to the diphenylnitrogen radical, $(Ph)_2\dot{N}$. The formation of transient species is excellently correlated with fluorescence quenching, thus suggesting the existence and essential role in the photochemical reaction of an exiplex.

The electron-accepting power of CBr_4 is higher than that of CCl_4. The reduction potentials of CBr_4 and CCl_4 are repoted to be -0.30 and -0.78 V, respectively; of course, the similar relation exists also between their electron affinities which were estimated from the absorption spectra of their complexes with hexametylbenzene (1.2 and 0.12 eV, respectively) (ref. 77). For this reason, the ground-state charge-transfer complex amine-tetrabromomethane may dominate in free-radical systems because of its high concentration in solid films.

In order to elucidate reaction mechanisms of photosensitive materials based on CBr_4, flash photolytic investigations in solutions were made by Miwa and co-workers (refs. 78-81) on the various systems consisting of diphenylamine, N-phenyl-1-naphthylamine, indole, N,N-dimethylaniline, and CBr_4, $CHBr_3$, CH_2Br_2, CCl_4, $CHCl_3$, 1,2-dibromoethane, benzylbromide.

In all systems, the absorption spectra and decay of the corresponding amine cation radical were observed. As an example, for the system DPA-CBr_4 it was obtained that cation radicals of DPA have an absorption maximum at 640, 650 and 670 nm in n-hexane, benzene and methanol, respectively. In benzene the transient absorption spectrum decays within time interval about 20 msec in accordance with second order kinetics and gives the very unstable compound with absorption centered about 430 nm. In DPA-CCl_4 system the similar process results in quantative formation of diphenylnitrogen radical from DPA cation radical. Further, it was found that linear relationship exists between the yield of cation radicals and the light intensity of flash. Based on these findings and on the results of the investigation of quenching effects of bromoalkanes on the fluorescence of amines, it seems reasonable to assume that the cation radicals of arylamines are produced by one photon process and from excited singlet state.

According to these results, it was proposed that the common photochemical process via exciplex formation, charge separation and dissociation to amine cation radical and anion radical of corresponding halogenated compound takes place on every system.

As noted above, in solid films a direct electron transfer in charge-transfer complex may occur as a primary process.

It is well known that the anion radicals of a haloderivatives are unstable and that electron transfer to these molecules results in an irreversible heterolytic dissociation of the C-Hal bond. Thus, in the systems considered, products of the primary photochemical reaction in "cage" of the solvent comprise a cation radical of the amine, a radical formed upon a dissociative reduction of a halo-derivative, and a halide ion.

$$[Am\cdots CBr_4] \xrightarrow{h\nu} [Am^{+}\cdots CBr_4^{-}]^{*} \longrightarrow \{Am^{+} + \dot{C}Br_3 + Br^{-}\} \quad (15)$$

In the case of CBr_4, the course of further reactions is determined by chemical reactivity of the cation radical of amine and of the tribromomethyl radical. Because of the appearance of the nonionic diphenylnitrogen radical in model DPA-CCl_4 system, it seems likely that further sequence of reactions may include only neutral radicals. Such mechanism was proposed by Sprague (ref. 72) and Miwa (refs. 78-81). More recently, these assumptions were seemingly confirmed (ref. 82). The photochemical processes of CBr_4-DPA and CBr_4-leuco crystal violet systems were studied with the combination of ESR spectra and spin trapping technique. From the hyperfine structures of the ESR spectra and their variation varying with the concentration ratio of CBr_4 to amines free radical such as

$$(C_6H_5NHC_6H_5-)_3C^{\bullet} \quad \text{and} \quad ((CH_3)_2NC_6H_5)_2\dot{C}-C_6H_5$$

were identified , in addition to bromine radical, $Br^{\bullet}$.

On the other hand, Latowski et al. (ref. 83) for the photochemical formation of crystal violet in the analogous system N,N-dimethylaniline-CCl_4 on the ground of spectrophotometric investigations suggested a mechnism in which the recombination of the trichloromethyl radical with the cation radical has been taken into account.

It should still be noted that the free-radical materials may also be sensitized by dyes. Vannikov and Grishina (ref. 11, ch. 5) have demonstrated that effective sensitization of diphenylbenzylamine-CBr_4 system occurs only if E_{ox}(cyanine dye) is higher than E_{ox}(amine). This fact seems to indicate that the excited dye transfer an electron from amine to CBr_4.

3.3 Photoinduced electron transfer from organic ligands to metal ions

It is well known that certain compounds of transition metals may be reduced when exposed to light. For example, ferric chloride is reduced to ferrous chloride in the presence of an electron donors. Similarly, ferric salts of organic acids, e.g. oxalic, citric, etc., are reduced to ferrous salt on exposure to actinic radiation.

The salts of such transition metals as Fe, Co, Cr posess comparatively high light sensitivity and have therefore found applications in many photographic processes.

Thus, for instance, it is known (ref. 84) that in complexes of the cobalt (III) ion with a ligand such as a Lewis base the ligands are relatively tenaciously held in complexes, but released when the cobalt is reduced to the (II) state. Hence, the photoreduction of cobalt (III) amines liberates ammonia as well as forms cobalt (II) compounds; evolved ammonia is capable of initiating a variety of colour forming or colour bleaching reactions, whilst cobalt (II) can form a coloured image on reaction with appropriate cheleting agents. Such systems have been studied extensively (ref. 10). The spectral sensitizers (such as cyanines, merocyanines, acridines, etc.) may be used in these systems.

Let us now consider two most thoroughly studied reactions: the photochemical reduction of dichromate ion to a compound of lower valency and the photochemical decomposition of the ferric complexes.

3.3.1 Dichromates in the presence of gelatin and polyvinylalcohol

At the present time the photochemical reactions between dichromate or chromate ion and gelatin or polyvinylalcohol (PVA) are widely used in photoresists and for holographic purposes (refs. 4,9,85).

Films of polymers containing a small amount of dichromates become insoluble in water upon exposure to ultraviolet or blue light. A chromium (III) coordinated complex having a polymeric structure is believed to be responsible for this insolubilization of the photoresists. It is assumed that during exposure Cr(VI) is reduced to Cr(III) which then react with the gelatin. All unexposed Cr(VI) is removed during the development.

The phase hologram formation in dichromated gelatin is based on modulation of the refractive index. It is assumed that the primary cause of such modulation may be the formation in the gelatin structure of the Cr(III)-gelatin complex.

Datta and Soller (ref. 86) have studied the photochemical reactions in dry dichromated photoresist using ultraviolet and electron paramagnetic resonance (EPR) spectroscopy. EPR is the most useful technique to detect intermediate chromium species because the chromium photoreaction intermediates have small extinction coefficients. On the other hand, these intermediates, Cr(V) d_1, Cr(IV) d_2, and Cr(III) d_3, have one, two, and three unpaired "d" electrons respectively, and would exhibit EPR spectra. The EPR technique is capable of giving a quantative measure of concentration of the chromium species and is extremely sensitive.

Dry films of $K_2Cr_2O_7$-PVA on quartz plates were exposed to a high-pressure mercury lamp light. Irradiation causes an increase in optical density in the 350-400 nm and 440 nm regions. A new broad absorption band was also observed at 520-600 nm after iradiation. The increasing of the intensity of the EPR spectrum of Cr(V) with peak at g value of 1.98 was also found. Since it was known that CrO_4^{-3}, a form of Cr(V), has two absorption bands at wavelengths λ_{max}= 625 nm and λ_{max}= 355 nm and since the optical density at the 370 mn region increases with short exposure and begins to decrease after prolonged exposure similar to the observed changes in the EPR spectra of Cr(V), it was concluded that pentavalent chromium is formed under illumination of the films.

The reaction path for the reduction of Cr(V) to Cr(III) is not yet clear. The authors failed to detect Cr(IV) in any of the polymer matrices.

The similar results were obtained for dichromated gelatin by Sherstyuk et al. (ref. 9, ch. 6). It was obtained that the photoreduction of Cr(VI) compounds in photoresists with poly(vinylchloride) or gelatin matrices produces paramagnetic Cr(V) complexes which are stable in the absence of H_2O. Transformation of Cr(V) to Cr(III) in the presence of H_2O controls the secondary relief images and the resistance of the photoresists to acids.

Thus, it is at present definitely established that Cr(V) is an important intermediate for both dark and photochemical reactions. This means that photoinduced electron transfer from the organic ligands to chromium ion is a primary photochemical act.

The spectral sensitivity of this material is generally limited to wavelengths of 520 nm or less. The discovery that dichromated gelatin is a very efficient phase holographic material give rise to a new interest in extending the spectral sensitivity of the DCG to red light wavelengths (refs. 87-89). For example, it was found by Graube (ref. 89) that thiazine dyes (e.g., methylene blue and methylene green) produce the most sensitive films which need about 200 mJ/cm^2 exposure energy at 633 nm for 80% diffraction efficiency.

3.3.2 Blueprint process

The blueprint materials (the colour of the image is generaly blue) are based on the light-sensitivity of ferric oxalate (refs. 4,6,9). They have been widely used for reproduction of engineering and architectural drawings.

. Blue colour is formed as a result of a rection between the ferrous salt and the ferricyanide. The composition of the sensitized solution used to impregnate the paper is based on an aqueous solution of potassium ferricyanide and ferric complex with oxalic acid. After exposure the print is washed with water and thus unreduced salts dissolve away in the unexposed areas of the coating.

The main reaction which takes place during the exposure may be written as:

$$Fe_2(C_2O_4)_3 \xrightarrow{h\nu} 2Fe(C_2O_4) + 2CO_2 \tag{16}$$

During a subsequent washing with water this ferrous salt react with ferricyanide to produce a blue-colored insoluble ferrous ferricyanide:

$$2K_3[Fe(CN)_6] + 3Fe(C_2O_4) \longrightarrow Fe_3[Fe(CN)_6]_2 + 3K_2C_2O_4 \tag{17}$$

An electron transfer mechanism of the blueprint processs was proposed in 1968 by Suzuki et al. (ref. 90). Their suggestion was based on the following observations for the system consisting of ammonium ferrioxalate and potassium ferricyanide:
(i) The absorption of ammonium ferrioxalate was changed by exposure to actinic radiation;
(ii) Clear photoconductive phenomenon was observed in ferrioxalate when it was irradiated;
(iii) The decrease of paramagnetism in irradiated ammonium ferrioxalate was found by the Guoy magnetic balance method;
(iv) The oxidation-reduction potential of ammonium ferrioxalate solution was decreased by irradiation in an atmosphere of nitrogen;
(v) The generation of carbon dioxide from irradiated ammonium ferrioxalate was detected by formation of white precipitates with barium salts.

All these facts are consistent with the assumption that upon excitation of the ferrioxalate, electrons of $(C_2O_4)^{2-}$ are transferred to adjacent Fe^{3+} thus producing Fe^{2+} and CO_2.

It should be noted that these findings are also in agreement with the results of more earlier works (refs. 91-93). On the one hand, the photoinduced formation of Fe^{2+} in solutions of ferrioxalates was used by Parker (ref. 91) for creation of sensitive chemical actinometer. On the other hand, Parker and Hatchard (ref. 92) have studied the photodecomposition of the ferrioxalate ions by flash photolysis and postulated that the primary photochemical process involved an electron transfer from the occupied π-orbital of a oxalate ligand to the empty antibonding orbital of a central Fe^{3+} ion; they have also observed that irradiation of a ferrioxalate solution frozen in liquid nitrogen produces an orange-brown substance identified as free radical $C_2O_4^-$ or CO_2^-. It was therefore concluded that photodecompocition of the ferrioxalate ion can take place in the following stages:

$$Fe^{3+}(C_2O_4)^{2-} \xrightarrow{h\nu} Fe^{2+}(C_2O_4)^{-} \longrightarrow Fe^{2+} + (C_2O_4)^{-} \quad (18)$$

$$(C_2O_4)^{-} + (C_2O_4)^{-} \longrightarrow 2CO_2 + (C_2O_4)^{2-} \quad (19)$$

Clearly, in the general case the free anion radical $C_2O_4^{-}$ may undergo other modes of reaction. For instance, Parker (ref. 93) has suggested that bimolecular reaction between oxalate radical and excess ferrioxalate ion may occur:

$$(C_2O_4)^- \; + \; Fe^{3+}(C_2O_4)^{2-} \; \longrightarrow \; Fe^{2+} \; + \; (C_2O_4)^{2-} \; + \; 2CO_2 \qquad (20)$$

3.4 The use of photoinduced electron transfer in photographic systems based on physical development, photopolymerization, and photochromism

The primary photochemical reactions in such systems may be not only the reaction of photoinduced electron transfer, but also such reaction as cis-trans isomerization, tautomerization, homolytic bond cleavage, etc. (refs. 2,3). Let us consider those systems that involve the photoinduced electron transfer as a primary act.

3.4.1 Physical development systems

The term "physical development" is used to denote process in which the reduction of solvated metal ions and the deposition of resultant atoms are triggered by photographically produced nuclei (refs. 4-10, 94). For example, it is known the physical development (PD) process based on the photosensitivity of potassium palladous oxalate in which the palladium nuclei are formed on exposure:

$$Pd^{2+} \; + \; 2e^- \; \longrightarrow \; Pd(0) \qquad (21)$$

$$(C_2O_4)^{2-} \; + \; 2h^+ \; \longrightarrow \; 2CO_2 \qquad (22)$$

The latent palladium image may be then amplified by nickel, cobalt, iron, etc., deposition.

Classical wet physical developers are silver developers. Their solutions contain silver ions together with a photographic reducing agent, such as hydroquinone, metol, etc. Wet processing is avoided in solid state silver physical development system, so-called the 3M Dry Silver process. In this process the coated layer contains a small quantity of silver halide as the photosensitive component, silver behenate (silver soap) as a supply of silver ions and a reducing agent. The latent image acts as a catalyst for reduction of the silver soap by the reducing agent when the layer is heated to ≈120°C. Because of its low sensitivity this system is widely used in laser recording and microfilm duplication.

In general, the metal salt which participates in the nucleation reaction can be introduced into the photographic layer after exposure, or it can already be present in the photosensitizing solution.

As an example, let us consider a recently proposed system with dry tellurium physical development (TPD) (refs. 95,96). A variety of tellurium (II) and (IV) compounds were employed as oxidants in tellurium physical developers in which the tellurium compounds are incorporated in a polymeric matrix with a reducing agents. The TPD process:

$$\text{Te oxidant + reducing agent} \dashrightarrow \text{Te(0) + oxidized developer} \quad (23)$$

is initiated at ≈ 150°C by photographically produced Pd(0) nuclei. The minimum size of Pd nuclei was determined by use of the technique of vacuum deposition. It was found that the thermal TPD process can be catalized by 6-atom Pd(0) centers. The 10^6-10^8 amplification factor and these minimum-size data indicate that TPD has the characteristics necessary for an efficient photographic system. A wide variety of reducing agents may be incorporated in TPD process, including conventional silver halide developers.

The Pd(0) nuclei may be generated by photolysis of transition metal compounds or an equivalent process with high quantum efficiency.

3.4.2 Photopolymerization imaging

Photopolymerization has the potential for high quantum yields due to involvement of chain reactions and is at present a well accepted technique of photographic image formation (refs. 4-10, 97,98). An image results by readout based on some property difference between the starting monomer and the photoproduced polymer (such as solubility, light scattering, etc.). There are various means to initiate polymer chain formation and propagation by free radical and ionic mechanisms depending on the type of initiator, monomer and reaction conditions. Let us consider some examples.

A high speed system based on N-vinylcarbasole (NVC) and carbon tetrabromide (CTB) was described by Notley (ref. 99). Droplets of

NVC and CTB in gelatin result in blue-green images on exposure and heating. This is particular case of so-called charge-transfer polymerization of NVC in the presence of various electron acceptors (ref. 100). The photochemical electron transfer from NVC to the electron acceptor (which occurs by selective excitation of the charge-transfer complex or via the formation of exciplex) leads to the formation of NVC cation radical which initiates the photosensitized ionic polymerization of NVC. For example, in the system NVC-chloranil by means of flash spectroscopy was demonstrated the photochemical formation of corresponding ion radical intermediates (ref. 101).

The dye-sensitized photolysis of diazonium salts may induce the polymerization of acrylamide with a very high rate (ref. 102). In this system an electron transfer to diazonium cation from unprotonated semiquinone methylene blue formed by a photoredox reaction with activator (this is a supersensitizing act because the day hole is trapped at the reducing agent) results in the formation of a phenyl radical. This phenyl radical induces the polymerization.

Similar results were reported in more earlier publication by Margerum et al. (ref. 103). They found that polymerization of barium acrylate may be initiated by reactive free radicals of activator (e.g., sodium p-toluensulfinate) formed upon oxidation by dye (methylene blue or thionine). In general, it has been established that xanthene, thiazine, oxazine and acridine dyes may all be reduced on irradiation in presence of electron donors such as secondary or tertially aliphatic amines, allyl thiourea, ascorbic acid, etc. These electron donors active as reducing agents may also be often the chain transfer agents in photopolymerization (dye redox initiation).

Many inorganic systems have been investigated as initiators of photopolymerization (ref. 9, ch. 3,5). The ions of heavy metals, e.g., Fe, Cr, Ti, Mo, may be oxidized or reduced on excitation by electron transfer.

3.4.3 Photochromic materials

Photochromic systems (refs. 4-8,10,11,104-106) are based on the ability of certain compounds to change reversibly their colour (i.e. their electronic absorption spectra) on exposure to visible or ultra-violet wavelengths. In the dark or upon excitation in other wavelength region these systems may revert to their original condition. In principle, such materials may be used as optical filters of controllable transmission, or as reversible photographic media. The promise of photochromics lies in a large possible information capacity and an ability to add on and erase images.

Many hundreds of photochromic compounds based on various reversible reactions are known for a long time. For example, the photochromic polymer systems based on an electron transfer between excited dye and corresponding donor or acceptor have been proposed (see ref. 106, ch. 1) It seems, however, that despite the extensive interest in photochromism, photochromic imaging systems have not achieved any significant commercial use. This is due to their low sensitivity and limited spectral sensitivity range.

Among organic substrates, the spiropyrans are a main class of photochromic compounds which are very interesting for various applications. The molecules of spiripyrans are made of two conjugated heterocucles bonded by a saturated carbon atom and located in two orthogonal planes. The spiropyran closed form absorbs in the UV region and the excitation in the range of 320-390 nm leads by breaking the C-O bond (and a subsequent 90 rotation of one side of the molecule in relation to the other) to the open form (its structure is like merocyanine dye) whose absorption lies in the visible region. It is not improbable that this photonic excitation may be regarded as an intramolecular charge transfer in benzopyran ring. Holmanski et al. (ref. 107) have investigated the solvatochromic shift of the fluorescence spectrum of the 1,3,3-trimethyl-6'-nitrospiro[indoline-2,2'[2H]-benzopyran]. The estimated change in dipole moment upon electronic excitation was 11 Debye. This change was interpreted as a consequence of an intramolecular charge transfer to the nitro group and a corresponding increase of positive charge on the oxygen atom.

Recently, the thin crystalline film of organometallic charge-transfer complex has been used to develop an erasable optical disk (refs. 108). The organometallic medium is layered on the suppoting base material with a transparent protective coating. A complex consists of a metallic donor (Na or Cu) and an organic acceptor (tetracyanoquinodimethane, TCNQ). Optical information processing is carried out when the TCNQ film is irradiated with a narrow laser light. This changes the chemical state of the TCNQ, which is seen as a color change at each data point. Since the color contrast represents a change in absorption and reflection of light, the methods of reading may be the same as well-known methods used currently in optical disk technology. To erase this color contrast - representing a stored bit of information - a higher-powered laser (e.g., a carbon dioxide laser) may be used in order to heat the complex and thus to convert it back to its original state. At present, the disc can be written on and erased 10 times.

4 CONCLUDING REMARKS

As one can see from the above discussion, the primary photochemical reaction in conventional and in various unconventional imaging systems actually involves very rapid irreversible photon-induced electron transfer between some donor and acceptor entities, which in general case may be separated from each other by some distance. This light-driven electron transfer in different imaging systems results in formation of atoms, ion radicals, free radical, etc., which are used for initiating further chemical reactions.

It is well known that the major drawback of the non-conventional photochemical imaging systems lies in their low energetic sensitivity as compared to the conventional AgHal systems and that the application of amplification processes may increase their sensitivity. In this connection, it is to be emphasized that the most important unconventional amplification photographic systems known today (physical development, polymerization, chain decomposition of diazonium salts, etc.) (ref. 2,3,7) are based on the reaction of the photoinduced electron transfer. Furthermore, the highest values of sensitivity may seemingly be obtained for layers in which the different elements of such systems are used in some combinations (refs. 109-111, for example).

The need for recording, storing, and reproducing large quantities of data will multiply the requirements for new, more effective photo processes. It is possible, in principle, that the most important new imaging applications - holography, microfilm, high-density optically accessed computer memories, and fabrication of printed circuits - will induce through the coming years the appearance of new imaging chemistry, outside the scope of well-known processes. There is little doubt, however, that such new systems will be based on the photoinduced electron transfer as a primary act.

5 REFERENCES

1 T.H. James, Ed., The theory of photographic process, 4th edn., Macmillan, New York, 1977.
2 M.V. Alfimov and O.B. Yakusheva, The photochemical methods of information recording. The primary photoprocesses, Usp. Chim., 48(4) (1979) 585-611 (Russ.).
3 M.V. Alfimov, Photochemistry as the basis of information phototechnology, J. Inf. Rec. Mater. 13(6) (1985) 387-399.
4 J. Kosar, Light-Sensitive Systems, Wiley, New York, 1965.
5 M.R.V. Sahyun, Non-conventional photochemical imaging process, J. Chem. Educ., 50(2) (1973) 88-93.
6 E. Brinkman, G. Delzenne, A. Poot, and J. Willems, Unconventional Imaging Processes, Focal Press, London, 1976.
7 L.H.G. Leenders, Amplification process in photochemical unconventional systems, J. Photogr. Sci., 26 (1978) 234- 240.
8 R.E. Jacobson, Photochemical imaging systems, Chemistry in Britain, 16 (1980) 468-473.
9 A.I. Kryukov, V.P. Sherstyuk, and I.I. Dilung, Electron phototransfer and its applied aspects, Naukova Dumka, Kiev, 1982 (Russ.).
10 R.E. Jacobson, A review of the principles and performances of photochemical imagingm processes, J. Photogr. Sci., 31 (1983) 1-12.
11 A.V. Vannikov and A.D. Grishina, Photochemistry of polymeric donor-acceptor complexes, Nauka, Moskow, 1984 (Russ.).
12 G.J. Kavarnos and N.J. Turro, Photosensitization by reversible electron transfer: theories, experimental evidence, and examples, Chem. Rev., 86 (1986) 401-449.
13 M. Julliard and M.Chanon, Photoelectron-transfer catalysis: its connections with thermal and electrochemical analogues, Chem. Rev., 83 (1983) 425-506.
14 V.A. Zasukha, Theory of electron phototransfer in a condense medium, Opt. Spectrosk., 61(1) (1986) 68-72. (Russ.).
15 R.W. Gerney and N.F. Mott, The theory of the photolysis of silver bromide and the photographic latent image, Proc. Roy. Soc., 164 (1938) 151-167.
16 C.R. Barry, Determination of energy levels in AgBr emulsion crystals from optical measurements, J. Photogr. Sci., 18(5) (1970) 169-175.
17 J.F. Hamilton, F.A. Hamm, and L.E. Brady, Motion of electrons and holes in photographic emulsion grains, J. Appl. Phys., 27(8) (1956) 874-875.
18 J.F. Hamilton and L.E. Brady, Electrical measurements on photographic emulsion grains. I. Dark conductivity, J. Appl. Phys., 30(12) (1959) 1893-1901.
19 J.F. Hamilton and L.E. Brady, Electrical measurements on photographic emulsion grains. II. Photoelectron carriers, J. Appl. Phys., 30(12) (1959) 1902-1913
20 J.F. Hamilton and L.E. Brady, The role of mobile silver ions in latent image formation, J. Phys. Chem., 66 (1962) 2384-2396.
21 L.E. Brady and J.F. Hamilton, Electrical measurements on photographic grains containing cadmium, J. Appl. Chem., 35(5) (1964) 1565-1569.
22 L.E. Brady and J.F. Hamilton, Mechanism of electron trapping in silver bromide photographic grains, J. Appl. Phys., 37(6) (1967) 2268-2271.
23 L.M. Kellog, Measurements of photoelectron lifetimes in silver halide microcrystals using microwave techniques, Photogr. Sci. Eng., 18(4) (1974) 378-382.
24 A. Hasegawa and T. Sakaguchi, Photoconductivity measurement in photographic silver halide microcrystals by microwave technique, Photogr. Sci. Eng., 24(2) (1980) 74-76.

25 R.J. Dery and J. P. Spoonhower, Fast photoelectron kinetics in silver bromide emulsions, Photogr. Sci. Eng., 28(3) (1984) 92-98.

26 H. Hada, M. Kawasaki, and T. Yoshida, Observation of latent image formation by intermittent Dember effect measurement, Phot. Sci. Eng., 25(5) (1981) 201-208.

27 H. Hada, M. Kawasaki, and T. Fukaya, The measurement of the time of flash-light induced latent image formation, J. Photogr. Sci., 33 (1985) 45-48.

28 V.I. Saunders, R.W. Tyler, and W.Rest, Mobility of electrons and positive holes in spectral sensitization and in desensitization of the photographic process by dyes, J. Chem. Phys., 46(1) (1967) 199-205.

29 B. Levy, M. Lindsey, and C.R. Dickson, The direction of photocurrents in silver chloride with adsorbed spectrally sensitizing and desensitizing dyes, Photogr. Sci. Eng., 17(1) (1973) 115-127.

30 L. Costa, F. Grum, and P.B. Gilman, Jr., A comparison of the photographic and luminescent properties of spectral sensitizing dyes adsorbed to AgBrI emulsions, Photogr. Sci. Eng., 18(3) (1974) 261-275.

31 P.B. Gilman, Jr., Use of spectral sensitizing dyes to estimate effective energy levels of silver halide substrates, Photogr. Sci. Eng., 18(5) (1974) 475-485.

32 T.L. Penner and P.B. Gilman, Jr., Influence of dye energy levels on the spectrally sensitized luminescence from silver bromoiodide, Photogr. Sci. Eng., 19(2) (1975) 102-114.

33 A.A. Muenter, Fluorescence lifetimes and sensitization rate constants for dyes adsorbed to silver halide microcrystals, J. Phys. Chem., 80(20) (1976) 2178-2183.

34 S.H. Erlich, Transient absorptions of cyanine dyes in films of gelatin and silver halide emulsions, Photogr. Sci. Eng., 18(2) (1974) 179-186.

35 S.H. Erlich, Influence of dye energy levels on the spectral sensitization of silver bromoiodide and the initial optical transient response of dye bleaching: a conversion efficiency study, Photogr. Sci. Eng., 20(1) (1976) 5-14.

36 E.E. Hilinski, J.M. Masnovi, C. Amatore, J.K. Kochi, and P.M. Rentzepis, Charge-transfer excitation of electron donor-acceptor complexes. Direct observation of ion pairs by time-resolved picosecond spectroscopy, J.Am. Chem. Soc., 105( ) (1983) 6167- 6168.

37 T. Tani, Light-induced ESR spectra of dyes adsorbed by AgBr emulsion grains, Photogr. Sci. Eng., 19(6) (1975) 356-363.

38 T. Tani and Y. Sano, Factors influencing light-induced ESR spectra of dyes adsorbed by AgBr emulsion grains, J. Photogr. Sci., 27(6) (1979) 231-240.

39 J. Siegel, M. Friedrich, V. Kuhn, J. Grossman, and D. Fassler, UV-VIS-und ESR-spectroskopische Untersuchungen an Polymethinfarbstoffradicales, J. Inf. Rec. Mater., 14(3) (1986) 215-227.

40 K. Kuwata, S Kominami, K. Hayachi, and K. Kobayashi, Time resolved ESR spectroscopy in radiation chemistry and photochemistry, Mem. Inst. Sci. Ind. Res., Osaka univ., 43 (1986) 59-67.

41 R. Steiger, H. Hediger, and P. Junod, Studies of the mechanisms of spectral sensitization of silver halides, Photogr. Sci. Eng., 24(4) (1980) 185-195.

42 R. Steiger and J.F. Reber, Gelatin-substituted cyanine dyes as spectral sensitizers for silver halide emulsions, Photogr. Sci. Eng., 27(2) (1983) 59-65.

43 R. Steiger, Spectral sensitization by energy transfer, Photogr. Sci. Eng., 28(2) (1984) 35-43.

44 J.E. Jones and P.B. Gilman, Jr., Study of the mechanism of supersensitization of 1,1'-diethyl-2,2'-cyanine, Photogr. Sci. Eng., 17(4) (1973) 367-372.

45 T. Iwasaki, S. Sumi, A. Fujishima, and K. Honda, Photoelectrochemical study of supersensitization with AgCl crystal, Photogr. Sci. Eng., 23(1) (1979) 17-19.

46 T.L. Penner and D. Mobius, Photoinduced electron-transfer processes in monolayer assemblies. Supersensitization in a three-component system, J. Am. Chem. Soc, 104(26) (1982) 7407-7413.

47 T.A. Babcock, B.P. Michrina, P.A. McCue, and T.H. James, Effects of moisture on photographic sensitivity, Photogr. Sci. Eng., 17(4) (1973) 373-381.

48 V.M. Fridkin, The physics of the electrophotographic process, Focal Press, London, 1973.

49 W.F. Berg and K. Hauffe, Eds., Current problems in electrophotography, W. de Gruyter, Berlin, 1972.

50 W.H. Hirschwald, Zinc oxide: an outstanding example of a binary compound semiconductor, Acc. Chem. Res., 18 (1985) 228-234.

51 I.A. Akimov, Yu.A. Cherkasov and M.I. Cherkashin, The sensitized photoeffect, Nauka, Moskow, 1980 (Russ.).

52 E. Daltrozzo and H. Tributsch, On the mechanism of spectral sensitization: Rhodamin B sensitized electron transfer to zinc oxide, Photogr. Sci. Eng., 19(6) (1975) 308-314.

53 H. Tributsch and H. Gerischer, The use of semiconductor electrodes in the study of photochemical reactions, Ber. Bunsenges. physik. Chem., 73(8/9) (1969) 850-854.

54 T. Tani, Modified electron transfer mechanism for spectral sensitization. VII. Spectral sensitization of photoelectric effects in various photoconductors, Photogr. Sci. Eng., 17(1) (1973) 11-20.

55 S. Dahne, Spectral sensitization and electronic structure of organic dyes, Photogr. Sci. Eng., 23(4) (1979) 219-239.

56 T. Watanabe, T. Takizawa, O. Kawanami, K. Kurita, and K. Honda, Quantum efficiencies for dye decomposition and silver formation in spectral sensitization of silver bromide particles. A geterogeneous photochemical approach, Nippon Shashin Gakkaishi, 49(1) (1986) 13-20.

57 A.B. Fisher and I. Bronstein-Bonte, Photoinduced electron transfer quenching of rhodamine B in polymer films, J.Photochem., 30 (1985) 475-485.

58 K.H. Brown and J.P. Montillier, Equilibrium constants for charge transfer complexes between TNF and various carbazolyl compounds, Photogr. Sci. Eng., 22(1) (1978) 22-28.

59 W.D. Gill, Drift mobilities in amorphous charge-transfer complexes of trinitrofluorenone and poly-N-vinylcarbazole, J. Appl. Phys., 43(12) (1972) 5033-5040.

60 M. Yokoyama, Y. Endo, A. Matsubara, and H. Mikawa, Mechanism of intrinsic carrier photogeneration in poly-N-vinylcarbazole. II. Quenching of exciplex fluorescence by electric field, J. Chem. Phys., 75(6) (1981) 3006-3011

61 M. Yokoyma and H. Mikawa, Extrinsic carrier photogeneration mechanism in photoconducting polymers - exciplex and charge transfer (CT) complex sensitization, Photogr. Sci. Eng., 26(3) (1982) 143-147.

62 J. Marx, J. Epperlein, and F. Walkow, Diazosysteme in der Reprographie - Eine Ubersicht, J.Signal AM, 11(2) (1983) 83-108.

63 Wai-Hon Lee and Paul-Henry Voisin, Diazotype materials for laser recording, Appl. Optics 22(5) (1983) 671-675.

64 R. Landau, P. Pinot de Moira, and A.S. Tanenbaum, J. Photogr. Sci., Diazo compounds in the photocopying industry, 13(3) (1965) 144-151.

65 M. Sukigara, K. Honda and S. Kikuchi, Photochemistry of aromatic diazonium salts, J. Photogr. Sci., 18 (1970) 38-40.

66 H. Mustroph, J. Epperlein, and H. Pietsch, An outline and interpretation of the spectral behavior of diazonium compounds, Photogr. Sci. Eng., 24(4) (1980) 196-200.

67 M. Tsuda and S. Oikawa, Theoretical research on the initial stage of photodecomposition of aromatic diazo compound. II p-Diazoquinone, Photogr. Sci. Eng., 23 (3) (1979) 177-181.

68 A. Cox, T.J. Kemp, D.R. Payne and P. Pinot de Moira, The photolysis of arenediazonium salts in solutions and films. Quantum efficiencies and reactive intermediates, J. Photogr. Sci., 25(5) (1977) 208-215.

69 A. Cox, T.J. Kemp, D.R. Payne, M.C.R. Symons and P. Pinot de Moira, ESR characterization of ground state triplet aryl cations substituted at the 4 position by dialkylamino groups, J. Am. Chem. Soc., 100(15) (1978) 4779-478

70 H.G.O. Becker, Grundlagen der Lichtempflindlichkeit und spektralen Sensibilisierung von Diazosystemen, J. Signal AM, 3(5) (1975) 381-394.

71 T. Yamase, T. Ikawa, H. Kokado, and E. Inoue, Dye-sensitized photolysis of diazonium compounds - Part II, Photogr., Sci. Eng., 17(3) (1973) 268-271.

72 R.H. Sprague, H.L. Fichter, and E.Wainer, New photographic processes. I. The arylamine-carbon tetrabromide system giving print-out dye images, Phot. Sci. Eng., 5(2) (1961) 98-103.
73 V.A. Sazhnikov, A.G. Strukov, M.G. Stunzhas, S.P. Efimov, O.M. Andreev, and M.V. Alfimov, Photoinduced formation of 9-substituted acridine from diphenylamine and tetrabromomethane, Dokl. Acad. Nauk SSSR 288(1) (1986) 172-176 (Russ.).
74 R.H. Sprague, J. Urbancik, and J. Stewart, New photographic processes. IV. The cyanine-base free-radical photosystem. A panchromatic black-image high-resolution print-out film, Photogr. Sci. Eng., 9(2) (1965) 133-137.
75 S. Matsuda, H. Kokado, and E. Inoue, The photoconductivity in a CCl4 solution of N,N-dimethylaniline, Bull. Chem. Soc. Japan, 43(9) (1970) 2994-2995.
76 T. Iwasaki, T. Sawada, M. Okuyama, and H. Kamada, The photochemical reaction of aromatic amines with carbon tetrachloride, J. Phys. Chem., 82(3) (1978) 371-372.
77 Y. Kuwae, M. Kamachi, K. Hayashi, and H.G. Viehe, ESR study of radical halogen abstractions. Participation of electron-transfer mechanism in the radical halogen abstractions, Bull. Chem. Soc. Japan, 59(7) (1986) 2325-2330.
78 T. Miva and T. Ishikawa, Reaction mechanism of free radical photosensitive materials.(I), Nippon Shashin Gakkaishi, 38(4) (1875) 360-370.
79 T. Miwa, Reaction mechanism of free radical photosensitive materials. (II). On systems of various aromatic amines-tetrabromomethane, Nippon Shashin Gakkaishi, 42(2) (1979) 105-111.
80 T. Miwa, M. Koizumi, and H. Genda, Reaction mechanism of free radical photosensitive materials. (III). On diphenylamine and N-phenyl-1-naphthylamine - various halogenated organic compound systems, Nippon Shashin Gakkaishi, 42(3) (1979) 174-186.
81 T, Miwa, M. Koizumi, and H. Genda, Reaction mechanism of free radical photosensitive materials. (IV). On photoprimary process, Nippon Shashin Gakkaishi, 42(4) (1979) 246-256.
82 Y. Zhou, J. Kuo, G. Xu, and D. Shi, ESR spectra of some radical photographic systems, Gangnang Kexue Yu Knang Huaxue, (1984) 14-23 (Ch.), (CA 104: 119827f).
83 T. Latowski and B. Zelent, Photochemical studies on an aromatic amine-methane polychloroderivative system. Part VII. Further investigation on the photochemical reaction of N,N-dimethylaniline in carbon tetrabromide, Roczniki Chemii, 51 (1977) 1883-1893 (Eng.).
84 Anonym., Cobalt (III) complex imaging compositions having improved photographic properties, Research Disclosure, No. 194 (1980) 214-218, Item 19423.
85 S. Sjolinder, J. Imaging Sci., Dichromated gelatin and light sensitivity, 30(4) (1986) 151-154.
86 P. Datta and B.R. Soller, A study of photochemical reactions in a dichromated photoresist, Photogr. Sci. Eng., 23(3) (1979) 203-206.
87 M. Akagi, Spectral sensitization of dichromated gelatin, Photogr. Sci. Eng., 18(3) (1974) 248-250,
88 T. Cubota, T. Ose, M. Sasaki, and K. Honda, Hologram formation with red light in methylene blue sensitized dichromated gelatin, Appl. Opt., 15(2) (1976) 556-558.
89 A. Graube, Dye-sensitized dichromated gelatin for holographic optical element fabrication, Photogr. Sci. Eng., 22(1) (1978) 37-41.
90 S. Suzuki, K. Matsumoto, K. Harada, and E. Tsubura, Photo-physicochemical behavior of iron and chromium compounds, Photogr. Sci. Eng., 12(1) (1968) 2-16.
91 C.A. Parker, A new sensitive chemical actinometer. I. Some trials with potassium ferrioxalate, Proc. Roy. Soc., 220(1140) (1953) 104-116.
92 C.A. Parker and C.G. Hatchard, Photodecomposition of complex oxalates. Some preliminary experiments by flash photolysis, J. Phys. Chem., 63(1) (1959) 22-26.
93 C.A. Parker, Induced autoxidation of oxalate in relation to the photolysis of potassium ferrioxalate, Trans. Faraday Soc., 50(11) (1954) 1213-1221.
94 H. Jonker, L.K.H. van Beek, C.J. Dippel, C.J.G.F. Janssen, A. Molenaar, and E.J. Spiertz, Principles of PD recording systems and their use in photofabrication, J. Photogr. Sci., 19(4) (1971) 96-105.
95 Lelental and H.J. Gysling, Tellurium physical development: a dry non-silver amplification process, J. Photogr. Sci., 28(6) (1980) 209-218.

96 H.J. Gysling, M. Lelental, M.G. Mason and L.J. Gerenser, Non-silver amplification processes: Part 3. Catalytic thermal decomposition of Te(II) coordination complexes via internal redox reactions, J. Photogr. Sci., 30(1) (1982) 55-65.
97 D.W. Woodward, V.C. Chambers and A.B. Cohen, Image-forming systems based on photopolymerization, Phot. Sci. Eng., 7(6) (1963) 360-368.
98 P. Walker, V.J. Webers and G.A. Thommes, Photopolymerizable reproduction systems-chemistry and applications, J. Photogr. Sci., 18(4) (1970) 150-158.
99 N.T. Notley, Photopolymerization imaging and the reaction between 9-vinylcarbazole and carbon tetrabromide, Phot. Sci. Eng., 14(1) (1970) 97-100.
100 K. Tada, Y.Shirota and H. Mikawa, Mechanism of charge-transfer polymerization. IV. Multiplicity of the reaction course in the photosensitized reaction of N-vinylcarbazole in the presence of the organic electron acceptor, Macromolecules, 6(1) (1973) 9-17, and references cited therein..
101 K. Hayashi, M.Irie, and Y. Yamamoto, Photoinduced ionic polymerization, Photogr. Sci. Eng., 23(4) (1979) 193-193.
102 T. Yamase, T. Sumi, T. Ikawa, H. Kokado and E. Inoue, Dye-sensitized photolysis of diazonium compounds. Part III. Rapid photopolymerization of acrylamide, Phot. Sci. Eng., 18(1) (1974) 25-28.
103 J.D. Margerum, L.J. Miller and J.B. Rust, Photopolymerization studies: II. Imaging and optical fixing, Phot. Sci. Eng., 12(4) (1968) 177-184.
104 G. Jackson, The properties of photochromic materials, Optica acta, 16(1) (1969), 1-16.
105 J. Epperlein, B Hofmann, Anorganische lichtempfindliche und photochrome Systeme Teil 1, J. Signal AM, 1(6) (1973) 395-415; Teil 2, J Signal AM, 2(1) (1974) 5-24.
106 V.A. Barachevski, G.I. Lashkov, and V.A. Tsechomski, Photochromism and its application, Chimiya, Moscow, 1977 (Russ.).
107 A.S. Holmanski, A.V. Zubkov, and K.M. Dyumaev, The nature of the primary photochemical act in spiropyrans, Usp. Chim., 50(4) (1981) 569-589 (Russ.).
108 M. Rand, Molecular electronics research growing despite controversy, Electronics Week, 58(18-19) (1985) 36-43.
109 J. Hanna and E. Inoue, The design of an organic photoreceptor with a charge-acceptance memory. II. Poly N-vinylcarbazole, diazonium salt, and TNF system, Photogr. Sci. Eng., 26(2) (1982) 69-74.
110 N.C Khe, I. Shimizu, and E. Inoue, The electrophotographic properties of a photoreceptor using ZnO in PVK, Photogr. Sci. Eng., 26(5) (1982) 249-253.
111 J.M. Winslow, High-speed photothermographic diazo films containing leuco dyes and nitrate salts: Part A - Sensitometry, Photogr. Sci. Eng., 28(3) (1984) 125-128.

Chapter 6.6

State of the Art in Resists Used in Microlithography

F. Schue, L. Giral, C. Montginoul and B. Serre

1. GENERAL CONSIDERATIONS

1.1. INTRODUCTION

A tiny modular silicon "chip" contains all the elements of an electronic circuit such as transistors, capacitors, etc., associated and connected to each other by metallic connections on the surface of the chip. Today, upward of 1,000 circuits, each incorporating a dozen transistors or other devices, can be put onto a silicon chip a few millimeters square. Within a few years, up to 50,000 circuits will be integrated on a similar chip. The continued increase in functional complexity over the past ten years has been provided by improved device design and improvements in the photolithographic art. Further refinements of the techniques demand line resolution in the submicrometer range. It appears that this aim can only be approached with radiation techniques and the use of radiation-sensitive polymeric materials. While photolithography will continue to be the mainstay of the microelectronics industry for many years to come, there are a number of disadvantages inherent in it. Photolithography, using near-ultraviolet radiation in the range of 300-400nm for exposure, is limited to the replication of fine structures with dimensions of about 1-2 μm. Below approximately 1 μm, the resolution capability of photolithographic processing becomes increasingly limited by diffraction. As has been shown during recent years, improvement in the resolving power and reproducibility of patterns is possible by virtue of high energy radiation, especially electron beams and X-rays.

In order to appreciate the role polymers play in the fabrication of such devices and, in particular, the necessity for continued developments in device lithography, it will be instructive to review the current fabrication technology.

1.1.1. Microlithography

Fig. I illustrates a typical procedure employed in the production of integrated circuits.

1.1.1.1. Passivation (Fig. Ia)

In order to protect the surface of silicon and to doped it only at the desired positions, the silicon wafer is covered with a thin layer of SiO_2 (0.1 μm). This oxide is impermeable to boron atoms and to impurities like oxygen. Silicon dioxide can

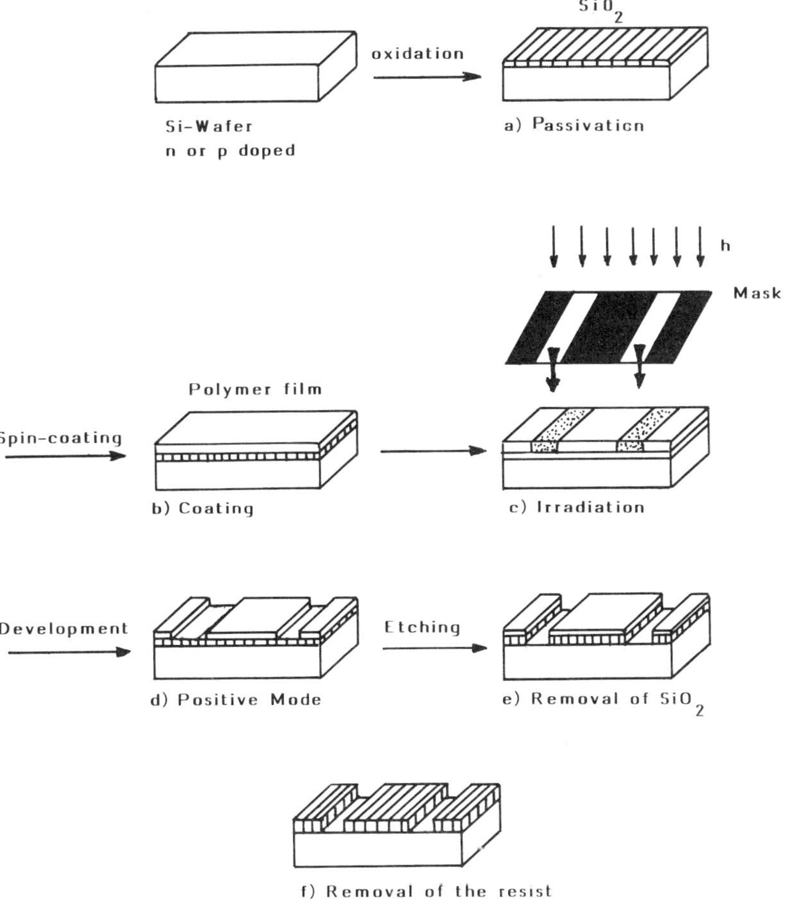

Fig. 1. Photolithography using a positive resist.

be stripped by hydrofluoric acid (Silicon itself is not attacked). Therefore, to strip selectively the oxide layer and obtain the design of a circuit, given positions have to be protected. It is at that level that the polymer intervenes, whereby it is supposed to exhibit a higher resistance to etchants than the substrate. Consequently, in this case, the polymers are denoted as "resists".

1.1.1.2. Coating (Fig. Ib)

The passivated silicon surface is spin-coated with a polymer of about one μm thickness. This step is followed by an annealing to remove the solvent and favor the adhesion of the resist on the substrate.

1.1.1.3. Irradiation (Fig. Ic)

Subsequently, the polymer film is irradiated using a mask. The irradiation induces a modification of the polymer linked to the energy absorbed by it.

1.1.1.4. Development (Fig. Id)

Areas of the resist layer which are exposed to light become detached in the subsequent development process.

Depending on whether the irradiated or the unirradiated regions of the polymer remain on the substrate after development, the polymers are classified as negative or positive working.

1.1.1.5. Removal of SiO_2 (Fig Ie)

In the non-protected zones, the silicon dioxide layer is etched away by hydrofluoric acid or plasma.

1.1.1.6. Removal of the resist (Fig. If)

The resist is removed and the substrate is only partially covered by the silicon dioxide layer. "Windows" have been opened which make the silicon substrate accessible for further treatment such as doping.

1.1.2. Manufacture of integrated circuits

The first steps for the production of a transistor may be illustrated by Fig. II.

The base material for a p-MOSFET (Metal Oxide Semiconductor Field Effect Transistor) is an n-doped silicon wafer. By oxidation, a thin silicon dioxide layer is grown on the substrate (Fig. IIa). This is followed by the steps described in Fig. I and which define the source and drain zones (Fig. IIb). The subsequent doping takes place only at these zones. All the other areas are protected by the silicon dioxide. If atoms, having only three valence electrons (e.g. boron) are diffusing into the source and drain zones, these zones become p-doped. The process is terminated by a slight oxidation of the p-doped areas (Fig. IIc). A new irradiation allows the opening of the contact windows and the gate zone is formed (Fig. IId). Another

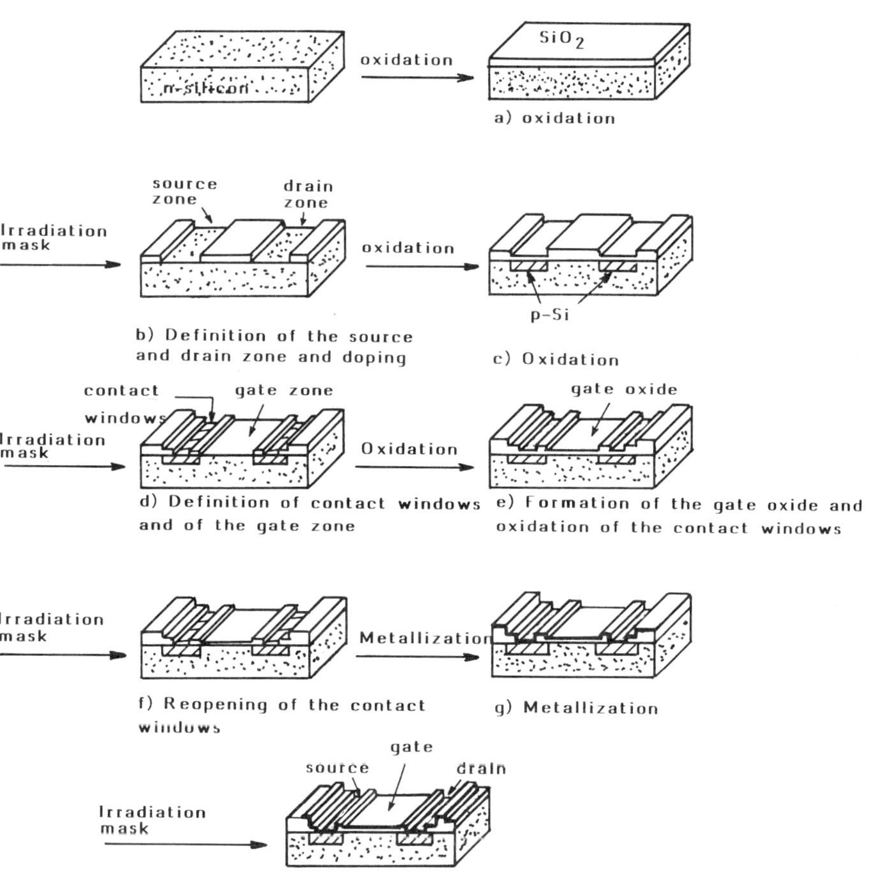

Fig. II. Manufacture for a transistor : p-MOSFET

oxidation produces a well-defined silicon dioxide layer called the gate oxide (Fig. IIe). During the step of oxidation the contact windows in the source and drain zones are closed and must be opened through a lithographic step (Fig. IIf). After metallization (Fig. IIg) and a lithographic step (Fig. IIh), the metal contacts between the source, the drain and the gate are established.

The manufacture of a transistor needs at least four lithographic steps. But often, up to twelve lithographic steps are necessary for more sophisticated circuits. Therefore, the precision of superposition is of great importance.

The resists used must have very particular properties because the sharpness of the designs and their sizes will be dependent on the quality of these resists and on the radiation used.

Before presenting in detail sections on different types of radiation, it is necessary to discuss a number of definitions constantly used.

1.2. RESISTS

1.2.1. Classification of resists

The resists used in microlithography to protect the surface of the substrate during the lithographic steps are divided into classes depending on their behavior under irradiation (Fig. III). Positive resists become more soluble in the irradiated area relative to the unexposed area, whereas negative resists become less soluble in the irradiated area than in the unexposed area. Differences in solubility of exposed and unexposed areas of polymer resists usually depend either on radiation-induced insolubilisation by cross-linking (negative resists) or radiation-induced polymer degradation leading to increased solubility (positive resists) (Fig. IV). Other methods of changing solubility by radiation induced changes in polarity are important, and these will be described later.

Positive resists, which require facile chain scission, are frequently vinylpolymers bearing tetrasubstituted carbons in the main chain :

$$\left(CH_2 - \underset{Y}{\overset{X}{\overset{|}{\underset{|}{C}}}} \right)_n$$

This is the case for poly(methylmethacrylate) : PMMA (X = CH_3, Y = $COOCH_3$). Negative resists, which require ease of crosslinking, are polymers bearing unsaturated double bonds or other reaction groups, e.g. :

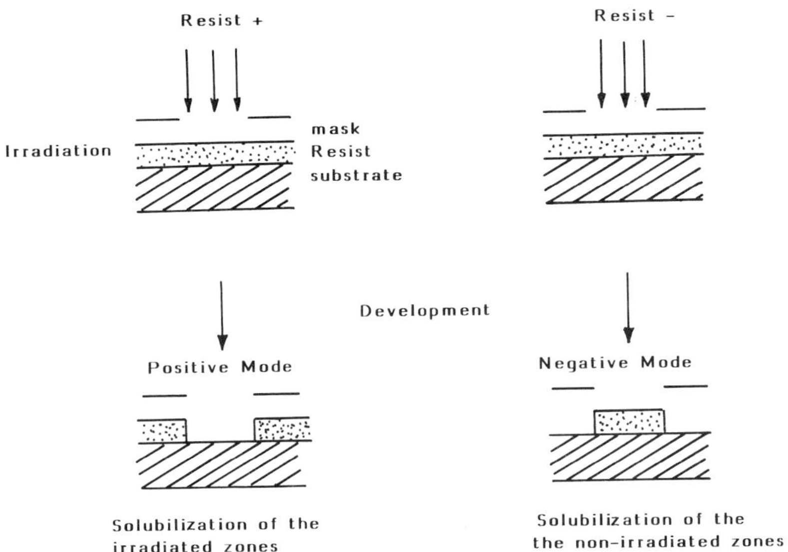

Fig. III. Exposure and development of positive and negative resists.

Resist +

Chain scission

Irradiation

Resist -

Chain crosslinking

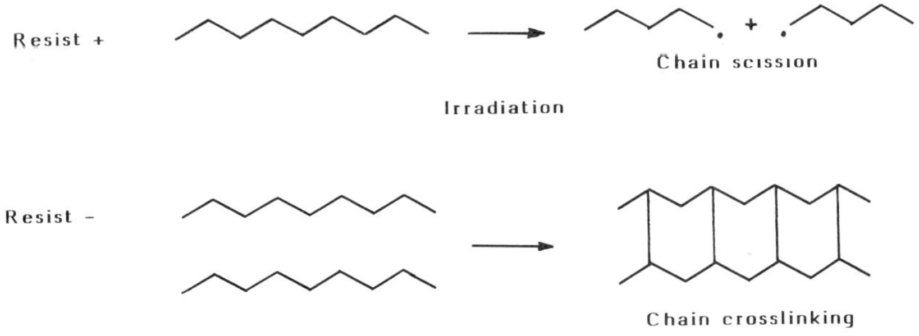

Fig. IV. Chemical reactions occuring during irradiation of positive and negative resists.

vinyl groups :

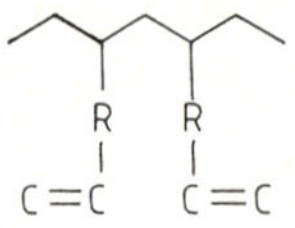

$$-R-C=C$$

episulfide groups :

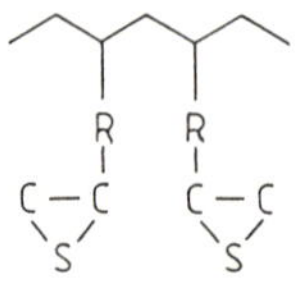

$$-R-\underset{\backslash S /}{C-C}$$

epoxide groups :

$$-R-\underset{\backslash O /}{C-C}$$

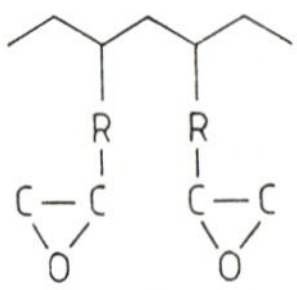

The same behavior is observed for polymers bearing trisubstituted carbons in the main chain, e.g. the polyacrylates :

$$\left(CH_2 - \underset{COOR}{\overset{H}{\underset{|}{\overset{|}{C}}}} \right)_n$$

In fact, both crosslinking and degradation may occur simultaneously in the same polymer, but one process usually predominates, giving rise to the classification of polymers as being of the degrading or cross-linking type.

Note :

The scission efficiency is described by a factor G(s) and the cross-linking efficiency is described by a factor G(x). G(s) and G(x) represent the number of main-chain scissions and intermolecular crosslinks, respectively, per 100 eV absorbed by the polymer.

1.2.2. CHARACTERISTIC PARAMETERS OF A RESIST

For each resist, negative or positive, two important parameters can be defined: the sensitivity and the contrast.

1.2.2.1. Sensitivity

The sensitivity expresses the aptitude of a resist to undergo structure modifications under irradiation. In order to define this parameter, it is necessary to introduce the incident dose "D". In the case of an electron beam exposure "D" is expressed as

$$D = \frac{I \times t}{S} \quad C/cm^2$$

where I is the beam current in Amperes, t is the exposure time in seconds, and S is the area exposed in square centimers.

For electron beam irradiations exposure doses are given in C/cm^2 (C: Coulombs) and for X-ray exposures in J/cm^2 (J: Joules).

The sensitivity of a polymer is determined by plotting the normalized thickness of resist layers after development (e_r/e_o = cross-linked thickness or residual thickness/ initial thickness, as a function of dose (Fig. V).

For a positive resist, the sensitivity is defined as being the dose D_o which corresponds to a normalized thickness equal to zero, in other words the dose D_o required to remove the irradiated polymer completely upon development.

For a negative resist, the ratio e_r/e_o increases with the dose up to a value of $i = 1$, for which the total thickness of the layer is cross-linked. The dose where the polymer commences to become insoluble is referred to as the gel dose D_g^i We may note here that the sensitivity of a negative resist is usually taken as the dose $D_g^{0.5}$ or $D_g^{0.7}$ required to cross-link the film so that respectively 50% or 70% of the initial thickness remains after development.

1.2.2.2. Contrast

The pattern resolution attainable with a given resist is determined by the resist contrast (Fig. V). It is defined as $\gamma = \frac{0.7}{\log \frac{D_{0.7}}{D_0}}$ for positive resists and as

$$\gamma_{0.7} = \frac{0.7}{\log \frac{D_g^{0.7}}{D_g^i}} \quad \text{or} \quad \gamma_{0.5} = \frac{0.5}{\log \frac{D_g^{0.5}}{D_g^i}} \quad \text{for negative resists.}$$

The contrast appears to be an important requirement for high resolution.

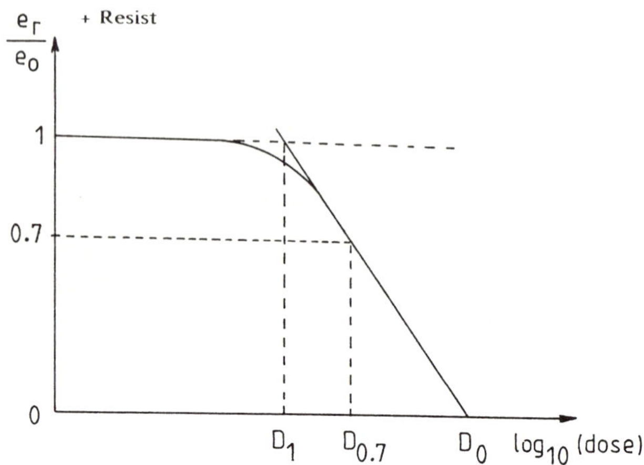

Sensitivity D = Do

Contrast = $\gamma = \dfrac{0.7}{\log \dfrac{D0.7}{Do}}$

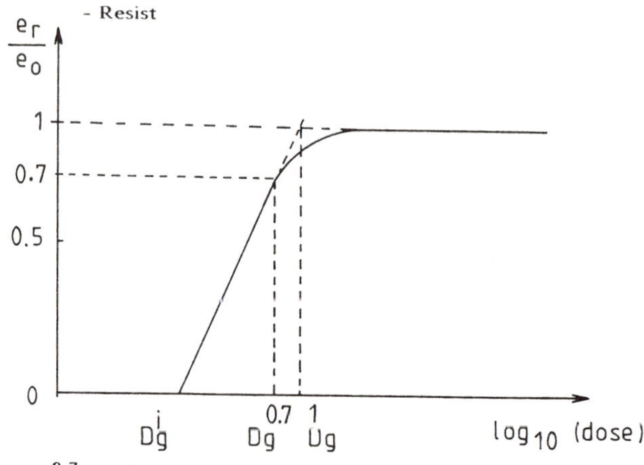

Sensitivity D = $D_g^{0.7}$ or $D_g^{0.5}$

Contrast = $\gamma_{0.7} = \dfrac{0.7}{\log \dfrac{D_g^{0.7}}{D_g^i}}$ or $\gamma_{0.5} = \dfrac{0.5}{\log \dfrac{D_g^{0.5}}{D_g^i}}$

Fig. V. Sensitivity and contrast.

In more general way, experience shows that negative resists possess a high sensitivity (10^{-6} to 10^{-7} C/cm²) but a low contrast ($\gamma \leqslant 1$). On the contrary, positive resists are characterized by a lower sensitivity but a good contrast.

1.2.3. Conditions required for a good resist

1.2.3.1. Spin-coating

It is generally accepted that a thickness of the resist layers between 0.6 and 1.5 µm shows a few defects. But a resolution of 0.1 µm needs a lower thickness. This is a limiting condition for negative resists. Finally, it will be shown later that the sensitivity is dependent on the molecular weight of the polymers. The higher the molecular weight the better the sensitivity will be. However a compromise has to be found because a high molecular weight induces a too high viscosity prejudicial to the formation of a film of good quality and uniform thickness.

1.2.3.2. Adhesion

The resist should have adequate adhesion during spin-coating, development and wet etching of SiO_2. Usually it seems to be beneficial to add an adhesion promoter.

Poor adhesion leads to marked undercutting of the patterns and loss of resolution.

If wet eching is replaced by a dry etching technique, less adhesion is required to maintain perfect image transfer.

1.2.3.3. Sensitivity

A high sensitivity allows a reduction in the time of irradiation. The following values have to be reached, in order to reduce the time of irradiation below one minute

- for electron beam irradiation : 10^{-6} to 10^{-7} C/cm² under 20 keV
- for X-ray irradiation : 2 to 7 mJ/cm²

1.2.3.4. Contrast

Resolution is partly linked to the contrast of the resist. A value at least equal to 2 seems necessary to obtain a 0.5 µm resolution.

1.2.3.5. Resolution

The resolution is characterized by the minimum width of the lines developed in the resist layer. It is measured by the number of lines per millimeter after irradiation and development.

The parameters which affect the contrast and therefore the resolution are:

- swelling during the development

- diffusion in the case of an electron beam irradiation.

During an X-ray exposure ($\lambda < 50$ Å) the diffraction and diffusion phenomena may be neglected and the resolution may be better. The present aim is to obtain features as fine as 0.1 μm.

The swelling of an organic polymer in a solvent proceeds in two steps :

- the swelling of the polymer when the solvent diffuses in the matrix (kinetic control)
- the transformation of the swollen polymer into a real solution.

The importance of the swelling is determined by :

- the molecular weight
- the structure of the polymer
- the glass transition temperature
- the solvent

1.2.3.6. Etch resistance

The requirements on the polymer depend on whether wet or dry etching processes are to be employed. The most common method by far is wet chemical etching. Buffered hydrofluoric acid solutions are used, for example, to etch SiO_2. Since lateral penetration of the chemical etchant can be significant for thick substrate films, new more unidirectional etching methods have been developed, including ion milling, plasma etching, and sputter etching. Each of these methods places its own requirement on the resist material, and while a material may be a perfectly adequate resist for one etching process, it may be completely unacceptable for another. As will also be shown, high radiation sensitivity, especially in the case of positive resists, is often coupled with low dry etch resistance, due to the fact that upon absorption of both high and low energy particles the same chemical reactions are initiated. Aromatic compounds being only weakly damaged by radiation exhibit high dry etch resistance.

1.2.3.7. Other properties

The synthesis of the polymers must be easy and reproductible and the monomers must have low toxicity.

1.2.4. Conclusion

The ideal resist, positive or negative, must possess a given number of properties summarized in Table I.

TABLE I

Conditions required for a resist

- Easy spin-coating
- Good adhesion
- High sensitivity electrons : 10^{-6} C/cm² (under 20 keV)
X-rays : 5 to 10 mJ/cm²
UV and visible : 5 to 20 mJ/cm²
- High contrast : $\gamma > 2$
- Good resolution : < 0.5 µm
- Mechanical resistance
- Chemical resistance to etching
- Easy removal after etching

2. ELECTRON SENSITIVE RESISTS

2.1. POSITIVE RESISTS

2.1.1. Introduction

The sensitivity of a positive resist is a function of a given number of factors which have been discussed by KU and SCALA (ref. 1). In the case of positive resists, the absorption of radiation causes scission of a certain fraction of the main-chain bonds, resulting in a lower value of the molecular weight. It has been shown that the incident dose "D" required for breaking a certain fraction "P_s" of the main-chain bonds is given by

$$D = \frac{P_s 100\, qz\rho N}{EG(s)\, M_o} \tag{1}$$

in which :

- D : total incident dose in C/cm²
- q : charge of the electron
- ρ : density of the polymer
- z : thickness of the resist layer
- N : Avogadro's number
- E : energy absorbed in the resist layer per incident electron (for a given resist this depends only on the accelerating voltage of the electron beam)
- M_o : molecular weight of a monomer unit
- G(s) : G (scission), number of main-chain bonds broken per 100 eV of energy absorbed
- P_s : probability of scission

CHARLESBY (ref. 2) has shown that the decrease in number average molecular weight ($\overline{M}'_n$) resulting from scission of a given fraction of bonds is related to the initial molecular weigh (M°_n) through the equation

$$\overline{M}'_n = \frac{\overline{M}^\circ_n}{1 + \dfrac{Ps\,\overline{M}^\circ_n}{M_o}} \quad (2)$$

This equation is only valid if no cross-linking of the polymer occurs during irradiation.

By combining Equations 1 and 2, the decreased molecular weight can be expressed as

$$\overline{M}'_n = \frac{\overline{M}^\circ_n}{1 + \dfrac{EG(s)\,\overline{M}^\circ_n\,D}{100\rho q\,z\,N}} \quad (3)$$

The formation of an image in the layer is obtained by differential solubility between the irradiated and the non-irradiated polymer. In order to appraise this property (refs. 3,4), UEBERREITER proposed an empirical relationship between the rate of dissolution "S" of a polymer and its number average molecular weight $\overline{M}_n$:

$$S = k\,\overline{M}_n^{-\alpha} \quad (4)$$

where k and α are constants for a particular polymer and development.

On this basis, the solubility ratio "S_r" is defined by the expression

$$S_r = \frac{S'}{S} = \left[\frac{\overline{M}'_n}{\overline{M}^\circ_n}\right]^{-\alpha} = \left[1 + \frac{EG(s)\overline{M}^\circ_n\,D}{100\rho q\,z\,N}\right]^{\alpha} \quad (5)$$

where S' and S are respectively the rate of dissolution of the irradiated and the non irradiated film.

It appears therefore that for constant values of the accelerating voltage, of the film thickness and of the solubility ratio, S_r, the dose D is lower the higher is G(s) $\overline{M}^\circ_n$.

Under these conditions, a positive resist will possess a good sensitivity if the values of G(s) and M°_n are high. Nevertheless, a compromise must be found because a high molecular weight induces a high viscosity prejudicial to the formation of a film of uniform thickness.

Moreover, the polydispersity I ($I = \overline{M}_w/\overline{M}_n$) of the polymer before irradiation has to be considered. Indeed, BOWDEN (ref. 5) described the effect of chain scission on polymers with different molecular weights and dispersities. The number average molecular weight as well as the polydisparity (which approaches two) resulting from scission of the main-chain bonds are independent of the initial molecular weight and distribution.

Consequently, it is necessary to maintain the distribution as narrow as possible in order to maximize sensitivity and contrast.

2.1.2. Poly(methyl Methacrylate) and derivatives

Poly (methyl methacrylate) (PMMA) was the first positive resist studied under high energy irradiation by HALLER et al (ref. 6).

$$\text{PMMA}: \left(CH_2 - \underset{COOCH_3}{\overset{CH_3}{\underset{|}{\overset{|}{C}}}} \right)_n$$

PMMA exhibits a high resolutive (< 0.8 µm) and a good adhesion to the substrate, but its sensitivity is moderate : $D_o = 5.10^{-5}$ C/cm² (10 keV).

2.1.2.1. Influence of the molecular weight (Table II)

As expected from theoretical considerations, high sensitivity requires a high molecular weight and a molecular weight distribution as narrow as possible. For example, M. GAZARD et al. (ref. 7) prepared high molecular weight PMMA (5.10^6 g/mole) by γ irradiation of the monomer obtaining thus a very high sensitivity: 8.10^{-7} C/cm² (20 keV).

TABLE II

Sensitivities of poly(methyl methacrylates) of high molecular weight (a)

a) For all the polymers, the polydispersity is close to 2.5

b) Elvacite 2041 : Commercial PMMA of DU PONT DE NEMOURS

c) IAA : Isoamyl acetate, poor solvent for PMMA

EA : Ethylacetate, good solvent for PMMA

Reference	Polymerisation	$\overline{M}_w \times 10^{-6}$	Development (c) % IAA/% EA	D_o (C/cm^2)(20keV)
Elvacite (b) 2041	-	0.37	90/10	$5x10^{-5}$
A	emulsion	1.1	80/20	$2x10^{-6}$
B	emulsion	1.6	80/20	10^{-6}
C	emulsion	1.7	80/20	10^{-6}
D	irradiation	2.9	75/25	$5x10^{-7}$
E	irradiation	5.4	70/30	$8x10^{-7}$
F	irradiation	6	70/30	$8x10^{-7}$

2.1.2.2. Influence of the developer

The choice of the developer is of prime importance because it affects directly the sensitivity and the contrast of the polymer. An example is given on Figure VI.

A study effected by E. GIPSTEIN et al (ref. 8) shows that G(s) and G(x) are independent of the molecular weight, the polydispersity and the stereoregularity of PMMA.

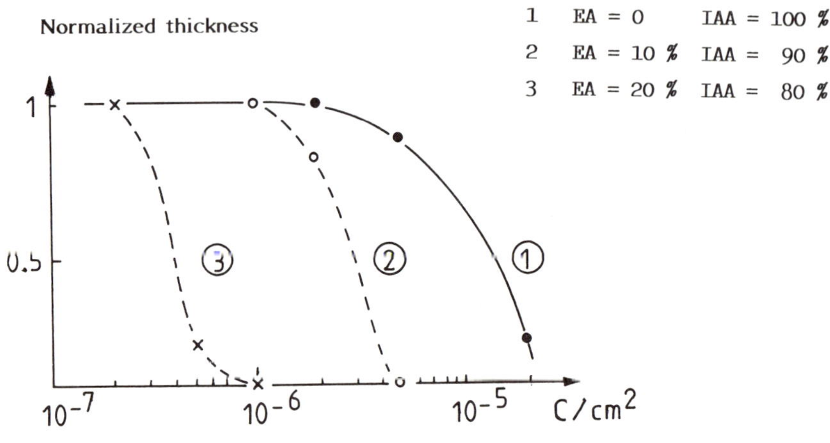

Fig. VI. Influence of the composition of the developer on the sensitivity of PMMA ($\overline{M}_w = 1.7 \times 10^6$)

EA : Ethyl acetate

IAI : Isoamyl acetate

2.1.2.3. Influence of the nature of the substituents

2.1.2.3.1. Polyalkylmethacrylates

Although PMMA is an excellent resist in many respects, it is deficient in one vital aspect : sensitivity. Therefore considerable effort has been expended on modifying the structure of PMMA to enhance its sensitivity.

T. AOKI et al. (ref 9), LAI and HELBERT (ref 10) and T. TADA (ref 21) have tried to correlate the steric hindrance of the substituent R_2 (R_1=CH_3) to G(s)-G(x), the glass transition temperature, the sensitivity and the contrast (Table III).

$$\left(CH_2 - \underset{\substack{| \\ C = O \\ | \\ O - R_2}}{\overset{\substack{R_1 \\ |}}{C}} \right)_n$$

TABLE III

G(s)-G(x) sensitivities and contrast factors of some poly(alkylmethacrylates)

$$\left(CH_2 - \underset{\substack{| \\ COOR_2}}{\overset{\substack{CH_3 \\ |}}{C}} \right)_n$$

Resist	R_2	$\bar{M}_n x10^{-5}$	I	Tg (°c)	G(s)-G(x)	Sensitivity D_o (C/cm²)	Contrast γ	Reference
Poly methyl methacrylate	$-CH_3$	1.59	2.5	105	1.3	$5x10^{-5}$	1.7-2	10
Poly n-butyl methacrylate	$-CH_2-CH_2-CH_2-CH_3$	-	-	40	0.6	$5x10^{-7}$	2.3	9
Poly isobutyl methacrylate	$-CH_2-CH<(CH_3)(CH_3)$	1.2	1.7	53	1.1	$8x10^{-6}$	0.9	9, 10
Poly tertiobutyl methacrylate	$-C(CH_3)_3$	0.65	-	118	1.28	-	-	9, 10
Poly cyclohexyl methacrylate	$CH<(CH_2-CH_2)(CH_2-CH_2)>CH_2$	1.73	2.4	90	0.44	$1.6x10^{-5}$	0.73	9, 10
Poly phenyl methacrylate	$-C<(CH-CH)(CH=CH)>CH$	4.09	2.8	54	0.29	$4.3x10^{-5}$	1.31	9, 10

According to the results obtained, it seems difficult to correlate G(s)-G(x) to the sensitivity. One notes that a lowering of the glass transition temperature induces an increase of the sensitivity. This is the case for poly(n-butyl methacrylate) for which the sensitivity is $5x10^{-7}$ C/cm² and Tg 40°c. But it seems that the higher sensitivity should be more correlated to some specific properties of solubility in relationship with the low glass transition temperature observed. Poly(phenyl methacrylate) is a particular case. Indeed, the phenyl ring present in the main chain protects the film against degradation by converting part of the incident energy absorbed into photonic emission.

2.1.2.3.2. Polyacrylates

HELBERT et al. (refs 11, 16, 17, 19, 20, 22) as well as B. SERRE et al. (ref 15) sought to enhance the sensitivity of PMMA via substitution of electron-withdrawing groups at the α -carbon atom. They found that poly(methyl- α -chloroacrylate) (PMCA), polymethacrylonitrile (PMCN) and poly(n-butyl- α -cyanoacrylate) (PBCyA) have substantially higher sensitivity. This can be seen from Table IV. We note that the substitution of a methyl group in PMMA by a chlorine or a cyano group enhances G(s) without changing G(x) too much. The highest sensitivity $2x10^{-6}$ C/cm² is observed with PBCyA.

Nevertheless, in the last case, the presence of a n-butyl group could possibly contribute to this high value by enhancing the solubility ratio Sr. We note also a higher plasma etch resistance for PMCN.

TABLE IV

Lithographic characteristics of some polyacrylates

$$\left(CH_2 - \underset{R_2}{\overset{R_1}{C}} \right)_n$$

Resist	R_1	R_2	$\overline{M}w$ $x10^{-5}$	I	Tg (°c)	G(s)	G(x)	Sensitivity D_o(C/cm²)	Contrast γ	Resolution (µm)	Etch resistance Å/min	References
PMMA Elvacite 2041	$-CH_3$	$COOCH_3$	4.8	3	109	1.3	0	$5.6x10^{-5}$(10keV)	1.7 - 2	≤ 1	100*	6,11,23
								$9.3x10^{-5}$(15keV)				20
								$20.6x10^{-5}$(20keV)				20
PMCA	-Cl	$COOCH_3$	1.25	1.7	151	6.0	0.7-0.9	$2.7x10^{-5}$(15keV)	-	-	186*	11,17,18,19
			16.0	2.6	130	6.0	0.7-0.9	$2.3x10^{5}$(10keV)	-	-		20
PMCN	-CN	$-CH_3$	1.33	2.6	118	3.3	0	$1.9x10^{-5}$(10keV)	-	1	36*	11,16
								$7.0x10^{-5}$(20keV)	-	1		20,22
PBCyA	-CN	$COOC_4H_9$	14.3	2.6	115	4	0.4	$2x10^{-6}$(20keV)	2.3	-	-	15

*plasma : $CF_4/4\%O_2$

2.1.2.3.3. Fluorinated polymethacrylates (table V)

The partial substitution of hydrogen on the ester group by fluorine enhances the sensitivity significantly. We note a value of $4x10^{-7}$ C/cm² for FBM_1. Not much change is observed for G(s) compared to PMMA. Nevertheless, some preliminary studies made by TADA (ref 21) suggest that the substitution of fluorine does not alter the C-C bond in the main chain but would modify significantly the solubility ratio Sr.

TABLE V

Lithographic characteristics of fluorinated polymethacrylates.

$$-\!\!\left[CH_2 - \underset{\underset{\displaystyle OR_2}{|}}{\underset{C=O}{\underset{|}{\overset{\overset{\displaystyle R_1}{|}}{C}}}} \right]\!\!-_n$$

Resist	R_1	R_2	$\bar{M}w$ $x10^{-5}$	I	Tg (°c)	G(s)	G(x)	Sensitivity $D_o(C/cm^2)$	Contrast γ	Resolution (μm)	Etch resistance Å/min	References
FBM-1	$-CH_3$	$-CH_2-CF_2-CFH-CF_3$	5	2.8	50	1.2	-	$4x10^{-7}$(20keV)	5	0.3	-	14,30
FBM-1	-	-	2.8	1.74	-	-	-	$1.2x10^{-6}$(20keV)	4	-	-	24
EBR-9	-Cl	$-CH_2-CF_3$	1.38	2.9	133	-	-	$1x10^{-6}$(20 keV)	2.3	0.2-0.5	-	13
PFEtMA	$-CH_3$	$-CH_2-CH_2F$	7.2	2.49	96	-	-	$1.5x10^{-5}$(20keV)	1	-	-	25
PTFEM	$-CH_3$	$-CH_2-CF_3$	9.25	1.35	69	2.3	0	$2-3x10^{-5}$(20keV)	1	-	230*	22,25
PHFIM	$-CH_3$	$-CH<^{CF_3}_{CF_3}$	-	-	80	2.6	0	$2.3x10^{-5}$(15keV)	-	-	230*	11

*plasma : $CF_4/4\%O_2$

2.1.3. Copolymers

2.1.3.1. Copolymers of methylmethacrylate

TABLE VI

Lithographic characteristics of copolymers of methylmethacrylate

$$-\!\!\left[CH_2 - \underset{\underset{\displaystyle OCH_3}{|}}{\underset{C=O}{\underset{|}{\overset{\overset{\displaystyle CH_3}{|}}{C}}}} \right]_m\!\!-\!\!\left[CH_2 - \underset{\underset{\displaystyle R_2}{|}}{\overset{\overset{\displaystyle R_1}{|}}{C}} \right]\!\!-_m$$

Resist	R_1	R_2	$\bar{M}w$ $x10^{-5}$	I	Tg (°c)	G(s)	G(x)	Sensitivity $D_o(C/cm^2)$	Contract γ	Resolution (μm)	Etch resistance Å/min	Reference
P(MMA-CO-MCA) (54/56)	-Cl	$-COOCH_3$	8.5	2.0	125	3.1	0.06	$2.5x10^{-5}$(15 keV)	-	0,4	59*	11,18,19,33, 17
P(MMA CO-HMA) (80/20)	$-CH_3$	$-COO(CH_2)_5CH_3$	1.76	1.8	67	-	-	$2.5x10^{-5}$(15keV)	-	-	-	28
P(MMA-CO-TFMAN) (68/32)	CH_3	-CN	3.1	2.0	98	3.1	0	$3x10^{-5}$(20keV)	-	-	79*	22
P(MMA-CO-TFMMA) (89/11)	$-CF_3$	$-C00CH_3$	3.7	2.8	125	2.4	0	$15x10^{-5}$(20keV)	-	-	100*	22
P(MMA-CO-IB) (75/25)	$-CH_3$	$-CH_3$	1.15	2.8	43 / 47	-	-	$4.5x10^{-6}$(15keV)	-	1.25	—	29
P(MMA-CO-MAA) (60/40)	$-CH_3$	-COOH	-	-	160	-	-	$2x10^{-5}$(20keV)	-	0.4	-	31
P(MMA-CO-AN) (93/7)	-H	-CN	-	-	-	-	-	$4x10^{-6}$(15keV)	2.5	0.5	—	34
P(MMA-CO-MCN)	$-CH_3$	-CN	-	-	-	-	-	$4-6x10^{-6}$(15keV)	-	0.5	—	35,36

*plasma : $CF_4/4\%O_2$

Considerable effort has been expended on modifying the structure of PMMA to enhance its sensitivity. Thus, the copolymerization of MMA with various monomers of which the scission factor G(s) and etch resistance are higher to those of MMA induces an important enhancement of the above-mentioned factors. Some of the must striking results are given in table VI. We note particularly a sensitivity of $4x10^{-6}$ C/cm² for the copolymers with isobutylene (IB), acrylonitrile (AN) and methacrylonitrile (MCN).

2.1.3.2. Copolymers of methacrylonitrile

Copolymers of methacrylonitrile and various halogenated monomers (MCA, FMA, TFEM and TFMAN) were investigated in order to enhance not only the sensitivity but mostly the etch resistance compared to MMA. The most striking results are given in Table VII.

TABLE VII

Lithographic characteristics of copolymers of methacrylonitrile

$$-\!\!\left[CH_2 - \underset{CN}{\overset{CH_3}{C}} \right]_m\!\!-\!\!\left[CH_2 - \underset{R_2}{\overset{R_1}{C}} \right]_n\!\!-$$

Resist	R_1	R_2	Mw $x10^{-5}$	I	Tg (°c)	G(s) — G(x)		Sensitivity D_o(C/cm²)	Contrast γ	Resolution (μm)	Etch resistance Å/min	References
P(MCN CO MCA) (50/50)	-Cl	$-COOCH_3$	10.5	2.3	120	3.9	0	$1.8x10^{-5}$(15 keV)	-	1	50*	11,20,23
P(MCN-CO-FMA) (80/20)	-F	$-COOCH_3$	8.3	2.8	-	2.0	0	$6x10^{-5}$ (20 keV)	-	-	30*	22
P(MCN-CO-TFEM) (31/69)	$-CH_3$	$-COOCH_2CF_3$	4.4	2.1	86	2.2	0	$3\text{-}4x10^{-5}$(20 keV)	-	-	93*	22
P(MCN CO TFMAN) (88/12)	$-CF_3$	-CN	16.8	2.5	123	3.3	0	$5x10^{-5}$(20 keV)	-	-	33*	22

*plasma : $CF_4/4\%O_2$

For instance, it is claimed that a copolymer containing 50 mol % of methyl-α-chloroacrylate has a sensitivity and an etch resistance of $1.8x10^{-5}$ C/cm² (15 keV) and 50 Å/min respectively.

2.1.4. **Cross-linked homo- and copolymers** (Table VIII)

PMMA is an excellent positive resist from the viewpoint of adhesion and resolution, but suffers from rather low sensitivity and poor resistance to dry-etching techniques. ROBERTS (refs 27, 37, 38, 39) has reported a very simple, but highly effective modification of PMMA, which affords higher thermal stability without loss of resolution. Overall, a small number of acid chloride and carboxylic acid

TABLE VIII

Lithographic characteristics of cross-linked homo- and copolymers

Resist	R_1	R_2 R_3	$\overline{M}_w$ $x10^{-5}$	I	Tg (°c)	Sensitivity D_o (C/cm²)	Contrast γ	Resolution (μm)	Etch resistance Å/min	References
EBR-1 cross-linked	$-CH_2CCl_3$		5.74	2.4	138	$1.25x10^{-6}$ (20 keV)	-	<1	160 (Ar ionic etching)	12
PMAAD	$-NH_2$	-	2.4		200	$6x10^{-7}$ (20keV)	2	0,2		26,40
PMMA cross-linked	-	R_2 OH R_3 Cl	-	-	-	8 to $40x10^{-6}$ 10 to 30 keV	-	<1	-	27,37,38,39
PCA FMR-E 101	-	-	-	-	125	$8x10^7$ (20 keV)	4	0.5	-	41
P(MMA-MAA) + 13% TPG	-	-	5.8	-	-	10^{-6} (20 keV)	-	0.5	-	52
PbBMA + 1% DABC	-	-	11.0	-	-	$2.3x10^{-6}$ (20 keV)	-	-	-	52
PMIPK + 1.9% DABC	-	-	1.8	-	-	$5.1x10^{-7}$ (20 keV)	-	-	-	52

are introduced into the PMMA backbone in order to allow for thermal crosslinking in the resist (via formation of anhydride linkages) after it has been coated onto the appropriate substrate (Fig. VII). Anhydride crosslinks in the exposed areas of the resist are broken during electron-beam irradiation so that there is no loss of sensitivity. Resistance to plasma and to developer solvent is much improved by the anhydride crosslinks.

Fig. VII Cross-linked homo- and copolymers

A similar approach was attempted by MATSUDA et al. (ref. 26). They studied polymethacrylamide (PMAAD). After it has been coated onto the substrate, thermal crosslinking occurs via formation of inter and intramolecular inside linkages (Fig. VII). A sensitivity of $6x10^{-7}$ C/cm² has thus been obtained.

FUJI Chemical Corp. has commercialized a resist which is a copolymer of ethyl α-amidoacrylate :

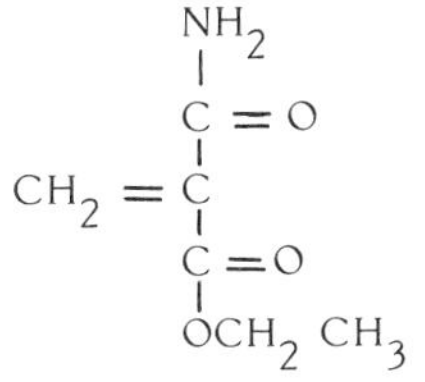

$$CH_2=C(C(=O)NH_2)(C(=O)OCH_2CH_3)$$

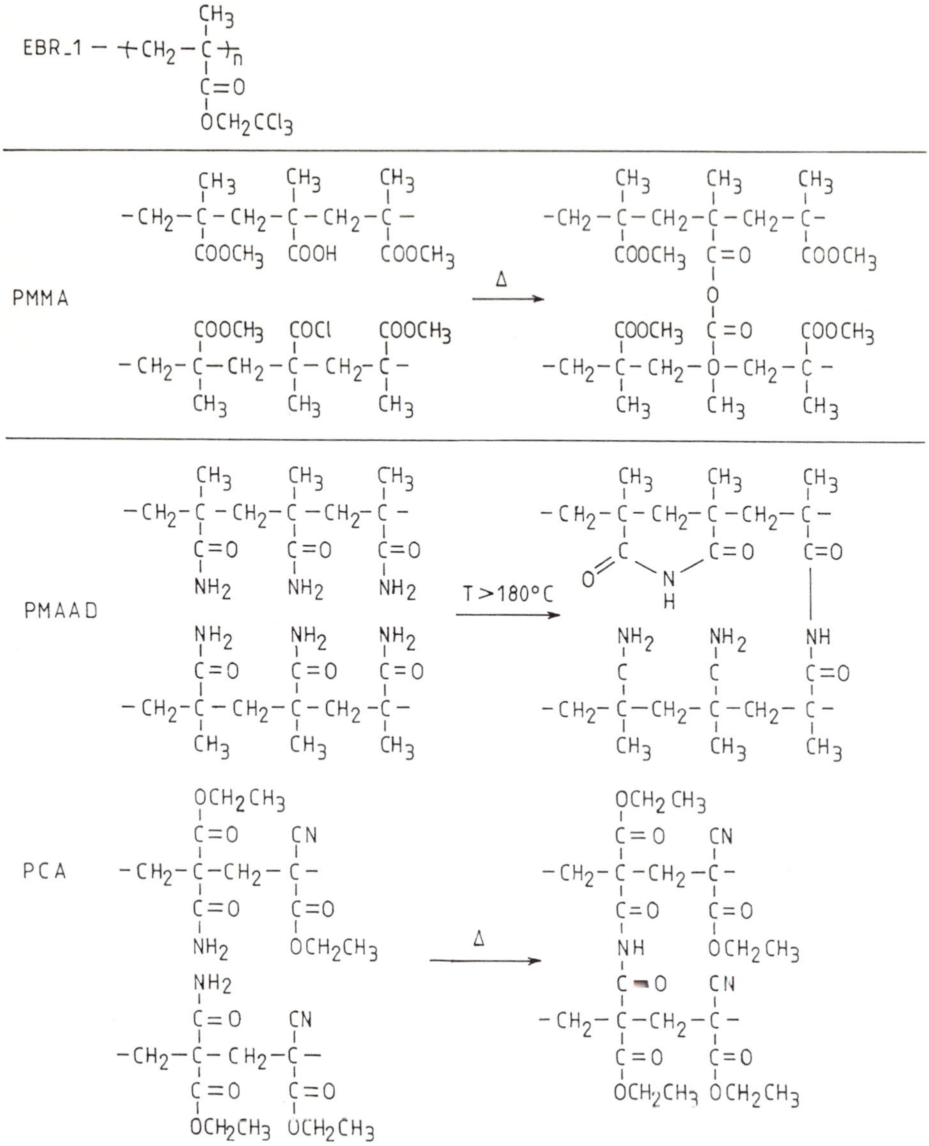

EBR-1 — $+CH_2-C(CH_3)(C{=}O\,OCH_2CCl_3)+_n$

PMMA

Δ

PMAAD

$T > 180°C$

PCA

Δ

Figure VII

and ethyl α -cyanoacrylate :

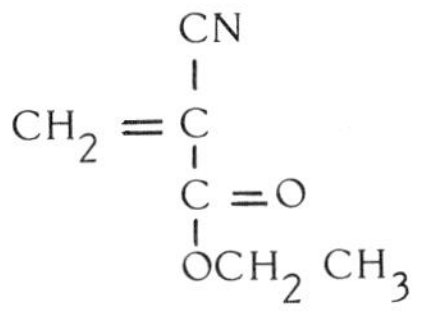

$$CH_2 = C(CN) - C(=O) - OCH_2CH_3$$

and of which the trade name is FMR-E 101. This resist has a sensitivity of 8×10^{-7} C/cm² with a contrast of 4 and a resolution of 0.5 μm.

2.1.5. Polysulfones

A new class of positive resists based on alternating copolymer of olefins with sulfur dioxide was reported by BOWDEN and THOMPSON (refs 4 3, 44). Their structure is as follow :

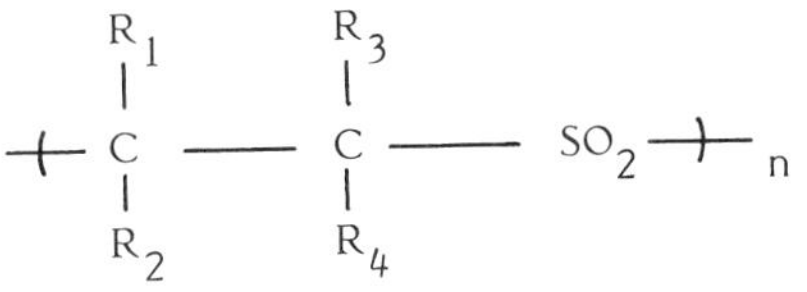

$$-\!\!\left[C(R_1)(R_2) - C(R_3)(R_4) - SO_2 \right]\!\!-_n$$

The high sensitivity observed results from the weakness of the C-S bond (ref. 42). THOMPSON and BOWDEN (refs 45, 47) investigated several poly (olefin sulfones) from aliphatic- and cyclic- substituted olefins. They confirmed their high sensitivity values of the order of 1 to 3×10^{-6} C/cm² as reflected in their high G (scission) (Table IX).

TABLE IX

Lithographic characteristics of some polysulfones

Resist	$\overline{M}w$ $\times 10^{5}$	I	Tg (°C)	G(s)	G(x)	Sensitivity D_o (C/cm²)	Contrast γ	Resolution (μm)	Etch resistance Å/min	References
PMMA Elvacite 2041	4.8	3	109	1.3	0	5.6×10^{-5}(10kev)	1.7-2	≤ 1	100*	6,11,23
PMCA	16.0	2.6	130	6.0	0.7-0.9	2.3×10^{-5}(10keV)	-	-	186*	11,17,18,19,20
PMCN	13.3	2.6	118	3.3	0	1.9×10^{-5}(10keV)	-	1	36*	11,16
Poly(Butene-1 Sulfone	30	3	-	11	-	8×10^{-7}(10keV)	-	0.5	-	46
Poly(Styrene Sulfone	2.7	-	180 200	-	-	10^{-5} (5keV)	-	-	300 (ionic etching)	47
Poly(methyl-1 cyclopentene-1 Sulfone	-	-	-	-	-	1.3×10^{-6} (10keV)	-	-	-	50,51

*plasma $CF_4/4\%O_2$

Poly (butene-1 sulfone) (refs 45, 46) (PBS) is considered to have the best properties and therefore was commercialized. It has a sensitivity of $8x10^{-7}$ C/cm².

Polysulfones allow two types of development :

- solvent development
- vapor development

Vapor development is related to the spontaneous decomposition of polysulfones upon irradiation with high energy radiation. The capability of decomposing into gaseous products is based on the fact that for several poly (olefin sulfones) the ceiling temperature T_c is rather low, in some cases lower than room temperature. The ceiling temperature phenomenon refers to the propagation-depropagation equilibrium which is formulated as shown below :

$$\text{Poly (olefin sulfone)} \rightleftharpoons \text{olefin} + SO_2$$

As the temperature increases, the equilibrium is shifted to the right side. Under these conditions, it is possible to realize a good development. The rate of this vapor development is mostly determined by the structure of the olefin, the temperature of irradiation (ref. 48) and the irradiation parameters (ref. 49). Extensive research (ref. 52) has been carried out on morphological features of various polysulfones. Indeed, many of these polymers tend to crack during the formation of the film and especially during solvent development following exposure. This phenomenon is attributed to the presence of microscopic crystalline region and hence swelling stresses at the crystalline-amorphous interface.

One limiting feature of the polysulfones is their poor resistance to dry processing techniques. This has been attributed to the depolymerization reaction which enhances the rate of film loss. BOWDEN and THOMPSON (ref. 47) circumvent this problem by using a 2:1 copolymer of styrene to SO_2. As expected, cleavage of the styrene-SO_2 bond produced a styryl radical with another styrene unit in the penultimate position. Therefore, the radical behaves like a polystyrene radical with a tendency to unzip. The presence of benzene rings results in a marked lowering of the sensitivity to 10^{-5} C/cm² but a good resistance to dry etching.

2.1.6. Use of photosensitive resists in electron beam irradiation

There is some interest in the use of positive photosensitive resists such as the AZ resists commercialized by SHIPLEY (refs 90, 92) for this purpose.

The basic polymeric components of optical positive resists are novolak resin precursors. These are relatively low-molecular-weight condensation products derived from phenols and formaldehyde. A simplified structure may be written as

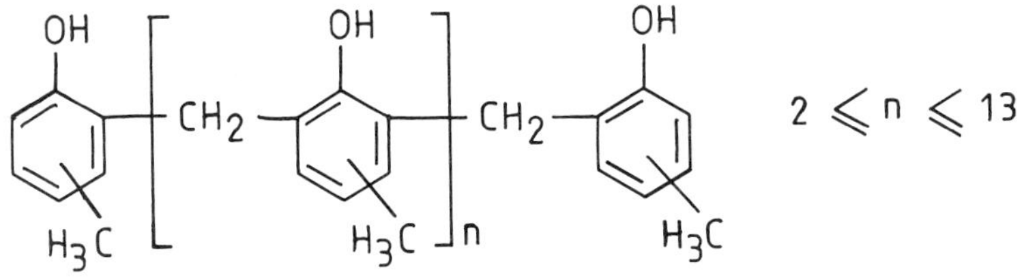

$2 \leqslant n \leqslant 13$

Such polyphenols are soluble in alkaline solutions. For resists applications, the alkali-solubility of the novolak is inhibited by the presence of a photo-active compound which is a derivative of a 1,2-naphthoquinone diazide.

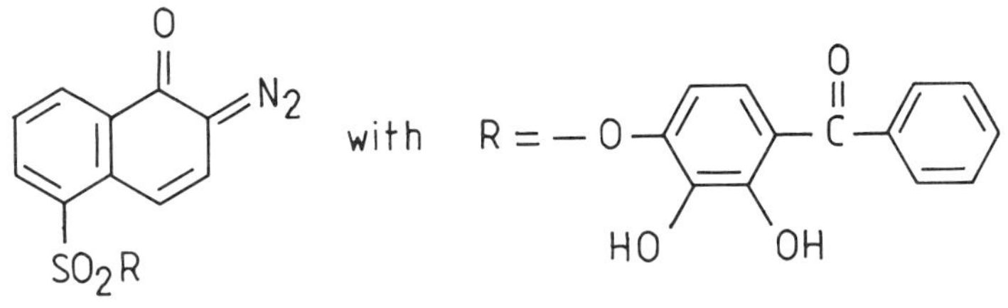

with

The later readily decomposes on photolysis to yield indene carboxylic acids via successive keto-carbene and ketene intermediates

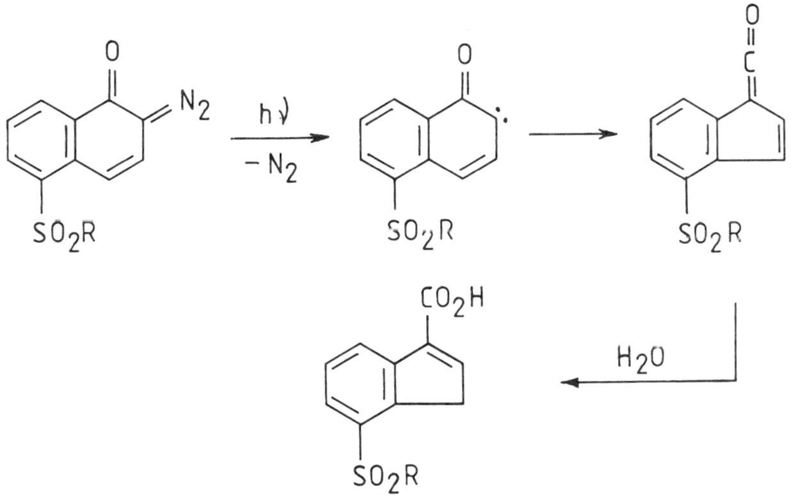

As is shown above, water is an essential reactant in the resist and is present as an impurity in the novolak. Controlled humidity is also essential for reproducible positive resists. Where the exposure produces carboxylic acid, the area readily

dissolves in solutions containing metallic or quaternary ammonium hydroxides. In the absence of photo-decomposition producing alkali-soluble carboxylic acids, the novolak is inhibited from dissolving in alkali by the hydrophobic character of the quinone diazide and the azo coupling induced by the alkali between the diazide and the novolak (ref. 91).

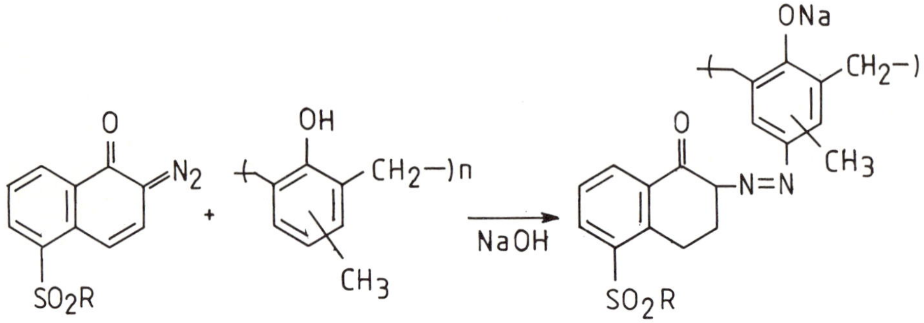

These resists have a good resolution although the sensitivity is low, varying from $2x10^{-5}$ to $1.5x10^{-4}$ C/cm² (25 keV) according to the chemical structure of the diazide.

A new formulation has been found by combining the high sensitivity of poly(olefin sulfones) with the high resistance to dry etching of photosensitive resists (ref. 94). The 1,2-naphthoquinone diazide, which inhibits the dissolution of the novolak, was replaced by a polymeric inhibitor which depolymerizes upon irradiation. Poly(methyl-2 pentene-1 sulfone) PMPS was choosen having a given number of desired properties

- it is compatible with the novolak and allows the formation of a homogeneous film by spin-coating
- it is chemically stable
- its rate of depolymerization is fast upon irradiation at room temperature and allowing a reasonable sensitivity to be reached ($< 10^{-5}$ C/cm², 20 keV)
- concentration lower than 30 % (in weight) in the resist are sufficient to inhibit the dissolution of the novolak.

In parallel, FAHRENHOLTZ (ref. 95) used the novolak resin alone as a positive resist under electron-beam irradiation. He obtained a resolution of 0.5 µm with an incident dose of $5x10^{-6}$ C/cm² (20 keV).

2.2. NEGATIVE RESISTS

2.2.1. Introduction

When a negative resist is irradiated by an electron-beam, crosslinking occurs causing the formation of a three-dimensional network, which is insoluble in the solvent for the original linear chain polymer. The minimum incident dose D^i_g (C/cm^2) required to produce this network is given by the relation of KU and SCALA (ref. 1)

$$D^i_g = \frac{50\ q\rho z\ N}{E\ G(x)\ \overline{M}_w} \qquad (6)$$

in which :

q : charge on electron

ρ : density of the polymer

z : thickness of the film

N : Avogadro's number

E : energy absorbed in the resist layer per incident electron

$\overline{M}_W$: weight average molecular weight of the polymer

G(x) : number of cross-links produced per 100 eV of absorbed energy.

Thus, a resist having a high molecular weight and G(x) will need a lower incident dose and therefore will be more sensitive.

The contrast of a negative resist will be better described by the two following parameters :

1) the molecular weight distribution

2) the effect of scission probability.

The rate of gel formation decreases as the molecular weight distribution widens. This will induce a decrease of the contrast. Therefore, high contrast demands a narrow molecular weight distribution.

We have already noted that both cross-linking and degradation occur simultaneously. A polymer behaves as a negative resist providing the ratio of cross-linking to degradation is greater than a critical value. The effect of a significant scission probability will reduce the amount of gel formed and decrease the contrast.

Compared to the number of studies made in the field of positive resists, much less has been done in the field of negative resists. The sensitivity of a cross-linkable polymer is enhanced by the introduction of chemical groups sensitive to cross-linking upon an electron beam irradiation. These groups are the following :

- double bond : $- CH = CH_2$
- epoxide group : $- CH - CH_2$ (bridged by O)
- epithio group : $- CH - CH_2$ (bridged by S)
- carbon-halogen bond : C - Cl, C - Br

2.2.2. Polymers containing epoxide groups

2.2.2.1. Homo- and copolymers of glycidylmethacrylate

A copolymer based on glycidylmethacrylate (GMA) and ethylacrylate (EA) known as COP is used extensively in many laboratories. The effects of the molecular weight and the polydispersity index on the sensitivity were thoroughly examined (refs 55, 63, 64). The glycidyl group is very sensitive to electron irradiation, whereas ethylacrylate improves the adhesion to the substrate. The most important results are given in Table X.

TABLE X

Lithographic characteristics of homo- and copolymers of glycidylmethacrylate.

Resist	R_1	R_2	Epoxide group (% in mole)	$\overline{M}_w$ x10^{-5}	I	$Dg^{0.5}$ C/cm² 10keV	Contrast γ	Resolution (µm)	Etch resistance Å/min	References
PGMA	$-CH_3$	$-CH_2-CH-CH_2$ (epoxide, O)	100	1.11	1.7	5.4x10^{-7}	1	-	33*	63, 57
P(GMA-CO-EA) (COP)	-H	$-CH_2-CH_3$	72	0.74	2.1	8.4x10^{-7}	1.2**	1-2	40*	64,57
P(GMA-CO-EA) (COP)***	-H	$-CH_2-CH_3$	68	5.52	2.4	2.5x10^{-7}	1.0**	1-2	40*	64,57
P(GMA-CO-EA) (COP)	-H	$-CH_2-CH_3$	70	21.6	4.5	1.7x10^{-7}	0.70**	1-2	40*	64-57
P(GMA-CO-PM)	CH_3	phenyl ring	74	3.8	2.1	3.2x10^{-7}	1.1	2	-	55

* plasma $CF_4/4\%O_2$

** γ_1

*** Tg=34°c

An increase of the molecular weight and a decrease of the molecular weight distribution induce respectively an enhancement of the sensitivity and of the contrast. Thus, a copolymer containing 68% in mole of GMA, with $\overline{M}_w$ = 5.5.x10^5 and I=2.4 possesses a sensitivity of 2.5x10^{-7} C/cm² and a contrast of only 1. Under these conditions, the resolution is the limiting factor when using negative resists. Generally, it is of the order of 1 to 2 µm. This limitation is due to swelling problems during liquid development. In order to reduce swelling, copolymers of GMA and respectively styrene, phenylmethacrylate, 3-chlorostyrene were synthesized (refs 55, 58). As foreseen, the sensitivity of these copolymers is decreased compared to the one of COP because of the presence of the phenyl ring, but seems to be sufficient for electron applications (Table XI).

TABLE XI

Lithographic characteristics of copolymers of styrene.

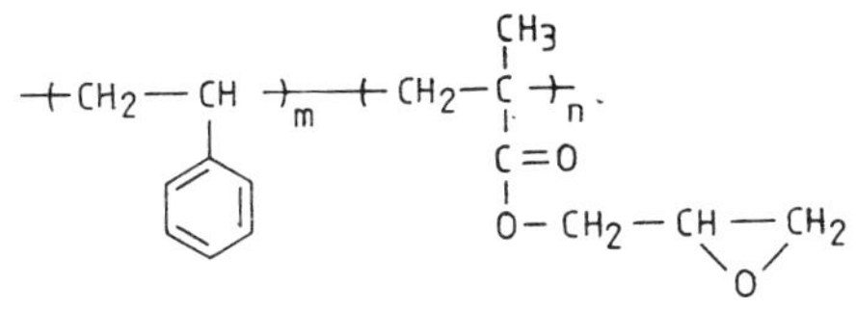

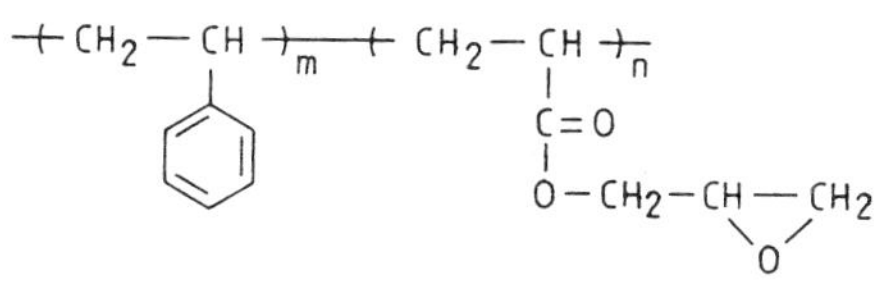

$$-\!\!\left(CH_2-\underset{C_6H_5}{CH}\right)_m\!\!-\!\!\left(CH_2-\underset{\substack{| \\ C=O \\ | \\ O-CH_2-CH-CH_2 \\ \backslash O /}}{\overset{CH_3}{C}}\right)_n \qquad -\!\!\left(CH_2-\underset{C_6H_5}{CH}\right)_m\!\!-\!\!\left(CH_2-\underset{\substack{| \\ C=O \\ | \\ O-CH_2-CH-CH_2 \\ \backslash O /}}{CH}\right)_n$$

Resist	Epoxide group in mole %	$\overline{M}w$ $x10^{-4}$	I	Tg (°c)	$Dg^{0.5}$ C/cm^2 10keV	Contrast γ	Resolution (µm)	Etch resistance	References
P(St-CO-GA)	22	5.7	1.6	53	$50x10^{-7}$	1.1	-	-	55
P(St-CO-GA)	44	5.02	2.0	-	$16x10^{-7}$	1.1	-	-	55
P(St-CO-GMA)	64	14.1	1.8	90	$20x10^{-7}$	1.3	-	-	55
P(St-CO-GMA)	79	21.8	1.9	-	$9.5x10^{-7}$	0.9	-	-	55
P(Cl-St-CO-GMA)	71	27	2.16	-	$**9x10^{-7}$	1.2*	1-1.5	equivalent to AZ 1350 and to HRP 204	58
P(Cl-St-CO-GMA)	50	29.3	2.16	-	$15x10^{-7}$**	1.5*	1-1.5		58

* $\gamma_{10.7}$
**Dg

Poly(3-chlorostyrene-co-glycidyle methacrylate)
[P(Cl-St-CO-GMA)]

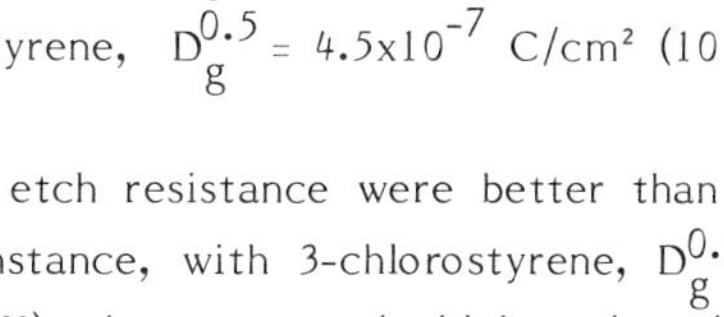

$$-\!\!\left(CH_2-\underset{C_6H_4Cl}{CH}\right)_m\!\!-\!\!\left(CH_2-\underset{\substack{| \\ C=O \\ | \\ O-CH_2-CH-CH_2 \\ \backslash O /}}{\overset{CH_3}{C}}\right)_n$$

Thus, for 26 mole % of phenylmethacrylate, $D_g^{0.5} = 3.2x10^{-7}$ C/cm² (10keV) and $\gamma = 1.1$ (Table X). For 21 mole % of styrene, $D_g^{0.5} = 4.5x10^{-7}$ C/cm² (10 keV) and $\gamma = 0.9$ (Table XI).

The thermal stability and the plasma etch resistance were better than that of COP due to the higher Tg value. For instance, with 3-chlorostyrene, $D_g^{0.5}$ is given between 9 and $15x10^{-7}$ C/cm² (10 keV), the contrast is higher than 1 and the resolution is of the order of 1 µm. The plasma etch resistance is equivalent to the onces of AZ 1350 and HPR 204.

2.2.2.2. Epoxide containing polybutadienes

NONOGAKI (refs 60, 62) and THOMPSON (ref. 61) reported on epoxide-containing polybutadienes. These polymers are synthetized by epoxidation of polybutadiene dissolved in an aromatic solvent, with peracetic acid. They possess reactive groups and crosslink under irradiation at doses as low as $5x10^{-8}$ C/cm² (5 keV), but the contrast is lower than 1 (Table XII).

TABLE XII

Lithographic characteristics of epoxide-containing polybutadienes

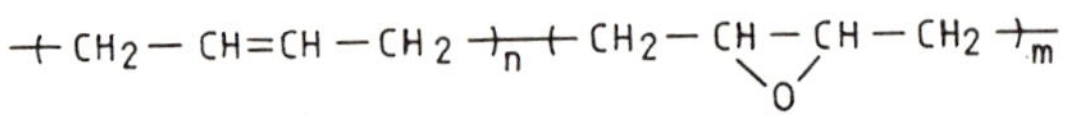

$$-\!\!\left[CH_2-CH=CH-CH_2 \right]_n\!\!-\!\!\left[CH_2-\underset{\diagdown O \diagup}{CH-CH}-CH_2 \right]_m \quad (EPB)$$

Epoxide group in mole %	$\overline{Mw}$ $x10^{-5}$	I	$Dg^{0,5}$ C/cm² 5keV	Contrast γ	Resolution (μm)	References
0	2.04	2.65	$1.5x10^{-7}$	0.93	-	60,61,62
23	1.74	3.45	$5X10^{-8}$	0.61	1.5	-
86	0.95	2.07	$5x10^{-8}$	0.62	1.5	-

2.2.3. Polystyrene and halogenated polystyrene

LAI (ref. 53), IMAMURA (ref. 54), THOMPSON (ref. 55) and ITAYA (ref. 56) have studied polystyrene as a negative resist. The presence of the phenyl ring decreases the sensitivity but the contrast and the plasma etch resistance are excellent (Table XIII).

TABLE XIII

Lithographic characteristics of polystyrene

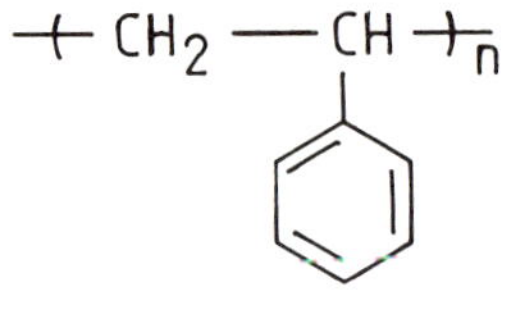

$$-\!\!\left[CH_2-\underset{C_6H_5}{CH} \right]_n$$

$\bar{M}w$ $\times10^{-5}$	I	$Dg^{0.5}$ ($10^{-6}C/cm^2$)	Contrast γ	Resolution (μm)	Etch resistance Å/min	Reference
14.4	1.56	11 (15keV)	1.5	⩽ 1	40*	53,54,55
6.03	1.07	22 (15keV)	2	⩽1	40*	53,54,55
0.821	1.09	80(15keV)	2	⩽1	40*	53,54,55
0.128	1.12	300 (15keV)	2	⩽1	40*	53,54,55
1.0	1.01	60 (20 keV)	2.3	0.25	-	56

*plasma $CF_4/4\%O_2$

In order to increase the sensitivity and maintain their other properties mentioned above, IMAMURA (ref. 54) and CHOONG (ref. 57) studied respectively chloromethylated polystyrene (CMPS) and poly(chloromethylstyrene) (PCMS). The results obtained are given in Table XIV.

TABLE XIV

Lithographic characteristics of halogenated polystyrenes

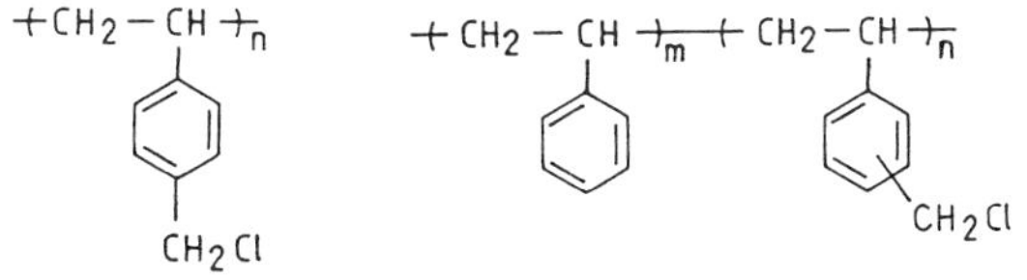

Resist	Cl (in weight %)	$\bar{M}w$ $\times10^{-5}$	I	$Dg^{0.5}$ C/cm^2 ***	Contrast γ	Resolution (μm)	Etch resistance Å/min **	References
CMS	22.0	27	3.0	1.4×10^{-7}	1	0.3	40	54
CMS	21.3	9.2	1.4	2.3×10^{-7}	1.3	0.3	40	54
CMS	16.0	5.6	1.1	4×10^{-7}	1.4	0.3	40	54
CMS	19.6	3	1.3	7.1×10^{-7}	1.5	0.3	40	54
CMS	18.9	1.8	1.1	11.4×10^{-7}	1.7	0,3	40	54
PCMS	-	0.22	1.32	67×10^{-7}	2.1*	0.5-1	-0	57
PCMS	-	0.26	2.08	61×10^{-7}	1.5*	0.5-1	-0	57
PCMS	-	1.12	1.95	15×10^{-7}	1.4*	0.5-1	-0	57
PCMS	-	1.75	1.85	12×10^{-7}	1.3*	0.5-1	-0	57
PCMS	-	3.71	2.10	5×10^{-7}	1.0*	0.5-1	-0	57
PCMS	-	3.81	1.56	3×10^{-7}	1.6*	0.5-1	-0	57
PCMS	-	4.20	1.76	3.5×10^{-7}	1.0*	0.5-1	-0	57
PCMS	-	4.52	2.28	3.5×10^{-7}	0.8*	0.5-1	-0	57

* $\gamma_{0.5}$ **plasma $CF_4/4\%O_2$ ***at 20keV

AZ 1350 10 Å/min

In the case of CMPs, the presence of a C-Cl bond (20 in weight % of chlorine) increases the sensitivity by a factor of 100. As expected for negative resists, the sensitivity increases with increasing molecular weight. The contrast increases with decreasing molecular weight distribution, the best value obtained being of the order of 1.7. The resolution observed is about 0.3 µm and the plasma etch resistance 40 Å/min. More recently CHOONG (ref. 57) thoroughly examined PCMS whose lithographic properties are better compared with CMS.

Indeed, at equal molecular weight, the higher sensitivity is directly linked to the higher percentage of C-Cl bonds. Likewise, the higher contrast of PCMS compared to PGMA and COP is ascribed respectively to a step by step crosslinking mechanism and to a chain reaction (Fig. VIII).

Fig. VIII - Influence of the nature of the crosslinking reaction on the sensitivity and the contrast

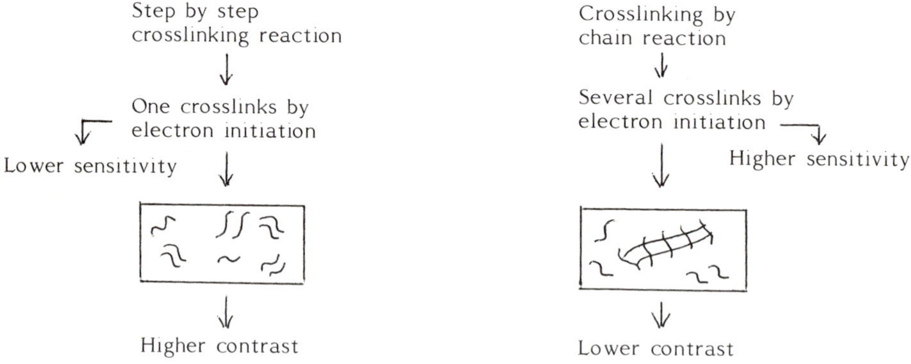

As observed for CMPS, the sensitivity of PCMS increases with increasing molecular weight. The contrast increases with decreasing molecular weight distribution. For the same molecular weight distribution, significantly higher contrast is found for lower molecular weights than for high molecular weights. High molecular weights result in a lower crosslinking density after exposure than for low molecular weights. As a consequence, high-sensitivity PCMS exhibits higher swelling and lower contrast than low-sensitivity.

Resolutions between 0.5 and 1.0 µm were obtained. Also a high dry-etch resistance comparable to AZ photoresists' is observed.

2.2.4. Poly(2-vinyl naphthalene) and derivatives

The results obtained for homo- and copolymers of 2-vinylnaphthalene are given in Table XV.

TABLE XV

Lithographic properties of poly(2-vinylnaphthalene) and derivatives

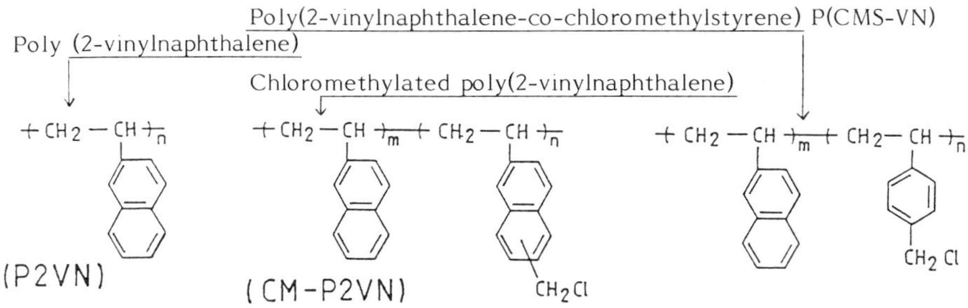

Resist	$\bar{M}w$ $x10^{-5}$	I	$Dg^{0.5}$ (C/cm^2)*	Contrast $\gamma_{0.5}$	Resolution (µm)	Etch resistance (A/min)	Reference
P2VN	0.39	2.3	$3x10^{-4}$	2.1	0.5	150	59
($CM_{10.5}P2VN$)	0.51	3.0	$1.3x10^{-5}$	1.3	0.5	180	59
P(CMS_{21} VN_{79})	0.186	1.6	$5x10^{-5}$	1.7	0.5	185	59

*at 20 keV **RIE (Ar)

If resolution and contrast seem to be acceptable, the sensitivity is relatively low ($D_g^{0.5}$: $3x10^{-4}$ to $1.3x10^{-5}$ C/cm^2 at 20 keV).

2.2.5. Various negative resists

In Table XVI, we have reported the lithographic properties of several negative resists whose structures are given in Fig. IX.

TABLE XVI

Lithographic characteristics of various negative resists.

Resist	$\overline{M}_w$ $x10^{-5}$	I	Dg (C/cm^2)	Contrast γ	Resolution (μm)	Etch resis- (Å/min)	References
P(ETMA-CO-MMA)	-	-	$6x10^{-7}$(20keV)	0.9	-	-	65
CER	-	-	$3x10^{-7}$(10keV*)	-	4	-	66
P DOP	-	-	$4x10^{-7}$(10keV**)	-	2.34	-	67
VBZ	1.69	1.87	$5x10^{-7}$(20keV***)	1.07	1-2	-	68
PVS	1350 ▲	-	$8x10^{-7}$(20keV**)	1	0.5	-	69
EK	0.35 à 0.7	⩽3	1 to $3x10^{-7}$ (20keV)	0.85 to 1.20 •	0.5	50 ▲▲	70
DCPA	5.24	2.00	$1.1x10^{-7}$(10keV*)	1.13 ••	-	350 RIE (Ar)	79
PSTTF	-	-	$5.6x10^{-6}$(20keV*)	3.3	0.1	equivalent to AZ	86

* $Dg^{0.5}$ ▲ $\overline{M}n$ ▲▲ plasma $CF_4/8\%O_2$

** D^1_g • $\gamma_{0.5}$

*** $Dg^{0.6}$: γ_1

Very high sensitivities are observed for VBZ, PVS, EK and DCPA.

Fig. IX : Structures of several negative resists.

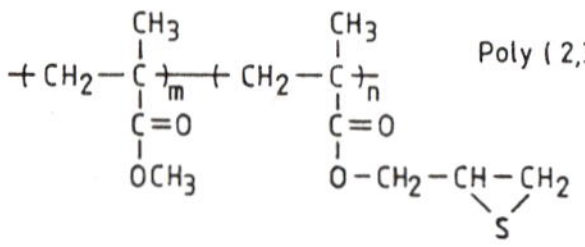

Poly (2,3 epithio propyl methacrylate-co-methyl methacrylate)
P(ETMA-CO-MMA)

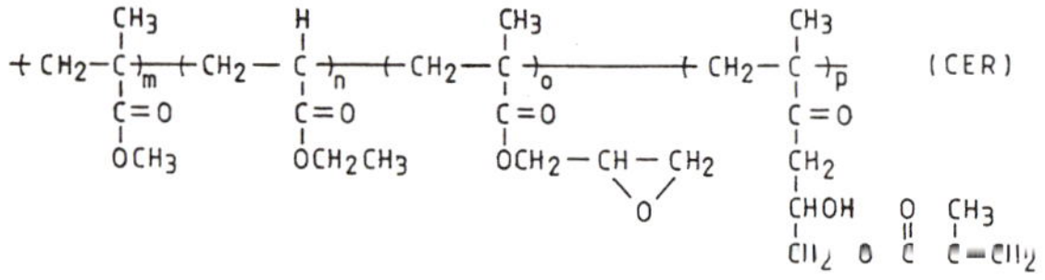

(CER)

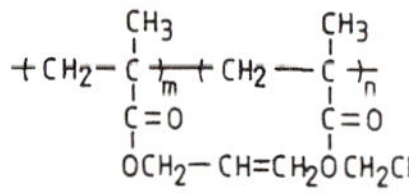

(EK)

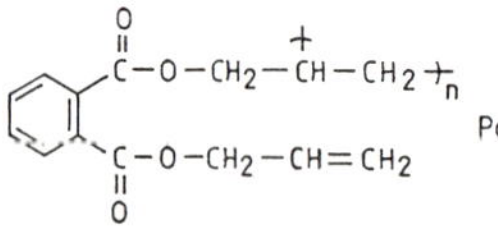

Poly (diallyl orthophtalate)
(PDOP)

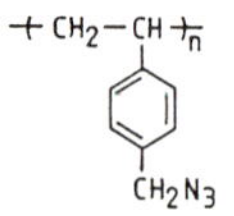

Poly (vinyl azidobenzyle)
(VBZ)

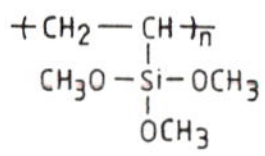

Poly (vinylsiloxane)
(PVS)

In the systems described until now, polymer chains crosslink to form a three-dimensional network upon irradiation. Development is carried out using an organic solvent which not only dissolves the unexposed material but also swells the cross-linked areas to a greater or lesser extent. The undesirable swelling can be suppressed by converting the non-polar crosslinked polymer into a more ionic one. These polymers are soluble in solvents of low polarity before irradiation but insoluble after irradiation. This aim can be attained by mixing a polystyrene (PS) bearing tetrathiafulvalene groups (TTF) tetrabromomethane as shown below.

PSTTF

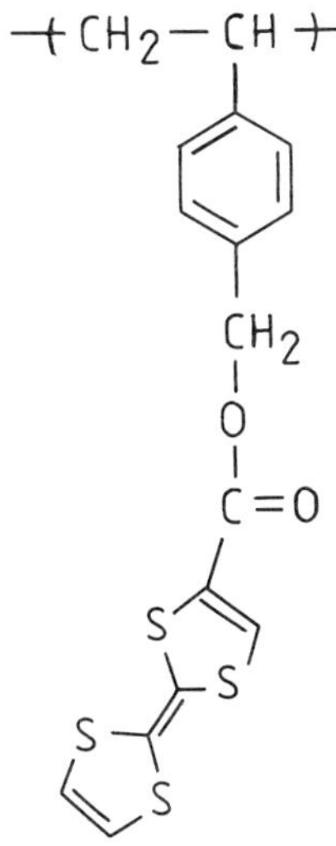

$$-\!\!\left(CH_2 - CH \right)\!\!-$$

$$PSTTF + CBr_4 \xrightarrow{\text{irradiation}} PSTTF^{+}Br^{-} + \cdots$$

Under irradiation, a reaction represented as a redox process, leads to polar salt structures in the polymers.

PSTTF is shown to have a desirable combination of properties, including high sensitivity, high dry-etch resistance, and high resolution. As a consequence, it is a very attractive candidate for high performance electron beam lithography.

3. X-RAY SENSITIVE RESISTS

3.1. GENERAL CONSIDERATIONS

The use of X-rays ($\lambda < 50$ Å) allows a decrease in the diffusion and diffraction phenomena. Indeed these phenomena limit the resolution in photo and electron micro-lithography.

It is generally supposed that organic polymers (those containing carbon, hydrogen, oxygen, and/or nitrogen) have the same sensitivities (G) to X-ray and electron irradiation. A correlation between their sensitivities has been made evident by THOMPSON (ref. 71). This is not surprising since the fundamental mechanism resulting from the absorption of X-rays by organic materials is the ejection of electrons whose energy has values between 1 and 3 keV depending of the wavelength of the X-rays. These electrons induce in the polymer, identical reactions to the ones observed where irradiating with an electron beam. The same chemical intermediates are obtained :

- secondary electrons
- cations, anions
- anion- and cation-radicals
- neutral radicals
- electronically excited species

Under these conditions it seems obvious that the resists used for electron beam irradiation are suitable for X-ray irradiation.

3.2. POSITIVE RESISTS

3.2.1. Introduction

As observed in the case of resists exposed to electron beam irradiation, the sensitivity of a positive resist is a function of a given number of parameters which have been discussed by KU and SCALA (ref. 1). It has been shown that the absorbed dose σ_{abs} required, in order to break a fraction p_s of bonds is given by the following equation :

$$p_s = \frac{\sigma_{abs}\ G(s)\ \ Mo}{100\ \rho\ N} \qquad (6)$$

where

σ_{abs} is the absorbed dose in eV/cm^3

G(s): G(scission), number of main-chain bonds broken per 100 eV of energy absorbed

N : Avogadro's number

ρ : density of the polymer

Mo : molecular weight of a monomer unit

p_s : probability of scission

SPEARS and SMITH (ref. 72) have linked the incident irradiation dose OT_{irrad} to σ_{abs} according to the following relationship :

$$\left[\phi T_{irrad}\right] = \frac{\sigma_{abs}}{(\frac{\mu}{\rho})\rho} \qquad (7)$$

in which

ϕ : incident X-ray flux

T_{irrad} : time of irradiation

ρ : density of the polymer

$\frac{\mu}{\rho}$: the mass absorption coefficient at a given wave length.

By combining equations 6 and 7, p_s can be expressed as :

$$p_s = \frac{\phi T_{irrad}(\frac{\mu}{\rho})\rho\, G(s)\ \bar{M}o}{100\ \rho\ N} \qquad (8)$$

By combining now equations 2, 4 and 8, the solubility ration Sr is defined by the expression

$$S_r = \frac{S'}{S} = \left[\frac{\bar{M}'_n}{\bar{M}^\circ_n}\right]^{-\alpha} = \left[1 + \frac{\left[\phi T_{irrad}\right](\frac{\mu}{\rho})\ G(s)\ \bar{M}^\circ_n}{100\quad N}\right]^{\alpha} \qquad (9)$$

It appears therefore that for constant values of the film thickness and the solubility ration Sr, the required dose σ_{abs} is lower, the higher is $(\frac{\mu}{\rho})G(s)\bar{M}^\circ_n$.

Under these conditions, a positive resist will possess a good sensitivity if the values of $(\frac{\mu}{\rho})$ G(s) and $\bar{M}^\circ_n$ are high.

The absorption of X-rays of incident intensity I_o follows an exponential law (ref. 73) :

$$I = I_o e^{-(\frac{\mu}{\rho})\rho z} \qquad (10)$$

Where I is the intensity after the passage through an absorbing layer of thickness z, $\frac{\mu}{\rho}$ is the mass absorption coefficient and ρ is the density.

For a given element and wavelength :

$$(\frac{\mu}{\rho}) = C\lambda^n \qquad (11)$$

where C and n are constants.

For a polymer consisting of i elements of mass absorption coefficient $(\frac{\mu}{\rho})_i$ and atomic weight A_i the resulting mass absorption coefficient $(\frac{\mu}{\rho})$ is given by equation 12 :

$$(\frac{\mu}{\rho}) = \frac{\sum_i (\frac{\mu}{\rho})_i \, A_i}{\sum A_i} \tag{12}$$

Under these conditions, the atomic composition and the weight percentage of each absorbing element in a polymer determine the mass absorption coefficient $\frac{\mu}{\rho}$.

3.2.2. Polymethylmethacrylate and halogenated derivatives

SPEARS and SMITH (ref. 72) were the first to use PMMA as a resist sensitive to X-ray irradiation. But the drawback of this material is its low sensitivity. It has been pointed out that chlorine-containing polymers appeared to be appropriate for X-ray lithography. Indeed, if halogen replaces hydrogen, the absorption is increased appreciably due to the higher atomic number.

As shown in table XVII, the sensitivities are directly connected to the mass absorption of the fluorine atoms at 8.34 A.

TABLE XVII

Lithographic characteristics of polymethylmethacrylate and halogenated derivatives.

$$-\!\!\left[CH_2 - \underset{\underset{OR}{|}\atop{C=O}}{\overset{CH_3}{\overset{|}{C}}} \right]\!\!-_n$$

Resist	R	(μ/ρ) (cm²/g)	[η]• 30°C (dl/g)	Sensitivity Do (mJ/cm²)(8.34 Å)	Contrast γ	Resolution (μm)	Etch resistance (Å/min)	References
PMMA Elvacite 2041	CH_3	941.5	1.11*	2,180	2.4	0.1	-	74,75
PFEMA	$-CH_2-CH_2F$	1071.5	0.46**	730	3.4	0.3	380***	74
			0.88**	730	4.2	0.3	380***	74
PF_3EMA	CH_2-CF_3	1302.5	0.21**	1255	1.8	-	-	74
			0.31**	730	2.4	-	-	74
			0.45*	730	3	-	-	74

•intrinsic viscosity *Benzene **Acetone ***plasma CCl_4

The gain in sensitivity compared to PMMA (Elvacite 2041) is about three. Moreover, most of the polymers possess a contrast at least comparable to that of PMMA, reaching 4.2 for PFEMA.

The incorporation of fluorine atoms in PMMA has also been studied by KAKUCHI et al. (refs 76, 77). The results obtained are reported in table XVIII.

TABLE XVIII
Lithographic characteristics of fluorinated polymethacrylates.

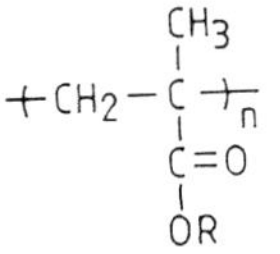

$$\left[CH_2 - \underset{\substack{| \\ C=O \\ | \\ OR}}{\overset{\substack{CH_3 \\ |}}{C}} \right]_n$$

Resist	R	[η]*at 30°c (dl/g)	Tg (°c)	Sensitivity Do (mJ/cm²)(at 5.4 Å)	Contrast γ	Resolution (µm)	References
FBM-1	CF_3-CFH-CF_2-CH_2-	1.0	50	52	4.5	0.3	76,77
FBM-2	CF_3-CHF-CF_2-CH(CH_3)	0.59	64	80	4.5		76,77
FBM-3	CF_3-CFH-CF_2-(C_2H_5)-	0.57	47	46	4.5		76,77
FBM-4	CF_3-CFH-CF_2-CH(C_3H_7)	0.59	45	70	2.5		76,77
FBM-5	CF_3-CFH-CF_2-C(CH_3)$_2$-	0.52	83	53	2.5		76,77
FBM-6	CF_3-CFH-CF_2-C(CH_3)(C_2H_5)	0.35	73	62	2.5		76,77

*methylethylketone

These resists show very high sensitivities. Thus FBM-1 has a sensitivity of 52 mJ/A/cm² (at 5.4 A), a contrast equal to 4.5 and a resolution of about 0.3 µm.

An increase of the absorption can also be obtained by doping PMMA with heavy elements like U, Er, Au, Pd and Pt. Therefore an increase in sensitivity will be observed (ref. 78).

3.2.3. Copolymers of Methylmethacrylate

HATZAKIS et al. (refs 31, 32) have shown that significant improvements in sensitivity of PMMA can be obtained by synthesizing a copolymer P(MMA-co -MAA)- heavy metal salts. Several results are reported in Table XIX.

TABLE XIX

Lithographic characteristics of copolymers of methylmethacrylate and heavy metal salts of a methacrylic acid

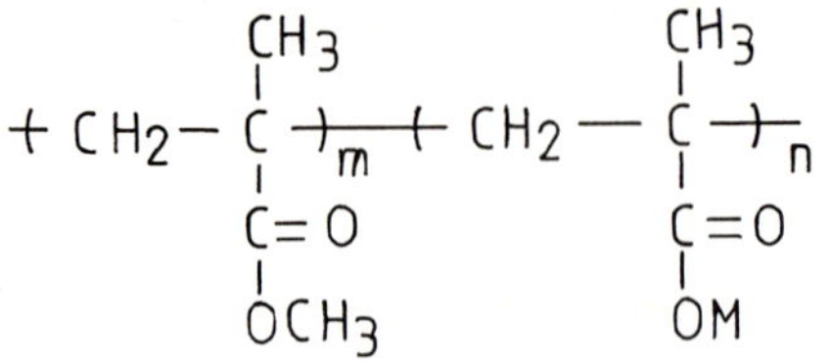

$$\left[CH_2 - \underset{\underset{\underset{OCH_3}{|}}{C=O}}{\overset{CH_3}{\overset{|}{\underset{|}{C}}}} \right]_m \left[CH_2 - \underset{\underset{\underset{OM}{|}}{C=O}}{\overset{CH_3}{\overset{|}{\underset{|}{C}}}} \right]_n$$

Resist	M	$\bar{M}w$ x10^{-5}	I	Sensitivity Do (mJ/cm²)(8,3Å)	Contrast γ	Resolution (μm)	References
P(MMA-Co-MAA)(75/25)	H	-	-	100-200	2	-	31
P(MMA-Co-MAATl)(75/25)	Tl 28 % in weight	-	-	24	2	0.1	31
P(MMA-Co-MAA Pb)	Pb 3 % in weight	1	2	200	2	1	32

The copolymer based on the thallium salt is ten times mores sensitive than the non doped copolymer.

3.2.4. Various positive resists

Some resists, already used in e-beam irradiation, habe been tested under X-ray irradiation.

The main results obtained are listed in Table XX. It is interesting to note that PiBA and PBCyA have sensitivities respectively equal to 25 and 20 mJ/cm². In a more general way, the existing resists have too low a sensitivity to be used in X-ray lithography.

TABLE XX

Lithographic characteristics of various positive resists.

Resist	Structure	$\overline{M}w \times 10^{-5}$	I	(μ/ρ) (cm^2/g)	Sensitivity Do (mJ/cm^2)	Contrast γ	Resolution (μm)	References
Poly(butene-1 sulfone) PBS	$(CH_2-CH(CH_2CH_3)-SO_2)_n$	3	3	553	94(4.37 A)	-	0.5	79,80,75
Poly(methacrylonitrile) PMCN	$(CH_2-C(CH_3)(CN))_n$	13.3	2.6		1800(8.3 A)	-	1	16
Poly(isobutyraldehyde) PIBA	$(CH(CH_2-CH(CH_3)_2)-O)_n$	-	-	-	25(5.41 A)	1	-	81
Poly(dichloro 2,3 acetaldehyde) PDCA	$(CH(CHCl_2)-O)_n$	-	-	-	140(5.41 A)	4.0	-	81
Poly(n-butyl cyanoacrylate) PBCyA	$(CH_2-C(CN)(C(=O)-O-(CH_2)_3-CH_3))$	14.3	2.6	-	20(13.34 A)	2.3	-	15

3.3. NEGATIVE RESISTS

3.3.1. Introduction

When a negative resist is irradiated by an X-ray beam, crosslinking occurs causing the formation of a three-dimensional network, which is insoluble in the solvent for the original linear chain polymer. The minimum absorbed dose $(\sigma_{abs})_i$ required to produce this network is given by the relation of KU and SCALA (ref. 1).

$$(\sigma_{abs})_i = \frac{50\,\rho N}{G(x)\,\overline{M}_w} \qquad (13)$$

where

ρ : density of the polymer

$G(x)$: number of crosslinks produced per 100eV of absorbed energy

N : Avogadro's number

$\overline{M}_w$: average molecular weight of the polymer

SPEARS and SMITH (ref. 72) having linked the incident irradiation dose ϕT_{irrad} to σ_{abs} (Equation 7), the minimum incident dose ϕT_{irrad} or D_g^i is given by the relation :

$$D_g^i = \frac{50\,N}{G(x)\,\overline{M}_w\,(\frac{\mu}{\rho})} \qquad (14)$$

Thus, a resist having a high molecular weight and G(x) will need a lower incident dose and therefore will be more sensitive. Moereover, as for e-beam resists, high contrast demands a narrow molecular weight distribution.

3.3.2. Acrylic homo- and copolymers

3.3.2.1. Homopolymers

As has been pointed out by TAYLOR et al. (ref. 79), chlorine containing polymers appear to be suitable for X-ray lithography. Indeed, if halogen replaces hydrogen the absorption is increased appreciably, due to the higher atomic number. Structures and sensitivity values of various chloroalkyl acrylates are presented in Table XXI. But according to us, it seems that the sensitivity of these resists is not really improved by an increase of the mass absorption coefficient.

TABLE XXI

Lithographic characteristics of chlorinated polyacrylates

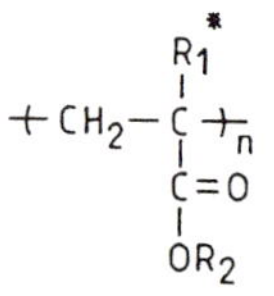

$$-\!\!\left[CH_2 - \underset{\underset{\underset{OR_2}{|}}{C=O}}{\overset{\overset{R_1^{*}}{|}}{C}} \right]\!\!-_n$$

Resist	R_2	(μ/ρ) cm^2/g	$\bar{M}w$ $x10^{-5}$	I	Tg (°c)	G(x)	Sensitivity $D_g^{0.5}$ (mJ/cm^2)**	Contrast γ	Resolution (μm)	Etch resistance (Å/min)
Poly-2 chloro ethyl acrylate	$-CH_2-CH_2Cl$	629	4.48	2.79	-1	2.6	27	0.75	-	-
Poly(-1,3 dichloro propyl-2 acrylate)	$-CH-(CH_2Cl)_2$	859	8.19	2.43	20	0.27	39	1.13	-	-
Poly(2,3 dichloro propyl-1 acrylate) (DCPA)	$-CH_2-CHCl-CH_2Cl$	859	5.24	2.00	6	1.8	13	1.12	1-2	308 ionic Ar etching
			12.4	2.20	-	-	5.3	0.73	—	—
Poly(trichloro neo-pentyle acrylate)	$-CH_2-C(CH_2Cl)_3$	941	3.37	2.26	75	-	260	2	—	—
Poly(2,2,2 trichloro ethyl acrylate)	$-CH_2\ C\ Cl_3$	1117	8.96	2.09	50	-	44	1.10	—	—
Poly(glycidile metacrylate) (GMA)	$-CH_2-CH-CH_2$ (epoxide, O bridging CH and CH_2)	136	0.65	2.18	-	-	67	—	—	—

*R_1=H except for (GMA) R_1=CH_3 **at 4.37 A

Indeed, the most sensitive resists studied (e.g. poly-2-chloroethyl acrylate and DCPA) are less absorbant. In contrast, we observe an increase of sensitivity with decreasing glass transition temperature. It seems that a lower T_g allows a higher mobility of the polymeric radicals and therefore favors the interchain coupling reactions.

Moreover, the sensitivity seems to be linked to the nature of the C-Cl bond in the polymer. This seems to be obvious for the two isomers of poly(dichloro-propyl- acrylate). The 1,3 isomer ($-CH-(CH_2Cl)_2$) possesses two primary C-Cl bonds and the 2,3 isomer ($-CH_2-CHCl-CH_2Cl$) shows a primary bond and a secondary one which is weaker. This secondary bond is responsible for the significant increase of the sensitivity.

Moreover, the contrast decreases with increasing weight average molecular weight. TAYLOR et al. (ref. 74) presume that an increase in molecular weight favors intramolecular crosslinking reactions over the intermolecular ones which reduce the solubility of the polymer.

3.3.2.2. Copolymers

TABLE XXII

Lithographic characteristics of acrylic copolymers

$$-\!\!\left[CH_2-\underset{\underset{OR_2}{|}}{\underset{C=O}{|}}\overset{R_1}{\overset{|}{C}}\right]_m\!\!-\!\!\left[CH_2-\underset{\underset{OR_4}{|}}{\underset{C=O}{|}}\overset{R_3}{\overset{|}{C}}\right]_n$$

Resist	epoxy group (% in mole)	$\overline{M}w$ $x10^{-5}$	I	Sensitivity $Dg^{0.5}$ (mJ/cm²)	contrast γ	resolution (μm)	etch resistance (Å/min)	references
GA-Co-DCPA	0	5.24	2.00	13	1.12	1	356*	79
	22	4.87	2.29	9.0	0.87	-	308*	79
	40.7	-	-	6.6	0.67	-	-	79
GMA-Co-DCPA	15.2	8.13	1.81	7.7	0.83	-	-	82
	32	4.66	1.96	23.0	0.97	-	-	82
	65.5	3.9	1.87	39.6	1.01			82
GMA-Co-EA (COP)	68	2.2	2.1	18.5	1.00	-	356*	82,79
EK	-	0.35to0.7	⩽3	9	1.3	<1	30**	70
DCOPA (DCPA+COP 7% in weight)	-	-	-	10,2	1.23	1	-	82,83

*ionic etching Ar **plasma $CF_4/8\%O_2$

Resist	R_1	R_2	R_3	R_4
GA-Co-DCPA	H	$-CH_2-CH-CH_2$ \/ O	H	$-CH_2CHCL$ | CH_2CL
GMA-Co-DPCA	CH_3	$-CH_2-CH-CH_2$ \/ O	H	$-CH_2-CHCl$ | CH_2Cl
GMA-Co-EA	CH_3	$-CH_2-CH-CH_2$ \\| O	H	$-CH_2-CH_3$
EK	CH_3	$CH_2-CH = CH_2$	$-CH_3$	$-CH_2-CH_2OH$

In order to improve the adhesion and the sensitivity of DCPA, TAYLOR et al. (refs 79, 82) copolymerized it with glycidyl acrylate and glycidyl methacrylate respectively (Table XXII). The introduction of glycidyl groups enhances the sensitivity of the resist but decreases its contrast. This behavior, according to TAYLOR (ref. 82), can be ascribed to the presence of HCl produced during the irradiation of the resist which allows ring opening polymerization of the epoxy groups and therefore the subsequent crosslinking of the resist.

Mixtures of DCPA and COP have been tested in order to increase the adherence of DCPA (ref. 82). Thus, a mixture containing 7% of COP (DCOPA) and irradiated under nitrogen has a lower sensitivity, but a contrast, resolution and wet etch resistance which are higher.

More recently, ZAN et al. (ref. 70) described a copolymer of allyl methacrylate and hydroxy-2-ethyl methacrylate (EK). It possesses a rather high sensitivity (9 mJ/cm² at 7 Å) and a contrast higher than 1. Its plasma etch resistance is good (30 Å/min) and the resolution is lower than 1 micron.

3.3.3. Various negative resists

Other negative resists have been listed in Table XXIII. All of them present a high sensitivity but their contrast is rather low ($\gamma < 1$) except for PSTTF for which γ equals 2.3 and the resolution 0.1 µm.

TABLE XXIII

Lithographic characteristics of various negative resists.

Resist	Structure	$\bar{M}_w$ $x10^{-5}$	I	(μ/ρ) (cm^2/g)	Sensitivity Dg (mJ/cm^2)	Contrast γ	Resolution (μm)	References
Poly(chloroprene)	$-(CH_2-CCl=CH-CH_2)_n-$	4.71	6.1	861	11*	0.58	-	79
Poly(epichlorhydrine)	$-(CH_2-CH-O)_n-$ with CH_2Cl side group	13.0	-	811	21*	<0.5	-	79
P(AM-Co-EVE)	AM:° EVE: $(CH_2-CH)_n$ with OCH_2-CH_3 side group	-	-	-	2.4**	0.47	-	84
P(TFE-Co-nBVE)	TFE: $-(CF_2-CF_2)_n-$ nBVE: $-(CH_2-CH)_n-$ with $O(CH_2)_3CH_3$ side group	-	-	-	11**	1.0	-	85
CER		-	-	-	25***	0.78	-	80
PDOP		-	-	-	14***	0.78	-	80
EPB		-	-	-	1.5***	0.56	-	80
PSTTF		-	-	-	44***	2.3	0.1	86

* $Dg^{0.5}$ at 4.37 Å ** Dg* at 5.41 Å ***$Dg^{0.5}$ at 8.34 Å ° $-(CH-CH)_n-$ with side groups $CO_2-CH_2-CH=CH_2$ on each carbon

3.4. PLASMA DEVELOPED RESISTS (93)

Quite recently, intense investigations directed at the improvement of the resolution of negative resists were devoted to the elaboration of methods to avoid the use of liquid development procedures which inevitably cause distortions of structures due to swelling and contraction of the polymers. An interesting method was proposed by TAYLOR and WOLF (refs 87- 89) which is illustrated in Fig. X.

Plasma resistance in organic polymers is enhanced by the presence of aromatic and heteroatomic groups. TAYLOR et al. have utilised this result in a very elegant resist technique referred to as "photolocking". In the original work, the resists consisted of a plasma degradable acrylic polymer intimately mixed with a volatile, but readily polymerized, solid monomer such as N-vinyl-carbazole. Highest sensitivity was achieved using poly(2,3-dichloropropyl acrylate) as the host polymer, and the optical-wave length sensitivity was adjusted by incorporation of sensitisers. As originally conceived, and recently extended, the photolocking technique may be used for optical, e-beam, and X-ray resists. It is particularly applicable to X-ray lithography, and, in a significant development, TAYLOR et

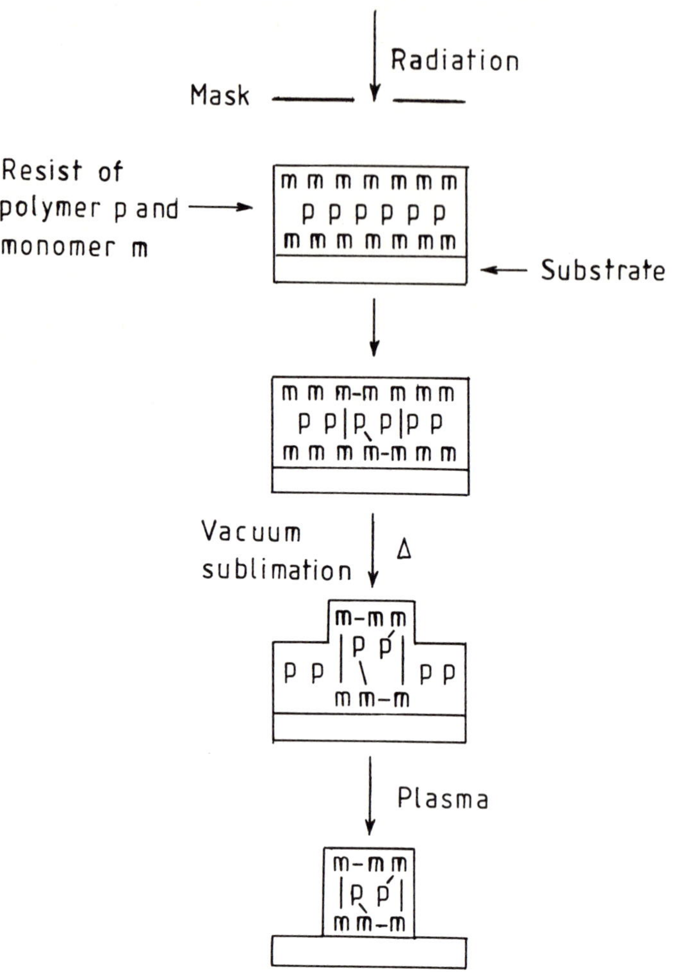

Fig. X - Schematic representation of the plasma development process involving irradiation with X-rays, fixation and plasma etching.

al. have shown that much improvement may be obtained by replacement of the rather toxic N-vinyl by a number of polymerizable silicon compounds, of which bi-acryloxy and methacryloxy derivations are best, e.g.

$$CH_2 = CHCOO(CH_2)_n - \underset{CH_3}{\overset{CH_3}{\underset{|}{\overset{|}{Si}}}} - O - \underset{CH_3}{\overset{CH_3}{\underset{|}{\overset{|}{Si}}}} -(CH_2)_n\ O\ CO\ H = CH_2$$

Values of n may be varied to improve compatibility with the host polymer. The polymer/monomer system is irradiated with X-rays through a mask. In the irradiated regions, the monomer guest forms a homo-polymer and is grafted to some extent to the polymer host. Unreacted monomer is removed by heat treatment under vacuum. Subsequently, the polymer in the non-irradiated regions is removed with the aid of an O_2-plasma which attacks the reacted regions only slightly. A sensitivity of about 1.5 mJ/cm^2 at 4.37 Å is reached and the resolution is lower than 0.5 μm.

4. PHOTOSENSITIVE RESISTS

4.1. INTRODUCTION

Optical lithographic methods are most important for the manufacture of integrated circuits on a world-wide basis. The photoresists used are mainly sensitive in the wavelength range from 300 to about 500 nm.

4.2. NEGATIVE-WORKING PHOTORESIST SYSTEMS

4.2.1. Photo-crosslinking systems

4.2.1.1. Diazide-Polydiene systems (refs 96-99)

Negative optical resists are derived mainly from light-induced crosslinking of cyclized (cis-1,4-polyisoprene with bis-azido compounds. Natural rubber and synthetic cis-1,4-polyisoprene are known to undergo a variety of internal cyclisations when heated with Friedel-Crafts catalysts. This produces a much tougher polymer with good adhesion and film-forming properties, and an abundance of sites for reactions with nitrenes.

The partially cyclized polyisoprene can be depicted as follow :

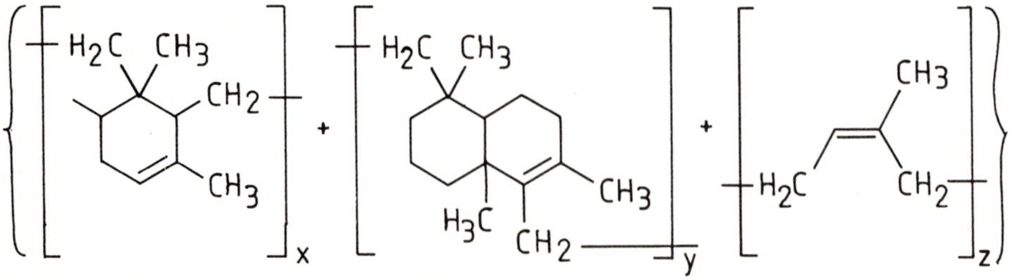

The photosensitive bifunctional crosslinking agent is a diazide of the following structure :

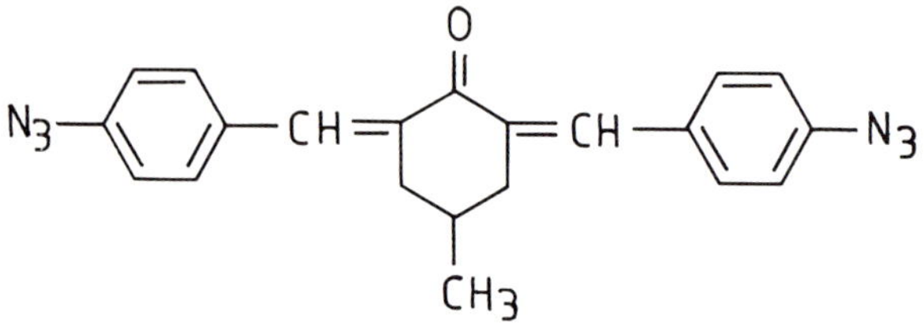

Upon irradiation, nitrogen cleaves off and the azide groups form very reactive nitrene (R-N).

$$N_3\text{-R-}N_3 \xrightarrow{h\nu} N_3\text{-R-}\dot{N}\bullet + N_2 \xrightarrow{h\nu} \bullet\dot{N}\text{-R-}\dot{N}\bullet + N_2$$

Each nitrene undergoes the following reactions :

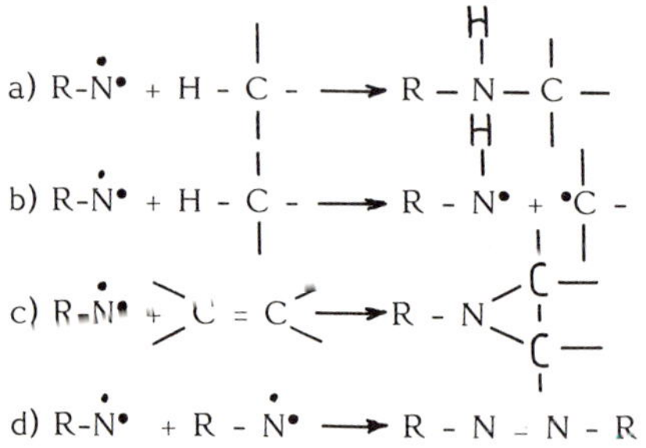

a) $\text{R-}\dot{N}\bullet + \text{H} - \overset{|}{\underset{|}{C}} - \longrightarrow \text{R} - \overset{H}{N} - \overset{|}{\underset{|}{C}} -$

b) $\text{R-}\dot{N}\bullet + \text{H} - \overset{|}{\underset{|}{C}} - \longrightarrow \text{R} - \overset{H}{N}\bullet + \bullet\overset{|}{\underset{|}{C}} -$

c) $\text{R-}\dot{N}\bullet + {>}C = C{<} \longrightarrow \text{R} - N{<}\begin{matrix} C- \\ | \\ C- \end{matrix}$

d) $\text{R-}\dot{N}\bullet + \text{R} - \dot{N}\bullet \longrightarrow \text{R} - N = N - \text{R}$

e) $\text{R} - \dot{N}\bullet + O_2 \longrightarrow \text{R} - N = O$ + other products

Nitrenes undergo insertion reactions with aliphatic carbon-hydrogen bonds to produce secondary amines and undergo addition reactions with carbon-carbon double bonds to yield aziridines. It is apparent, therefore that bis-azido compounds, when photolysed in a polymer matrix, readily produce crosslinks.

These negative photoresists are commercialized under the trade name KTFR (KODAK) and OMR-83 (TOKYO OHKA KOGYO C.). The sensitivity of the KTFR resist in the wavelength range from 300 to 480 nm is equal to 20 mJ/cm^2.

4.2.1.2. Photodimerization systems

A large number of photoresists use photodimerization for the crosslinking reaction. These include the classical cinnamic acid derivatives. The particularly extensively investigated crosslinking reaction of cinnamic acid derivatives proceeds essentially as a photoinduced cycloaddition, and leads to formation of cyclobutanes (refs 100, 101).

The systems are of industrial interest as photoresists when a polymer chain carries a large number of photodimerizable groups. For polyvinyl cinnamate, which is used industrially (trade name KPR by KODAK), the photodimerization is depicted as follow :

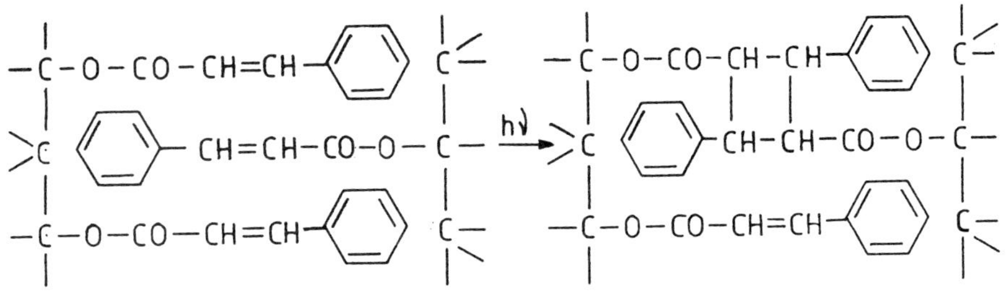

$$-C-O-CO-CH{=}CH-C_6H_5 \qquad C_6H_5-CH{=}CH-CO-O-C- \qquad -C-O-CO-CH{=}CH-C_6H_5 \xrightarrow{h\nu} -C-O-CO-CH-CH-C_6H_5 \;/\; C_6H_5-CH-CH-CO-O-C- \;/\; -C-O-CO-CH{=}CH-C_6H_5$$

The sensitivity attained is about 16 mJ/cm^2.

Recently, a type of photodimerizable epoxide resin, has also been used as a photoresist (ref. 102).

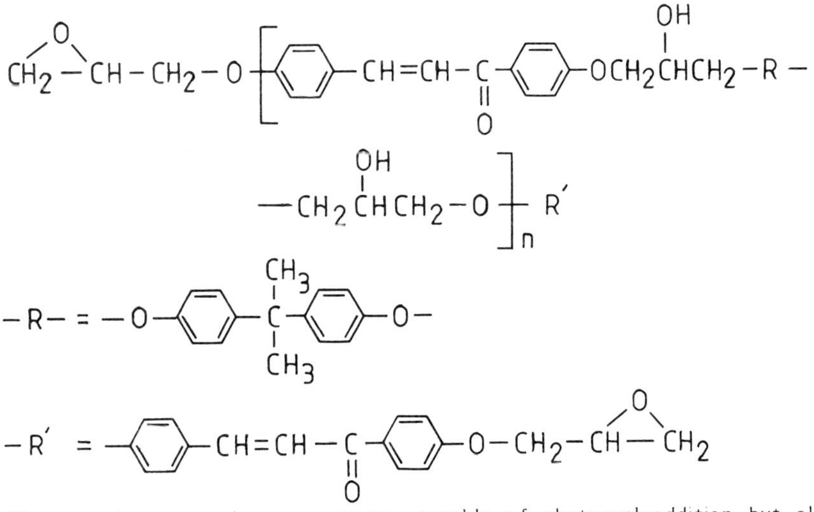

$$\overset{O}{CH_2-CH}-CH_2-O\left[-C_6H_4-CH{=}CH-\underset{\|}{\overset{}{C}}(=O)-C_6H_4-OCH_2\overset{OH}{CH}CH_2-R-CH_2\overset{OH}{CH}CH_2-O-\right]_n R'$$

$$-R- = -O-C_6H_4-C(CH_3)_2-C_6H_4-O-$$

$$-R' = -C_6H_4-CH{=}CH-C(=O)-C_6H_4-O-CH_2-\overset{O}{CH-CH_2}$$

These resins not only carry groups capable of photocycloaddition but also epoxy groups, which permit suitably catalyzed thermal crosslinking.

4.2.2. Photoinitiated systems

A very large number of photoinitiators or photoinitiator systems are known, which upon irradiation produce polymerization-initiating free radicals. Thus, TAYLOR et al. (ref. 103) used a resist consisting of a plasma degradable acrylic polymer intimately mixed with a volatile, but readily polymerized, solid monomer such as N-vinylcarbazole or acenapthylene. Highest sensitivity was achieved using poly(2,3-dichloropropyl acrylate) as the host polymer, and the optical-wavelength sensitivity was adjusted by incorporation of sensitizers, such as phenanthraquinone, which absorb in the 290-350 nm range. It is assumed that photolysis of N-vinylcarbazole/sensitizer pair leads to cationic and free-radical polymerization of the monomer with the resulting polymeric product "locked" into the original host-polymer matrix.

4.3. POSITIVE-WORKING PHOTORESIST SYSTEMS

4.3.1. Diazonaphthalenone systems

The basic polymeric components of optical positive resists are novolak resin precursors. These are relatively low molecular weight condensation products derived from phenols and formaldehyde. A simplified structure may be written as:

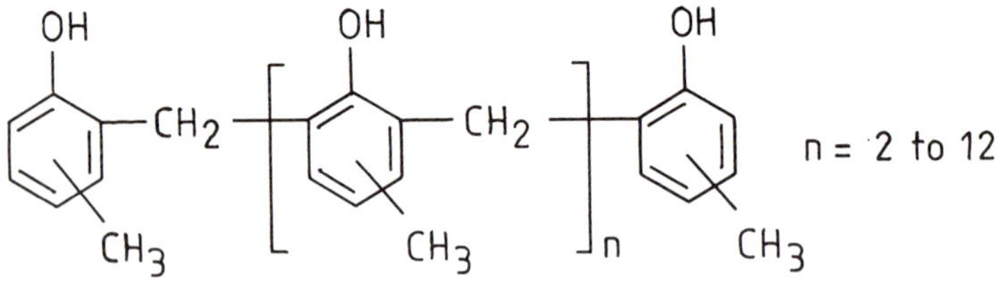

OH, CH_2, OH, CH_2, OH, CH_3, CH_3, CH_3, n

n = 2 to 12

and it must be remembered that such polyphenols are soluble in alkaline solutions. For resists applications, the alkali solubility of the novolak is inhibited by the presence of a so-called "photo-active compound" which is invariably a derivative of a 1,2-naphtoquinone diazide (refs 90-92, 104-106).

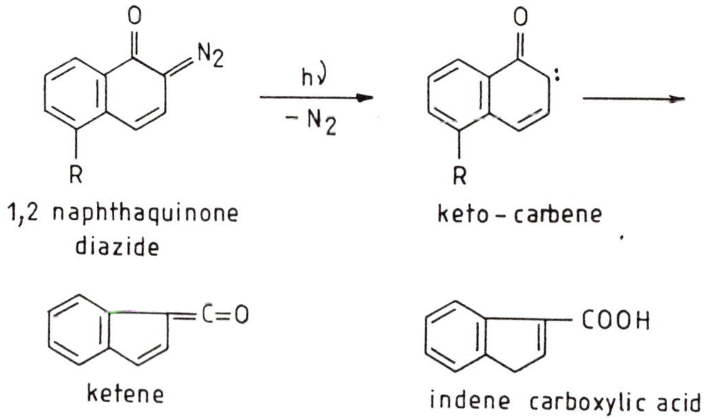

O, N_2, R — $h\nu$, $-N_2$ → O, R →

1,2 naphthaquinone diazide — keto-carbene

C=O — COOH

ketene — indene carboxylic acid

Where exposure produces carboxylic acid, the area readily dissolves in solutions containing metallic or quaternary ammonium hydroxides. In the absence of photo-decomposition to produce alkali-soluble carboxylic acids, the novolak is inhibited from dissolving in alkali by the hydrophobic character of the quinone diazide. The possibility of aqueous-alkaline development of photoresists of this type is one of the essential reasons for using this system to produce very fine structures on semiconductors surfaces and mask substrates. The hydrophobic nature of the unexposed resist elements prevents them from swelling during development.

4.3.2. O-Nitro-benzyl systems

The primary mechanism by which 9-nitrobenzyl cholate/P (MMA-MAA) works as a resist involves a photochemical reaction of the cholate ester solubility inhibitor. Upon irradiation, the nitrobenzyl ester undergoes a rearrangement to generate O-nitro-benzaldehyde plus a carboxylic acid. In this manner, an alkali-insoluble ester is converted into a free acid which is choosen so as to be readily soluble in alkaline developers (refs 107-109).

With R =

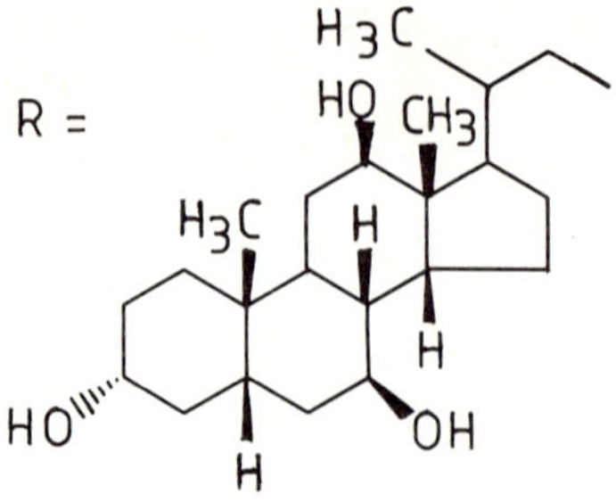

2-Nitrobenzyl cholate poly(methyl methacrylate-co-methacrylic acid) is a high contrast ($\gamma > 5$), high resolution photoresist which has a sensitivity of 150-200 mJ/cm² at 260±20 nm.

4.3.3. Photolysis of salts

WILLSON and his collaborators at IBM have demonstrated the feasibility of positive optical lithography via light-induced acid-catalysed hydrolysis (refs 110-112). It is well known that aryl iodonium and sulphonium salts decompose on photolysis to yield a mixture of products including the appropriate protonic acid e.g.

$$Ar_3S^+ \; AsF_6^- \xrightarrow[RH]{h\nu} Ar_2S + Ar^\bullet + R^\bullet + H^+AsF_6^-$$

Onimum salt of these types have maximum absorbance in the 250 nm region and hence offer ideal sensitisers for deep UV lithography, but may also be co-sensitised for photolysis at longer wavelenghts.

The polymer most useful for this type of lithography is the tert-butoxy-carbonyl ester of poly(4-vinylphenol) (ref. 113). Tert-butoxycarbonylesters of phenols are highly reactive towards acids, decomposing to give the appropriate phenol, carbon dioxide and gaseous isobutene:

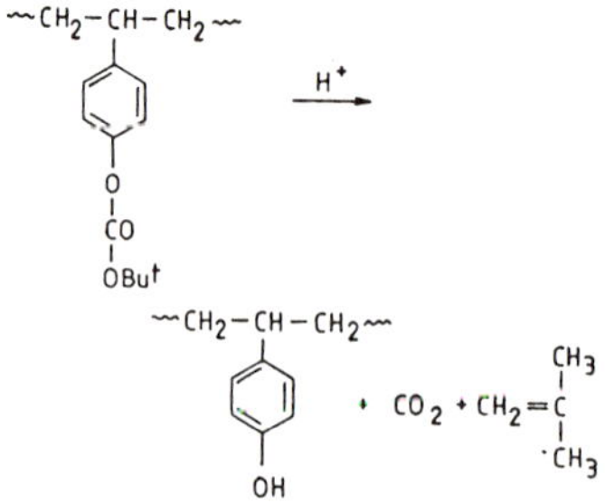

~CH2–CH–CH2~ (O–CO–OBu^t) $\xrightarrow{H^+}$ ~CH2–CH–CH2~ (OH) + CO_2 + $CH_2=C(CH_3)_2$

In this way, a neutral styrene-based polymer is converted into a poly(4-vinylphenol), which may, in principle, be extracted with alkali to give a positive resist. Such a system has a sensitivity of about 15 to 25 mJ/cm^2 (at 254 nm).

5. CONCLUSION

Chemists today are facing growing demands for supplying new compounds to the electronics industry. Perhaps the greatest opportunities lie in inventing resists for photolithographic techniques of ever higher resolution.

As makers of integrated circuits work to fit increasing numbers of components onto silicon chips, the sizes of features on a chip keep getting smaller. Indeed, some observers say that resolution of such minute features almost exceeds capabilities of present photolithographic methods.

The new lithographic processes that integrated-circuit makers are looking at include those based on deep-ultra-violet (less than 300 nm), electron beam, and x-ray irradiation. Until now, the near-UV region has been satisfactory, such as at the 436-, 404 and 366-nm lines of mercury sources.

For the electronics industry, resists are resins that can be applied to a substrate in a pattern of exposed and covered areas to leave some areas open for later etching or plating while protecting others. For many types of printed circuit boards, whose lines and spaces are 40 nm or more wide, screening is sufficient to give needed resolution. But integrated circuit chips need the finer resolution afforded by photolithography, which uses light-sensitive resists for imaging.

Such photoresists may be either positive or negative. Negative photoresists include resins such as cyclized polybutadiene or polyisoprene, mixed with a bis(azide), which become crosslinked and insolubilized by exposure to ultraviolet light. UV irradiation photolyzes bis(azides) to bis(nitrenes), which form aziridine groups with double bonds of neighboring chains. Developing exposed negative photoresists in a solvent dissolves away resin from unexposed areas.

Posivite resists are formulations like phenolic resins mixed with relatively large amounts of light-sensitive solubility inhibitors, such as diazonaphthoquinones. On exposure to light, the solubility inhibitors rearrange to acidic compounds that can be leached out by bases. Thus, exposed areas of positive resists are the ones dissolved away on development. Alternatively, positive resists can be resins that are depolymerized by irradiation.

In general, the electronics industry is moving toward increased use of positive resists at the expense of negative ones. Also, for finer resolution, the trend is toward resists that can be developed by plasma etching or other dry processing techniques in preference to solvent development. To date, negative resists have had advantages of good chemical etch resistance, light sensitivity, adherence to substrates, and latitude of processing conditions, in comparison with positive types. Positive resists are winning out, however, because they do not suffer in resolution from polymer swelling during development, can be imaged by projection rather than contact printing, and are less sensitive to oxygen.

REFERENCES

1 H.Y. Ku and L.C. Scala, J. Electrochem. Soc. 116 (1969) 980.
2 A. Charlesby, Atomic Radiation and Polymers. Pergamon Press, London (1960).
3 K. Ueberreiter, The solution process, in Diffusion in Polymers, Crank. J. and Park, G.S., Eds Academic Press, New York (1968).
4 K. Ueberreiter and F. Asmussen, Makromol. Chem. 46 (1961) 327.
5 M.J. Bowden, J. Polym. Sci. Polym. Symp. 49 (1974) 221.
6 I. Haller, M. Hatzakis and R. Srinivasan, I.B.M. J. Res. Devel. 12 (1968) 251.
7 J.C. Dubois, M. Gazard and C. Duchesne, Colloque international sur la Microlithographie (Paris) (1977) 289.
8 E. Gipstein, A.C. Ouano, D.E. Johnson and O.U. Need, I.B.M. J. Res. Devel. (1977) 143.
9 T. Aoki, T. Takeuchi and K. Sekikawa, Proc of Spring Meeting of Jap. Soc. of Precision Eng. (1972) 143.
10 J.H. Lai and J.H. Helbert, Macromolecules, 11 (1978) 617.
11 J.N. Helbert, C. Pittman Jr and J.H. Lai, Proceedings of Regional Technical Conference, "Photopolymers, Principles Processes and Materials". SPE Ellenville (N.Y.) (1979) 137.
12 T.Tada, J. Electrochem. Soc. 126 (1979) 1635.
13 T. Tada, J. Electrochem. Soc. 126 (1979) 1829.
14 M. Kakuchi, S. Sugawara, K. Murase, K. Matsuyama, J. Electrochem. Soc. 124, (1977) 1648.
15 A. Eranian, E. Datamanti, J.C. Dubois, B. Serre, M.J.M. Abadie, F. Schué, C. Montginoul and L. Giral, Int. Conf. on Microlithography, Lausanne (1981) p. 404.
16 J.N. Helbert, C.F. Cook Jr., C.Y. Chen and C.U. Pittman Jr., J. Electrochem. Soc., 126 (1979) 694.
17 J.N. Helbert, C.Y. Chen, C.U. Pittman Jr., G.L. Hagnauer, Macromolecules, 11 (1978) 1104.
18 J.H. Lai, L.T. Shepherd, R. Ulmer and C. Griep, Proceedings of the fourth photopolymer Conf. Soc. Plast. Eng. Ellenville (N.Y.) (1976) 259.
19 J.N. Helbert, P.J. Caplan and E.H. Poindexter, J. Appl. Polymer Sci., 21 (1977) 797.
20 J.H. Lai, J.N. Helbert, C.F. Cook Jr., and C.U. Pittman Jr., J. Vac. Sci. Technol., 16 (1979) 1992.
21 T.Tada, J. Electrochem. Soc., 128 (1981) 1791.
22 C.U. Pittman Jr., M. Ueda, C.Y. Chen, J.H. Kwiatkowski, C.F. Cook Jr., and J.N. Helbert, J. Electrochem. Soc., 128 (1981) 1758.
23 M. Hatzakis, Appl. Polym. Symp., 23 (1974) 73.
24 F. Biguez, J. Hartmann, P. Parrens, E. Tabouret and B. Serre, Note technique LETI Grenoble (1980).
25 A. Eranian, E. Datamanti, J.C. Dubois, B. Serre, F. Schué, C. Montginoul and L. Giral, Thomson CSF Orsay, Communication privée (1982).
26 S. Matsuda, S. Tsuchiya, M. Honma, K. Hasegawa, G. Nagamatsu and T. Asano, Proceedings 4th, Photopolym. Conf. Soc. Plast. Eng. Ellenville (N.Y.) (1976) 284.
27 E.D. Roberts, Applied Polymer Symposium, 87 (1974).
28 J.H. Lai and S. Shrawagi, J. of Applied Polym. Sci., 22 (1978) 53.
29 E. Gipstein, W. Moreau and O. Need, J. Electrochem. Soc., 123 (1976) 1105.
30 M. Kakuchi, Jpn Kokai Tokyo Koho, 80 (1980) 58.243.

31 I. Haller, R. Feder, M. Hatzakis and E. Spiller, J. Electronchem. Soc., 126 (1979) 154.
32 D.J. Webb and M. Hatzakis, J. Vacuum Science and Technology, 16 (1979) 2008.
33 C.U. Pittman Jr., C.Y. Chen and J.H. Helbert, A.C.S. Div. Polym. Chem. Preprints 19 (1978) 565.
34 Y.H. Atano, H. Morishita and S. Nonogaki, A.C.S. Div. Org. Coatings and Plast. Chem. Preprints, 35 (1975) 58.
35 L.E. Stillwagon, E.M. Doerries, L.F. Thompson and M.J. Bowden, A.C.S. Div. Org. Coatings and Plast. Chem. Preprints, 37 (1977) 38.
36 L.E. Stillwagon, E.M. Doerries, L.F. Thompson and M.J. Bowden, 8th Int. Conf. On Electron and Ion Beam Sci. and Technol. Electrochem. Soc. Meeting, May 21-26, Seattle (1978).
37 E.D. Roberts, ACS Div. Org. Coatings and Plast. Chem. Preprints, 33 (1973) 359.
38 E.D. ROberts, ACS Div. Org. Coatings and Plast. Chem. Preprints, 35 (1975) 281.
39 E.D. Roberts, ACS Div. Org. Coatings and Plast. Chem. Preprints, 37 (1977) 36.
40 S. Homma, S. Matsuda, S. Tsuchiya, K. Hasegawa, G. Nagamatsu and T. Asano, 1st Photopolymer Conf. (Tokyo, September, 1976).
41 S. Matsuda, S. Tsuchiya, M. Honma, K. Hasegawa, G. Nagamatsu and T. Asano, Colloque International sur la Microlithographie (Paris), 277 (1977).
42 J.R. Brown and J.H. O'Donnell, Macromolecules 5 (1972) 109.
43 M.J. Bowden and L.F. Thompson. J. Appl. Polym. Sci., 17 (1973) 3211.
44 L.F. Thompson and M.J. Bowden, J. Electrochem. Soc., 120 (1973) 1722.
45 M.J. Bowden and L.F. Thompson, Applied Polymer Symposium, 23 (1974) 99.
46 M.J. Bowden, L.F. Thompson and J.P. Ballantyne, J. Vac. Sci. Technol., 12 (1975) 1294.
47 M.J. Bowden and L.F. Thompson, J. Electrochem. Soc., 121 (1974) 1620.
48 M.J. Bowden and L.F. Thompson, Polym. Eng. Sci., 17 (1977) 269.
49 M.J. Bowden and L.F. Thompson, Polym. Eng. Sci., 14 (1974) 525.
50 M.J. Himics, N. Desai, M. Kaplan and E.S. Poliniak, ACS Div. Org. Coatings and Plast. Chem. Preprints, 35 (1975) 266.
51 R.J. Himics, N. Desai, M. Kaplan and E.S. Poliniak, ACS, Div. Org. Coatings and Plast. Preprints, 35 (1975) 273.
52 M. Suzuki and J. Ohnishi, J. Electrochem. Soc., 129 (1982) 402.
53 J.H. Lai and L.T. Sheperd, J. Electrochem. Soc., 126 (1979) 696.
54 S. Imamura, J. Electrochem. Soc., 126 (1979) 1628.
55 L.F. Thompson, L.E. Stillwagon and E.M. Doerris, J. Vac. Sci. Technol., 15 (1978) 938.
56 K. Itaya, K. Shibayama and T. Fujimoto, J. Electrochem. Soc., 129 (1982) 663.
57 H.S. Choong et F.J. Kahn, J. Vac. Sci. Technol., 19 (1981) 1121.
58 L.F. Thompson, L. Yau and E.M. Doerries, J. Electrochem. Soc., 126 (1979) 1703.
59 Y. Ohnishi, J. Vac. Sci. Technol., 19 (1981) 1136.
60 T. Hirai, Y. Hatano and S. Nonogaki, J. Electrochem. Soc., 118 (1971) 669.
61 E.D. Feit, R.D. Heidenreich and L.F. Thompson, Applied Polymer Symposium, 23 (1974) 125.
62 S. Nonogaki, H. Morishita and N. Saitou, Applied Polymer Symposium, 23 (1974) 117.
63 C. Vigot, These Thompson CSF 1979.
64 L.F. Thompson, J.P. Ballantyne and E.D. Feit, J. Vac. Sci. Technol., 12 (1975) 1280.
65 M. Gazard, J.C. Dubois and A. Eranian, Soc. Plast. Eng. Regional Techn. Conference Photopolymers, Ellenville (1976).
66 J.L. Bartelt and E.D. Feit, J. Electrochem. Soc., 122 (1975) 542.

67 J.L. Bartelt, Applied Polymer Symposium, 23 (1974) 139.
68 C.W. Wil-Kins Jr., E.D. Feit and M.E. Wurtz, 8th Int. Conf. Proceedings of the symposium on Electron and Ion Beam Science and Technology, May, 1978, 341.
69 M. Gazard, J.C. Dubois and C. Duchesne, Applied Polymer Symposium, 23 (1974) 107.
70 Z.C.H. Tan, C.C. Petropoulos and F.J. Rauner, J. Vac. Sci. Technol., 19 (1981) 1348.
71 L.F. Thompson, E.D. Feit, M.J. Bowden, P.V. Lenzo and E.G. Spencer, J. Electrochem. Soc., 121 (1974) 1500.
72 D.L. Spears and H.I. Smith, Solid State Techn., 15 (1972) 29.
73 The Encyclopedia of X-rays Gamma Rays, pp. 11-15. G.L. Clark, Ed., Reinhold Publishing Co., New York (1973).
74 A. Eranian, A. Coutet, E. Datamanti and J.C. Dubois, ACS Symposium, 151 (1980) 275.
75 G.N. Taylor, Solid State Technol., 23 (1980) 73.
76 M. Kakuchi, S. Sugawara, K. Murase and K. Matsuyama, J. Electrochem. Soc., 124 (1977) 1648.
77 K. Murase, M. Kakuchi and S. Sugawara, Colloque International sur la Microlithographie, Paris, p. 261 (1977).
78 R. Feder, I. Haller, M. Hatzakis, L.T. Romankiw and E. Spiller, IBM, Techn. Disclosure Bulletin, 18 (1975) 2343.
79 G.N. Taylor, G.A. Coquin and S. Somekh, Polym. Eng. Sci. 17 (1977) 420.
80 L.F. Thompson, E.D. Feit, M.J. Bowden, P.V. Lenzo and E.G. Spencer, J. Electrochem. Soc. 12 (1974) 1500.
81 K. Sukegawa and S. Sugawara, 24th Meeting of Japan Soc. of Appl. Phys. (Yokohama, March 1977).
82 J.M. Moran and G.N. Taylor, J. Vac. Sci. and Technol., 16 (1979) 1980.
83 J.M. Moran and G.N. Taylor, J. Vac. Sci. and Technol., 16 (1979) 2014.
84 K. Sukegawa, S. Sugawara and K. Matsuyama, 1977 Ann. Meeting of IECE Japan (Tokyo, March, 1977).
85 M. Kakuchi and S. Sugawara, 23rd Meeting of Japan Sci. of Appl. Phys. (Tokyo, March, 1976).
86 D.C. Chofer, F.B. Kaufman, S.R. Kramer and A. Aviram, Appl. Phys. Lett., 37 (1980) 314.
87 G.N. Taylor and T.M. Wolf, J. Electrochem. Soc., 127 (1980) 2665.
88 G.N. Taylor, T.M. Wolf and J.M. Moran, J. Vac. Sci. Technol., 19 (1981) 872.
89 G.N. Taylor and T.M. Wolf, International Conference on Microlithography p. 381 (Lausanne, 1981).
90 H. Hiraoka and A.R. Gutierrez, J. Electrochem. Soc., 126 (5) (1979) 860-865.
91 J. Pacansky and J.R. Lyeria, IBM, J. Res. Develop., 23 (1) (1979) 45-55.
92 B. Broyde, J. Electrochem. Soc., 117 (12) (1970) 1555 1556.
93 M. Tsuda, S. Oikawa, W. Kanai, A. Yokota, I. Hijikata, A. Uehara, H. Nakane, J. Vac. Sci. Technol., 19 (2) (1981) 259-261.
94 M.J. Bowden, L.F. Thompson, S.R. Fahrenholtz, E.M. Doerries, J. Electrochem. Soc., 128 (6) (1981) 1304-1313.
95 S.R. Fahrenholtz, J. Vac. Sci. Technol., 19 (4) (1981) 1111 1115.
96 K.G. Clark, Solid State Technology, 73 (1978).
97 G.A. Delzenne, Advances in Photochemistry II, 1 (1979).
98 T. Iwayanagi, T. Kohashi, S. Nonogaki, J. Electrochem. Soc., 2759 (1980).
99 Y. Harita, M. Ichigawa, K. Harada, T. Tsunoda, Polym. Eng. Sci., 17 (6) (1977) 372.
100 B.M. Gong, Y.D. Ye, M.Q. Gu, Q.B. Zhang, J. Vac. Sci. Technol., 16 (6) (1979) 1980.
101 T. Nishikubo, T. Iizawa, K. Tsuchiya, Makromol. Chem., 3 (1982) 377.
102 S.A. Zahir, J. Appl. Polym. Sci., 23 (1979) 1355.
103 G.N. Taylor, T.M. Wolf, M.R. Goldrick, J. Electrochem. Soc. 128 (2) (1981) 361.

104 L.F. Thompson and R.E. Kerwin, Annual Review of Materials Science, 6 (1976) 267.
105 F.A. Vollenbroek, E.J. Spiertz, H.J.J. Kroon, Photopolymer Conference, Ellenville, 1982.
106 H. Meier and K.P. Zeller, Angew. Chem., 87 (1975) 52.
107 E. Reichmanis, C.W. Wilkins Jr. and E.A. Chandross, J. Vac. Sci. Techn., 19 (4) (1981) 1338.
108 C.W. Wilkins Jr., E. Reichmanis and E.A. Chandross, J. Electrochem. Soc., 129 (11) (1982) 2552.
109 E. Reichmanis, C.W. Wilkins Jr., D.A. Price, and E.A. Chandross, J. Electrochem. Soc., 130 (6) (1983) 1433.
110 J.V. Crivello, and J.H.W. Law, Macromolecules 10 (1977) 1307.
111 J.V. Crivello and J.H.W. Law, J. Polym. Sci., Polym. Chem. Ed., 17 (1979) 977.
112 J.V. Crivello and J.H.W. Law, J. Polym. Sci., Polym. Chem. Ed., 17 (1979) 1059.
113 J.M.J. Frechet, H. Ito and C.G. Willson, Microcircuit Engineering 82, Grenoble (1982).

Chapter 6.7

Electron Transfer in Drug-Induced Photosensitization Processes

N. Paillous and M. Comtat

1 - INTRODUCTION

The beneficial effects of the sun on the human organism have been recognized since ancient times (ref. 1). A well known example is the conversion of dietary 7-dehydrocholesterol into vitamin D in the body under the influence of sunlight. This metabolically active form of vitamin D prevents rickets by increasing the retention of calcium (ref. 2).

In combination with drug treatment, exposure to the sun or U.V. radiation can also have beneficial cutaneous effects. The application of crude coal tar onto the skin, followed by exposure to UV radiation has been found to be a remarkably effective treatment for lupus vulgaris and psoriasis. This represents one of the first techniques of phototherapy and photosensitization (refs. 3,4). Both psoriasis (refs. 5,6) and vitiligo (refs. 7,8) are now treated using a combination of psoralens and exposure to radiation. Derivatives of hematoporphyrin in combination with irradiation can also be used in the treatment of some cancers (ref. 9). The photochemotherapy of cancer by porphyrins is now under intensive investigation, and holds considerable promise (ref. 10).

However, in many instances, adverse effects result from exposure to the sun. Apart from the delayed pigmentation reaction or sun-tan, which exposes 20th century holidaymakers to the risk of senescence of the skin from solar elastosis, it can lead to various skin reactions in patients treated with certain drugs.

Some drugs induce specific effects in irradiated tissue that would not have taken place in the absence of irradiation. Such compounds are referred to as photosensitizers, and their effects are generally divided into two main categories: phototoxicity and photoallergy (ref. 11).

Phototoxicity, which will be considered first, is the net result of chemical reactions induced in the organism by the photosensitizing drug (refs. 12,13). The photochemical reactions can cause damage to DNA, cytoplasmic organelles and cell membranes (ref. 14). In general, such reactions occur in all individuals who have absorbed relatively large amounts of the relevant drug, and have been exposed to adequate doses of light energy. Clinically these reactions are seen as an enhanced sunburn response with erythematous edema, and the occasional formation of blisters followed by pigmentation and desquamation (ref. 15). These manifestations are limited to the exposed areas of skin, and usually appear within a few minutes to a few days after the first exposure to the sun. More serious reactions such as angioneurotic edema and skin cancer may also occur. Some phototoxic compounds can affect the eye, and lead to the formation of cataracts (ref. 16).

Photoallergy results from a different process which involves the immune system (refs. 17-22). The sensitizing drug, or a metabolite, behaves either as a hapten, or transforms an endogenous protein into an antigen. The photoantigen then primes immunocompetent cells. An immunological reaction thus takes place after subsequent exposure of the photosensitizing substance to light of the appropriate wavelength. This cell-mediated hypersensitivity reaction involves the secretion of lymphokines by T-lymphocytes.

The photoallergic reaction requires prior sensitization, and is only observed in predisposed individuals. Once the allergy has been acquired, small amounts of the drug, and limited exposure to light can set off the reaction. Clinically it is observed as urticarial, or delayed papular to eczematous lesions (ref. 15). The eruption can extend to areas of the body that have not been exposed, and only regresses slowly after elimination of the agent responsible. It can lead to chronic invalidating disorders.

A number of systemic drugs are now recognized to be phototoxic or photoallergic. Some examples are listed in Table 1 (ref. 23). As can be seen from their diversity, the effects cannot be directly attributed to either a characteristic chemical structure or to any particular pharmacological activity. It is therefore difficult to predict which compounds might lead to such reactions, hence the interest in trying to elucidate the mechanisms involved in these phenomena.

This chapter emphasises the role of electron transfer in the mechanism of drug-induced phototoxicity. It is not meant to be exhaustive, and the few examples chosen have deliberately excluded dyes, perfumes, cosmetics and porphyrins. The place of electron transfer among the various mechanisms of phototoxicity is dealt with in the first section. Examples of electron transfer occurring in the absence of oxygen are discussed in the second section, while the last section is devoted to electron transfer under aerobic conditions.

2 - MECHANISMS OF PHOTOTOXICITY

The chemical and biological aspects of the reactions responsible for phototoxicity have been investigated in detail for several compounds. Table 2 shows the general chemical methodology and the corresponding biological investigations.

Irradiation-induced transformation of the drug can be studied *in vitro*,

TABLE 1 Drugs causing photosensitivity reactions

ANTICANCER DRUGS
Dacarbazine (DTIC-Dome)
Fluorouracil (Fluoroplex *)
Methotrexate (Mexate *)
Vinblastine (Velban)

ANTIDEPRESSANTS
Amitriptyline (Elavil *)
Desipramine (Norpramin; Pertofrane)
Doxepin (Adapin; Sinequan)
Imipramine (Tofranil *)
Nortriptyline (Aventyl)
Protriptyline (Vivactil)
Trimipramine (Surmontil)

ANTIHISTAMINES
Cyproheptadine (Periactin)
Diphenhydramine (Benadryl *)

ANTIMICROBIALS
Demeclocycline (Declomycin *)
Doxycycline (Vibramycin *)
Griseofulvin (Fulvicin U/F *)
Nalidixic acid (NegGram *)
Oxytetracycline (Terramycin *)
Sulfacytine (Renoquid)
Sulfamethazine (Neotrizine *)
Sulfamethizole (Thiosulfil *)
Sulfamethoxazole (Gantanol *)
Sulfamethoxazole-trimethoprim (Bactrim; Septra)
Sulfasalazine (Azulfidine *)
Sulfathiazole
Sulfisoxazole (Gantrisin *)
Tetracycline (Achromycin *)

ANTIPSYCHOTIC DRUGS
Chlorpromazine (Thorazine *)
Fluphenazine (Permitil; Prolixin)
Haloperidol (Haldol)
Perphenazine (Trilafon)
Piperacetazine (Quide)
Prochlorperazine (Compazine *)

Antipsychotics drugs (continued)
Promethazine (Phenergan *)
Thioridazine (Mellaril)
Trifluoperazine (Stelazine *)
Triflupromazine (Vesprin)
Trimeprazine (Temaril)

DIURETICS
Bendroflumethiazide (Naturetin *)
Benzthiazide (Exna *)
Chlorothiazide (Diuril *)
Cyclothiazide (Anhydron)
Furosemide (Lasix)
Hydrochlorothiazide (HydroDIURIL *)
Hydroflumethiazide (Diucardin *)
Methyclothiazide (Aquatensen;Enduron)
Metolazone (Diulo; Zaroxolyn)
Polythiazide (Renese)
Quinethazone (Hydromox)
Trichlormethiazide (Metahydrin *)

HYPOGLYCEMICS
Acetohexamide (Dymelor)
Chlorpropamide (Diabinese; Insulase)
Tolazamide (Tolinase)
Tolbutamide (Orinase *)

SUNSCREENS
Para-aminobenzoic acid (PABA-Pabagel; Pabanol; PreSun *)
PABA esters (Eclipse; Pabafilm *)

OTHER
Bergamot oil, oils of citron, lavender, lime, sandalwood, cedar (used in many perfumes and cosmetics)
Carbamazepine (Tegretol)
Coal tar (Zetar *)
Disopyramide (Norpace)
Gold salts (Myochrysine; Solganal)
Hexachlorophene (pHisoHex *)
Methoxsalen (Oxsoralen)
Oral contraceptives
Trioxsalen (Trisoralen)

* and others drugs of the same family.

Reproduced with permission from Medical Letter (ref. 23)

which may also help identify reactive intermediates, and analyze interactions with isolated biological molecules. Parallel studies on the response of biological systems can also be carried out *in vitro*, using techniques such as inactivation of viruses, lethality to bacteria and mammalian cells, mutation frequency in bacteria, erythematous reactions in human skin samples and opacification of lens tissue.

Mechanisms of action accounting for the observed biological effects can thus be deduced from the combined results of these two approaches.

TABLE 2 Experimental approach for the determination of phototoxicity mechanisms

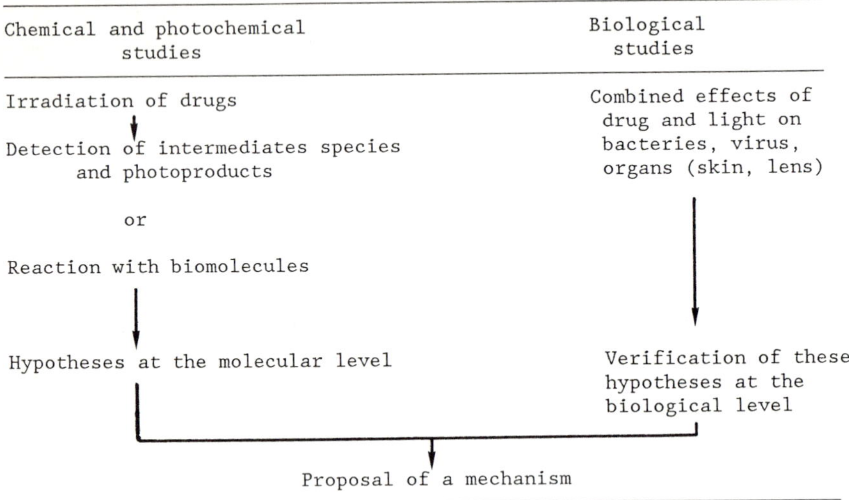

2.1 Classification of mechanisms

The first step in any photochemical process is the absorption of a photon by a molecule A, which goes from the fundamental to the excited singlet state. The molecule can then undergo various deactivation processes. These are summarised in Fig. 1:

The excited molecule may return to the ground state:

- by emission of fluorescence radiation (fluo)
- by internal conversion of electronic energy into vibrational energy (also leading to the formation of radicals ($R_1^{\bullet}$).

It can also:

- eject an electron giving rise to a cation ($A^{\bullet +}$ + e^-)
- lead to non-radical photochemical reactions by a concerted pathway
- go into the triplet state by intersystem crossing (isc). This may return to the ground state by thermal or radiative deactivation (phosphorescence), or

lead to photochemical reactions involving free radicals ($R_2^{\bullet}$). The longer lifetime and greater reactivity of the triplet state with respect to the singlet state is a determining factor in photosensitization processes.

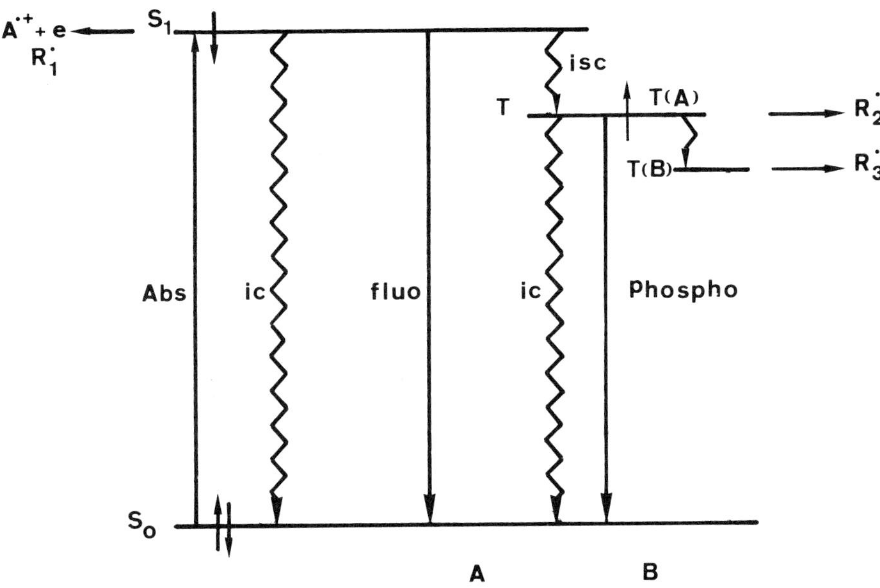

Fig. 1 Excitation and deactivation processes in an organic molecule

It can be seen from Fig. 1 that the excited molecule can lead to the formation of radicals either by photoionization ($A^{\dot{+}}$) or by cleavage from one or other excited state ($R_1^{\bullet}$)($R_2^{\bullet}$). The interaction of the excited molecule with another molecule B can also produce radicals either by transfer of hydrogen or electrons ($R_3^{\bullet}$). Such radicals can either react with substances in their immediate environment or with oxygen. These oxidation reactions are called Type I reactions. The Type II oxidation reactions involve direct interaction between the triplet state of the sensitizer and oxygen, forming singlet oxygen. These two types of reaction which both play an important part in phototoxic processes are illustrated in Table 3.

TABLE 3 Type I and type II processes

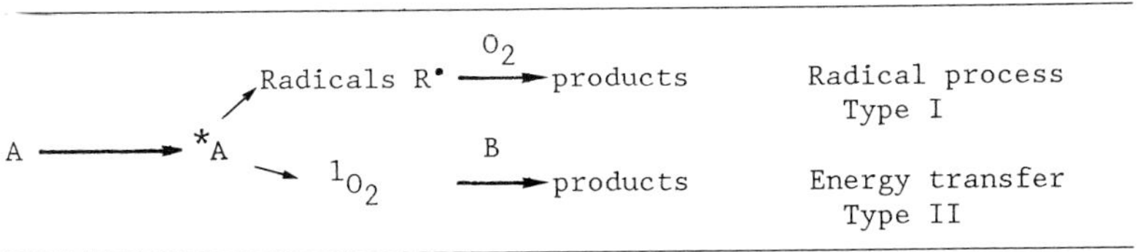

2.2 Electron transfer and phototoxicity

Electron transfer is also involved in many of the photochemical reactions mentioned above.

2.2.1 *In the absence of oxygen*

This type of electron transfer can be due to ejection of an electron, probably from the first excited state of the sensitizer. These solvated electrons are subsequently captured by a substrate

$$A^1 \longrightarrow A^{\dot{+}} + e$$

Direct transfer between the sensitizer and the substrate can also lead to a reduced form of the sensitizer and the oxidized form of the substrate:

$$*A + D \longrightarrow A^{\dot{-}} + D^{\dot{+}}$$

Less commonly, the sensitizer can transfer an electron to an oxidant:

$$*A + D \longrightarrow A^{\dot{+}} + D^{\dot{-}}$$

The types of structure which favor electron transfer are thus those that are easily oxidized or reduced.

2.2.2 *In the presence of oxygen*

Various mechanisms of photoinduced electron transfers involving oxygen can lead to the formation of the superoxide anion:

- photoionization of the sensitizer can liberate an electron which then reacts with molecular oxygen
- an electron transfer between the sensitizer and a substrate. The reduced forms of the sensitizer (or substrate) donates an electron to molecular oxygen
- the excited sensitizer reacts directly with oxygen forming a complex by charge transfer in which O_2 is the electron acceptor. In aqueous media, the solvation which stabilizes the ions also facilitates dissociation of the complex. However, this reaction is relatively slower than that producing singlet oxygen
- the singlet oxygen itself can act as oxidant, and remove an electron from a substrate

Table 4 summarises the different electron transfer reactions.

TABLE 4 Photoinduced electron transfer reactions in the presence and absence of oxygen

$^1A \longrightarrow A^{\dot{+}} + e^-$	$O_2 + e^- \longrightarrow O_2^{\dot{-}}$
$^*A + B \longrightarrow A^{\dot{+}}_{ox} + B^{\dot{-}}_{red}$	$B^{\dot{-}} + O_2 \longrightarrow O_2^{\dot{-}} + B$
$^*A + B \longrightarrow A^{\dot{-}}_{red} + B^{\dot{+}}_{ox}$	$A^{\dot{-}} + O_2 \longrightarrow O_2^{\dot{-}} + A$
$^*A + O_2 \longrightarrow$	$[A^{\dot{+}} - O_2^{\dot{-}}] \longrightarrow O_2^{\dot{-}} + A^{\dot{+}}$
$^*A + O_2 \longrightarrow$	$A + {}^1O_2; {}^1O_2 + B \longrightarrow O_2^{\dot{-}} + B^{\dot{+}}$

As soon as the ion $O_2^{\bar{\cdot}}$ appears, other intermediates involved in the reduction of oxygen can be produced. This is related to the ability of this ion to undergo dismutation:

$$O_2^{\bar{\cdot}} + O_2^{\bar{\cdot}} + 2\ H_2O \longrightarrow H_2O_2 + O_2 + 2\ OH^-$$

This reaction is fairly slow in neutral and alkaline media but can become fast when the pH favors formation of the protonated species HO_2.

$$O_2^{\bar{\cdot}} + H^+ \rightleftharpoons HO_2^{\cdot} \qquad pK = 4.69$$

The dismutation produces the more stable intermediate, hydrogen peroxide

TABLE 5 Dismutation of $O_2^{\bar{\cdot}}$ and its protonated form

Reactions	Rate constants
$O_2^{\bar{\cdot}} + O_2^{\bar{\cdot}} + 2H^+ \longrightarrow H_2O_2 + O_2$	$k < 100M^{-1}S^{-1}$
$HO_2^{\cdot} + HO_2^{\cdot} \longrightarrow H_2O_2 + O_2$	$k = 7.6\ 10^5 M^{-1}S^{-1}$
$HO_2^{\cdot} + O_2^{\bar{\cdot}} + H^+ \longrightarrow H_2O_2 + O_2$	$k = 8.5\ 10^7 M^{-1}S^{-1}$

The hydrogen peroxide can react with the $O_2^{\bar{\cdot}}$ ion in a Haber_Weiss reaction catalyzed by Fe ions. The hydroxyl ion can thus appear in any system in which $O_2^{\bar{\cdot}}$ is produced. Although the $O_2^{\bar{\cdot}}$ ion can diffuse, the highly reactive hydroxyl ion only interacts with substances in its immediate environment. The chemical and toxic reactions of these species are in fact general to all biological systems under aerobic conditions.

Type I reactions involving neutral radicals and Type II reactions involving singlet oxygen are currently thought to play the major roles in the processes underlying phototoxicity. The role of electron transfer processes has been less well studied. However, photoionization reactions producing radical cations, or superoxide anions in the presence of oxygen, have been described for some drugs. The interactions between radical cations or superoxide anions with biological substrates and their possible role in phototoxicity have also been investigated in some cases. A few examples will be mentioned with particular attention to promazine and its derivatives which have been the most extensively studied of these drugs.Anaerobic processes will be discussed first.

3 - ELECTRON TRANSFER NOT INVOLVING OXYGEN IN DRUG-INDUCED PHOTOTOXICITY

3.1 Chlorpromazine and related phenothiazines

Chlorpromazine (CPZ) and other related phenothiazines are employed as neuroleptics and anti-inflammatory agents (ref. 24). They are both phototoxic and photoallergic (refs. 25-27,11), and can lead to ocular opacity in some patients. Their chemical structures are shown in Table 6.

TABLE 6 Structure of the phenothiazines

Compound	R_1	R_2
Phenothiazine (PMZ)	H	H
Promazine	$(CH_2)_3N(CH_3)_2$	H
Chlorpromazine (CPZ)	$(CH_2)_3N(CH_3)_2$	Cl
Triflupromazine	$(CH_2)_3N(CH_3)_2$	CF_3
Methoxypromazine ($POCH_3$)	$(CH_2)_3N(CH_3)_2$	OCH_3
Acepromazine	$(CH_2)_3N(CH_3)_2$	$COCH_3$
Compazine	$(CH_2)_3N\quad N-CH_3$	Cl
Perphenazine	$(CH_2)_3-N\quad N-CH_2OH$	Cl
Stelazine	$(CH_2)_3N\quad N-CH_3$	CF_3
Fluphenazine	$(CH_2)_3-N\quad N-CH_2OH$	CF_3
Promethazine	$CH_2CHN(CH_3)_2$ with CH_3 on CH	H
Thioridiazine	$(CH_2)_2-$ ring, N, CH_3	SCH_3
Trifluoperazine	$(CH_2)_3-N\quad N-CH_3$	CF_3

As for all nitrogen-containing heterocyclic compounds, these compounds are good electron donors (ref. 28), and they ionize readily. Their ionization potentials, derived from their charge transfer bands, are all close to 6.96 eV (ref. 29), the value for chlorpromazine. In deaerated solutions, photoionization occurs at a much lower potential (around 4.1 eV).

Formation of such radical cations appears to play an important part in the pharmacological action of these compounds (ref. 30). Such intermediates are thought to be formed *in vivo* (ref. 31). For chlorpromazine, it has been suggested that this radical cation might be responsible for its psychotropic action (ref. 32), possibly due to its planar configuration which could intercalate between the strands of DNA, or more probably due to its reactivity towards nucleophiles on nerve cell membranes (ref. 33).

These radical cations may be produced chemically (refs. 34,35), enzymatically (refs. 32,36) or photochemically. Many studies both *in vivo* and *in vitro* have been devoted to the implication of radical cations of phenothiazine in phototoxic processes (for reviews see refs. 37 and 38).

3.1.1 *Photoionization*

The photochemical transformations of phenothiazine have now been investigated *in vitro* (refs.39,40). In nitrogen saturated aqueous or alcoholic solution, chlorpromazine undergoes cleavage of the C-Cl bond at position 2 producing promazine and 2-substituted hydroxy or alkoxy derivatives (refs. 39,41,42), as well as higher molecular weight substances such as a dimer (ref. 43) and various multimers.

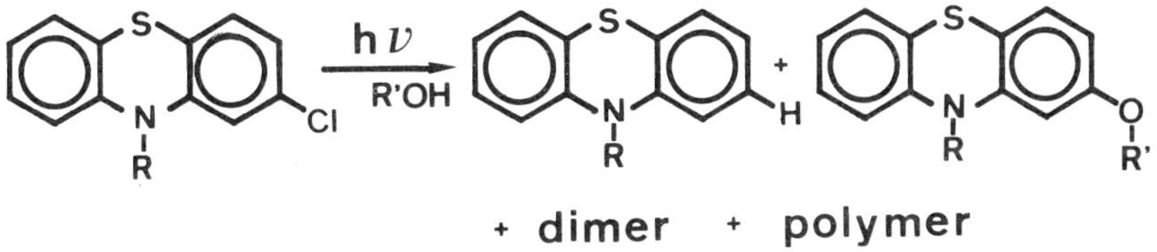

S N R Cl $\xrightarrow[R'OH]{h\nu}$ S N R H + S N R O R'

+ dimer + polymer

In an inert atmosphere, this cleavage of the C-Cl bond may involve neutral radicals as proposed by Grant and Green (ref. 42). This mechanism has been confirmed by ESR studies (ref. 44).

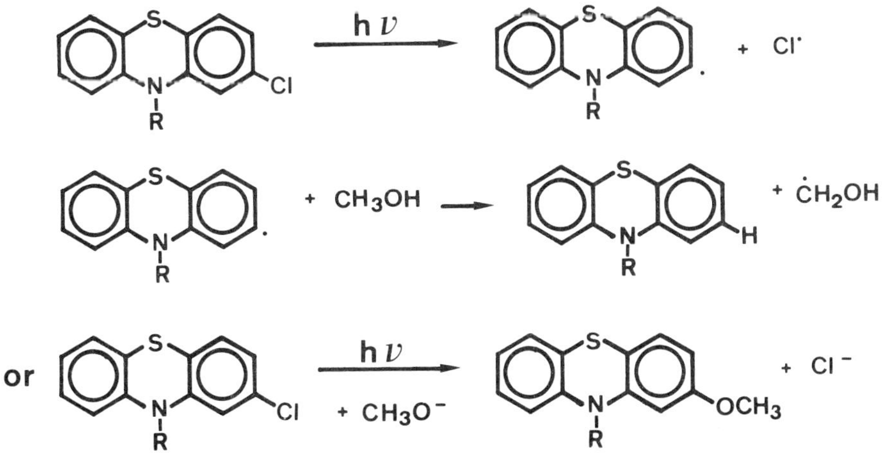

S N R Cl $\xrightarrow{h\nu}$ S N R $\cdot$ + $Cl^{\cdot}$

S N R $\cdot$ + CH_3OH $\longrightarrow$ S N R H + $\dot{C}H_2OH$

or S N R Cl $\xrightarrow[+ CH_3O^-]{h\nu}$ S N R OCH_3 + Cl^-

It has also been suggested that radical cations are involved during the degradation of chlorpromazine in propanol, giving rise to the same products (ref. 45).

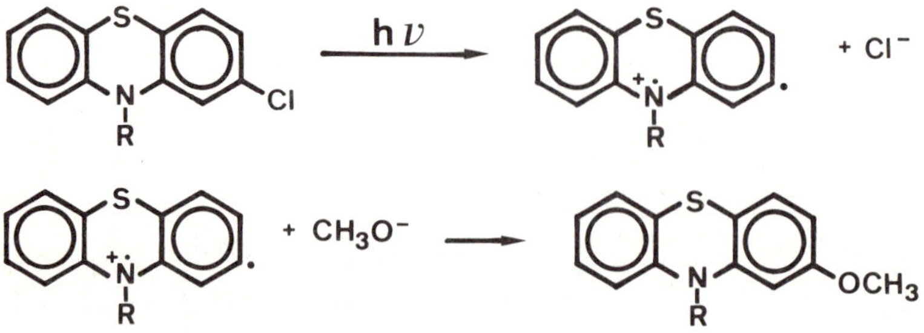

In fact, formation of this radical cation under irradiation has been recognized since 1962 (refs. 34,35,46). Its properties have been studied under different conditions and in various media (refs. 39,47-49), as well as for other phenothiazines (refs. 50-53). It should be mentioned that these electron transfers are favored in micellar solutions. In the case of the photoionization of the triplet state, the yield is observed to be higher in micelles (ϕ = 0.5) than in methanol (ϕ = 0.1) (ref. 47). The rate of dechlorination to produce chloride ions is also higher in micelles. Navaratnam (ref. 54), Moore (ref. 55) and Epling (ref. 56) have suggested that this enhanced dehalogenation is due to capture of solvated electrons by chlorpromazine in the ground state.

The electron spin resonance studies published recently by Chignell *et al.* (ref. 44) have led to a reconsideration of the role of the radical cation during photolysis of chlorpromazine or promazine (ref. 37). In contrast to the scheme proposed by Navaratnam *et al.* (ref. 54) photoionization of chlorpromazine only takes place for wavelengths of less than 270 nm which correspond to the S_2 absorption band (ref. 44). At longer wavelengths, irradiation of CPZ leads only to radicals, forming peroxides in the presence of oxygen.

3.1.2 *Intramolecular electron transfer*

Intramolecular electron transfer between the nitrogen atom on the side chain and the aromatic ring has also been demonstrated. In an analogous way to the reaction in micelles, a radical anion could be involved in this dehalogenation. It is known that amines can act as electron donors to halogenated aromatic compounds in the excited state, leading to more rapid formation of Cl^- ions via the radical anion $ArCl^{\overline{\cdot}}$.

In an investigation of intramolecular electron transfer, Bunce compared the reactivity of unsubstituted chlorpromazine to that of its N-substituted derivatives (ref. 57). Although the presence of an amino group on the side chain or the addition of triethylamine led to a low fluorescence quenching, they markedly increased the reactivity of the compounds in degassed media. Electron transfer from the side chain to the aromatic nucleus does therefore contribute to the instability of these drugs under irradiation. However, the side chain is not strictly required for the lability of chlorpromazine since the unsubstituted compound decomposes in the absence of amines with a measurable quantum yield.

3.1.3 *Electron transfer in photoinduced interactions between promazines and biological media*

Radical cations have often been invoked to account for interactions between phenothiazines and biological macromolecules (membranes, soluble proteins, DNA, RNA), and in a more general way in the phototoxic responses elicited in red blood cells, skin, bacteria, mammalian cells, bacteriophages and viruses (Table 7).

TABLE 7 Involvement of cation radicals in the phototoxicity of phenothiazines

Phenothiazine	Biological systems	Phototoxic effect	Reference
Methoxypromazine, promethazine, triflupromazine, acepromazine	Human erythrocyte membranes	Crosslinking	82
Promethazine, Chlorpromazine	BSA	Binding	45,70-71
Chlorpromazine	DNA	Intercalation	33,65
Chlorpromazine	DNA	Damage	65
Chlorpromazine, promethazine	DNA	Binding	70,71
Promazine, triflupromazine methoxypromazine	DNA	Strand breakage	66
Chlorpromazine	$(Na^{+} + K')$-ATPase	Inhibition	78,79,80
Thioridazine triflupromazine, trifluperazine	$(Na^{+} + K')$-ATPase	Inhibition	81

3.1.3a D N A

Studies on the photochemical interactions between phenothiazine-type drugs and DNA have led to conflicting results. Some authors have shown that chlorpromazine can photosensitize DNA strand breaks in viruses (ref. 58) and mammalian cells (refs. 59,60), although other workers have not observed DNA damage in dead cells after irradiation in the presence of CPZ (refs. 61-64).

The DNA strand breaks observed have been attributed to photoionization of chlorpromazine either via hydrated electrons which are known to produce species that are able to break DNA chains, or via radical cations from chlorpromazine which is stabilized by intercalation in the the DNA chains (refs. 33,65). In a similar way to mutagens like the acridine dyes, the plane of the molecule lies perpendicular to the axis of the DNA helix.

Decuyper *et al.* have shown that radical cations from promazine, methoxypromazine, triflupromazine, produced either chemically or enzymatically (peroxidase, H_2O_2) lead to similar strand breakage in ϕX174 DNA as that following irradiation in the presence of promazines (ref. 66). This is not observed for chlorpromazine, for which a neutral radical mechanism has been proposed. However, apart from the $POCH_3$ radical cation (ref. 67), these radical cations are generally poor DNA breakers.

These compounds can also react photochemically with DNA to form covalently linked adducts (refs. 45,60,68). In a similar method to that used for chain breakage, De Mol has shown that the enzymatically produced radical cation of promethazine, of known reactivity (ref. 69), is irreversibly bound to DNA (refs. 70,71). This binding is similar to that seen for photoactivated drugs. Comparison of the results obtained with ^{35}S labeled phenothiazines has shown that the covalent DNA binding is significantly less for PMZ^+ than for CPZ^+. This is in agreement with the results of bacterial mutagenicity tests, which have shown that photoactivated CPZ is mutagenic, although photoactivated PMZ has not been found to be genotoxic.

The radical cations of phenothiazines can also be formed metabolically *in vivo*. Genotoxic effects could thus be due to interactions with DNA, although there is no clear-cut evidence for this. On the other hand, there is evidence that phenothiazines have anti-tumor activity (refs. 72,73).
The site of interaction of promazine with DNA under irradiation has been investigated using ^1H and ^{13}C photo-CIDNP by Lablache-Combier *et al.* (ref. 74). They employed test solutions containing phenothiazines together with purines, pyrimidines or free nucleotides. Polarization behavior was detected from the NMR spectra of such irradiated mixtures, except for those containing guanine and its nucleotide. These polarizations indicate the formation of a pair of biradicals. In this case, the radical ion pair results from an electron transfer between CPZ and the DNA bases.

$$CPZ \xrightarrow{h\nu} {}^1CPZ \longrightarrow {}^3CPZ$$

$${}^3CPZ + B \longrightarrow \overline{CPZ^{\dot{+}}\ B^{\dot{-}}} \qquad \text{radical pair}$$

$$\overline{CPZ^{\dot{+}}\ B^{\dot{-}}} \longrightarrow CPZ^* + B^*$$

$$\overline{CPZ^{\dot{+}}\ B^{\dot{-}}} \longrightarrow CPZ^{\dot{+}} + B^{\bullet} \qquad \text{polarisation due to}$$

$$CPZ^{\dot{+}} + B^{\dot{-}} \longrightarrow CPZ^* + B^* \qquad \text{a back electron transfer}$$

In the case of guanine, the base with the highest affinity, there is no electron transfer as shown by the absence of polarization, and an adduct is formed. Van der Vorst *et al.* have shown that the photochemical interaction between promazine and single stranded DNA is selective for guanine residues (ref.75). Kochevar *et al.* have isolated an adduct between CPZ and guanosine-5'monophosphate with the same emission characteristics as the compound described by Van der Vorst (ref. 76). Recently, the transcription by *E. coli* RNA polymerase of SV40 DNA *in vitro* was found to be inhibited after photoreaction with promazine derivatives. Promazine adducts on guanine moieties seem to be bypassed by the RNA polymerase (ref. 77).

At present there is a lack of general agreement about the mechanism of these photoadditions. Kochevar has suggested that a neutral radical pathway is more likely to be involved than a radical cation mechanism (ref. 76), and the recent results of Chignell tend to support this view (ref. 44).

3.1.3b Proteins and Enzymes

Chlorpromazine photosensitizes inhibition of enzymes (refs. 34,78-81), binding to proteins (refs. 45,70,71) and cross-linking of membrane proteins.

An inhibitory effect of various phenothiazines on $Na^{+}K^{+}$ activated adenosine triphosphatase has also been demonstrated *in vitro*. The involvement of a radical cation (refs. 78-81) has been suggested by analogy with the inhibitory effect observed after treatment with peroxidase and H_2O_2. It is known that radical cations react with sulfhydryl compounds, and so this effect may be due to oxidation of the sulfhydryl groups on the enzyme.

Merville *et al.* have shown that promazine derivatives can induce cross-linking of membrane proteins in red blood cell ghosts under irradiation (ref. 82). The aggregation observed in the two spectrin subunits in the presence of oxygen was attributed to the formation of a radical cation. Similar results have been obtained when the radical is produced by purely chemical means. This rather unstable radical cation can either react directly with nucleophiles in the binding sites, or is deactivated and leads to the formation of products which can react with amino acid residues in membrane proteins. The binding of a radical cation of promethazine or CPZ to bovine serum albumin (BSA) has been demonstrated using radiotracers (refs. 45,70,71). This may account for the photoallergic properties of these drugs.

3.1.4 *Conclusion*

Photoionization has long been considered as an important factor in the phototoxic and photoallergic reactions of chlorpromazine. Many studies have shown that the radical cations of CPZ can bind to DNA or induce DNA chain scission. It can also inhibit enzymes and induce cross-linking of membrane proteins. However, its action towards biological molecules has usually only been demonstrated indirectly, by comparing the results of photochemical studies with those obtained by production of the radical cation by chemical or enzymatic means.

Recent studies have shown that the radical cation is only observed after irradiation at 270 nm rather than 330 nm. This would tend to rule out a role for the radical cation in phototoxic reactions at wavelengths above 330 nm. The mechanism of the phototoxicity of chlorpromazine is thus largely an open question.

3.2 Furocoumarins

The furocoumarins are a family of compounds derived from the condensation of a coumarin with a furan ring. Natural products with a linear structure such as the psoralens, or with a non-linear structure, like angelicin belong to this group (Table 8) (refs. 83,84)

TABLE 8 Structures of various furocoumarins

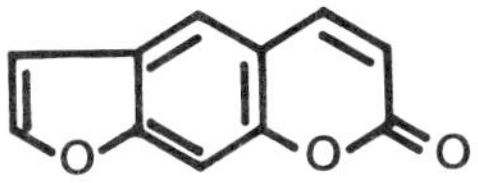

psoralen

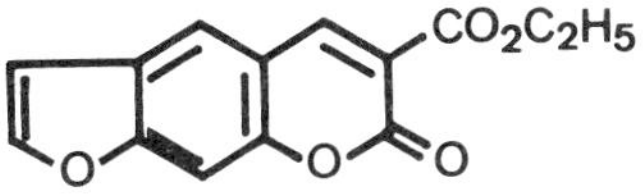

3-carbethoxypsoralen

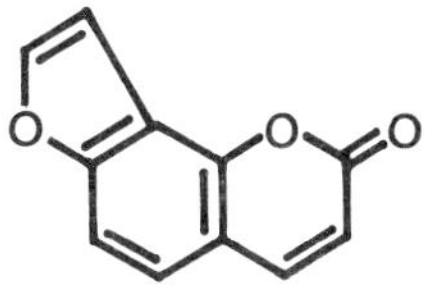

Angelicin

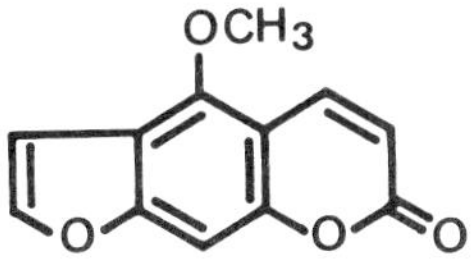

5-methoxypsoralen

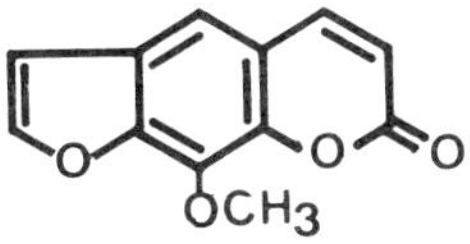

8-Methoxypsoralen

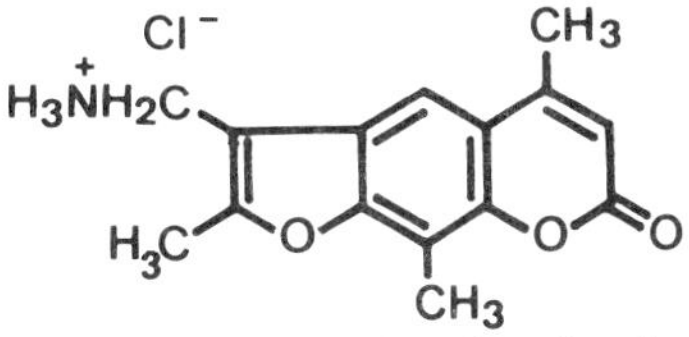

4'-aminomethyl-4,5',8 trimethyl psoralen hydrochloride

These compounds whether natural or synthetic can act as photosensitizers of biological systems (for reviews see refs. 85-87). These properties are exploited in the therapy of skin disorders such as psoriasis, vitiligo, mycosis and other fungal infections (for reviews see refs. 88,89).

A large number of studies have been devoted to their mode of action. It is generally thought that their antiproliferative effect, and part of their phototoxicity is due to their ability to bind to DNA. The mechanism of their photoaddition to DNA has now been well worked out. The first step is a cycloaddition between the 3,4 or 4',5' double bonds of psoralen and the 5,6 double bond in the pyrimidine rings of the DNA bases. Absorption of a second photon leads to a further cycloaddition between the 3,4 double bond of the 4',5' monoadduct of psoralen, and the double bond of the pyrimidine in the complementary strand of the macromolecule (for reviews see refs. 88,90). 3,4 cycloadducts of psoralens are not converted into cross-linkages by further irradiation at 365 nm as they do not absorb at this wavelength.

These additions are generally considered to proceed via a concerted pathway. However, the triplet state is normally involved, although the singlet state is thought to play a role for some furocoumarins (refs. 91,92). A stepwise mechanism (ref.93) via a charge transfer has been proposed on the basis of flash photolysis experiments carried out by Bensasson *et al.* (ref. 94). These workers demonstrated a charge transfer between the triplet state of the furocoumarins and the bases of the nucleic acid (ref. 95):

$^{3}FC + Thy \longrightarrow FC^{\overline{\cdot}} + Thy^{\stackrel{+}{\cdot}}$

The determination of the quenching constants of the triplet states of the furocoumarin derivatives by purines and pyrimidines (Table 9) showed that thymine (and its derivatives) is more reactive than uracil or cytosine to the psoralens.

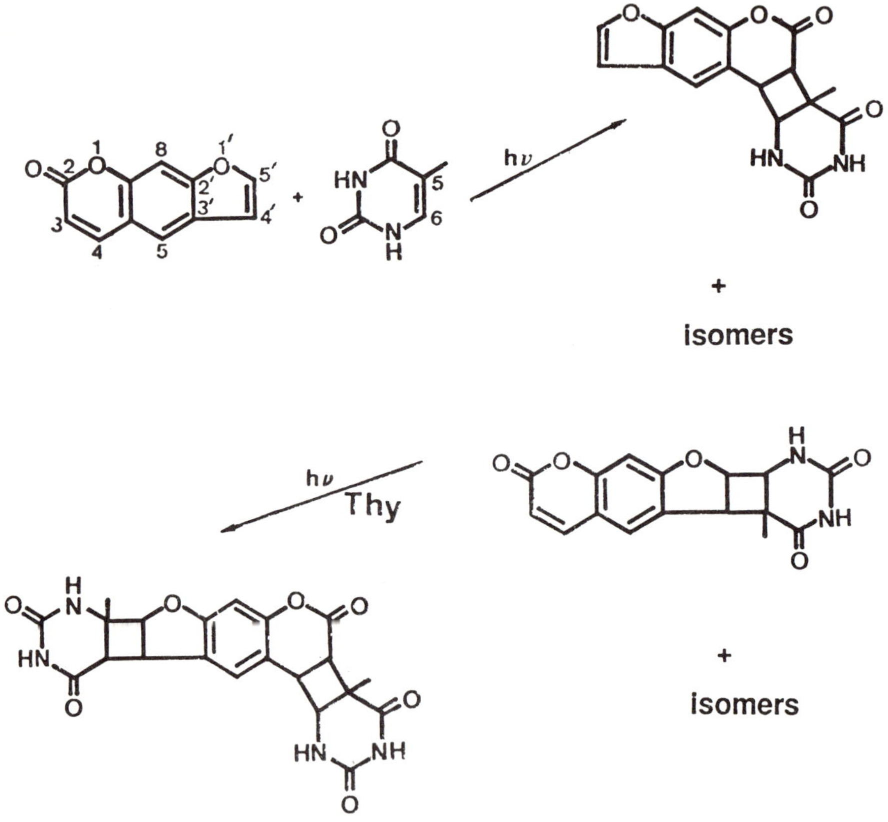

Fig. 2 Photoaddition reaction of psoralen to thymine

The formation of a radical anion of psoralen with a long half-lifetime could thus represent the first step in the photochemical generation of DNA-psoralen adducts. Although the escape yield of the charged species is relatively low, the electron transfer is reversible in the absence of oxygen (ref. 96).

TABLE 9 Rate constants for psoralen and angelicin triplet quenching by nucleic acid bases (ref. 94)

Quencher	Rate constant $M^{-1}S^{-1}$ Psoralen	Angelicin
Thymine	7.5×10^8	1.1×10^9
Thymidine	1.9×10^8	3.5×10^8
Uracil	1.1×10^8	2.1×10^8
Uridine	1.4×10^8	2.0×10^8
Cytosine	1.7×10^8	2.6×10^7
Cytidine	3.4×10^7	5.5×10^7
Adenine	1.0×10^8	1.6×10^8
Adenosine	4.1×10^7	4.7×10^7
Guanine	$< 10^9$	$< 10^9$
Guanosine	7.3×10^8	7.2×10^8

Other highly efficient reactions have been demonstrated to occur between the triplet of the furocoumarins and amino acids, especially tryptophan and other aromatic amino acids (refs. 94,97). Tryptophan has been shown to quench the triplets of psoralen and angelicin at much higher rates than those found with other additives.

$$^3FC + Try \longrightarrow FC^{\overline{\cdot}} + Try^{\dot{+}}$$

These interactions could be responsible for the covalent photobinding of furocoumarins to proteins in the absence of oxygen (refs. 98-101), although these reactions have usually been attributed to a radical mechanism (ref. 102).

In conclusion, the strong electron affinity of the triplet of the furocoumarins which is readily quenched by amino acids, pyrimidines and purines, indicates a charge transfer mechanism for the photoaddition reactions of these compounds with DNA or proteins. These relatively little studied processes may play a role in the photosensitizing properties of these drugs.

3.3 Anthracene

Certain polycyclic aromatic hydrocarbons have been found to have powerful photosensitizing properties (refs. 103-106). Frequently they are also carcinogenic (refs. 107,108) which has been related to their ability to bind to DNA *in vivo* (ref. 109). The particular size and flatness of these molecules probably enables them to intercalate into the DNA helices. Like many chemical carcinogens anthracene has been found to bind covalently to DNA *in vitro* under irradiation (refs. 110-112). It also leads to the formation of interstrand crosslinks with DNA (ref. 113), and inhibits DNA synthesis in mammalian epidermis (ref. 114). Addition of anthracene to DNA has been thought to occur via a radical cation mechanism (ref. 111). The photoionization of anthracene, mainly observed in the gas and solid phases, can also take place in liquid media at room temperature under certain conditions (refs. 115,116). However, anionic micellar solutions would appear to be more suitable media for studies on the reactivity and role of the radical cation in photosensitization (refs. 55,117). Interactions of radical cations with biomolecules have been investigated in such media by Putna, who has demonstrated the involvement of electron transfer between anthracene (or other polycyclic hydrocarbons) and tryptophan (ref. 117). However, it seems that there is no relationship between the carcinogenic and photodynamic activities of the hydrocarbons and the electron transfer behavior of their radical cations (ref. 117).

4 - ELECTRON TRANSFER INVOLVING OXYGEN IN DRUG-INDUCED PHOTOTOXICITY

Irradiation of drugs in the presence of oxygen can often produce superoxide anions by electron transfer. We will only discuss this mechanism for the compound groups which have been mentioned above. However, it should be noted that electron transfer to oxygen is much easier to detect than electron transfer under anaerobic conditions, and has therefore been more extensively investigated. Many other drugs including adriamycin, daunomycin, benoxaprofen, amiodarone, etc. have also been shown to produce superoxide anions.

The superoxide anion can be detected by conductivity, spectroscopy, mass spectrometry, and electron spin resonance spectroscopy. However, these direct physical methods are not particularly sensitive especially in neutral and basic media due to the rapidity of the dismutation reaction. The presence of the $O_2^{\bullet-}$ ion can also be deduced by its reaction with various compounds which can be detected spectroscopically. For example, this ion will reduce ferricytochrome to ferrocytochrome, or tetranitromethane to the nitroformate anion. One of the most widely employed reactions is the reduction of nitro blue tetrazolium to formazan blue.

The $O_2^{\bullet-}$ ion can also oxidize ascorbic acid, collagen, glutathione, hydroquinone, NADH lactate dehydrogenase, and adrenaline to adenochrome. Most of the systems used are non-specific, and therefore require the presence of superoxide dismutase (SOD) to discriminate between the effects of the superoxide anion and those of other reductive or oxidative species (for reviews see refs. 118-123).

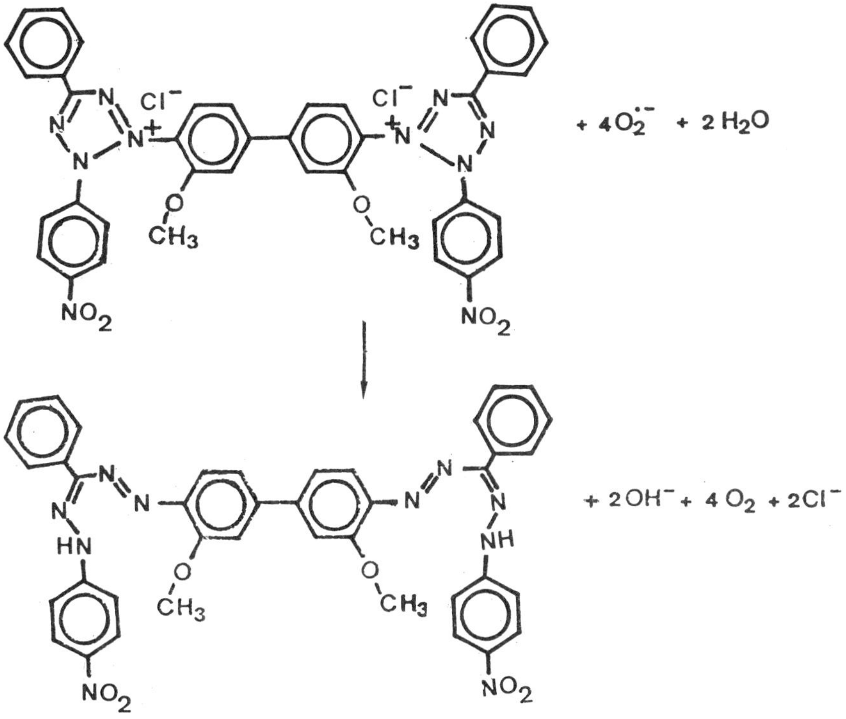

Fig. 3 Action of $O_2^{\bar{\cdot}}$ on nitro blue tetrazolium

4.1 Phototoxicity of some drugs involving superoxide anions

4.1.1 *Chlorpromazine and phenothiazines*

For chlorpromazine, flash photolysis experiments have demonstrated that under irradiation there is an electron transfer between CPZ and oxygen leading to the formation of a radical cation and a superoxide anion. The radical cation then reacts slowly with oxygen to produce the sulfoxide (refs. 41,124,125).

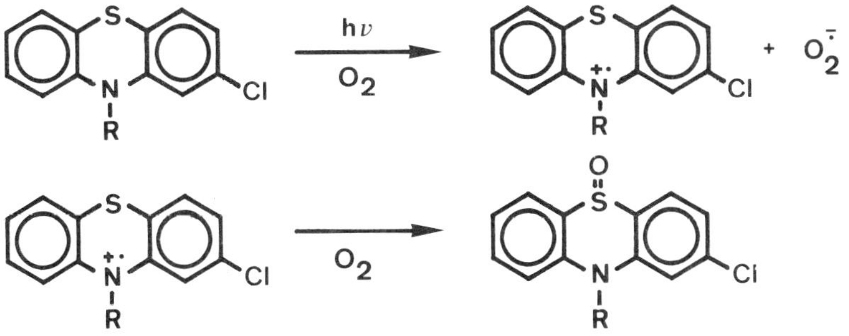

ESR studies using 5,5 dimethylpyrroline N-oxide (DMPO) as spin trap have, however, indicated another mechanism (ref. 44). Irradiation of chlorpromazine in the presence of DMPO leads to the formation of a peroxy radical which decays to the hydroxyl radical adduct DMPO-OH, although the signal intensity of this adduct is not affected by addition of superoxide dismutase.

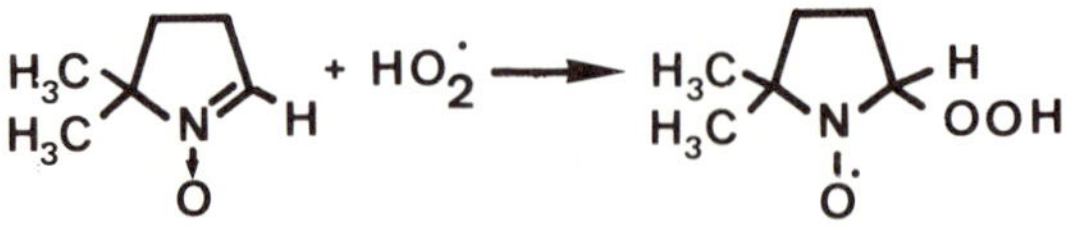

It thus appears that superoxide anions are not produced, and that the sulfoxide derives from an intermediate sulfur peroxy radical CPZSOO.

On the other hand, during irradiation of promazine in the presence of DMPO, production of the DMPO-OH adduct is reduced by about 50% in the presence of superoxide dismutase. In this case it would seem that there is an electron transfer from promazine to oxygen-producing superoxide anions.

4.1.2 *Psoralens*

In parallel with their action on DNA, photoexcited psoralens react with molecular oxygen to form $^1O_2$ and $O_2^{\bar{\cdot}}$ (ref. 126). The reduction of nitro blue tetrazolium to blue diformazan with a characteristic absorption at 560 nm has been used to compare the production of $O_2^{\bar{\cdot}}$ by various psoralens (ref. 127).

TABLE 10 Production of superoxide anions by psoralens photosensitized reactions and its quenching by SOD (ref. 127)

Compounds	$O_2^{\bar{\cdot}}$ (μM)	% quenching by SOD
Bifunctional psoralens		
Psoralen	73.3	91
4,8-Dimethyl, 5'-carboxypsoralen	44.0	94
4,5', 8-Trimethylpsoralen	32.6	100
5-Diethylamino butoxypsoralen	29.7	88
8-Methoxypsoralen	23.0	97
3,4'-Dimethyl, 8-methoxypsoralen	19.1	97.5
5-Methoxypsoralen	10.6	97.5
Monofunctional psoralens		
Angelicin	38.8	88
5-Methylangelicin	32.0	93.5
4,5'-Dimethylangelicin	25.3	100
3-Carbethoxypsoralen	24.5	76

Additional evidence for the presence of $O_2^{\cdot-}$ is derived from examination of the quenching of this reaction by superoxide dismutase. The results show that monofunctional psoralens produce $O_2^{\cdot-}$ at a similar rate to that from the bifunctional compounds. Since the monofunctional psoralens, which do not form interstrand crosslinks are more carcinogenic than the bifunctional compounds, it was suggested that the reactive oxygen species may play a role in their carcinogenicity and cell membrane damaging properties (ref. 127).

Visnagin and khellin have similar structures to the furocoumarins, and are also used in the treatment of psoriasis and vitiligo. Their ability to form $O_2^{\cdot-}$ has also been determined using the nitro blue tetrazolium reaction (ref. 128).

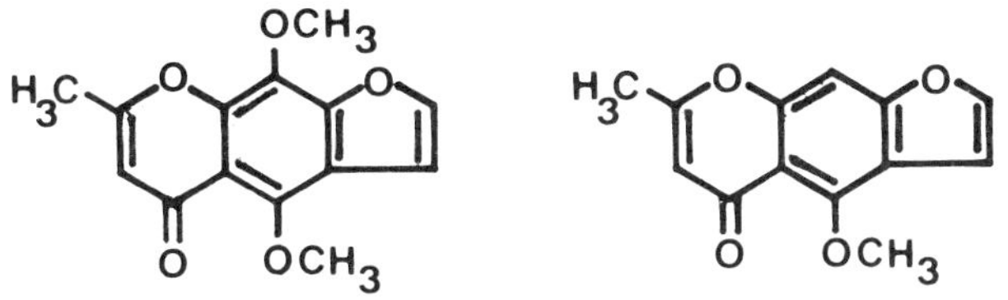

Fig. 4 Structures of visnagin and khellin

The production of $O_2^{\cdot-}$ by visnagin is much higher that from khellin, which has been related to differences in their phototoxicity and genotoxicity (ref. 129).

4.1.3 *Crude coal tar*

Coal tar preparations from various sources (CCT) have been used in combination with UV irradiation in the treatment of various skin disorders (ref. 130). These preparations contain at least 10,000 different compounds of which only 400 or so (60% of the total) have been isolated. Polycyclic aromatic compounds some of which may be photoreactive (e.g. anthracene, phenanthrene, pyrene, fluoranthene, benzanthracene, chrysene, benzopyrene) are well represented in this mixture.

Irradiation of these compounds in the presence of oxygen leads to the formation of reactive oxygen species. Positive evidence for the formation of $O_2^{\cdot-}$ from CCT and it constituents was obtained by spectroscopic analysis of the reduction of nitro blue tetrazolium. Production of $O_2^{\cdot-}$ was the same order of magnitude for constituents such as acridine or phenanthridine which are potent producers of $O_2^{\cdot-}$, although less than that for anthracene which is one of the most active producers of superoxide anions. This anion would thus appear to play a role not only in the therapeutic effectiveness of CCT but also in its photosensitizing and carcinogenic properties.

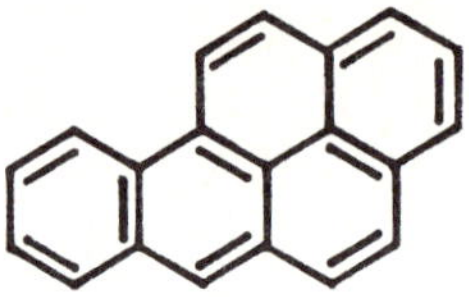

Benzo[*a*]pyrene

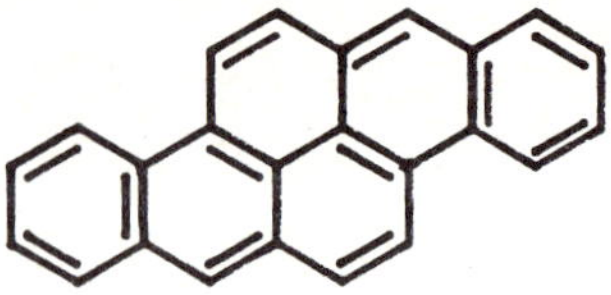

Dibenzo[*a,h*]pyrene

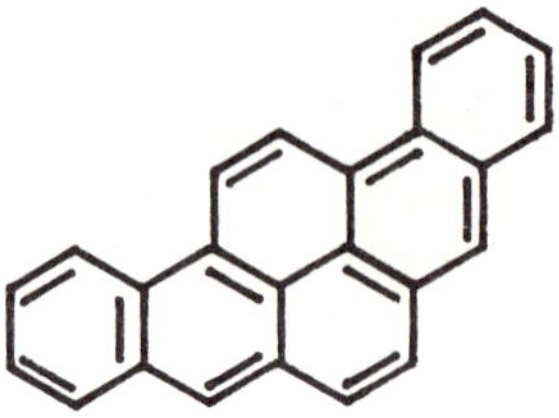

Dibenzo[*a,i*]pyrene

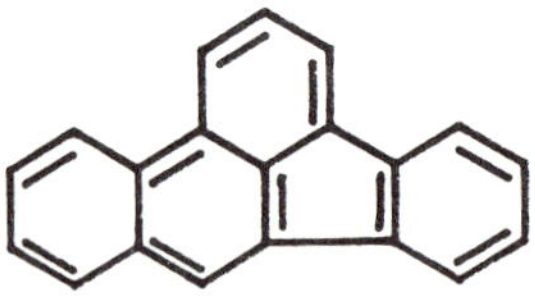

Benzo[*b*]fluoranthene

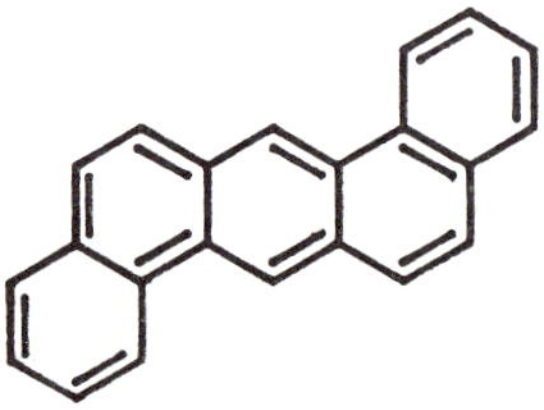

Dibenzo[*a,h*]anthracene

Fig. 5 Hydrocarbons isolated from coal tar pitch (ref. 131)
(Reproduced with permission of the W.B.Saunders Company)

4.2 Biological effects

The production of superoxide anions and its reduction products leads to a variety of little recognized biological effects. We will mention a few observations on the effects on biomolecules and whole or isolated biosystems of superoxide anions derived from exogenous (drugs) or endogenous compounds. For example, tryptophan or oxyhemoglobin can undergo autosensitizing oxidation reactions under irradiation.

4.2.1 *Tryptophan*

UV irradiation of tryptophan leads to the formation of N-formylkynurine which is a good photosensitizer of free tryptophan (ref. 132). This compound can initiate oxidation of tryptophan itself at wavelengths of 320 nm and above (wavelengths which have no effect on other amino acids in proteins).

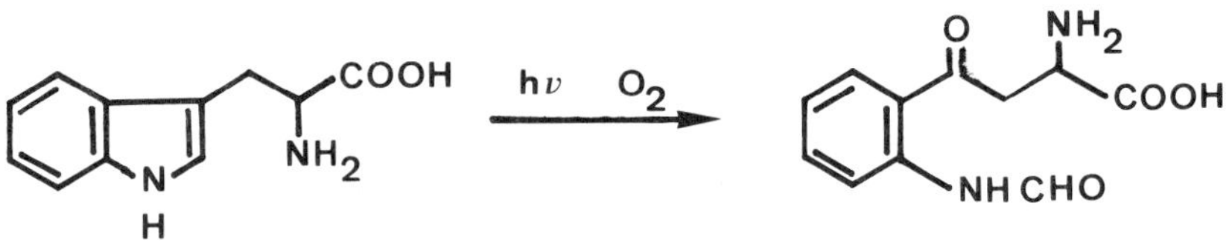

That this oxidation of tryptophan at pH 7.6 is partly due to $^1O_2$(50%) is shown by the effect of D_2O which increases the lifetime of $^1O_2$ while NaN_3 reduces it. However, the kinetics and the influence of pH on the rate of oxidation also tend to indicate formation of $^1O_2$ (ref. 133). The increase in rate of production of hydrogen peroxide at the start of the photolysis and the subsequent fall demonstrate that the formation of $^1O_2$ is probably mediated by N-formylkynurenine (ref. 134).

4.2.2 *Methemoglobin*

The formation of superoxide anions from oxyhemoglobin plays a major role in the production of methemoglobin MetHb in erythrocytes (refs. 135-137). The photolytic production of MetHb can intervene directly or by a process of oxidative attack by the $O_2^{\bar{\cdot}}$ ion formed in the medium or produced at the same time as MetHb. Most of the MetHB is produced in the β chains (ref. 138).

$$(HbO_2)_4 + 4O_2^{\bar{\cdot}} + 8H^+ \longrightarrow MetHb + 4O_2 + 4H_2O_2$$

4.2.3 *Lipids*

The superoxide anion can also initiate the peroxidation of unsaturated lipids which can lead to alterations in their fluidity, inactivation of enzymes and transport molecules, and cell lysis (ref. 139). The reaction mechanism for the peroxidative degradation of lipids in membranes may be illustrated as as follows:

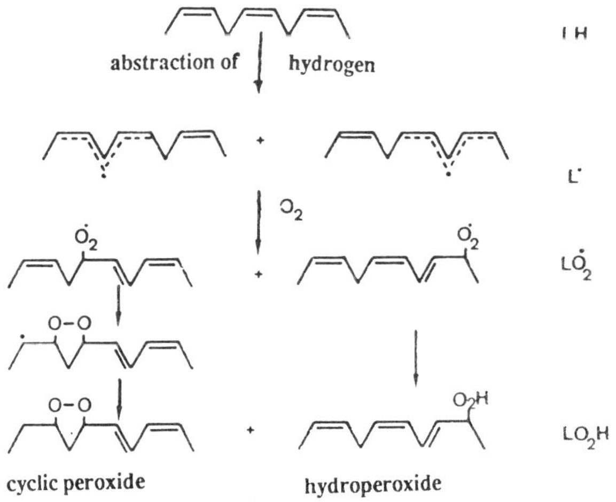

LH

abstraction of hydrogen

L·

O_2

$LO_2^{\cdot}$

LO_2H

cyclic peroxide

hydroperoxide

This type of degradation is indicated by the presence of malonaldehyde. The superoxide anion itself is not directly responsible, but rather its conjugate acid $HO_2^{\bullet}$ or the $OH^{\bullet}$ radical resulting from the redox metal-mediated reduction of H_2O_2 by $O_2^{\bar{\bullet}}$. It should be noted, however, that superoxide anion has both the ability to decompose the hydroperoxides and to promote the peroxidation of unsaturated fatty acids (ref. 140)

4.2.4 *DNA*

Exposure of DNA to producers of superoxide anions can lead to extensive strand breakage (ref. 141) and degradation of desoxyribose. Thus the $O_2^{\bar{\bullet}}$ ion, or in fact any radical able to react with DNA is potentially carcinogenic. However, at present there is no direct evidence linking cancer with superoxide anion production *in vivo*.

4.2.5 *Synovial fluid*

The formation of the $O_2^{\bar{\bullet}}$ ion and its transformation into $OH^{\bullet}$ radicals by a Haber_Weiss reaction can lead to breakage of the hyaluronic acid polymer leading to an alteration in the viscosity of synovial fluid (ref. 142).

4.2.6 *Eye*

The photosensitization reactions involving electron transfer can also have deleterious effects in both the lens and cornea (ref. 16). The phototoxic effects are seen as fine deposits in deep corneal stroma and corneal epithelium. The mechanism of the loss of transparency of the lens is not well understood but the loss of solubility of proteins after disulfide bridge formation has often been proposed. Furthermore, oxygenated radicals can induce loss of structural homogeneity of the lens tissue by rupturing epithelial cells which reduces light transmission (refs. 143,144).

4.3 Detoxication

Although superoxide anions can be formed *in vivo*, organisms also have regulating systems for their removal preventing their phototoxic effects. These defense systems are based on metallo-enzymes called superoxide dismutases (ref. 118). They are found in almost all organisms. These enzymes catalyze the reaction of $O_2^{\bar{\bullet}}$ into O_2 and H_2O_2. Many of them contain zinc and copper complexes, and they give rise to reactions that can be schematized as follows

$$-Zn^{2+}-Im^{-}-Cu^{2+}(Im)_3 \xrightarrow[H^+]{O_2^{\cdot -}} -Zn^{2+}-ImH-Cu^{+}(Im)_3 + O_2$$

$$-Zn^{2+}-ImH-Cu^{+}(Im)_3 \xrightarrow{O_2^{\cdot -}} -Zn^{2+}-Im^{-}-Cu^{2+} + HO_2^{-}$$

where Im refers to imidazolate.

Since $O_2^{\cdot -}$ is one of the main toxic species of oxygen, and superoxide dismutase is the natural defense against this ion, it follows that cells which use oxygen should also contain SOD. The toxicity of $O_2^{\cdot -}$ can thus be due either to an excess of this ion or to a defect in the catalytic activity of SOD.

5 - CONCLUSION

Electron transfer processes have received relatively little attention in attempts to account for the phototoxicity of drugs. In general, the effects of neutral radicals and singlet oxygen have been more extensively investigated. This is perhaps due to our greater understanding of the chemical reactivity and biological effects of these species, which can be formed by irradiation of compounds of pharmacological importance both *in vitro* and *in vivo*.

The various examples described here show that electron transfer is nevertheless involved in phototoxicity, especially for some drugs. However, the relevant studies are of recent origin and the results are not yet unambiguous. Further work will be required to rationalize the experimental findings in the light of thermodynamic and kinetic considerations. Research will need to be focussed on studies which discriminate between the effects of electron transfer and other photoinduced processes.

6 REFERENCES

1 L.C. Harber and D.R. Bickers, Photosensitivity diseases, W.B. Saunders Company, Philadelphia, 1981, pp 3-4.
2 A.C. Giese, in Living with our sun's ultraviolet rays, Plenum Press, New York, 1976, chap. 3 and 5.
3 N.R. Finsen, Phototherapy, London, 1901.
4 W.H. Goeckerman, Northwest Med., 24 (1925) 229-231.
5 J.F.Walter, J.J. Voorhees, W.H. Kelsey and E.A. Duell, Arch. Dermatol., 107 (1973) 861-865.
6 J.A. Parrish, T.B. Fitzpatrick, L. Tanenbaum and M. A. Pathak, New Engl. J Med., 291 (23) (1974) 1207-1211.
7 A.M. El Mofty, J. Roy. Egypt. Med. Assoc., 31 (1947) 651-655.
8 A.B. Lerner, C.R. Denton and T.B. Fitzpatrick, J. Invest. Dermatol., 20 (1953) 299-308.
9 A. Policard, C.R.Soc.Biol. Paris 91(1924) 1423.
10 For a recent review see : H.E. Van der Bergh, Chemistry in Britain, (1986) 430-439 and references therein.
11 H.E. Kirschbaum and H. Beerman., Am. J. Med. Sci., 248 (1964) 445-468.
12 J.H. Epstein in : Year book of dermatology, Malkinson FD, Pearson R.W. (Eds.), Chicago, 1971, Year Book Publishers Inc., pp 5-43.
13 H.F. Blum, Photodynamic action and diseases caused by light, New York, 1941, Reinhold Publishing Corp.
14 L.C. Harber, R.L. Baer and D.R. Bickers in Sunlight and Man. T.B. Fitzpatrick (Ed.) University of Tokyo Press, Tokyo, 1974, pp 319-333.
15 J.H. Epstein, J. Am. Acad. Dermatol., 8 (2) (1983), 141-147.
16 S. Lerman, Ophtalm., 93 (1986) 304-318.
17 S. Epstein, in Dermatology due to environmental and physical factors. Springfield I.L., Rees B.B. (Ed.). Charles C. Thomas Publisher, 1962, pp 119-137.
18 M. Stork, Arch. Dermatol., 91 (1965) 469-482.
19 J.H. Epstein, Arch. Dermatol., 106 (1972) 741-748.
20 W.L. Morrison, J.A. Parrish and J.H. Epstein, Arch. Dermatol., 115 (1979) 350-356.
21 I.E. Kochevar, Photochem. Photobiol., 30 (1979) 437-442.
22 E.A. Emmett, Int. J. Dermatol., 17 (1979) 370-379.
23 Medical Letter, 22 (1980) 64.
24 E. Hadin, Critical reviews in Clinical Laboratories Sciences, 2 (1971) 347-391.
25 C.L. Huang and F.L. Sands, J. Pharm. Sc., 56 (1967) 259-264.
26 A. Satanove and J.S. Mc Intosh, J. Am. Med. Assoc., 200 (1967) 209-212.
27 J.H. Epstein, L.A. Brunsting, M.C. Peterson and B.E. Schwarz, J. Invest. Dermatol., 28 (1957) 329-338.
28 J.P. Billion, Bull. Soc. Chim. Fr., (1961), 1923-1929.
29 A. Fulton and L.E. Lyons, Austr. J. Chem., 21 (1968) 873-882.
30 For a review see : I.S. Forrest, C.J. Carr and E. Lesdin, (Eds) Phenothiazines and structurally related drugs, Raven Press, New York 1974.
31 P.H. Sackett, J.S. Mayausky, T. Smith, S. Kalus and R.L. Mc Creery, J. Med Chem., 24 (1981) 1342 1349 and references therein.
32 L.H. Piette and G. Bulow and I. Yamazaki, Biochem. Biophys. Acta, 88 (1964) 120-129.
33 S-I. Ohnishi and H.M. Mc Connell, J. Amer. Chem. Soc., 87 (1965) 2293.
34 D.C. Borg and G.C. Cotzias, Proc. Natl. Acad. Sci., 48 (1962) 623-642.
35 L.H. Piette and I.S. Forrest, Biochem. Biophys. Acta, 57 (1962) 419-420.
36 D.J. Cavanauch, Science, 125 (1957) 1040-1041.
37 C.F. Chignell, A.G. Motten and G.R. Buettner, Environmental Health Perspect., 64 (1985) 103-110.
38 J. Piette, J. Decuyper, M.P. Merville-Louis and A. Van de Vorst, Biochimie, 68 (1986) 835-842.

39 A.K. Davies, S. Navaratnam and G.O. Philips, J. Chem. Soc. Perkin II, (1976) 25-29.
40 S. Hashiba, M. Tatsuzawa and A. Ejima, Bull. Nat. Inst. Hygienic Science., (Eisei Shisensho Hokoku), 97 (1979) 73-78.
41 C.L. Huang and F.L. Sands, J.Chromatog., 13 (1964), 246-249.
42 F.W. Grant and J. Greene, Toxicol. and Appl. Pharmacol., 23 (1972) 71-74.
43 I.E. Kochevar and J. Hom, Photochem. Photobiol., 27 (1983) 163-168.
44 A.G. Motten, G.R. Buettner and C.F. Chignell, Photochem. Photobiol., 42 (1985) 9-15.
45 I. Rosenthal, E. Ben-hur, A. Prager and E. Riklis, Photochem. Photobiol., 28 (1978) 591-594.
46 C. Lagercrants, Psychopharmacol. Serv. Centr. Bull., 2 (1962) 53.
47 S.A. Alkaitis, G. Beck and M. Grätzel, J. Am. Chem. Soc., 97 (20) (1975) 5723-5729.
48 T. Iwaoka and M. Kondo, Chem. Lett., (1978), 731-734.
49 T. Iwaoka and M. Kondo, Bull. Chem. Soc. Japn., 50 (1) (1977) 1-5.
50 H.J. Shine and E.E. Mach. J. Org. Chem., 30 (1965) 2130-2139.
51 B.R. Henry and M. Kasha, J. Chem. Phys., 47 (1967) 3319-3327.
52 T. Iwaoka, H. Kokubun and M. Koizumi, Bull. Chem. Soc. Japn., 44 (1971), 341-347.
53 H.D. Burrows, T.J. Kemp and M.J. Welbourn, J. Chem. Soc. Perkin Trans II, (1973) 969-974.
54 S. Navaratnam, B.J. Parsons, G.O. Philips and A.K. Davies, J. Chem. Soc. Faraday Soc. I, (1978) 1811-1819.
55 D.E. Moore and C.D. Burt, Photochem. Photobiol., 34 (1981) 431-439.
56 G.A. Epling, C. Black and V. Rawal, J. Chem. Soc. Perkin Trans II, (1983) 1313-1320.
57 N.J. Bunce, Y. Kumar and L. Ravanal, J. Med. Chem., 22 (2) (1979) 202-204.
58 R.S. Day and M. Dimattina, Chem. Biol. Interact., 17 (1977) 89-97.
59 B. Ljunggren, S.R. Cohen, D.M. Carter and S.I. Wayne, J. Invest. Dermatol. 75 (1980) 253-256.
60 H.F. Fujita, F. Yanagisawa, A. Endo and K. Suzuki, J. Radiat. Res., 21 (1980) 279-287.
61 I.E. Kochevar and A.A. Lamola, Photochem. Photobiol., 29 (1979) 791-796.
62 I. Matsuo, M. Ohkido, H. Fujita and K. Suzuki, Photochem. Photobiol., 31 (1980) 175-178.
63 Y. Kanako, T. Sato and T. Fujui, Mol. Pharmacol., 20 (1981) 704-708.
64 S. Yonei and T. Todo, Photochem. Photobiol., 35 (1982) 591-592.
65 N.J. de Mol, R.H. Posthuma and G.R. Mohn, Chem. Biol. Interact., 47 (1983) 223-237.
66 J. Decuyper, J. Piette, M. Lopez, M.P. Merville and A. Van de Vorst, Biochem. Pharmacol., 33 (1985) 4025-4031.
67 H. Fujita, A. Endo and K. Suzuki, Photochem Photobiol., 33 (1981) 215-222.
68 G. Kahn and B.P. Davis, J. Invest. Dermatol., 55 (1970) 47-51.
69 N.J. de Mol and J. Koenen, Pharm. Weekblad Sci. Ed., 7 (1985), 121.
70 N.J. de Mol, A.B.C. Becht, J. Koenen and G. Lodder, Chem. Biol. Interact., 57 (1986) 73-83.
71 N.J. de Mol and R.W. Busker, Chem. Biol. Interact., 52 (1984) 79-92.
72 J. Katz, S. Kunofsky, R.E. Patton and N.C. Allaway, Cancer (NY), 20 (1967) 2194.
73 N. Motohashi, Yakugaku Zasshi, 103 (1983) 364-371.
74 J. Marko, G. Vermeesch, N. Febvay-Garot and A. Lablache-Combier, Photochem. Photobiol., 42 (1985) 213-221.
75 M.P. Merville, J. Piette, M. Lopez, J. Decuyper and A. Van de Vorst, J. Biol. Chem., 259 (1984) 15069-15077.
76 T.A. Ciulla, G.A. Epling and I.E. Kochevar, Photochem. Photobiol., 43 (1986) 607-613.
77 J. Decuyper, J. Piette, M.P. Merville-Louis and A. Van de Vorst, Biochem. Pharmacol., 36 (1987) 1069-1076.
78 T. Akera and T.M. Brody, Mol. Pharmacol., 5 (1968) 600-612.

79 T. Akera and T.M. Brody, Mol. Pharmacol., 5 (1969) 605-617.
80 C.Y. Lee, T. Akera and T.M. Brody, Biochem. Pharmacol., 25 (1976) 1751-1756.
81 R.H. Gubitz, T. Akera and T.M. Brody, Biochem. Pharmacol., 22 (1973) 1229-1235.
82 M.P. Merville, J. Piette, J. Decuyper, C.M. Calberg-Bacq and A. Van de Vorst,Chem. Biol. Interact., 44 (1983) 275-287.
83 L. Musajo and G. Rodighiero, Experientia, 18 (1962) 153-162.
84 M.A. Pathak and T.B. Fitzpatrick, J. Invest. Dermatol., 32 (1959) 509-518.
85 L. Musajo and G. Rodighiero, Photophysiology, A.C. Giese (ed.), Academic Press, New York, 1972, Vol. VII. pp 115-147.
86 P.S. Song and J.K. Tapley, Photochem. Photobiol., 29 (1979) 1177-1197 and references therein.
87 G. Rodighiero, F. Dall'Acqua and M. Pathak, Topics in photomedecine, K.C. Smith (Ed.), Plenum Press, New York, 1984, pp 319-398.
88 J.A. Parrish, R.S. Stern, M.A. Pathak and T.B. Fitzpatrick : The Science of Photomedecine, J.D. Regan and J.A. Parrish (Eds.) Plenum Press, New York, 1982, pp 595-624.
89 J.A. Parrish, The Science of Photomedecine, J.D. Regan and J.A. Parrish (Eds.) Plenum Press, New York, 1982, pp 511-531.
90 F. Dall'Acqua, Research in Photobiology, A. Castellani (Ed.) Plenum Press, New York, 1976, pp 245-254.
91 R. Bevilacqua and F. Bordin, Photochem. Photobiol., 29 (1978) 1115-1118.
92 H. Fujita, M. Sano and K. Suzuki, Photochem. Photobiol., 29 (1979) 71-74.
93 I.E. Kochevar, J. Invest. Dermatol., 76 (1981) 59-64.
94 R.V. Bensasson, E.J. Land and C. Salet, Photochem. Photobiol., 27 (1978) 273-280.
95 R.V. Bensasson, E.J. Land and T.G. Truscott, in Flash photolysis and pulse radiolysis, Contributions to the chemistry of Biology and Medecine, Pergamon (Ed.) 1983.
96 R.V. Bensasson., NATO ASI. Ser., Ser. A. (Primary Photoprocesses Biol. Med) (85) ,1985,241-254.
97 E.J. Land and T.G. Truscott, Photochem. Photobiol., 29 (1979) 861-866.
98 F.M. Veronese, R. Bevilacqua and O. Schiavon, Farmaco., 34 (1979) 3-12.
99 F.M. Veronese, O. Shiavon, R. Bevilacqua, F. Bordin and G. Rodighiero, Photochem. Photobiol., 34 (1981) 351-354.
100 K. Yoshikawa, N. Nori, S. Sakakibara, N. Mizuno and P.S. Song, Photochem. Photobiol., 29 (1979) 1127-1133.
101 M. Granger, F. Toulmé and C. Helene, Photochem. Photobiol., 36 (1982) 175-180.
102 M. Granger and C. Helene, Photochem. Photobiol., 38 (1983) 563-568.
103 K.D. Crow, E. Alexander, W.H.L. Buck, B.E. Johnson, I.A. Magnus and A.D. Porter, Br. J. Dermatol., 73 (1961) 220-232.
104 K.H. Kaidbey and A.M. Kligman, Arch. Dermatol., 113 (1977) 592-595.
105 L. Schwartz, L. Tulipan and D.G. Birmingham, in Occupational Diseases of the skin, Lea and Febiger (Eds.) Philadelphia, 1957 pp 301-307.
106 K.H. Kaidbey and S. Nonaka, Photochem. Photobiol., 39 (1984) 375-378.
107 S.S. Epstein, M. Small, H.L. Falk and N. Mantel, Cancer Res., 24 (1964) 855-862.
108 D.D. Morgan and D. Warshawky, Photochem. Photobiol., 25(1977) 39-46.
109 P. Brooks and P.D. Lawley, Nature (London) 202 (1964) 781-784.
110 J. A. Miller, Cancer Res., 30 (1970) 559-576.
111 G.M. Blackburn, J. Buckingham, R.G. Fenwick, P. Taussig and M.H. Thompson, J. Chem. Soc. Perkin Trans. I, (1973) 2809-2813.
112 G.M. Blackburn and P.E. Taussig, Biochem. J., 149 (1975) 289-291.
113 M. A. Pathak and R.K. Biswas, J. Invest. Dermatol., 68 (1977) 236.
114 J.F. Walter and P.R. DeQuoy, Arch. Dermatol., 114 (1978) 1463-1465.
115 G. K. Oster and N-L. Yang, J. Phys. Chem. 77 (1973) 2159-2160.
116 G. Beck and J.K. Thomas, Chem. Phys. Lett., 13, (3) (1972) 295-297.

117 L. Putna, G. Reske and H. Schmidt, Photochem. Photobiol., 30 (1979) 723-725.
118 B. Halliwell in Copper Proteins and Copper enzymes, R. Lontie (Ed.), CRC Press, Florida, 1983, Vol II, chap 3 pp 63-102.
119 C. Ferradini, Biochimie, 68 (1986) 779-785.
120 B. Halliwell and J.M.C. Gutteridge, Biochem. J., 219 (1984) 1-14.
121 I. Fridovich, Free radicals in biology, W. A. Pryor (Ed.) Academic Press New York, 1976, Vol 1, Chap 6 pp 239-277.
122 C.S. Foote, Pathology of oxygen, A.P. Autor (Ed.) Academic Press, New York 1982, chap 2,pp 21-42.
123 W. Bors, M. Saran, E. Lengfelder, M.C. Fuchs and C. Frenzel, Photochem. Photobiol., 28 (1978) 629-638.
124 A. Felmeister and C.A. Discher, J. Pharm. Sci., 53 (1964) 756-762.
125 T. Iwaoka and M. Kondo, Bull. Chem. Soc. Japn., 47 (1974) 1980-1986.
126 P.C. Joshi and M.A. Pathak, Biochem. Biophys. Res. Comm., 112 (1983) 638-646.
127 M.A. Pathak and P.C. Joshi, Biochem. Biophys. Acta, 798 (1984) 115-126.
128 P. Martelli, L. Bovalini, S. Ferri, G.G. Franchi and M. Bari, FEBS, 189 (1985) 255-257.
129 B.F. Abeysekera, Z. Abramowski and G.H.N. Towers, Photochem. Photobiol., 38 (1983) 311-315.
130 P.C. Joshi and M.A. Pathak, J. Invest. Dermatol., 82 (1984) 67-73.
131 Photosensitivity Diseases, Harber and Bickers (Eds.) W.B. Saunders Company, Philadelphia, 1981, pp 59.
132 K. Inoue, T. Matsuura and I. Saito, Photochem. Photobiol., 35 (1982) 133-139.
133 P. Walrant and Santus, Photochem. Photobiol., 19 (1974) 411-417.
134 J.P. Mc Cormick and Thomason J. Am. Chem. Soc., 100 (1978) 312-313.
135 I. Fridovich, Adv. Enzymol. Relat. Areas. Mol. Biol., 41 (1974) 35-97.
136 R.E. Lynch, G.R. Lee and G.E. Cartwight, J. Biol. Chem., 251 (1976) 1015-1019.
137 C.C. Winterbourn, B.M. Mc Grath and R.W. Carrell, Biochem. J., 155 (1976) 483-502.
138 L.S. Demma and J.M. Salhany. J. Biol. Chem., 254 (1979) 4532-4535.
139 B.H. Bielski, R.L. Arudi and M.W. Sutherland, J. Biol. Chem., 258 (1983) 4759-4761.
140 M.J. Thomas, K.S. Mehl and W.A. Pryor, Biochem. Biophys. Res. Commun., 83 (1979) 927-932.
141 J.R. White, T.O. Vaughan and Wei-Shiang Yeh, Fed. Proc., Fed. Amer. Soc. Exp.Biol., 30 (1971) 1145.
142 J.M. McCord, Science, 1985 (1974) 529-531.
143 D.S. Hull, S. Csukas and K. Green, Invest. Ophtalmol. Vis. Sci., 22 (1982) 502-508.
144 S.D. Varma, D. Chand, Y.R. Sharma, J.F. Kuck and R.D. Richards, Current Eye Res., 3 (1984) 35-57.

Cumulative Subject Index, Parts A–D

D

E

F

G

H

I

J

K

L

M

N

Q

R

S

T

Cumulative Chemical Index, Parts A–D

B

E

N

O

Q

T

U

V

W

X

Z

Cumulative Author Index, Parts A–D

A program working on HP1000 System and written by Dr. J.L. Larice (same address as M. Chanon) has made possible the direct printing of this table of authors. Persons interested by this program may write to J.L. Larice.